Microbial bioprotectants for plant disease management

It is widely recognised that agriculture is a significant contributor to global warming and climate change. Agriculture needs to reduce its environmental impact and adapt to current climate change whilst still feeding a growing population, i.e. become more 'climate-smart'. Burleigh Dodds Science Publishing is playing its part in achieving this by bringing together key research on making the production of the world's most important crops and livestock products more sustainable.

Based on extensive research, our publications specifically target the challenge of climate-smart agriculture. In this way we are using 'smart publishing' to help achieve climate-smart agriculture.

Burleigh Dodds Science Publishing is an independent and innovative publisher delivering high quality customer-focused agricultural science content in both print and online formats for the academic and research communities. Our aim is to build a foundation of knowledge on which researchers can build to meet the challenge of climate-smart agriculture.

For more information about Burleigh Dodds Science Publishing simply call us on +44 (0) 1223 839365, email info@bdspublishing.com or alternatively please visit our website at www.bdspublishing.com.

Related titles:

Biostimulants for sustainable agriculture
Print (ISBN 978-1-78676-336-5); Online (ISBN 978-1-78676-338-9, 978-1-78676-339-6)

Achieving sustainable crop nutrition
Print (ISBN 978-1-78676-312-9); Online (ISBN 978-1-78676-314-3, 978-1-78676-315-0)

Pesticides and agriculture Profit, politics and policy
Print (ISBN 978-1-78676-276-4); Online (ISBN 978-1-78676-279-5, 978-1-78676-278-8)

Chapters are available individually from our online bookshop:
https://shop.bdspublishing.com

climate•SMART•publishing
IN AGRICULTURAL SCIENCE

BURLEIGH DODDS SERIES IN AGRICULTURAL SCIENCE
NUMBER 108

Microbial bioprotectants for plant disease management

Edited by Dr Jürgen Köhl, Wageningen University & Research, The Netherlands and Dr Willem J. Ravensberg, Koppert Biological Systems, The Netherlands

burleigh dodds
SCIENCE PUBLISHING

Published by Burleigh Dodds Science Publishing Limited
82 High Street, Sawston, Cambridge CB22 3HJ, UK
www.bdspublishing.com

Burleigh Dodds Science Publishing, 1518 Walnut Street, Suite 900, Philadelphia, PA 19102-3406, USA

First published 2021 by Burleigh Dodds Science Publishing Limited

Library of Congress Control Number: 2021940792

British Library Cataloguing in Publication Data
A catalogue record for this book is available from the British Library

ISBN 978-1-78676-813-1 (Print)
ISBN 978-1-78676-816-2 (PDF)
ISBN 978-1-78676-815-5 (ePub)
ISSN 2059-6936 (print)
ISSN 2059-6944 (online)

DOI 10.19103/AS.2021.0093

Typeset by Deanta Global Publishing Services, Dublin, Ireland

Contents

Series list

Title	Series number
Achieving sustainable cultivation of maize - Vol 1 From improved varieties to local applications Edited by: Dr Dave Watson, CGIAR Maize Research Program Manager, CIMMYT, Mexico	001
Achieving sustainable cultivation of maize - Vol 2 Cultivation techniques, pest and disease control Edited by: Dr Dave Watson, CGIAR Maize Research Program Manager, CIMMYT, Mexico	002
Achieving sustainable cultivation of rice - Vol 1 Breeding for higher yield and quality Edited by: Prof. Takuji Sasaki, Tokyo University of Agriculture, Japan	003
Achieving sustainable cultivation of rice - Vol 2 Cultivation, pest and disease management Edited by: Prof. Takuji Sasaki, Tokyo University of Agriculture, Japan	004
Achieving sustainable cultivation of wheat - Vol 1 Breeding, quality traits, pests and diseases Edited by: Prof. Peter Langridge, The University of Adelaide, Australia	005
Achieving sustainable cultivation of wheat - Vol 2 Cultivation techniques Edited by: Prof. Peter Langridge, The University of Adelaide, Australia	006
Achieving sustainable cultivation of tomatoes Edited by: Dr Autar Mattoo, USDA-ARS, USA; and Prof. Avtar Handa, Purdue University, USA	007
Achieving sustainable production of milk - Vol 1 Milk composition, genetics and breeding Edited by: Dr Nico van Belzen, International Dairy Federation (IDF), Belgium	008
Achieving sustainable production of milk - Vol 2 Safety, quality and sustainability Edited by: Dr Nico van Belzen, International Dairy Federation (IDF), Belgium	009
Achieving sustainable production of milk - Vol 3 Dairy herd management and welfare Edited by: Prof. John Webster, University of Bristol, UK	010
Ensuring safety and quality in the production of beef - Vol 1 Safety Edited by: Prof. Gary Acuff, Texas A&M University, USA; and Prof. James Dickson, Iowa State University, USA	011
Ensuring safety and quality in the production of beef - Vol 2 Quality Edited by: Prof. Michael Dikeman, Kansas State University, USA	012
Achieving sustainable production of poultry meat - Vol 1 Safety, quality and sustainability Edited by: Prof. Steven C. Ricke, University of Arkansas, USA	013
Achieving sustainable production of poultry meat - Vol 2 Breeding and nutrition Edited by: Prof. Todd Applegate, University of Georgia, USA	014

Acknowledgements

Cover image acknowledgements

- Top left = © Copyright Koppert Biological Systems. All Rights Reserved.
- Top right = Wageningen Plant Research, The Netherlands.
- Bottom left = Cover image: courtesy of Stéphane Compant, AIT Austrian Institute of Technology, Austria.
- Bottom right = © Copyright Koppert Biological Systems. All Rights Reserved.

Introduction

With growing concerns about the environmental impact of synthetic fungicides, increasing levels of fungicide resistance and increasing regulatory restrictions on fungicide use, the crop protection sector and farmers face mounting pressure to replace synthetic fungicides with environmentally friendly biological alternatives for disease control. This volume examines the recent advances in the development of ecologically balanced biological methods using bioprotectants based on beneficial microorganisms to control plant diseases.

Part 1 of the book focuses on product development of microbials, discussing aspects such as modes of action of microbial bioprotectants and developments in screening approaches. Chapters in Part 1 also examine methods to visualise and determine microbials, durability of efficacy of microbial bioprotectants, the commercialisation of microbial bioprotectants, advances in production of commercial microbials and the regulation of microbials in the European Union. Part 1 concludes with a chapter on microbial bioprotectants and the marketplace. Part 2 examines the use of biological control agents to control plant disease, focusing on bacterial, fungal and viral agents. Part 3 provides examples of the use of bioprotectants, specifically highlighting the use of bioprotectants to protect food from aflatoxin contamination and the use of *Verticillium albo-atrum* WCS850 to control Dutch elm disease. Part 4 of the book concludes with a discussion of the future of bioprotectants, focusing on their role in integrated crop protection approaches.

Part 1 Product development of microbials

Part 1 opens with a chapter that focuses on the advances in understanding modes of action of microbial bioprotectants. Plant-associated microorganisms are involved in important functions related to growth, performance and health of their hosts. Understanding their modes of action is pivotal for the development and application of microbial bioprotectants and biostimulants. Chapter 1 summarises current knowledge about beneficial plant-microbe interactions, discusses recent insights into the functioning of the plant microbiome and beneficial plant-microbe networks. It shows that the use of microorganisms and the exploitation of beneficial plant-microbe interactions offer promising and environmentally-friendly strategies to achieve sustainable agriculture on a global scale.

Chapter 2 examines advances in screening approaches for the development of microbial bioprotectants to control plant diseases. The chapter begins by discussing screening of microorganisms for biological control of plant diseases, specifically focusing on exclusive and inclusive approaches. It then goes on to discuss the nine-step approach to screening for biocontrol agents, which is then followed by an analysis of non-traditional biocontrol agents of plant diseases such as entomopathogenic fungi, bacteria, bacteriophages and mycoviruses. Sections on niche and regional markets for biocontrol agents are also included. The chapter also provides an overview of the formulation of biocontrol agents, the role of the microbiome in biocontrol and the use of microbiome engineering for disease control.

The subject of Chapter 3 is visualising plant colonisation by beneficial bacteria, which is considered a key step to improve the understanding of plant-microbe interactions. Plants contain diverse microorganisms that interact with their hosts and with each other. Beneficial bacteria can be utilised on crops to protect plants against biotic and abiotic stresses and to stimulate plant growth. However, the behaviour of specific microorganisms on and within plants is still underexplored. The chapter begins by highlighting the methods that can be used to visualise and determine microbial colonisation of plants. The chapter then goes on to examine colonisation of beneficial bacteria from the soil to the root surface and to root internal tissues. A section on the colonisation of the aerial parts of plants is also provided, which is then followed by a discussion on improving applications by understanding colonisation processes. These techniques are crucial in the selection process of biocontrol agents and the development of a product formulation and method of application.

Chapter 4 addresses the durability of efficacy of microbial bioprotectants against plant diseases. The chapter highlights current knowledge concerning erosion of efficacy of microbial bioprotectants against plant pathogens and its possible consequences for field applications. It starts by reviewing the durability of plant protection against bio-aggressors, then goes on to discuss the main modes of action of microbial bioprotectants against plant pathogens in relation to potential mechanisms of resistance developed by plant pathogens. The chapter also reviews the diversity of susceptibility within plant pathogen populations to microbial bioprotectants and the related risk of emergence of resistance to microbial bioprotectants in plant pathogens as well.

The next chapter reviews the recent commercialisation of microbial bioprotectant products containing bacteria, fungi, yeast and bacteriophages for the control of plant diseases. Chapter 5 also summarises recent development activities of new bioprotectant products based on microorganisms or their metabolites, including induced resistance products, single domain antibody proteins produced by microorganisms, and protozoans (amoebae). Production, mainly by submerged fermentation, and formulation processes of microbial bioprotectants are discussed. Key factors influencing the fermentation, formulation and the scale up for industrial production of such microorganism as bioprotectant products are also addressed, including stability and viability of the active substances produced by liquid fermentation processes.

Chapter 6 examines the key issues in the regulation of microbial bioprotectants in the European Union, focusing specifically on the challenges and solutions to achieve more sustainable crop protection. Microbial bioprotectants, like chemical pesticides, are required to pass a risk assessment and risk management procedure prior to use in plant protection, which in many countries is an obstacle for market access, in particular, the European Union. Administrative issues and data requirements, adapted from those used for chemicals, cause issues for both applicants and evaluators. These issues are reviewed and improvements are proposed. Biology should be the basis of the evaluation and data requirements for microorganisms, with an emphasis in this chapter on microbial compounds and testing methods. Political actions involving the use of pesticides are reviewed and recommendations are made on how to improve the system for microbial bioprotectants, including new uses of microorganisms. New legislation is suggested for all microorganisms used in agriculture and related uses based on the assumption that well-known microorganisms are of low risk to human health and the environment.

The final chapter of Part 1 reviews microbial bioprotectants and the marketplace. Chapter 7 begins by discussing the latest figures in terms of microbial bioprotectants global market value. It also highlights the different types and ways these bioprotectants can be used. The chapter also discusses the trends and drivers in the microbial market, focusing specifically on why microbials dominate and the factors that drive bioprotectant adoption. A section on the myths about the bioprotectant market is also included, which is then followed by a discussion of the limitations for using microbial bioprotectants. Future opportunities and threats for microbial bioprotectants are also highlighted.

Part 2 Biological control agents

The first chapter of Part 2 focuses on the use of *Bacillus* spp. as bacterial control agents to control plant diseases. Biocontrol agents (BCAs) based on plant growth promoting rhizobacteria have recently been developed as alternatives to chemical pesticides. Among those beneficial bacteria, *Bacillus* spp. are one of the most promising BCAs. A wide range of bioactive secondary metabolites (BSMs) are involved in biocontrol via antibiosis to plant pathogens and/or via elicitation of systemic resistance in their host plants. Chapter 8 illustrates the diversity of pathosystems in which BCA based on *Bacillus* spp. have proved effective. It describes the mechanisms underpinning this biocontrol activity via production of a wide range of enzymes, proteins and small-size BSMs. As these BSMs are clearly involved in pathogen control, the chapter emphasises the importance of understanding the ecological factors influencing their production. The final section of the chapter highlights the potential interactions between *Bacillus* spp. and other soil microorganisms in developing consortia of biocontrol agents combining species with synergistic activities for plant health improvement.

Chapter 9 examines the use of *Pseudomonas* spp. as bacterial biocontrol agents to control plant diseases. The chapter starts by highlighting recent advances in *Pseudomonas* taxonomy and a summary of its most important biocontrol traits. It then provides an overview of the most important *Pseudomonas* groups and subgroups harbouring biocontrol strains. Based on their activity, *Pseudomonas* biocontrol strains come in three types. The first type, represented by *P. chlororaphis*, *P. protegens*, *P. corrugata* and *P. aeruginosa* (sub)group strains, produces an arsenal of secondary metabolites with broad antimicrobial activity. The second type is found in the *P. putida*, *P. fluorescens*, *P. koreensis*, *P. mandelii*, and *P. gessardii* (sub)group. The spectrum of biocontrol properties of these strains is less diverse and involves siderophores and cyclic lipopeptides. The third type colonizes above-ground plant parts. Strains from this type mainly belong to the *P. syringae* group and are used to control postharvest pathogens. Examples of well-characterized and representative biocontrol strains show the links between phylogeny, ecology and biocontrol traits. The chapter concludes by reviewing commercially-available biocontrol strains.

Moving on from Chapters 8 and 9, Chapter 10 focuses on bacterial bioprotectants besides *Bacillus* and *Pseudomonas* species. The chapter begins by discussing the taxonomy of non-*Bacillus* and *Pseudomonas* (NBP) bioprotectant strains, including enterobacteria, actinomycetes, *Sphingomonas, Methylobacterium, Agrobacterium-Rhizobium* and *Lactobacillus*. The chapter reviews their mechanisms of action against plant pathogens. Sources of isolates and methods of isolation are discussed in building strain collections. The chapter then reviews procedures for screening antagonistic bacteria candidates as bioprotectants using biochemical and molecular markers, including the example of lactic acid bacteria. The chapter also covers strain improvement to increase fitness and efficacy in the field through physiological and genetic manipulation. Since they are essential for commercial development, biosafety issues are discussed, followed by an overview of patented substances and commercialized products. The chapter concludes with a summary and future trends in research section on non-*Bacillus* and *Pseudomonas* species.

The next chapter addresses the use of *Trichoderma* spp. to control plant diseases. *Trichoderma* is one of the most studied genera of ascomycetous fungi due to the beneficial effects it has on plants. *Trichoderma* spp. are involved in the production of cell wall-degrading enzymes and metabolites with antimicrobial activity. It also produces volatile compounds that act together as direct biocontrol agents to protect plants against phytopathogenic fungi, oomycetes, nematodes and bacteria. Chapter 11 first reviews the various biocontrol mechanisms that can be used, specifically focusing on mycoparasitism, antibiosis and competition between the soil microbiota and plants for resources such as nutrients. The chapter also examines systemic defence and growth promotion in plants and how these can both have beneficial effects on the plant. A section on the importance of registration and commercialisation of *Trichoderma* strains is also provided as well as an overview of successful uses.

The subject of Chapter 12 is the use of *Clonostachys rosea* to control plant diseases. The fungus *C. rosea* was recognized as an aggressive parasite on other fungi already in the late 1950s. Research into its potential use in biological control of plant diseases soon followed, resulting in several commercial products. The chapter begins by reviewing the taxonomy and sources of *C. rosea*, then goes on to discuss the range of mechanisms of action involved in combating *C. rosea*. A section on the lessons that can be learnt from genomics and transcriptomics and *C. rosea* is also provided, which is then followed by a review of *C. rosea* product development and commercialisation. The chapter also analyses the delivery and action of *C. rosea* as a biocontrol agent, before concluding with an overview of why the use of *C. rosea* for plant disease control is a promising prospect.

Chapter 13 looks at the use of bacteriophages to control plant diseases. Bacteriophages are viruses that kill target bacteria without affecting another microorganism and environment. Bacteriophage efficiency on the phyllosphere is mainly affected by ultraviolet (UV) light. The chapter first discusses the advantages of using bacteriophages for biocontrol over other strategies, then goes on to examine the early research on bacteriophages. The chapter also reviews the sources of bacteriophages and the interactions they can have with host cells. A section on the concerns in using bacteriophages as biocontrol agents is also provided which is followed by an analysis of the facts that can affect the efficacy of bacteriophages for plant disease control. Approaches for optimum efficacy of bacteriophage for plant disease control are discussed and the bacteriophage therapy challenges that need to be addressed are also included. A list of applications of bacteriophages is also provided.

The next chapter focuses on the use of mild viruses for the control of plant pathogenic viruses. Plant virus management strategies have largely been limited to the application of hygiene protocols, the control of viral vectors such as insects and nematodes and the use of resistant varieties. However, these approaches are often insufficient to prevent infections. The rapid control of newly emerging viral diseases remains challenging. Chapter 14 focuses on cross-protection using mild viruses as active substances in biocontrol. The chapter begins by describing the theoretical modes of action of cross-protection. It then goes on to discuss crucial elements in the development of a cross-protection strategy, taking into account new insights based on commercial application of cross-protection. The chapter also provides case studies in which cross-protection has been applied in commercial crops, in particular vaccination strategies to control Pepino mosaic virus (PepMV) in greenhouse tomato. Finally, developments that may impact future research into the control of emerging viral pathogens are discussed.

The final chapter of Part 2 looks at the possibility of mycoviruses as a means of biocontrol. Chapter 15 discusses the characteristics of mycoviruses and the conditions that need to be met to make a successful virocontrol agent. Virocontrol involves biocontrol of pathogens via hypovirulence-causing mycoviruses. The chapter describes one of the success stories so far, that of virocontrol of *Cryphonectria parasitica*, the chestnut blight pathogen. The chapter ends by reviewing future trends and where to find more information on mycoviruses and virocontrol.

Part 3 Examples of use of microbial bioprotectants

Part 3 opens with a chapter that focuses on the development and scale-up of bioprotectants to keep staple foods safe from aflatoxin contamination in Africa. Aflatoxins pose a significant public health risk, decrease productivity and profitability and hamper trade. To minimise aflatoxin contamination a biocontrol technology based on atoxigenic strains of *Aspergillus flavus* that do not produce aflatoxin is used widely in the United States. The technology, with the generic name Aflasafe, has been improved and adapted for use in Africa. Aflasafe products have been developed or are currently being developed in 20 African countries. Aflatoxin biocontrol is being scaled up for use in several African countries through a mix of public, private, and public-private interventions. Farmers in several countries have commercially treated nearly 400 000 ha of maize and groundnut achieving >90% reduction in aflatoxin contamination. Chapter 16 summarises the biology of aflatoxin-producing fungi and various factors affecting their occurrence, including climate change. Various management practices for aflatoxin mitigation are then discussed. These include biological control, which is increasingly being adopted by farmers in several countries. The chapter also discusses biocontrol product development and commercialization in various African countries. Subsequently, the chapter highlights some barriers to adoption and other challenges.

Chapter 17 highlights the potential of using *Verticillium albo-atrum* WCS850 to control Dutch elm disease, specifically focusing on its prominence in Europe. The chapter first describes the development of the *Verticillium albo-atrum* WCS850 strain, which first started in the 1980s. The chapter then goes on to review the production, method of application, registration and use of the *Verticillium albo-atrum* WCS850 strain as a biocontrol agent. A section on testing long-term efficacy of the biocontrol strain is also provided, before concluding with an analysis of how Dutch elm disease treatment has changed since it was first identified.

Part 4 Future outlook on microbial bioprotectants

Chapter 18 addresses the role of bioprotectants for disease control in integrated crop protection approaches. Bioprotectants have the potential to replace chemical pesticides in agricultural cropping systems and crop protection approaches. Development of new bioprotectants in combination with more restricted use of chemical crop protection will result in their much stronger market position in the future. Bioprotectants fulfil particular roles in current and future crop protection approaches, primarily reducing pesticide residues in harvested products in conventional systems, as well as being the first and preferred control option in integrated pest management programs and organic farming, and complementing resident microbiomes in future resilient cropping systems. The process of developing bioprotectants until market access can take from up to 10 to 15 years. The chapter gives a brief overview of the role of bioprotectants in current and future crop protection approaches to stimulate discussion within the biocontrol industries, and amongst scientists and funding agencies on the need for new generations of bioprotectants for an agriculture industry undergoing transition.

The final chapter of the book highlights the future outlook on microbial bioprotectants in agriculture. Microbial bioprotectants have the potential to play a major role in the future of crop protection. Agriculture needs to become more sustainable and still provide food security within planetary borders. New technologies and scientific discoveries can unravel the interactions between the plant, the microbiome and the soil and provide new opportunities for crop protection and more resilient cropping systems. Regulatory issues delay and hamper exploitation and research of genetic resources. Chapter 19 describes the factors that promote the use of microbial bioprotectants as well as those that hamper their further adoption. A sustainable and resilient agriculture depends on the microbial interactions between plants in promoting plant growth and combatting biotic and abiotic threats. The transition to a resilient agriculture requires big changes in policy, regulation and farming practices. The chapter assesses the future outlook for the methods for controlling plant diseases described in this book as well as the factors determining their uptake and success.

Part 1

Product development of microbials

Chapter 1

Advances in understanding modes of action of microbial bioprotectants

Gabriele Berg, Graz University of Technology and Austrian Centre of Industrial Biotechnology, Austria; Peter Kusstatscher, Franz Stocker and Ahmed Abdelfattah, Graz University of Technology, Austria; and Tomislav Cernava, Graz University of Technology and Austrian Centre of Industrial Biotechnology, Austria

1 Introduction

Human activities substantially affect the environment. According to the Intergovernmental Science-Policy Platform on Biodiversity and Ecosystem Services (IPBES), for example, species extinctions are currently up to 1000 times higher than normal (ipbes.net). Intensive agriculture has contributed significantly to these long-lasting anthropogenic impacts (Crutzen, 2002). The conversion of atmospheric nitrogen to ammonia by the Haber–Bosch process, for efficient production of fertilizers, has fundamentally altered the global nitrogen cycle. Overuse of chemicals leads to the development of resistant plant and human pathogens, which are difficult or impossible to suppress. Current agricultural practices focus on increased productivity to ensure high crop yields with maximized profitability. Growing levels of fertilizer and pesticide inputs globally have led to the contamination of the environment as well as of food, resulting in negative impacts on so-called *One health* or planetary health (Flandroy et al.,

http://dx.doi.org/10.19103/AS.2021.0093.01

2018). We have already exceeded planetary boundaries for nitrogen supply, land use, as well as biodiversity loss (Rockström et al., 2009).

Microbial bioprotectants and biostimulants could provide a solution to these problems (Köhl et al., 2019). Plants and soil naturally harbor specific microorganisms that fulfill functions, such as nutrient, mineral, and vitamin supply, and protect against biotic and abiotic stress (Bakker et al., 2020). Such beneficial microbes can be categorized according to their modes of action into bioprotectants and biostimulants (Lugtenberg and Kamilova, 2009; Dunham Trimmer, 2017). In this chapter, we follow the definition of bioprotectants provided by the International Biocontrol Manufacturers Association (IBMA) (https://ibma-global.org/what-is-biocontrol), in order to avoid the ambiguity that might be caused by different national and continental definitions.

Bioprotectants based on beneficial microbes can potentially replace chemical pesticides. Gram-positive bacteria with spore-forming ability are currently the main group being commercialized; however, they represent only a small proportion of the potential microbial community. The commercial use of bioprotectants depends on the reliability of microbial formulations, their modes of action, as well as their safety. A low shelf life and inconsistent effects in the field characterize many of the products that have been commercialized so far. There have been recent developments in methods for the analysis of microbial communities based on next-generation sequencing (Jansson and Baker, 2016; Berg et al., 2020). They have resulted in an improved knowledge base in terms of understanding the modes of action of biocontrol agents (Spadaro and Droby, 2016). This knowledge is important during all stages of the development of microbial products, from their discovery to risk and efficacy assessment.

Microbiome research has revolutionized our understanding of the way plants function and opened new possibilities to develop more sustainable agriculture (Berg et al., 2020). This new field of research has contributed to our understanding of the modes of action of binary plant-microbe interactions as well as the interplay within the whole plant microbiome. Promising results from microbiome research have provided a boost for the whole 'microbiome market' as well as private investment in companies and startups (global-engage.com). Advances in the engineering of environmental microbiomes are predicted to replace potentially toxic chemicals in agriculture in the future, stimulate a more sustainable use of environmental resources, and improve food processing. Currently, agricultural products based on the microbiota are one of the fastest-growing sectors in agronomy, with a compound annual growth rate (CAGR) of 15–18% and a predicted value of over $10 billion US dollars by 2025 for the whole biocontrol sector (Dunham Trimmer, 2017).

This chapter summarizes current knowledge about beneficial plant-microbe interactions and discusses recent insights into the functioning of the plant microbiome and beneficial plant-microbe networks. It shows that the use of microorganisms and the exploitation of beneficial plant-microbe interactions offer promising and environmental-friendly strategies to achieve sustainable agriculture on a global scale.

2 Modes of action of plant-associated microorganisms with their host

Over the past 125 years, research and practical applications have both repeatedly demonstrated that microorganisms can have an intimate interaction with their host plants and are able to promote plant growth as well as suppress plant pathogens (Hiltner, 1902; Whipps, 2001; Berg, 2009). Plant-beneficial effects can result either from direct interactions of microorganisms with their plant hosts or indirectly by the suppression of plant pathogens. Under *in vitro* conditions, the principles of these interactions, their regulation, and stimulation are well understood at the molecular level. However, in nature, they often occur in combination and direct mechanisms are difficult to differentiate from indirect disease suppression (Haas and Défago, 2005). The ability of a microorganism to colonize plant habitats is essential for all successful plant-microbe interactions (Lugtenberg et al., 2002). Distinct steps in the initiation of such interplay include recognition, adherence, invasion (only endophytes and pathogens), colonization, and growth to establish interactions (Hardoim et al., 2015). Figure 1 provides an overview of tripartite interactions between beneficial microorganisms and plants, including the native microbiome as well as pathogens. It also illustrates the modes of action that are involved. Box 1 includes examples of studies on modes of action of plant-beneficial microorganisms with their plant hosts.

Box 1 Selected reviews of the modes of action of plant-associated microorganisms with their host and pathogens

Reviews describing modes of action in general were published by Whipps (1988, 2001), Weller et al. (2002), Berg (2009), and Lugtenberg and Kamilova (2009). Moreover, reviews focusing on modes of action of specific genera are also available, for example, for *Pseudomonas* (Haas and Défago, 2005), *Burkholderia* (Compant et al., 2008), *Bacillus* (Santoyo et al., 2012), and *Paenibacillus* (Rybakova et al., 2016); all of them are representatives of common plant-beneficial genera. Köhl et al. (2019) discussed the modes of action especially with regard to the potential risk for human and environmental health.

There are three general types of direct beneficial interaction of microorganisms with their plant hosts:

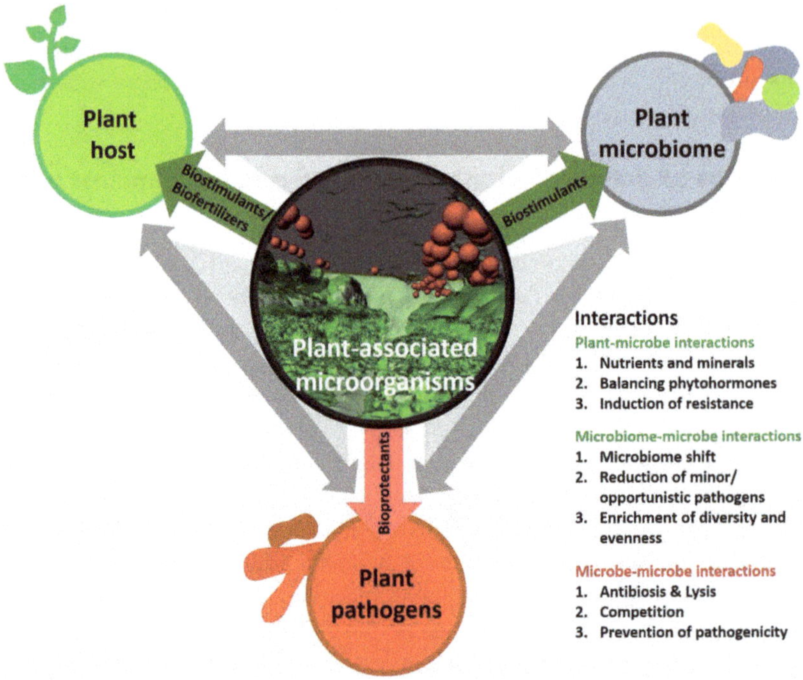

Figure 1 Plant-microbe interactions promoting plant growth and health: mode of action and potential use in biotechnological applications as bioprotectants, biofertilizers, and biostimulants.

- Provision of nutrients and minerals;
- Balancing of the hormonal status in plants; and
- Induction of resistance by activating the indigenous plant defense system.

All of them contribute to enhanced plant health and fitness. They are briefly discussed in the following subsections in order to provide an overview of the current knowledge base for classical approaches used to discover modes of action of beneficial microorganisms.

2.1 Provision of nutrients and minerals

Plant-associated microorganisms can directly supply nutrients and minerals for plant growth and health. The symbiosis between rhizobia and legume host plants is one of the oldest descriptions of such beneficial interactions but remains a very important example of the role of plant growth-promoting rhizobacteria (PGPR). Bacteria of this group metabolize root exudates (carbohydrates) and in turn provide nitrogen to the plant for amino acid synthesis. The ability to fix nitrogen also occurs in free-living bacteria like *Azospirillum*, *Burkholderia*,

Stenotrophomonas, and *Gluconacetobacter diazotrophicus* (Dobbelaere et al., 2003; Nascimento et al., 2020). Another important nutrient is sulfate, which can be provided to the plant following oxidation by bacteria (Banerjee and Yesmin, 2009), or phosphorous derived from organic compounds such as phytates, thus indirectly involving bacteria in the promotion of plant growth (Unno et al., 2005). When plants are treated with *Azospirillum*, it often results in the enhancement of root growth and increased uptake of phosphorous and other macro- and microelements (Dobbelaere and Okon, 2007). Mineral supply is also important for plant growth promotion and includes microbial synthesis of siderophores and complementary siderophore uptake systems (Katiyar and Goel, 2004). In addition, other poorly soluble inorganic nutrients can be made available by solubilization with bacterial siderophores and the secretion of organic acids. It has also been shown that organic fertilizers containing Bacillus *amyloliquefaciens* cause disease suppression by influencing resident soil microbial communities, specifically by promoting specific *Pseudomonas* spp. (Tao et al., 2020).

2.2 Balancing the hormonal status of plants

Ethylene illustrates the critical importance of hormonal balance in plants. At low levels, ethylene can promote plant growth in several plant species including *Arabidopsis thaliana*, while at higher levels, it is an inhibitor of plant growth and is known as a senescence hormone (Pierik et al., 2006). The hormonal status in plants can be stabilized by the production and degradation of phytohormones, for example, ethylene, indole-3-acetic acid (IAA), cytokinins, and gibberellins. They can be synthesized by the plant itself or by associated microorganisms. Bacteria are, for example, able to reduce ethylene levels by interfering with its biosynthesis. The compound 1-aminocyclopropane-1-carboyclic acid (ACC) is a precursor of ethylene in plants. ACC deaminase-producing bacteria are able to degrade this substance, which results in reduced uptake by the plant and lowered levels in the roots. Bacteria can thus increase root growth by lowering endogenous ACC levels (Glick, 2005). Due to an additional role of ethylene as a stress hormone, ACC deaminase-producing bacteria can also protect plants against biotic and abiotic stress (Saleem et al., 2007).

2.3 Induction of resistance

Induction of resistance by the activation of the indigenous plant defense system is another type of direct interaction of microorganisms with plant hosts (Bakker et al., 2007). Defense mechanisms can be activated via stimuli recognized by specific pattern-recognition receptors. These stimuli, such as pathogen-associated molecular patterns or microbe-associated molecular patterns (MAMPs), require only milli- to sub-nanomolar levels to induce resistance. Systemic acquired resistance describes the direct reaction of the plant against plant pathogens, while induced systemic resistance describes the role of beneficial microorganisms (Conrath et al., 2015). Antibiotics, biosurfactants, volatile organic compounds, lipopolysaccharides, and many other components of microbial origin can serve as MAMPs (Pieterse et al., 2014). Defense mechanisms are induced by a mechanism termed 'priming' (Conrath et al., 2015). After induction of priming, reactive oxygen species, phytoalexins, phenolic compounds, and pathogenesis-related proteins are produced, resulting in the formation of physical barriers by the plant (Wiesel et al., 2014). Plant hormones and volatile organic compounds have been identified as important signaling molecules in local and systemic-induced defense responses to pathogen or insect attack (Pieterse et al., 2013).

3 Modes of action between plant-associated microorganisms and pathogens

There are four general types of direct inhibitory interaction of microorganisms with plant pathogens:

- Antibiosis: the inhibition of microbial growth by diffusible antibiotics and volatile organic compounds (VOCs), bioactive compounds, and biosurfactants;
- Competition for colonization sites and nutrients and minerals such as iron through the production of siderophores or efficient siderophore uptake systems;
- Degradation of pathogenicity factors of pathogens; and
- Parasitism that may involve the production of extracellular cell wall-degrading enzymes such as chitinases and β-1,3-glucanase.

Each of these is briefly described in the following subsections.

3.1 Antibiosis: inhibition of microbial growth by bioactive substances

Antibiosis is one of the most commonly described modes of action of microbial antagonism, occurring via diffusible antibiotics, VOCs, and biosurfactants. Antimicrobial secondary metabolites are produced by a great number of microbes and released into the environment in small amounts. These heterogeneous groups of organic low-molecular-weight compounds can have adverse effects on microbial growth and metabolic activity (Bérdy, 2005; Thomashow et al., 2008). In resource-limited environments, antimicrobial metabolites are effective against competitors because a significant competitive advantage is conferred to microbes via such metabolites (Raaijmakers and Mazzola, 2012). Since most bacteria cannot be cultured in usual media (Amann et al., 1995), many antibiotics produced *in situ* are still unknown. Microbial genome analyses also reveal a high quantity of cryptic antibiotic gene clusters that potentially encode unknown antibiotics (Raaijmakers and Mazzola, 2012). A highly potent mode of action is also known to be exerted by the secretion of biosurfactants, a specific group of organic compounds (Burch et al., 2014).

Despite many studies of antimicrobial compounds, we still do not know their exact function in nature. Potential functions include signaling and competition (Linares et al., 2006). The expression of many bioactive compounds can be confirmed by transcriptomic studies under *in vitro* conditions; however, the same compounds were not produced or play a minor role *in situ* (Koch et al., 2018). This raises the question of whether commonly known compounds are the main drivers in the inhibition of phytopathogens. Under *in situ* conditions, transcriptomic studies often reveal novel molecules such as spermidine, produced as a general stress response by beneficial microorganisms (Alavi et al., 2014).

3.2 Competition and modification of microenvironments

Growth and germination of plant pathogens always depend on the availability of nutrients (Köhl and Fokkema, 1998). Microorganisms can locally induce various environmental stresses including drought, moisture fluctuations, limited nutrient availability, pH shifts, salt enrichment, and enrichment of bioactive compounds to gain significant advantages against a pathogen (Kunz et al., 2010). An important advantage is low selection pressure, which limits the development of resistance by pathogens (Köhl et al., 2019). An indirect mode of bacterial antagonism against pathogens is competition for nutrients and space (Spadaro and Droby, 2016). Siderophores and efficient uptake systems, for example, offer a powerful tool to compete for iron (Raaijmakers et al., 1995; Jan et al., 2011).

3.3 Interference with pathogenicity

Reduction or complete elimination of pathogenicity is an additional mechanism that can be utilized for biocontrol. Microbes employ bioactive compounds or antibiotics to survive in competing environments (Molina-Favero et al., 2008). Potential bioprotectants can degrade such virulent compounds and thus subdue the growth of the pathogen (Duffy et al., 2003). Virulence factor genes can also be induced by quorum sensing: beneficial microorganisms are able to inhibit gene expression in pathogens through degradation of autoinducer signals (Von Bodman et al., 2003). The degradation efficiency of virulence factors decreases when microorganisms colonize interior parts of plants (Sturz et al., 2000).

3.4 Parasitism and lysis

Parasitism, hyperparasitism, and lysis are direct types of interaction between microorganisms (Jeffries, 1995). Bioprotectants can secrete cell wall-degrading enzymes such as chitinases, β-1,3-glucanases, proteases, or cellulases, in combination with secondary metabolites, which invade and kill pathogenic cells (Jeffries, 1995). Such enzymes are substrate specific and highly sensitive in terms of environmental denaturation, resulting in low toxicological and ecotoxicological risks. The risk of resistance formation by a plant pathogen is low, though pathogens are able to evolve resting structures like endospores, chlamydospores, and melanized sclerotia, or suppress the synthesis of antagonistic enzymes promoting hyperparasitic interactions (Bardin et al., 2015).

4 Understanding the structure and function of the plant microbiome

4.1 Discovery and definition

Microbiome research is a new research field in understanding microbial communities (Berg et al., 2020). It has evolved into an interdisciplinary field combining microbiology, bioinformatics, biotechnology, agriculture, food safety, plant pathology, and human medicine. The field has been revolutionized by advanced molecular and microscopic tools, especially by various next-generation sequencing technologies. These methods now allow us to obtain a full picture of the structure and function of any microbiome (Fig. 2).

In order to understand plant-microbe interactions, it is important to understand the plant microbiome and its drivers and to know how the 'microbiota' and 'microbiome' are defined. The first definition of the term 'microbiome' in a biocontrol context was provided in 1988 by Whipps and colleagues. They described the 'microbiome' as a 'characteristic microbial community' in a 'reasonably well-defined habitat which has distinct physio-chemical properties' as their 'theatre of activity' (Whipps et al., 1988). The definition by Whipps et al. (1988) has been recently updated by Berg et al. (2020). The microbiota are a part of the microbiome and comprise bacteria, fungi, algae, and archaea, as well as small protists. The microbiota are embedded in an ecosystem that is also termed as the 'theatre of activity'. It involves the whole spectrum of molecules produced by the microorganisms and their hosts, including their structural elements (proteins and peptides, lipids, polysaccharides, nucleic acids), metabolites (signaling molecules, bioactive substances, organic and inorganic molecules), as well as their surrounding environmental conditions (Berg et al., 2020). All mobile genetic elements, such as phages, viruses, plasmids, prions, viroids, and free DNA are not considered as living microorganisms (Dupré and O'Malley, 2013). They do not belong to the microbiota but are a part of the microbiome (Berg et al., 2020).

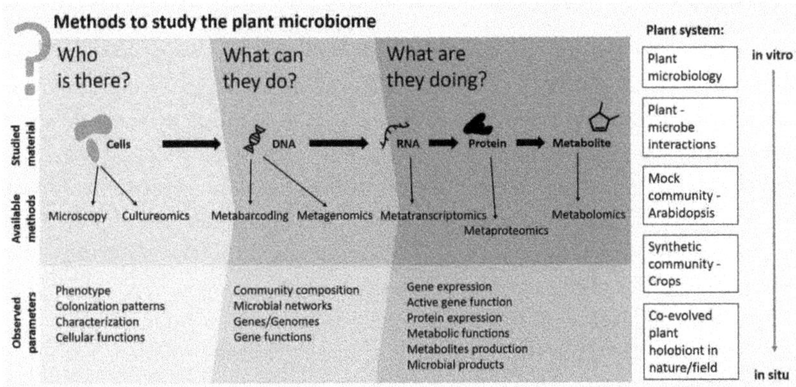

Figure 2 Methods for plant microbiome analyses and modes of action at different levels and in different systems.

Plants are naturally colonized by a variety of microorganisms and are able to engage in complex and dynamic interactions with their associated microbes. Such microbes associated with the plant are termed the plant's microbiota (also described as the plant's second genome) and contribute a broad range of beneficial traits (Mendes et al., 2011; Turner et al., 2013). The plant and its associated microbiota can also be regarded as a meta-organism, the so-called holobiont (Vandenkoornhuyse et al., 2015). Many components of such holobionts are not cultivable; the proportion of those that can be cultured is under 10%. There are, for example, no plant-associated archaea in culture, although their proportion in the rhizosphere can be up to 30% (Taffner et al., 2019). Novel tools make it possible to assemble genomes of plant-associated microorganisms (metagenome-assembled genomes (MAGs)) from metagenomic data, which makes it possible to study them in detail without the need for cultivation (Wicaksono et al., 2020).

4.2 Plant microbiome structure

Plants harbor distinct habitat-specific microbial signatures, which are mainly shaped by abiotic factors. The phyllosphere microbiome is located in all above-ground plant organs. These are typically exposed to the surrounding air and permanently changing abiotic factors such as ultraviolet (UV) radiation, temperature, and water, and generally low nutrient availability (Remus-Emsermann and Schlechter, 2018). The phyllosphere can be further subdivided into the caulosphere (stems), phylloplane (leaves), anthosphere (flowers), and carposphere (fruits). Endophyte communities are a key part of the plant microbiota, and their role in connecting the different plant microhabitats during development stages is of particular importance for plant health (Hardoim et al., 2015). The rhizosphere is an important reservoir for plant endophytes; it represents the below-ground interface with highly diverse, local soil microbiota (Philippot et al., 2013). The spermosphere is an important microhabitat during early plant development. It is the zone surrounding seeds where interactions between the soil, microbial communities, and germinating seeds take place (Schiltz et al., 2015). In addition, the airborne microbiome represents a further source of microorganisms for plant recruitment and a pathway for microbial dissemination (Abdelfattah et al., 2019). This microenvironment links the above- and below-ground microbiome of plants.

The factors that shape the plant microbiome have been studied for a long time. The plant genotype and soil quality have been identified as crucial factors influencing the composition of the rhizosphere microbiota (Berg and Smalla, 2009). Their impact depends on many factors (plant morphology, secondary metabolism, and soil type) and is triggered by plant root exudates and signaling (Doornbos et al., 2012). Other factors such as biotic interactions, plant diversity, climate, and agricultural practices influence the plant microbiota as well (Philippot et al., 2013). Exposure to pathogens as well as beneficial microorganisms can act as an important driver of microbiome composition (Rezki et al., 2016).

4.3 Plant microbiome assembly

For a long time, it was assumed that emerging plant seedlings were colonized by microbes from their surrounding environment. Soil was seen as the main source reinforced by plant-controlled enrichment through, for example, root exudates and the plant defense system (Sánchez-Cañizares et al., 2017). In the last decade, seeds have been discovered as a major source for transmitting plant-specific core microbiota (Adam et al., 2018; Johnston-Monje et al., 2016; Shade et al., 2017). Studies have revealed an unexpectedly high diversity and abundance of seed-associated microbiota (Berg and Raaijmakers, 2018). The current model of plant microbiome assembly suggests that both the seed and soil microbiota are able to colonize the plant seedling, promoting microbial diversity and function as well as adaptation to the local environment (Bergna et al., 2019). Plant-beneficial bacteria are enriched in the core microbiome, which is vertically transmitted to the next generation (Bergna et al., 2019). Recent research on the seed's role in vertical transmission across generations has shown high spatial partitioning of the fungal and bacterial community within both seed and seedling, indicating inheritance, niche differentiation, and divergent transmission routes for the establishment of root and phyllosphere communities (Abdelfattah et al., 2020; Wassermann et al., 2019). However, the mechanisms of species-specific plant microbiome assembly remain a significant knowledge gap.

4.4 Plant microbiome functioning

Plants and their associated microbes have interacted with each other for a long time, forming assemblages of species that are referred to as holobionts (Vandenkoornhuyse et al., 2015; Berg et al., 2016; Turner et al., 2013). The plant-associated microbiota supply multiple functions to the holobiont, including:

- Seed germination and growth;
- Nutrient supply;
- Resistance against biotic and abiotic stress factors; and
- Production of bioactive metabolites.

New functions of plant microbiota have recently been identified. The plant-associated microbiota are able to alter above-ground herbivory (Hol et al., 2010) and flowering time (Wagner et al., 2014), and are involved in cold acclimation of plants (Etemadi et al., 2018), as well as the attraction of predators and parasites of herbivores (Pieterse et al., 2013).

4.5 Interactions within a healthy plant microbiome

Microbes are embedded in complex networks, where symbiotic interactions have diverse consequences for microbial fitness, population dynamics, and functional capacities within each microbiome. These interactions can either be between microorganisms of the same species or between different species, genera, families, and domains. The interactive patterns within these webs may be positive, negative, or neutral. Antagonistic interactions are classified as negative microbial antagonists and are able to suppress pathogens. They are naturally occurring in the plant microbiome, and their proportion can account for up to 80% in natural environments (Berg 2009). Healthy plant microbiomes are characterized by a high microbial diversity and evenness of plant-associated microbial communities. All healthy plant microbiomes naturally contain potential pathogens (Wassermann et al., 2019). Manzotti et al. (2020) showed that after isolation and re-inoculation, they could exert pathogenicity on the host plant. This suggests that balance within the plant microbiome is crucial for plant health. Each healthy plant microbiome is characterized by keystone species within the core microbiome and a dense network of positive interactions among the constituents (Berg et al., 2020).

4.6 Communication within the plant microbiome

The cooperation between plants and millions of microbes as well as intra-microbiome interplay requires intense communication (Venturi and Keel, 2016). A high number of signatures encoding interactions via quorum sensing and other signaling molecules have been found in plant-associated microbial metagenomes (Bragina et al., 2014). However, the mechanism by which microorganisms interact as a community to benefit plants is still poorly understood. VOCs are responsible for 'microbial small talk' but can also act as long-distance messengers for interactions with the plant host (Schmidt et al., 2016). A review by Rosier et al. (2016) highlights the importance of chemical signaling and biochemical and genetic events which determine the efficacy of benign microbes in promoting the development of beneficial traits in plants. Middleton et al. (2020) suggested miRNAs as a key mediator for plant-microbe interactions. They suggest, as an alternative to microbial products, miRNAs as a promising tool for novel plant holobiont engineering.

5 The role of the plant microbiome in plant health and biocontrol

Biocontrol agents have been used to control plant diseases for decades (Wilson and Wisniewski, 1989). However, the host microbiome was often not considered in understanding the interaction between biocontrol agents and the native community (Droby et al., 2016; Droby and Wisniewski, 2018). This may be one reason for the inconsistent effect of many biocontrol agents. Studies conducted during the last years have provided the basis to change our perception of plants as independent organisms to a more complex model that includes their microbiota. When plants are considered as holobionts, it becomes clear that the health and fitness of the whole system rely on each of its components. Pathogens as well as beneficial microorganisms induce changes in the holobiont when they are introduced. These changes remain to be studied in more detail in order to make use of bioprotectants that exert their function at the holobiont level, that is, interacting with plants as well as their microbiota. Success in introducing allochthonous microorganisms in such complex systems depends on many factors. The autochthonous microbiome can be a barrier or even 'protection shield' of the holobiont (Weller et al., 2002). Microbial diversity is a key factor (Van Elsas et al., 2012). Adam et al. (2016) suggest the idea of a minor disturbance combined with timely application of bioprotectants to improve their establishment.

5.1 All plant diseases are connected to a microbiome shift commonly termed as dysbiosis

The classification of microorganisms as beneficial, pathogenic, and neutral (based on microbial interactions with their hosts) is a part of the widely cited microbiome definition by Lederberg and McCray (2001) and Whipps et al. (1988). Categorizing pathogenic, beneficial, and neutral microorganisms according to their interaction with their host may be useful for studies where microbe-host interactions play a central role in mediating host fitness, such as plant pathology studies. However, the interpretation of pathogenicity should always be done with caution. Recent studies on soil-borne and opportunistic pathogens showed that host-microbe interactions not only depend on the status of the host but also on the entire microbiome (Berg et al., 2017; Lee et al., 2020). The functioning of naturally disease-suppressive soils may act in a similar way (Mendes et al., 2011). Instead of solely focusing on the interaction of one specific microorganism with its host, holistic approaches based on the holobiont theory should be taken into consideration.

The beneficial interplay of the host and its microbiome is essential for maintaining the health of the holobiont, while diseases are often correlated with microbial dysbiosis (Fig. 3). The concept of 'pathobiome', integrating pathogenic agents within biotic environments, has been applied to multiple pathosystems (Vayssier-Taussat et al., 2014). Analyses of pathobiomes have often revealed multiple pathogens embedded in a severe dysbiosis. Another interesting interpretation of microbial-host interactions is a so-called Anna Karenina principle (Zaneveld et al., 2017). It states that, paralleling Leo Tolstoy's dictum, 'all happy families look alike; each unhappy family is unhappy in its own way', dysbiotic individuals vary more in microbial community composition than

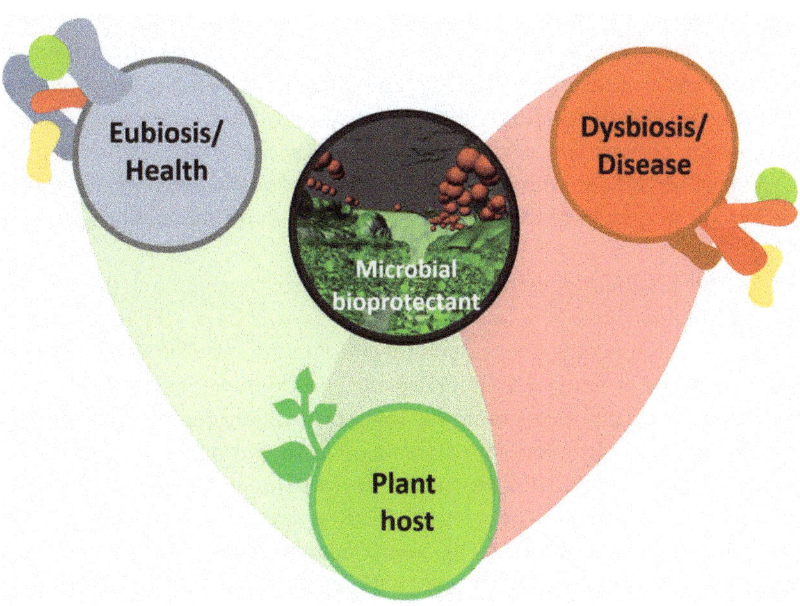

Figure 3 Differences in the plant microbiome between health and disease. Plant diseases are always connected to a dysbiosis of the microbiome.

healthy individuals. For instance, infestations by *Bactrocera oleae* in olives were found to alter the natural fungal community into a dysbiotic, unique in every geographical location (Abdelfattah et al., 2018). These results also suggest that microbial diversity is a key factor in preventing diseases in plants because it avoids outbreaks as well as invasions of pathogens (Berg et al., 2017). Despite numerous studies, defining a 'healthy microbiota' and the difference between eubiosis and dysbiosis remains a major challenge.

5.2 The plant microbiome is affected by the application of bioprotectants

Next-generation sequencing techniques have shown that microbial treatments are able to shape the plant microbiome. This has provided new insights into the modes of action of various products, highlighting the importance of microbiome-bioprotectant interactions. Recent findings have shown that both agrochemicals and bioprotectants induce changes in the plant microbiome (Cernava et al., 2019a; Guimarães et al., 2020). Examples of interactions of bioprotectants with the plant microbiome are listed in Table 1.

Microbiome modulation by bioprotectants affects the holobiont, as recently reviewed by Berg et al. (2021). One fragile phase during plant development is the flowering period (Berg, 2015). The stabilization of the microbiome during the flowering period was observed after an inoculation of cucumber with *Bacillus subtilis* B579 (Chen et al., 2013). In this study, *B. subtilis* increased the microbial diversity transiently by promoting the secretion of root exudates during flowering. This resulted in enhanced protection against the pathogen *Fusarium oxysporum* f. sp. *cucumeris*. Disease suppression of *Fusarium wilt* caused by *F. oxysporum* f. sp. *cubense* (FOC) was also observed after inoculation with *B. amyloliquefaciens*, which indirectly supported the growth of the indigenous *Pseudomonas* population (Tao et al., 2020).

Another holobiont-level mode of action protecting cucumber is the induction of a microbiome shift by *Pseudomonas fluorescens* 2P24 or CPF10 that modify the bacterial community composition toward *Bacillus* spp. (Yin et al., 2013). An additional mode of action was observed with *Pseudomonas jessenii* RU47, which stabilized the host plant's microbiome. The bioprotectant counteracted diversity losses that were observed in combination with *Rhizoctonia solani* AG1-IB infection (Adesina et al., 2009). A similar effect was found in combination with *R. solani* AG1-IB following co-inoculation of lettuce with *Serratia plymuthica* 3Re4-18 and *Trichoderma viride* GB7. The combination of these two strains enhanced biocontrol efficacy and balance within the plant microbiome (Grosch et al., 2012). This was observed in tomato plants infected with *Ralstonia solanacearum* B3B, following inoculation with *Bacillus velezensis* B63 or *P. fluorescens* P142.

Here, a microbiome shift occurred toward plant growth-promoting bacterial phyla (*Actinobacteria*, *Verrucomicrobia*) (Elsayed et al., 2020).

Fusarium stalk rot in maize caused by *Fusarium graminearum* can be controlled via application of *Trichoderma harzianum* CCTCC-RW0024, which resulted in an increase of plant growth-promoting *Acidobacteria* (Saravanakumar et al., 2017). A significant increase in microbial activity was also observed following the inoculation of pepper plants infected with *Phytophthora capsici* with *Pseudomonas corrugata* CCR80 or *Chryseobacterium indologenes* ISE14. Both biocontrol agents increased microbial diversity and induced microbial shifts toward *Pseudomonads* and *Actinomycetes* that are well-known for their antagonistic potential (Sang and Kim, 2012). When a bacterial consortium consisting of the PGPR strains *Bacillus cereus* AR156, *B. subtilis* SM21, and *Serratia* sp. XY21 was used for the inoculation of sweet pepper infected with *P. capsici*, disease symptoms were significantly reduced. The consortium significantly increased the abundance of *Burkholderia*, *Comamonas*, and *Ramlibacter*. *Burkholderia* spp. are common antagonists of various plant pathogens, *Comamonas* spp. commonly have antifungal effects, and *Ramlibacter* spp. potentially improve plant growth under adverse conditions (Zhang et al., 2019).

The extent to which bioprotectants affect the plant microbiome can also differ. While *Stenotrophomonas rhizophila* SPA P69 is only able to suppress minor fungal pathogens within the rhizosphere microbiome (Schmidt et al., 2012), *B. velezenzis* FZB42 was observed to generally increase the microbial diversity of the holobiont, which led to the identification of an advanced mode of action (Erlacher et al., 2014). Similar results were obtained by Chowdhury et al. (2013) who showed that the treatment with FZB42 resulted in a shift of the indigenous rhizosphere bacterial community, although the inoculant was identified only temporarily. Lee et al. (2020) were the first to demonstrate that dysbiosis of the protective gram-positive bacterial community in diseased tomato plants promotes the incidence of disease. This underlines the importance of microbiome modulation by bioprotectants as a novel mode of action.

5.3 The cultivar-specific plant microbiome and bioprotectants

Plant and cultivar-specific signatures in the microbiota were discovered 20 years ago (Berg and Smalla, 2009). However, the full impact of breeding on the plant microbiome was only recently discovered (Pérez-Jaramillo et al., 2018). Breeding not only changes plant metabolism and morphology; it also changes the composition of the microbiota by affecting up to 50% of the naturally occurring members of the community (Cardinale et al., 2012). Breeding is often directed at increasing yield or plant resistance to pathogens and pests. Both

Table 1 Bioprotectants that were shown to interact with the plant microbiome

Bioprotectant	Pathogen	Impact on the microbiome	References
Pseudomonas jessenii RU47	*R. solani* AG1-IB	Stabilization of microbial diversity	Adesina et al. (2009)
Stenotrophomonas rhizophila P69	*no*	Microbiome shift, reduction of potential fungal pathogens	Schmidt et al. (2012)
Serratia plymuthica 3RE4-18 *and Trichoderma viride* GB7	*R. solani* AG1-IB	Stabilization of microbial diversity	Grosch et al. (2012)
Chryseobacterium indologenes SE14	*Phytophthora capsici*	Increasing microbial activity and diversity, modification interaction with the microbiome of bacterial community toward *Pseudomonads* and *Actinomycetes*	Sang and Kim (2012)
Pseudomonas corrugata CCR80	*P. capsici*	Increasing microbial activity and diversity, modification of bacterial community toward *Pseudomonads* and *Actinomycetes*	Sang and Kim (2012)
Bacillus subtilis B579	*Fusarium oxysporum* f. sp. *cucumerinum*	Increasing microbial biodiversity during flowering	Chen et al. (2013)
Pseudomonas fluorescens 2P24	*Ralstonia solanacearum*, *F. oxysporum*, *Rhizoctonia solani*	Modification of bacterial community toward *Bacillus* spp.	Yin et al. (2013)
P. fluorescens CPF10	*Rastonia solanacearum*, *F. oxysporum*, *R. solani*	Modification of bacterial community toward *Bacillus* spp.	Yin et al. (2013)
Bacillus amyloliquefaciens FZB42	*R. solani* AG1-IB	Microbiome shift	Chowdhury et al. (2013;Kröber et al. (2014)
B. amyloliquefaciens FZB42	*R. solani* AG1-IB	Increase microbial diversity, modification of bacterial community toward *Acinetobacter* spp. and *Alkanindiges* spp.	Erlacher et al. (2014)

Trichoderma harzianum CCTCC-RW0024	*Fusarium graminearum*	Modification of bacterial community toward *Acidobacteria*	Saravanakumar et al. (2017)
Bacillus cereus AR156, *B. subtilis* SM21, *Serratia sp.* XY21	*P. capsici*	Modification of bacterial community toward *Burkholderia* spp., *Comamonas* spp., *Ramlibacter* spp.	Zhang et al. (2019)
Bacillus velezensis B63	*R. solanacearum* B3B	Modification of bacterial community toward *Actinobacteria*, *Verrucomicrobia*	Elsayed et al. (2020)
P. fluorescens P142	*R. solanacearum* B3B	Modification of bacterial community toward *Actinobacteria*, *Verrucomicrobia*	Elsayed et al. (2020)
B. amyloliquefaciens W19	*Fusarium wilt* pathogen *F. oxysporum f.* sp. *cubense (FOC)*	Increase of indigenous *Pseudomonas* spp.	Tao et al. (2020)
Brevibacterium frigoritolerans HRS1, *Bacillus niacini* HRS2, *Solibacillus silvestris* HRS3, *Bacillus luciferensis* HRS4	*R. solanacearum*	Consortium of functional guilds of healthy plants restoring the dysbiosis	Lee et al. (2020)

yield and resistance can be mediated by plant-associated microorganisms. There are several studies showing the correlation between cultivars and the resistance toward specific pathogens and the microbiota, for example in the pathosystems of pumpkin-*Erwinia* (Adam et al., 2018), beans-*Fusarium* (Mendes et al., 2019), or sugar beet-*Rhizoctonia* (Wolfgang et al., 2020). *Rhizoctonia*-tolerant cultivars of sugar beet, mainly mediated by Fort Collins Resistance, resulted in an enrichment of bioactive *Pseudomonas* strains in the rhizosphere. Strains of *Pseudomonas* are also able to directly induce this resistance. This strongly suggests integrating microbiome research into breeding strategies. This will facilitate the development of synergistic strategies that will address the plant holobiont as a whole.

6 Consequences for discovery and application of bioprotectants

The plant microbiome represents species- and habitat-specific microbial ecosystems containing thousands of microbial species. Intensive agriculture, especially through breeding and chemical use, has changed microbial populations and reduced microbial diversity. Microbial diversity can be restored with microbial products. Microbiome shifts, and enhancing microbial diversity in particular, have been identified as a new mode of action of bioprotectants. Understanding the impacts of bioprotectants on the plant holobiont has highlighted various crucial factors that must be considered in the future to improve the functionality and reliability of new products. Box 2 summarizes key factors in the discovery, application, and risk assessment of potential bioprotectants.

Box 2 Consequences for discovery, application, and risk assessment of bioprotectants

Discovery and evaluation of new bioprotectants

New insights into the plant microbiome have revealed an impressive microbial diversity among plant species. This provides new strategies to isolate and enrich bioprotectants from microbial hotspots identified in the plant kingdom, for example, mosses (Bragina et al., 2012), medicinal plants (Köberl et al., 2013), suppressive soils, or wild plant relatives (Pérez-Jaramillo et al., 2018). The oldest technique for studying microorganisms by cultivation should be extended (Berg et al., 2020).

Integrating microbiome-based concepts, for example, (i) identifying keystone species and using them for microbiome modulation, (ii) exploiting network analysis and co-occurrence patterns, (iii) using the core microbiome for identification of beneficials, or (iv) comparing healthy and diseased plant microbiomes and developing methods for microbiome tracking (Erlacher et al., 2014; Kusstatscher et al., 2019) offer novel possibilities for bioprotectant discovery.

Toju et al. (2020) proposed a framework for exploring 'functional core' species that have a substantial impact on processes and functions in the whole community. They identified functional core microbes to modulate rhizosphere microbiomes by suppressing pathogens.

Microbial diversity can be implemented as a biomarker in discovery strategies (Berg et al., 2017). Due to the fact that plant-microbial diversity is associated with plant health by providing a defense and interaction network, this parameter can be easily measured in all biocontrol experiments by analyzing the microbiome.

Most of the potential biocontrol agents are still screened in terms of *in vitro* antagonism against the target pathogen. Despite the fact that this led to the detection of promising biocontrol agents in the past, the majority of them were not successful *in situ* (Adesina et al., 2009; Schreiter et al., 2014; Rybakova et al., 2016). Evaluation of potential plant growth-promoting and biocontrol strains must be conducted in natural pathosystems.

The mode of action should be analyzed by combining different available tools. While *in vitro* assays often indicate antibiosis effects, (meta)transcriptomic analyses can reveal more complex interactions and general stress responses of the holobiont. *In vitro* results often do not correlate with the *in situ* production of distinct effectors.

Application of bioprotectants

Microbial consortia based on different combinations of modes of action are a promising approach for plant disease management in the future (Massart et al., 2015). Unraveling complex modes of action helps the development of appropriate formulations and methods of application (Spadaro and Droby, 2016).

Seed application is a promising strategy to apply bioprotectants; after germination, they can sustainably colonize the plant's microhabitats (Berg and Raaijmakers, 2018).

Long-term establishment of plant-associated microorganisms depends on concepts from invasion ecology, which suggest that survival rates of invaders are inversely related to the diversity of the native microbiome (Adam et al., 2016).

Evaluation of bioprotectants

Most genomes and MAGs (Wicaksono et al., 2020) of plant-associated microorganisms analyzed so far contain antibiotic gene clusters as well as genes encoding antimicrobial resistance (Raaijmakers and Mazzola, 2012; Obermeier et al., 2020). This requires re-thinking the importance of antibiosis which, from the current scientific point of view, seems to be overestimated.

In contrast, the induction of shifts in the plant or even soil microbiota seems to be an underestimated mode of action. It can be found in the most beneficial plant-microbe interactions based on the introduction of beneficial microorganisms.

The current classification of bioprotectants and biostimulants according to their mode of action is not reflected by genomic, MAG, and metagenomic studies. These indicate a more complex interplay that includes different modes of action. Depending on environmental conditions, different genes and traits can be expressed, enabling a flexible response of the plant microbiome to environmental conditions.

Analyses of the modes of action should consider microbial bioprotectants in combination with the microbiome, rather than as an isolated microorganism or its genome.

Inter-connected microbiomes, the role of microbial diversity, as well as the *One health* or planetary health approach should be included in the evaluation of bioprotectants.

7 Conclusion and future trends

Targeted manipulation of the plant microbiome with bioprotectants will provide the key to increasing the sustainability of agricultural practices in the future. Advances will rely on technological developments, especially in next-generation sequencing and other high-throughput analyses for *in situ* studies of plant holobionts. Various molecular techniques have recently become established, resulting in substantial improvement in understanding of modes of action of bioprotectants (Spadaro and Droby, 2016). They have also uncovered various intrinsic characteristics of plant-associated microorganisms that remained mostly hidden until the advent of metagenomic analyses.

Natural plant microbiomes typically harbor a highly diverse repertoire of antibiotic resistance genes (ARGs) typically present in low numbers (Martínez, 2008; Cernava et al., 2019b). Such resistomes can harbor ARGs which act as an 'insurance system' of the microbiome. The presence and activity of such resistomes has been confirmed by isolate screening and metagenomic clone libraries, with the latter specifically implemented for uncultivable microorganisms (Obermeier et al., 2020). It remains unknown if naturally occurring resistomes can be triggered by agricultural inputs. However, it can be assumed that conventional chemicals have a higher potential to evoke new multi-resistant pathogens than bioprotectants that rely on more complex modes of action, rather than long-time exposure to a single bioactive compound. Another important challenge in the development of bioprotectants is their consistency and efficacy under field conditions. A better understanding of the plant holobiont and its interaction with distinct inoculants will facilitate the establishment of bioprotectants that complement the functioning of the native microbiome. They will induce targeted shifts in the holobiont, with the aim of enhancing indigenous microbial diversity, reducing pathogen impact, and promoting indigenous beneficial microorganisms (Berg, 2015).

8 Where to look for further information

Research on the plant microbiota and their internal and external networking will influence our understanding of the mode of action of bioprotectants. The future challenge will be to assess these results and transform them into science-based regulations and registration procedures. This needs interdisciplinary work and involvement of the society as well. International networks such as the International Organisation for Biological Control (IOBC; https://www.iobc-wprs.org/) and the IBMA, the International Biocontrol Manufacturers Association (https://ibma-global.org/), provide further information.

9 Acknowledgement

The authors received funding from the European Union's Horizon 2020 research and innovation program under grant agreement No. 817946 (Excalibur) [G.B.] and by European Union through Marie Skłodowska-Curie Actions - Individual Fellowships (MSCA-IF) grant agreement 844114 [A.A.]. We have received support from the ERA-NET SusCrop project 'potatoMETAbiome' funded by the Federal Ministry Republic of Austria for Agriculture, Regions and Tourism (project No. 101392) and from The COMET center acib: Next Generation Bioproduction funded by BMK, BMDW, SFG, Standortagentur Tirol, Government of Lower Austria und Vienna Business Agency in the framework of COMET - Competence Centers for Excellent Technologies, both to [G.B.]. The COMET-Funding Program is managed by the Austrian Research Promotion Agency FFG.

10 References

Abdelfattah, A., Ruano-Rosa, D., Cacciola, S. O., Li Destri Nicosia, M. G. and Schena, L. (2018). Impact of *Bactrocera oleae* on the fungal microbiota of ripe olive drupes, *PLoS ONE* 13(11), e0199403.

Abdelfattah, A., Sanzani, S. M., Wisniewski, M., Berg, G., Cacciola, S. O. and Schena, L. (2019). Revealing cues for fungal interplay in the plant-air interface in vineyards, *Frontiers in Plant Science* 10, 922.

Abdelfattah, A., Wisniewski, M., Schena, L. and Tack, A. J. M. (2020). Experimental evidence of microbial inheritance in plants and transmission routes from seed to phyllosphere and root, *Environmental Microbiology* 23(4), 2199-2214. doi: 10.1111/1462-2920.15392.

Adam, E., Groenenboom, A. E., Kurm, V., Rajewska, M., Schmidt, R., Tyc, O., Weidner, S., Berg, G., de Boer, W. and Falcão Salles, J. (2016). Controlling the microbiome: microhabitat adjustments for successful biocontrol strategies in soil and human gut, *Frontiers in Microbiology* 7, 1079.

Adam, E., Bernhart, M., Müller, H., Winkler, J. and Berg, G. (2018). The *Cucurbita pepo* seed microbiome: genotype-specific composition and implications for breeding, *Plant and Soil* 422(1-2), 35-49.

Adesina, M. F., Grosch, R., Lembke, A., Vatchev, T. D. and Smalla, K. (2009). *In vitro* antagonists of *Rhizoctonia solani* tested on lettuce: rhizosphere competence, biocontrol efficiency and rhizosphere microbial community response, *FEMS Microbiology Ecology* 69(1), 62-74.

Alavi, P., Starcher, M. R., Thallinger, G. G., Zachow, C., Müller, H. and Berg, G. (2014). *Stenotrophomonas* comparative genomics reveals genes and functions that differentiate beneficial and pathogenic bacteria, *BMC Genomics* 15(1), 482.

Amann, R. I., Ludwig, W. and Schleifer, K. H. (1995). Phylogenetic identification and *in situ* detection of individual microbial cells without cultivation, *Microbiological Reviews* 59(1), 143-169.

Bakker, P. A. H. M., Pieterse, C. M. J. and van Loon, L. C. (2007). Induced systemic resistance by fluorescent *Pseudomonas* spp., *Phytopathology* 97(2), 239-243.

Bakker, P. A. H. M., Berendsen, R. L., van Pelt, J. A., Vismans, G., Yu, K., Li, E., Van Bentum, S., Poppeliers, S. W. M., Sanchez Gil, J. J., Zhang, H., Goossens, P., Stringlis, I. A., Song, Y., de Jonge, R. and Pieterse, C. M. J. (2020). The soil-borne identity and microbiome-assisted agriculture: looking back to the future, *Molecular Plant* 13(10), 1394-1401.

Banerjee, M. R. and Yesmin, L. (2009). Sulfur-oxidizing plant growth promoting rhizobacteria for enhanced canola performance, U.S. Patent No. 7,491,535, Washington, DC: U.S. Patent and Trademark Office.

Bardin, M., Ajouz, S., Comby, M., Lopez-Ferber, M., Graillot, B., Siegwart, M. and Nicot, P. C. (2015). Is the efficacy of biological control against plant diseases likely to be more durable than that of chemical pesticides?, *Frontiers in Plant Science* 6, 566.

Bérdy, J. (2005). Bioactive microbial metabolites, *The Journal of Antibiotics* 58(1), 1-26.

Berg, G. (2009). Plant-microbe interactions promoting plant growth and health: perspectives for controlled use of microorganisms in agriculture, *Applied Microbiology and Biotechnology* 84(1), 11-18.

Berg, G. and Smalla, K. (2009). Plant species and soil type cooperatively shape the structure and function of microbial communities in the rhizosphere, *FEMS Microbiology Ecology* 68(1), 1-13.

Berg, G. (2015). Beyond borders: investigating microbiome interactivity and diversity for advanced biocontrol technologies, *Microbial Biotechnology* 8(1), 5-7.

Berg, G., Rybakova, D., Grube, M. and Köberl, M. (2016). The plant microbiome explored: implications for experimental botany, *Journal of Experimental Botany* 67(4), 995-1002.

Berg, G., Köberl, M., Rybakova, D., Müller, H., Grosch, R. and Smalla, K. (2017). Plant microbial diversity is suggested as the key to future biocontrol and health trends, *FEMS Microbiology Ecology* 93(5), fix050.

Berg, G. and Raaijmakers, J. M. (2018). Saving seed microbiomes, *The ISME Journal* 12(5), 1167-1170.

Berg, G., Rybakova, D., Fischer, D., Cernava, T., Vergès, M. C., Charles, T., Chen, X., Cocolin, L., Eversole, K., Corral, G. H., Kazou, M., Kinkel, L., Lange, L., Lima, N., Loy, A., Macklin, J. A., Maguin, E., Mauchline, T., McClure, R., Mitter, B., Ryan, M., Sarand, I., Smidt, H., Schelkle, B., Roume, H., Kiran, G. S., Selvin, J., Souza, R. S. C., van Overbeek, L., Singh, B. K., Wagner, M., Walsh, A., Sessitsch, A. and Schloter, M. (2020). Microbiome definition re-visited: old concepts and new challenges, *Microbiome* 8(1), 103.

Berg, G., Kusstatscher, P., Abdelfattah, A., Cernava, T. and Smalla, K. (2021). Microbiome modulation–toward a better understanding of plant microbiome response to microbial inoculants, *Frontiers in Microbiology* 12, 650610.

Bergna, A., Cernava, T., Rändler, M., Grosch, R., Zachow, C. and Berg, G. (2019). Tomato seeds preferably transmit plant beneficial endophytes, *Phytobiomes Journal* 2(4), 183–193.

Bragina, A., Berg, C., Cardinale, M., Shcherbakov, A., Chebotar, V. and Berg, G. (2012). *Sphagnum* mosses harbour highly specific bacterial diversity during their whole lifecycle, *The ISME Journal* 6(4), 802–813.

Bragina, A., Oberauner-Wappis, L., Zachow, C., Halwachs, B., Thallinger, G. G., Müller, H. and Berg, G. (2014). The *Sphagnum* microbiome supports bog ecosystem functioning under extreme conditions, *Molecular Ecology* 23(18), 4498–4510.

Burch, A. Y., Zeisler, V., Yokota, K., Schreiber, L. and Lindow, S. E. (2014). The hygroscopic biosurfactant syringafactin produced by *Pseudomonas syringae* enhances fitness on leaf surfaces during fluctuating humidity, *Environmental Microbiology* 16(7), 2086–2098.

Cardinale, M., Grube, M., Castro Jr., J. V., Müller, H. and Berg, G. (2012). Bacterial taxa associated with the lung lichen *Lobaria pulmonaria* are differentially shaped by geography and habitat, *FEMS Microbiology Letters* 329(2), 111–115.

Cernava, T., Chen, X., Krug, L., Li, H., Yang, M. and Berg, G. (2019a). The tea leaf microbiome shows specific responses to chemical pesticides and biocontrol applications, *Science of the Total Environment* 667, 33–40.

Cernava, T., Erlacher, A., Soh, J., Sensen, C. W., Grube, M. and Berg, G. (2019b). *Enterobacteriaceae* dominate the core microbiome and contribute to the resistome of arugula (*Eruca sativa* Mill.), *Microbiome* 7(1), 13.

Chen, F., Wang, M., Zheng, Y., Li, S., Wang, H., Han, D. and Guo, S. (2013). The effect of biocontrol bacteria on rhizosphere bacterial communities analyzed by plating and PCR-DGGE, *Current Microbiology* 67(2), 177–182.

Chowdhury, S. P., Dietel, K., Rändler, M., Schmid, M., Junge, H., Borriss, R., Hartmann, A. and Grosch, R. (2013). Effects of *Bacillus amyloliquefaciens* FZB42 on lettuce growth and health under pathogen pressure and its impact on the rhizosphere bacterial community, *PLoS ONE* 8(7), e68818.

Compant, S., Nowak, J., Coenye, T., Clèment, C. and Barka, E. A. (2008). Diversity and occurrence of *Burkholderia* spp. in the natural environment, *FEMS Microbiology Reviews* 32(4), 607–626.

Conrath, U., Beckers, G. J. M., Langenbach, C. J. G. and Jaskiewicz, M. R. (2015). Priming for enhanced defense, *Annual Review of Phytopathology* 53, 97–119.

Crutzen, P. J. (2002). The "Anthropocene", *Journal de Physique IV* 12(10), 1–5.

Dobbelaere, S., Vanderleyden, J. and Okon, Y. (2003). Plant growth-promoting effects of diazotrophs in the rhizosphere, *Critical Reviews in Plant Sciences* 22(2), 107–149.

Dobbelaere, S. and Okon, Y. (2007). The plant growth-promoting effect and plant responses. In: Elmerich, C. and Newton, W. E. (Eds), *Associative and Endophytic Nitrogen-Fixing Bacteria and Cyanobacterial Associations*, Dordrecht: Springer, pp. 145–170.

Doornbos, R. F., van Loon, L. C. and Bakker, P. A. H. M. (2012). Impact of root exudates and plant defense signaling on bacterial communities in the rhizosphere: a review, *Agronomy for Sustainable Development* 32(1), 227–243.

Droby, S., Wisniewski, M., Teixidó, N., Spadaro, D. and Jijakli, M. H. (2016). The science, development, and commercialization of postharvest biocontrol products, *Postharvest Biology and Technology* 122, 22–29.

Droby, S. and Wisniewski, M. (2018). The fruit microbiome: a new frontier for postharvest biocontrol and postharvest biology, *Postharvest Biology and Technology* 140, 107–112.

Duffy, B., Schouten, A. and Raaijmakers, J. M. (2003). Pathogen self-defense: mechanisms to counteract microbial antagonism, *Annual Review of Phytopathology* 41(1), 501–538.

Dunham Trimmer, L. L. C. (2017). Biological control global market overview. Available at: http://wrir4.ucdavis.edu/events/2017_SLR_Meeting/Presentations/GeneralPresentations/1%20Trimmer%20-%20Global%20Biocontrol%20Market%202017.pdf; accessed 12 August 2019.

Dupré, J. and O'Malley, M. A. (2013). Varieties of living things: life at the intersection of lineage and metabolism. In: Normandin, S. and Wolfe, C. T. (Eds), *Vitalism and the Scientific Image in Post-Enlightenment Life Science, 1800–2010*, Dordrecht: Springer, pp. 311–343.

Elsayed, T. R., Jacquiod, S., Nour, E. H., Sørensen, S. J. and Smalla, K. (2020). Biocontrol of bacterial wilt disease through complex interaction between tomato plant, antagonists, the indigenous rhizosphere microbiota and *Ralstonia solanacearum*, *Frontiers in Microbiology* 10, 2835.

Erlacher, A., Cardinale, M., Grosch, R., Grube, M. and Berg, G. (2014). The impact of the pathogen *Rhizoctonia solani* and its beneficial counterpart *Bacillus amyloliquefaciens* on the indigenous lettuce microbiome, *Frontiers in Microbiology* 5, 175.

Etemadi, M., Zuther, E., Müller, H., Hincha, D. K. and Berg, G. (2018). Ecotype-dependent response of bacterial communities associated with *Arabidopsis* to cold acclimation, *Phytobiomes Journal* 2(1), 3–13.

Flandroy, L., Poutahidis, T., Berg, G., Clarke, G., Dao, M. C., Decaestecker, E., Furman, E., Haahtela, T., Massart, S., Plovier, H., Sanz, Y. and Rook, G. (2018). The impact of human activities and lifestyles on the interlinked microbiota and health of humans and of ecosystems, *Science of the Total Environment* 627, 1018–1038.

Glick, B. R. (2005). Modulation of plant ethylene levels by the bacterial enzyme ACC deaminase, *FEMS Microbiology Letters* 251(1), 1–7.

Grosch, R., Dealtry, S., Schreiter, S., Berg, G., Mendonça-Hagler, L. and Smalla, K. (2012). Biocontrol of *Rhizoctonia solani*: complex interaction of biocontrol strains, pathogen and indigenous microbial community in the rhizosphere of lettuce shown by molecular methods, *Plant and Soil* 361(1–2), 343–357.

Guimarães, R. A., Pherez-Perrony, P. E., Müller, H., Berg, G., Medeiros, F. H. V. and Cernava, T. (2020). Microbiome-guided evaluation of *Bacillus subtilis* BIOUFLA2 application to reduce mycotoxins in maize kernels, *Biological Control* 150, 104370.

Haas, D. and Défago, G. (2005). Biological control of soil-borne pathogens by fluorescent pseudomonads, *Nature Reviews. Microbiology* 3(4), 307–319.

Hardoim, P. R., van Overbeek, L. S., Berg, G., Pirttilä, A. M., Compant, S., Campisano, A., Döring, M. and Sessitsch, A. (2015). The hidden world within plants: ecological and evolutionary considerations for defining functioning of microbial endophytes, *Microbiology and Molecular Biology Reviews* 79(3), 293–320.

Hiltner, L. (1902). *Die Keimungsverhältnisse der Leguminosensamen und ihre Beeinflussung durch Organismenwirkung*, Heidelberg: Springer-Verlag.

Hol, W. H. G., de Boer, W., Termorshuizen, A. J., Meyer, K. M., Schneider, J. H. M., van Dam, N. M., van Veen, J. A. and van der Putten, W. H. (2010). Reduction of rare soil microbes modifies plant-herbivore interactions, *Ecology Letters* 13(3), 292–301.

Jan, A. T., Azam, M., Ali, A. and Mohd, Q. M. R. (2011). Novel approaches of beneficial *Pseudomonas* in mitigation of plant diseases–an appraisal, *Journal of Plant Interactions* 6(4), 195–205.

Jansson, J. K. and Baker, E. S. (2016). A multi-omic future for microbiome studies, *Nature Microbiology* 1(5), 16049.

Jeffries, P. (1995). Biology and ecology of mycoparasitism, *Canadian Journal of Botany* 73(S1), 1284–1290.

Johnston-Monje, D., Lundberg, D. S., Lazarovits, G., Reis, V. M. and Raizada, M. N. (2016). Bacterial populations in juvenile maize rhizospheres originate from both seed and soil, *Plant and Soil* 405(1–2), 337–355.

Katiyar, V. and Goel, R. (2004). Siderophore mediated plant growth promotion at low temperature by mutant of fluorescent pseudomonads, *Plant Growth Regulation* 42(3), 239–244.

Köberl, M., Ramadan, E. M., Adam, M., Cardinale, M., Hallmann, J., Heuer, H., Smalla, K. and Berg, G. (2013). *Bacillus* and *Streptomyces* were selected as broad-spectrum antagonists against soilborne pathogens from arid areas in Egypt, *FEMS Microbiology Letters* 342(2), 168–178.

Koch, E., Becker, J. O., Berg, G., Hauschild, R., Jehle, J., Köhl, J. and Smalla, K. (2018). Biocontrol of plant diseases is not an unsafe technology!, *Journal of Plant Diseases and Protection* 125(2), 121–125.

Köhl, J. and Fokkema, N. J. (1998). Strategies for biological control of necrotrophic fungal foliar pathogens. In: Boland, G. L. and Kuykendall, D. L. (Eds), *Plant–Microbe Interactions and Biological Control*, New York: Marcell Dekker Inc., pp. 49–88.

Köhl, J., Kolnaar, R. and Ravensberg, W. J. (2019). Mode of action of microbial biological control agents against plant diseases: relevance beyond efficacy, *Frontiers in Plant Science* 10, 845.

Kröber, M., Wibberg, D., Grosch, R., Eikmeyer, F., Verwaaijen, B., Chowdhury, S. P., Hartmann, A., Pühler, A. and Schlüter, A. (2014). Effect of the strain *Bacillus amyloliquefaciens* FZB42 on the microbial community in the rhizosphere of lettuce under field conditions analyzed by whole metagenome sequencing, *Frontiers in Microbiology* 5, 252.

Kunz, S., Schmitt, A. and Haug, P. (2010). Fire blight control strategies in organic fruit growing, Fördergemeinschaft Ökologischer Obstbau e.V. (FÖKO), Growing, Germany: 14th International Conference on cultivation technique and phytopathological problems in organic fruit, pp. 118–125.

Kusstatscher, P., Zachow, C., Harms, K., Maier, J., Eigner, H., Berg, G. and Cernava, T. (2019). Microbiome-driven identification of microbial indicators for postharvest diseases of sugar beets, *Microbiome* 7(1), 112.

Lederberg, J. and McCray, A. T. (2001). Ome sweet. *Omics–A Genealogical Treasury of Words, The Scientist* 15(7), 8–8.

Lee, S. M., Kong, H. G., Song, G. C. and Ryu, C. M. (2020). Disruption of Firmicutes and Actinobacteria abundance in tomato rhizosphere causes the incidence of bacterial wilt disease, *The ISME Journal* 15(1), 330–347.

Linares, J. F., Gustafsson, I., Baquero, F. and Martinez, J. L. (2006). Antibiotics as intermicrobial signaling agents instead of weapons, *Proceedings of the National Academy of Sciences of the United States of America* 103(51), 19484 –19489.

Lugtenberg, B. J., Chin-A-Woeng, T. F. and Bloemberg, G. V. (2002) Microbe-plant interactions: principles and mechanisms, *Antonie van Leeuwenhoek* 81(1-4), 373-383.

Lugtenberg, B. and Kamilova, F. (2009). Plant-growth-promoting rhizobacteria, *Annual Review of Microbiology* 63, 541-556.

Manzotti, A., Bergna, A., Burow, M., Jørgensen, H. J. L., Cernava, T., Berg, G., Collinge, D. B. and Jensen, B. (2020). Insights into the community structure and lifestyle of the fungal root endophytes of tomato by combining amplicon sequencing and isolation approaches with phytohormone profiling, *FEMS Microbiology Ecology* 96(5), fiaa052.

Martínez, J. L. (2008). Antibiotics and antibiotic resistance genes in natural environments, *Science* 321(5887), 365-367.

Massart, S., Margarita, M. M. and Jijakli, M. H. (2015). Biological control in the microbiome era: challenges and opportunities, *Biological Control* 89, 98-108.

Mendes, R., Kruijt, M., de Bruijn, I., Dekkers, E., van der Voort, M., Schneider, J. H., Piceno, Y. M., DeSantis, T. Z., Andersen, G. L., Bakker, P. A. and Raaijmakers, J. M. (2011). Deciphering the rhizosphere microbiome for disease-suppressive bacteria, *Science* 332(6033), 1097-1100.

Mendes, L. W., de Chaves, M. G., Fonseca, M. C., Mendes, R., Raaijmakers, J. M. and Tsai, S. M. (2019). Resistance breeding of common bean shapes the physiology of the rhizosphere microbiome, *Frontiers in Microbiology* 10, 2252.

Middleton, H., Yergeau, É., Monard, C., Combier, J. P. and El Amrani, A. (2020). Rhizospheric plant-microbe interactions: miRNAs as a key mediator, *Trends in Plant Science* 26(2), 132-141. doi: 10.1016/j.tplants.2020.09.005.

Molina-Favero, C., Creus, C. M., Simontacchi, M., Puntarulo, S. and Lamattina, L. (2008). Aerobic nitric oxide production by *Azospirillum brasilense* Sp245 and its influence on root architecture in tomato, *Molecular Plant-Microbe Interactions* 21(7), 1001-1009.

Nascimento, F. C., Kandasamy, S., Lazarovits, G. and Rigobelo, E. C. (2020). Effect of chemical fertilization on the impacts of plant growth-promoting rhizobacteria in maize crops, *Current Microbiology* 9, 1-10.

Obermeier, M. M., Wicaksono, W. A., Taffner, J., Bergna, A., Poehlein, A., Cernava, T., Lindstaedt, S., Lovric, M., Müller Bogotá, C. A. and Berg, G. (2020). Plant resistome profiling in evolutionary old bog vegetation provides new clues to understand emergence of multi-resistance, *The ISME Journal* 15(3), 921-937. doi: 10.1038/s41396-020-00822-9.

Pérez-Jaramillo, J. E., Carrión, V. J., de Hollander, M. and Raaijmakers, J. M. (2018). The wild side of plant microbiomes, *Microbiome* 6(1), 143.

Philippot, L., Raaijmakers, J. M., Lemanceau, P. and van der Putten, W. H. (2013). Going back to the roots: the microbial ecology of the rhizosphere, *Nature Reviews. Microbiology* 11(11), 789-799.

Pierik, R., Tholen, D., Poorter, H., Visser, E. J. and Voesenek, L. A. (2006). The Janus face of ethylene: growth inhibition and stimulation, *Trends in Plant Science* 11(4), 176-183.

Pieterse, C. M., Poelman, E. H., van Wees, S. C. and Dicke, M. (2013). Induced plant responses to microbes and insects, *Frontiers in Plant Science* 4, 475.

Pieterse, C. M., Zamioudis, C., Berendsen, R. L., Weller, D. M., van Wees, S. C. and Bakker, P. A. (2014). Induced systemic resistance by beneficial microbes, *Annual Review of Phytopathology* 52, 347-375.

Raaijmakers, J. M., van der Sluis, Lvd, Bakker, P. A. H. M., Schippers, B., Koster, M. and Weisbeek, P. J. (1995). Utilization of heterologous siderophores and rhizosphere competence of fluorescent *Pseudomonas* spp., *Canadian Journal of Microbiology* 41(2), 126–135.

Raaijmakers, J. M. and Mazzola, M. (2012). Diversity and natural functions of antibiotics produced by beneficial and plant pathogenic bacteria, *Annual Review of Phytopathology* 50, 403–424.

Remus-Emsermann, M. N. P. and Schlechter, R. O. (2018). Phyllosphere microbiology: at the interface between microbial individuals and the plant host, *New Phytologist* 218(4), 1327–1333.

Rezki, S., Campion, C., Iacomi-Vasilescu, B., Preveaux, A., Toualbia, Y., Bonneau, S., Briand, M., Laurent, E., Hunault, G., Simoneau, P., Jacques, M. A. and Barret, M. (2016). Differences in stability of seed-associated microbial assemblages in response to invasion by phytopathogenic microorganisms, *PeerJ* 4, e1923.

Rockström, J., Steffen, W., Noone, K., Persson, Å, Chapin, F. S. I., Lambin, E., Lenton, T. M., Scheffer, M., Folke, C., Schellnhuber, H. J., Nykvist, B., de Wit, C. A., Hughes, T., van der Leeuw, S., Rodhe, H., Sörlin, S., Snyder, P. K., Costanza, R., Svedin, U., Falkenmark, M., Karlberg, L., Corell, R. W., Fabry, V. J., Hansen, J., Walker, B., Liverman, D., Richardson, K., Crutzen, P. and Foley, J. (2009). Planetary boundaries: exploring the safe operating space for humanity, *Ecology and Society* 14(2), 32.

Rosier, A., Bishnoi, U., Lakshmanan, V., Sherrier, D. J. and Bais, H. P. (2016). A perspective on inter-kingdom signaling in plant–beneficial microbe interactions, *Plant Molecular Biology* 90(6), 537–548.

Rybakova, D., Cernava, T., Köberl, M., Liebminger, S., Etemadi, M. and Berg, G. (2016). Endophytes-assisted biocontrol: novel insights in ecology and the mode of action of *Paenibacillus*, *Plant and Soil* 405(1–2), 125–140.

Saleem, M., Arshad, M., Hussain, S. and Bhatti, A. S. (2007). Perspective of plant growth promoting rhizobacteria (PGPR) containing ACC deaminase in stress agriculture, *Journal of Industrial Microbiology and Biotechnology* 34(10), 635–648.

Sánchez-Cañizares, C., Jorrín, B., Poole, P. S. and Tkacz, A. (2017). Understanding the holobiont: the interdependence of plants and their microbiome, *Current Opinion in Microbiology* 38, 188–196.

Sang, M. K. and Kim, K. D. (2012). Plant growth-promoting rhizobacteria suppressive to Phytophthora blight affect microbial activities and communities in the rhizosphere of pepper (*Capsicum annuum* L.) in the field, *Applied Soil Ecology* 62, 88–97.

Santoyo, G., del Carmen Orozco-Mosqueda, MdC. and Govindappa, M. (2012). Mechanisms of biocontrol and plant growth-promoting activity in soil bacterial species of *Bacillus* and *Pseudomonas*: a review, *Biocontrol Science and Technology* 22(8), 855–872.

Saravanakumar, K., Li, Y., Yu, C., Wang, Q. Q., Wang, M., Sun, J., Gao, J. X. and Chen, J. (2017). Effect of *Trichoderma harzianum* on maize rhizosphere microbiome and biocontrol of *Fusarium* Stalk rot, *Scientific Reports* 7(1), 1771.

Schiltz, S., Gaillard, I., Pawlicki-Jullian, N., Thiombiano, B., Mesnard, F. and Gontier, E. (2015). A review: what is the spermosphere and how can it be studied?, *Journal of Applied Microbiology* 119(6), 1467–1481.

Schmidt, C. S., Alavi, M., Cardinale, M., Müller, H. and Berg, G. (2012). *Stenotrophomonas rhizophila* DSM14405 T promotes plant growth probably by altering fungal communities in the rhizosphere, *Biology and Fertility of Soils* 48(8), 947-960.

Schmidt, R., Etalo, D. W., de Jager, V., Gerards, S., Zweers, H., de Boer, W. and Garbeva, P. (2016). Microbial small talk: volatiles in fungal-bacterial interactions, *Frontiers in Microbiology* 6, 1495.

Schreiter, S., Ding, G. C., Grosch, R., Kropf, S., Antweiler, K. and Smalla, K. (2014). Soil type-dependent effects of a potential biocontrol inoculant on indigenous bacterial communities in the rhizosphere of field-grown lettuce, *FEMS Microbiology Ecology* 90(3), 718–730.

Shade, A., Jacques, M. A. and Barret, M. (2017). Ecological patterns of seed microbiome diversity, transmission, and assembly, *Current Opinion in Microbiology* 37, 15–22.

Spadaro, D. and Droby, S. (2016). Development of biocontrol products for postharvest diseases of fruit: the importance of elucidating the mechanisms of action of yeast antagonists, *Trends in Food Science and Technology* 47, 39–49.

Sturz, A. V., Christie, B. R. and Nowak, J. (2000). Bacterial endophytes: potential role in developing sustainable systems of crop production, *Critical Reviews in Plant Sciences* 19(1), 1–30.

Taffner, J., Cernava, T., Erlacher, A. and Berg, G. (2019). Novel insights into plant-associated archaea and their functioning in arugula (*Eruca sativa* Mill.), *Journal of Advanced Research* 19, 39–48.

Tao, C., Li, R., Xiong, W., Shen, Z., Liu, S., Wang, B., Ruan, Y., Geisen, S., Shen, Q. and Kowalchuk, G. A. (2020). Bio-organic fertilizers stimulate indigenous soil *Pseudomonas* populations to enhance plant disease suppression, *Microbiome* 8(1), 137.

Thomashow, L. S., Bonsall, R. F. and Weller, D. M. (2008). Detection of antibiotics produced by soil and rhizosphere microbes *in situ*. In: Karlovsky, P. (Ed), *Secondary Metabolites in Soil Ecology. Soil Biology*, Germany: Springer, pp. 23–36.

Toju, H., Abe, M. S., Ishii, C., Hori, Y., Fujita, H. and Fukuda, S. (2020). Scoring species for synthetic community design: network analyses of functional core microbiomes, *Frontiers in Microbiology* 11, 1361.

Turner, T. R., James, E. K. and Poole, P. S. (2013). The plant microbiome, *Genome Biology* 14(6), 209.

Unno, Y., Okubo, K., Wasaki, J., Shinano, T. and Osaki, M. (2005). Plant growth promotion abilities and microscale bacterial dynamics in the rhizosphere of *Lupin* analysed by phytate utilization ability, *Environmental Microbiology* 7(3), 396–404.

Vandenkoornhuyse, P., Quaiser, A., Duhamel, M., Le Van, A. and Dufresne, A. (2015). The importance of the microbiome of the plant holobiont, *New Phytologist* 206(4), 1196–1206.

Van Elsas, J. D., Chiurazzi, M., Mallon, C. A., Elhottovā, D., Krištůfek, V. and Salles, J. F. (2012). Microbial diversity determines the invasion of soil by a bacterial pathogen, *Proceedings of the National Academy of Sciences of the United States of America* 109(4), 1159–1164.

Vayssier-Taussat, M., Albina, E., Citti, C., Cosson, J. F., Jacques, M. A., Lebrun, M. H., Le Loir, Y., Ogliastro, M., Petit, M. A., Roumagnac, P. and Candresse, T. (2014). Shifting the paradigm from pathogens to pathobiome: new concepts in the light of meta-omics, *Frontiers in Cellular and Infection Microbiology* 4, 29.

Venturi, V. and Keel, C. (2016). Signaling in the rhizosphere, *Trends in Plant Science* 21(3), 187–198.

Von Bodman, S. B., Bauer, W. D. and Coplin, D. L. (2003). Quorum sensing in plant-pathogenic bacteria, *Annual Review of Phytopathology* 41(1), 455–482.

Wagner, M. R., Lundberg, D. S., Coleman-Derr, D., Tringe, S. G., Dangl, J. L. and Mitchell-Olds, T. (2014). Natural soil microbes alter flowering phenology and the intensity of selection on flowering time in a wild Arabidopsis relative, *Ecology Letters* 17(6), 717–726.

Wassermann, B., Cernava, T., Müller, H., Berg, C. and Berg, G. (2019). Seeds of native alpine plants host unique microbial communities embedded in cross-kingdom networks, *Microbiome* 7(1), 108.

Weller, D. M., Raaijmakers, J. M., Mc Spadden Gardender, B. B. and Thomashow, L. S. (2002). Microbial populations responsible for specific soil suppressiveness to plant pathogens, *Annual Review of Phytopathology* 40(1), 309–348.

Whipps, J. M., Karen, L. and Cooke, R. C. (1988). Mycoparasitism and plant disease control. In: Burge, M. N. (Ed), *Fungi in Biological Control Systems*, Manchester: Manchester University Press, pp. 161–187.

Whipps, J. M. (2001). Microbial interactions and biocontrol in the rhizosphere, *Journal of Experimental Botany* 52(Spec Issue), 487–511.

Wicaksono, W. A., Cernava, T., Berg, C. and Berg, G. (2020). Reconstruction of bacterial metagenome-assembled genome sequences from Alpine bog vegetation, *Microbiology Resource Announcements* 9(35), e00821-20.

Wiesel, L., Newton, A. C., Elliott, I., Booty, D., Gilroy, E. M., Birch, P. R. and Hein, I. (2014). Molecular effects of resistance elicitors from biological origin and their potential for crop protection, *Frontiers in Plant Science* 5, 655.

Wilson, C. L. and Wisniewski, M. E. (1989). Biological control of postharvest diseases of fruits and vegetables: an emerging technology, *Annual Review of Phytopathology* 27(1), 425–441.

Wolfgang, A., Zachow, C., Müller, H., Grand, A., Temme, N., Tilcher, R. and Berg, G. (2020). Understanding the impact of cultivar, seed origin, and substrate on bacterial diversity of the sugar beet rhizosphere and suppression of soil-borne pathogens, *Frontiers in Plant Science* 11, 560869.

Yin, D., Wang, N., Xia, F., Li, Q. and Wang, W. (2013). Impact of biocontrol agents *Pseudomonas fluorescens* 2P24 and CPF10 on the bacterial community in the cucumber rhizosphere, *European Journal of Soil Biology* 59, 36–42.

Zaneveld, J. R., Mc Minds, R. and Thurber, R. V. (2017). Stress and stability: applying the Anna Karenina principle to animal microbiomes, *Nature Microbiology* 2(9), 1–8.

Zhang, L. N., Wang, D. C., Hu, Q., Dai, X. Q., Xie, Y. S., Li, Q., Liu, H. M. and Guo, J. H. (2019). Consortium of plant growth-promoting rhizobacteria strains suppresses sweet pepper disease by altering the rhizosphere microbiota, *Frontiers in Microbiology* 10, 1668.

Chapter 2

Advances in screening approaches for the development of microbial bioprotectants to control plant diseases

Wagner Bettiol, Embrapa (Brazilian Agricultural Research Corporation), Brazil; Flávio Henrique Vasconcelos de Medeiros, Universidade Federal de Lavras, Brazil; Josiane Barros Chiaramonte, Vittia Fertilizantes e Biológicos SA, Brazil; and Rodrigo Mendes, Embrapa (Brazilian Agricultural Research Corporation), Brazil

1 Introduction

The ultimate goal in developing a microbial bioprotectant against pathogens (including nematodes) is large-scale commercialisation and use by growers around the world. The global bioprotectants market was estimated at US$4.3 billion in 2020 and is expected to reach US$8.5 billion by 2025, with a projected

http://dx.doi.org/10.19103/AS.2021.0093.02

compound annual growth rate (CAGR) of 14.7% over that period (Markets and Markets, 2020). However, this sustained growth will depend on the availability of new bioprotectant products. Consequently, it is necessary to isolate and select novel microorganisms with the potential to efficiently control plant diseases, as well as to develop marketable products (Cotes et al., 2019). Currently, a limited number of microbial species have been commercially exploited in this way. For example, although 254 species of *Trichoderma* are known (Bissett et al., 2015), only 10 of them have been commercialised as biocontrol agents (Bettiol et al., 2019b).

The isolation and selection of antagonistic microorganisms support the entire programme of biological control (Andrews, 1985). In determining the chances of success in developing commercial bioprotectants, the environmental conditions where these organisms will be obtained must be carefully observed, as well as the methods of isolation and selection. Given the number of species of microorganisms and diversity of environments, such as soil, rhizosphere, phylloplane, spermosphere and carposphere, this is a formidable task (Andrews, 1985). In addition, due to the microbial diversity characterised by new molecular tools, isolation and screening of antagonists has proved extremely challenging, requiring the efforts of a large number of specialists. Success in the development of new products also requires interdisciplinary teamwork. The presence of experienced field technicians on teams with knowledge of integrated pest management (IPM) (such as farmers and agronomists) is of utmost importance, as they are able to visualise how new technology can be integrated into current cultivation systems and assess how farmers can adopt them.

This chapter discusses ways of surmounting key challenges in getting bioprotectants onto the market. Firstly, it focuses on the screening strategies for microbial agents which increase the likelihood of success, then discusses how alternative control agents other than fungi and bacteria may be explored, the possibilities of targeting specific niches and regional markets and the importance of formulation during bioprotectant development. Finally, the possibilities of engineering soil and host microbiomes to benefit plant health are considered, taking into account recent advances in understanding the underlying mechanisms of microbiome assembly in plant environments and the role of the microbiome in plant defence, especially in the rhizosphere.

2 Screening microorganisms for biological control of plant diseases: exclusive and inclusive approaches

Over the last five decades, detailed methodologies for isolation and selection of antagonists have been presented by Morgan (1963), Leben (1964), Jenkins (1968), Fokkema (1976), Fokkema and Van der Meulen (1976), Fravel and Spurr

Jr (1977), Thirumalachar and O'Brien (1977), Trutmann et al. (1980), Utkhede and Rahe (1980/1983), Adams and Ayers (1981), Andrews et al. (1983), Baker et al. (1983), Weller and Cook (1983), Pusey and Wilson (1984), Weller et al. (1985), Garibaldi et al. (1987), Janisiewicz (1987/1988), Janisiewicz and Roitman (1988), Clarkson and Lucas (1993), Mariano (1993), Landa et al. (1997), Montesinos (2003), Shi and Sun (2017), Cotes et al. (2019) and García et al. (2019). Given the wealth of existing literature, these will not be addressed in this chapter.

Thousands of molecules are evaluated during the development of new chemical pesticides. Similarly, when developing a new bioprotectant, the initial selection must be based on a process of isolating a wide range of potential candidates. Following this, a thorough selection process is required. Detailed screening systems have been proposed and used (Andrews, 1985; Clarkson and Lucas, 1993; Montesinos, 2003; Köhl et al., 2011). However, success is not guaranteed by relying on artificial, simplified testing systems. There is also no guarantee that the efficacy of antagonists against a specific disease can be extrapolated to other diseases (Köhl et al., 2020). The chances of success have been shown to increase with a well-trained team (Köhl et al., 2011, 2019). Currently, efforts are being directed mainly at the isolation and screening of a large number of species. However, there have been several successful cases in which the in-depth knowledge of researchers achieved a marketable product after screening only a few isolates (Table 1).

The natural occurrence of mycoparasitism on the structures of pathogens as a potential source of antagonists has often been neglected during the development of bioprotectants. Observations of mycoparasitic interactions in the environment have made it possible to obtain biocontrol agents such as *Dycima pulvinata* for the control of South American leaf blight in rubber trees, *Acremonium* for the control of tar spot on coconuts and *Trichoderma stromaticum* for the control of cacao witches' broom (*Moniliophthora perniciosa*) (see case studies later in this chapter). Field observations have also identified the mild virus Tristeza for the control of Tristeza in citrus (Müller and Costa, 1991), and mild strains of the papaya ringspot virus-type W for control of zucchini squash mosaic (Rezende and Pacheco, 1998). The natural biological control that occurs in virtually all environments should not be overlooked in programmes for the isolation and selection of biocontrol agents.

Culturable microorganisms are currently used for biological control of plant diseases. However, the vast majority of microorganisms are not culturable and thus are not used in biological control. The development of technologies to exploit non-culturable organisms depends on knowledge of the microbiome. The growing knowledge of the soil microbiome enables understanding of the interaction of rhizosphere microorganisms with plants in the control of soil-borne plant pathogens. Additionally, the phyllosphere microbiome harbours

Table 1 Screening approaches for the selection of biological control agents against plant diseases

Year/Region	Pathosystem	Parameter evaluated	Antagonists	Approach and results	Commercial product/Country	Reference
1963 USA	*Colletotrichum lagenarium* and *Erysiphe cichoracearum* x cucumber	Control of cucumber anthracnose and powdery mildew in greenhouse	Bacteria	From 235 isolates of bacteria obtained from cucumber leaves, only the isolate A180 consistently decreased anthracnose. Powdery mildew was not controlled by the bacteria isolates		Leben (1964)
1970 New Zealand	Crown gall (*Agrobacterium tumefaciens*) x apricot, nectarine, peach, prune, plum, rose, and ornamentals, etc	Control of crown gall in plants	*Agrobacterium radiobacter* (now *Rhizobium rhizogenes* strain K 84 and K1026)	According to Kerr (2016), "Not all non-pathogen had this effect but fortunately, the first non-pathogen we tried was strain K84, a non-pathogenic strain of *R. rhizogenes* and the 84th strain in my culture collection"	Nogall™ (K1026) (K84 - the first registered bioprotectant for biocontrol of soil-borne plant disease)	Jones and Kerr (1989), Keane et al. (1970), Kerr (2016), Kerr and Bullard (2020), Kerr and Htay (1974)
1982 USA	*Rhizoctonia, Sclerotinia, Sclerotium, Fusarium, Pythium, Botrytis* x vegetables, ornamentals, soybean, maize	Rhizosphere competence, control of several soilborne plant pathogens and plant growth promotion (PGP)	*Trichoderma harzianum* T22 (derived from the protoplast fusion of strains T-95 and T-12)	T-22 was selected based on control of several soilborne plant pathogen, colonisation and protection of the root system, and conidia production	Bio-Trek 22G® F-Stop® Plantshield® Rootshield® Triaction® Trianum®	Hadar et al. (1984), Harman (n.d.), Harman et al. (1989), Woo et al. (2014)

1983 Canada	*Pythium* in several crops	Growth promotion and disease control *in vitro*, in greenhouse and field	*Pseudomonas chlororaphys* 63-28 (before *Pseudomonas fluorescens)*	More than 4000 bacterial strains were isolated in Canada from rhizosphere. In greenhouse conditions, 887 isolates were tested for growth-promoting activity on canola, and 35 strains were selected for enhanced growth in at least two tests. Among them, isolate 63-28 of *P. fluorescens* now *P. chlororaphys* was selected. *P. chloraraphys* 63-28 was registered for the control of *Pythium* spp., *Rhizoctonia solani* and *Fusarium oxysporum* in 2001	AtEze®	Kloepper et al. (1988), Paulitz and Bélanger (2001), EPA (2001)
1983 USA	*Pseudomonas avenae* x sweet corn	*In vitro* antagonism, in greenhouse and field disease control	*Pseudomonas fluorescens* F-11 and F-24 and *Bacillus* sp. U-24.	From leaves of corn, 179 bacteria isolates were obtained, 34 were antagonistic to *P. avenae* (29 fluorescent pseudomonads). In greenhouse trials, two *P. fluorescens* (F-11 and F-24) strains and one *Bacillus* sp. (U-46) were selected. *P. fluorescens* F-11 reduced the disease under field conditions, but *Bacillus* did not affect disease development		Lopes and Stall (1990)
1984 Brazil	*Pyricularia oryzae* x rice	*In vitro* antagonism and in the field disease control	Several isolate of *Bacillus subtilis*	From 472 isolates of bacteria obtained from rice plants and soil, 348 showed antagonism against *P. oryzae in vitro*. Twenty-seven were selected, and 10 top isolates were selected for field evaluations to control rice blast		Bettiol (1988), Bettio and Kimati (1989, 1990)

(*Continued*)

Table 1 (*Continued*)

Year/Region	Pathosystem	Parameter evaluated	Antagonists	Approach and results	Commercial product/Country	Reference
1988 Canada	*Botrytis cinerea* x strawberry	Pathogen control in leaf flower tissues, whole plants in greenhouse and field. Disease control and PGP	*Gliocladium roseum* 88-710 (*Clonostachys rosea* since 1999)	*C. rosea* isolate PG-A-Fr-88-710 was selected from 1400 isolates of bacteria and fungi from strawberry plants based on its effectiveness against *Botrytis cinerea* in strawberries	EndoFine®	Peng and Sutton (1991), Sutton et al. (1997, 2002), Canadian (2007)
1989 Israel	*Ampelomyces quisqualis* x powdery mildew	Parasitism *in vitro* and plants	*A. quisqualis*	*A. quisqualis* M-10 was isolated from powdery mildew infecting the slow-growing flowering shrub of *Catha edulis*	AQ10®	Sztejnberg et al. (1989), AQ10 (1991), AQ10 (2020)
1990 Canada	*Pythium aphanidermatum* x hydroponic cucumber	Inhibition of zoospore germination and motility	*P. fluorescens* Pf15 and *Pseudomonas corrugata* Pc13	From cucumber rhizosphere and forest soils, 604 bacteria strains were isolated. Thirty-five isolates were selected based on inhibition of zoospore germination and motility of *Pythium*, and five were selected based on assays on cucumber. From these, three also increased plant growth in the absence of *Pythium*. *P. fluorescens* Pf15 and *P. corrugata* Pc13 increased fruit production		Paulitz et al. (1992), Rankin and Paulitz (1994)

1990 Cuba	*Phytophthora, Rhizoctonia, Pythium, Sclerotium, Fusarium* x tobacco and vegetables	Parasitism, inhibition of mycelial growth; control of the diseases in greenhouse and field	*T. harzianum* A-34 and *Trichoderma viride* TS-3	*T. harzianum* A-34 and *T. viride* TS-3 were selected from several *Trichoderma* strains isolated from soil	Tricosave A-34 Tricosave TS-3	Stefanova (2007), Stefanova et al. (1999/2014)
1995 Germany	*Sclerotinia* spp. x sclerotial mycoparasitism	Sclerotial mycoparasitism in soil	*Coniothyrium minitans* CON/M/91-08 (DSM 9660)	*C. minitans* was isolated directly from sclerotia of *Sclerotinia*	Contans®	Luth (2001), Patents (1996), Whipps et al. (2008), Whipps and Gerlagh (1992)
1996 Colombia	*Pythium splendens* x common bean (first step). Thereafter, *R. solani, S. sclerotiorum* and *Fusarium oxysporum*	Inhibition of conidia germination and mycelial growth; control of the diseases in greenhouse and field	*Trichoderma koningiopsis* Th003	Th003 (*T. koningiopsis)* was selected amongst 14 *Trichoderma* isolates. Th003 was registered for control of *F. oxysporum, R. solani, S. minor* and *S. sclerotiorum* in tomato, lettuce and rice	Tricotec®	Beltrán et al. (2007, 2012), Bettiol et al. (2019), Cotes et al. (1996, 2001, 2007, 52019)
1997 Spain	*R. solani, S. sclerotiorum, Fusarium* sp., *Pythium* sp. and *Phytophthora x* tomato, sugar beet, lettuce, strawberry, cucumber, and pepper	Parasitism, antibiosis, induction of resistance in plants, and PGP.	*Trichoderma atroviride* NBT-T11 and *Trichoderma asperellum* NBT-T25	*T. atroviride* T11 was isolated from soil cultivated with sugar beet. *T. asperellum* T-25 was isolated from soil cultivated with citrus for 30 years	Tusal®	Grandona et al. (1997, 2004), Hermosa et al. (2000, 2001), Sanz et al. (2004)

(Continued)

Table 1 (*Continued*)

Year/Region	Pathosystem	Parameter evaluated	Antagonists	Approach and results	Commercial product/Country	Reference
1998 Brazil	*Fusarium solani, R. solani, Sclerotinia sclerotiorum, Thielaviopsis paradoxa* x several crops	Parasitism of pathogen structure and control of the diseases in greenhouse and field	*T. harzianum*	Considering the parasitism, growth inhibition of plant pathogens, and conidia production on rice, the isolate ESALQ 1306 (*T. harzianum)* was selected from a few numbers of *Trichoderma* isolated from soil, and plant litter (Lopes, pers. comm.)	Trichodermil® (first registered bioprotectant for the biocontrol plant disease in Brazil in 2006)	Bettiol et al. (2019a), Lopes, Rogerio Baiggioni (pers. comm.)
1998 Costa Rica	*Rhizoctonia, Pythium, Phytophthora, Fusarium, Botrytis* and *Colletotrichum* x several crops	Parasitism of pathogen structure and control of the diseases in greenhouse and field	*T. asperellum*	Forty-two *Trichoderma* isolates were obtained from soil. All isolates were evaluated for *in vitro*, inhibition of the pathogens, control of the diseases in field, and mass production. Three isolates controlled Fusarium in pineapple	Bioprotection TR on farm production.	Alfaro and Obregón-Gómez (2015), Mata and Obregón (2014), Mata et al. (2012), Obregon-Gómez (2010), Obregón and Arias (2008), Obregón (2012)
2001 Israel	Postharvest disease (B. *cinerea* on pepper and tomatoes; *Rhizopus stolonifera* on potatoes and peaches)	Control incidence of postharvest in fruits	*Metschnikowia fructicola*	*M. fructicola* was isolated from grapes. It was found to be effective for the control of *B. cinerea* in grapes. The mode of action of this bioagent is through competition for space and nutrients	Shemer®	Blachinsky et al. (2007), Droby et al. (2009), Kurtzman and Droby (2001)

2003 Brazil	*Hemileia vastatrix* x coffee	Control of coffee leaf rust in coffee seedling in greenhouse and field	Coffee phyllosphere bacteria (*Bacillus* spp. and *Pseudomonas* sp.) and fungi	393 isolates were obtained from coffee leaves, crop residues and soil. On infected coffee plants, 17 isolates were effective. The efficiency of these isolates was further evaluated, and seven isolates were selected: six of *Bacillus* spp. and one of *Pseudomonas* sp. *Bacillus* B157 reduced rust intensity in field assays	Haddad et al. (2007, 2009)
2004 Brazil	*H. vastatrix* x coffee	Inhibition of urediniospores germination, and control of rust on detached leaves and plants	Endophytic bacteria from *Coffea arabica* and *Coffea robusta*	Forty-four endophytic bacteria isolates were evaluated to inhibit urediniospores germination and control rust on coffee leaf discs. Ten were selected and evaluated on detached leaves and coffee plants. *Bacillus cereus* and *Bacillus lentimorbus* were identified	Shiomi et al. (2006)
2004 Chile	*B. cinerea* x *Eucalyptus* *Fusarium circinatum* x *Pinus*	Parasitism *in vitro*, leaf colonization, greenhouse and nursery assays. Induction of resistance and PGP	*C. rosea* UDC-32 and UDC-222; *Trichoderma* UDC-23 and UDC-408	Seventy-one *Trichoderma* and 10 *Clonostachys* strains from *Eucalyptus* and *Pinus* nurseries were selected. Recently, other isolates were selected from forest nursery. Two *C. rosea* strains (UDC-32 and UDC-222) and two *Trichoderma* (UDC-23 and UDC-408) were selected *in vitro*. These isolates reduced the disease on *Pinus radiata*	Moraga-Suazo et al. (2011a,b-2016), Zaldúa and Sanfuentes (2010), Moraga-Suazo and Sanfuentes (2017), Sanfuentes and Zaldúa (2006), Sanfuentes et al. (2006)

(*Continued*)

Table 1 (*Continued*)

Year/Region	Pathosystem	Parameter evaluated	Antagonists	Approach and results	Commercial product/Country	Reference
2005 Brazil	*B. cinerea* x strawberry	Inhibition of sporulation on leaf discs and control of gray mold under field conditions	*C. rosea*	From 189 samples from soil and dead leaves, 22 isolates of *Clonostachys* were obtained, 16 of them were identified as *C. rosea*. Four *C. rosea* isolates were selected based on ability to control *Botrytis*. In field, these isolates required spraying twice a week for successful control of gray mold		Nobre et al. (2005), Cota et al. (2008a,b)
2006 Brazil	*H. vastatrix* x coffee	Growth promotion of coffee seedlings, urediniospore reduction of germination, and control of coffee rust on detached leaf discs and coffee seedlings	Endophytic bacteria and fungi from *C. arabica* and *C. robusta*	A total of 234 strains of endophytic bacteria (217) and fungi (17) from coffee tissues were observed: no fungi strains induced PGP or reduced disease severity; six bacteria isolates induced PGP; and six bacteria isolates reduced the severity of rust. There was no correspondence between isolates causing PGP and isolates reducing coffee leaf rust		Silva et al. (2012)
2009 Canada	*B. cinerea*, *S. sclerotiorum*, *Monilinia*, *Pythium*, *Rhizoctonia* x several crops	Plant colonisation, disease control in greenhouse and field, PGP	*C. rosea f. rosea* BVT- Cr-7	Based on abundant spore production and superior suppression of various pathogens, *C. rosea f. rosea* BVT- Cr-7 was selected from 25 *C. rosea* isolates obtained in several crops	Vectorite®	US Patents (2017), Sutton, John (pers. comm.) Vectorite (2020)

2010 Chile	*F. circinatum x P. radiata*	Antagonism *in vitro* (substrate competition, mycoparasitism and antibiosis). *In vivo* assays for disease incidence, mortality and asymptomatic infections in *P. radiata* seedlings	Mixture of *Bionectria ochroleuca* CBF2120 (RGM-2147), *T. harzianum* CBF84 (RGM2148) and *Gliocladium penicillioides* CBF 1774 (RGM2258)	More than 3000 isolates of fungi and bacteria from different organs, species and environments were selected. *In vitro,* 300 isolates were selected. In greenhouse, 90 isolates reduced seedlings mortality (eight of them reduced > 90%). Also, in greenhouse, 16 fungal isolates were evaluated individually or in mixtures in addition to two trials in forest nurseries. The best results were obtained by a mixture of three species, which generally maintained a level of biocontrol of *F. circinatum* > 85%	Bioproduct M3. Does not have registration until now	Sanfuentes (2017).
2010 Uruguay	*B. cinerea* and *Penicillium expansum* causal agents of postharvest diseases of apple	Control the pathogens in apple and *in vitro* antagonism	*Leucosporidium scottii*	From 21 soil samples collected in Antarctica (King George Island), five yeast isolates were obtained. *L. scottii* At17 was the most efficient to control postharvest diseases on apples		Vero et al. (2013)
2011 Colombia	*B. cinerea x* blackberry	Control of *B. cinerea* in rose petals and blackberry	*Rhodotorula glutinis* Lv316	From 265 yeast isolates obtained from the blackberry phyllosphere and rhizosphere, 100 were pre-selected *in vitro* and *in vivo*. These isolates were evaluated on rose petals for the control of *Botrytis* (with 50 isolates selected). *R. glutinis* Lv316 was selected for the development of a product		Cotes et al. (2011), Zapata et al. (2011, 2013, 2016)

(*Continued*)

Table 1 (*Continued*)

Year/Region	Pathosystem	Parameter evaluated	Antagonists	Approach and results	Commercial product/Country	Reference
2014 Spain	*Monilinia* spp. x stone fruit in pre-harvest and postharvest	Control incidence of the pathogens in stone fruit	*Penicillium frequentans* Pf909 and *Bacillus amyloliquefaciens* CPA-8	*P. frequentans* Pf909 (a common colonizer of peach twigs and flowers) and *B. amyloliquefaciens* CPA-8 (isolated from nectarine) were characterized. The multiplication of these isolates was optimized, and a formulation was developed for both antagonists. These antagonists are now at commercial development phase		Gotor-Vila et al. (2017a,b), Biocomes (2016)
2014 Netherlands Germany Sweden	*Blumeria graminis* x cereal crops	Control of powdery mildew on wheat plants in greenhouse conditions	*Tiletiopsis pallescens*	From 504 samples of powdery mildew pustules on cereal, grasses and other plant leaves, 1237 fungi isolates were obtained. From them, 862 were pre-selected (based on mass production, safety, cold and drought tolerance and UV-resistance), which resulted in 185 strains selected assays on wheat seedling inoculated with powdery mildew. Four strains of *T. pallescens* were selected for field assays and two strains showed higher efficacy in the reduction of powdery mildew. These antagonists are now in the commercial development phase		Biocomes (2014, 2015c, 2016a, 2017a,b), Cotes et al. (2019), Köhl and Rijk (2016), Köhl et al. (2019)

2014 Austria The Netherlands Sweden	*Verticillium longisporum* x brassicaceae	Anti-fungal activities, plant colonization, PGP and disease control	*Serratia plymuthica* HRO-C48; *S. plymuthica* F20; *Paenibacillus polymixa* SB3-1 and endophytes	*Serratia plymuthica* HRO-C48 and *P. polymixa* SB3-1 were selected based on the ability to PGP of oilseed rape and cauliflower and to suppress *Verticillium*. The effects on PGP, and disease control were dependent on grown conditions, cultivar and disease pressure. After these observations, new endophytes were selected to co-formulate with *Serratia* and *Paenibacillus*. From over 550 strains, *S. plymuthica* F20 and five isolates that showed antagonistic effects against the pathogen were selected	Biocomes (2015a,b, 2016b, 2017c) Rybakova et al. (2016a,b, 2017, 2020)

PGP = plant growth promotion.

several bacterial communities with distinct and diverse metabolisms which may be explored as a source of natural products, including those suitable for biocontrol (Moitinho et al., 2020). Perazzolli et al. (2020) showed that applications of a rare sugar (tagatose) significantly reshaped the microbiome community by stimulating beneficial microorganisms over pathogens. This suggests that the possibility of developing new biological control strategies through modulation of resident microbiomes, for example the stimulation of certain microbial populations, is a new option in biocontrol which has great potential.

During the search for biocontrol agents over many years, scientists have proposed the isolation and screening of these agents. However, even when the most effective microorganisms for disease control have been found, the lack of stability under field conditions or the lack of economic feasibility in mass production have precluded obtaining marketable products as discussed by Ravensberg (2011).

A combination of exclusive and inclusive methods is more likely to result in economically feasible and agronomically effective biocontrol agents to control plant diseases. By excluding potentially undesirable strains and including effective strains, the screening of effective strains may not only be accelerated but will increase the chance of obtaining an effective and lower-cost biocontrol product.

Both inclusive and exclusive approaches are intended to exclude undesirable strains. When applying an exclusive test, a higher number of (undesirable) strains will be excluded. Therefore, the more exclusive the initial applied test, the more likely it is to be an effective part of stepwise screening. Obviously, such tests should always result in the elimination of strains which are undesirable in a biocontrol product. Such preliminary tests should be undertaken before isolating strains and should take into account the relevance of the target disease in terms of its impact on the crop. The industry should try to isolate as many strains as possible from the diverse niches and locations where the disease normally occurs. At this point, screening based on exclusive methods, evaluating strains based on their potential as a biocontrol agent, should commence. Those strains passing these tests are then identified at the species level. Any which potentially represent a threat to either plants or animals are eliminated.

The more inclusive steps are then started. Inclusive tests are repeated under the same conditions and with the same set of strains to check for consistent results whilst eliminating those which do not display a consistent performance. These inclusive steps are:

- efficacy testing of pathogen control in plants;
- preliminary assessment of mass production potential;

- pilot formulation;
- upscaling of mass production;
- full field testing; and
- integration into cropping systems management.

Checking for consistent results in these inclusive steps is of the greatest importance. The same strains should therefore be checked more than once under the same conditions (Fig. 1).

3 The nine-step approach to screening for biocontrol agents

Köhl et al. (2011) proposed a step-by-step screening approach in which various aspects of the biocontrol agent are taken into account in order to bring a successful biocontrol agent to the market. In summary, the authors proposed a nine-step procedure:

1 Assessment of the targeted crop, disease and market
2 Origin and isolation of candidate antagonists
3 Rapid throughput screening
4 Database mining
5 Efficacy testing in bioassays
6 Preliminary assessment of mass production
7 Pilot formulation and registration costs
8 Upscaling mass production and full field testing
9 Integration into cropping systems

We have followed their approach, encompassing particular niche and regional differences, as well as changes in testing technologies and costs.

3.1 Relevance of the target disease in terms of loss and relevance of the crop

Investment in product development needs to be recovered and therefore the adoption of a product capable of managing a particular crop and disease is essential. Once a product for such a pathosystem has been developed, it can be adapted to other crops and diseases in terms of formulation, dose or other types of customisation to make it suitable for new targets. This process of adaptation may be assisted by legislation. In Brazil, bioprotectants are registered for a given target disease and not for the crop. For example, once a product for white mould (*Sclerotinia sclerotiorum*) has been registered in soybean, it can be recommended for any other crop in which this disease is a threat without the need to submit a new dossier for registration.

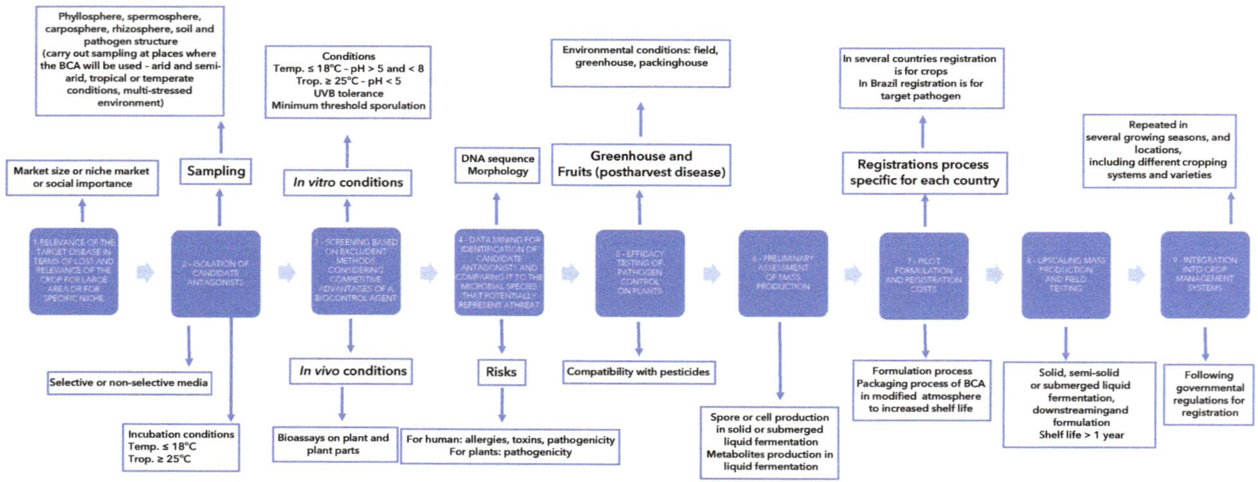

Figure 1 Stepwise screening of microorganisms for development of traditional biocontrol agents. Steps 4, 5, 6, 8, and 9 are similar for all environmental conditions. Temp = temperature, Trop. = tropical conditions. pH for or soilborne plant pathogens (adapted from Kohl et al., 2011; Ravensberg, 2011).

3.2 Isolation of candidate antagonists

This step is of great importance as it will determine the ultimate success of biocontrol. The search for biocontrol agents must be based on finding the most promising antagonist when the life cycle of the pathogen is taken into account. In some situations, the most promising biocontrol agent will be isolated from a plant in which disease is already established (Köhl et al., 2011). In others, healthy plants will be investigated (Medeiros et al., 2010). There is no absolute rule as to which direction will be most successful: a combination of both strategies increases the likelihood of obtaining the most promising biocontrol agent. The last step of the screening approach is integration of the biocontrol product with IPM approaches. The initial search should be undertaken in the context of the relevant crop management practice(s). For instance, in the search for a biocontrol agent for foliar disease control in conventionally managed soybean, it is less likely that the most promising biocontrol agent will be obtained from an organically managed crop. In order to assure foliar protection, particularly against Asian soybean rust, fungicides from different modes of action are sprayed throughout the growing cycle of the crop (Godoy et al., 2016). Even where a single mode of action is used, the sprays will alter the composition of the microbiome (Batzer and Mueller, 2020). Putative biocontrol agents such as *Mortierella* may be considered regardless of the soybean management system (Longley et al., 2020). However, it is probable that these populations will have originated from having withstood fungicide exposure in the conventionally managed crop.

3.3 Exclusive screening methods based on competitive advantages of a biocontrol agent

Ecological, regulatory and culturable traits are fundamental for a microorganism used in disease control. Traits include:

- tolerance to low temperature (4°C) and water availability (-10 MPa);
- no growth at human body temperature (37°C);
- tolerance to ultraviolet light B; and
- growth at pH 5 to 8.

Testing for as many of these traits as possible should be considered in product development. Other tests which are of particular importance in the targeted disease should also be carried out. In potato production, for example, soft rot during storage is a significant problem caused by *Pectobacterium carotovorum* even under a high CO_2 environment to improve shelf life. The ability of the biocontrol agent to grow as a facultative anaerobe should be considered in testing. *Bacillus* spp. may meet this requirement (Clements et al., 2002). This implies both that screening tests will be successful in recovering bacterial strains, and that an approach considering strains already identified from deposited collections could direct attention towards anaerobe bacteria such as *Pantoea, Erwinia, Salmonella, Enterobacter, Citrobacter* and *Klebsiella* (Wang et al., 2006).

3.4 Data mining for identification of candidate antagonists and comparison with microbial species presenting a potential threat to plants or animals

Although only proposed at Step 4, high throughput DNA extraction and barcode gene sequencing should be performed at the initial screening step. This reduces the cost of testing microbial strains and speeds up the screening procedure, reducing the risk of exposing workers to potentially harmful microorganisms. For instance, plant-associated bacteria such as *Salmonella, Lysteria* or even *Bacillus anthrax* (Wardeh et al., 2015) would be excluded from all screening approaches at this step, or possibly earlier.

3.5 Efficacy testing of pathogen control on plants

Microbial strains which pass exclusive tests are evaluated for their efficacy against the target disease. Various *in vitro* based tests have been proposed to speed up such steps. Some of these have recently been evaluated for grey mould (*Botrytis cinerea*), bacterial spot (*Pseudomonas syringae* pv. *tomato*) and growth promotion in tomatoes and do not always correlate (Köhl et al., 2020). Once a screening approach has been attempted for a target disease, a second screening needs to be performed for a second disease or desired microbial trait. High throughput screening approaches rely heavily on a given trait or mode of action and good potential antagonists may be lost through such a narrow approach. For instance, evaluating the ability of an antagonist to displace the pathogen from a piece of tomato stem, used in bioassays as a model for stem wounds, reinforces the importance of competition for space and nutrients or even antibiosis. However, the capacity for induced resistance may be neglected because, even when the plant stem shows an immune response, it does not occur in a similar pattern to that of an unwounded plant. Conversely, for better performance of biocontrol agents, different antagonists are often necessary for each target pathogen within the same crop. Companies should work with products which combine more than one microorganism. Further details about this topic are discussed later in the chapter.

3.6 Preliminary assessment of mass production

Depending on the number of strains which pass Steps 3 and 4, mass production tests may be performed before or in parallel for evaluation for disease control.

Minimum threshold sporulation of 10^5 cells/mL is proposed to exclude strains which have sporulation less than that obtained in the industry (>10^9), but excludes microorganisms which are difficult to mass-produce economically.

3.7 Pilot formulation

This is of primary importance in making an economically viable product. Formulation must assure a good shelf life and the dispersal of propagules in the spray tank. This step should be taken before field trials in order to discard unsuitable microorganisms and to incorporate traits which may assure better performance. These include surfactants, anti-humectants, adhesive or even UV protectants, depending on the intended technical positioning of the final product.

3.8 Upscaling mass production and full field testing

At this stage, there should be one microbial strain which fulfils all the previous criteria. This strain should work by itself with *in situ* production of the metabolites involved in biocontrol. Once biocontrol has been proved effective in disease control, fermentation can optimise production of the metabolites in order to improve efficacy. These metabolites are implicated in initial disease control in both fungi (Reino et al., 2007) and bacterial biocontrol agents (Compant et al., 2005). This essential function can be confirmed by further evaluating defective mutant lines through application of the metabolite (Zeriouh et al., 2011).

3.9 Integration into cropping systems management

Despite being proposed as the final step, this process of integration should be considered throughout the screening process. Once validated by the previous steps, integration is achieved through fine-tuning the technical specification of the product. This process must consider its efficacy in differing regions and disease pressures, delivery options, etc. It is important to consider the terms of the product recommendation if the biocontrol product is to be widely adopted. For instance, in recommending products for white mould (*S. sclerotiorum*) management (see later case study), the purpose of the biocontrol products is presented as being for sclerotia parasitism. Therefore, they should be sprayed early in the crop season and also at the end of the season in order to parasitise the sclerotia as they are produced. This should also be done in a non-host crop rotated with the susceptible crop. A product sprayed at the beginning of the season should be combined with a compatible herbicide before or after sowing. Since a particular level of soil humidity is required for parasitism to occur, performance guidelines are required which should describe the timing of product usage and the most conditions required for its optimum performance.

3.10 *Case study: searching* Trichoderma *spp. for control of white mould caused by* Sclerotinia sclerotiorum

White mould is the most significant disease of legume field crops such as soybean, bean and many other horticultural crops. Biocontrol relies on parasitism of sclerotia and therefore reduction of apothecia formation and ascospore release. The biocontrol product is applied at the initial phenological growth stages of the crop (e.g. V2 and V4 of soybean) and is normally sprayed in combination with a compatible herbicide. Efficacy depends on the presence of infested crop stubble in the soil which not only offers an alternative nutrient source for the initial establishment of the applied microorganism, but also an environmentally buffered system to assure sclerotia parasitism (Görgen et al., 2009). The search for suitable biocontrol agents for white mould and other sclerotia-forming pathogens should be attempted in the overwintering stage as this is the weakest niche and moment for the pathogens. More importantly, the pathogens present throughout the year should be considered, especially where two or three crops are cultivated in 1 year, as in tropical regions. The biocontrol effect here should also be considered at the most unfavourable moment for the pathogen. In the case of *S. sclerotiorum,* recent evidence shows that the time of optimum performance for the parasitism of sclerotia is not in the cropping season of the most susceptible crop planted in the most favourable moment for white mould epidemics (an average temperature below 20°C). It will be more effective in the summer when a less susceptible or even non-host crop such as maize is planted.

4 Non-traditional biocontrol agents of plant diseases: entomopathogenic fungi and bacteria

When screening bioprotectant agents for plant disease pathogens, non-traditional options should also be considered. These non-traditional agents are rarely considered or even analysed in the selection process. However, in recent years the potential of entomopathogenic fungi (*Beauveria bassiana, Metarhizium anisopliae, M. robertsii*) and the bacteria *Bacillus thuringiensis* in controlling plant diseases and promoting growth has been demonstrated (Ownley et al., 2008; Zhou et al., 2008; Ownley et al., 2010; Sasan, 2012; Sasan and Bidochka, 2012; Ravindran et al., 2014; Greenfield et al., 2016; Qi et al., 2016; Liu et al., 2017; Raad et al., 2019; Sarven et al., 2020; Tomilova et al., 2020).

There is an extensive body of knowledge related to the fermentation and formulation of *Metarhizium* and *Beauveria* in regard to their efficiency in controlling various pests. There are also a large number of commercial products available on the market. It is therefore necessary to assess the effect of these specific fungi-based products on plant pathogens. The potential for controlling plant diseases should also be considered in the process of selecting microbial agents. This will reduce the cost of obtaining a bioprotectant and will increase its market potential. Considering that *B. thuringiensis* has been successfully used for some decades (Cawoy et al., 2012), the potential of this organism as biological control agent of plant diseases and as plant growth promoter should be explored.

4.1 Powdery mildew

The suppression of powdery mildew in grapevine by mycophagous mites has been reported (English-Loeb et al., 1999; Melidossian et al., 2005; English-Loeb et al., 2007; Weber et al., 2016). There are programmes for developing biocontrol agents for this group of pathogens, for example, those causing powdery mildew in cereals (*Blumeria graminis*) (Köhl et al., 2019). However, these programmes are based only on the selection of microorganisms for controlling the causal agents of powdery mildew.

Powdery mildew caused by *Erysiphe necator* is a serious worldwide pathogen in grapevine plants (Gadoury et al., 2012). Its control is based on chemical fungicide sprays which often give rise to the development of resistance. English-Loeb et al. (1999) first reported the potential of *Orthotydeus lambi,* a tydeid mite, for controlling powdery mildew in riverbank grape (*Vitis riparia*) and cultivated grape (*Vitis vinifera*) under greenhouse and vineyard conditions. They also reported that the fungicides used to control the pathogen suppress this mite. Melidossian et al. (2005) and English-Loeb et al. (2007), using wild populations of *O. lambi* obtained from *V. riparia,* confirmed its potential to control grape powdery mildew. According to English-Loeb et al. (2007), the combination of the two tactics (mites and selective fungicides) provided a significant increase in yield over either of these tactics used alone. Thus, for an IPM programme for powdery mildew, it is important to select fungicides which do not affect the population of mites.

Rearing *O. lambi* in the laboratory is a prerequisite to the release of these biocontrol agents in the vineyard. However, according to Gadoury et al. (2012), for *O. lambi* to be commercially viable, fungicides which control powdery mildew and other diseases without harming mites are required. This requirement has been a concern during the last decade in developing strategies to control diseases and pests and has lead to an increase in the use of new organisms.

4.2 Coffee leaf rust

Coffee leaf rust caused by *Hemileia vastatrix* is the most significant disease in *Coffea arabica* and *C. canephora* in Brazil, causing losses ranging from 30%

to 50% if no disease control measure is implemented (Zambolim, 2016). Despite the availability of resistant varieties, susceptible varieties are still the most widely grown, which has led to widespread use of chemical fungicides (Zambolim, 2016). However, with the need to reduce pesticide use in coffee plantations, the use of key fungicides such as triazoles in rust management has been restricted. In consequence, the adoption of biological control has grown along with the need for research into biocontrol products and their development.

While only *Bacillus*-based products are available for coffee leaf rust control, other biocontrol agents have already been proposed, such as *Lecanicillium* (Eskes et al., 1991; Bettiol et al., 1994; Haddad et al., 2009) which act synergistically in disease control alongside *Bacillus* products. While the bacterium protects the leaf tissue through antibiosis and induced resistance, the hyperparasite reduces inoculum viability. The lower growth rate, reduced sporulation and dependence on environmental conditions (Martins et al., 2015) may have caused a lack of commercial interest in developing a *Lecanicillium*-based product. However, there is clear evidence of the role played by involving ant-hemipteran mutualism which allows *Lecanicillium* to become locally abundant, thus contributing to the increase in sporulation to the parasite *Hemilia vastatrix* urediniospores (Vandermeer et al., 2009). Such ecological relationships in naturally occurring biocontrol may enhance performance of biological control. In this particular case, the amendment of chitin to the medium/formulation boosts chitinase production (Villa-Verma et al., 2013), which plays an important role in the efficacy of the biological control (Nguyen et al., 2015).

Another promising biocontrol approach for coffee leaf rust has been proposed by Oliveira et al. (2014) who highlighted *Ricoseius loxocheles*, a phytoseiid mite which feeds on the urediniospores of coffee leaf rust. Oliveira et al. (2014) concluded that some species of mites may control plant diseases caused by fungi and suggest that *R. loxocheles* might be considered as a biological control agent for coffee rust.

5 Non-traditional biocontrol agents of plant diseases: bacteriophages and mycoviruses

Several mites have been commercialised as biocontrol agents for pests (van Lenteren et al., 2018). The use of this group of organisms as more widely marketable biological control agents for plant diseases should be considered.

The management of plant diseases caused by bacteria is basically achieved through the use of antibiotics and copper compounds. This control model has caused problems such as the emergence of antibiotic-resistant strains. Bacteriophages have been posited as an alternative in controlling bacterial plant diseases (Jones et al., 2007, 2012; Frampton et al., 2012; Buttimer et al.,

2017; Harada et al., 2018; Vu and Oh, 2020). They comprise a group of viruses which infect and kill only targeted bacteria with no negative effects on human or animal cells (Jones et al., 2012; Principi et al., 2019; Vu and Oh, 2020).

According to Jones et al. (2007, 2012), the use of bacteriophages to control plant diseases presents significant challenges. In addition to the correct selection of bacteriophages, application strategies and the development of formulations are crucial for success. Plant bioassays are indispensable for the evaluation of biocontrol activities with bacteriophages because *in vitro* selection provides optimal conditions for phage infections which do not represent environmental conditions (Jones et al., 2012). García et al. (2019) have reviewed models of bacteriophage production. The efficiency of commercial bacteriophage use depends on environmental conditions, although formulation is also very important for commercial success (Frampton et al., 2012; Jones et al., 2012; Vu and Oh, 2020). AgriPhage-CMM for canker, AgriPhage-Fire Blight, AgriPhage-Citrus canker and AgriPhage XCV and PST which respectively control bacterial spot and bacterial speck are, among others, examples of very selective biocontrol agents which are commercially available (https://www.agriphage.com/product-info/; Jones et al., 2012; Buttimer et al., 2017; Harada et al., 2018; García et al., 2019).

Mycoviruses (fungal viruses) which induce hypovirulence offer another non-traditional biocontrol agent and have been receiving considerable attention (Picarelli et al., 2017; Wang et al., 2017; García-Pedrajas et al., 2019; Jiao, 2020). Cryphonectria parasitica hypovirus 1 (CHV1) is the best-known mycovirus and also the first to be used as an agent for the control of chestnut blight caused by *Cryphonectria parasitica* (Heiniger and Rigling, 1994; Jiao, 2020).

Mycoviruses are associated with inducing suppression of a large number of plant pathogenic fungi by hypovirulence. Mycoviruses involved in suppression include:

- CHV1, CHV2 and CHV3 on *C. parasitica*;
- HetRVP on *Heterobasidion annosum*;
- SsHV1, SsHV2, SsDRV, SsBRV2, SsHADV1 and SsPV1 on *S. sclerotiorum*;
- SmEV1 on *Sclerotonia minor*;
- RsEV1 and RsPV2 on *Rhizoctonia solani*;
- FgHV2 and FgV-ch9 on *Fusarium graminearum*;
- FodV1 on *Fusarium oxysporum* f. sp. *dianthi*;
- FsV1 on *Fusarium solani*;
- VdCV1 on *Verticillium dahliae*;
- RnMBV1 and RnMBV2 on *Roselinia necatrix*;
- BcMV1 and BcPV2 on *B. cinerea*;
- BpRV1 on *Botrytis porri*;
- AsHV1 on *Alternaria alternata*;

- BdCV1 on *Botryosphaeria dothidea*;
- MoCV1-A on *Magnaporthe oryzae*; and
- EcEV on *Erysiphe cichoracearum*.

These and others have been extensively researched (Yu et al., 2013; Picarelli et al., 2017; Wang et al., 2017; García-Pedrajas et al., 2019; Lemus-Minor et al., 2019).

García-Pedrajas et al. (2019) have suggested four steps in the implementation of control strategies based on the use of mycoviruses:

1 Identification of the mycovirus infecting the pathogen to be controlled through the production of isogenic infected and virus-free strains.
2 Analysis of the potential for hypovirulence-inducing mycovirus to be efficiently spread in the pathogen populations.
3 Molecular manipulation of the mycovirus and/or its hosts to enhance its biological control potential.
4 Development of biocontrol formulations and deployment methods.

These steps provide a framework for commercial development.

6 Niche markets for biocontrol agents

Waage (2007) has augmentation biological control as a sustainable method of pest control that promotes biodiversity. This has found its niche in crops where chemical pesticides are undesirable, difficult to use and/or relatively ineffective. However, the market has undergone major changes during the last 15 years, and major crops, such as soybean, cotton, corn and coffee, among others, are also being targeted by companies for the development of bioprotectants. In general, large companies marketing chemical pesticides and bioprotectants focus on the largest world markets to optimise income and cover significant development costs. However, the development of bioproducts for certain niche markets may be commercially viable for small- to medium-sized companies and start-ups.

Rezende and Pacheco (1998) demonstrated the effectiveness of cross-protection with two mild strains of papaya ringspot virus-type W (PRSV-W-1 and PRSV-W2) for the control of mosaic disease caused by PRSV-W in zucchini squash. The authors observed that the production of marketable fruits increased by more than 327% in assays conducted under field conditions for 2 years. According to the authors, 'some small growers and one nursery company are already producing zucchini squash seedlings pre-immunised with PRSV-W-1 for commercial use, and some growers have expressed satisfaction with the result they have obtained with their protected crop'. Due to the limited market of zucchini growers, PRSV-W1 or PRSV-W2 is only commercially viable for small companies and there is a need to develop technologies that will deliver a stable and easily used product.

Other examples of niche markets are:

- *Acremonium altematum* and *A. persicinum* for the control of tar spot in coconut (*Catacauma torrendiella* and *Coccostroma palmicola*) (Sudo, 1989; Bettiol, 1996);
- *Dycima pulvinata* (syn. *Hansfordia pulvinata)* for the control of South American leaf blight (*Microcyclus ulei*) in rubber tree (*Hevea brasiliensis*) associated with polyclone plantation (Junqueira et al., 1989/1993/1996; Melo et al., 2008);
- *Trichoderma stromaticum* for the control of witches' broom in cacao caused by *Moniliophthora perniciosa* (Costa et al., 2009); and
- mild strains of Citrus Tristeza virus for the control of Citrus tristeza through preimmunisation or cross-protection (Costa and Müller, 1980; Müller and Costa, 1991).

Some of these are discussed in the following case studies.

6.1 Case study: mycoparasite isolation directly from the pathogen structure in tropical conditions

Acremonium altematum and *Acremonium persicinum* are used for the control of tar spot of coconut (*Catacauma torrendiella* and *Coccostroma palmicola*). Tar spot disease in coconut causes premature senescence of the leaves which bend at the axilla, resulting in the loss of fruit bunches. The mycoparasites *A. alternatum* and *A. persicinum* were isolated directly from stromas of *C. palmicola* and *C. torrendiella* in a commercial orchard in Paraiba, Brazil, in 1985. These antagonists were sprayed on cultivated areas where there was no natural occurrence of the mycoparasites. Three weeks after application, 80% of the stromas were colonised by the mycoparasites which reduced development of the disease. In spite of the pathogen being present, no death of leaves and bunches occurred and satisfactory yields were obtained. The application of mycoparasites must be carried out in the rainy season and at the end of the day. This is an example of successful control in a niche market (Sudo, 1989; Bettiol, 1996).

6.2 Case study:* Dicyma pulvinata *(syn.* Hansfordia pulvinata*) for the control of South American leaf blight (*Microcyclus ulei*) in rubber trees associated with polyclone plantations

South American rubber tree leaf blight has caused drastic economic losses in Brazil. This pathogen infects young leaves, causing senescence and weakening of the plants. Following the observation of *D. pulvinata* colonising stromas of *M. ulei,* a programme for the application of this parasite to monoclonal and polyclonal rubber tree plantations began in 1985. The fungus was isolated directly from the parasitised stromas. The number of stromas of *M. ulei* on the leaf surface, the number of stromas colonised by the mycoparasite, and leaf retention were evaluated at six monthly intervals after applications of *D. pulvinata* between 1986 and 1989. *Dycima pulvinata* efficiently colonised stromas on the susceptible clone cultivated in a polyclonal rubber tree plantation, though not in monoclonal plantations. *D. pulvinata* was also effective in commercial rubber tree plantations. *D. pulvinata* was used along with *Sporotrix insectorum* and *Hirsutela verticillioides* for the control of South American leaf blight, hevea lace bug and *Calacarus hevea* in an IPM programme (Junqueira et al., 1989, 1993, 1996; Junqueira and Gasparatto, 1991; Bettiol, 2003; Melo et al., 2008).

6.3 Case study: **Trichoderma stromaticum** *for the control of witches' broom in cacao caused by* **Moniliophthora perniciosa**

Witches' broom poses a major threat to cacao plantations in Brazil. *Trichoderma* was isolated directly from dry brooms of cacao collected in Belém in the Brazilian Amazon. Initially, the isolate was misidentified as *Trichoderma viride*, but was later correctly identified as *T. stromaticum,* a new parasitic species of the pathogen. A commercial product (Tricovab®) based on *T. stromaticum* has been registered. The product is recommended for application on dry brooms located on the ground or in the crown of cacao trees and inhibits the formation of basidiomas of the pathogen. It has been used successfully to reduce the disease in cocoa in Bahia, Brazil since 1999 although the hyperparasite does not assure a 100% reduction of basidiocarp formation. In IPM, the spraying of fungicides, such as copper oxychloride, is essential. The biocontrol agent and the fungicide need to be sprayed in alternation every 15 days. This has proved a successful control for countering witches' broom in cacao (Bastos, 1988, 1996; Bezerra et al. (2000); Samuels et al., 2000; Costa et al., 2009; Medeiros et al., 2010).

6.4 Mild strains of Citrus tristeza virus for the control of Citrus tristeza through preimmunisation or cross-protection in São Paulo State, Brazil

During the 1940s it was found that plants of Pera sweet orange, even when budded on Tristeza tolerant rootstocks, were susceptible to the Tristeza virus, thus reducing their productivity. However, observation showed that citrus trees infected with mild strains of the Tristeza virus were protected against infection by severe strains of the virus and a programme of pre-immunisation or cross-protection was started in 1961. A mild virus strain obtained from badly infected orchards was selected and used to infect healthy plants. In 1971, the widespread use of pre-immunised plants was begun. Currently, almost all the trees of Pera sweet orange in production in the State of São Paulo are pre-immunised. The cost of pre-immunised and non-pre-immunised plants is similar. Plants cross-protected with these mild strains grew well, had practically no symptoms of tristeza or stem pitting, and produced good yields when compared with plants from the same clones which had not been cross-protected. Weak strains are not commercialised but are made available to seedling producers free of charge. This is perhaps one of the most successful examples of a biocontrol agent for the control of plant diseases, having been effective since its launch in 1971 and at practically no cost to the farmer (Meneghini, 1946; Grant and Costa, 1951; Moreira, 1959; Costa and Müller, 1980; Müller and Costa, 1991).

6.5 Niche markets: mycotoxin control

Public awareness of the chemical contamination of food by pesticides has also aided the global promotion of biocontrol and led to further investigation into other contaminants which could harm animal and human health. Among those food contaminants, harmful metabolites produced by fungi, also known as mycotoxins, pose the greatest threat. Although there is extensive literature addressing the biocontrol of mycotoxin, few of these examples have been turned into commercial products available to growers.

In order to prevent mycotoxin contamination of food and feed, management of the fungus which produces this toxin is essential. Aflatoxin, fumonisin, deoxynivalenol, ochratoxin and patulin are considered the most significant toxins due to the frequency with which they are found in higher than acceptable concentrations. However, many other mycotoxins are reported which pose local threats to food and feed production. While very successful biocontrol strategies have been proposed, none of them are effective against all mycotoxins.

The most important strategy for controlling mycotoxins is control of the mycotoxigenic fungus population. The most successful biocontrol strategy relies on competitive advantage over mycotoxin-producing fungi. The most widely known method of biocontrol of mycotoxins is inundative delivery of defective strains of *Aspergillus flavus* in the production of aflatoxin and cyclopiazonic acid produced by *Aspergillus flavus* and *Penicillium cyclopium,* respectively (King et al., 2011). Such mycotoxin-producing defective strains occur naturally and belong to the same vegetative-compatibility group (VCG).

The success of these methods of biocontrol relies on their distribution in the field in reducing aflatoxin accumulation or colonisation of the grain by aflatoxin-producing fungi (Grubisha and Cotty, 2015). An example may be found in the United States where AF36, one of the USEPA licensed atoxigenic strains which bears a single nucleotide polymorphism (SNP), causes an early stop codon in the polyketide synthase gene (*aflC*) (Ehrlich and Cotty, 2004). This belongs to the Y36 VCG widespread in the USA and Mexico (Ortega-Beltran et al., 2016). A product based on this strain can therefore be deployed throughout the country, wherever the mycotoxin problem occurs (Grubisha and Cotty, 2015).

However, the isolation of locally occurring atoxigenic VCGs is more likely to result in a greater reduction of aflatoxin contamination than that achieved by the use of exotic strains (Mehl et al., 2012). Governments should invest in technologies for isolation and mass production of the local atoxigenic-*A. flavus* VCG (Bandyopadhyay et al., 2016). This approach has been proposed for the development of Aflasafe®, a biocontrol product which utilises the active ingredient of the dominant strain of each region/country in the African continent (Moral et al., 2020).

Both fungi and bacteria have been proposed for biocontrol of *Fusarium* spp. within the *Fusarium fujikuroi* species complex, particularly *Fusarium verticillioides,* and the *Fusarium graminearum* species complex, particularly *Fusarium graminearum* (Nguyen et al., 2017). While *A. flavus* colonises the grain after it has been formed, *Fusarium* spp. may infect the plant systemically or through the stigma-stylet, resulting in early season infection which is difficult to control (Medeiros et al., 2012). Consideration should therefore be given to microorganisms which rapidly colonise flowers, bacteria and yeasts. There is extensive literature on the role of fungi such as *Trichoderma* spp. (Ferrigo et al., 2014), *Clonostachys* spp. (Palazzini et al., 2013) and even *Fusarium* spp. (Luongo et al., 2005) in the control of *Fusarium verticillioides* and *F. graminearum* and in the reduction of mycotoxin production. However, these putatively beneficial fungi may also colonise maize kernels, thus reducing the marketability of the product (Agrios, 2005). The longer-lasting benefit of colonisation by fungi (Ferrigo et al., 2014), and the degree of resistance induced (Saravanakumar et al., 2018), has encouraged their adoption in the management of mycotoxins, particularly through the use of *Trichoderma* spp. The role of cellulase produced by *Trichoderma harzianum* T22 in the activation of defence-related genes is significant; notably, 2,4-dihydroxy-7-methoxy-2H-1,4-benzoxazin-3(4H)-one has been shown to be directly toxic to *F. graminearum* (Saravanakumar et al., 2018)

Similar issues in the selection of biocontrol agents also apply to other important mycotoxins such as patulin, which is significant in fruit contamination and is produced by *Penicillium expansum* (Morales et al., 2008) and ochratoxin in crops such as coffee (Souza et al., 2021) and grapes (Zhang et al., 2016), produced by *Aspergillus ochraceus* (Nguyen et al., 2017). There is significant research on the breakdown of mycotoxins by microorganisms (Medeiros et al., 2012). However, these are limited due to the specific requirements of individual mycotoxin-degrading enzyme activity (Ji et al., 2016) and also because of the wide availability of mycotoxin scavengers on the market. Another strategy may be to combine both a good competitor strain, which hampers the mycotoxin-producing fungus in colonising the plant tissue, with a microorganism having a powerful break-down effect on the mycotoxin. It should be noted that many mycotoxins also have an antimicrobial role (Burmeister and Hesseltine, 1966; Ali-Vehmas et al., 1998; Burel et al., 2013; Waché et al., 2009). Competition with the introduced beneficial microbe may foster mycotoxin production during competition, whereas the introduction of a mycotoxin-degrading microorganism may assist the competitor microbial strain.

7 Regional markets for biocontrol agents

Many of the biocontrol agents developed in temperate and tropical conditions are not efficient in regions with arid or semi-arid climates as high temperatures and drought are inhospitable to microorganisms. The isolation and selection of potential biocontrol agents from organisms originating in the target environment are thus essential for the development of products to be used in these regions. However, as this market is limited in many pathosystems, these agents may not be financially viable for large companies. It would therefore be appropriate for public institutions to develop biocontrol agents for these pathosystems and transfer them to small or regional enterprises to meet the demand. This has become more important due to the restrictions the bigger agri-food market chains place on the levels of permissible pesticide residue in fruits and vegetables. Papaya, cape gooseberry, mango, melon, pineapple, date and others in specific climatic conditions are therefore potential niche markets for bioprotectants.

7.1 Temperature

Several biocontrol agents developed in a temperate climate may not be adaptable to the tropics. The average maximum and mean annual air temperatures in the most important cereal production regions in Brazil are >30°C and >24°C, respectively (Alvares et al., 2013/2014). Assessment of the soil temperature in Irecê, Bahia, Brazil, Farias et al. (2018) showed that January, March, September and October had the highest mean soil temperature readings (reaching 46.2°C, 37.3°C and 32.3°C at depths of 0.05 m, 0.10 m and 0.2 m depths respectively, in hourly data, and 41.5°C, 34.0°C and 32.5°C in daily readings). The authors observed that the thermal amplitude of the soil at a depth of 5 cm was 24.4°C and 18.1°C in September and October, respectively. Antagonists are therefore required to survive these conditions. These climatic data suggest that air and soil temperatures for tropical climates should be considered in the process of isolating and selecting antagonists. Isolation and selection under extreme conditions may be employed, for example, as described by Vero et al. (2013) when isolating yeasts from Antarctic soils for control of pathogens in post-harvest apple fruits stored in cold chambers (Table 1).

7.2 Climate change and other stressors

Desertification, salinisation, extreme temperatures and drought all impact agricultural production in many regions of the world (Ilangumaran and Smith, 2017; Ray et al., 2018; WMO, n.d.). Microorganisms may be used to mitigate the negative effects of abiotic stresses by increasing plant tolerance under these conditions (Medeiros and Bettiol, 2021; Ilangumaran and Smith, 2017; Numan et al., 2018), Vardharajula et al. (2011). The combination of protective beneficial microbes with biological control agents protects plants against abiotic and biotic stresses and promotes growth, enabling continuity of food production in regions affected by adverse environmental conditions. The search for antagonists with these characteristics could be carried out by public institutions in public-private collaboration. This would minimise development costs by making products available for these growing markets which use organisms selected for their regional conditions. Climate change will affect the occurrence of plant diseases and control methods. Although the effects of climate change on plant diseases have been studied for the past two decades, there is limited information as to the effects of biocontrol agents. It is therefore important to consider climate change during the process of isolation and selection of antagonists.

8 Formulation of biocontrol agents

The most important function in formulating a microorganism is to maintain propagules with low or no metabolic activity, whilst keeping it viable and virulent for the longest possible storage time (Ravensberg, 2011). Bioprotectants are composed of an active ingredient (microbial structure and/or metabolite) which exerts biological activity, and excipients or additives (inert substances) which protect and release the active ingredient. These additives or excipients must achieve a number of objectives (Montesinos, 2003; Ravensberg, 2011; Corrêa et al., 2015; Segarra et al., 2015; García et al., 2019):

- ensure the stability of the product;
- protect the active principle from ultraviolet radiation and desiccation;
- retain moisture;
- promote the dispersion of the biocontrol product;

- facilitate its application; and
- ensure it continues to act effectively on its target.

According to Garcia et al. (2019), a formulation should improve the stability of a product in storage to ensure a shelf life of more than 1 year. Formulation must also support the consistent effectiveness of a product in the field, preferably in several crops and against various pests under different climatic conditions.

Segarra et al. (2015) proposed six steps in developing a formulation for *Lysobacter capsici* AZ78, a non-spore-forming bacterium:

1. Selection of growth conditions and medium
2. Assessment of the shelf life of the harvested cells
3. Selection of additives to protect *L. capsici* AZ78 against desiccation, UV light, wash off and other environmental stresses
4. Evaluation of the side effects of additives on *L. capsici* AZ78 cells, *P. viticola* and leaf discs from grapevine plants
5. Validation of the combined additives on the efficacy and persistence of *L. capsici* AZ78 against *P. viticola* under field conditions
6. Choosing whether to use additives in tank mixes or as co-formulants

García et al. (2019) divided the development of formulation into two phases:

- pre-formulation; and
- formulation.

In the pre-formulation step, the active ingredients and additives of the formulation prototypes are designed and their compatibility is tested. After these steps, the formulation itself is prepared. Next, stability and efficiency studies for the selection of the possible formulation are carried out. These studies cover the composition of the formulation, the manufacturing process, the type of packaging and stability studies.

9 The role of the microbiome in biocontrol

Recent developments in sequencing technology have advanced knowledge of the structure and function of plant-associated microbial communities. Several studies have described the composition of microbial communities associated with different cropping systems (Hamonts et al., 2018), their occurrence in natural environments (e.g. Moitinho et al., 2020; Viana et al., 2011) and the changes arising from different strategies in management of the soil (Hartmann et al., 2015; Zhu et al., 2016). Other studies have highlighted the role of the plant microbiome on host nutrition (Singh and Trivedi, 2017), resilience to abiotic stress and protection against diseases (Carrión et al., 2019; Mendes et al., 2011). These studies have highlighted the importance of the plant microbiome and its role in sustainable agriculture (Busby et al., 2017; Singh and Trivedi, 2017).

9.1 Microbiome composition and biotic interaction

The term microbiome refers to the microbial community and the structural elements, molecules and genes present in the environment (Berg et al., 2020). The composition of the plant microbiome relies on biotic interactions with the host which continually releases exudates through its roots which select resident microorganisms from the soil (Bais et al., 2006). Colonisation is frequently established by evolution and shows the positive effects of plant fitness in the environment (Bulgarelli et al., 2013; Lemanceau et al., 2017). Changes during the plant's developmental stages also cause significant alteration in exudation patterns, thus altering the microbial community according to substrate preferences in the rhizosphere (Zhalnina et al., 2018).

9.2 Microbiome studies

Despite the variation in microbiome composition among different crops, plant developmental stages and soil's physical and chemical characteristics, some bacterial phyla have been observed to predominate in rhizosphere studies. Proteobacteria, Actinobacteria, Firmicutes, Acidobacteria and Bacteroidetes are frequently found both in the rhizosphere and other areas, including the phyllosphere and endosphere (Rossmann et al., 2017). However, due to the functional redundancy observed in complex ecological systems such as soil, it is expected that the same plant species growing in different soil types may develop associations with a different bacterial group having a common function. It is therefore a sound strategy to target these functions by identifying the genes associated with them and then extrapolate to the bacterial taxa which enable these functions in the plant (Lemanceau et al., 2017). This suggests the possibility of targeting a minimally functional microbiome composed of the minimum set of microbial traits necessary for a given function in the ecosystem (Mendes et al., 2013).

Microbiome studies enable the understanding of mechanisms involved in plant-pathogen-microbe interactions in the context of biological control. The basic mechanism for controlling pathogen invasion in the rhizosphere is niche occupation by a structured and diverse microbial community. Invasion by soil-borne pathogens may be inhibited by assembling an intricate network of microorganisms which use plant exudates as resources. In common bean, for instance, plant breeding may have unintentionally selected plants which are able to recruit a well-structured community of bacteria involved in their defence (Mendes et al., 2018). Chapelle et al. (2016) described how sugarbeet seedlings, when attacked by a soil-borne pathogen, release exudates which promote the proliferation of specific bacterial taxa and the activation of microbial functions which protect their roots. This mechanism makes the rhizosphere microbiome the first line of defence against soil-borne pathogens (Berendsen et al., 2012).

9.3 The rhizosphere microbiome and control of above-ground pathogens and root parasitic weeds

The rhizosphere microbiome is also important in the biological control of above-ground pathogens or insect pests through induced systemic resistance (ISR). When a plant is injured above or below ground, several chemicals and signalling hormones in its exudation profile will change. These changes select specific microorganisms in the rhizosphere or in different organs of the plant which are able to trigger ISR, thus protecting the host (Pieterse et al., 2014; Pascale et al., 2020).

The plant microbiome also has the potential to control root parasitic weeds through various mechanisms. Microorganisms can show specific pathogenicity to these weeds, by limiting their growth (Chen et al., 2016), killing the endosperm of seeds in the soil (Thomas et al., 1999) and interfering with signals recognised by the parasite. These responses limit pathogen invasion and may even induce systemic resistance in the plant (Ali et al., 2013; Yoder and Scholes, 2010). However, many questions still remain unanswered on this subject (Masteling et al., 2019).

10 Microbiome engineering for disease control

The understanding of plant microbiome assembly and function has offered possible new strategies for biological control. While traditional strategies are based on a single or a small number of microbial isolates (Kaminsky et al., 2019), microbiome-based manipulation or engineering allows for the use of complex microbial communities in supporting plant development and health. Some of these strategies are shown in Fig. 2 and are listed in the following sections.

10.1 Plant breeding for beneficial microbiomes

Based on tomato plants, Smith et al. (1999) were among the first to highlight the potential of exploring relationships between crops and their associated microbiome for the purpose of biological control. This strategy of rhizosphere

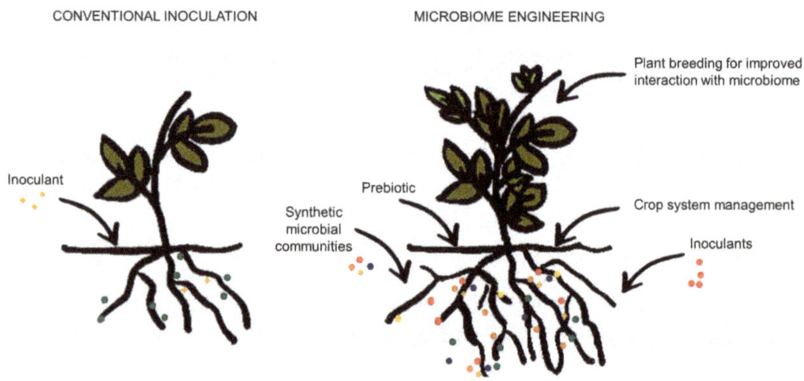

Figure 2 Comparison between conventional inoculant approach and strategies for microbiome engineering aiming for biological control in plants. Microbiome engineering consists of promoting richness and diversity in the microbial communities associated with the plant in order to modulate the host resulting in protection and growth promotion. This can be achieved by different strategies, including the use of plants improved to recruit beneficial microbes, crop system management, use of prebiotics, use of synthetic microbial communities and use of inoculants aiming for changes in microbiome composition and functions.

engineering consists in changing plant metabolism through breeding to improve a beneficial microbiome (Ahkami et al., 2017).

10.2 Management of cropping systems

Interactions between microbes determine rhizosphere assembly and pathogen control. When using this strategy, the concepts of microbial ecology are important for understanding the success of a given crop management system. For example, microbial interactions such as competition and cooperation promote a diversity of microorganisms and niche occupation that can also determine the fate of a pathogen invasion (Mendes et al., 2018). In this case, system management exerts a strong effect on disease control. Crop rotation and intercropping potentially promote soil diversity and are currently applied to manage pathogen propagules and decrease the incidence of disease (Hwang et al., 2009). Microbial ecology concepts are important both for the improvement of microbial communities as a whole and for the promotion of the viability and persistence of biological control agents (Massart et al., 2015).

10.3 Exploitation of suppressive soils

Soil-borne pathogens are frequently present as part of soil microbiota and may eventually invade plant tissues, potentially causing outbreaks of disease. Soils in which the pathogen is present but unable to establish disease in a susceptible host are known as suppressive soils (Baker and Cook, 1974). The suppressive feature occurs as a response of the plant to the presence of the pathogen in the soil. Disease suppressive soils are a consequence of natural enrichment-driven selection pressure imposed by the pathogen on specific microbial consortia (Mendes et al., 2011).

10.4 Prebiotics

As all plants are naturally associated with a complex and diverse microbiome, promoting specific members of this microbiome can lead to a desired phenotypic alteration in the host without the need to introduce exogenous inoculants. While the application of a single inoculant would act as a probiotic, the stimulation of specific members of the resident microbial community through the addition of a substrate may be considered to be a prebiotic approach. Despite developments in the understanding of microbiome assembly, knowledge of correlations between chemical molecules and the introduction of the beneficial microbiome is limited. Understanding the underlying mechanisms of microbiome assembly aids the identification of chemical compounds which can be used as prebiotics to modulate the microbial community when introduced into the system. Studying sugarbeet infection caused by the soil-borne pathogen *Rhizoctonia solani*, Chapelle et al. (2016) showed that the plant responds to oxalic acid produced by the pathogen during rhizosphere invasion by triggering microbiome activation to protect the host. This suggests that the introduction of oxalic acid can act as a prebiotic.

10.5 Synthetic microbial communities

A combination of traditional and microbiome approaches for rhizosphere modulation can be achieved by using a microbial consortium. The application of multiple microorganisms to manipulate the plant microbiome and promote phenotypical traits in the host must be informed by microbiome studies after a collection of culturable microbial isolates has been obtained (Niu et al., 2017). A minimal synthetic community able to colonise maize roots has been obtained by combining information from the maize microbiome with traditional culture-dependent approaches. By removing one microorganism at a time, it was found that interactions between the microorganisms from the synthetic community were also important in sustaining the community structure and biological control of *Fusarium verticillioides*. This illustrates the importance of making innovative use of traditional inoculants by using consortia, and by considering the ecological concept of keystone species (van der Heijden and Hartmann, 2016).

11 Future trends

The manipulation of the microbiome for disease control depends upon advances in several areas. Firstly, an extensive knowledge of the plant-associated microbiome in different soil types. NGS technologies are providing considerable information about crop-associated microbial communities (Gloor et al., 2017). The availability of information from metagenomic data on metabolic pathways is also increasing rapidly (Medema and Fischbach, 2015). A second area is the growing development of bioinformatic and statistical sources for analysing microbiome data. However, critical analysis and interpretation may still prove to be a limitation in applying microbiome data in biocontrol (Massart et al., 2015). Thirdly, accessing and isolating microorganisms by means of:

a taxonomical identification
b using a selective method to test for desired function
c searching by enzymatic activity and antagonistic chemical compounds
d developing strategies for microbiome engineering

In the latter case, there is an urgent need to elucidate the underlying mechanisms involved in the complex interactions of the 'disease pyramid' (Chiaramonte et al., 2021). This includes the host, the pathogen, the environment and the associated microbiome.

12 Where to look for further information

There are a number of articles, books, professional organisations, and research projects to support discussions about screening of antagonists for the development of microbial bioprotectants for control of plant diseases, including:

12.1 Books

- Cotes, A. M. (2018). Control biológico de fitopatógenos, insectos y ácaros. V. 1: Agentes de control biológico. Mosquera, Colombia: AGROSAVIA. 568 p.
- Cotes, A. M. 2018. Control biológico de fitopatógenos, insectos y ácaros. V. 2: Aplicaciones y perspectivas. Mosquera, Colombia: AGROSAVIA. 512 p.
- Van Lenteren, J. C., Bueno, V. H. P., Luna, M. G. and Colmenarez, Y. C. 2020. *Biological Control in Latin America and the Caribbean: Its Rich History and Bright Future*. CABI. 522p.

12.2 Articles

- Berg, G., Rybakova, D., Fischer, D., Cernava, T., Vergès, M. C. C., Charles, T. and Schloter, M. (2020). Microbiome definition re-visited: old concepts and new challenges, *Microbiome*, 8(1), 1–22.
- Carrión, V. J., Perez-Jaramillo, J., Cordovez, V., Tracanna, V., De Hollander, M., Ruiz-Buck, D. and Raaijmakers, J. M. (2019). Pathogen-induced activation of disease-suppressive functions in the endophytic root microbiome, *Science* 366(6465), 606–612.
- Cernava, T. (2021). How microbiome studies could further improve biological control, *Biological Control* 160, 104669.
- Frampton, R. A., Pitman, A. R. and Fineran, P. C. (2012). Advances in bacteriophage-mediated control of plant pathogens, *International Journal of Microbiology* 2012, ID326452.
- García-Pedrajas, M. D., Cañizares, M. C., Sarmiento-Villamil, J. L., Jacquat, A. G. and Dambolena, J. S. (2019). Mycoviruses in biological control: from basic research to field implementation, *Phytopathology* 109, 1828–1839.
- Köhl, J., Postma, J., Nicot, P., Ruocco, M. and Blum, B. (2011). Stepwise screening of mi- croorganisms for commercial use in biological control of plant pathogenic fungi and bacteria, *Biological Control* 57, 1–12.
- Köhl, J., Medeiros, F. H. V., Lombaers-van der Plas, C., Groenenboom-de Haas, L. and van den Bosch, T. (2020). Efficacies of bacterial and fungal isolates in biocontrol of *Botrytis cinerea* and *Pseudomonas syringae* pv. *tomato* and growth promotion in tomato do not correlate, *Biological Control* 150, 104375.
- Müller, C. A., Obermeier, M. M. and Berg, G. (2016). Bioprospecting plant-associated microbiomes. *Journal of Biotechnology* 235, 171–180.

12.3 Professional organisations

- IBMA: International Biocontrol Manufactures Association (https://ibma-global.org/).
- IOBC-WPRS: International Organization for Biological Control (https://www.iobc-wprs.org/).

12.4 Research project

- Biocomes. New biological control products for sustainable farming and forestry (http://www.biocomes.eu/).
- Back to the Roots (https://back2roots.nioo.knaw.nl/project-1-going-back-roots).

12.5 Centres of expertise include

- Embrapa - Brazilian Agricultural Research Corporation (https://www.embrapa.br/).
- Institute of Environmental Biotechnology/Graz University of Technology (https://www.tugraz.at/en/institutes/ubt/home/).
- Aflasafe Technology Transfer and Commercialization – ATCC (https://www.iita.org/iita-project/aflasafe-technical-transfer-commercialization/).
- Istituto Agrario San Michele All'Adige – (https://www.fmach.it/).
- Netherlands Institute of Ecology (https://nioo.knaw.nl/en).
- Wageningen University & Research (https://www.wur.nl/).

13 Acknowledgements

Wagner Bettiol (CNPq 307855/2019-8), Flavio H. V. Medeiros (CNPq 309307/2017-1) and Rodrigo Mendes (CNPq 302591/2019-2) acknowledge Conselho Nacional de Desenvolvimento Científico e Tecnológico – CNPq for the productivity fellowship. Rodrigo Mendes acknowledges Fundação de Amparo à Pesquisa do Estado de São Paulo for the research support (FAPESP 2020/00469-2).

14 References

Adams, P. B. and Ayers, W. A. (1981). Sporidesmium sclerotiorum: distribution and function in natural biological control of sclerotial fungi, *Phytopathology* 71, 90–93.

Agrios, G. N. (2005). *Plant Pathology* (5th edition). San Diego, CA: Elsevier, Academic Press.

Ahkami, A. H., Allen White, R., Handakumbura, P. P. and Jansson, C. (2017). Rhizosphere engineering: enhancing sustainable plant ecosystem productivity, *Rhizosphere* 3(4), 233–243.

Alfaro, E. E. and Obregón-Gómez, M. (2015). *Trichoderma* spp. y su rol en la agricultura moderna de Costa Rica, *Fitosanidad* 19(2), 102.

Ali, H. A., Elamin, H. B. and Dirar, H. A. (2013). Biological control of *Striga hermonthica* Del. Bendth : screening for bacteria scavenging strigol, *Am. J. Biochem.* 3(4), 89–92.

Ali-Vehmas, T., Rizzo, A., Westermarck, T. and Atroshi, F. (1998). Measurement of antibacterial activities of T-2 toxin, deoxynivalenol, ochratoxin A, aflatoxin b1 and fumonisin B1 using microtitration tray-based turbidimetric techniques, *Zentralbl. Veterinarmed. A* 45(8), 453–458.

Alvares, C. A., Stape, J. L., Sentelhas, P. C. and Gonçalves, J. L. M. (2013). Modeling monthly mean air temperature for Brazil, *Theor. Appl. Climatol.* 113(3–4), 407–427.

Alvares, C. A., Stape, J. L., Sentelhas, P. C., Gonçalves, J. L. M. and Sparovek, G. (2014). Köppen's climate classification map for Brazil, *Meteorol. Z.* 22(6), 711–728.

Andrews, J. H. (1985). Strategies for selecting antagonistic microorganisms from the phylloplane. In: Windels, C. L. and Lindon, S. E. (Eds), *Biological Control on the Phylloplane*, St Paul: American Phytopathological Society, pp. 31–44.

Andrews, J. H., Berbee, F. M. and Nordhein, E. U. (1983). Microbial antagonism to the imperfect stage of the apple scab pathogen, *Venturia inaequalis*, *Phytopathology* 73(2), 228–234.

AQ10. (1991). *Ampelomyces quisqualis* AQ10, CNCM I-807, for biological control of powdery mildew. Available at: https://patents.google.com/patent/US5190754A/en (accessed 10 December 2020).

AQ10WG Biofungicide. (2020). Available at: https://www.dejex.co.uk/PDFs/Dejex%20Supplies%20-%20AQ10%20fact%20sheet.pdf (accessed 10 December 2020).

Bais, H. P., Weir, T. L., Perry, L. G., Gilroy, S. and Vivanco, J. M. (2006). The role of root exudates in rhizosphere interactions with plants and other organisms, *Annu. Rev. Plant Biol.* 57(9), 233–266.

Baker, K. F. and Cook, R. J. (1974). *Biological Control of Plant Pathogens*, San Francisco: W. H. Freeman.

Baker, C. J., Stavely, J. R., Thomas, C. A., Sasser, M. and MacFall, J. S. (1983). Inhibitory effect of *Bacillus subtilis* on *Uromyces phaseoli* and on development of rust pustules on bean leaves, *Phytopathology* 73(8), 1148–1152.

Bandyopadhyay, R., Ortega-Beltran, A., Akande, A., Mutegi, C., Atehnkeng, J., Kaptoge, L., Senghor, A. L., Adhikari, B. N. and Cotty, P. J. (2016). Biological control of aflatoxins in Africa: current status and potential challenges in the face of climate change, W. Mycotoxin J. 9, 771–789.

Bastos, C. N. (1988). Resultados preliminares sobre a eficácia de *Trichoderma viride* no controle da vassoura-de-bruxa (*Crinipellis perniciosa*) do cacaueiro, *Fitopatol. Bras.* 13, 340–342.

Bastos, C. N. (1996). Mycoparasitic nature of the antagonism between *Trichoderma viride* and *Crinipellis perniciosa*, *Fitopatol. Bras.* 21, 50–54.

Batzer, J. C. and Mueller, D. S. (2020). Soybean fungal endophytes *Alternaria* and *Diaphorte* spp. are differentially impacted by fungicide application, *Plant Dis.* 104(1), 52–59.

Beltrán, C., Cotes, A. M. and París, A. (2007). Selection of isolates of *Trichoderma* spp. with biocontrol activity over *Rhizoctonia solani* in potato, *IOBC WPRS Bull.* 30, 55–58.

Beltrán, C., Moreno, C. A., Blanco, P., Villamizar, L. and Cotes, A. M. (2012). Biological control of *Rhizoctonia solani* and growth promotion activity of *Trichoderma koningiopsis* Th003 and *Trichoderma asperellum* Th034 formulations in potato (*Solanum tuberosum*), *IOBC WPRS Bull.* 78, 223–227.

Berendsen, R. L., Pieterse, C. M. J. and Bakker, P. A. (2012). The rhizosphere microbiome and plant health, *Trends Plant Sci.* 17(8), 478–486.

Berg, G., Rybakova, D., Fischer, D., Cernava, T., Vergès, M. C. C., Charles, T., Chen, X., Cocolin, L., Eversole, K., Corral, G. H., Kazou, M., Kinkel, L., Lange, L., Lima, N., Loy, A., Macklin, J. A., Maguin, E., Mauchline, T., McClure, R., Mitter, B., Ryan, M., Sarand, I., Smidt, H., Schelkle, B., Roume, H., Kiran, G. S., Selvin, J., Souza, R. S. C., van Overbeek, L., Singh, B. K., Wagner, M., Walsh, A., Sessitsch, A. and Schloter, M. (2020). Microbiome definition re-visited: old concepts and new challenges, *Microbiome* 8(1), 103.

Bettiol, W. (1988). Seleção de microrganismos antagônicos a Pyricularia oryzae Cav. para o controle da brusone do arroz (Oryza sativa L.), Ph.D. Thesis, Piracicaba: ESALQ.

Bettiol, W. (1996). Biological control of plant pathogens in Brazil: application and current research, *World J. Microbiol. Biotechnol.* 12(5), 505–510.

Bettiol, W. (2003). Controle de doenças de plantas com agentes de controle biológico e outras tecnologias alternativas. In: Campanhola, C. and Bettiol, W. (Eds), *Métodos alternativos de controle fitossanitário*, Jaguariúna: Embrapa Meio Ambiente, pp. 191–215.
Bettiol, W. and Kimati, H. (1989). Seleção de microrganismos antagônicos a *Pyricularia oryzae* Cav. para o controle do brusone do arroz (*Oryza sativa* L.), *Summa Phytopathol.* 15, 257–266.
Bettiol, W. and Kimati, H. (1990). Efeito de *Bacillus subtilis* sobre *Pyricularia oryzae* agente causal da brusone do arroz, *Pesq. Agropecu. Bras* 25(8), 1165–1174.
Bettiol, W., Saito, M. L. and Brandão, M. S. B. (1994). Controle da ferrugem do cafeeiro com produtos a base de *Bacillus subtilis*, *Summa Phytopathol.* 20, 119–122.
Bettiol, W., Silva, J. C. and Castro, M. L. M. P. (2019a). Uso atual e perspectivas do *Trichoderma* no Brasil. In: Meyer, M. C., Mazaro, S. M. and Silva, J. C. (Eds) *Trichoderma uso na agricultura*, Brasília: Embrapa, pp. 21–43.
Bettiol, W., Pinto, Z. V., Silva, J. C., Forner, C., Faria, M. R., Pacifico, M. G. and Costa, L. S. A. S. (2019b). Produtos comerciais à base de *Trichoderma*. In: Meyer, M. C., Mazaro, S. M. and Silva, J. C. (Eds) *Trichoderma uso na agricultura*, Brasília: Embrapa, pp. 45–179.
Bezerra, J. L., Costa, J. C. B., Pomella, A. W. V. and Almeida, O. C. (2000). Como produzir Tricovab para controlar a vassoura-de-bruxa do cacaueiro, *Fitopatol. Bras.* 25, 359.
Biocomes. (2014). Powdery mildew update, December 2014. Available at: http://www.biocomes.eu/update/powdery-mildew-update-december-2014/ (accessed 11 December 2020).
Biocomes. (2015a). New biological control products for sustainable farming and forestry–soil-borne Verticillium wilt. Available at: http://www.biocomes.eu/update/soil-borne-verticillium-wilt-update-january-2015/ (accessed 4 December 2020).
Biocomes. (2015b). New biological control products for sustainable farming and forestry–soil-borne Verticillium wilt. Available at: http://www.biocomes.eu/update/soil-borne-verticillium-wilt-update-july-2015/ (accessed 4 December 2020).
Biocomes. (2015c). Powdery mildew update, June 2015. Available at: http://www.biocomes.eu/update/powdery-mildew-update-june-2015/ (accessed 11 December 2020).
Biocomes. (2016a). Powdery mildew update, January 2016. Available at: http://www.biocomes.eu/update/powdery-mildew-update-january-2016/ (accessed 11 December 2020).
Biocomes. (2016b). New biological control products for sustainable farming and forestry–soil-borne Verticillium wilt. Available at: http://www.biocomes.eu/update/soil-borne-verticillium-wilt-update-november-2016/ (accessed 4 December 2020).
Biocomes. (2017a). Powdery mildew update, March 2017. Available at: http://www.biocomes.eu/update/powdery-mildew-update-march-2017/ (accessed 11 December 2020).
Biocomes. (2017b). Available at: http://www.biocomes.eu/wp-content/uploads/2017/11/WP8-powdery-mildew.pdf (accessed 11 December 2020).
Biocomes. (2017c). New biological control products for sustainable farming and forestry–soil-borne Verticillium wilt. Available at: http://www.biocomes.eu/update/bacterial-seed-treatments-brassica-diseases/ (accessed 4 December 2020).
Bissett, J., Gams, W., Jaklitsch, W. and Samuels, G. J. (2015). Accepted *Trichoderma* names in the year 2015, *IMA Fungus* 6(2), 263–295.

Blachinsky, D., Antonov, J., Bercovitz, A., Elad, B., Feldman, K., Husid, A., Lazare, M., Marcov, N., Shamai, I., Keren-Zur, M. and Droby, S. (2007). Commercial applications of "Shemer" for the control of pre- and postharvest diseases, *IOBCWPRS Bull* 30, 75–78.

Bulgarelli, D., Schlaeppi, K., Spaepen, S., van Themaat, E. V. L. and Schulze-Lefert, P. (2013). Structure and functions of the bacterial microbiota of plants, *Annu. Rev. Plant Biol.* 64(1), 807–838.

Burel, C., Tanguy, M., Guerre, P., Boilletot, E., Cariolet, R., Queguiner, M., Postollec, G., Pinton, P., Salvat, G., Oswald, I. P. and Fravalo, P. (2013). Effect of low dose of fumonisins on pig health: immune status, intestinal microbiota and sensitivity to *Salmonella*, *Toxins* 5(4), 841–864.

Burmeister, H. R. and Hesseltine, C. W. (1966). Survey of the sensitivity of microorganisms to aflatoxin, *Appl. Microbiol.* 14(3), 403–404.

Busby, P. E., Soman, C., Wagner, M. R., Friesen, M. L., Kremer, J., Bennett, A. and Dangl, J. L. (2017). Research priorities for harnessing plant microbiomes in sustainable agriculture, *PLoS Biol.* 15(3), 1–14.

Buttimer, C., McAuliffe, O., Ross, R. P., Hill, C., O'Mahony, J. and Coffey, A. (2017). Bacteriophages and bacterial plant diseases, *Front. Microbiol.* 8, 34.

Canadian Patent Application (2007). Use of *Clonostachys rosea* strain 88-710 and natural emulsifiers as an inoculant for promoting enhanced plant vigor. Available at: https://patentimages.storage.googleapis.com/3f/df/a1/534c7c23eb3b4d/CA2646428A1.pdf; https://patents.google.com/patent/CA2646428A1/en (accessed 15 December 2020).

Carrión, V. J., Perez-Jaramillo, J., Cordovez, V., Tracanna, V., De Hollander, M., Ruiz-Buck, D., Mendes, L. W., van Ijcken, W. F. J., Gomez-Exposito, R., Elsayed, S. S., Mohanraju, P., Arifah, A., van der Oost, J., Paulson, J. N., Mendes, R., van Wezel, G. P., Medema, M. H. and Raaijmakers, J. M. (2019). Pathogen-induced activation of disease-suppressive functions in the endophytic root microbiome, *Science* 366(6465), 606–612.

Cawoy, H., Bettiol, W., Fickers, P. and Ongena, M. (2012). *Bacillus*-based biological control of plant diseases. In: Stoytcheva, M. (Ed), *Pesticides in the Modern World–Pesticides Use and Management*, Rijeka: InTech Europe, pp. 273302.

Chapelle, E., Mendes, R., Bakker, P. A. and Raaijmakers, J. M. (2016). Fungal invasion of the rhizosphere microbiome, *ISME J.* 10(1), 265–268.

Clarkson, J. P. and Lucas, J. A. (1993). Screening for potential antagonists of *Pseudocercosporella herpotrichoides*, the causal agent of eyespot disease of cereals 1. Bacteria, *Plant Pathol.* 42(4), 543–551.

Clements, L. D., Miller, B. S. and Streips, U. N. (2002). 'Comparative growth analysis of the facultative anaerobes *Bacillus subtilis*, *Bacillus licheniformis*, and *Escherichia coli*', *System, App. Microbiol.* 25(2), 284–286.

Chen, J., Xue, Q. H., McErlean, C. S. P., Zhi, J. H., Ma, Y. Q., Jia, X. T., Zhang, M. and Ye, X. X. (2016). Biocontrol potential of the antagonistic microorganism *Streptomyces enissocaesilis* against *Orobanche cumana*, *BioControl* 61(6), 781–791.

Chiaramonte, J. B., Mendes, L. W. and Mendes, R. (2021). Rhizosphere microbiome and soil-borne diseases. In: Gupta, V. V. S. R., Sharma, A. K. (Eds), *Rhizosphere Biology: Interactions Between Microbes and Plants*, Singapore: Springer, pp. 155–168.

Compant, S., Duffy, B., Nowak, J., Clément, C. and Barka, E. A. (2005). Use of plant growth-promoting bacteria for biocontrol of plant diseases: principles, mechanisms of action, and future prospects, *Appl. Environ. Microbiol.* 71(9), 4951–4959.

Corrêa, E. B., Sutton, J. C. and Bettiol, W. (2015). Formulation of *Pseudomonas chlororaphis* strains for improved shelf life, *Biol. Control* 80, 50–55.

Costa, A. S. and Müller, G. W. (1980). Tristeza control by cross protection, *Plant Dis.* 64(6), 538–541.

Costa, J. C. B., Bezerra, J. L., Santos Filho, L. P. S., Alves, M. C. and Moura, E. M. (2009). Controle biológico da vassoura-de-bruxa do cacaueiro na Bahia, Brasil. In: Bettiol, W. and Morandi, M. A. B. (Eds), *Biocontrole de doenças de plantas: uso e perspectivas*, Jaguariúna: Embrapa, pp. 245–266.

Cota, L. V., Maffia, L. A., Mizubuti, E. S. G., Macedo, P. E. F. and Antunes, R. F. (2008a). Biological control of strawberry gray mold by *Clonostachys rosea* under field conditions, *Biol. Control* 46(3), 515–522.

Cota, L. V., Maffia, L. A. and Mizubuti, E. S. G. (2008b). Brazilian isolates of *Clonostachys rosea*: colonization under different temperature and moisture conditions and temporal dynamics on strawberry leaves, Lett, *Lett. Appl. Microbiol.* 46(3), 312–317.

Cotes, A. M., Cárdenas, A. and Pinzón, H. (2001). Effect of seed priming in the presence of *Trichoderma koningii* on seed and seedling disease induced in tomato by *Rhizoctonia solani* and *Fusarium oxysporum* f. sp. *lycopersici*, *IOBC WPRS Bull.* 24, 259–264.

Cotes, A. M., Fargetton, X. and Köhn, J. (2019). Diseño conceptual, selección y prueba de concepto de microorganismos biocontroladores. In: Cotes, A. M. (Ed), *Control biológico de fitopatógenos, insectos y ácaro* (vol. 2), Mosquera: AGROSAVIA, pp. 598–627.

Cotes, A. M., Lepoivre, P. and Semal, J. (1996). Correlation between hydrolytic enzyme activities measured in bean seedlings after *Trichoderma koningii* treatment combined with pregermination and the protective effect against *Pythium splendens*, *Eur. J. Plant Pathol.* 102(5), 497–506.

Cotes, A. M., Moreno, C. A., Molano, L. F., Villamizar, L. and Piedrahita, W. (2007). Prospects for integrated management of *Sclerotinia sclerotiorum* in lettuce, *IOBC WPRS Bull.* 30, 391–394.

Cotes, A. M., Zapata, J., Díaz, A., García, M., Medina, C., Cristancho, D., Rodríguez, S. and Uribe, D. (2011). Selección de levaduras filosféricas con potencial para el control biológico de *Botrytis cinerea*. *Fitopatol. Colomb.* 35(2), 51–56.

Droby, S., Wisniewski, M., Macarisin, D. and Wilson, C. (2009). Twenty years of postharvest biocontrol research: is it time for a new paradigm?, *Postharvest Biol. Technol.* 52(2), 137–145.

Ehrlich, K. C. and Cotty, P. J. (2004). An isolate of *Aspergillus flavus* used to reduce contamination in cottonseed has a defective polyketide synthase gene, *Appl. Microbiol. Biotechnol.* 65, 473–478.

English-Loeb, G., Norton, A. P., Gadoury, D. M., Seem, R. C. and Wilcox, W. F. (1999). Control of powdery mildew in wild and cultivated grapes by a Tydeid mite, *Biol. Control* 14(2), 97–103.

English-Loeb, G., Norton, A. P., Gadoury, D., Seem, R. and Wilcox, W. (2007). Biological control of grape powdery mildew using mycophagous mites, *Plant Dis.* 91(4), 421–429.

EPA. (2001). *Pseudomonas chlororaphys* strain 63-28 (006478) fact sheet. Available at: https://www3.epa.gov/pesticides/chem_search/reg_actions/registration/fs_PC-006478_01-Apr-01.pdf (accessed 4 December 2020).

Eskes, A. B., Mendes, M. D. L. and Robbs, C. F. (1991). Laboratory and field studies on parasitism of *Hemileia vastatrix* with *Lecanicillium lecanii* and *V. leptobactrum*, *Café Cacao Thé* 35, 275–282.

Farias, PdSd, Souza, LdS., Paiva, AdQ., Oliveira, ÁSd, Souza, L. D. and Ledo, C. AdS. (2018). Hourly, daily, and monthly soil temperature fluctuations in a drought tolerant crop, *Rev. Bras. Cienc. Solo* 42, e0170221.

Ferrigo, D., Raiola, A., Rasera, R. and Causin, R. (2014). *Trichoderma harzianum* seed treatment controls *Fusarium verticillioides* colonization and fumonisin contamination in maize under field conditions, *Crop Prot.* 65, 51–56.

Fokkema, N. J. (1976). Antagonism between fungal saprophytes and pathogens and aerial plant surfaces. In Dickinson, C. H. and Preece, T. F. (Eds.), *Microbiology of Aerial Plant Surfaces*, London: Academic Press, pp. 487–506.

Fokkema, N. J. and Van der Meulen, F. (1976). Antagonism of yeast like phyllosphere fungi against *Septoria nodurum* on wheat leaves, *Neth. J. Plant Pathol.* 82(1), 13–16.

Frampton, R. A., Pitman, A. R. and Fineran, P. C. (2012). Advances in bacteriophage-mediated control of plant pathogens, *Int. J. Microbiol.* 2012, ID326452.

Fravel, D. R. and Spurr Jr., H. W. (1977). Biocontrol of tobacco brown spot disease by *Bacillus cereus* subsp. *mycoides* in a controlled environment, *Phytopathology* 77(7), 930–932.

Gadoury, D. M., CAdle-Davidson, L., Wilcox, W. F., Dry, I. B., Seem, R. C. and Milgroom, M. G. (2012). Grapevine powdery mildew (*Erysiphe necator*): a fascinating system for the study of the biology, ecology and epidemiology of an obligate biotroph, *Mol. Plant Pathol.* 13(1), 1–16.

García, R., Latz, S., Romero, J., Higuera, G., García, K. and Bastías, R. (2019). Bacteriophage production models: an overview, *Front. Microbiol.* 10, 1187.

García-Pedrajas, M. D., Cañizares, M. C., Sarmiento-Villamil, J. L., Jacquat, A. G. and Dambolena, J. S. (2019). Mycoviruses in biological control: from basic research to field implementation, *Phytopathology* 109(11), 1828–1839.

García, A. D., Álvarez, M. I. G., Bernal, E. P. G., Diaz, A. M. S., Barrera, F. M. C., Moreno, D. M. L., Torres, E. A. and Cotes, A. M. (2019). Development and scaling of biopesticides. In: Cotes, A. M. (Ed), *Control biológico de fitopatógenos, insectos y ácaros* (vol. 2), Mosquera: AGROSAVIA, pp. 628–691.

Garibaldi, A., Brunatti, F. and Gullino, M. L. (1987). Evaluation of several antagonistics and different methods of applications against Fusarium wilt of carnation, *EPPO Bull.* 17(4), 625–629.

Gloor, G. B., Macklaim, J. M., Pawlowsky-Glahn, V. and Egozcue, J. J. (2017). Microbiome datasets are compositional: and this is not optional, *Front. Microbiol.* 8, 2224.

Godoy, C. V., Seixas, C. D. S., Soares, R. M., Marcelino-Guimarães, F. C., Meyer, M. C. and Costamilan, L. M. (2016). Asian soybean rust in Brazil: past, present, and future, *Pesq. agropec bras* 51(5), 407–421.

Görgen, C. A., Silveira-Neto, A. Nd, Carneiro, L. C., Ragagnin, V. and Lobo Junior, M. (2009). Controle do mofo-Branco com palhada e *Trichoderma harzianum* 1306 em soja, *Pesq. agropec bras* 44(12), 1583–1590.

Gotor-Vila, A., Usall, J., Torres, R., Abadias, M. and Teixidó, N. (2017a). Formulation of the biocontrol agent *Bacillus amyloliquefaciens* CPA-8 using different approaches: liquid, freeze-drying and fluid-bed spray-drying, *BioControl* 62(4), 545 -555.

Gotor-Vila, A., Teixidó, N., Di Francesco, A., Usall, J., Ugolini, L., Torres, R. and Mari, M. (2017b). Antifungal effect of volatile organic compounds produced by *Bacillus amyloliquefaciens* CPA-8 against fruit pathogen decays of cherry, *Food Microbiol.* 64, 219–225.

Grant, T. J. and Costa, A. S. (1951). A mild strain of the tristeza virus of citrus, *Phytopathology* 41, 114–122.

Greenfield, M., Gómez-Jiménez, M. I., Ortiz, V., Vegha, F. E., Kramer, M. and Parsa, S. (2016). *Beauveria bassiana* and *Metarhizium anisopliae* endophytically colonize cassava roots following soil drench inoculation. *Biol. Control* 95, 40–48.

Grondona, I., Hermosa, R., Tejada, M., Gomis, M. D., Mateos, P. F., Bridge, P. D., Monte, E. and García-Acha, I. (1997). Physiological and biochemical characterization of *Trichoderma harzianum*, a biological control agent against soilborne fungal plant pathogens, *Appl. Environ. Microbiol.* 63(8), 3189-3198.

Grondona, I., Rodríguez, A., Gómez, M. I., Camacho, R., Llobell, A. and Monte, E. (2004). TUSAL®, a commercial biocontrol formulation based on *Trichoderma*, *IOBC WPRS Bull.* 27(8), 285–288.

Grubisha, L. C. and Cotty, P. J. (2015). Genetic analysis of the *Aspergillus flavus* vegetative compatibility group to which a biological control agent that limits aflatoxin contamination in USA crops belongs, *Appl. Environ. Microbiol.* 81(17), 5889-5899.

Hadar, Y., Harman, G. E. and Taylor, A. G. (1984). Evaluation of *Trichoderma koningii* and *T. harzianum* from New York soils for biological control of seed rot caused by *Pythium* spp., *Phytopathology* 74(1), 106–110.

Haddad, F., Maffia, L. A., Mizubuti, E. S. G. and Teixeira, H. (2007). Biocontrol as an alternative for leaf rust management in organically-grown coffee, *IOBC WPRS Bull.* 30, 17.

Haddad, F., Maffia, L. A., Mizubuti, E. S. G. and Teixeira, H. (2009). Biological control of coffee rust by antagonistic bacteria under field conditions in Brazil, *Biol. Control* 49(2), 114–119.

Hamonts, K., Trivedi, P., Garg, A., Janitz, C., Grinyer, J., Holford, P., Botha, F. C., Anderson, I. C. and Singh, B. K. (2018). Field study reveals core plant microbiota and relative importance of their drivers, *Environ. Microbiol.* 20(1), 124–140.

Harada, L. K., Silva, E. C., Campos, W. F., Fiol, F. S. D., Vila, M., Dabrowska, K., Krylov, V. N. and Balcão, V. M. (2018). Biotechnological applications of bacteriophages: state of the art, *Microbiol. Res.* 212-213, 38–58.

Harman, G. E. (n.d.). *Trichoderma for Biocontrol of Plant Pathogens: from Basic Research to Commercialized Products'.* Available at: http://web.entomology.cornell.edu/shelton/cornell-biocontrol-conf/talks/harman.html (accessed 16 December 2020).

Harman, G. E., Taylor, A. G. and Stasz, T. E. (1989). Combining effective strains of *Trichoderma harzianum* and solid matrix priming to provide improved biological seed treatment systems, *Plant Dis.* 72, 631-637.

Hartmann, M., Frey, B., Mayer, J., Mäder, P. and Widmer, F. (2015). Distinct soil microbial diversity under long-term organic and conventional farming, *ISME J.* 9(5), 1177 -1194.

Heiniger, U. and Rigling, D. (1994). Biological control of chestnut blight in Europe, *Annu. Rev. Phytopathol.* 32(1), 581-599.

Hermosa, M. R., Grondona, I., Iturriaga, E. A., Díaz-Mínguez, J. M., Castro, C., Monte, E. and García-Acha, I. (2000). Molecular characterization and identification of biocontrol isolates of *Trichoderma* spp., *Appl. Environ. Microbiol.* 66(5), 1890–1898.

Hermosa, M. R., Grondona, I., Díaz-Mínguez, J. M., Iturriaga, E. A. and Monte, E. (2001). Development of a strain-specific SCAR marker for the detection of *Trichoderma atroviride* 11, a biological control agent against soilborne fungal plant pathogens, *Curr. Genet.* 38(6), 343–350.

Hwang, S., Ahmed, H., Gossen, B., Kutcher, H., Brandt, S., Strelkov, S. and Turnbull, G. (2009). Effect of crop rotation on soil pathogen population dynamics and canola seedling establishment, *Plant Pathol. J.* 8(3), 106-112.

Ilangumaran, G. and Smith, D. L. (2017). Plant growth promoting rhizobacteria in amelioration of salinity stress: a systems biology perspective, *Front. Plant Sci.* 8, 1768.

Janisiewicz, W. J. (1987). Post harvest biological control of blue mold on apples, *Phytopathology* 77(3), 481-485.

Janisiewicz, W. J. (1988). Biocontrol of postharvest diseases of apples with antagonist mixtures, *Phytopathology* 78(2), 194-198.

Janisiewicz, W. J. and Roitman, J. (1988). Biological control of blue mold and gray mold on apple and pear with *Pseudomonas cepacia*, *Phytopathology* 78(12), 1697-1700.

Jenkins, P. T. (1968). A method to determine the frequency of airborne bacteria antagonistic to *Sclerotinia fructicola*, *Aust. J. Exp. Agric.* 8(33), 434-435.

Ji, C., Fan, Y. and Zhao, L. (2016). Review on biological degradation of aflatoxin, zearalenone and deoxynivalenol, *Anim. Nutr.* 2(3), 127-133.

Jiao, J. (2020). Research and development status of mycovirus. *6th International Conference on Agricultural and Biological Sciences, IOP Conference Series: Earth and Environmental Science* 594, p. 012025.

Jones, D. A. and Kerr, A. (1989). *Agrobacterium radiobacter* strain K1026, a genetically engineered derivative of strain K84, for biological control of crown gall, *Plant Dis.* 73(1), 15-18.

Jones, J. B., Jackson, L. E., Balogh, B., Obradovic, A., Iriarte, F. B. and Momol, M. T. (2007). Bacteriophages for plant disease control, *Annu. Rev. Phytopathol.* 45, 245-262.

Jones, J. B., Vallad, G. E., Iriarte, F. B., Obradovic, A., Wernsing, M. H., Jackson, L. E., Baloggh, B., Hong, J. C. and Momol, M. T. (2012). Considerations for using bacteriophages for plant disease control, *Bacteriophage* 2(4), 208-214.

Junqueira, N. T. V. and Gasparatto, L. (1991). Controle biológico de fungos estromáticos causadores de doenças de foliares em seringueira. In: Bettiol, W. (Ed), *Controle biológico de doenças de plantas*, Jaguariúna: Embrapa/CNPMA, pp. 307-331.

Junqueira, N. T. V., Gasparatto, L., Lima, M. I. P. M., Leiberei, R. and Normando, M. C. S. (1989). Potential do fungo *Hansfordia pulvinata* no controle biológico do mal das folhas da seringueira, *Fitopatol. Bras.* 14, 158.

Junqueira, N. T. V., Moraes, V. H. F., Lieberei, R. and Gasparotto, L. (1993). Induced polyploidy potential for improving resistance in *Hevea* clones to rubber tree leaf blight, *Fitopatol. Bras.* 18, 12-18.

Kaminsky, L. M., Trexler, R. V., Malik, R. J., Hockett, K. L. and Bell, T. H. (2019). The inherent conflicts in developing soil microbial inoculants, *Trends Biotechnol.* 37(2), 140-151.

Keane, P. J., Kerr, A. and New, P. B. (1970). Crown gall of stone fruit II. *Identification and nomenclature of *Agrobacterium* isolates, *Aust. J. Biol. Sci.* 23(3), 585-595.

Kerr, A. (2016). Biological control of crown gall, *Australas. Plant Pathol.* 45(1), 15-18.

Kerr, A. and Htay, K. (1974). Biological control of crown gall through bacteriocin production, *Physiol. Plant Pathol.* 4(1), 37-44.

Kerr, A. and Bullard, G. (2020). Biocontrol of crown gall by *Rhzobium rhizogenes*: challenges in biopesticide commercialisation, *Agronomy* 10, 1026.

King, E. D., Bassi Jr., A. B., Ross, D. C. and Druebbisch, B. (2011). An industry perspective on the use of "atoxigenic" strains of *Aspergillus flavus* as biological control agents and the significance of cyclopiazonic acid, *Toxin Rev.* 30(2-3), 33-41.

Kloepper, J. W., Hume, D. J., Scher, F. M., Singleton, C., Tipping, B., Laliberté, M., Frauley, K., Kuthaw, T., Simonson, C., Lifshitz, R., Zaleska, I. and Lee, L. (1988). Plant growth-promoting rhizobacteria on canola (rapeseed), *Plant Dis.* 72(1), 42–46.

Köhl, J. and Rijk, N. (2016). The Biocomes consortium, a model private-public partnership project. In: Medeiros, et al. (Eds). *Proceedings of the 12th Brazilian Meeting of Biological Control of Plant Disease*, Lavras: Universidade Federal de Lavras, pp. 63–75.

Köhl, J., Postma, J., Nicot, P., Ruocco, M. and Blum, B. (2011). Stepwise screening of microorganisms for commercial use in biological control of plant pathogenic fungi and bacteria, *Biol. Control* 57(1), 1–12.

Köhl, J., de Geijn, H. G., Haas, L. G., Henken, B., Hauschild, R., Hilscher, U., der Plas, C. L., van den Bosch, T. and Wikström, M. (2019). Stepwise screening of candidate antagonists for biological control of *Blumeria graminis* f. sp. *tritici*, *Biol. Control* 136, 104008.

Köhl, J., Medeiros, F. H. V., Lombaers-van der Plas, C., Groenenboom-de Haas, L. and van den Bosch, T. (2020). Efficacies of bacterial and fungal isolates in biocontrol of *Botrytis cinerea* and *Pseudomonas syringae* pv. *tomato* and growth promotion in tomato do not correlate, *Biol. Control* 150, 104375.

Kurtzman, C. P. and Droby, S. (2001). *Metschnikowia fructicola*, a new ascosporic yeast effective for biocontrol of postharvest fruit rots, *Syst. Appl. Microbiol.* 24(3), 395–399.

Landa, B. B., Hervás, A., Bettiol, W. and Jiménez-Díaz, R. M. (1997). Antagonistic activity of bacteria from the chickpea rhizosphere against *Fusarium oxysporum* f. sp. *ciceris*, *Phytoparasitica* 25(4), 305–318.

Leben, C. (1964). Influence of bacteria isolated from healthy cucumber leaves on two leaf diseases of cucumber, *Phytopathology* 54, 405–408.

Lemanceau, P., Blouin, M., Muller, D. and Moënne-Loccoz, Y. (2017). Let the core microbiota be functional, *Trends Plant Sci.* 22(7), 583–595.

Lemus-Minor, C. G., Cañizares, M. C., García-Pedrajas, M. D. and Pérez-Artés, E. (2019). Horizontal and vertical transmission of the hypovirulence-associated mycovirus Fusarium oxysporum f. sp. dianthi virus, *Eur. J. Plant Pathol.* 153(2), 645–650.

Liu, S.-F., Wang, G.-J., Nong, X.-Q., Liu, B., Wang, M.-M., Li, S.-L., Cao, G.-C. and Zhang, Z.-H. (2017). Entomopathogen *Metarhizium anisopliae* promotes the early development of peanut root, *Plant Prot. Sci.* 53, 101–107.

Longley, R., Noel, Z. A., Benucci, G. M. N., Chilvers, M. I., Trail, F. and Bonito, G. (2020). Crop management impacts the soybean (*Glycine max*) microbiome, *Front. Microbiol.* 11, 1116.

Lopes, C. A. and Stall, R. E. (1990). Biological control of *Pseudomonas avenae* with epiphytici bacteria isolated from corn, *Pesqui. Agropecu Bras.* 25, 1133 -1141.

Luongo, L., Galli, M., Corazza, L., Meekes, E., Haas, L. D., Van Der Plas, C. L. and Köhl, J. (2005). Potential of fungal antagonists for biocontrol of *Fusarium* spp. in wheat and maize through competition in crop debris. *Biocontrol Sci. Technol.* 15(3), 229–242.

Luth, P. (2001). The control of *Sclerotinia* sp. and *Sclerotium cepivorum* with the biological fungicide Contans® WG–experiences from field trials and commercial use. In: Young, C. S. and Hughes, K. J. D. (Eds), *Proceedings of the XI International Sclerotinia Workshop*, York: Central Science Laboratory, pp. 37–38.

Mariano, R. L. R. (1993). Métodos de seleção in vitro para o controle microbiológico de patógenos de plantas, *Rev. An. Patol. Plant* 1, 369–409.

Markets and Markets (2020). Biopesticides market by type (bioinsecticides, biofungicides, bionematicides, and bioherbicides), source (microbials, biochemicals, and beneficial insects), mode of application, formulation, crop application, and region–global forecast to 2025. Available at: https://www.marketsandmarkets.com/Market-Reports/biopesticides-267.html?gclid=CjwKCAiA17P9BRB2EiwAMvwNyN1J3lIlag6ewvyHPSndQLQhV5r5rkpbhZIGCAR6Ju61h366uRJ0EhoCov0QAvD_BwE (accessed 12 November 2020).

Martins, S. J., Soares, A. C., Medeiros, F. H. V., Santos, D. B. C. and Pozza, E. A. (2015). Contribution of host and environmental factors to the hyperparasitism of coffee rust under field conditions, *Australas. Plant Pathol.* 44(6), 605–610.

Massart, S., Martinez-Medina, M. and Jijakli, M. H. (2015). Biological control in the microbiome era: challenges and opportunities, *Biol. Control* 89, 98–108.

Masteling, R., Lombard, L., de Boer, W., Raaijmakers, J. M. and Dini-Andreote, F. (2019). Harnessing the microbiome to control plant parasitic weeds, *Curr. Opin. Microbiol.* 49, 26–33.

Mata, G. S. V. and Obregón, G.. M. A. (2014). Control biológico de enfermidades de plantas em Costa Rica. In: Bettiol, W., Rivera, M. C., Mondino, P., Montealegre, J. R. and Colmenárez, Y. C. (Eds) *Control biológico de enfermedades de plantas en América Latina y el Caribe*, Montevideo: Universidad de la Republica, pp. 182-199.

Mata, X., Obregón, M., Lorito, M. and Monte, E. (2012). Antagonism of local isolates of Trichoderma spp. on pineapple heart rot disease caused by Phytophthora cinnamomi var. cinnamomi in Costa Rica, *The 12th International Trichoderma and Gliocladium Workshop*. Abstract. New Zealand, pp. 107.

Medeiros, C. A. A. and Bettiol, W. (2021). Multifaceted intervention of *Bacillus* spp. against salinity stress and *Fusarium wilt* in tomato. *J. Appl. Microbiol.* doi: 10.1111/jam.15095. Epub ahead of print. PMID: 33817910.

Medeiros, F. H. V., Pomella, A. W. V., de Souza, J. T., Niella, G. R., Valle, R., Bateman, R. P., Fravel, D., Vinyard, B. and Hebbar, P. K. (2010). A novel, integrated method for management of witches' broom disease in Cacao in Bahia, Brazil, *Crop Prot.* 29(7), 704–711.

Medeiros, F. H. Vd, Martins, S. J., Zucchi, T. D., Melo, I. Sd, Batista, L. R. and Machado, JdC. (2012). Biological control of mycotoxin-producing molds, *Ciênc. agrotec* 36(5), 483–497.

Medema, M. H. and Fischbach, M. A. (2015). Computational approaches to natural product discovery, *Nat. Chem. Biol.* 11(9), 639–648.

Mehl, H. L., Jaime, R., Callicott, K. A., Probst, C., Garber, N. P., Ortega-Beltran, A., Grubisha, L. C. and Cotty, P. J. (2012). *Aspergillus flavus* diversity on crops and in the environment can be exploited to reduce aflatoxin exposure and improve health. *Ann. N. Y. Acad. Sci.* 1273(1), 7–17.

Melidossian, H. S., Seem, R. C., English-Loeb, G., Wilcox, W. F. and Gadoury, D. M. (2005). Suppression of grapevine powdery mildew by a mycophagous mite, *Plant Dis.* 89(12), 1331–1338.

Melo, D. F., Mello, S. C. M., Mattos, C. R. R. and Cardoso, S. E. A. (2008). Compatibilidadae de *Dycima pulvinata* com defensivos agrícolas e eficiência do biocontrole do mal-das-folhas da seringueira em campo, *Pesqui. Agropecu Bras.* 43, 179–185.

Mendes, L. W., Raaijmakers, J. M., Hollander, M. de, Mendes, R. and Tsai, S. M. (2018). Influence of resistance breeding in common bean on rhizosphere microbiome composition and function, *ISME J.* 12(1), 212–224.

Mendes, R., Garbeva, P. and Raaijmakers, J. M. (2013). The rhizosphere microbiome: significance of plant beneficial, plant pathogenic, and human pathogenic microorganisms, *FEMS Microbiol. Rev.* 37(5), 634–663.

Mendes, R., Kruijt, M., De Bruijn, I., Dekkers, E., Van Der Voort, M., Schneider, J. H. M., Piceno, Y. M., DeSantis, T. Z., Andersen, G. L., Bakker, P. A. and Raaijmakers, J. M. (2011). Deciphering the rhizosphere microbiome for disease-suppressive bacteria, *Science* 332(6033), 1097–1100.

Meneghini, M. (1946). 'Sobre a natureza e transmissibilidade da doença 'tristeza' dos citros', *Biológico* 12, 285–287.

Moitinho, M. A., Souza, D. T., Chiaramonte, J. B., Bononi, L., Melo, I. S. and Taketani, R. G. (2020). The unexplored bacterial lifestyle on leaf surface, *Braz. J. Microbiol.* 51(3), 1233–1240.

Montesinos, E. (2003). Development, registration and commercialization of microbial pesticides for plant protection, *Int. Microbiol.* 6(4), 245–252.

Moraga-Suazo, P. and Sanfuentes, E. (2017). Growth promotion of *Pinus radiata* seedlings by soil inoculation and seed pretreatment with the biological control agent *Clonostachys rosea*, *Gayana Bot.* 74(ahead), 0.

Moraga-Suazo, P., Sanfuentes, E. and Le-Feuvre, R. (2011a). Induced systemic resistance triggered by *Clonostachys rosea* against *Fusarium circinatum* in *Pinus radiata*, *Forest Res.* 5(2), 1000174.

Moraga-Suazo, P., Opazo, A., Zaldúa, S., Gonzáles, G. and Sanfuentes, E. (2011b). Evaluation of *Trichoderma* spp. and *Clonostachys* spp. strains to control *Fusarium circinatum* in *Pinus radiata* seedlings, *Chil. J. Agric. Res.* 71(3), 412–417.

Moral, J., Garcia-Lopez, M. T., Camiletti, B. X., Jaime, R., Michailides, T. J., Bandyopadhyay, R. and Ortega-Beltran, A. (2020). Present status and perspective on the future use of aflatoxin biocontrol products, *Agronomy* 10(4), 491.

Morales, H., Sanchis, V., Usall, J., Ramos, A. J. and Marín, S. (2008). Effect of biocontrol agents *Candida sake* and *Pantoea agglomerans* on *Penicillium expansum* growth and patulin accumulation in apples. *Int. J. Food Microbiol.* 122(1-2), 61–67.

Moreira, S. (1959). Tristeza na laranja Pera, *Bol. Campo* 15, 3–4.

Morgan, F. L. (1963). Infection inhibition and germtube lysis of three cereal rust by *Bacillus pumilus*, *Phytopathology* 53, 1346–1348.

Müller, G. W. and Costa, A. S. (1991). Premunização de plantas cítricas'. In: Bettiol, W. (Ed), *Controle biológico de doenças de plantas*, Jaguariúna: Embrapa/CNPMA, pp. 285–293.

Nguyen, H. Q., Quyen, D. T., Nguyen, S. L. T. and Vu, V. H. (2015). An extracellular antifungal chitinase from *Lecanicillium lecanii*: purification, properties, and application in biocontrol against plant pathogenic fungi, *Turk. J. Biol.* 39, 6–14.

Nguyen, P. A., Strub, C., Fontana, A. and Schorr-Galindo, S. (2017). Crop molds and mycotoxins: alternative management using biocontrol, *Biol. Control* 104, 10–27.

Niu, B., Paulson, J. N., Zheng, X. and Kolter, R. (2017). Simplified and representative bacterial community of maize roots, *Proc. Natl Acad. Sci. U. S. A.* 114(12), E2450–E2459.

Nobre, S. A. M., Maffia, L. A., Mizubuti, E. S. G., Cota, L. V. and Dias, A. P. S. (2005). Selection of *Clonostachys rosea* isolates from Brazilian ecosystems effective in controlling *Botrytis cinerea*, *Biol. Control* 34(2), 132–143.

Numan, M., Bashir, S., Kan, Y., Mumtaz, R., Shinwari, Z. K., Khan, A. L., Khan, A. and Al-Harrasi, A. (2018). Plant growth promoting bacteria as an alternative strategy for salt tolerance in plants: a review, *Microbiol. Res.* 209, 21-32.

Obregón, M. and Arias, M. (2008). Evaluación *in vitro* de la capacidad antagonista de 42 cepas de *Trichoderma* spp. para el combate de *Fusarium* spp., causante de la pudrición basal en plantas de Piña, In *X Congreso Mundial de Trichoderma y Gliocladium, VII Congreso Nacional de Fitopatología, II Congreso Nacional de Fitoprotección*. Resúmenes. San José, pp. 67.

Obregón-Gómez, M. (2010). *Trichoderma* a beneficial microorganism used in integrated production systems of Costa Rica, Book of Abstracts: *Trichoderma*: molecular mechanisms and application of biocontrol in agriculture. Haifa, pp. 54.

Obregón, M. (2012). Application of *Trichoderma asperellum* in the control of pineapple caused by *Fusarium oxysporum* in the field in Costa Rica, *12th International Trichoderma and Gliocladium Workshop*. Abstract. New Zealand, pp. 65.

Oliveira, C. M., Ferreira, J. A. M., Oliveria, R. M., Santos, F. O. and Pallini, A. (2014). Ricoseius loxocheles, a phytoseiid mite that feeds on coffee leaf rust, *Exp. Appl. Acarol.* 64(2), 223-233, DOI: 10.1007/s10493-014-9814-y.

Ortega-Beltran, A., Grubisha, L. C., Callicott, K. A. and Cotty, P. J. (2016). The vegetative compatibility group to which the US biocontrol agent *Aspergillus flavus* AF 36 belongs is also endemic to Mexico, *J. Appl. Microbiol.* 120(4), 986-998.

Ownley, B. H., Griffin, M. R., Klingeman, W. E., Gwinn, K. D., Moulton, J. K. and Pereira, R. M. (2008). *Beauveria bassiana*: endophytic colonization and plant disease control, *J. Invertebr. Pathol.* 98(3), 267-270.

Ownley, B. H., Gwinn, K. D. and Veja, F. E. (2010). Endophytic fungal entomopathogens with activity against plant pathogens: ecology and evolution, *BioControl* 55(1), 113-128.

Palazzini, J. M., Groenenboom-de Haas, B. H., Torres, A. M., Köhl, J. and Chulze, S. N. (2013). Biocontrol and population dynamics of *Fusarium* spp. on wheat stubble in Argentina. *Plant Pathol.* 62(4), 859-866.

Pascale, A., Proietti, S., Pantelides, I. S. and Stringlis, I. A. (2020). Modulation of the root microbiome by plant molecules: the basis for targeted disease suppression and plant growth promotion, *Front. Plant Sci.* 10, 1-23.

Patents (1996). WO/1996/021358–fungus isolate, preparation for combatting plant-pathogenic fungi. Process for Producing It and Its Use. Available at: https://patentscope.wipo.int/search/en/detail.jsf?docId=WO1996021358 (accessed 10 December 2020) (vol. 1).

Paulitz, T. C. and Bélanger, R. R. (2001). Biological control in greenhouse systems, *Annu. Rev. Phytopathol.* 39, 103-133.

Paulitz, T. C., Zhou, T. and Rankin, L. (1992). Selection of rhizosphere bacteria for biological control of *Pythium aphanidermatum* on hydroponically grown cucumber, *Biol. Control* 2(3), 226-237.

Peng, G. and Sutton, J. C. (1991). Evaluation of microorganisms for biocontrol of *Botrytis cinerea* in strawberry, *Can. J. Plant Pathol.* 131, 247-257.

Perazzolli, M., Nesler, A., Giovannini, O., Antonielli, L., Puopolo, G. and Pertot, I. (2020). Ecological impact of a rare sugar on grapevine phyllosphere microbial communities, *Microbiol. Res.* 232, 126387.

Picarelli, M. A. S. C., Gobatto, D., Patrício, F., Rivas, E. B. and Colariccio, A. (2017). Vírus que infectam fungos fitopatogênicos, *Arq. Inst. Biol.* 84, e0162016.

Pieterse, C. M. J., Zamioudis, C., Berendsen, R. L., Weller, D. M., Van Wees, S. C. M. and Bakker, P. A. H. M. (2014). Induced systemic resistance by beneficial microbes, *Annu. Rev. Phytopathol.* 52, 347–375.

Principi, N., Silvestri, E. and Esposito, S. (2019). Advantages and limitations of bacteriophages for the treatment of bacterial infections, *Front. Pharmacol.* 10, 513.

Pusey, P. L. and Wilson, C. L. (1984). Postharvest biological control of stone fruit brown rot by *Bacillus subtilis*, *Plant Dis.* 68, 753–756.

Qi, J., Aiuchi, D., Tani, M., Asano, S.-I. and Koike, M. (2016). Potential of entomopathogenic *Bacillus thuringiensis* as plant growth promoting rhizobacteria and biological control agents for tomato *Fusarium* wilt, *Int. J. Environ. Agric. Res.* 2, 55–63.

Raad, M., Glare, T. R., Brochero, H. L., Müller, C. and Rostás, M. (2019). Transcriptional reprogramming of *Arabidopsis thaliana* defence pathways by the entomopathogen *Beauveria bassiana* correlates with resistance against a fungal pathogen but not against insects, *Front. Microbiol.* 10, 615.

Rankin, L. and Paulitz, T. C. (1994). Evaluation of rhizosphere bacteria for biological control of *Pythium* root rot of greenhouse cucumber in hydroponic culture, *Plant Dis.* 78(5), 447–451.

Ravensberg, W. J. (2011). *A Roadmap to the Successful Development and Commercialization of Microbial Pest Control Products for Control of Arthropods*, Dordrecht: Springer Science & Business Media.

Ravindran, K., Chitra, S., Wilson, A. and Sivaramakrishnan, S. (2014). Evaluation of antifungal activity of *Metarhizium anisopliae* against plant phytopathogenic fungi. In: Krarwar, R. N., Upadhyay, R. S., Dubey, N. K. and Raghuwasnshi, R. (Eds), *Microbial Diversity and Biotechnology in Food Security*, Tamil Nadu: Springer India, pp. 251–255.

Ray, R. L., Fares, A. and Risch, E. (2018). Effects of drought on crop production and cropping areas in Texas, *Agric. Environ. Lett.* 3(1), 170037.

Reino, J. L., Guerrero, R. F., Hernández-Galán, R. and Collado, I. G. (2007). Secondary metabolites from species of the biocontrol agent *Trichoderma*, *Phytochem. Rev.* 7(1), 89–123.

Rezende, J. A. M. and Pacheco, D. A. (1998). Control of papaya ringspot virus-type W in zucchini squash by cross-protection in Brazil, *Plant Dis.* 82(2), 171–175.

Rossmann, M., Sarango-Flores, S. S., Chiaramonte, J. B., Kmit, M. C. P. and Mendes, R. (2017). Plant microbiome: composition and functions in plant compartments. In: Pylro, V. and Roesch, L. (Eds), *The Brazilian Microbiome: Current Status and Perspectives*, Cham: Springer International Publishing, pp. 1–123.

Rybakova, D., Cernava, T., Köberl, M., Liebminger, S., Etemadi, M. and Berg, G. (2016a). Endophytes-assisted biocontrol: novel insights in ecology and the mode of action of *Paenibacillus*, *Plant Soil* 405(1–2), 125–140.

Rybakova, D., Schmuck, M., Wetzlinger, U., Varo Suarez, A., Murgu, O., Müller, H. and Berg, G. (2016b). Kill or cure? The interaction between endophytic *Paenibacillus* and *Serratia* strains and the host plant is shaped by plant growth conditions, *Plant Soil* 405(1–2), 65–79.

Rybakova, D., Mancinelli, R., Wikström, M., Birch-Jensen, A. S., Postma, J., Ehlers, R. U., Goertz, S. and Berg, G. (2017). The structure of the *Brassica napus* seed microbiome is cultivar-dependent and affects the interactions of symbionts and pathogens, *Microbiome* 5(1), 104.

Rybakova, D., Wikström, M., Birch-Jensen, F., Postma, J., Ehlers, R. U., Schmuck, M., Kollmann, R., Köhl, J. and Berg, G. (2020). Verticillium wilt in oilseed rape–the microbiome is crucial for disease outbreaks as well as for efficient suppression, *Plants (Basel)* 9(7), 866.

Samuels, G. J., Pardo-Schultheiss, R., Hebbar, K. P., Lumsden, R. D., Bastos, C. N., Costa, J. C. B. and Bezerra, J. L. (2000). *Trichoderma stromaticum* sp. nov., a parasite of the cacao witches' broom pathogen, *Mycol. Res.* 104(6), 760–764.

Sanfuentes, E. (2017). Desarrollo de un bioproducto para el control de *Fusarium circinatum* en viveros de *Pinus radiata,* FONDEF IDeA IT13I10057. 102p. Final Report.

Sanfuentes, E., Molina, G., Zaldúa, S. and Gonzáles, G. (2006). Selección de hongos antagonistas para el control biológico de *Botrytis cinerea* en viveros forestales de Chile, *Bosque* 27, 126–134.

Sanz, L., Montero, M., Grondona, I., Vizcaíno, J. A., Llobell, A., Hermosa, R. and Monte, E. (2004). Cell wall degrading isoenzyme profiles of *Trichoderma* biocontrol strains show correlation with rDNA taxonomic species, *Curr. Genet.* 46(5), 277–286.

Saravanakumar, K., Dou, K., Lu, Z., Wang, X., Li, Y. and Chen, J. (2018). Enhanced biocontrol activity of cellulase from *Trichoderma harzianum* against *Fusarium graminearum* through activation of defense-related genes in maize, *Physiol. Mol. Plant Pathol.* 103, 130–136.

Sarven, M. S., Hao, Q., Deng, J., Yang, F., Wang, G., Xiao, Y. and Xiao, S. (2020). Biological control of tomato gray mold caused by *Botrytis cinerea* with entomopathogenic fungus *Metarhizium anisopliae*, *Pathogens* 9, 213.

Sasan, R. K. (2012). The analysis of *Metharizium robertsii* potential as endophytic plant root colonizer, plant growth enhancer and antagonist to bean root pathogen *Fusarium solani* f. sp. *phaseolis*, Thesis, Brock University, Ontario. 84 p.

Sasan, R. K. and Bidochka, M. J. (2012). The insect-pathogenic fungus *Metarhizium robertsii* (Clavicipitaceae) is also an endophyte that stimulates plant root development, *Am. J. Bot.* 99(1), 101–107.

Segarra, G., Puopolo, G., Giovannini, O. and Pertot, I. (2015). Stepwise flow diagram for the development of formulations of non spore-forming bacteria against foliar pathogens: the case of *Lysobacter capsica* AZ78. *J. Biotechnol.* 216, 56–64.

Shi, J. F.. and Sun, C. Q. (2017). Isolation, identification, and biocontrol of antagonistic bacterium against *Botrytis cinerea* after tomato harvest, *Braz. J. Microbiol.* 48(4), 706–714.

Shiomi, H. F., Silva, H. S. A., Melo, I. Sd, Nunes, F. V. and Bettiol, W. (2006). Bioprospecting endophytic bacteria for biological control of coffee leaf rust, *Sci. Agric.* 63(1), 32–39.

Silva, H. S. A., Tozzi, J. P. L., Terrasan, C. R. F. and Bettiol, W. (2012). Endophytic microorganisms from coffee tissues as plant growth promoters and biocontrol agents of coffee leaf rust, *Biol. Control* 63(1), 62–67.

Singh, B. K. and Trivedi, P. (2017). Microbiome and the future for food and nutrient security, *Microb. Biotechnol.* 10(1), 50–53.

Smith, K. P., Handelsman, J. and Goodman, R. M. (1999). Genetic basis in plants for interactions with disease-suppressive bacteria, *Proc. Natl Acad. Sci. U. S. A.* 96(9), 4786–4790.

Souza, M. L., Ribeiro, L. S., Miguel, M. G. C. P., Batista, L. R., Schwan, R. F., Medeiros, F. H. and Silva, C. F. (2021). Yeasts prevent ochratoxin A contamination in coffee by displacing *Aspergellus carbonarius*, *Biol. Control* 155, 104512.

Stefanova, M. (2007). Introducción y eficacia técnica del biocontrol de fitopatógenos con *Trichoderma* spp. en Cuba, *Fitosanidad* 11(3), 75–79.

Stefanova, M., Leiva, A., Larriganaga, L. and Coronado, M. F. (1999), Actividad metabólica de cepas de *Trichoderma* spp. para el control de hongos fitopatógenos del suelo, *Rev. Fac. Agron.* 16, 509–516.

Stefanova, M., Villegas, M. E. D. and Campos, J. M. (2014). Control biológico de enfermedades de plantas en Cuba. In: Bettiol, W., Rivera, M. C., Mondino, P., Montealegre, J. R. and Colmenárez, Y. C. (Eds) *Control biológico de enfermedades de plantas en América Latina y el Caribe*, Montevideo: Unviersidad de la Republica, pp. 201–218.

Sudo, S. (1989). Biocontrole de *Catacauma torrendiella* e *Coccostroma palmicola*, agentes causadores da lixa-preta do coqueiro. In: Anais (Translator). *Reunião Brasileira sobre controle biológico de doenças de plantas*, Piracicaba: U.S. pharmacopeia/ EMBRAPA, pp. 57–59.

Sutton, J. C., Li, D. W., Peng, G., Yu, H., Zhang, P.. and Valdebenito-Sanhueza, R. M. (1997). Gliocladium Roseum a Versatile Adversary of Botrytis Cinerea in Crops, *Plant Dis.* 81(4), 316–328.

Sutton, J. C., Liu, W., Huang, R. and Owen-Going, N. (2002). Ability of *Clonostachys rosea* to establish and suppress sporulation potential of *Botrytis cinerea* in deleafed stems of hydroponic greenhouse tomatoes, *Biocontrol Sci. Technol.* 12(4), 413–425.

Sztejnberg, A., Galper, S., Mazar, S. and Lisker, N. (1989). *Ampelomyces quisqualis* for biological and integrated control of powdery mildews in Israel, *J. Phytopathol.* 124(4), 285–295.

Thirumalachar, M. J. and O'Brien, M. J. (1977). Suppression of charcoal rot in potato with a bacterial antagonist. *Plant Dis. Rep.* 61, pp. 543–546.

Thomas, H., Heller, A., Sauerborn, J. and Müller-Stöver, D. (1999). *Fusarium oxysporum* f. sp. *orthoceras*, a potential mycoherbicide, parasitizes seeds of *Orobanche cumana* (sunflower broomrape): a cytological study, *Ann. Bot.* 83(4), 453–458.

Trutmann, P., Keang, P. J. and Merriman, P. R. (1980). Reduction of sclerotial inoculum of *Sclerotinia sclerotiorum* with *Coniothyrium minitams*, *Soil Biol. Biochem.* 12(5), 461–465.

Tomilova, O. G., Shaldyaeva, E. M., Kryukova, N. A., Pilipova, Y. V., Schmidt, N. S., Danilov, V. P., Kryukov, V. Y. and Glupov, V. V. (2020). Entomopathogenic fungi decrease rhizoctonia disease in potato in field conditions, *PeerJ* 8, e9895.

US Patent (2017). Isolated strain of *Clonostachys rosea* for use as a biological control agent. Available at: https://patents.google.com/patent/US9758758B2/en (accessed 16 December 2020).

Utkhede, R. S. and Rahe, J. E. (1980). Biological control of onion white rot, *Soil Biol. Biochem.* 12(2), 101–104.

Utkhede, R. S. and Rahe, J. E. (1983). Interactions of antagonist and pathogen in biological control of onion white rot, *Phytopathology* 73(6), 890–893.

van der Heijden, M. G. A. and Hartmann, M. (2016). Networking in the plant microbiome, *PLoS Biol.* 14(2), e1002378.

van Lenteren, J. C., Bolckmans, K., Köhl, J., Ravensberg, W. J. and Urbaneja, A. (2018). Biological control using invertebrates and microorganisms: plenty of new opportunities, *BioControl* 63(1), 39–59.

Vandermeer, J., Perfecto, I. and Liere, H. (2009). Evidence for hyperparasitism of coffee rust (*Hemileia vastatrix*) by the entomogenous funguss, *Lecanicillium lecanii*, through a complex ecological web, *Plant Pathol.* 58(4), 636–641.

Vardharajula, S., Ali, S. Z., Grover, M., Reddy, G. and Bandi, V. (2011). Drought-tolerant plant growth promoting *Bacillus* spp.: effect on growth, osmolytes, and antioxidant status of maize under drought stress, *J. Plant Inter.* 6(1), 1–14.

Vectorite (2020). Available at: https://www.vectoritebvt.com/wp-content/uploads/2020/08/BVT-Vectorite-with-CR-7-Technical-Data-Sheet-for-Berries-20200826-General-Contact.pdf (accessed 16 December 2020).

Vero, S., Garmendia, G., Gonzáles, M. B., Bentancur, O. and Wisniewski, M. (2013). Evaluation of yeasts obtained from Antarctic soil samples as biocontrol agents for the management of postharvest diseases of apple (*Malus x domestica*), *FEMS Yeast Res.* 13(2), 189–199.

Viana, L. T., Bustamante, M. M. da C., Molina, M., Pinto, A. de S., Kisselle, K., Zepp, R. and Burke, R. A. (2011). Microbial communities in cerrado soils under native vegetation subjected to prescribed fire and under pasture, *Pesqu Agropecu Bras.* 46(12), 1665–1672.

Villa-Lerma, G., Gonzáles-Márquez, H., Gimeno, M., López-Luna, A., Bárzana, E. and Shirai, K. (2013). Ultrasonication and steam-explosion as chitin pretreatments for chitin oligosaccharide production by chitinases of *Lecanicillium lecanii*, *Biores. Technol.* 146, 794–798.

Vu, N. T. and Oh, C. S. (2020). Bacteriophage usage for bacterial disease management and diagnosis in plants, *Plant Pathol. J.* 36(3), 204–217.

Waage, J. (2007). The sustainable management of biodiversity for biological control in food and agriculture: status and needs, *FAO: Background Study Paper No. 38*, pp. 54.

Waché, Y. J., Valat, C., Postollec, G., Bougeard, S., Burel, C., Oswald, I. P. and Fravalo, P. (2009). Impact of deoxynivalenol on the intestinal microflora of pigs, *Int. J. Mol. Sci.* 10(1), 1–17.

Wang, S., Ongena, M., Qiu, D. and Guo, L. (2017). Fungal viruses: promising fundamental research and biological control of fungi, *SM Virol.* 2(1), 1011.

Wang, E. T., Tan, Z. Y., Guo, X. W., Rodriguez-Duran, R., Boll, G. and Martinez-Romero, E. (2006). Diverse endophytic bacteria isolated from a leguminous tree *Conzattia multiflora* grown in Mexico, *Arch. Microbiol.* 186(4), 251–259.

Wardeh, M., Risley, C., McIntyre, M. K., Setzkorn, C. and Baylis, M. (2015). Database of host-pathogen and related species interactions, and their global distribution, *Sci. Data* 2, 150049.

Weber, M. G., Porturas, L. D. and Taylor, S. A. (2016). Foliar nectar enhances plant-mite mutualisms: the effect of leaf sugar on the control of powdery mildew by domatia-inhabiting mites, *Ann. Bot.* 118(3), 459–466.

Weller, D. M. and Cook, R. J. (1983). Suppression of take-all of wheat by seed treatments with fluorescent *pseudomonads*, *Phytopathology* 73(3), 463–469.

Weller, D. M., Zhang, B. X. and Cook, R. J. (1985). Application of a rapid screening test for selection of bacteria suppressive to take-all of wheat, *Plant Dis.* 69, 710–713.

Whipps, J. M. and Gerlagh, M. (1992). Biology of *Coniothyrium minitans* and its potential for use in disease biocontrol, *Mycol. Res.* 96(11), 897–907.

Whipps, J. M., Sreenivasaprasad, S., Muthumeenakshi, S., Rogers, C. W. and Challen, M. P. (2008). Use of *Coniothyrium minitans* as a biocontrol agent and some molecular aspects of sclerotial mycoparasitism. *Eur. J. Plant Pathol.* 121, 323–330.

WMO. (n.d.), *Clim. Change Desertification*. Available at: https://www.wmo.int/pages/prog/wcp/agm/publications/documents/wmo_cc_desertif_foldout_en.pdf (accessed 4 January 2021).

Woo, S. L., Ruocco, M., Vinale, F., Nigro, M., Marra, M., Lombardi, N., Pascale, A., Lanzuise, S., Manganiello, G. and Lorito, M. (2014). Trichoderma-based products and their widespread use in agriculture. *Open Mycol. J.* 8, 71-126.

Yoder, J. I. and Scholes, J. D. (2010). Host plant resistance to parasitic weeds; recent progress and bottlenecks, *Curr. Opin. Plant Biol.* 13(4), 478-484.

Yu, S., LI, B., Fu, Y., Xie, J., Cheng, J., Ghabrial, S. A., Li, G., Yi, X.. and Jiang, D. (2013). Extracellular transmission of a DNA mycovirus and its use as a natural fungicide, *PNAS* 110(4), 1452-1457.

Zaldúa, S. and Sanfuentes, E. (2010). Control of *Botrytis cinerea* in *Eucalyptus globulus* mini-cuttings using *Clonostachys* and *Trichoderma* strains, *Chil. J. Agric. Res.* 70(4), 576-582.

Zambolim, L. (2016). Current status and management of coffee leaf rust in Brazil, *Trop. Plant Pathol.* 41(1), 1-8.

Zapata, J., Acosta, C., Díaz, A., Villamizar, L. and Cotes, A. M. (2011). Characterization of *Rhodotorula glutinis* and *Pichia onychis* isolates with potential as biopesticides for controlling *Botrytis cinerea*, *Acta Hortic.* 905(905), 155-160.

Zapata, J., Villamizar, L., Díaz, A., Uribe, L., Bolaños, C., Gómez, M. and Cotes, A. M. (2013). Development of a biopesticide prototype based on the yeast *Rhodotorula glutinis* Lv316 for controlling *Botrytis cinerea* in blackberry, *IOBC WPRS Bull.* 86, 263-269.

Zapata, Y., Díaz, A., Grijalba, E., Rodríguez, F., Elad, Y. and Cotes, A. M. (2016). Phyllosphere yeasts with potential for biological control of *Botrytis cinerea* in rose, *Acta Hortic.* 1144(1144), 77-84.

Zeriouh, H., Romero, D., García-Gutiérrez, L., Cazorla, F. M., de Vicente, A. and Pérez-García, A. (2011). 'The iturin-like lipopeptides are essential components in the biological control arsenal of *Bacillus subtilis* against bacterial diseases of cucurbits', *Molec, Plant Microbe Interac.* 24(12), 1540-1552.

Zhalnina, K., Louie, K. B., Hao, Z., Mansoori, N., Da Rocha, U. N., Shi, S., Cho, H., Karaoz, U., Loqué, D., Bowen, B. P., Firestone, M. K., Northen, T. R. and Brodie, E. L. (2018). Dynamic root exudate chemistry and microbial substrate preferences drive patterns in rhizosphere microbial community assembly, *Nat. Microbiol.* 3(4), 470-480.

Zhang, H., Apaliya, M. T., Mahunu, G. K., Chen, L. and Li, W. (2016). Control of ochratoxin A-producing fungi in grape berry by microbial antagonists: a review, *Trends Food Sci. Technol.* 51, 88-97.

Zhou, Y., Choi, Y. L., Sun, M. and Yu, Z. (2008). Novel roles of *Bacillus thuringiensis* to control plant diseases, *Appl. Microbiol. Biotechnol.* 80(4), 563-572.

Zhu, S., Vivanco, J. M. and Manter, D. K. (2016). Nitrogen fertilizer rate affects root exudation, the rhizosphere microbiome and nitrogen-use-efficiency of maize, *Appl. Soil Ecol.* 107, 324-333.

Chapter 3

Visualising plant colonisation by beneficial bacteria: a key step to improve the understanding of plant-microbe interactions

Stéphane Compant, Günter Brader and Angela Sessitsch, AIT Austrian Institute of Technology GmbH, Austria

1 Introduction

Plants host diverse microbial complex assemblages in and on various vegetative and reproductive or disseminative organs (Brader et al., 2017; Papik et al., 2020). These microbial assemblages comprise bacteria, fungi, oomycetes, viruses and archaea, which may be pathogenic, neutral or beneficial to their hosts (Compant et al., 2010; Hardoim et al., 2015; Afzal et al., 2019). Starting in the early nineteenth century (1809, with the work of Heinrich Friedrich Link on fungi), pioneering investigations into plant-associated microorganisms have shown their dynamic nature and ability to interact closely with their hosts. In the last few decades, we have obtained an advanced understanding of the mechanisms involved in plant-microbe interactions as well as the ecology of plant-associated microbial consortia (Compant et al., 2020; Harrison and Griffin, 2020). Microorganisms exhibiting plant-beneficial properties can be utilised to improve plant health, growth and yields (Trivedi et al., 2017). However, the current application of microbials in the field faces multiple challenges, ranging

http://dx.doi.org/10.19103/AS.2021.0093.04

from compatibility issues between plants and microorganisms, competition with highly complex native microbiota, to challenges in formulation development (Sessitsch et al., 2019). There is a need to improve the delivery of the target microorganism(s) to the receiving (plant) environment, in terms of dosage, timing and target tissue. Timing and required colonisation behaviour will depend on the relevant modes of action. For example, a strain inducing systemic resistance against a phytopathogen has to colonise the plant before pathogen infection, but often systemic colonisation of the whole plant is not required. Furthermore, if a pathogen colonises a particular plant tissue and the mode of action is antibiosis, the inoculant strain should colonise the same niche at the same time as the pathogen (Compant et al., 2019).

An understanding of the colonisation characteristics of a commercial strain (i.e., the distinct niches a strain colonises, the timing of spread within and on the plant and the biotic and abiotic factors influencing colonisation) is needed to assess the suitability of strains for certain applications and to further improve the delivery and application approach (Rilling et al., 2019; Compant et al., 2020). Knowledge of colonisation characteristics is also useful for assessing any non-target impacts associated with the application (Compant et al., 2019). In this chapter, we set out how beneficial bacteria can colonise their host plants under various conditions and demonstrate how an understanding of plant colonisation can be used to improve bacterial application approaches.

2 Methods to visualise and determine microbial colonisation of plants

Different tools are available to monitor microbial colonisation in plants. Monitoring can be undertaken using direct fluorescent antibodies, although the development of specific antibodies is a potential limitation for the technique as it is laborious and difficult to find specific antibodies. Other methods to detect bacteria include counting colony-forming units (CFUs) and the use of spontaneous mutants, being resistant to antibiotics, and selective plating techniques (Romano et al., 2020), but this does not identify the specific location of bacterial colonisation (only cell densities are obtained). Molecular methods such as quantitative PCR of specific marker genes, as well as analysing the whole microbiota abundance, are also used (discussed in Romano et al., 2020, for rhizosphere bacteria). Although considerable information has been obtained on the abundance of bacterial cells, these methods do not provide insight into specific colonisation niches. Colonisation behaviour of specific strains has been studied using techniques such as quantitative PCR and CFU counting as shown, for example, with *Bacillus firmus* I-1582 and *B. amyloliquefaciens* QST713 (present in VOTiVO® and Serenade®, respectively) in non-sterile soil and under field conditions (Mendis et al., 2018) or with *Streptomyces lydicus* WYEC108

(Actinovate®) in sterile and non-sterile soils (Yuan and Crawford, 1995). This has allowed the determination of the abundance of bacterial cells on and inside the roots. However, this methodology (i.e. targeting microbial DNA followed by PCR or targeting culturable cells) provides results only on the entire bacterial or strain population in the whole plant, without knowing the plant compartments or niches that are colonised. Moreover, it is important to consider surface-colonising bacteria and the need to treat samples appropriately if endophytic colonisation is specifically addressed, that is, by removing bacterial DNA from the plant surfaces. A combination of the approaches, including microscopy analyses, can provide specific information on bacterial colonisation of different plant environments, allowing better monitoring of the colonisation characteristics of specific strains.

Some approaches are useful to monitor a target strain in different plant niches. Strains labelled with marker genes (e.g. *gfp*, *dsred* or *gus*A), the use of fluorescence *in situ* hybridisation (FISH) analysis (with taxa-targeting DNA probes), or general staining methods (involving, e.g. syto9® or acridine orange targeting DNA) have been shown to allow *in planta* visualisation using light microscopy, epifluorescence, confocal or live imaging microscopy (Ahmad et al., 2011; Cardinale, 2014; Massalha et al., 2017). These methods have provided considerable information on the niches and plant compartments that bacteria can colonise.

3 Colonisation of beneficial bacteria from the soil to the root surface

The soil is the most important reservoir of plant-associated microorganisms, including beneficial strains (Benizri et al., 2001; Lugtenberg and Kamilova, 2009). Diverse microorganisms colonise and proliferate in the rhizosphere (the narrow soil region surrounding the roots), which is influenced by the plant and microorganisms and was first described by Lorenz Hiltner (1904; Lemanceau et al., 2017). Additionally, when seeds germinate, the surrounding environment (termed the spermosphere) allows the attachment of rhizosphere microorganisms to the surface of the roots, that is, the rhizoplane, for instance, by entrapment on biofilms (Nelson, 2018). Sugars, amino acids, organic acids, vitamins, phytohormones, flavonoids, terpenes and several other compounds are typically released by the plant into the rhizosphere/spermosphere and act together with dead plant cells as attractants for microorganisms (Bais et al., 2006; Musilova et al., 2016; Backer et al., 2018). There is fierce competition for access to the released nutrients. Microorganisms have a number of features to increase their competitiveness, such as the formation of pili, flagella, fimbriae, nod factors, lipopolysaccharides, use of quorum sensing and twitching motility (Lugtenberg and Kamilova, 2009; Compant et al., 2010).

Different microorganisms have been shown to preferentially colonise different parts of the roots (Mercado-Blanco, 2015). Some are attracted by the root cap and mucilage and colonise the root tip, while others colonise the root elongation zone as well as the root hair zone (Parke, 1991; Fig. 1). Other microorganisms colonise different root zones more uniformly. Furthermore, emerging secondary roots are hotspots for microorganisms. Microorganisms

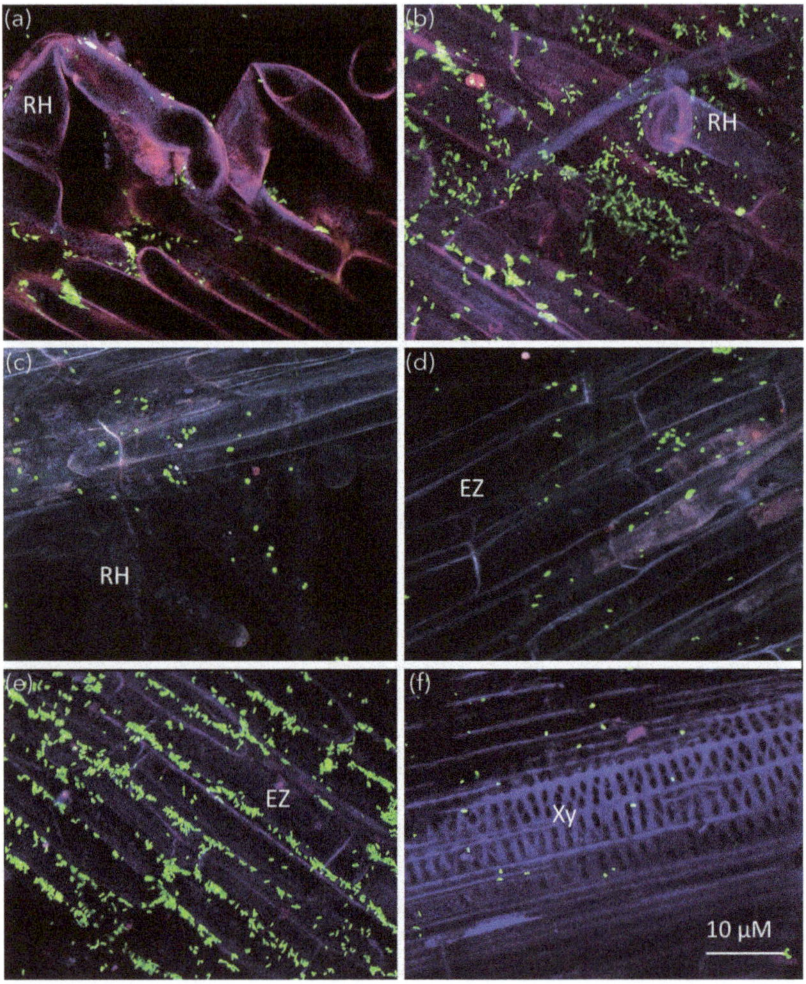

Figure 1 Confocal microscopic images of maize (a, b and e, f) and tomato (c, d) roots colonised by *Paraburkholderia phytofirmans* strain PsJN::*gfp*2x, 5 days post seed inoculation, showing colonisation of the surface of the root hair zone (a–c), root elongation (d, e) and inside xylem vessels (f). Bacteria were single (c, d and f), clustered in groups (a and b) or forming lines (e). EZ: elongation zone, RH: root hair, Xy: xylem. Green: bacteria, Purple/blue: plant. Size of a bacterium: 1.5 μm.

can not only occur as single cells but also multiply by making doublets of cells, creating colonies, lines of bacteria or biofilms, in zones where they can use the nutrients released by the plant (Fig. 1; Compant et al., 2005, 2010). Inoculated microorganisms need to establish and successfully compete against other microorganisms (Mendes et al., 2013). In addition to the mechanisms employed by the microorganisms to outcompete their neighbours, the soil condition, the plant species and genotype, growth stage and physiological conditions of the plant as well as environmental conditions are known to influence the colonisation processes (Parke, 1991; Mendes et al., 2013).

As an example, *Paraburkholderia phytofirmans* strain PsJN (previously known as a *Burkholderia* strain), which confers induced resistance to pathogens and tolerance to abiotic stresses (frost and drought) as well as stimulation of plant growth, has been seen on grapevine roots (root hair zone, root elongation zone and root tip) using *gfp*- and *gusA*-marked strains (Compant et al., 2005, 2008; Mitter et al., 2013a,b). Figure 1 shows the colonisation of the *gfp* transformed strain on the root surface, particularly at the root hair zone and the elongation zone, on maize and tomato plants, with different densities following seed inoculation and plant growth. A strain of *Trinickia* sp. has also been observed colonising roots of tomatoes using a *Dsred* transformed strain (del Barrio-Duque et al., 2020), especially in the root hair zone. Strains of *Pseudomonas* spp. have been seen colonising the root hair zone as well as penetrating inside the hairs using transformed strains (Mercado-Blanco and Prieto, 2012). Many other examples targeting different microbial taxa can be found in the literature. Strains of *Streptomyces* spp. were labelled with *egfp* (Chen et al., 2016), and spores, single hyphae, spore chains and/or mycelium were observed on the root surface, mostly in the zone of cellular maturation of the main and lateral roots and at the root hair zone (Chen et al., 2016).

In the case of commercial strains, plant colonisation by *B. velezensis* strain FZB42 (RhizoVital®42) has been thoroughly investigated (Fan et al., 2011, 2012). Transformed strains have been seen on corn and *Arabidopsis thaliana* roots. On Arabidopsis roots, strain FZB42 preferentially colonised the root tips, whereas on maize, it colonised the root hair zone. Biofilms consisting of extracellular matrix and cells with altered morphology were also observed on the rhizoplane (Fan et al., 2012). Dietel et al. (2013) further analysed, using mutants, which bacterial genes have an influence on colonisation processes of strain FZB42. Important factors include swarming motility and biofilm formation, and knock-out mutant strains affected by these features have reduced ability to colonize plant roots. However, this ability also depends on the additional genes as six knock-out mutants without phenotype in biofilm formation still displayed reduced root colonisation (Dietel et al., 2013). While FZB42 has been well investigated, comparable information is still missing for most commercial strains.

4 Colonisation of beneficial bacteria to root internal tissues

Upon colonising the root surfaces, some bacteria have the capability to enter the roots (Reinhold-Hurek and Hurek, 2011; Mengistu, 2020). This behaviour (known as endophytism) is common under natural conditions (Rosenblueth and Martínez-Romero, 2006; Liu et al., 2017; Kandel et al., 2017; Rana et al., 2020). Endophytic colonisation has been frequently addressed by academic research not only following the application of beneficial plant-associated microorganisms but also in samples from diverse plant environments. Interestingly, it has been shown that beneficial microorganisms showing an endophytic lifestyle have a greater impact on plant growth and health than strains colonising the rhizosphere or rhizoplane only (Ali et al., 2012; Coutinho et al., 2015).

Endophytic microorganisms can migrate passively to the inner root tissues through openings of the plant (e.g. cracks in the secondary emergence root zone and holes made by insects or pathogenic microorganisms), whereas others actively colonise the inner plant tissues by secreting plant cell wall-degrading enzymes (Sørensen and Sessitsch, 2015; Santoyo et al., 2016; Mengistu, 2020). Typically, only small amounts of plant cell wall-degrading enzymes such as cellulases or polygalacturonases are produced to prevent plant damage. Beneficial endophytes are usually recognised as non-pathogenic strains and evade plant defence (Hardoim et al., 2015); however, the mechanisms are not clearly understood. The root epidermal cells (with hairs or not) and the exodermis beneath them can be colonised. However, some strains can migrate up to several cortical cell layers, going deeper and deeper inside the roots (Hallmann, 2001). They can be intercellular but may also be found inside the cells. As an example, scanning electron microscopic observations showed endophytic colonisation of lettuce root cortex tissues by a *Streptomyces exfoliatus* strain (Chen et al., 2016). Some strains can also pass the endodermis barrier, which is difficult to pass due to its thick cell wall. This type of entry occurs before the cells reach the central cylinder containing the vascular system and therefore colonise the xylem vessels (Compant and Vacher, 2019). The *P. phytofirmans* strain PsJN was found to cross from the root surface to the cortical cell layer and then to the endodermis and the xylem vessels as shown in Fig. 1 with maize, but this can also be found in other plants such as, for instance, grapevines (Compant et al., 2005, 2008).

It has been hypothesised that entering roots and subsequent endophytic colonisation confers protection to the bacterium, compared to the rhizosphere colonisation where there is tremendous competition (Santoyo et al., 2016). Bacterial traits required for endophytic competence include chemotaxis, flagella formation, carbohydrate utilisation, secretion factors and detoxification of plant stress compounds, for example, reactive oxygen species, among others, enabling them to move up to the xylem vessels. Furthermore, microbial compounds may be released inside roots. Plant defence may facilitate or restrict endophytic progression inside root tissues (Miché et al., 2006; Lòpez-Fernàndez et al., 2016). On the other hand, colonising root tissues internally might be just a consequence of high competition on the root surfaces, especially in the case of passive entrance (Compant and Vacher, 2019). However, some strains use smart tools, described above, enabling endophytic colonisation up to the xylem, which cannot be achieved without employing active mechanisms.

Upon entering the xylem vessels, some bacteria can colonise the upper plant parts (James et al., 1997, 2002; Hallmann, 2001) and systemically colonise the plant, moving slowly from one xylem element to another. Colonising a plant part of just 20 cm may take a month or more (Compant et al., 2008, 2010). The holes of the perforated plates of the xylem vessels allow the passage of these microorganisms (Compant et al., 2010). This has been well described using the *gfp* or *gus*A labelled *P. phytofirmans* strain PsJN in grapevine (Compant et al., 2005, 2008). Why these microbes colonise xylem vessels is still not well explored. However, it has been shown that xylem vessels, usually known to contain water and minerals, also contain traces of specific sugars such as L-arabinose, and chemotaxis of bacteria towards these sugars may play a role (Malfanova et al., 2013). Water availability inside roots can also explain why microorganisms colonise xylem vessels. The hydraulic pressure due to the transpiration stream might allow bacteria to move faster from below to the upper plant parts (Compant et al., 2010). However, bacteria still colonise the plant from one xylem element to another and sometimes stay at the junction between these two elements. Evidence of these colonisation routes has been obtained by microscopic analyses with strains of *Herbaspirillum seropedicae* and *P. phytofirmans*, among others, but only a few case studies reported systemic colonisation of beneficial bacteria (see section 'Where to look for further information'). Another possible spread of bacteria throughout the plant can be related to intercellular colonisation of root tissues, although this has not been shown to be the main bacterial route of colonisation. Using the *P. phytofirmans* strain PsJN and its *gfp*-marked derivative, we demonstrated colonisation from the xylem vessels of the roots to the aerial plant parts (Compant et al., 2005, 2008; Mitter et al., 2013b).

Most of the beneficial properties mediated by endophytes can be associated with their surface colonisation, their microbial-associated molecular patterns and their secreted secondary metabolites (Hardoim et al., 2015; Brader et al., 2017). Bacterial hormones may be secreted inside plant tissues, although proof of this is still lacking. Plant defence reactions also occur within or between cells due to bacterial colonisation. Generally, research should not only address colonisation but also should be complemented by studies addressing microbial activities. For instance, microbial gene expression inside plant tissues and organs or even single-cell analysis could pinpoint the role of endophytes in certain plant (or plant-microbe) traits. Sessitsch et al. (2012) showed that different N-cycling genes of bacterial endophytes are expressed in rice roots, therefore inside the plants, and thus highlighted the importance of endophytes in N uptake and metabolism. Carrión et al. (2019) further studied the role and disease suppression of secondary metabolites produced by the endophytic microbiome in the plant-pathogen interaction of sugar beet and the wilt fungus *Rhizoctonia solani*.

Some endophytes may colonise even the lumen of the xylem of leaves or fruits but with reduced subpopulations compared to the roots (Compant et al., 2010; Hardoim et al., 2015). They may protect the host plant from pathogens colonising aerial parts, induce stress resistance on leaves and protect their host from abiotic stresses. Some endophytes can also be horizontally transmitted to finally re-colonise the soil (Compant et al., 2016). However, this is not yet fully explored. Considerable information should be obtained by tracking microorganisms inside plants, in addition to analysing functionalities with techniques such as laser microdissection, *in situ* metabolite analysis (Geier et al., 2020) and transcriptome analysis.

5 Colonisation of the aerial plant parts

Following the application of bacteria on leaves, fruits or flowers, plant-beneficial bacteria may colonise different organs, migrate and proliferate (Brader et al., 2017; Afzal et al., 2019). Compared to the rhizosphere and the rhizoplane, the surfaces of above-ground tissues have different water content and exudates, and UV and temperature conditions can be detrimental for bacteria (Brader et al., 2017; Vacher et al., 2016). Bacteria have to adapt to these specific microclimates to survive. On leaves, they form dynamic populations, assemble between the cell interstices, colonise hydathodes and stomata and enter inside plants. Colonisation depends on the environment, the plant's physiological status, species and genotype as well as the genomic properties of bacterial strains (Vacher et al., 2016; Vorholt, 2012). Wounds caused by pathogens or animals can also impact the colonisation of these bacteria (Compant and Vacher, 2019). Colonisation from the surface to the inner tissues can protect microorganisms from the external environment and defend the plant from biotic and abiotic stressors. Peredo and Simmons (2018) demonstrated colonisation of *Methylobacterium extorquens* PA1 cells in the substomatal chambers of *Arabidopsis thaliana*, following a phase of leaf surface colonisation. Generally, whether a strain can colonise leaves or other tissues internally will depend on the genetic repertoire of the microorganism, the plant species and cultivar and the environmental conditions. The *P. phytofirmans* strain PsJN is known to reduce grapevine Pierce disease infection by the pathogen *Xylella fastidiosa* (Baccari et al., 2019). This deleterious pathogen colonises xylem vessels, blocks water supply, induces leaf scorches and can kill the plant. As the PsJN strain is known to systemically colonise plants by spreading through the xylem vessels, Baccari et al. (2019) inoculated strain PsJN into a branch using a syringe and thereby reduced *Xylella* pathogen loads. Knowing the niches of colonisation and the routes used by beneficial bacteria can provide information regarding new strategies for crop protection.

On flowers, some bacteria can successfully establish microcolonies, depending on the stress factors and external conditions. It has been demonstrated that flowers contain, under natural conditions, various bacterial assemblages featuring different composition, diversity and structure. All flower tissues can be colonised (Compant et al., 2011; Franck et al., 2017). Bacteria can colonise pollen, stamina and anthers as well as sepals, petals, flower receptacle and the stigma, depending on the compounds secreted by flowers. The *Pseudomonas fluorescens* A506 strain of the commercial product Blightban® A506, utilised against fire blight and severity of fruit russet, has been shown to colonise flowers, leading to a significant reduction in populations of the fire blight pathogen *Erwinia amylovora* (Lindow and Suslow, 2003). Immature flower buds contain stomata that can be used by the bacteria to enter the flower tissues. Pollen germinating on stigma might allow further penetration of the bacteria into flowers. Wounds caused by pathogens or animals and wind can also allow bacteria to travel from the surface of the flowers to the inner plant tissues. Pollinators can also help bacteria to colonise the surface of flowers (Compant and Vacher, 2019).

Inside flowers, bacteria can be present in the epidermis and the ovary, and later inside the fruits and seeds, as well as in the embryonic parts of seeds (Escobar-Rodríguez et al., 2018). While this was demonstrated for plants and their natural microflora using FISH analyses (Compant et al., 2011), flower colonisation has also been used as a delivery strategy to stimulate plant growth. Mitter et al. (2017) developed an approach based on flower inoculation to provide seeds containing beneficial endophytes internally. The inoculated strain, that is strain PsJN, which served as a model strain, emerged with the germinating seed and efficiently colonised the progeny plant (Mitter et al., 2017). The niches of colonisation have been demonstrated by visualising the strain inside the seed embryo using a FISH probe targeting the bacterial species (Mitter et al., 2017).

Once applied to fruits, bacteria can also penetrate inside plant tissues and be present in the flesh parts, such as the exo-, meso- and endocarp (Compant et al., 2019). Colonisation of the interior of seeds inside fruits following fruit surface inoculation has not been demonstrated to date, to our knowledge. This is, however, not surprising as the seeds are already mature, and the seed coat may not allow the bacteria to enter inside mature seeds (Escobar-Rodríguez et al., 2018). If bacteria are also sprayed on seed surfaces, they usually do not colonise the plant's inner tissues due to the barrier of the seed coat. However, once these seeds germinate, bacteria will colonise the spermosphere, as described at the beginning of this chapter, attracted by seed exudates, and will colonise the new plant surface, both at root and short shoot level (Nelson, 2018; Nelson et al., 2018).

6 Improving applications by understanding the colonisation processes

There is little knowledge of the colonisation characteristics of commercial strains. Generally, industrial research has focused on developing the application and on assessing its efficacy in the field, often without considering *in planta* colonisation and establishment. However, understanding the requirements of a commercial strain to colonise the target plant niche at the right time as well as to express the desired activities under field conditions can be of great help to further improve a microbial application (Compant et al., 2019, 2020; Rana et al., 2020), especially in the case of controlling a pathogen colonising specific niches such as xylem vessels inside plants. If a biocontrol strain has to colonise the root hairs to encounter the target pathogen but only colonises the root tip, its efficiency will be reduced, even if the responsible antimicrobial metabolites can be translocated to other tissues. Furthermore, if a biocontrol strain colonises the rhizosphere but not the plant endosphere and is used to control a pathogen colonising the plant internally, it might reduce pathogen load, but the strain will not affect most of the pathogen cells that have entered inside the plant tissues. Even if a biocontrol strain triggers induced systemic resistance in addition to antimicrobial activities, efficient colonisation of the target niches will be necessary. Moreover, if a pathogen is present inside seeds prior to germination, a biocontrol strain has to colonise the same niche. To prevent infestation of seeds with a pathogen, biocontrol strains colonising flowers or immature seeds may be applied, where colonisation can occur in mature seeds. A detailed understanding of colonisation process as regards time and space will lead to tailored applications of beneficial microorganisms, particularly, biocontrol strains. In addition, the efficiency and application range of already developed microbial products can be further improved based on the colonisation characteristics of the responsible microorganism in the target plant environments. These characteristics might be different depending on the host plant species, as it has been shown that colonisation of the commercial biocontrol strain *B. velezensis* FZB42 shows distinct patterns in three plant species (Fan et al., 2012).

With the rapidly growing knowledge of the importance of plant microbiomes, scientists are increasingly aware of microorganisms colonising the interior of plants. However, the development of microbial applications has initially focused on plant growth-promoting rhizobacteria, that is, strains largely colonising the rhizosphere. There is substantial evidence that endophytes are ubiquitous and may confer several benefits to their plant hosts (Hardoim et al., 2015; Compant et al., 2020). Understanding the colonisation characteristics and ecology of microorganisms on and inside the plants has not yet been fully explored for commercial applications.

7 Conclusion and future trends in research

In this chapter, we have described the niches and routes of colonisation of plant-associated bacteria. Although considerable research has been carried out on plant microbiomes, few studies have investigated the colonisation processes by bacteria. This is particularly the case for microbial products. However, understanding the colonisation characteristics of an applied microorganism as well as the potential failures in the application is of tremendous value in terms of tailored application design. Understanding the colonisation characteristics of a given strain will determine the right application time as well as the application mode to allow the most efficient colonisation of the target plant compartment. Colonisation should be an important aspect of application development. New strategies may be developed, and the consideration of colonisation characteristics might offer new opportunities for strain delivery and targets, as shown by the case study in this chapter involving *P. phytofirmans* strain PsJN. Additional research should be carried out on the relevant mechanisms and on elucidating the factors (plant, environment or microorganisms) that are involved in the colonisation process. Such fundamental understanding, coupled with tailored and knowledge-based microbial application strategies, will result in microbial products demonstrating improved field success and will lead to fewer toxic chemicals in agriculture.

8 Where to look for further information

Readers can discover more information regarding the ecology, role, mechanisms and future prospects related to bacterial endophytes in the following reviews:

- Brader, G., Compant, S., Vescio, K., Mitter, B., Trognitz, F., Ma, L.-J., et al. (2017), 'Ecology and genomic insights into plant-pathogenic and plant-nonpathogenic endophytes', *Annu. Rev. Phytopathol.*, 55, 61–83.
- Compant, S., Clément, C. and Sessitsch, A. (2010), 'Plant growth-promoting bacteria in the rhizo- and endosphere of plants: their role, colonization,

mechanisms involved and prospects for utilization', *Soil Biol. Biochem.*, 42, 669–678.

- Hardoim, P. R., Van Overbeek, L. S., Berg, G., Pirttilä, A. M., Compant, S., Campisano, A., et al. (2015), 'The hidden world within plants: ecological and evolutionary considerations for defining functioning of microbial endophytes', *Microbiol. Mol. Biol. Rev.*, 79, 293–320.

Readers can also find more information on emerging trends to increase microbial application in the field in the following reviews:

- Compant, S., Samad, A., Faist, H. and Sessitsch, A. (2019), 'A review on the plant microbiome: ecology, functions, and emerging trends in microbial application', *J. Adv. Res.*, 19, 29–37.
- Sessitsch, A., Pfaffenbichler, N. and Mitter, B. (2019), 'Microbiome applications from lab to field: facing complexity', *Trends Plant Sci.*, 24, 194–198.

9 References

Afzal, I., Shinwari, Z. K., Sikandar, S. and Shahzad, S. (2019). Plant beneficial endophytic bacteria: mechanisms, diversity, host range and genetic determinants, *Microbiol. Res.* 221, 36–49.

Ahmad, I., Ahmad, F. and Pichtel, J. (2011). Rhizosphere and root colonization by bacterial inoculants and their monitoring methods: a critical area in PGPR research. In: Ahmad, I., Ahmad, F. and Pichtel, J. (Eds), *Microbes and Microbial Technology: Agricultural and Environmental Applications*, Springer, New York, pp. 363–391.

Ali, S., Charles, T. C. and Glick, B. R. (2012). Delay of flower senescence by bacterial endophytes expressing 1-aminocyclopropane-1-carboxylate deaminase, *J. Appl. Microbiol.* 113(5), 1139–1144.

Baccari, C., Antonova, E. and Lindow, S. (2019). Biological control of Pierce's disease of grape by an endophytic bacterium, *Phytopathology* 109(2), 248–256.

Backer, R., Rokem, J. S., Ilangumaran, G., Lamont, J., Praslickova, D., Ricci, E., Subramanian, S. and Smith, D. L. (2018). Plant growth-promoting rhizobacteria: context, mechanisms of action, and roadmap to commercialization of biostimulants for sustainable agriculture, *Front. Plant Sci.* 9, 1473.

Bais, H. P., Weir, T. L., Perry, L. G., Gilroy, S. and Vivanco, J. M. (2006). The role of root exudates in rhizosphere interactions with plants and other organisms, *Annu. Rev. Plant Biol.* 57, 233–266.

Benizri, E., Baudoin, E. and Guckert, A. (2001). Root colonization by inoculated plant growth rhizobacteria, *Biocontrol Sci. Technol.* 11(5), 557–574.

Brader, G., Compant, S., Vescio, K., Mitter, B., Trognitz, F., Ma, L. J. and Sessitsch, A. (2017). Ecology and genomic insights into plant-pathogenic and plant-nonpathogenic endophytes, *Annu. Rev. Phytopathol.* 55, 61–83.

Cardinale, M. (2014). Scanning a microhabitat: plant-microbe interactions revealed by confocal laser microscopy, *Front. Microbiol.* 5, 94.

Carrión, V. J., Perez-Jaramillo, J., Cordovez, V., Tracanna, V., de Hollander, M., Ruiz-Buck, D., Mendes, L. W., van Ijcken, W. F. J., Gomez-Exposito, R., Elsayed, S. S., Mohanraju, P., Arifah, A., van der Oost, J., Paulson, J. N., Mendes, R., van Wezel, G. P., Medema, M. H. and Raaijmakers, J. M. (2019). Pathogen-induced activation of disease-suppressive functions in the endophytic root microbiome, *Science* 366(6465), 606–612.

Chen, X., Pizzatti, C., Bonaldi, M., Saracchi, M., Erlacher, A., Kunova, A., Berg, G. and Cortesi, P. (2016). Biological control of lettuce drop and host plant colonization by rhizospheric and endophytic Streptomycetes, *Front. Microbiol.* 7, 714.

Compant, S. and Vacher, C. (2019). Sources, niches and routes of colonization by beneficial bacterial endophytes. In: Schouten, A. (Ed). *Endophyte Biotechnology: Potential for Agriculture and Pharmacology*, CABI, UK, pp. 32–41.

Compant, S., Cambon, M. C., Vacher, C., Mitter, B., Samad, A. and Sessitsch, A. (2020). The plant endosphere world–bacterial life within plants, *Environ. Microbiol.* 23(4), 1812–1829, https://doi.org/10.1111/1462-2920.15240.

Compant, S., Clément, C. and Sessitsch, A. (2010). Plant growth-promoting bacteria in the rhizo- and endosphere of plants: their role, colonization, mechanisms involved and prospects for utilization, *Soil Biol. Biochem.* 42(5), 669–678.

Compant, S., Kaplan, H., Sessitsch, A., Nowak, J., Ait Barka, E. and Clément, C. (2008). Endophytic colonization of *Vitis vinifera* L. by *Burkholderia phytofirmans* strain PsJN: from the rhizosphere to inflorescence tissues, *FEMS Microbiol. Ecol.* 63(1), 84–93.

Compant, S., Mitter, B., Coli-Mull, J. G., Gangl, H. and Sessitsch, A. (2011). Endophytes of grapevine flowers, berries, and seeds: identification of cultivable bacteria, comparison with other plant parts, and visualization of niches of colonization, *Microb. Ecol.* 62(1), 188–197.

Compant, S., Reiter, B., Sessitsch, A., Nowak, J., Clément, C. and Ait Barka, E. (2005). Endophytic colonization of *Vitis vinifera* L. by plant growth-promoting bacterium *Burkholderia* sp. strain PsJN, *Appl. Environ. Microbiol.* 71(4), 1685–1693.

Compant, S., Saikkonen, K., Mitter, B., Campisano, A. and Mercado-Blanco, J. (2016). Editorial special issue: soil, plants and endophytes, *Plant Soil* 405(1–2), 1–11.

Compant, S., Samad, A., Faist, H. and Sessitsch, A. (2019). A review on the plant microbiome: ecology, functions, and emerging trends in microbial application, *J. Adv. Res.* 19, 29–37.

Coutinho, B. G., Licastro, D., Mendonça-Previato, L., Cámara, M. and Venturi, V. (2015). Plant-influenced gene expression in the rice endophyte *Burkholderia kururiensis* M130, *Mol. Plant Microbe Interact.* 28(1), 10–21.

del Barrio-Duque, A., Samad, A., Nybroe, O., Antonielli, L., Sessitsch, A. and Compant, S. (2020). Interaction between endophytic Proteobacteria strains and *Serendipita indica* enhances biocontrol activity against fungal pathogens, *Plant Soil* 451(1–2), 277–305.

Dietel, K., Beator, B., Budiharjo, A., Fan, B. and Borriss, R. (2013). Bacterial traits involved in colonization of *Arabidopsis thaliana* roots by *Bacillus amyloliquefaciens* FZB42, *Plant Pathol. J.* 29(1), 59–66.

Escobar-Rodríguez, C. E., Mitter, B., Barret, M., Sessitsch, A. and Compant, S. (2018). Commentary: seed bacterial inhabitants and their routes of colonization, *Plant Soil* 422(1–2), 129–134.

Fan, B., Borriss, R., Bleiss, W. and Wu, X. (2012). Gram-positive rhizobacterium *Bacillus amyloliquefaciens* FZB42 colonizes three types of plants in different patterns, *J. Microbiol.* 50(1), 38–44.

Fan, B., Chen, X. H., Budiharjo, A., Bleiss, W., Vater, J. and Borriss, R. (2011). Efficient colonization of plant roots by the plant growth promoting bacterium *Bacillus amyloliquefaciens* FZB42, engineered to express green fluorescent protein, *J. Biotechnol.* 151(4), 303–311.

Franck, A. C., Saldierna Guzmán, J. P. and Shay, J. E. (2017). Transmission of bacterial endophytes, *Microorganisms* 5(4), 70.

Geier, B., Sogin, E. M., Michellod, D., Janda, M., Kompauer, M., Spengler, B., Dubilier, N. and Liebeke, M. (2020). Spatial metabolomics of *in situ* host-microbe interactions at the micrometre scale, *Nat. Microbiol.* 5(3), 498–510.

Hallmann, J. (2001). Plant interactions with endophytic bacteria. In: Jeger, M. J. and Spence, N. J. (Eds), *Biotic Interactions in Plant-Pathogen Associations*, CABI Publishing, New York, pp. 87-119.

Hardoim, P. R., Van Overbeek, L. S., Berg, G., Pirttilä, A. M., Compant, S., Campisano, A., Döring, M. and Sessitsch, A. (2015). The hidden world within plants: ecological and evolutionary considerations for defining functioning of microbial endophytes, *Microbiol. Mol. Biol. Rev.* 79(3), 293–320.

Harrison, J. G. and Griffin, E. A. (2020). The diversity and distribution of endophytes across biomes, plant phylogeny and host tissues: how far have we come and where do we go from here?, *Environ. Microbiol.* 22(6), 2107–2123.

Hiltner, L. (1904). Über neuere Erfahrungen und Probleme auf dem Gebiete der Bodenbakteriologie unter besonderer Berücksichtigung der Gründüngung und Brache, *Arb. D.L.G.* 98, 59–78.

James, E. K., Gyaneshwar, P., Manthan, N., Barraquio, W. L., Reddy, P. M., Iannetta, P. P., Olivares, F. L. and Ladha, J. K. (2002). Infection and colonization of rice seedlings by the plant growth-promoting bacterium *Herbaspirillum seropedicae* Z67, *Mol. Plant Microbe Interact.* 15(9), 894–906.

James, E. K., Olivares, F. L., Baldani, J. I. and Döbereiner, J. (1997). *Herbaspirillum*, an endophytic diazotroph colonizing vascular tissue in the leaves of *Sorghum bicolor* L. Moench, *J. Exp. Bot.* 48, 785–797.

Kandel, S. L., Joubert, P. M. and Doty, S. L. (2017). Bacterial endophyte colonization and distribution within plants, *Microorganisms* 5(4), 77.

Lemanceau, P., Blouin, M., Muller, D. and Moënne-Loccoz, Y. (2017). Let the core microbiota be functional, *Trends Plant Sci.* 22(7), 583–595.

Lindow, S. E. and Suslow, T. V. (2003). Temporal dynamics of the biocontrol agent *Pseudomonas fluorescens* strain A506 in flowers in inoculated pear trees, *Phytopathology* 93(6), 727–737.

Liu, K., Newman, M., McInroy, J. A., Hu, C. H. and Kloepper, J. W. (2017). Selection and assessment of plant growth-promoting rhizobacteria for biological control of multiple plant diseases, *Phytopathology* 107(8), 928–936.

Lòpez-Fernàndez, S., Compant, S., Vrhovsek, U., Bianchedi, P. L., Sessitsch, A., Pertot, I. and Campisano, A. (2016). Grapevine colonization by endophytic bacteria shifts secondary metabolism and suggests activation of defense pathways, *Plant Soil* 405(1-2), 155–175.

Lugtenberg, B. and Kamilova, F. (2009). Plant-growth-promoting rhizobacteria, *Annu. Rev. Microbiol.* 63, 541–556.

Malfanova, N., Kamilova, F., Validov, S., Chebotar, V. and Lugtenberg, B. (2013). Is L-arabinose important for the endophytic lifestyle of *Pseudomonas* spp.?, *Arch. Microbiol.* 195(1), 9-17.

Massalha, H., Korenblum, E., Malitsky, S., Shapiro, O. H. and Aharoni, A. (2017). Live imaging of root-bacteria interactions in a microfluidics setup, *Proc. Natl Acad. Sci. U. S. A.* 114(17), 4549-4554.

Mendes, R., Garbeva, P. and Raaijmakers, J. M. (2013). The rhizosphere microbiome: significance of plant beneficial, plant pathogenic, and human pathogenic microorganisms, *FEMS Microbiol. Rev.* 37(5), 634-663.

Mendis, H. C., Thomas, V. P., Schwientek, P., Salamzade, R., Chien, J. T., Waidyarathne, P., Kloepper, J. and De La Fuente, L. (2018). Strain-specific quantification of root colonization by plant growth promoting rhizobacteria *Bacillus firmus* I-1582 and *Bacillus amyloliquefaciens* QST713 in non-sterile soil and field conditions, *PLoS ONE* 13(2), e0193119.

Mengistu, A. A. (2020). Endophytes: colonization, behaviour, and their role in defense mechanism, Int. J. Microbiol., 30, 6927219.

Mercado-Blanco, J. (2015), Life of microbes inside the plant, in Lugtenberg, B. J. J. (Ed), Principles of Plant-Microbe Interactions, Springer International Publishing, Berlin, pp. 25-23.

Mercado-Blanco, J. and Prieto, P. (2012). Bacterial endophytes and root hairs, *Plant Soil* 361(1-2), 301-306.

Miché, L., Battistoni, F., Gemmer, S., Belghazi, M. and Reinhold-Hurek, B. (2006). Upregulation of jasmonate-inducible defense proteins and differential colonization of roots of *Oryza sativa* cultivars with the endophyte *Azoarcus* sp., *Mol. Plant Microbe Interact.* 19(5), 502-511.

Mitter, B., Brader, G., Afzal, M., Compant, S., Naveed, M., Trognitz, F. and Sessitsch, A. (2013a). Advances in elucidating beneficial interactions between plants, soil and bacteria, *Adv. Agron.* 121, 381-445.

Mitter, B., Petric, A., Chain, P. S., Trognitz, F., Nowak, J., Compant, S. and Sessitsch, A. (2013b). Genome analysis, ecology, and plant growth promotion of the endophyte *Burkholderia phytofirmans* strain PsJN. In: Bruijn, F. J. (Ed), *Molecular Microbial Ecology of the Rhizosphere* (1st edn.), John Wiley & Sons, Inc., pp. 865-874.

Mitter, B., Pfaffenbichler, N., Flavell, R., Compant, S., Antonielli, L., Petric, A., Berninger, T., Naveed, M., Sheibani-Tezerji, R., von Maltzahn, G. and Sessitsch, A. (2017). A new approach to modify plant microbiomes and traits by introducing beneficial bacteria at flowering into progeny seeds, *Front. Microbiol.* 8, 11.

Musilova, L., Ridl, J., Polivkova, M., Macek, T. and Uhlik, O. (2016). Effects of secondary plant metabolites on microbial populations: changes in community structure and metabolic activity in contaminated environments, *Int. J. Mol. Sci.* 17(8), 1205 -1236.

Nelson, E. B. (2018). The seed microbiome: origins, interactions, and impacts, *Plant Soil* 422(1-2), 7-34.

Nelson, E. B., Simoneau, P., Barret, M., Mitter, B. and Compant, S. (2018). Editorial special issue: the soil, the seed, the microbes and the plant, *Plant Soil* 422(1-2), 1-5.

Papik, J., Folkmanova, M., Polivkova, M., Suman, J. and Uhlik, O. (2020). The invisible life inside plants: deciphering the riddles of endophytic bacterial diversity, *Biotechnol. Adv.* 44, 107614.

Parke, J. L. (1991). Root colonization by indigenous and introduced microorganisms. In: Keister, D. L. and Cregan, P. B. (Eds), *The Rhizosphere and Plant Growth*, Beltsville Symposia in Agricultural Research, Vol. 14, Springer, Dordrecht, pp. 33-42.

Peredo, E. L. and Simmons, S. L. (2018). Leaf-FISH: microscale imaging of bacterial taxa on phyllosphere, *Front. Microbiol.* 8, 2669.

Rana, K. L., Kour, D., Kaur, T., Devi, R., Yadav, A. N., Yadav, N., Dhaliwal, H. S. and Saxena, A. K. (2020). Endophytic microbes: biodiversity, plant growth-promoting mechanisms and potential applications for agricultural sustainability, *Antonie Leeuwenhoek* 113(8), 1075–1107.

Reinhold-Hurek, B. and Hurek, T. (2011). Living inside plants: bacterial endophytes, *Curr. Opin. Plant Biol.* 14(4), 435–443.

Rilling, J. I., Acuña, J. J., Nannipieri, P., Cassan, F., Maruyama, F. and Jorquera, M. A. (2019). Current opinion and perspectives on the methods for tracking and monitoring plant growth-promoting bacteria, *Soil Biol. Biochem.* 130, 205–219.

Romano, I., Ventorino, V. and Pepe, O. (2020). Effectiveness of plant beneficial microbes: overview of the methodological approaches for the assessment of root colonization and persistence, *Front. Plant Sci.* 11, 6.

Rosenblueth, M. and Martínez-Romero, E. (2006). Bacterial endophytes and their interactions with hosts, *Mol. Plant Microbe Interact.* 19(8), 827–837.

Santoyo, G., Moreno-Hagelsieb, G., del Carmen Orozco-Mosqueda, Mdel C. and Glick, B. R. (2016). Plant growth-promoting bacterial endophytes, *Microbiol. Res.* 183, 92–99.

Sessitsch, A., Hardoim, P., Döring, J., Weilharter, A., Krause, A., Woyke, T., Mitter, B., Hauberg-Lotte, L., Friedrich, F., Rahalkar, M., Hurek, T., Sarkar, A., Bodrossy, L., van Overbeek, L., Brar, D., van Elsas, J. D. and Reinhold-Hurek, B. (2012). Functional characteristics of an endophyte community colonizing rice roots as revealed by metagenomic analysis, *Mol. Plant Microbe Interact.* 25(1), 28–36.

Sessitsch, A., Pfaffenbichler, N. and Mitter, B. (2019). Microbiome applications from lab to field: facing complexity, *Trends Plant Sci.* 24(3), 194–198.

Sørensen, J. and Sessitsch, A. (2015). Plant-associated bacteria lifestyle and molecular interactions. In: van Elsas, J. D., Trevors, J. T., Jansson, J. K. and Nannipieri, P. (Eds), *Modern Soil Microbiology* (2nd edn.), CRC Press, Boca Raton, pp. 211–236.

Trivedi, P., Schenk, P. M., Wallenstein, M. D. and Singh, B. K. (2017). Tiny microbes, big yields: enhancing food crop production with biological solutions, *Microb. Biotechnol.* 10(5), 999–1003.

Vacher, C., Hampe, A., Porte, A. J., Sauer, U., Compant, S. and Morris, C. E. (2016). The phyllosphere: microbial jungle at the plant-climate interface, *Annu. Rev. Ecol. Evol. Syst.* 47(1), 1–24.

Vorholt, J. A. (2012). Microbial life in the phyllosphere, *Nat. Rev. Microbiol.* 10(12), 828-840.

Yuan, W. M. and Crawford, D. L. (1995). Characterization of *Streptomyces lydicus* WYEC108 as a potential biocontrol agent against fungal root and seed rots, *Appl. Environ. Microbiol.* 61(8), 3119–3128.

Chapter 4

Durability of efficacy of microbial bioprotectants against plant diseases

Marc Bardin, Thomas Pressecq and Philippe C. Nicot, INRAE, Pathologie Végétale, Avignon, France; and Yousra Bouaoud, University of Bejaia, Algeria

1 Introduction

Much research effort has been dedicated in the last 50 years to the development of microbial bioprotectants against plant pathogens, resulting in the registration of an increasing number of commercial products. A recent study by van Lenteren et al. (van Lenteren et al., 2018) has identified 91 microbial bioprotectants (49 filamentous fungi, yeasts or oomycetes, 37 bacteria and 5 viruses including bacteriophages) registered against plant pathogens in Australia, Brazil, Canada, the European Union, Japan, New Zealand and the United States of America. This number of microbial bioprotectants is probably underestimated due to the lack of data for some countries where biocontrol is experiencing increasing use, such as China, Russia or India. Moreover, a similar strain of a given microbe can be commercialized with different names, either because it is formulated differently or because it is distributed by different companies. The development of a microbial bioprotectant product involves the

http://dx.doi.org/10.19103/AS.2021.0093.05

screening of a large number of microbial strains on the basis of different criteria (Köhl et al., 2011). A number of issues must be considered before a product is registered and placed on the market (Glare et al., 2012). They include technical issues, such as the study of environmental factors on the efficacy against the targeted plant pathogen, the development of methods for mass production of the microorganism, the elaboration of a formulation of the product and the validation of its effect under commercial production conditions of the crop. They also include regulatory issues, related to toxicology and ecotoxicology and economic issues (Ravensberg, 2011).

A recurring problem encountered with microbial bioprotectants is their occasional inconsistent efficacy in the field (Nicot et al., 2011). This situation is generally attributed to microclimatic variations encountered in agricultural production conditions (e.g. temperature, relative humidity and radiation), a lack of ecological competence of the microbial bioprotectant (e.g. survival and colonization capacity), intrinsic characteristics of the product (e.g. variable production of metabolites or enzymes involved in the efficacy of the control) and/or the unstable quality of the bioprotectant product (e.g. shelf life).

However, possible loss of efficacy of a biocontrol agent may arise if isolates with low susceptibility occur in natural populations of plant pathogens and become abundant as a result of the selection pressure exerted by the products applied by farmers. Though the resistance of plant pathogens to microbial bioprotectants is still understudied, recent work shows that the susceptibility of plant pathogens to microbial bioprotectants can be highly variable (Bardin et al., 2015). Moreover, cases of resistance (or reduced sensitivity) to microbial bioprotectants have been observed under commercial growing conditions among invertebrate pests (Siegwart et al., 2015). Thus, the diversity in the sensitivity of various strains of a pathogen to a microbial bioprotectant, and its ability to evolve towards resistance, must be considered to ensure effective and durable protection. This review highlights current knowledge concerning the erosion of microbial protection against plant pathogens and its possible consequences for field efficacy.

2 Durability of plant protection against pests

The durability of a control method for plant protection is defined as the persistence of its efficacy in space and time. It depends on the selection pressure exerted by the control method itself on populations of pests, as well as on the ability of these pests to adapt to this method.

Erosion of efficacy has been widely studied in the past for several plant protection methods. The durability of chemical control has, for instance, long been challenged by the frequent and recurrent emergence of resistance to fungicides in populations of various major plant pathogenic fungi (Brent and Hollomon, 2007b;

REX-Consortium, 2013). Likewise, the breakdown of plant varietal resistances, in particular those linked to major resistance genes, is also well documented for plant pathogens and insect pests (McDonald and Linde, 2002; REX-Consortium, 2016). Cases of resistance (or reduced sensitivity) of plant pathogens to physical control methods have also been described. Thus, the phytopathogenic fungus *Botrytis cinerea* is able to become gradually accustomed to the inhibitory effect of UV-filtering films (Nicot et al., 2001). Filtering out the small amounts of near-UV radiation provided by natural light causes a sharp reduction in spore production by this fungus. After several successive generations under selection pressure (absence of near-UV radiation), sporulation of the fungus was shown to increase again. Although many pests are known for their ability to develop resistance to conventional methods of plant protection, only a few studies have explored their ability to overcome the effect of microbial bioprotectants.

Resistance (or lower susceptibility) in the field to microbial biocontrol agents or to metabolites produced by microbials has already been observed in arthropod pests (Siegwart et al., 2015). Resistance to one or several toxins of the most widely used microbial bioprotectant in the world, *Bacillus thuringiensis*, has, for example, been described several years after its market approval in three species of pests (*Plutella xylostella*, *Trichoplusia ni*, *Plodia interpunctella*). Under laboratory conditions, at least 27 insect species are capable of developing resistance to *B. thuringiensis* (Tabashnik, 1994). But resistance of pests to other microbial bioprotectants has also been pointed out. A striking example concerns the resistance of codling moth (*Cydia pomonella*) to the bioprotectant *C. pomonella* granulovirus. This resistance was detected jointly in France and in Germany in pome fruit orchards, after more than 10 years of intensive use of commercial products derived from an identical entomopathogenic viral isolate (Asser-Kaiser et al., 2007; Sauphanor et al., 2006). A very high resistance factor of 13 000 was observed (i.e. it took 13 000 times more virus particles to eradicate a population of resistant insects than to eradicate a population of susceptible insects). Under laboratory conditions, repeated treatments with this virus on the resistant population of *C. pomonella* made it possible to select an even higher resistance level, with a resistance factor of 60000 (Berling et al., 2009).

To our knowledge, most of the proven cases of resistance of pests to microbial bioprotectants concern harmful insects. Although biological control against plant pathogens (bacteria, fungi, oomycetes, nematodes, viruses) has been widely studied in the past years leading to the marketing of numerous microbial bioprotectants, the durability of their efficacy is still poorly studied, and no case of resistance in the field has yet been described (Bardin et al., 2015). This lack of documented cases of resistance of plant pathogen to microbial bioprotectants may be explained by their still limited use in agriculture compared to conventional protection tools.

3 Main modes of action of microbial bioprotectants against plant pathogens and potential mechanisms of resistance developed by plant pathogens

The mechanisms underlying the protective effect of a given microbial bioprotectant are often only partially understood, although the number of scientific studies done on this subject is quite important. However, data available to date suggest that the protection of plants against harmful microorganisms by microbial bioprotectants is based on mechanisms linked to direct (antibiosis, hyperparasitism) and/or indirect (competition, induced resistance, interference with the mechanisms of pathogenesis) interactions with the pathogens (Alabouvette et al., 2009; Bardin and Pugliese, 2020; Haidar et al., 2016; Jacobsen, 2006; Köhl et al., 2019).

In the case of antibiosis, the antagonistic microorganism produces secondary metabolites that are toxic to the targeted plant pathogen. Metabolites produced at low concentrations may, for instance, inhibit spore germination, mycelial growth or the sporulation of fungal plant pathogens. There are many examples of bacteria and fungi producing compounds with an antimicrobial effect, including *Bacillus* sp., *Pseudomonas* sp., *Streptomyces* sp., *Trichoderma* sp. Toxic substances responsible for antibiosis have been characterized in different strains belonging to various microbial species, and in some cases, genes involved in the synthesis of some of these substances have been identified (Raaijmakers et al., 2002). Even if this mode of action is easily revealed in the laboratory by confrontation tests on a nutrient medium, knowledge of the conditions for the synthesis of antibiotic metabolites by the bioprotectant and of the quantities produced *in situ* is often not available. And finally, the ability of microbial bioprotectants to synthesize these molecules under the conditions of their use on crops is rarely demonstrated (Raaijmakers and Mazzola, 2012). The production of antibiotic compounds may depend on many characteristics of the environment in which the microorganism develops, such as the water potential, the pH of the medium, the temperature or the nutrient availability (Whipps, 1987). This mode of action is the closest to that of molecules used in chemical control. It can, therefore, be very effective in inhibiting the development of a plant pathogen but can raise concerns about the emergence of resistant strains. Plant pathogens are, for instance, known to have developed mechanisms of resistance to various toxic compounds produced by beneficial microorganisms (Duffy et al., 2003), particularly during their interaction in the soil (Raaijmakers et al., 2009). These mechanisms include excretion by active efflux and degradation of toxic compounds, interference with the regulation and biosynthesis of enzymes and metabolites produced by microbial bioprotectants.

In the case of hyperparasitism, the microbial bioprotectant parasitizes the plant pathogen. It specifically recognizes its target, enters these cells (e.g. bacterial cells, mycelium, spores or resting structures for fungal pathogens) and causes its destruction through the colonization of its organs. *Paraphaeosphaeria* (formerly *Coniothyrium*) *minitans* is, for example, a hyperparasitic fungus of the sclerotia of *Sclerotinia* sp., able to excrete enzymes degrading cell walls of the fungal pathogen (Whipps and Gerlagh, 1992). The fungus *Ampelomyces quisqualis* parasitizes specifically the plant pathogenic fungi that cause powdery mildew on several crops (Kiss et al., 2004). Bacteriophages can be considered as hyperparasites of bacteria as they proliferate in and eventually destroy the cells of their bacterial host (lytic phages). Phage-based commercial products are available worldwide against specific plant-pathogenic bacterial species (Zaczek et al., 2014; Buttimer et al., 2017). Direct contact of the microbial bioprotectant acting through hyperparasitism with the plant pathogen and the need for an adequate timing of the treatment are necessary to ensure satisfactory protection against the disease before too much damage is caused to the crop. Some plant pathogens have developed the ability to resist or repress the synthesis of cell wall-degrading enzymes required by microbial protectants to parasitize them (Bardin et al., 2015).

Competition for nutrients and space is another mode of action by which microbial bioprotectants can limit or even suppress the development of plant pathogens. For instance, competition is effective against phytopathogenic fungi whose spores need a source of nutrients to initiate their germination. In this case, the reduction in the nutrient concentration in the medium leads to a reduced rate of germinated spores of the pathogen and a slow-down of mycelial growth, thus decreasing the number of infections and the expansion of lesions (Blakeman and Fokkema, 1982). Some microorganisms (bacteria, yeasts, filamentous fungi) can inhibit the germination of conidia from fungal plant pathogens through competition for nutrients such as nitrogen, carbon or macro- or microelements present in the medium. For instance, the fungus *Trichoderma harzianum* inhibits the germination of *B. cinerea* conidia by competing for nutrients at the early stages of interaction (Zimand et al., 1996). Competition is not restricted to interactions during spore germination, it can also occur later in the pathogen growth process if the biocontrol agent is present in sufficient quantity. Competition for nutrients or for space has, for instance, been demonstrated for different yeast species against post-harvest fungal pathogens, including *B. cinerea* or *Penicillium. expansum* (Mercier and Wilson, 1994; Spadaro and Droby, 2016).

In addition to competition for carbohydrates or nitrogen, a well-studied case concerns competition for iron as well. Restricted iron availability can be a limiting factor for microbial growth. Various bacteria and fungi can synthesize high-affinity chelators named siderophores (Köhl et al., 2019). For example, isolates of *Pseudomonas* spp. compete for iron via siderophores, resulting in a reduction of populations of pathogenic *Fusarium* in the rhizosphere (Lemanceau and Alabouvette, 1993; Lemanceau et al., 1992).

The resistance of pathogens to competitive mechanisms seems unlikely (Köhl et al., 2019). However, plant pathogens are continually subjected to stress related to their environment, such as nutrient limitation, and some strains may have developed the ability to rapidly exploit the nutritional resources available in their ecological niche. This type of ability could slow down the development of microbial bioprotectants by using more efficiently the resources and ultimately, influence the efficacy of biocontrol. For instance, in a study by Mazzola and Cook (1991), root infections by the plant pathogen *Pythium irregulare*, which causes root rot of wheat, significantly reduced the population of *Pseudomonas fluorescens* strains introduced in the soil for its control. The question of the ecological competence of biocontrol agents (ability to survive and develop in a given ecological niche) is, therefore, essential to ensure good efficacy against the targeted pathogen.

Many microorganisms have the ability to induce or to prime defense mechanisms in plants after a signal recognition phase (Pieterse et al., 2014). Induced resistance can result from various modifications in the plant: thickening of the parietal structures reinforcing their role of physical barrier, stimulation of secondary metabolic pathways allowing the synthesis of antimicrobial substances (such as phytoalexins) and accumulation of defense proteins (or PR proteins). This topic is currently the subject of a lot of research. To our knowledge, resistance to the mechanism of induced resistance has never been formally demonstrated, and due to the involvement of several defense compounds, it is considered unlikely to occur (Romanazzi et al., 2016). However, to survive and thrive in their hosts, plant pathogens have developed various strategies (Morrissey and Osbourn, 1999). These strategies include the excretion of toxic metabolites produced by the plants via efflux pumps, the degradation of defense compounds and the protection against oxidative damages (Bardin et al., 2015). These mechanisms of resistance to plant defense molecules might eventually lead to the selection of strains resistant to plant-inducted resistance.

Finally, during the first stage of host-pathogen interaction, the synthesis of hydrolytic enzymes (cutinases, pectinases, etc.) by plant pathogens and, in particular, fungi is crucial for the infection process. Microbial bioprotectants can interfere with pathogenicity factors of the pathogens by inhibiting or degrading certain hydrolytic enzymes (Duffy et al., 2003). Additionally, some microorganisms can have an indirect effect on the establishment of pathogenesis, for example, by modifying the pH of the medium (Swadling and Jeffries, 1998). For example, an alkaline pH may reduce the efficiency of the enzymes necessary for a plant pathogen to degrade the tissues of the plants (Manteau et al., 2003). Microbial bioprotectants degrading oxalic acid, a molecule produced by *Sclerotinia sclerotiorum* or *B. cinerea* during their interaction with the plant, can protect plants against attacks by these pathogens (Schoonbeek et al., 2007). Some bacteria have the ability to change the surface characteristics of plant leaves by synthesizing compounds with a surface-active effect, such as lipopeptides (Ongena and Jacques, 2008). This has the effect of hampering the process of attachment and growth of several plant pathogens on the leaves. In addition to this effect, cyclic lipopeptides are known to have a direct antimicrobial effect on plant pathogens, and they can elicit induced systemic resistance in certain host plants (Falardeau et al., 2013). According to several authors, the development of resistance to lipopeptides in microorganisms seems unlikely because of their biosurfactant properties which tend to disrupt or inhibit microbial biofilms (Zhao et al., 2017), and because their mode of action directly targets membranes, whose repair implicates a costly mechanism at the cellular level for microorganisms (Falardeau et al., 2013). However, very different levels of sensitivity to the lipopeptides Fengycin and Surfactin, both produced by *Bacillus subtilis*, were observed in two strains of *Venturia inaequalis* (Desmyttere et al., 2019). But to our knowledge, mechanisms of resistance to lipopeptides have not been described in plant pathogens.

These modes of action are not exclusive and a combination of mechanisms has been demonstrated for various bioprotectant microorganisms. The most studied case concerns the fungal genus *Trichoderma*, for which different modes of action have been demonstrated for the same strain (Elad, 2000; Harman, 2006; Howell, 2003; Karlsson et al., 2017), combining, for example, hyperparasitism and antibiosis (Lorito et al., 1993). According to available references, the strain QST713 of *Bacillus subtilis* acts by secretion of metabolites with antimicrobial activities including lipopeptides, by nutrient and spatial competition, and by induction of systemic plant resistance resulting in increased synthesis of peroxidases. This multitude of modes of action allows this bacterium to control various plant pathogens such as *B. cinerea, Pythium ultimum, Rhizoctonia solani, Plasmodiophora brassicae,* or the fungal agents responsible for powdery mildew (Lahlali et al., 2013; Levasseur et al., 2005; Paulitz and Belanger, 2001). In the case of the combination of several modes of action for a given microbial bioprotectant, the role and precise importance of each of them and the conditions under which the microorganisms switch from one mode of action to another in the control of the disease are not yet clear. The different modes of action may be expressed successively, simultaneously or synergistically, depending on the environmental conditions encountered.

The durability of biocontrol may depend on the mode of action involved in the effect of microbial bioprotectant on pathogens. All microbial bioprotectants can generate selection pressure on populations of plant pathogens when applied for crop treatments. However, certain modes of action are probably more conducive to the development of resistance than others. This crucial question will have to be tackled by research in the future, in order to select modes of action with the most durable efficacy possible.

4 Diversity of susceptibility of plant pathogens to microbial bioprotectants

The diversity in the level of susceptibility to microbial bioprotectants within plant pathogen populations may reduce their efficacy in the field. Due to the existence (even before the use of bioprotectants) of less sensitive phenotypes in natural populations of plant pathogens, the widespread use of bioprotectants could lead to rapid selection for resistance. Testing a sufficiently high number of isolates of a target plant pathogen for its level of resistance to a microbial bioprotectant is, therefore, the first step to estimate the risk of emergence of resistance. In addition, monitoring the level of resistance in natural populations of a plant pathogen before and after treatment can provide information on the distribution of resistance in the field, its impact on the efficacy of protection and its evolution over time.

A rapid assessment of resistance to fungicides in large collections of strains of fungal plant pathogens is commonly carried out by plant protection companies. This may be done, for example, according to the recommendations of the Fungicide Resistance Action Committee (FRAC), to establish a baseline for these products (https://www.frac.info/).

This type of study is increasingly carried out for microbial bioprotectants used against plant diseases. As shown in the review by Bardin et al. (Bardin et al., 2015), the diversity of the efficacy of microbial bioprotectants against plant pathogens has mainly been studied for microbial bioprotectants that produce metabolites with a direct effect on plant pathogens. Several studies have established a baseline for molecules produced by microbial bioprotectants. For instance, various levels of susceptibility to phenazine-1-carboxylic acid (PCA) and 2,4-diacetylphloroglucinol (2,4-DAPG), two compounds synthesized by antagonistic *Pseudomonads* present in the wheat rhizosphere, were detected among 66 strains of *Gaeumannomyces graminis* var. *tritici*, with one strain being resistant to both compounds (Mazzola et al., 1995). Various levels of susceptibility to 2.4-DAPG were also observed among 117 strains of *Fusarium oxysporum* (Schouten et al., 2004). A wide range of sensitivity to pyrrolnitrin, an antibiotic produced by microbial bioprotectants including *Pseudomonas chlororaphis*, was detected among 204 strains of *B. cinerea* (Ajouz et al., 2011b). Several studies have also been carried out on antibiotic-producing bioprotectants themselves. A wide spectrum of sensitivity to 19 antagonistic strains of *Streptomyces* sp. was, for example, identified among 15 strains of the plant pathogenic bacterium *S. scabies*, the causal agent of potato scab (Otto-Hanson et al., 2013). Identically, an *in vitro* study revealed that 3 out of

41 strains of *B. cinerea* were not fully inhibited by any of the 3 antagonistic strains of *Streptomyces* sp. tested (Boukaew et al., 2017). The results of these two studies suggest the need to use a mixture of antagonistic *Streptomyces* sp. strains to effectively control the targeted diseases.

Scientific information available on the diversity of susceptibility of populations of plant pathogens to the protective effect of microbial bioprotectants that do not produce toxic metabolites or antibiotics against pathogens is scarce. However, available data show that, even in this case, a large diversity of susceptibility to microbial bioprotectants in populations of plant pathogens is generally observed. For example, in a study by Schnabel and Jones (Schnabel and Jones, 2001), considerable diversity in the susceptibility of 52 strains of *Erwinia amylovora*, the causal agent of fire blight, towards five specific bacteriophages was observed. According to Jones et al. (2012), bacterial strains within a species vary in their susceptibility to bacteriophages, suggesting that a mixture of bacteriophages is generally preferable for controlling bacterial strains present in the field. In order to ensure a durable efficacy of this method of protection, the authors recommend that a survey of resistance of bacterial strains should be carried out periodically by testing the sensitivity to virulent phages that have been deployed in the field (Jones et al., 2012). Another study has shown differences in susceptibility among 18 strains of *Erysiphe necator*, the causal agent of powdery mildew on grape, to the fungal hyperparasite *A. quisqualis* (Falk et al., 1995). Recently, the diversity for susceptibility to the fungus *Paraphaeosphaeria minitans* has been evaluated in the plant pathogenic fungus *S. sclerotiorum* (Nicot et al., 2019). This plant pathogen is responsible for white rot, which causes significant economic damage to a wide range of crops in most parts of the world (Bardin and Gullino, 2020; Boland and Hall, 1994). This fungus can attack more than 400 plant species, many of which are used in rotation cycles in arable cropping systems. In plant tissues contaminated with *S. sclerotiorum*, sclerotia, the resting structure of the fungus, are formed. Once incorporated in the soil of a field, the sclerotia constitute a stock of inoculum for the following years, and thus potentially contaminate subsequent crops. Sclerotia can survive for years in the soil (Bolton et al., 2006) and can produce either ascospores or mycelium, both of which can infect healthy plants at varying distances. Therefore, sclerotia constitute a source of primary inoculum playing a major role in the disease cycle. This disease is difficult to control by conventional methods, and currently, various microbial biocontrol products are approved worldwide (Bardin and Pugliese, 2020). The commercial bioprotectant Contans® contains as an ingredient the fungus *Paraphaeosphaeria minitans*, a hyperparasite that reduces the infectious potential of a field by parasitizing and destroying the sclerotia of *S. sclerotiorum*. Substantial variability in susceptibility to the commercial bioprotectant has been observed among 75 strains of *S. sclerotiorum*, regardless of their geographic area and host of isolation (Nicot et al., 2019). This study suggests that the efficacy of the bioprotectant may vary from field to field depending on the presence of less sensitive strains. It also raises the question that selection pressure could possibly lead to a gradual selection of less susceptible strains, as this bioprotectant becomes widely adopted by farmers.

Other examples of diversity in susceptibility concern microbial bioprotectants combining several modes of action (Bardin et al., 2015). A recent study concerns the development of a new microbial bioprotectant adapted to tomato growing conditions in Algeria against *B. cinerea* that constitutes one of the major phytosanitary problems of greenhouse tomato crops in most parts of the world (Bardin and Gullino, 2020), including Algeria (Aissat et al., 2008). This disease is difficult to control by conventional control methods, and currently, several microbial biocontrol products are approved worldwide (Bardin and Pugliese, 2020; Nicot et al., 2016). However, the selection of new microbial bioprotectants is necessary in order to broaden the range of biocontrol tools for farmers and to offer solutions adapted to the climatic and cropping specificities encountered in certain countries of the world. To this end, a high potential microorganism, *Pseudomonas helmanticensis* CT22, has been selected to control grey mold in tomato greenhouses encountered in Northern Algeria (Bouaoud et al., 2018b). In order to evaluate the stability of efficacy of this potential new microbial bioprotectant towards *B. cinerea*, the susceptibility of various strains of the pathogen to two doses of the bioprotectant (10^7 CFU/mL and 10^6 CFU/mL) has been tested (Bouaoud et al., 2018a). The efficacy of the bacterium was significantly influenced by the strains of *B. cinerea* tested at both doses of the product (ANOVA, $P = 0.03$ and $P < 0.001$ for high dose and low dose, respectively). At the 10-fold reduced concentration, a wide range of susceptibility was observed among the 62 strains of the pathogen tested with protection levels ranging from 24% to 100% (mean = 80 ± 3%; median = 87%; see Fig. 1). It reveals the importance of considering several strains of the pathogen when screening for new microbial bioprotectants, to obtain a good representation of the pathogen population and thus consider the future protective efficacy of the product in the field and its potential durability. The probable combined modes of action of this microbial bioprotectant (Bouaoud, unpublished data) and its high level of protection provided may reduce the risk of resistance development of *B. cinerea* strains.

Another example concerns the commercial product Serenade Max® (whose active ingredient is the bacterial strain *B. subtilis* QST713) that proved to have a significant protective efficacy against *B. cinerea* on various plant species including tomato (Bardin et al., 2013; Nicot et al., 2016). To assess the possible presence of low susceptibility to this microbial bioprotectant in populations of *B. cinerea*, the protective efficacy of this product was evaluated at two doses on whole tomato plants against 50 strains of the plant pathogen differing in their

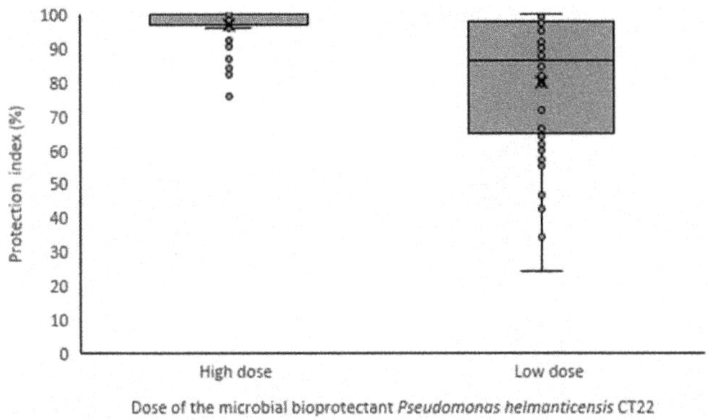

Figure 1 Box plot representing the diversity of protection triggered by the microbial bioprotectant *Pseudomonas helmanticensis* at high dose (10^7 CFU/mL) and low dose (10^6 CFU/mL) against 62 strains of *Botrytis cinerea* on tomato plants. The protection index was computed as 100 * ($AUDPC_{BC}$ - $AUDPC_{CT22}$)/$AUDPC_{BC}$, where $AUDPC_{BC}$ was the average AUDPC for a given strain of *Botrytis cinerea* and $AUDPC_{CT22}$ was the average AUDPC for the same strain in presence of the microbial bioprotectant. For each boxplot, percent of protection for each strain (O), mean (x), median (horizontal line), 25–75% values (grey box), and maximum and minimum values (vertical line) are represented.

geographic area and host of isolation. At both concentrations of the product (low dose = 2 g/L and high dose = 8 g/L), a wide range of susceptibility was observed among the 50 strains, with protection levels significantly different according to the dose of product used (24 ± 3% at low dose and 44 ± 3% at high dose, ANOVA, $P < 0.001$). The efficacy of the product was significantly influenced by the strains of *B. cinerea* tested at both doses of the product (ANOVA, $P < 0.001$), with some of the strains not controlled by the microbial bioprotectant (protection index less than 0) in the conditions of test on tomato plants (see Fig. 2).

There are other examples of diversity studies of susceptibility of plant pathogens to microbial bioprotectants (Bardin et al., 2015). All these studies clearly show that the protective efficacy of a microbial bioprotectant, whatever its mode of action, can vary depending on the strain of the plant pathogen. This observed diversity may contribute to the reduced efficacy of microbial bioprotectants in the field, by means of the existence of less susceptible phenotypes in natural plant pathogen populations and by means of the possible emergence of new resistant phenotypes from these less susceptible isolates due to the selection pressure exerted by the microbial bioprotectant used by farmers.

5 Risk of emergence of resistance to microbial bioprotectants in plant pathogens

An important issue concerns whether the widespread use of microbial bioprotectants in commercial fields or greenhouses could lead to the selection

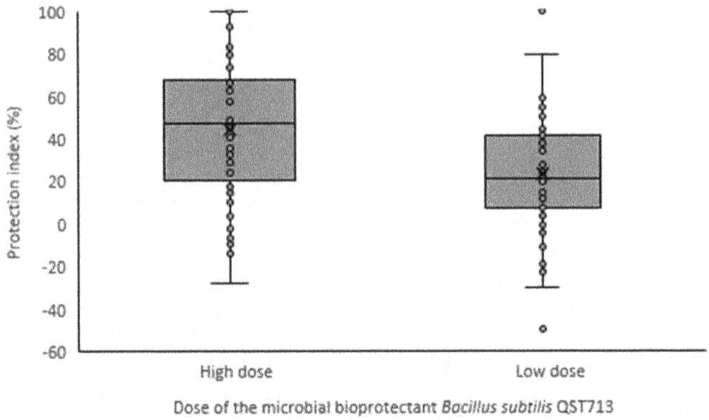

Figure 2 Box plot representing the diversity of protection triggered by the microbial bioprotectant *Bacillus subtilis* QST713 at high dose (8 g/L) and low dose (2 g/L) of the commercial product Serenade Max® against 50 strains of *Botrytis cinerea* on whole tomato plants. The protection index was computed as 100 * ($AUDPC_{BC}$ - $AUDPC_{ser}$)/ $AUDPC_{BC}$, where $AUDPC_{BC}$ was the average AUDPC for a given strain of *Botrytis cinerea* and $AUDPC_{ser}$ was the average AUDPC for the same strain in presence of the microbial bioprotectant. For each boxplot, percent of protection for each strain (O), mean (x), median (horizontal line), 25-75% values (grey box), and maximum and minimum values (vertical line) are represented.

of resistant strains which may eventually jeopardize the efficacy of this control method, as it did in the case of chemical pesticide use. To elucidate this potential risk, repeated exposure of successive generations of a plant pathogen to compounds produced by a microbial bioprotectant or to a microbial bioprotectant itself needs to be realized. This procedure of experimental evolution is commonly carried out to evaluate the durability of the efficacy of antimicrobial agents in human pathology (Cowen, 2001) or to assess the capacity of fungal plant pathogens to adapt to fungicides (Brent and Hollomon, 2007a,b). A first experiment conducted with the microbial bioprotectant *B. subtilis* CL27 in an *Astilbe hybrida* greenhouse revealed that after ten successive treatments the microbial bioprotectant had become ineffective against *B. cinerea* (Li and Leifert, 1994). Another documented example describes a procedure of experimental evolution done *in vitro* to test the hypothesis of an adaptation of a plant pathogen to the effect of a microbial bioprotectant in the presence of continuous selection pressure (Ajouz et al., 2010). This study reveals that *B. cinerea* was able to adapt to the effect of pyrrolnitrin, an antibiotic synthesized by several microbial bioprotectants. Additionally, cross-resistance was observed between pyrrolnitrin and iprodione, a commonly used anti-*Botrytis* fungicide (Fillinger et al., 2012). It suggests a potential risk of development of resistance to this antibiotic and to the pyrrolnitrin-producing microbial bioprotectants in the field, even in the absence of treatment by the microbial bioprotectant itself. However, the fitness cost associated with this resistance was significant for the pathogenic fungus (Ajouz et al., 2011a). Thus, mutants of *B. cinerea* resistant to pyrrolnitrin showed reduced mycelial growth, a low level of aggressiveness on the host plant (tomato and apple) and a reduction in the capacity to produce conidia compared to sensitive wild-strains. This, therefore, suggests a limited capacity for epidemic dissemination of strains of *B. cinerea* resistant to this antibiotic.

Knowledge concerning the potential erosion of efficacy of microbial bioprotectants is essential to ensure a durable use of these products on target plant pathogens. This knowledge will help to identify risk factors that can foster the selection of resistant strains of plant pathogens to microbial bioprotectants. It will also help identify the types of microbial bioprotectants with a lower risk of loss of efficacy.

6 Future trends in research and recommendations for a durable use of microbial bioprotectants

The durability of the efficacy of microbial bioprotectants may be linked to the characteristics of plant pathogen populations, such as genetic diversity and the ability to evolve in response to selection pressure. The selection pressure exerted by field use of microbial bioprotectant depends on the surfaces treated, the application rates or treatment cadences. The outcome of this pressure for specific microorganisms ultimately depends on evolutionary mechanisms such as mutation, recombination, migration and selection. These mechanisms have been the subject of numerous studies for microorganisms in other contexts, for example, to assess the durability of plant varietal resistance to plant diseases. Thus, McDonald and Linde (McDonald and Linde, 2002) postulated that populations of pathogens with high evolutionary potential were more likely to overcome varietal resistance. The same hypothesis can be proposed for the development of resistance of plant pathogens to microbial bioprotectants. Although many plant pathogens are known for their capacity to develop resistance to conventional plant protection methods, only a few studies have explored their ability for resistance to microbial bioprotectants. Available studies highlight differences in susceptibility of various strains of plant pathogens to microbial bioprotectants or to compounds that are produced or synthesized as the result of their interaction with the host plant.

The durability of the efficacy of biocontrol may also be linked to intrinsic characteristics of the microbial bioprotectant such as its mode of action. Significant research efforts are, therefore, still necessary to acquire precise knowledge on the modes of action of microbial bioprotectants. This knowledge will make it possible to optimize their use and anticipate the potential failures of efficacy, for example, by recommending the use of bioprotectants combining several modes of action or of complex formulations (mixtures of metabolites or of microorganisms with different modes of action) or alternating products. It will also help to integrate durability concerns in the screening procedure for new microbial bioprotectants, aiming at the selection of the more durable modes of action and avoiding modes of action that can easily lead to the selection of resistant isolates in natural populations of plant pathogens.

If the commercialization of microbial bioprotectants became widespread in the world and their use increased in cropping systems (expansion of the treated surfaces, increase of the applied dose or in the frequency of applications), the emergence of new phenotypes resistant to these products could not be excluded. Data are still too scarce to allow the development of specific principles on necessary precautions for the use of microbial bioprotectants in practice. However, this chapter underlines the need to develop now the appropriate management schemes for these new products, in order to avoid repeating errors made in the past when deploying chemical fungicides or varietal resistance. One possible strategy could be to start applying for the deployment of commercial microbial bioprotectants some of the FRAC recommendations (https://www.frac.info/). Such recommendations may, for instance, concern the alternation of products with known differences in modes of action, the establishment of sensitivity baselines and the survey of evolution in populations of the plant pathogens and the development of risk prediction models. Another strategy can be suggested for the selection of new microbial bioprotectants, such as the use of more than one strain of the target plant pathogen during the screening procedure, the use of criteria to maximize durability (e.g. avoid certain modes of action, privilege combined modes of action) and promote the development of mixtures of strains (assembled consortia) with complementary modes of action. However, on this last point, regulatory, technical and cost issues may complicate the commercialization of assembled consortia.

7 Where to look for further information

The following articles provide a good overview of key aspects of the subject:

- Bardin, M., Ajouz, S., Comby, M., Lopez-Ferber, M., Graillot, B., Siegwart, M., and Nicot, P. C. (2015). Is the efficacy of biological control against plant diseases likely to be more durable than that of chemical pesticides? Frontiers in Plant Science 6, 566.
- Bardin, M. and Nicot, P. (2020). Chapter 11. Using microorganisms as biocontrol products. In: Fauvergue et al. eds. Biocontrôle. Eléments pour une protection agroécologique des cultures (p. 151-160). Quæ. (in French).
- Bardin, M. and Pugliese, M. (2020), Biocontrol agents against diseases. In: Gullino, M., Albajes, R. and Nicot, P. (eds.) Integrated Pest and Disease Management in Greenhouse Crops. Springer.

- Bardin, M and Siegwart, M. (2020). Chapter 23. Can pests develop resistance to biocontrol products? In : Fauvergue et al. eds. Biocontrôle. Eléments pour une protection agroécologique des cultures (p. 299-305). Quæ. (in French).
- Köhl, J., Kolnaar, R. and Ravensberg, W. J. (2019). Mode of action of microbial biological control agents against plant diseases: relevance beyond efficacy. Frontiers in Plant Science, 10, 845. doi: 10.3389/fpls.2019.00845.
- Duffy, B., Schouten, A., and Raaijmakers, J. M. (2003). Pathogen self-defense: mechanisms to counteract microbial antagonism. Annual Review of Phytopathology 41, 501–538.
- Raaijmakers, J. M., Paulitz, T. C., Steinberg, C., Alabouvette, C., and Moënne-Loccoz, Y. (2009). The rhizosphere: a playground and battlefield for soilborne pathogens and beneficial microorganisms. Plant and Soil 321(1–2), 341–361.
- van Lenteren, J. C., Bolckmans, K., Köhl, J., Ravensberg, W. J., and Urbaneja, A. (2018). Biological control using invertebrates and microorganisms: plenty of new opportunities. BioControl 63(1), 39–59.

Research in this area is conducted by member institutions of the International Organization of Biological Control (https://www.iobc-global.org/) and in particular by scientists participating in the activities of the following Working Groups of the West Palaearctic Regional Section:

- Biological and integrated control of Plant Pathogens (https://www.iobc-wprs.org/expert_groups/06_wg_plant_pathogens.html).
- Induced resistance in plants against insects and diseases (https://www.iobc-wprs.org/expert_groups/17_wg_induced_resistance.html).
- Integrated control in Protected crops, Temperate climate (https://www.iobc-wprs.org/expert_groups/08_wg_protected_crops_temperate.html) and Mediterranean climate (https://www.iobc-wprs.org/expert_groups/09_wg_protected_crops_mediterranean.html).
- Integrated control in Citrus fruit crops (https://www.iobc-wprs.org/expert_groups/16_wg_citrus_crops.html).
- Integrated protection of Field vegetable Crops (https://www.iobc-wprs.org/expert_groups/12_wg_vegetables.html).
- Integrated protection of Fruit crops (https://www.iobc-wprs.org/expert_groups/02_wg_fruit_crops.html).
- Integrated protection of Olive crops (https://www.iobc-wprs.org/expert_groups/15_wg_olive_crops.html).

Recommendations and strategy suggested for the deployment of chemical fungicides in Agriculture can be found online at the Fungicide Resistance Action Committee: https://www.frac.info/.

8 Acknowledgements

This work was partly supported by funding from the European Union's Horizon 2020 research and innovation programme under grant agreement N° 633184 ('EUCLID' project - http://www.euclidipm.org/).

9 References

Aissat, K., Nicot, P. C., Guechi, A., Bardin, M. and Chibane, M. (2008). Grey mould development in greenhouse tomatoes under drip and furrow irrigation. *Agronomy for Sustainable Development* 28(3), 403–409.

Ajouz, S., Bardin, M., Nicot, P. C. and El Maataoui, M. (2011a). Comparison of the development *in planta* of a pyrrolnitrin-resistant mutant of *Botrytis cinerea* and its sensitive wild-type parent isolate. *European Journal of Plant Pathology* 129(1), 31–42.

Ajouz, S., Nicot, P. C. and Bardin, M. (2010). Adaptation to pyrrolnitrin in *Botrytis cinerea* and cost of resistance. *Plant Pathology* 59(3), 556–566.

Ajouz, S., Walker, A. S., Fabre, F., Leroux, P., Nicot, P. C. and Bardin, M. (2011b). Variability of *Botrytis cinerea* sensitivity to pyrrolnitrin, an antibiotic produced by biological control agents. *BioControl* 56(3), 353–363.

Alabouvette, C., Olivain, C., Migheli, Q. and Steinberg, C. (2009). Microbiological control of soil-borne phytopathogenic fungi with special emphasis on wilt-inducing *Fusarium oxysporum*. *New Phytologist* 184(3), 529–544.

Asser-Kaiser, S., Fritsch, E., Undorf-Spahn, K., Kienzle, J., Eberle, K. E., Gund, N. A., Reineke, A., Zebitz, C. P. W., Heckel, D. G., Huber, J. and Jehle, J. A. (2007). Rapid emergence of baculovirus resistance in codling moth due to dominant, sex-linked inheritance. *Science* 317(5846), 1916–1918.

Bardin, M., Ajouz, S., Comby, M., Lopez-Ferber, M., Graillot, B., Siegwart, M. and Nicot, P. C. (2015). Is the efficacy of biological control against plant diseases likely to be more durable than that of chemical pesticides? *Frontiers in Plant Science* 6, 566.

Bardin, M., Comby, M., Lenaerts, R. and Nicot, P. C. (2013). Diversity in susceptibility of *Botrytis cinerea* to biocontrol products inducing plant defence mechanisms. *IOBC/WPRS Bulletin* 88, 45–49.

Bardin, M. and Gullino, L. M. (2020). Fungal diseases. In: *Integrated Pest and Disease Management in Greenhouse Crops* (vol. 9) Gullino, L. M., Albajes, R. and Nicot, P. C. (Eds), pp. 55–100. Springer Nature Switzerland AG, Switzerland.

Bardin, M. and Pugliese, M. (2020). Biocontrol agents against diseases. In: *Integrated Pest and Disease Management in Greenhouse Crops* (vol. 9) Gullino, L. M., Albajes, R. and Nicot, P. C. (Eds), pp. 385–407. Springer Nature Switzerland AG, Switzerland.

Berling, M., Blachere-Lopez, C., Soubabere, O., Lery, X., Bonhomme, A., Sauphanor, B. and Lopez-Ferber, M. (2009). *Cydia pomonella* granulovirus genotypes overcome virus resistance in the codling moth and improve virus efficiency by selection against resistant hosts. *Applied and Environmental Microbiology* 75(4), 925–930.

Blakeman, J. P. and Fokkema, N. J. (1982). Potential for biological control of plant diseases on the phylloplane. *Annual Review of Phytopathology* 20(1), 167–190.

Boland, G. J. and Hall, R. (1994). Index of plant hosts of *Sclerotinia sclerotiorum*. *Canadian Journal of Plant Pathology* 16(2), 93–108.

Bolton, M. D., Thomma, B. P. H. J. and Nelson, B. D. (2006). *Sclerotinia sclerotiorum* (Lib.) de Bary: biology and molecular traits of a cosmopolitan pathogen. *Molecular Plant Pathology* 7(1), 1-16.

Bouaoud, Y., Troulet, C., Berge, O., Aissat, K. and Bardin, M. (2018a). Diversity in the susceptibility of *Botrytis cinerea* strains to the biological control agent *Pseudomonas helmanticensis*. *IOBC WPRS Bulletin* 133, 100-104.

Bouaoud, Y., Troulet, C., Foughalia, A., Berge, O., Aissat, K. and Bardin, M. (2018b). A multi-criteria approach for the selection of efficient biocontrol agents against *Botrytis cinerea* on tomato in Algeria. *BioControl* 63(2), 299-311.

Boukaew, S., Prasertsan, P., Troulet, C. and Bardin, M. (2017). Biological control of tomato gray mold caused by *Botrytis cinerea* by using *Streptomyces* spp. *BioControl* 62(6), 793-803.

Brent, K. J. and Hollomon, D. W. (2007a). Fungicide resistance: the assessment of risk. In: *FRAC Monograph No.2* 2nd, revised edition, pp. 1-53. CropLife International, Brussels, Belgium.

Brent, K. J. and Hollomon, D. W. (2007b). Fungicide resistance in Crop Pathogens: how can it be managed. In: *FRAC Monograph No.1* 2nd revised edition. CropLife International, Brussels, Belgium.

Buttimer, C., McAuliffe, O., Ross, R. P., Hill, C., O'Mahony, J. and Coffey, A. (2017) Bacteriophages and Bacterial Plant Diseases. *Front. Microbiol.* 8, 34. doi: 10.3389/fmicb.2017.00034

Cowen, L. E. (2001). Predicting the emergence of resistance to antifungal drugs. *FEMS Microbiology Letters* 204(1), 1-7.

Desmyttere, H., Deweer, C., Muchembled, J., Sahmer, K., Jacquin, J., Coutte, F. and Jacques, P. (2019) Antifungal Activities of *Bacillus subtilis* Lipopeptides to Two *Venturia inaequalis* Strains Possessing Different Tebuconazole Sensitivity. *Front. Microbiol.* 10, 2327. doi: 10.3389/fmicb.2019.02327

Duffy, B., Schouten, A. and Raaijmakers, J. M. (2003). Pathogen self-defense: mechanisms to counteract microbial antagonism,. *Annual Review of Phytopathology* 41, 501-538.

Elad, Y. (2000). Biological control of foliar pathogens by means of *Trichoderma harzianum* and potential modes of action. *Crop Protection* 19(8-10), 709-714.

Falardeau, J., Wise, C., Novitsky, L. and Avis, T. J. (2013). Ecological and mechanistic insights Into the direct and indirect antimicrobial properties of *Bacillus subtilis* lipopeptides on plant pathogens. *Journal of Chemical Ecology* 39(7), 869-878.

Falk, S. P., Gadoury, D. M., Pearson, R. C. and Seem, R. C. (1995). Partial control of grape powdery mildew by the mycoparasite *Ampelomyces quisqualis*. *Plant Disease* 79(5), 483-490.

Fillinger, S., Ajouz, S., Nicot, P. C., Leroux, P. and Bardin, M. (2012). Functional and structural comparison of pyrrolnitrin- and iprodione-induced modifications in the Class III histidine-kinase Bos1 of *Botrytis cinerea*. *PLoS ONE* 7(8), e42520.

Glare, T., Caradus, J., Gelernter, W., Jackson, T., Keyhani, N., Köhl, J., Marrone, P., Morin, L. and Stewart, A. (2012). Have biopesticides come of age ? *Trends in Biotechnology* 30(5), 250-258.

Haidar, R., Fermaud, M., Calvo-Garrido, C., Roudet, J. and Deschamp, A. (2016). Modes of action for biological control of *Botrytis cinerea* by antagonistic bacteria. *Phytopathologia Mediterranea* 55, 13-34.

Harman, G. E. (2006). Overview of mechanisms and uses of *Trichoderma* spp. *Phytopathology* 96(2), 190-194.

Howell, C. R. (2003). Mechanisms employed by *Trichoderma* species in the biological control of plant diseases: the history and evolution of current concepts. *Plant Disease* 87(1), 4–10.

Jacobsen, B. J. (2006). Biological control of plant diseases by phyllosphere applied biological control agents. In: *Microbial Ecology of Aerial Plant Surfaces* Bailey, M. J., Lilley, A. K., Timms-Wilson, T. M. and Spencer-Phillips, P. T. N. (Eds), pp. 133–147. Athenaeum Press, Gateshead, UK.

Jones, J. B., Vallad, G. E., Iriarte, F. B., Obradovic, A., Wernsing, M. H., Jackson, L. E., Balogh, B., Hong, J. C. and Momol, M. T. (2012). Considerations for using bacteriophages for plant disease control. *Bacteriophage* 2(4), 208–214.

Karlsson, M., Atanasova, L., Jensen, D. F. and Zeilinger, S. (2017). Necrotrophic mycoparasites and their genomes. *Microbiology Spectrum* 5(2), FUNK-0016.

Kiss, L., Russell, J. C., Szentivanyi, O., Xu, X. and Jeffries, P. (2004). Biology and biocontrol potential of *Ampelomyces* mycoparasites, natural antagonists of powdery mildew fungi. *Biocontrol Science and Technology* 14(7), 635–651.

Köhl, J., Kolnaar, R. and Ravensberg, W. J. (2019). Mode of action of microbial biological control agents against plant diseases: relevance beyond efficacy. *Frontiers in Plant Science* 10, 845.

Köhl, J., Postma, J., Nicot, P., Ruocco, M. and Blum, B. (2011). Stepwise screening of microorganisms for commercial use in biological control of plant-pathogenic fungi and bacteria. *Biological Control* 57(1), 1–12.

Lahlali, R., Peng, G., Gossen, B. D., McGregor, L., Yu, F. Q., Hynes, R. K., Hwang, S. F., McDonald, M. R. and Boyetchko, S. M. (2013). Evidence that the biofungicide serenade (*Bacillus subtilis*) suppresses clubroot on canola via antibiosis and induced host resistance. *Phytopathology* 103(3), 245–254.

Lemanceau, P. and Alabouvette, C. (1993). Suppression of *Fusarium* wilt by fluorescent *Pseudomonas*. Mechanisms and applications. *Biocontrol Science and Technology* 3(3), 219–234.

Lemanceau, P., Bakker, P. A. H. M., Dekogel, W. J., Alabouvette, C. and Schippers, B. (1992). Effect of pseudobactin 358 production by *Pseudomonas putida* WCS358 on suppression of *Fusarium* wilt of carnations by nonpathogenic *Fusarium oxysporum* Fo47. *Applied and Environmental Microbiology* 58(9), 2978–2982.

Levasseur, M., Tombu, B. and Zambaux, C. (2005). Mieux connaitre Serenade Biofungicide. Fongicide biologique contre le *Botrytis* de la Vigne Phytoma 588, 52–53.

Li, H. and Leifert, C. (1994). Development of resistance in *Botryotinia fuckeliana* (de Barry) Whetzel against the biological control agent *Bacillus subtilis* CL27. *Zeitschrift fur Pflanzenkrankheiten und Pflanzenschutz* 101, 414–418.

Lorito, M., Harman, G. E., Hayes, C. K., Broadway, R. M., Tronsmo, A., Woo, S. L. and Pietro, A. d. (1993). Chitinolytic enzymes produced by *Trichoderma harzianum*: antifungal activity of purified endochitinase and chitobiosidase. *Phytopathology* 83(3), 302–307.

Manteau, S., Abouna, S., Lambert, B. and Legendre, L. (2003). Differential regulation by ambient pH of putative virulence factor secretion by the phytopathogenic fungus *Botrytis cinerea*. *FEMS Microbiology Ecology* 43(3), 359–366.

Mazzola, M. and Cook, R. J. (1991). Effects of fungal root pathogens on the population dynamics of biocontrol strains of fluorescent pseudomonads in the wheat rhizosphere. *Applied and Environmental Microbiology* 57(8), 2171–2178.

Mazzola, M., Fujimoto, D. K., Thomashow, L. S. and Cook, R. J. (1995). Variation in sensitivity of *Gaeumannomyces graminis* to antibiotics produced by fluorescent *Pseudomonas* spp. and effect on biological control of take-all of wheat. *Applied and Environmental Microbiology* 61(7), 2554–2559.

McDonald, B. A. and Linde, C. (2002). Pathogen population genetics, evolutionary potential, and durable resistance. *Annual Review of Phytopathology* 40, 349–379.

Mercier, J. and Wilson, C. L. (1994). Colonization of apple wounds by naturally occuring microflora and introduced *Candida oleophila* and their effect on infection by *Botrytis cinerea* during storage. *Biological Control* 4(2), 138–144.

Morrissey, J. P. and Osbourn, A. E. (1999). Fungal resistance to plant antibiotics as a mechanism of pathogenesis. *Microbiology and Molecular Biology Reviews* 63(3), 708–724.

Nicot, P. C., Avril, F., Duffaud, M., Leyronas, C., Troulet, C., Villeneuve, F. and Bardin, M. (2019). Differential susceptibility to the mycoparasite *Paraphaeosphaeria minitans* among *Sclerotinia sclerotiorum* isolates. *Tropical Plant Pathology* 44(1), 82–93.

Nicot, P. C., Bardin, M., Alabouvette, C., Köhl, J. and Ruocco, M. (2011). Potential of biological control based on published research. 1. Protection against plant pathogens of selected crops. In: *Classical and Augmentative Biological Control against Diseases and Pests: Critical Status Analysis and Review of Factors Influencing Their Success* Nicot, P. C. (Ed), pp. 1–11. IOBC-wprs, Zurich.

Nicot, P. C., Morison, N. and Mermier, M. (2001). Optical filters against grey mould of greenhouse crops. In: *Physical Control Methods in Plant Protection* Vincent, C., Panneton, B. and Fleurat-Lessard, F. (Eds), pp. 134–145. Springer Verlag, Paris.

Nicot, P. C., Stewart, A., Bardin, M. and Elad, Y. (2016). Biological control and biopesticide suppression of *Botrytis*-incited diseases. In: *Botrytis–The Fungus, the Pathogen and Its Management in Agricultural Systems* Fillinger, S. and Elad, Y. (Eds), pp. 165-187. Springer International Publishing, Switzerland.

Ongena, M. and Jacques, P. (2008). *Bacillus* lipopeptides: versatile weapons for plant disease biocontrol. *Trends in Microbiology* 16(3), 115-125.

Otto-Hanson, L. K., Grabau, Z., Rosen, C., Salomon, C. E. and Kinkel, L. L. (2013). Pathogen variation and urea influence selection and success of *Streptomyces* mixtures in biological control. *Phytopathology* 103(1), 34–42.

Paulitz, T. C. and Belanger, R. R. (2001). Biological control in greenhouse systems. *Annual Review of Phytopathology* 39, 103–133.

Pieterse, C. M. J., Zamioudis, C., Berendsen, R. L., Weller, D. M., van Wees, S. C. M. and Bakker, P. A. H. M. (2014). Induced systemic resistance by beneficial microbes. *Annual Review of Phytopathology* 52, 347–375.

Raaijmakers, J. M. and Mazzola, M. (2012). Diversity and natural functions of antibiotics produced by beneficial and plant pathogenic bacteria. *Annual Review of Phytopathology* 50, 403–424.

Raaijmakers, J. M., Paulitz, T. C., Steinberg, C., Alabouvette, C. and Moënne-Loccoz, Y. (2009). The rhizosphere: a playground and battlefield for soilborne pathogens and beneficial microorganisms. *Plant and Soil* 321(1–2), 341–361.

Raaijmakers, J. M., Vlami, M. and Souza, J. T. d. (2002). Antibiotic production by bacterial biocontrol agents. *Antonie van Leeuwenhoek* 81(1–4), 537–547.

Ravensberg, W. J. (2011). *Roadmap to the Successful Development and Commercialization of Microbial Pest Control Products for Control of Arthropods*. Springer, Dordrecht, The Netherlands.

REX Consortium (2013). Heterogeneity of selection and the evolution of resistance. *Trends in Ecology and Evolution* 28(2), 110–118.

REX Consortium (2016). Combining selective pressures to enhance the durability of disease resistance genes. *Frontiers in Plant Science* 7, 1916.

Romanazzi, G., Sanzani, S. M., Bi, Y., Tian, S., Gutiérrez Martinez, P. and Alkan, N. (2016). Induced resistance to control postharvest decay of fruit and vegetables. *Postharvest Biology and Technology* 122, 82–94.

Sauphanor, B., Berling, M., Toubon, J. F., Reyes, M., Delantte, J. and Allemoz, P. (2006). Carpocapse des pommes: cas de résistance au virus de la granulose en vergers biologiques. *Phytoma- La Défense des végétaux* 590, 24–27.

Schnabel, E. L. and Jones, A. L. (2001). Isolation and characterization of five *Erwinia amylovora* bacteriophages and assessment of phage resistance in strains of *Erwinia amylovora*. *Applied and Environmental Microbiology* 67(1), 59–64.

Schoonbeek, H. J., Jacquat-Bovet, A. C., Mascher, F. and Metraux, J. P. (2007). Oxalate-degrading bacteria can protect *Arabidopsis thaliana* and crop plants against *Botrytis cinerea*. *Molecular Plant-Microbe Interactions* 20(12), 1535–1544.

Schouten, A., van den Berg, G., Edel-Hermann, V., Steinberg, C., Gautheron, N., Alabouvette, C., de Vos, C. H., Lemanceau, P. and Raaijmakers, J. M. (2004). Defense responses of *Fusarium oxysporum* to 2,4-diacetylphloroglucinol, a broad-spectrum antibiotic produced by *Pseudomonas fluorescens*. *Molecular Plant-Microbe Interactions* 17(11), 1201–1211.

Siegwart, M., Graillot, B., Blachere, C., Besse, S., Bardin, M., Nicot, P. C. and López-Ferber, M. (2015). Resistance to bio-insecticides or how to enhance their sustainability: a review. *Frontiers in Plant Science* 6, 381.

Spadaro, D. and Droby, S. (2016). Development of biocontrol products for postharvest diseases of fruit: the importance of elucidating the mechanisms of action of yeast antagonists. *Trends in Food Science and Technology* 47, 39–49.

Swadling, I. R. and Jeffries, P. (1998). Antagonistic properties of two bacterial biocontrol agents of grey mould disease. *Biocontrol Science and Technology* 8(3), 439–448.

Tabashnik, B. E. (1994). Evolution of resistance to *Bacillus thuringiensis*. *Annual Review of Entomology* 39(1), 47–79.

van Lenteren, J. C., Bolckmans, K., Köhl, J., Ravensberg, W. J. and Urbaneja, A. (2018). Biological control using invertebrates and microorganisms: plenty of new opportunities. *BioControl* 63(1), 39–59.

Whipps, J. M. (1987). Effect of media on growth and interactions between a range of soil-borne glasshouse pathogens and antagonistic fungi. *New Phytologist* 107(1), 127–142.

Whipps, J. M. and Gerlagh, M. (1992). Biology of *Coniothyrium minitans* and its potential for use in disease biocontrol. *Mycological Research* 96(11), 897–907.

Zaczek, M., Weber-Dabrowska, B. and Gorski, A. (2014). Phages in the global fruit and vegetable industry. *Journal of Applied Microbiology* 118(3), 537–556.

Zhao, H., Shao, D., Jiang, C., Shi, J., Li, Q., Huang, Q., Riaz Rajoka, M. S. R., Yang, H. and Jin, M. (2017). Biological activity of lipopeptides from *Bacillus*. *Applied Microbiology and Biotechnology* 101(15), 5951–5960.

Zimand, G., Elad, Y. and Chet, I. (1996). Effect of *Trichoderma harzianum* on *Botrytis cinerea* pathogenicity. *Phytopathology* 86(11), 1255–1260.

Chapter 5

Advances in production and formulation of commercial microbial bioprotectant products

Jacob Eyal, Michael Dimock, and José João Carvalho, Certis USA LLC, USA

1 Introduction

Plant pathogenic fungi, bacteria and viruses cause significant economic damage to many crops during propagation, cultivation and postharvest storage. Over the last decade, significant advances have been made in the use and commercialization of microorganisms to improve plant health and to control pathogens causing crop damage. This chapter reviews such developments.

Advances in developing new bioprotectants are mainly driven by food safety concerns and stricter regulatory requirements around environmental protection. Increasing regulatory pressure has resulted in the loss or restriction of the use of synthetic chemical fungicides, copper compounds and antibiotics for the treatment of plant disease. The European Regulation (EC) No 1107/2009, which regulates plant protection products, has led to the withdrawal of more than 200 chemical substances, including many fungicides, from the European market in the last decade. Microbial control of pests and plant diseases offers

http://dx.doi.org/10.19103/AS.2021.0093.06

alternative products with no toxic residues, minimum pre-harvest and re-entry intervals and reduced risk of resistance development. This has led in recent years to a resurgence of interest by crop protection companies to fill gaps in their product portfolios with microbial bioprotectant products and accelerated the pace of development and application of microbial bioprotectant products (Bullock and Calvert 2020).

Progress in commercialization and application of biological control agents for plant disease was relatively slow prior to 2008 (Kiewnick 2007, Köhl et al. 2011). Most of the commercial products developed for control of plant disease during this period contain gram-negative, non-spore-forming bacteria such as *Pseudomonas* spp. or *Trichoderma* spp. and related fungi, targeting seed- and soilborne fungal pathogens through application to the soil (Montesinos 2003). Recently there has been renewed interest in the production and formulation of fungal propagules such as conidia and microsclerotia produced in liquid fermentation for broader application against plant disease (Agosin et al. 1997, Mancera-López et al. 2018, Kobori et al. 2014, Jackson and Payne 2016).

During the last decade, many commercial bioprotectant products based on gram-positive bacteria in the genus *Bacillus* have been introduced into the market as biofungicides (Zaki et al. 2020). The predominance of *Bacillus* spp. as microbial bioprotectants is due to their ability to produce spores that survive unfavorable conditions. These spores are relatively easy to produce by liquid fermentation and they can be dried and formulated into products having longer shelf life and more consistent performance in the field. The mechanisms by which *Bacillus* spp. affect plants and plant pathogens remain the subject of ongoing research and debate, but it is clear that the strains present in bioprotectant products produce many bioactive metabolites and are reported to be involved in the plant–microbial interactions (Ongena and Jacques 2007, Mnif and Ghribi 2015).

Many other microorganisms are under investigation for potential commercial development as inducers of plant resistance to fungi, bacteria and viruses, with two such microbial bioprotectant products recently introduced into the crop protection market as microbial induced resistance (IR) products to control fungi, bacteria and viruses. Other microorganisms include yeasts such as *Pseudozyma churashimaensis* (Lee et al. 2017) and *P. flocculosa* (Bélanger et al. 2012) and bacteria such as *Paenibacillus alvei* (Tjamos et al. 2005) and *Pseudomonas fluorescens* (Van de Mortel et al. 2012) are in the development stage. In addition to live microorganisms, several unique new induced-resistance (IR) bioprotectant products are being developed based on bacterial and fungal fermentation for large-scale production of phytohormones such as jasmonate fatty acid derivatives, which have an important role in regulating plant defense responses against pathogens (Pieterse et al. 2009, Eng. 2018).

Other new developments include the production and formulation of bacteriophages and protozoans for plant protection, the use of microorganisms to produce metabolites, and polypeptides such as single-domain antibodies (sdAb) to treat plant disease, as well as novel microbial formulations - encapsulation, nanoencapsulation and nanoparticles.

As a final point, considerations about fermentation processes and product formulation are also included here because they are a fundamental part of the bioprotectant products' development cycle. Fermentation and formulation processes directly affect the performance (efficacy) of bioprotectant products in the field, through effects on stability (shelf life) and viability of the microbial bioprotectant active ingredients (Fravel et al. 1998, Schisler et al. 2004). Hence, it is critical to establish the optimal fermentation conditions and suitable downstream processes for the proper development, scale-up and manufacturing of a bioprotectant product. Examples of such conditions and manufacturing parameters, and their effect on the product stability and viability of the microorganisms, are addressed at the end of this chapter.

2 Bioprotectant products containing bacteria

Various types of bacteria are used as microbial bioprotectant products for the control of plant disease, and they can be divided into two main groups: gram-negative and gram-positive bacteria. Gram-negative bacteria can produce biomass containing viable cells, while gram-positive bacteria produce biomass mainly containing spores, a more stable form of the bacterial life cycle. For the production of biomass by fermentation, selection of processing technology and equipment must take into account the effects on the survival of each bacterial group in the environmental conditions created during the downstream process to obtain the active ingredient (e.g. physical stress and pressure during the concentration of the active ingredient and temperature used for spray drying). Likewise for the formulation of the final product, where for each group, the manufacturing conditions need to be adapted to ensure that the key chemical and physical properties remain stable during product storage and handling (e.g. active ingredient content, no microbial contamination, formulation pH, appearance of the product, homogeneity, wettability, dispersibility, suspensibility, dust content and other properties depending on the formulation type).

2.1 Gram-negative bacteria bioprotectants

Few commercial products containing gram-negative bacteria are available for the control of plant disease. *Pseudomonas* spp. have been the most extensively studied gram-negative bacteria and for decades have been used as a model for biological control of plant disease as well as plant growth promotion (Emmert and Handelsman 1999, Sivasakthi et al. 2014).

Many product registrations containing *Pseudomonas* spp. have been withdrawn from the commercial market in the past 20 years (EPA 2020). Products such as 'BioSave' (Gustafson, USA), 'Dagger G' (Ecogen, USA), 'Blue Circle' (CTT Crop, USA), 'Blight Ban' (Plant Health Technologies, USA), and 'Intercept' (Soil Technologies Corp, USA), to cite some examples, have been withdrawn or discontinued due to difficulties in achieving an adequate shelf life for the formulation (Lisansky and Coombs 1993). Today, few *Pseudomonas* spp. products are available in the United States and Europe for the control of plant disease (Table 1). One single product 'Howler fungicide', from AgBiome (NC, USA), was commercially introduced in 2017, for the prevention and control of several plant diseases caused by species of *Rhizoctonia, Pythium, Colletotrichum, Phytophthora* and *Botrytis*, and it is available as a water-dispersible granule (WG or WDG) formulation (AgBiome 2020, Hammer et al. 2017).

Pseudomonas spp. whole cells can be easily produced by submerged fermentation followed by harvest of the whole cell biomass, which is then mixed with a suitable carrier and/or additives to provide a wet or dry formulation suitable for application and having an acceptable shelf life. The process starts with a high titer of whole cells through the batch or fed-batch fermentation process. Media for such fermentation can be simple carbon sources such as glucose, glycerol or molasses, with a nitrogen source such as ammonia or soy flour, and the addition of trace minerals containing K_2PO_4, $FeSO_4$, $ZnSO_4$, $MgSO_4$ and others. Under such conditions, a high titer of up to 10^{11} cells/ml can be achieved in a 24–48 h fermentation (Hammer et al. 2017).

Whole cells of *Pseudomonas* spp. can be concentrated using membrane adsorption/elution techniques, centrifugation or evaporation to obtain either a concentrated suspension or a paste of whole cells. It is critical to maintain the temperature below 10°C throughout the entire concentration process. Formulations of whole cells or cell paste with or without co-formulants, such as whey protein, corn starch, trehalose or more inexpensive materials such as starch, peat, calcium carbonate, talc or clay, have been extensively investigated by many researchers. Different drying methods have also been examined, including low-temperature low-humidity drying, spray drying, freeze drying or simple air drying in rotary drum dryers or tray dryers (Umashankar et al. 2019, Waffa et al. 2013, Nam et al. 2018, Vidhyasekaran et al. 1997, Ganeshan and Kumar 2005, Weller 2007, Chakravarty and Kalita 2011, Kumar et al. 2017). Despite these extensive efforts, the shelf life reported for such formulations remains very short, spanning from days to a maximum of 6 months.

A high titer biopolymer formulation of *P. fluorescens* strain A506 demonstrated shelf life of several months but only when stored under refrigeration (Stockwell and Stack 2007). A product based on *P. syringae* has been

Table 1 *Pseudomonas* spp. commercial products registered in the United States and Europe

Product name	Strain/composition	Company	Uses	Formulation	References
Cerall®	*Pseudomonas chlororaphis* strain MA 342	Koppert Biological Systems (The Netherlands)	Seed treatment against seedborne diseases (cereals). Another product trade name (Cedress®) is registered for use on carrot and peas (seed treatment)	FC	Tombolini et al. (1999), Johnsson et al. (1998)
Cedomon®	*Pseudomonas chlororaphis* strain MA 342	Koppert Biological Systems (The Netherlands)	Seed treatment (cereals)	EC	Tombolini et al. (1999), Johnsson et al. (1998)
Proradix®	*Pseudomonas* spp. strain DSMZ 13134	Sourcon-Padena (Italy)	Control of *Rhizoctonia* for potato planting	WP	Yusran et al. (2009)
Spot Less®	*Pseudomonas aureofaciens* strain Tx-1	Eco Soils Systems (USA)	Control of Turf diseases	Liquid (*)	Sigler et al. (2001)
Howler®	*Pseudomonas chlororaphis strain AFS009*	AgBiome (USA)	Soil and foliar plant fungal diseases	WG	www.agbiome.com (accessed 5 October 2020)

(*) whole fermentation broth.
Note: WP: wettable powder; WG: water dispersible granule; FC: flowable concentrate, liquid water-based formulation; EC: emulsifiable concentrate.

reformulated several times to obtain the most stable formulation (Janisiewicz and Korsten 2002). Initially the product was formulated as wettable powder (WP) or frozen pellets (Lindow et al. 1996), then reformulated as frozen liquid concentrate (LC) packed in containers similar to milk cartons, and finally, as lyophilized cells frozen for storage (Stockwell and Stack 2007).

More advanced formulations of *Pseudomonas* spp. include WPs with polysaccharides such as methylcellulose (CMC) or dextran starch with 0.01% glycerol, which still exhibits a shelf life of only 3 months at 4°C (Hagag et al. 2013). Shelf life of up to 6 months at 25°C has been achieved, by mixing wet cells of *P. chlororaphis* with additives comprising 5% skimmed milk, 7.5% monosodium glutamate and 0.5% of ascorbic acid, prior to freeze drying. The resulting pulverized powder was then mixed at a ratio of 1:9 by weight with diatomaceous earth to obtain a WP formulation (Nam et al. 2018).

Various monosaccharides (glucose, fructose), disaccharides (trehalose, sucrose, lactose) and trisaccharides (melezitose, raffinose, stachyose) have been used to protect cells and proteins against damage due to desiccation stress (Carpenter and Crowe 1988). Trehalose (also known as alpha-D-trehalose) was considered the most effective in maintaining the conformation of proteins, cells and other biomolecules when water is removed during the drying process and especially during the freeze-drying process (Crowe et al. 1987, 1988, Miao et al. 2016). Recently, isomelezitose has been proposed as the superior protectant for the preservation of *P. fluorescens* during drying and storage (Slininger et al. 2020).

In recent years, several US companies have invested heavily in the so-called '-omics' technologies to identify links between the plant and soil microbiomes and its application to crop disease control and plant health (Berg et al. 2017). One application of this approach is the use of gram-negative bacteria isolated from soils to stimulate crop yields, improve plant nutrition and health and protect against abiotic stress such as drought. Crop protection is also an objective of the microbiome approach, controlling pests and diseases through the incorporation of such microorganisms into soil, spraying them onto plants or by using seed treatment technologies to inoculate seeds with different consortia of gram-negative bacteria. To date, the mode of action of such products remains unclear. Whether the observed yield increase in the field is resulting from the ability of the microorganism to control plant soilborne diseases, the production of antibiotics, metabolites or enzymes or a combination of those remains open for further investigations.

Several US-based companies have been exploring the use of microbiome technologies for crop protection and plant health: NewLeaf Symbiotics, Indigo Agriculture and BioConsortia are three examples.

NewLeaf Symbiotics (St. Louis, Missouri) is developing products for yield increase and pest control based on a group of gram-negative, pink-pigmented bacteria called Methylotrophs, abbreviated as M-trophs. These bacteria can utilize reduced forms of one-carbon compounds, notably methane and methanol, as energy sources for growth (Chistoserdova and Kalyuzhnaya 2018, Bennett et al. 2018). However, published reports on the biocontrol potential of M-trophs are still scarce. Recent studies showed direct antagonistic effect of three facultative methylotrophic isolates against fungal pathogens, such as *Fusarium oxysporium*, *Sclerotium rolfsii*, *Pythium ultimum* and *Rhizoctonia solani*, through direct inhibition of mycelial growth and production of volatile inhibitory compounds (Janahiraman et al. 2016). Under greenhouse conditions, the M-troph isolates have shown to significantly reduce incidence of wilt and root rot diseases of tomato (Janahiraman et al. 2016). Large-scale production of M-trophs can be achieved using fermentation processes under special safety conditions allowing for use of methane gas or methanol as a carbon source to obtain a high titer of cells (Trotsenko and Torgonskaya 2018, Schrader et al. 2009). Other carbon sources have also been described, including unsaturated carbon hydrocarbons, to produce dry or liquid formulation of M-trophs (Trotsenko and Torgonskaya 2018).

Indigo Agriculture (Boston, Massachusetts), another US-based company, is using endophytic gram-negative bacterial strains such as *Burkholderia phytofirmans* strain PsJN, *Enterobacter* and *Pseudomonas* spp. (US patent no.10,362,787 B2. 2019), as part of its program to introduce gram-negative endophytic strains for the propagation and treatment of seeds in order to improve crop yield.

Notwithstanding the good efficacy, ease of liquid fermentation and the desire by many researchers to use gram-negative bacteria in biological disease control programs due to good performance, product shelf life remains a pitfall for widespread commercialization. The inability of these bacteria to produce spores (more stable than whole cells) and the fact that some gram-negative bacteria are associated with an overproduction of metabolites, which can impact regulatory approvals or cause crop phytotoxicity, also contribute to the dearth of products available to farmers in the commercial market. Overproduction of specific metabolites can actually be beneficial for efficacy, as shown for *P. fluorescens* strain ACK55, recently registered by the USEPA as a bioherbicide for control of downy brome (cheatgrass, *Bromus tectorum* L.). ACK55 colonizes intercellular spaces in grass roots and produces weed-suppressive metabolites that can inhibit cell membrane expansion at the seedling stage (US patent no. US9578884B).

Pseudomonas strains are known to produce other potentially relevant metabolites (phenazines, 2*R*- + 3*R*-butanediol, 4-ACPA, HCN, Prn, HPR, HHL, fluorescens insecticidal toxin (Fit)), which can be a barrier for registration (Chin-A-Woeng et al. 2003, Han et al. 2006, Péchy-Tarr et al. 2008, EFSA 2017a).

Without a breakthrough in formulation technology to overcome the aforementioned pitfalls and extend product shelf life, the commercialization and use of gram-negative bacteria formulations is expected to remain limited.

2.2 Gram-positive bacteria bioprotectants

Gram-positive bacteria have not raised much interest in the past because their mode of action to control plant disease was not well understood (Handelsman and Stabb 1996). The first commercially successful product in this group for the control of plant disease was introduced in the 1990s as Kodiak® (Gustafson, Inc., Plano. TX, USA), containing *Bacillus subtilis* strain GB03 (Brannen and Kenney 1997). This product proved highly effective as a seed treatment for control of *Fusarium* and *Rhizoctonia*, with the additional benefit of stimulating plant growth (Turner and Backman 1991), resulting in increased yield due to combined fungicidal and biostimulant effects (Beckman et al. 1994). The success of this first product in the market was in part due to the straightforward production, formulation technology and product storage stability. It was the first *Bacillus*-based formulation produced by liquid fermentation followed by centrifugation and spray drying, resulting in a stable technical powder used to formulate the end-user product. Following the success of Kodiak, the number of *Bacillus* spp. bioprotectants in the market for plant disease control has increased. As an example, Serenade® was introduced in 2007 by AgraQuest (later purchased by Bayer CropSciences AG). The active ingredient in Serenade® is strain QST713 of *B. subtilis*, later reclassified as *B. amyloliquefaciens* ssp. *plantarum* and more recently as *B. velezensis* strain QST 713 (Dunlap 2019, EFSA 2021). The taxonomy of the *B. subtilis* group is dynamically changing due to the availability of new genetic data, yet *B. velezensis* and *B. amyloliquefaciens* ssp. *plantarum* are considered synonymous taxonomic designations (EFSA 2021). This product has been commercialized in the United States, Europe and other world markets as both liquid and solid (WP) formulations. Serenade® was followed by the introduction, in 2010, of Amylo-X® by Certis USA, with a solid (WG) and an LC formulation, containing *B. amyloliquefaciens* var. *plantarum* strain D747, marketed in the United States, Europe and other markets.

The commercial success of these products resulted in the development and commercialization of large numbers of *Bacillus* spp.-based technologies for plant disease control, including other *B. subtilis* (Schisler et al. 2004, Stein 2005, Sandra et al. 2019) and *B. amyloliquefaciens* (Rotolo et al. 2015, Hanif et al. 2019) strains, *B. licheniformis* (Won et al. 2019), *B. mojavensis* (Hinton 2001, László 2012, Blacutt et al. 2016), *B. velezensis* (Jiang et al. 2018), *B. atrophaeus* (Rajaofera et al. 2020), *B. nakamurai* (Leathers et al. 2020), *B. polymyxa* (Grady et al. 2016), *B. pumilus* (Serrano et al. 2013) and *Lactobacillus plantarum* (Daranas et al. 2019). A summary of the key commercial products based on different *Bacillus* spp. strains is presented in Table 2.

The development of such *Bacillus* spp. received much attention due to their ability to produce various non-ribosomal cyclic lipopeptides (Stein 2005), which have an extraordinary bio-surfactant activity through hydrophobic and hydrophylic properties, low toxicity to mammals and high environmental compatibility, including low toxicity for non-target organisms. These lipopeptides have specific bioactivity against certain plant pathogens and particularly against fungal growth (Cochrane and Vederas 2016, Yang et al. 2020). The group of lipopeptides produced by different *Bacillus* spp. include more than 30 different peptides, linked to various fatty acid chains, and can be classified into four main families: surfactins, iturins, fengycins and kurstakins (Jacques 2011, Ye et al. 2012, Dunlap et al. 2013, Jamil Shafi et al. 2017, Yang et al. 2020). Their mode of action leads to fungal death through blockage, disruption and formation of holes in the cell wall and cell membrane (Dunlap et al. 2013). Surfactins are important antimicrobial metabolites, regarded as the most powerful bio-surfactants in the lipopeptide family (Pagadoy et al. 2005). Despite being discovered about 40 years ago, there is today a great revival of interest in surfactins triggered by an increasing demand for effective bio-surfactants and 'new' antimicrobial agents. Today, surfactins are produced commercially using *B. subtilis* fermentation and are being registered for cosmetic applications worldwide, for example, sodium surfactin for cosmetic products, produced by Kaneka, Japan (Kaneka 2020). Efforts and investments are rising in this area to produce and include these lipopeptides in diverse applications for the food industry, therapeutics and agriculture as new properties of these surfactants are discovered, driving several companies to explore novel methods for synthesis and purification of lipopeptides derived from *Bacillus* spp. (e.g. Eurogentec of Belgium and Lipopeptide AB of Sweden). In recent years, the area of lipopeptide production has attracted considerable investment, including efforts to register isolated lipopeptide(s) as a natural bioprotectant fungicide (Lipofabrik, France, http://www.lipofabrik.com/). Thus far lipopeptide production is still at an early stage of its potential. High costs and low yields of large-scale production are still limiting factors to the successful commercialization of lipopeptide products. A wide range of fermentation studies have been reported, utilizing different carbon sources, low-priced raw material, different culture conditions and the use of nanoparticles to improve lipopeptide production of *Bacillus* metabolites such as iterins, surfactins and fengycins from *Bacillus* spp. (Beltran-Gracia et al. 2017, Yang et al. 2020). Divalent cations have an important influence on lipopeptide production by *Bacillus* cell cultures, especially the addition of divalent manganese (Mn^{2+}) and ferrous iron (Fe^{2+}). Cationic manganese (Mg^{2+}) and trace minerals such as zinc and iron – in their salt forms of zinc sulfate ($ZnSO_4$), ferric chloride ($FeCl_3$) and manganese sulfate ($MnSO_4$) – when added into the growth media have shown to increase lipopeptide yield and spore production (Wei and Chu 2002, Abdel Mawgoud

Table 2 Commercial biofungicide products containing *Bacillus* spp in North America, Europe, Japan and Brazil

Product name	Species and strain (*)	Company	Formulation	Market status (**)
Serenade®	*B. amyloliquefaciens QST 713* (*)	Bayer CropSciences (Germany)	WG, LC, WP	Commercial
Sonata®	*B. pumilus QST 2808*	Bayer CropScience (Germany)	SC	Commercial
Amylo-X®	*B. amyloliquefaciens* ssp. *plantarum* D747	Certis USA (USA)	WG, LC, SC	Commercial
Serifel®	*B. amyloliquefaciens* MBI 600	BASF (Germany)	WP	Commercial
Taegro®	*B. amyloliquefaciens* FZB24	Novozymes (Denmark)	WP	Commercial
Mildore®	*B. subtilis* IAB/BS03	Investigaciones y Aplicaciones Biotecnológicas S.L. (Spain)	WP	Commercial
Velondis®	*B. subtilis* BU1814	BASF (Germany)	LC	Commercial
(AH2 strain) (#)	*B. amyloliquefaciens* AH2	Probelte (Spain)	SC	Under registration
(IT-45 strain) (#)	*B. amyloliquefaciens* IT-45	Danstar Ferment (Switzerland); Comercial Quimica Masso (Spain)	Data not available	Under registration
Avoda®	*B. subtilis* FMCH001 and *B. licheniformis* FMCH001	FMC (USA)	WP	Commercial
AmyProtect42®	*B. amyloliquefaciens* FZB42	Andermatt (Switzerland)	SC	Commercial
Biobac®	*B. subtilis* Y1336	UPL Brazil (India)	WP	Commercial
Biobaci®	*B. subtilis* BV09	Biovalens (Brazil)	WP	Commercial
Botkiller®	*B. subtilis* AT-332	Idemitsu Kosan (Japan)	WP	Commercial
Ecoshot®	*B. amyloliquefaciens* D747	Kumiai chemical industry (Japan)	WG	Commercial
Rizo®	*B. subtilis* UFPEDA 764	Lallemand Brazil (USA)	SC	Commercial
Amplitude™ Stargus™	*B. amyloliquefaciens* strain F727	Marrone Bio Innovations (USA)	SC	Commercial

(*) Recent advances in genomics have resulted in taxonomic revision of many *Bacillus* spp. and strains used as plant bioprotectants. For example, some *B. subtilis* and *B. amyloliquefaciens* strains have recently been reclassified as *B. velezensis* (Dunlap 2019, EFSA 2020).
(**) Commercial: at least one product is commercially available in one of the markets: the United States, Canada, Brazil, Japan or European Union.
(#) Trade name not available in the public domain (October 2020).
Note: SC: suspension concentrate, WP: wettable powder, WG (or WDG): water dispersible granule, LC: liquid concentrate.

et al. 2008). Still, the separation of pure lipopeptides from the fermentation broth is a difficult task, requiring sophisticated separation processes such as membrane adsorption technology, ultra-filtration chromatographic techniques and the use of resin columns (Beltran-Gracia et al. 2017). New research on the genetics of hyper-producing strains and improved fermentation processes, followed by fit-for-purpose extraction techniques in purification, are the next development steps in this emergent field of bioprotectants.

Another group of gram-positive bacteria investigated in the past as promising biocontrol agents is the *Streptomyces* spp., which grow in various environments and resemble filamentous fungi in appearance. The morphological differentiation of *Streptomyces* involves the formation of a layer of hyphae that can differentiate into a chain of spores. The most interesting property of *Streptomyces* is their ability to produce bioactive secondary metabolites, which have antifungal, antiviral and especially antibacterial activity. Production of most antibiotics by microorganisms is species specific, and these secondary metabolites are important for *Streptomyces* spp. in order to outcompete other microorganisms for nutrients and growth, even within the same genus (Emmert and Handelsman 1999). *Streptomyces* products were extensively investigated as potential bioprotectants; however, the commercial development has not progressed as expected due to overproduction of antibiotics and other metabolites by *Streptomyces* strains (Levy 1978). Currently, only two commercial *Streptomyces* products are available: Mycostop® by Verdera / Lallemand (formerly Kemira Aro Oy, Helsinki, Finland: www.lallemand.com), based on *Streptomyces griseoviridis*, and Actinovate® by Novozymes based on *Streptomyces lydicus* WYEC 108. Mycostop® is listed as biofungicide for the control or suppression of seed and soilborne damping off and root diseases caused by *Alternaria* spp., *Phytophthora* spp, *Pythium* spp and *Rhizoctonia* spp. in seedling production of vegetables, herbs and ornamentals as well as trees and forest seedlings, and Actinovate® is used as a soil drench or seed treatment, for control of soilborne fungi like *Fusarium*, *Rhizoctonia*, *Pythium*, *Phytophthora* and others.

The accelerated commercial interest in developing gram-positive bacteria products, especially *Bacillus* spp., is mainly due to their stability, which plays an important role in achieving a consistent efficacy on the field (Zaki et al. 2018). Fermentation processes for *Bacillus* spp. and other gram-positive bacteria can be optimized to produce maximum number of spores rather than producing numerous vegetative cells by the end of the process (Schisler et al. 2004). As a result, a stable (liquid suspension) concentrate of spores is obtained which can be further spray dried to obtain a stable technical powder containing heat- and desiccation-resistant spores. Production of *Bacillus* spp. spores via liquid fermentation is reproducible, scalable and cost-effective (Sandra et al. 2019), and the spores can be easily formulated into different stable formulation types (Schisler et al. 2004) including: aqueous suspension concentrate (SC), dry WP, WG or WDG, oil flowable and oil dispersion (OF/OD). Large-scale fermentation (up to 200 000 L) and downstream processes are currently used to produce gram-positive spore formulations. Information on specific commercial product composition, formulation, production protocols, process/equipment used in production, stabilization and drying of biomass of *Bacillus* products are proprietary information of commercial companies. Nonetheless, publicly available data describing the production process can be found in production manuals, the published scientific literature (Moazami 2002, Yezza et al. 2004, Kumar et al. 2019) and in various patents (Zhang 2013). From the public information, a generic schematic diagram for the production of *Bacillus* spp. bioprotectant products is presented in Fig. 1. The diagram displays the different methods and the unit operation equipment employed in a generic manufacturing process for *Bacillus* products.

An enormous number of formulation agents (co-formulants) are available for development products based on gram-negative bacteria, from companies specialized in the formulation of biocontrol products. Such formulation agents can be grouped by their function in the formulation as follows: stabilizer, binder, surfactant, dispersant, UV protectant, sticker, filler, buffer and more. Formulation additives can be found in many published catalogs (Bernhard et al. 1998, Manjula and Podile 2001, Schisler et al. 2004) and in experimental formulations of *Bacillus* spp., which are also published in the literature (Schmidt et al. 2001). In recent years, researchers and companies are developing products using mixtures of different *Bacillus* spp. in one formulation, or premixing *Bacillus* spp. with other microorganisms, or even in combination with synthetic chemicals. Such formulation combinations aim at obtaining several modes of action in a single product – for example, biofungicide and bioinsecticide or biofungicide and bionematicide – and to combine the benefit of plant disease control with crop yield increase. Examples of such approaches of pre-mixture are the commercial products: Ethos 3D™ by FMC corporation (Philadelphia, PA) and Trunemco™ by Nufarm America Inc. (Alsip, IL). Ethos 3D is a combination insecticide, biofungicide and growth promotor product, in a two-part formulation containing the active ingredients bifenthrin (15.67%) and *B. amyloliqufaciens* strain D-747 (5.5%). Trunemco™ was launched as a bionematicide and growth promotor product containing *B. amyloliquefaciens* strain MBI 600 (1.00%) and cis-jasmone (0.88%) (2-Cyclopenten-1-one, 3-methyl-2-(2Z)-2-pentenyl), used as seed treatment for protection against nematodes on corn, cotton and soybeans.

With growing research interest and commercial development activities, improvements in production processes, better formulations and an increasing market uptake by users, we envision that gram-positive bacterial products, led by *Bacillus* spp., will soon become mainstream bioprotectants for controlling plant disease pathogens in agriculture.

3 Bioprotectant products containing fungi propagules produced by fermentation

Most of the early work on fungal bioprotectants for the control of plant disease was focused on the development of *Trichoderma* spp. products containing conidia produced by solid-state fermentation (SSF) (Chet and Sivas 1988, Harman et al. 1991, Lumsden et al. 1995, Fravel et al. 1998) and few by liquid submerged fermentation (Eyal et al. 1997, Lumsden et al. 1995). Several reviews have been published dealing with *Trichoderma* spp. products, their formulation, registrations and uses (Fravel et al. 1998, Heydari and Pessarakli 2010, De los Santos-Villalobos et al. 2013, Sheridan et al. 2014, Rajesh et al. 2016). Interest in *Trichoderma* spp. as a bioprotectant for control of plant disease continues to grow with a focus on improved formulation for use as a standalone product or in combination with other microbial agents (*Bacillus* spp. and *Pseudomonas fluorescens*) or with chemical pesticides (Lutz et al. 2004, Rajesh et al. 2016, Fuga et al. 2016, Hernández-Castillo et al. 2020, and Zaki et al. 2020), seed treatment for yield enhancement and disease control (De los Santos-Villalobos et al. 2013), control of plant pathogens by competing for both rhizosphere colonization sites and nutrients (Rajesh et al. 2016), induction of systemic resistance in plants (Martínez-Medina et al. 2013, Ferrigo et al. 2014, Hernández-Castillo et al. 2020) and protection of grapevine pruning wounds from infection (John et al. 2005, Mondello et al. 2018). Today, only a few fungal bioprotectant products are registered and used commercially which are not based on *Trichoderma* spp. (a summary is presented in Table 3). An example of a bioprotectant product is based on the fungi *Ampelomyces quisqualis* strain AQ10, a naturally occurring fungus isolated from powdery mildew parasitizing the plant *Catha edulis* in Israel (Angeli et al. 2017). The product is commercialized under the name AQ 10® WG by the company CBC (Italy), for the control of powdery mildew in grapes, tomato and pepper, both in the United States and the European Union (EU). According to the EU approval of AQ 10® WG, the formulation contains 5 × 10^9 viable spores/g (EFSA 2017b).

A unique group of bioprotectants are fungal products which are intented to displace *Aspergillus flavus* strains, a major source of production of Aflatoxins (AFs), which are carcinogenic and immunosuppressive mycotoxins that threaten the global food safety, especially in agricultural crops in Africa and Asia (Alshannaq et al. 2018, Mauro et al. 2018). One approach to reduce AF levels in agricultural commodities is to apply non-aflatoxigenic competitor fungi. This concept was employed by the US Department of Agrigulture (USDA-ARS) in 2003 with the first introduction of an atoxigenic biocontrol product based on *Aspergillus flavus* AF36 for use in cotton (USEPA 2003). Further development of this initiative resulted in the isolation of new atoxigenic strains, adapted to local growing conditions, for use in maize and groundnut in sub-Saharan

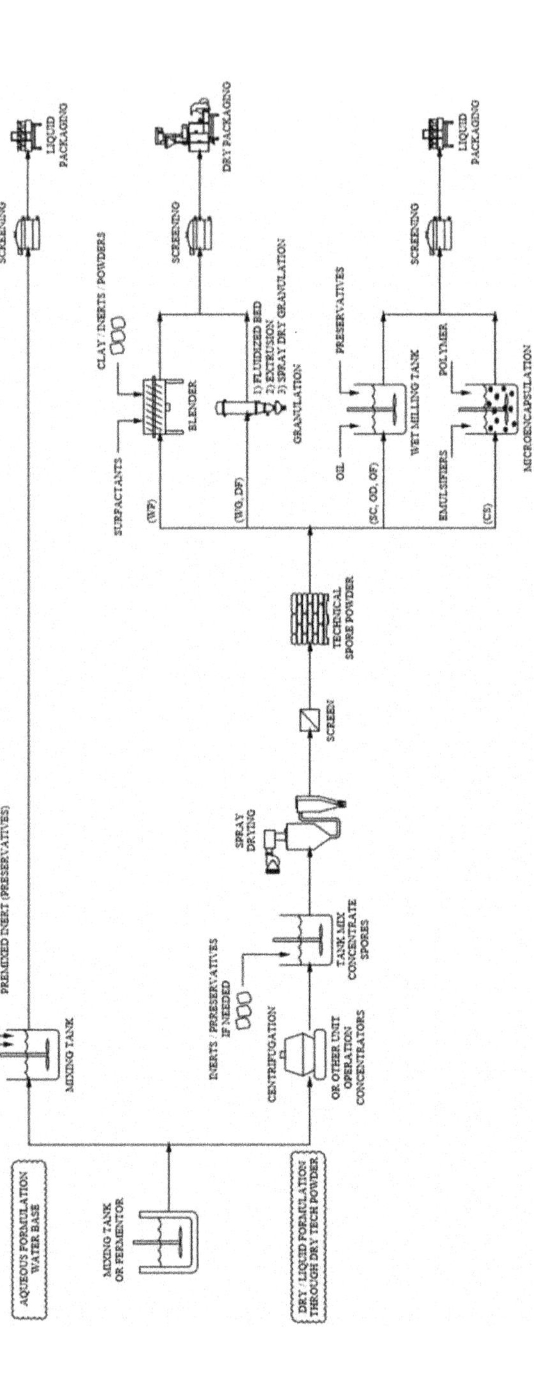

Figure 1 Schematic diagram illustrating a generic large-scale production and downstream options for the formulation of *Bacillus* spp. Note: EC: emulsifiable concentrate, SC: suspension concentrate, OD: oil dispersion, CS: capsule suspension, WP: wettable powder, WG (or WDG): water dispersible granule, DF: dry flowable.

Africa, under a joint development collaboration between USDA-ARS and the International Institute of Tropical Agriculture (IITA) (Atehnkeng et al. 2016). The resulting product is commercially available in several African countries under the trade name Aflasafe (details in Table 3), and efforts are underway to introduce it to other countries (www.aflasafe.com, Moral et al. 2020, Sarrocco et al. 2019).

Another example of an *A. flavus*-based product is Afla-Guard® GR, a naturally occurring and nontoxigenic strain of *A. flavus* (NRRL 21882). Afla-Guard® GR is marketed by Syngenta (Abbas et al. 2011a,b) as a granular formulation for use on peanuts and corn by ground and aerial application.

Currently no such product is registered in the commercial market in the EU, although, *Aspergillus flavus* strain MUCL 54911 is under assessment as a new active substance at the EU level, submitted by Pioneer Hi-Bred Italia Srl, a subsidiary of Corteva Agriscience (EFSA 2020). EU approval of the active substance is anticipated in 2021, but the timing of the commercial launch will also depend on authorizations by EU Member States, which typically requires 2–3 years after the approval of the active substance. The formulation under assessment is a solid granule (GR), code name AF-X1 (EFSA 2020). In Italy, the product AF-X1 can be used on maize for a period of 120 days per year, under an emergency authorization derogation procedure (Sarrocco et al. 2019).

Recently, the US company Marrone Bio Innovations (MBI) was granted registration in the United States for a bioprotectant product containing *Muscodor albus* SA-13, an endophytic non-spore-forming fungus which produces a complex of volatile compounds with broad-spectrum antimicrobial activity. The company claims that the product will also control plant-parasitic nematodes and soil-dwelling pests in addition to soilborne diseases such as root rot, damping off and blight. The active ingredient is produced by 13-day SSF on hulled barley grains. The finished formulation contains spent and unspent fermentation media and a minimum of 10^3 cfu/g of hyphae (Mercier et al. 2007, Strobel et al. 2013, Strobel et al. 2020).

Another interesting fungal bioprotectant product is based on the fungus *Coniothyrium minitans*, a mycoparasite of both hyphae and sclerotia of *Sclerotinia* spp. (Li et al. 2006). *C. minitans* has been proven to control *Sclerotinia* rots and white mold of vegetable crops, stem rot of rapeseed (*Brassica napus*) and head rot of sunflower (Whipps and Gerlagh 1992, Gerlagh et al. 1999, Budge and Whipps 2001, Chitrampalam et al. 2008, Zeng et al. 2012). *C. minitans* has been developed as a commercial biological agent and has been widely used in the United States and the EU under the name Contans® WG, commercialized by Bayer AG (Table 3). *C. minitans* is a coelomycete, which grows on sclerotia and produces pycnidia with numerous conidia on or in the sclerotia (Whipps et al. 2008).

Table 3 Commercial fungal bioprotectant products (other than *Trichoderma* spp.) registered in the United States and the EU for the control of plant diseases

Product name	Strain	Company	Main target	Formulation	References
AQ-10®	*Ampelomyces quisqualis* strain AQ10	CBC (Italy)	Control of powdery mildew	WG	Fondevilla and Rubiales (2012)
Contans®	*Coniothyrium minitans* CON/M/91-08 (DSM 9660)	Bayer CropScience (Germany)	Control of *Sclerotinia sclerotium* (white mold)	WG	Kiewnick (2001), Bennett et al. (2003)
Polygandron®	*Pythium oligandrum* M1	Biopreparáty (Czech Republic)	Control of *Pythium ultimum* (dumping off and root rot disease)	WP	Gerbore et al. (2014), Vesely et al. (1989).
Afla-Guard® GR	*Aspergillus flavus* NRRL 21882	Syngenta (Switzerland)	Displacing pathogenic *Aspergillus flavus* strain through competition	GR	Abbas et al. (2011a,b)
Aflasafe (*)	*Aspergillus flavus, several strains*	IITA (International Institute of Tropical Agriculture) (*)	Displacing pathogenic *Aspergillus flavus* strain through competition (maize and groundnut)	Solid*	Moral et al. (2020)
AF-X1(**)	*Aspergillus flavus,* MUCL54911 (**)	Pioneer (Italy), Corteva Agriscience (USA)	Displacing pathogenic *Aspergillus flavus* strain through competition (maize)	GR	Sarrocco et al. (2019)
Ennoble™	*Muscodor albus* SA-13	Marrone Bio Innovations (USA)	Control and inhibit plant parasitic nematodes, soil pests and soilborne diseases	GR	Strobel et al. (2013)

Note: WP: wettable powder, WG: water dispersible granule, GR: granules.
(*) IITA: IITA licenses biocontrol manufacturing and commercialization responsibilities to private companies or the public sector in several African countries. The formulation type may vary in the different countries but it is a grain-based solid formulation (Moral et al. 2020), most likely a GR formulation.
(**) under approval March 2021.

Such fungal bioprotectant products are produced by solid-state fermentation (SSF) methods. The economic feasibility, mass production methods and equipment used in SSF have been subjects of previous reviews (Ravensberg 2011, Chen et al. 2013). In the last 5 years, there has been growing interest in this area, as shown by the number of publications and patents covering bioprocessing, new equipment design and modeling of SSF for the production of bioprotectants, based on fungi and/or their secondary metabolites (Hongzhang 2013, Vandenberghe et al. 2020, Kumar et al. 2020). Such activities are mainly taking place in fast-developing countries such as China, India and Brazil, which is likely linked to the availability of cheap but high-quality substrate, the growing interest in bioprotectants in these markets (Kumar et al. 2020, Vandenberghe et al. 2020) and a more favorable regulatory context for registration of such technologies than Europe.

Recent years have seen an increase in research and development about the production of different propagules of fungi, including *Trichoderma* spp., using submerged culture fermentation, in which the fungus undergoes a dimorphic transformation from hyphae to submerged conidia or submerged sporidia (Zaki et al. 2018), to microsclerotia (Kobori et al. 2014) or to chlamydospores (Eyal et al. 1997). In all cases, these processes are mediated by environmental and nutritional cues to produce different propagules (Papavizas 1985, Mascarin et al. 2014), as further discussed in the next sections.

3.1 Production and formulation of chlamydospores

Chlamydospores are large resting spores with thick walls that are produced by many species of fungi (Ainsworth et al. 2011). Optimization of chlamydospore production by submerged liquid fermentation has been investigated for several fungi to produce stable commercial formulations based on this spore type. These include *Fusarium oxysporum* as a bioherbicide against *Striga hermonathica* (Ciotola et al. 2010), *Clonostachys rosea* (syn. *Gliocladium roseum*) (Sun et al. 2013, 2019) and *Trichoderma* spp. (Lewis and Papavizas 1983, Papavizas 1985, Herman et al. 1991, Jin et al. 1996, Eyal et al. 1997, Kobori et al. 2014) as biofungicides.

The kinetics and stability of chlamydospore formation by *Trichoderma* (formerly *Gliocladium*) *virens* strain GL-21 was investigated in large-scale (up to 60 000 L) submerged fermentation (Eyal et al. 1997). Only during the stationary phase (ca.160 h) did the morphology of the fermentation biomass change from hyphae to mature, thick-walled chlamydospores, which are considered stable and suitable for harvesting and formulation through drying (Fig. 2). Transmission electron microscopy (TEM) was used to examine different stages of maturity of chlamydospores versus immature or broken mature chlamydospores, as a function of fermentation harvest time (Fig. 3). Chlamydospores at varying

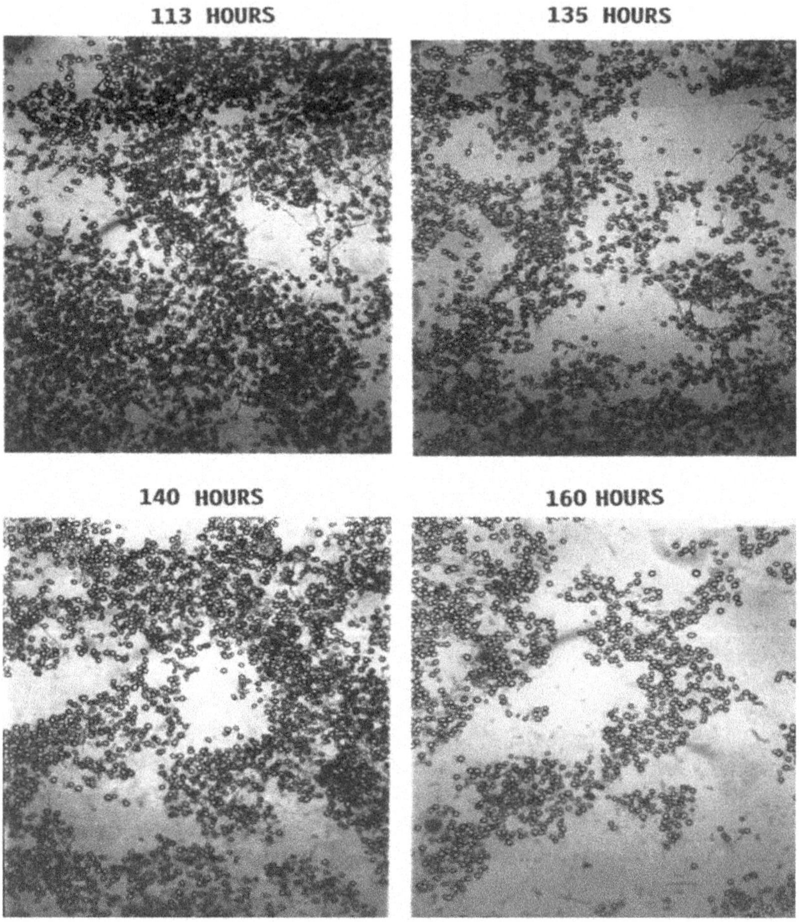

Figure 2 Fermentation broth of *Trichoderma virens* GL-21 at different stages of fermentation (×100). After 160 h, all mycelium are lysed to chlamydospores and ready to be harvested and formulated.

stages of development were harvested from the fermenter between 40 h and 160 h, concentrated and formulated into dry alginate granules (Fig. 4). Viability of chlamydospores within granules for each harvest time was evaluated after 300 days of storage at 4°C and 25°C. The results clearly indicated a correlation between stability (shelf life) and chlamydospore maturity stage (Fig. 5). Commercial fungal products based on chlamydospores, produced by submerged fermentation for control of plant disease, are mainly based on *Trichoderma* spp. Examples of such products include SoilGard® WG (Certis USA), Tricoguard® (Anu Bio-tech International, India) and Tricho-X® (Excel Industries Ltd., India) (Kumar 2013, Li et al. 2016).

3.2 Production and formulation of conidia

Most fungal bioprotectant products for the control of insects or plant disease (Table 3), and especially *Trichoderma* spp. products, contain aerial conidia produced by SSF on moistened grain (Woo et al. 2014). Several reports describe submerged fermentation methods for the production of conidia of various fungi including *Trichoderma viride*, *Trichoderma harzianum* (Said 2007, Lewis and Papavizas 1983, Papavizas 1985, Tabachnik 1989, Harman et al. 1991, Jin et al 1996, Kobori et al. 2014), *Ampelomyces quisqualis* (Angeli et al. 2017, Carbó et al. 2020) and *Colletotrichum truncatum* (Schisler et al. 1991). Conidia produced in different liquid media containing molasses, corn steep liquor (M-CSL), sucrose nitrate (SN), glucose, casein hydrolyzed using various trace minerals, vitamins and amino acids were evaluated for bioefficacy, storage stability and desiccation tolerance. Yet, there are conflicting reports about storage stability and desiccation tolerance of conidia produced by liquid fermentation: some researchers report good storage stability and good bioefficacy (Harman et al. 1991, Schisler et al. 1991) while others report very poor storage stability (Kobori et al. 2014).

Some biopesticide companies have adopted liquid fermentation for the production of conidia of *Trichoderma* spp., but details of fermentation and formulation procedures and composition have not been published. Few or no reports or patents are publicly available to compare yields, fermentation times, production costs or product stability obtained from liquid fermentation with aerial conidia production in solid culture. Such information is held as proprietary by the companies active in this area.

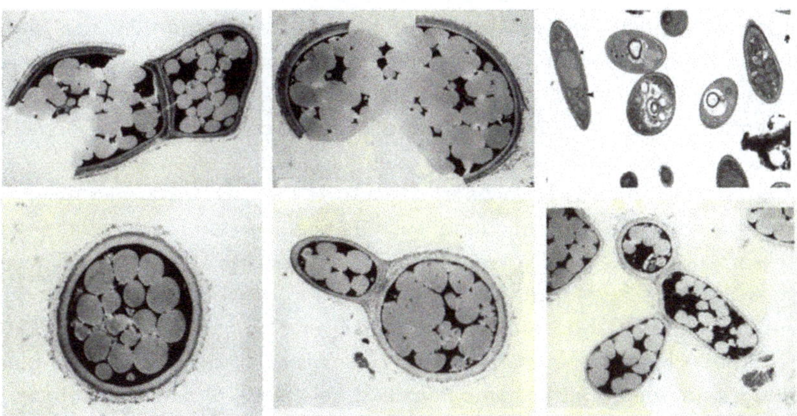

Figure 3 Transmission electron micrographs of chlamydospores of *Trichoderma virens* GL-21. The top images represent immature and broken chlamydospores, the bottom images represent mature chlamydospores after 160-h fermentation.

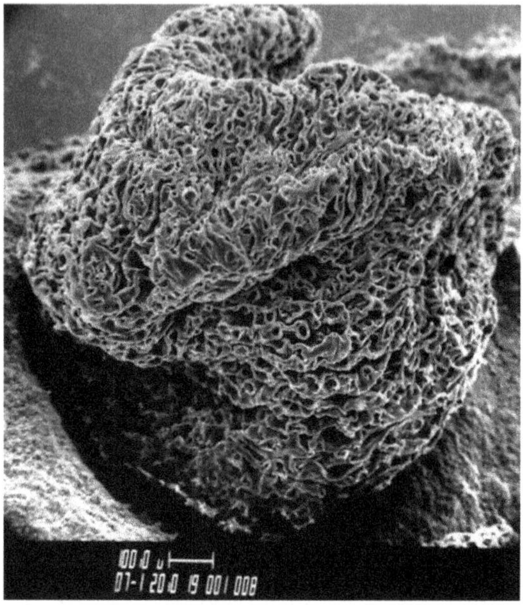

Figure 4 Scanning electron photomicrograph of a dry alginate bead containing chlamydospores of *Trichoderma virens* GL-21.

3.3 Production and formulation of fungal microsclerotia

Microsclerotia are aggregate structures of melanized hyphae produced naturally by many fungi as survival structures in soil (Coley-Smith and Cooke 1971, Cooke 1983). They are typically 50–600 μm in diameter, are often melanized and desiccated to some degree and contain endogenous nutritional reserves for use when favorable conditions return. Microsclerotia can also be formed during liquid fermentation under unfavorable nutritional and/or environmental conditions. Dry microsclerotia produced in submerged fermentation, upon rehydration after application to moist soil, produce conidia capable of infecting insect pests or controlling fungal diseases. Submerged liquid culture methods have been developed for producing microsclerotia of several entomopathogenic fungi such as *Beauveria* and *Metarhizium* spp. (Jackson and Jaronski 2009, Wang et al. 2011, Mascarin et al. 2014, Villamizar et al. 2018, 2020) and for various *Trichoderma* spp. (Kobori et al. 2014). Studies were also conducted on various pathogenic fungi to understand their mode of action, including *Gaeumannomyces graminis* var. *tritici*, a soilborne plant pathogen of wheat (Sivasithamparam and Parker 1976) and *Colletotrichum truncatum* as bioherbicide (Jackson and Schisler 1995, Jackson and Payne 2016). In these studies, culture conditions such as nutritional requirements, aeration C:N ratio, nitrogen source and culture age were optimized in relation

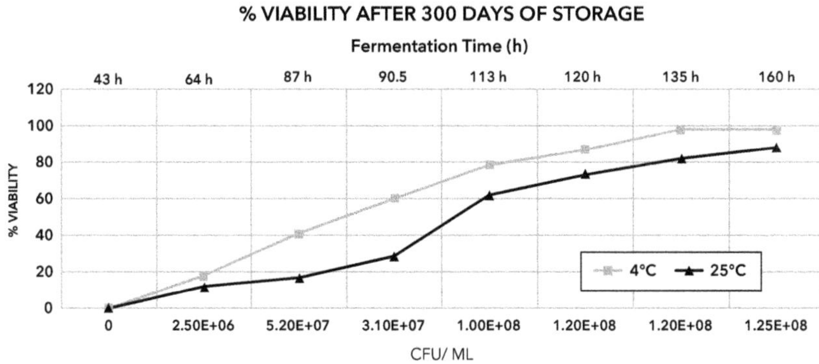

Figure 5 Effect of chlamydospore maturity (fermentation time) on storage stability of *Trichoderma virens* GL-21. Stability was evaluated by formulating chlamydospores from different stages of the fermentation broth, storing samples at 4°C and 25°C and assessing the remaining viability (CFU/ml) of each sample after 300 days (compared to time 0).

to desiccation tolerance, melanization and shelf life of harvested microsclerotia after drying and formulation. Sclerotia melanization was related to nitrogen source and culture age (Kobori et al. 2014). Melanization has direct influence on the persistence of microsclerotia in soil through tolerance of desiccation and other abiotic stress (Shearer and Jackson 2006, Jackson and Jaronski 2009). Therefore it was hypothesized that microsclerotia would be more suitable than conidia or chlamydospores for application against soilborne diseases (US patent no. US20160081351A1). A granular formulation containing air-dried microsclerotia of *Trichoderma harzianum* T22 (a strain used in some current biofungicide products) was stable for at least 12 months in both refrigerated and unrefrigerated storage (Kobori et al. 2014).

Currently there are no commercial biofungicides for the control of plant disease based on the production of fungal microsclerotia in liquid fermentation. Field application of microsclerotia will likely be limited to soil incorporation of granular formulations, since they suspend poorly in water and are too large to pass through most spray equipment. Additionally, separation of microsclerotia from other propagules (conidia, hyphae and chlamydospores) produced during fermentation can be difficult and may result in dry biomass containing these less stable spores, which can affect the product shelf life.

4 Bioprotectant products containing yeast

The possibility of using yeasts as biocontrol agents has been recognized, since the early 1960s, when it was found that some *Saccharomyces cerevisiae* strains secreted microbial compounds that killed other yeast strains but are immune to their own toxins (Bevan and Malkower 1963). Several studies detailed a long list of candidate yeast strains that can be used as bioprotectants or applied as postharvest treatment to prevent spoilage of fruits by several pathogenic fungi (Wisniewski et al. 2007, 2016, Wisniewski and Droby 2012, Ferraz et al. 2019). However, only a few yeast species are registered and used commercially today as bioprotectants for the control of plant disease (Table 4).

Candida oleophila was the first yeast to be developed into a commercial bioprotectant product. Its antifungal activity arises from the production of hydrolytic enzymes and volatile compounds (Huang et al. 2011). Biofilm formation, high osmotic tolerance, induction of systemic resistance and direct parasitism of fungal hyphae have also been associated with the biocontrol activity of *Candida* species (Droby et al. 2009, Ballet et al. 2016).

Another interesting yeast used as a bioprotectant product is the saprophytic ascomycete *Aureobasidium pullulans*, frequently isolated from leaf, flower or soil samples worldwide. This species exhibits an atypical polymorphism for a yeast and is more similar morphologically to other fungi. Antagonism against fungi has been documented for many *A. pullulans* strains, yet only one commercial product, containing a mixture of strains DSM 14940 (CF 10) and DSM 14941 (CF 40), is approved as a bioprotectant product, the first to consist of a mixture of two conspecific strains. The product is marketed under the trade name Blossom Protect® to control fire blight in pome fruits through competitive exclusion of the causal bacterium *Erwinia amylovora* from the floral hypanthium. The same product is used under a different tradename (Botector®) to control *Botrytis* on grapevine and soft fruits by the same mechanism (Seibold et al. 2004, Kunz et al. 2008, Weiss et al. 2006, Bio-ferm 2017).

Another bioprotectant product is based on nectar yeast of the genus *Metschnikowia* (Chappell and Fukami 2018). *Metschnikowia fructicola* has been studied extensively for its ability to inhibit a wide range of fungi causing postharvest and plant rots, and it is considered the most potent antagonistic yeast identified to date (Hilber-Bodmer et al. 2017). The strong antifungal activity of *Metschnikowia* spp. is ascribed to a variety of mechanisms including competition for nutrients (amino acids, iron, etc.), secretion of glucanases and chitinases and the production of volatile organic compounds (Hershkovitz et al. 2012, Banani et al. 2015, Gore-Lloyd et al. 2019, Zajc et al. 2019).

M. fructicola isolate NRRL Y-30752 was discovered in Israel, developed and registered by the Israeli company AgroGreen as a biocontrol product for the prevention of postharvest disease in sweet potato and carrot (Eshel et al. 2009). The product was sold to Bayer CropScience AG, then to Koppert Biological Systems and, in 2017, was granted regulatory approval in Europe as a plant protection product against fungal diseases in stone fruits, soft fruits and grapes under the trade name Noli®.

Various basidiomycete yeasts of the genus *Cryptococcus,* such as *C. albidus*, *C. laurentii* and *C. flavus*, have been shown to protect peach,

Table 4 Commercial yeast bioprotectants products for control of plant diseases.

Product name	Strain/composition	Company	Uses	Formulation	References	Market status 2020
Aspire®	*Candida oleophila* I-182	Ecogen (USA)	Postharvest diseases	WG	Droby et al. (1998)	Withdrawn
Nexy®	*Candida oleophila* strain O	Agrauxine (France)	Postharvest pome fruits	WG	Massart and Jijakli (2014)	Commercial
Blossom protect®	*Aureobasidium pullulans* mixture of strains DSM 14940/14941	Bio-ferm (Austria)	Fire blight diseases, postharvest pome fruit	WG	Kunz et al. (2008), Weiss et al. (2006)	Commercial
Botector®	*Aureobasidium pullulans* mixture of strains DSM 14940/14941	Bio-ferm (Austria)	*Botrytis* on grapevine, soft fruits and tomato	WG	Bio-ferm (2017)	Commercial
Noli®	*Metscnikowia fructicola* NRRL Y-30752	Koppert Biological Systems (The Netherlands)	Fungal diseases, stone fruits, soft fruits, grapevine	WG	Wisnieswski et al. (2012)	Commercial
Yield plus®	*Naganishia albida* (formerly *Cryptococcus albidus*)	Anchor Bio Technologies (South Africa)	*Botrytis* and blue mold (*P. expansum*), pome and citrus fruit	WP	Mbiti (2012), Kowalska et al. (2012)	Withdrawn
Romeo®	*Saccharomyces cerevisiae* cell wall strain LAS 117	Agrauxine (France)	Powdery mildew, downy mildew, *Botrytis* on grapevine and vegetables	WP	De Miccolis Angelini et al. (2019)	Commercial
Julietta®	*Saccharomyces cerevisiae* strain LAS 02	Agrauxine (France)	*Botrytis* on grapevine, vegetables, strawberries, berries; monilinia on fruit trees	WG	EFSA (2015)	Commercial

Note: WP: wettable powder; WG: water dispersible granule.

cherry, strawberry, tomato, citrus and pome fruits against postharvest decay (Zhang et al. 2007a,b). *C. albidus* was commercially used as a biocontrol agent in the product YieldPlus®, registered in 1997 and marketed by Anchor Biotechnologies in South Africa. This product was recently withdrawn from the market after being sold for over 15 years (Mbili 2012).

Romeo® and Julietta® are recently introduced yeast-based products from Agrauxine, a subsidiary of the French yeast manufacturer Lesaffre. Romeo® contains fractions of the cell wall of the yeast *Saccharomyces cerevisiae* strain LAS117 (De Miccolis Angelini et al. 2019) and is currently marketed in the United States and Europe as a systemic resistance inducer (SRI) product against powdery and downy mildew in grapes, fruits and vegetables. Julietta® is a biofungicide product based on live cells of *Saccharomyces cerevisiae* LAS02 used for the control of gray mold (*Botrytis cinerea*) on vegetables, grapes and other fruits and brown rot (*Monilinia* spp.) in stone fruits (Freimoser et al. 2019).

The epiphytic yeast *Pseudozyma* spp. has been the subject of much recent attention as a potential biocontrol agent, by colonizing different plant surfaces, thus providing a natural barrier against plant pathogens (Avis and Bélanger 2001, Spadaro and Gullino 2004). *Pseudozyma* spp. are a small group of yeasts, mostly epiphytic, that is, deriving moisture and nutrients from the air and rain, or saprophytic, and they are non-pathogenic to plants and animals (Avis and Bélanger 2002). Several *Pseudozyma* spp. have recently been found to exhibit biological activity against different powdery mildews with which they are associated, including *P. rugulosa, P. flocculosa*, *P. aphidis* and *P. churashimaensis* (Dik et al. 1998). For example, *P. flocculosa* is reported to be a powerful antagonist of powdery mildew (Bélanger et al. 2012) and also has antimicrobial activity (Avis and Bélanger 2001), while *P. churashimaensis* has been investigated as a potential IR agent for the control of various viruses and plant diseases, with foliar sprays of this species conferring significant protection under field conditions (Lee et al. 2017). Several attempts were made to develop commercial bioprotectant products based on *P. flocculosa* (Konstantinidou-Doltsinis et al. 2007). Sporodex® L, based on *P. flocculosa* strain PF-A22 UL (Plant Products Co. Ltd., Canada), is used for the control of powdery mildew on roses and cucumber in Canada and the United States. The product was temporarily registered in the United States in 2003 and field trials were conducted in Europe in 2007 for controlling powdery mildew on grape (Konstantinidou-Doltsinis et al. 2007). A more recent strain of *P. flocculosa* strain ATTC 64874 has been under assessment for several years in Europe but is yet to be approved by the EU Commission (EFSA 2015).

Stability and viability of yeasts in bioprotectant products are key for the successful commercialization of such products. Different methods employed for drying the yeast biomass after the fermentation process can lead to variation in viability between 10% and 80%, with a product shelf life of just a few months regardless of the drying method used (Yuan et al. 2015). For example, semi-commercial production and formulation of *A. pullulans strain* Ach 1-1 as a postharvest microbial control agent against *Penicillium expansum* in apples, stored in 5°C and 25°C, is well documented (Mounir et al. 2007). *A. pullulans* was grown in a glucose-fed batch fermenter for 48 h to obtain 106 g (dry weight) per liter. Cells were then harvested by centrifugation and dried in a fluidized bed dryer to obtain a viability of 62%. After 7 months at 4°C the viability dropped to 28%. This biomass preparation achieved 89% efficacy against *P. expansum* after 28 days and 7 days in apples stored at 5°C and 25°C, respectively. These results clearly demonstrate that shelf life remains the main bottleneck for successful commercialization of yeast bioprotectants.

Extensive studies on drying methods and formulation, mainly by modifying water activity or by adding protectants, were carried out for the yeast *Candida sake* (Torres et al. 2003, Abadias et al. 2003). Stability of up to 7 months was obtained with a liquid formulation containing cells grown in a sorbitol-modified medium and stored in an isotonic solution of trehalose, or polyols such as glycerol or polyethylene glycol under refrigeration of 4–10°C. Freeze drying (lyophilization) and spray drying with the addition of 10% skimmed milk and 1% peptone as protectants have not resulted in good viability of the yeast (Abadias et al. 2001, 2005). Such results show that drying and formulation processes are not yet economically feasible for large-scale production of *C. sake* due to the instability of the formulated product.

Similar formulation studies compared freeze drying, vacuum drying, fluidized bed drying and liquid formulation of the yeast *Pichia anomala* J121 (Melin et al. 2007). Viability of 80% (compared to initial level) after 6 months at room temperature was achieved by freeze drying and vacuum drying, adding trehalose and the polymer polyvinylpyrrolidone as a protectant during the drying process (Melin et al. 2007). Freeze-drying methods were also tested for other yeast species such as *Cryptococcus laurentii* and *Rhodatorula glutinis*. Formulation studies with both species showed that the addition of trehalose plays a critical role in enhancing cell viability of yeast (Li et al. 2006, 2008). Other liquid formulations containing glycerol (20%), xanthan (0.5%) and phosphate buffer solution were evaluated to formulate *Rhodatorula minuta* as a potential biocontrol product for mango anthracnose disease (Patiño-Vera et al. 2005). Such formulation agents greatly inhibited contamination of the final product, prevented cell sedimentation and maintained stability up to 6 months when the product was stored at 4°C. The same approach of testing other formulation additives was employed for other yeasts such as *P. anomala* J121, *P. membranaefaciens* and *Cryptococcus laurentii,* by adding different additives such as L-ascorbic acid, lactose, starch, trehalose or galactose (Melin et al. 2006, Liu et al. 2009, Li and Tian 2006). The main conclusion that can be drawn from the published data on viability and the shelf-life studies with various yeasts species is that reduction of the oxidative stress may be improved by supplementing the formulation with a sugar source such as trehalose or lactose. Yet, there is a great need for additional in-depth research in formulation and drying of yeast to obtain stable bioprotectant products for the control of plant disease.

5 Production and formulation of bacteriophage for controlling plant disease

Plant pathogenic bacteria cause significant crop losses throughout the world (Borkar and Yumlembam 2017). They can move undetected across borders, including in seeds, introducing crop diseases to regions where they were not present. Bacterial diseases represent about 15% of all damaging plant pathogens (Vidhyasekaran 2002), yet there are few plant protection products on the market to control bacterial disease, and even fewer bioprotectant products.

Bacteriophage are viruses that specifically infect and destroy only certain bacteria. Bacteriophage are ubiquitous in soil, water, plants and animals. They are highly specific antibacterial agents, with one phage typically being specific to one or a few strains of a particular bacteria species. At least one bacteriophage exists that can specifically infect one or more strains within almost every known species of plant pathogenic bacteria. Despite this great potential as biocontrol agents against bacterial plant pathogens, surprisingly, few products are available for this use.

The concept of bacteriophage as antibacterial agents for crop protection is not new, having been first proposed nearly a century ago. However, interest in phage therapy waned in most Western countries after the discovery of penicillin and other antibiotics. In recent years, widespread resistance to antibiotics has led to a renewed interest in the use of bacteriophage against pathogenic bacteria in different fields, including human health, food safety, aquaculture and crop protection (Doss et al. 2017, Buttimer et al. 2017, Garcia et al. 2019). Many bacteriophage products are in commercial use to reduce bacterial contamination in food and feed processing, but there are only five examples of bacteriophage approved for commercial use in plant protection, all of them in the United States and Canada. Currently no bacteriophage is approved in the EU for plant protection uses. The main driving forces in the development of phage for control of bacterial plant diseases are the restrictions imposed on the agriculture use of copper and antibiotics, as well as the development of resistance to these bactericides in field populations of bacterial plant pathogens (Marco and Stail 1983). Thus, phage therapy is now emerging as a serious commercial alternative to the use of antibiotics and copper for the treatment of bacterial diseases in plants (Frampton et al. 2012, Svircev et al. 2018, Vu and Oh 2020).

The application of phage as a disinfectant against highly pathogenic bacteria contaminating human and pet foods, food processing facilities and hides of living animals has been commercialized by several companies, including Intralytix, Inc. (USA), with products under the brand name Shield™; OmniLytics, Inc. (USA) with the AgriPhage™ product line; Micreos BV (The Netherlands) under the brand name PhageGuard™; and FINK TEC GmbH (Germany) with the product Applied™ Phage. Several companies in the United States, Japan and Europe are currently developing phage-based products for control of plant pathogenic bacteria in crop protection, as listed in Table 5.

A major impediment in the development and commercialization of bacteriophage is the availability of a cost-effective and controllable manufacturing process. Bacteriophage production is often assumed to be a routine batch fermentation process using conventional stirred backmixing bioreactors (fermenters). Indeed, existing fermentation equipment used for the growth of other microbial products can be easily adopted for phage production. Nonetheless, bacteriophage production processes are not easily scaled up, as there is a need for a system that accommodates biological interaction between the phage and its host bacterium, the growth of which is crucial to obtain the large titer of phage required. Additional drawbacks result from the inherently high rate of phage mutations during manufacturing, as well as the propensity of the host bacteria to evolve genetic resistance to the highly specific phage. Such genetic plasticity of both phage and target host complicates manufacturing, requiring the use of mixtures or phage 'cocktails' to overcome this mutation rate and maintain efficacy against genetically heterogeneous field populations of the host bacteria, that is, the target plant pathogen. Occasionally, there is a need to isolate new host strains from the field in order to adjust the phage cocktail to reflect the genetic composition of the target pathogen population. The composition of such cocktail can vary from season to season, requiring serious consideration of the dynamics between the phage and the host population genetics during the manufacturing process (Levin et al. 2013).

Since bacteriophage replication requires the presence of the host bacteria in the fermenter, the manufacturing of bacteriophage involves a production process with separate operations for the growth of the host bacteria and bacteriophage propagation (Fig. 6). To obtain a high phage titer during this process, optimization of basic fermentation parameters is essential. These parameters include the selection of nutritional media or substrate for both bacterial growth and phage infection (Tokman et al. 2016), temperatures for bacterial and phage growth, amount of phage inoculum, agitation and oxygen demand (dissolved oxygen or DO level) (Ali et al. 2019, Mutti and Corsini 2019).

Industrial-scale fermentation can be carried out in regular fermenters by fed batch, continuous (Mancuso et al. 2018) or semi-batch (Sauvageau

Table 5 Bacteriophage products for controlling of bacterial plat diseases

Country	Product name	Company	Target diseases	Target pathogen	Status 2020
USA, Canada	AgriPhage	OmniLytics Inc. (USA)	Bacterial spot of tomato and pepper	*Xanthomonas* spp.	Commercial
			Bacterial speck of tomato	*Pseudomonas syringae*	Commercial
	AgriPhage CMM		Bacterial canker of tomato	*Clavibacter michiganensis* ssp. *michiganensis*	Commercial
	AgriPhage CC		Citrus canker	*Xanthomonas citri* ssp. *citri*	Commercial
	AgriPhage FB		Fire blight	*Erwinia amylovora*	Commercial
USA	XylPhi-PD	Otsuka Pharmaceutical Co. (Japan)	Pierce's disease of grape	*Xylella fastidiosa*	Commercial
Hungary	Erwiphage	Enviroinvest Zrt. (Hungary)	Fire blight	*Erwinia amylovora*	Development (Dömötör et al. 2012)
USA	Fire Quencher	Brigham Young University (USA)	Fire blight	*Erwinia amylovora*	Development (Grose 2018)

and Cooper 2010) modes of operation. Semi-batch mode (Fig. 6) allows for continuous production while avoiding co-evolution of phages and bacteria and allowing removal of bacteria cell debris after phage amplification throughout the process (Jurac et al. 2019, Yuan et al. 2019). It is important to know well the biology of the phage to be produced and ensure the final product contains no living pathogenic bacteria (Santos et al. 2014, Garcia et al. 2019).

Process data obtained from laboratory studies is not always reliable for industrial scaling up of biological processes, and bacteriophage production is no exception (Kwok 2010). Researchers have been trying to fill up this gap, mainly through studies on bacteriophage production based on computer simulations, some of them validated experimentally (Garcia et al. 2019).

Bacteriophage formulation has been recently reviewed by Malik et al. (2017). Formulation type depends on the application, ranging from sterile liquids to non-sterile liquids, solids and semisolids or even encapsulated. Despite these examples, published data regarding formulation for phage-based products is scarce. Liquid formulations are predominant for the control of bacterial plant disease, but a key issue is the stability of the phages in liquid suspension without the presence of the biological host. Poor stability and decreasing potency of viruses can be the result of oxidation, aggregation and adsorption to the primary container or chemical degradation. Aggregation is mainly induced by electrostatic attraction, and it can be prevented with charged excipients, certain non-ionic surfactants or pH stabilization. Oxidation can be reduced by using antioxidants (Rodrigues et al. 2019). A typical phage formulation contains a buffer, antioxidants, surfactants, tonicity agents and cryoprotectants if the product is to be frozen during storage (Rodrigues et al. 2019).

As the final product may contain several phages, the identity of each phage and phage contaminants can be verified by genomic sequencing or PCR-based genotyping (Pirnay et al. 2015). While random mutations during the phage propagation stage are unavoidable, the production process should be designed to avoid significant phenotype changes, with functional properties tested under validated quality control (QC) protocols before production batches are released.

Notwithstanding the renewed interest in bacteriophage as a highly specific method for removing undesired bacteria, there is still much uncertainty about how to achieve rapid methods for analyzing the quality of bacteriophage products. Lysis performance of each individual bacteriophage is typically assessed by comparing the size of plaques formed on a lawn of the host bacterium growing on solid media (Okman et al. 2016, Ekwu et al. 2020), comparison of lysis kinetics (Fig. 7), or by using the Double Agar Layer (DAL) method (Rajnovic et al. 2019). An alternative to the DAL method is time-kill assays, where the kinetics of the phage-induced lysis is assessed by measuring the optical density of a suspension of host

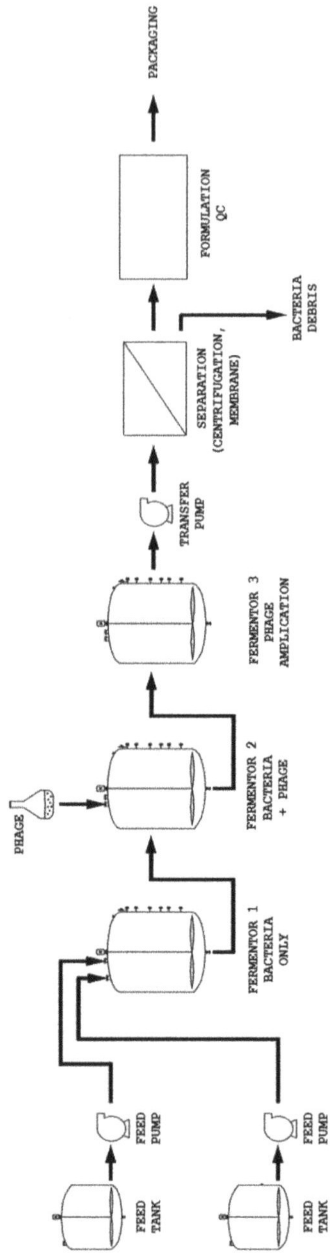

Figure 6 Schematic diagram for semi-continuous batch fermentation mode of operation and the removal of bacteria cell debris.

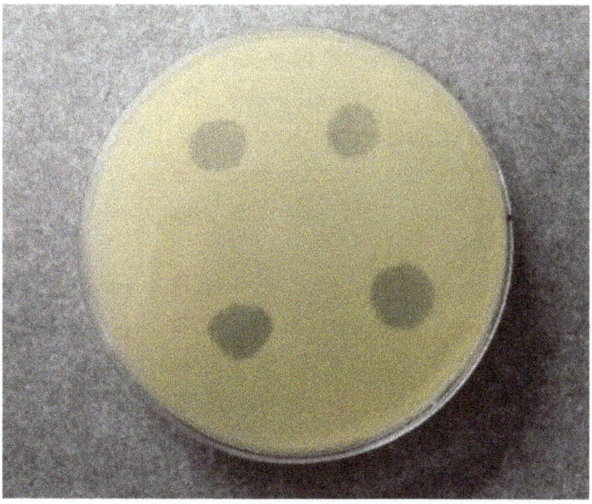

Figure 7 Plaques (cleared circles) formed on a 'lawn' culture of host bacteria at points of introduction of a lytic bacteriophage.

bacterial cells (Xie et al. 2018, Storms et al. 2019). Other methods, such as qPCR and ELISA, can also be used to determine the final phage titer (Pirnay et al. 2015).

Purity and composition of the finished product can be assessed by high-performance liquid chromatography (HPLC), coupled with mass spectrometry (Rogers et al. 2017). These methods can also be used to identify phage capsid proteins, toxins or other bacterial proteins (Rodrigues et al. 2019). Since the production of phages is specific to a plant pathogenic bacteria (i.e. they do not pose a safety risk to humans), QC requirements are less strict than in pharmaceutical applications, where maximum acceptable levels of specific contaminants such as endotoxins, enterotoxins or bacterial DNA are established and must not be exceeded.

Further research is needed in the field of phage production and formulation to adequately support any future use of phage for the control of bacterial plant disease. Even with these remaining challenges, the rediscovery of bacteriophage as bioprotectant products for control of bacterial diseases in agriculture is very exciting and promising.

6 Innovative microbial bioprotectant products for controlling plant disease

6.1 Induced resistance (IR) by microbial bioprotectant products

Treatment of plants with various agents (e.g. avirulent strains of pathogenic microorganisms, non-pathogenic microorganism, fungal cell wall fragments, plant extracts and synthetic chemicals) can induce resistance to subsequent pathogen attack, both locally and systemically (Kloepper et al. 2004, Walters et al. 2005, Choudhary et al. 2009, Conrath 2009, Pieterese et al. 2014, Tripathi et al. 2019). There is still much to be learned about the roles and interactions between different plant signaling pathways which govern this phenomenon through salicylic acid (SA), jasmonic acid (JA) and ethylene (Li et al. 2019). Nonetheless, recent advances in the development of microbial elicitors of IR, as bioprotectant products, reflect a new direction to combat plant disease.

The first living microbial product that acts entirely through activation of plant defense pathways was introduced in 2017 as LifeGard® biological plant activator (Certis USA, Columbia MD, www.certisusa.com), containing *Bacillus mycoides* isolate J (BmJ). This phyllosphere bacterium triggers plant resistance by activating the key IR regulatory protein NPR1 (Pieterse and van Loon 2004), independently of SA signaling (Bargabus-Larson and Jacobsen 2007), and is available as WG and aqueous LC formulations. Both formulations are labeled for the control of a variety of plant diseases caused by fungi: powdery mildew, anthracnose, white mold, early blight and *Cercospora* (leaf spot) (Bargabus et al. 2002, Neher et al. 2009). LifeGard has also proven efficacy against some diseases caused by oomycetes (grape downy mildew) and viruses (*Potato virus Y*).

Bacillus mycoides spp. is distinguished from other *Bacillus* spp. by the unusual colony morphology on agar plates. *B. mycoides* forms white opaque colonies that are characteristically hairy in appearance, often referred to as 'rhizoid' (Di Franco et al. 2002). The submerged fermentation process and the formulation of *B. mycoides* are much alike the production of any other commercial *Bacillus* spp. (Yanglei et al. 2018).

Another IR product recently introduced in Europe and the United States is Romeo® (Agrauxine, France) claimed to be the first SRI based on a non-living microbe. The active ingredient cerevisane, consisting of fractionated cell walls of the yeast *Saccharomyces cerevisiae* strain LAS117, induces the expression of genes involved in the plant response to fungal attack (De Miccolis Angelini et al. 2019). The product is commercially available as a WP formulation for protection of grapevine against powdery and downy mildews (De Miccolis Angelini et al. 2019).

Many other microorganisms are currently being investigated for prospective commercial development as inducer of plant resistance. Examples include the yeast *Pseudozyma churashimaensis*, where foliar spray of this yeast conferred significant protection against *Xanthomonas axonopodis* and against *Cucumber mosaic virus*, *Pepper mottle virus*, and *Broad bean wilt virus* under field conditions (Lee et al. 2017), as well as bacteria such as *Paenibacillus alvei* (Tjamos et al. 2005) and *Pseudomonas fluorescens* (Van de Mortel et al. 2012), and several fungi in the genus *Trichoderma* (Martínez-Medina et al. 2013).

6.2 Jasmonates as induced resistance bioprotectant products

Several research institutes and private companies are developing new IR products based on oxidized lipid signaling molecules, the so-called oxylipins (Blee 2002, Wasternack 2006). These efforts concentrate mainly on the fatty acid oxylipins molecule moiety: Jasmonic acid (JA), and its derivatives which are collectively referred to as jasmonates (Browse 2009, Wasternack and Feussner 2018, Ali and Baek 2020).

Jasmonate phytohormones play important roles in plant development and responses to physical damage caused by insects and other herbivores and attacks by pathogenic microorganisms (Ryan and Moura 2002, Howe and Jander 2008, Browse 2009, Wu and Baldwin 2010). They are also widely distributed in algae (Andreou et al. 2009), fungi and bacteria (Miersch et al. 1993, Miersch et al. 1999a,b, Abdala et al. 1999, Forchetti et al. 2007, Hause et al. 2007).

Jasmonates can be produced by liquid fermentation of fungi, bacteria or yeast. The highest yields of JA have been obtained from Ascomycete fungi in the genera *Diplodia* and *Lasiodiplodia* (Eng et al. 2018). Jasmonates have been investigated for their use as fragrances and flavoring agents in the food industry (Farbood et al. 2001), in medicine as therapeutic agents (Zhang et al. 2015) and in crop protection as inducers of plant resistance (Wasternack 2014). JA is produced when culturing bacteria or fungi strains in submerged culture with oleic or linoleic acid as fermentation substrates (Eng et al. 2018). The strains differ in their ability to produce JA with productivity as high as 1 g/L (Dhandhukia and Thakkar 2007). JA production by fermentation has been reported up to 150-L scale using *D. gossypina* (Farbood et al. 2001). Optimization studies of inoculum type and other fermentation parameters, using *Lasiodiplodia theobromae*, have shown that highest JA yields are obtained by using homogenized mycelium rather than spores; it is the inoculum and not the spores that improves JA yields (Dos Santos et al. 2014a,b). JA can also be produced by SSF (solid-state fermentation) using the same fungi strains with a similar culture medium to the one used for liquid fermentation. Output on solid media was up twofold, higher compared to liquid fermentation using similar growing media (Eng 1996, Eng et al. 2018). Other throughput experiments found nearly a fivefold higher titer of *Botryodiplodia theobromae* in SSF versus liquid fermentation (Laredo-Alcala et al. 2016).

A growing number of patents and scientific papers describe the production, formulation and application of JA and other jasmonates as IR products to defend crops against herbivores and pathogens, but also to improve crop yields (Ryan and Farmer 1999, Omer et al. 2000, Sanches et al. 2017). These patents include new formulations, combining JA with chemical and biological protectants, biological seed treatments or semiochemicals for enhanced performance (Ghasemi et al. 2014, Marks 2012).

At present, extraction of jasmonates from plant sources is superior in yield and purity to chemical synthesis and microbial fermentation. In fact, microbial fermentation results in jasmonate mixtures that require elaborate purification strategies during the implementation of an industrial process for jasmonate production. Jasmonates produced by microorganisms remain a promising field within bioprotectant products for controlling plant diseases through IR; however, additional research and production strategies for JA biosynthesis by microbial fermentation are needed.

6.3 Peptides for plant disease control

Plants defend themselves against fungal pathogen by producing several types of compounds, including (lipo)peptides and proteins with antimicrobial activity (Terras et al. 1995, Wright and Beer 1995, De Coninck et al. 2013). Various antifungal peptides produced by plants have been identified (Vriens et al. 2016). A key fungal pathogen target of such plant-produced peptides is the metabolic pathway leading to the production of fungal membrane compounds containing the sphingolipid glucosylceramide, or GlcCer (Gonçalves et al. 2012). GlcCer is a sugar sphingolipid, composed of a sphingoid backbone, a fatty acid and a glucose moiety (Fig. 8), which has recently emerged as a key regulator of pathogenicity in a variety of fungi (De-Connick et al. 2013, 2017, Thevissen et al. 2012). Notably, GlcCer is found in plants, fungi and animals but it is absent in bacteria and in some other eukaryotes such as baker yeasts.

GlcCer is essential for fungal cell division, hyphal formation, spore germination and for the regulation of fungal virulence. Hence, GlcCer is an important fungal target for compounds with specific and broad-spectrum activity, and much research interest is focused on inhibition of GlcCer synthesis for the development of new antifungal compounds for human therapeutic and agricultural uses (Del Poeta et al. 2014, Visesato et al. 2015).

By targeting the GlcCer production pathway, it was demonstrated that proteins derived from camelid single domain antibodies (sdAb), also known as VHH[1] act specifically against fungal glucosylceramides (De Coninck et al. 2017). This antibody technology involves the immunization of llamas with a purified fungal GlcCer and a mixture of mycelium and spores of fungus such as *Botrytis cinerea*. The llama immune repertoire is subsequently cloned in a phage-display vector to generate a library with a diversity of different clones, encoding for several antibodies with different specificities. This antibody library

[1] VHH: single domain variable region of the heavy chain of an antibody.

is screened for the desired specificity toward the target fungus by incubation with fungal GlcCer. This identifies phages producing antibodies which bind specifically to the fungal GlcCer, but not to mammalian or plant-derived GlcCer. Such specific sdAb showed high activity in inhibiting the growth of *B. cinerea in vitro* and reduced disease symptoms induced by *B. cinerea* when sprayed on tomato leaves (De Coninck et al. 2013).

The main advantages of sdAb are linked to their low molecular weight (2–15 kDa), that is, much smaller than a whole antibody (150–160 kDa), making sdAb highly stable in aqueous solutions, including at high temperatures, and they can refold spontaneously following denaturation. Their relatively small size makes sdAb attractive from a therapeutic standpoint, as they can penetrate tissues and cross natural barriers more easily than whole antibodies, therefore reaching hidden epitopes (the molecular target) located deeper in protein structures. Such penetrability is essential in the development of therapeutic antibodies as well as bioprotectant antibodies for the control of plant disease. An additional advantage is associated with their production because sdAb expresses extremely well in many systems, including bacterial hosts. The development of sdAb for inhibiting fungal growth led to the commercializing of this technology by Biotalys NV (formally known as Agrosavfe) (www.biotalys.com), which is developing a variety of products based on the use of sdAb library. For commercialization purposes, the sdAb technology was branded by the company as 'Agrobodies' which were screened using *in vitro* fungicidal assays against different plant pathogens. The first bioprotectant product based on sdAb, announced by Biotalys, is BioFun-1 for the control of *Botrytis cinerea* and powdery mildew. The product BioFun-1 is expected to be launched in the United States in the coming years, pending EPA regulatory approval (Biotalys 2020).

From a manufacturing point of view, sdAb are easy to clone, engineer and incorporate or fuse with other antibody's domains, and they are relatively easy to produce in microbial platforms. Most large-scale microbial production systems of sdAb use *E. coli* (Rahbarizadeh et al. 2005) or the yeast *Saccharomyces cerevisiae* (Van der Vaart 2002), through a regular fermentation process where the expressed sdAb can be harvested, homogenized and spray dried to be thereafter formulated into the final product. sdAb have a

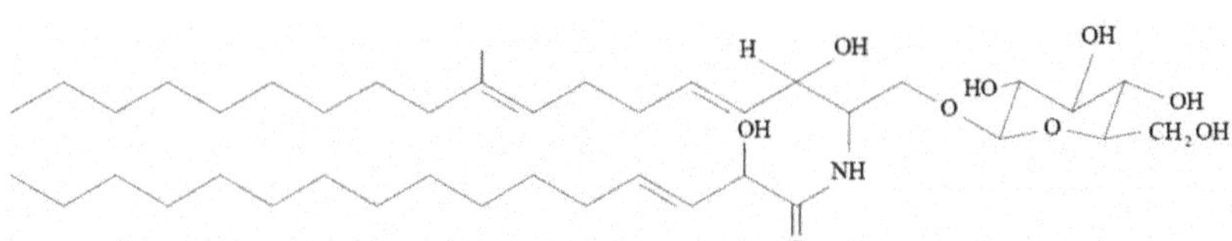

Figure 8 Fungal glucosylceramide (GlcCer) structure.

relatively short development time, from gene to product and require simple yet well-established fermentation conditions that can be adapted to large-scale production, resulting in relatively low costs of the final protein (Estell 2006).

6.4 Protozoans for plant disease control

Protozoans occur widely in soil and aquatic environments. Amoebae are members of a single clade within the Protozoa known to prey on bacteria or fungi (Heidel et al. 2011). Amoebae have been found in close association with natural biofilms where they feed on biofilm-enmeshed bacteria, disrupt biofilm structure and cause biofilm sloughing (Huws et al. 2004, Weitere et al. 2005). Research on amoebae as predators of other microorganisms has fostered new strategies to control bacteria and fungi in commercial applications, such as to reduce biofilms in cooling towers, destroy biofilms of gram-negative bacteria species in pharmaceutical applications and for the control of bacteria and fungi that cause plant diseases. At present, two companies are employing these strategies for plant protection: the French company Amoéba (www.amoeba-biocide.com, Amoeba 2020) and the US company AmebaGone, Inc.

Amoéba is developing biocide and biofungicide products based on the thermophilic amoeba *Willaertia magna* C2c Maky (De Jonckheere et al. 1984, 2018, Demaneche et al. 2020, Amoeba 2020). This free-living amoeba strain is present in aquatic environments and its use as a natural biocide to control the amoeba-resistant bacterium *Legionella pneumophila* in cooling towers was proposed by Hasni et al. 2019. The company also claims that a liquid lysate (homogenized cells) of the amoeba strain has fungicidal activity and is testing several formulations in field trials against downy mildew of grapevines, late blight of potatoes and bean rust. Recent transcriptomic and proteomic studies have defined the mode of action of *W. magna*, identifying enzymes associated with the production of secondary metabolites with antimicrobial activity (Hasni et al. 2019, Mameri et al. 2020).

The commercialization of live amoebae or lysate as a plant protection product can be a challenge. For the development of such products it is necessary first to produce a large number of amoeba cells economically and within a short time window. Traditional production methods based on axenic amoeba cultivation by adhesion to surfaces result in low cell density (Neff 1957). However, a recent study has demonstrated that *W. magna* C2c Maky can grow in axenic liquid medium. And under such conditions, the detection of transcripts and proteins related to the degradation and phagocytosis of bacteria suggests that *W. magna* retains the ability to feed on bacteria despite growing without them (Hasni et al. 2020). This study suggests that *W. magna* could be produced in large-scale fermentation (bioreactors) to yield higher cell titer than attainable by surface adhesion. On the other hand, little information on large-scale production of amoebae has been published, particularly for *W. magna* C2c Maky, to draw a conclusion on the efficiency of the manufacturing process. Available data indicate that production at 10-L scale is feasible, using special fermentation medium CGYEM at pH 7.0 and a temperature of 30–45°C, under constant agitation (200 rpm) and aeration (De Jonckheere 1977, Hasni et al. 2020). For inoculation purposes, the same medium was supplemented with 10% calf serum. Cells were harvested through centrifugation at 2000 rpm and washed by three steps of Page's modified Neff's Amoeba Saline medium (2 mM NaCl, 16 μM MgSO4, 27.2 μM CaCl2, 1 mM Na2HPO4, 1 mM KH2PO4). Concentrated cells of 5×10^8 cells/ml were lysed using 1× phosphate buffered saline (pH 8.0) and 1% sodium dodecyl sulfate (SDS) per gram of pellet for 2 h at 4°C under agitation to obtain cell lysate for future formulation of the finished product (Hasni et al. 2020).

The registration of *W. magna* C2c Maky as a biocidal product or biofungicide product will be a demanding task. The application submitted in 2018 for its registration as a biocidal product was denied by the Biocidal Product Committee (BPC), of the European Chemicals Agency (ECHA), due to concern over its risks to human health (BBPC report 2018). In addition, the development of resistance to amoeba by microorganisms can evolve as in some cases some of amoeba-resistant bacteria has become lytic for their amoeba host (Greub and Raoult 2004).

A similar approach to develop bioprotectant products against plant disease based on amoebae has been taken by AmebaGone, Inc. of Madison, Wisconsin (USA). This company has developed a group of social amoebae (family Dictyostelidae, nicknamed 'Dicty'), also known as cellular slime molds, which prey on pathogenic bacteria (Sanders et al. 2017). The company is developing Dicty strains for the control of fire blight in pome fruits by breaking biofilms caused by the causal pathogen (*Erwinia amylovora*) and through the production of secondary metabolites that inhibit further bacterial growth. Other products in development target the bacterial pathogens causing tomato canker (*Clavibacter michiganensis*) and potato soft rot (*Dickeya* and *Pectobacterium* spp.). Methods for commercial-scale spore production are under investigation, using probiotic bacteria as hosts to amplify Dicty spp. and producing high titer spores for spore preservation and application to crops (www.AmebaGone.com).

The dictyostelid amoeba *Dictyostelium discoideum* has been used as a model organism in molecular biology and genetics for studies on cell communication, differentiation and programmed cell death (Hickoop et al. 2001, Bozzaro 2019). A large body of knowledge around this organism is publicly available online (dictybase.org). Dicty strains produce large numbers of semi-dormant spores highly resistant to environmental stress and therefore stable in long-term storage. Spore germination releases metabolically active Dicty cells that immediately begin feeding on surrounding bacteria. These spores are being developed as future products, with efforts to develop a large-scale fermentation method for culturing such spores. Still, the key challenge is the development of economical large-scale processes for culture and propagation of Dicty spores or cells in large-scale fermentation.

Under laboratory conditions, *D. discoideum* cells grow readily at room temperature but they proliferate only in association with bacteria on solid substrate, or in a rich liquid medium, and require sterile conditions to prevent contamination with yeast, fungi or non-prey bacteria (Fey et al. 2007). The growth rate observed on bacteria was double the one in axenic medium (Fey et al. 2007, Sanders et al. 2017). The bacteria most commonly used to support rapid growth of *D. discoideum* are *Aerobacter (Klebsella) aerogenes* and *Escherichia coli*. A promising method for optimal growth of *Dictyostelium* spp. has been developed using synthetic medium immobilized in an inorganic porous support formed of ceramic catalyst carriers. This method achieves the highest cell densities observed so far in cultivation systems, up to 4.5×10^7 cells per mL (Lu et al. 2004).

In summary, the mass production of dictyostelids will require further development and the scale-up of a challenging semi-continuous fermentation process (similar to the aforementioned for bacteriophage production) in order to remove bacteria cells from the harvested amoeba product.

7 Innovative formulations for bioprotectants

The challenge of stabilizing bioprotectants during storage and after application has limited commercial development of bioprotectants, and some products have been withdrawn from the market due to stability issues. Formulation development must play a central role in addressing stability of bioprotectants, with the key goals of: (1) stabilizing the microbial active ingredient during harvesting and drying (manufacturing), (2) protecting the microbial bioprotectant from adverse environmental conditions during distribution and storage and (3) enhancing the protective activity of the microbial agent(s) after application. Examining the current commercial bioprotectant products for the control of plant disease, presented in this chapter, it can be perceived that the microbial bioprotectant industry still lacks the formulation knowledge and technology required for a successful commercialization of many of the fungi, yeast, virus and gram-negative bacteria with potential as bioprotection products (Leggett et al. 2011). Formulation development starts with understating the active ingredient and optimizing the fermentation process to maximize the production of stable cells or spores. Determining optimal fermentation media components, C:N ratio, pH, growth conditions (temperature, aeration, inoculum size) and the correct harvest time are all essential parameters in the production of stable spores or cells (Eyal et al. 1997, Schisler et al. 2004).

Much has been written in the last 20 years on formulation of microbial agents, including the effects of different formulation components, the relationship between the microbial control agent and the target host, the effect of the environment and the importance of understanding the needs of the product end-user in the success of the final product (Russell and Boyetchko 2006, Ravensberg 2011, Gasic and Brankica 2013, Keswani Cheta and Vivek 2016). In reality, most formulations used today for microbial agents are the same type as the ones used for chemical pesticides. The most common liquid formulations are emulsifiable concentrates (EC), SC (sometimes called flowable or FL), and OD. The most common solid formulations are WP and WG, sometimes called dry flowable or DF. WP formulations are blends of one or more solid active ingredients with solid formulation agents such as solid surfactants. WG formulations can be made in several different ways including extrusion granulation, spray-drying granulation or fluid bed spray granulation. WG formulations can be very suitable for all types of bioactive ingredients, including living microorganisms produced via either liquid or solid-state fermentation.

Significant progress has been made in the development of formulation and application methods but much work remains, such as improvements in product stability. Recent years have seen increased effort to develop more stable formulations of microbial active ingredients using new formulation technologies. These include bioencapsulation of living biological control agents (Vemmer and Patel 2013, Chen et al. 2013, Maruyama et al. 2020) and bio-nanoencapsulation techniques which are being explored for plant hormone delivery, seed germination, transfer of target genes and controlled release of agrochemical and microbial agent for the control of insects and plant diseases (Hayles et al. 2017, Pisoschi et al. 2020, Mancera-López et al. 2018). Several early studies employed polysaccharide microencapsulation of fungal spores such as *Trichoderma* and *Metarhizium* followed by spray drying and storage under vacuum to maintain viability for 1 year at 20°C (Goettel et al. 1990, Jin and Custis 2011). However, none of these techniques has been successfully scaled up for commercial use.

Nanoparticle encapsulation, with particle size ranging between 10 nm and 100 nm, has recently been explored as a method to formulate plant protectants (Worrall et al. 2018, Ramezani et al. 2019). Nanoparticles can function via two different mechanisms: (a) the nanoparticles themselves providing crop protection or (b) nanoparticles as carriers to entrap, encapsulate, absorb or attach microbial active ingredients such as the use of silver/titanium dioxide nanoparticles with the supernatant of bacteria or fungi (Ramezani et al. 2019). Research at several institutions is aimed at developing biodegradable nanoparticles as carriers, providing several potential benefits: (1) enhanced product shelf life and protection from UV radiation, (2) improved solubility of less water-soluble actives such as proteins, (3) reduced toxicity to non-target organisms and (4) boosting site-specific uptake into the target pest (Hayles et al. 2017).

New porous and biodegradable matrix materials for adherence of microorganisms or encapsulation such as *Bacillus* and *Pseudomonas* spp. are being developed using nanofibrillated cellulose, starch, alginate, chitosan, microcrystalline cellulose, cyclodextrin and lignin (Pinho et al. 2014, Pirzada et al. 2020). Such biodegradable matrices containing bioactive ingredients can then be applied directly to plants. Several of these matrix formulations have been developed and patented (Wicks and Pinchuk 2015), as encapsulation gains momentum and considerable commercial interest.

Dispersible concentrate (DC) formulation techniques have rarely been used for microbial active ingredients. Such techniques are more commonly used to formulate chemical active ingredients in chemical solvents with surfactant/dispersing agents. All the same, DC formulations have been recently reported to be suitable for biological products, especially aerial conidia produced by entomopathogenic fungi, to obtain a water-free formulation with an increased shelf life of up to 2 years at 20°C (Eiben et al. 2019).

8 Conclusion and future trends

The low stability of bioprotectant products containing fungi, yeast and gram-negative bacteria has largely hindered their commercial use to control plant disease and some products have been withdrawn from the market because of inadequate stability. Conversely, bioprotectant products containing the gram-positive bacteria *Bacillus* spp. have seen great progress during the last 10 years, resulting in the production of numerous, effective and stable commercial formulations for the control of various plant diseases.

Recent years have seen much interest in the development of bioprotectant products containing bacteria, yeast, fungi and their metabolites, for the control of plant disease through induced resistance (IR). We expect to see much more of such new bioprotectant products reaching the market in the coming years.

Growing research interest and new capital investment is directed toward the development and production of novel bioprotectant products for the control of plant disease, including bacteriophage and protozoans for the control of plant pathogenic bacteria, and single domain antibodies (sdAb) targeting specific plant pathogens. Yet, improvements are needed in submerged fermentation technologies to produce such active substances and products. Large-scale demonstration of production methodologies is to be attested along with a careful optimization of the key fermentation parameters and the downstream processing used for drying such active ingredients.

The pace at which new patents have been issued in recent years - focused on nanoencapsulation/bioencapsulation, to improve shelf life and performance of living microbial active ingredients - suggests that this fast progress will continue in this thriving area. Nonetheless, the feasibility of using such technologies in large-scale manufacturing is yet to be demonstrated alongside the economics and commercial viability of such approaches, which are still unknown. Training programs for distributors and the product end-users should also be promoted, along with investments throughout the supply chains to facilitate proper storage and transportation of bioprotectants.

9 Where to look for further information

Information on manufacturing and commercialization of bioprotectant products is very rarely available in public databases. New active ingredients and products close to reaching the market can be tracked via registration databases, such as:

- United States EPA:
  https://www.epa.gov/pesticides/biopesticides.
- European Union:
  https://ec.europa.eu/food/plants/pesticides/eu-pesticides-database_en.
  https://www.efsa.europa.eu/en/applications/pesticides.
- EPPO Database:
  https://www.eppo.int/ACTIVITIES/plant_protection_products/registe red_products.
- (EPPO database provides the links to several countries national databases).

Additional information about biocontrol technologies can be found in the website of the International Biocontrol Manufacturers Association (IBMA), which represents the biocontrol industry: https://ibma-global.org/.

10 References

Abadias, M., Benabarre, A., Teixidó, N., Usall, J. and Viñas, I. (2001). Effect of freeze drying and protectant on viability of the biocontrol yeast Candida sake. *International Journal of Food Microbiology* 65(3):173–182. 10.1016/s0168-1605(00)00513-4.

Abadias, M., Usall, J., Teixidó, N. and Viñas, I. (2003). Liquid formulation of the postharvest biocontrol agent Candida sake CPA-1 in isotonic solutions. *Phytopathology* 93(4):436–442. 10.1094/PHYTO.2003.93.4.436.

Abadias, M., Teixidó, N., Usall, J., Solsona, C. and Viñas, I. (2005). Survival of the postharvest biocontrol yeast Candida sake CPA-1 after dehydration by spray-drying. *Biocontrol Science and Technology* 15(8):835–846. 10.1080/09583150500187041.

Abbas, H. K., Zablotowicz, R. M., Horn, B. W., Phillips, N. A., Johnson, B. J., Jin, X. and Abel, C. A. (2011a). Comparison of major biocontrol strains of non-aflatoxigenic *Aspergillus flavus* for the reduction of aflatoxins and cyclopiazonic acid in maize. *Food Additives and Contaminants. Part A, Chemistry, Analysis, Control, Exposure and Risk Assessment* 28(2):198–208, 10.1080/19440049.2010.544680.

Abbas, H. K., Weaver, M. A., Horn, B. W., Carbone, I., Monacell, J. T. and Shier, W. T. (2011b). Selection of *Aspergillus flavus* isolates for biological control of aflatoxins in corn. *Toxin Reviews* 30(2–3):59–70. 10.3109/15569543.2011.591539.

Abdala, G., Miersch, O., Correa, N. and Rosas, S. (1999). Detection of jasmonic acid in cultures of *Escherichia coli* and *Saccharomyces cerevisiae*. *Natural Product Letters* 14(1):55–63 10.1080/10575639908045435.

Abdel Mawgoud, A. M., Aboulwafa, M. M. and Abdel Haleen, H. N. (2008). Optimization of surfactin production by *Bacillus subtilis* isolate BS5. *Applied Biochemistry and Biotechnology* 3:305–325. 10.1007/s12010=008-8155.

AgBiome (2020). Howler fungicide product brochure. Available at: https://www.agbiome.com/howler/, accessed 4 October 2020.

Agosin, E., Cotoras, M., Muñoz, G., Martin, R. S. and Volpe, D. (1997). Comparative properties of *Trichoderma harzianum* spore produced under solid state and submerged culture conditions, book chapter, 463–473. In: Roussos, S., Lonsane, B. K., Raimbault, M. and Viniegra-Gonzalez, G. (Eds) *Advances in Solid State Fermentation*. Springer, Dordrecht. 10.1007/978-94-017-0661-2_38.

Ainsworth, G. C., Kirk, P. M. and Bisby, G. R. (2011). *Ainsworth & Bisby's Dictionary of the Fungi*. CAB International, Wallingford, UK, 771.

Ali, J., Rafiq, Q. and Ratcliffe, E. (2019). A scaled-down model for the translation of bacteriophage culture to manufacturing scale. *Biotechnology and Bioengineering* 116(5):972–984. 10.1002/bit.26911.

Ali, M. S. and Baek, K. H. (2020). Jasmonic acid signaling pathway in response to abiotic stresses in plants. *International Journal of Molecular Sciences* 21(2):621. 10.3390/ijms21020621.

Alshannaq, A. F., Gibbons, J. G., Lee, M. K., Han, K. H., Hong, S. B. and Yu, J. H. (2018). Controlling aflatoxin contamination and propagation of *Aspergillus flavus* by a soy-fermenting *Aspergillus oryzae* strain. *Scientific Reports* 8(1):16871. 10.1038/s41598-018-35246-1.

Amoeba (2020). Amoéba a revolutionary biocide. Available at: https://www.amoeba-biocide.com/en, accessed 7 October 2020.

Andreou, A., Brodhun, F. and Feussner, I. (2009). Biosynthesis of oxylipins in non-mammals. *Progress in Lipid Research* 48(3–4):148–170. 10.1016/j.plipres.2009.02.002.

Angeli, D., Saharan, K., Segarra, G., Sicher, C. and Pertot, I. (2017). Production of *Ampelomyces quisqualis* conidia in submerged fermentation and improvements in the formulation for increased shelf-life. *Crop Protection* 97:135–144, 10.1016/j.cropro.2016.11.012.

Atehnkeng, J., Donner, M., Ojiambo, P. S., Ikotun, B., Augusto, J., Cotty, P. J. and Bandyopadhyay, R. (2016). Environmental distribution and genetic diversity of vegetative compatibility groups determine biocontrol strategies to mitigate aflatoxin contamination of maize by *Aspergillus flavus*. *Microbial Biotechnology* 9(1):75–88. 10.1111/1751-7915.12324.

Avis, T. J. and Bélanger, R. R. (2001). Specificity and mode of action of the antifungal fatty acid cis-9-heptadecenoic acid produced by *Pseudozyma flocculosa*. *Applied and Environmental Microbiology* 67(2):956–960. 10.1128/AEM.67.2.956-960.2001.

Avis, T. J. and Bélanger, R. R. (2002). Mechanisms and means of detection of biocontrol activity of *Pseudozyma* yeasts against plant-pathogenic fungi. *FEMS Yeast Research* 2(1, March):5–8. 10.1111/j.1567-1364.2002.tb00062.x.

Ballet, N., Souche, J. L. and Vandekerckove, P. (2016). Efficacy of *Candida oleophila* strain O in preventing postharvest diseases of fruit. *Acta Horticulturae* 1144:105–112. 10.17660/ActaHortic.2016.1144.15.

Banani, H., Spadaro, D., Zhang, D., Matic, S., Garibaldi, A. and Gullino, M. L. (2015). Postharvest application of a novel chitinase cloned from Metschnikowia fructicola and overexpressed in *Pichia pastoris* to control brown rot of peaches. *International Journal of Food Microbiology* 199:54–61 10.1016/j.ijfoodmicro.2015.01.002.

Bargabus, R. L., Zidack, N. K., Sherwood, J. E. and Jacobsen, B. J. (2002). Characterization of systemic resistance in sugar beet elicited by a non-pathogenic, phyllosphere-colonizing *Bacillus mycoides*, biological control agent. *Physiological and Molecular Plant Pathology* 61(5, November):289–298. 10.1006/pmpp.2003.0443.

Bargabus-Larson, R. L. and Jacobsen, B. J. (2007). Biocontrol elicited systemic resistance in sugar beet is salicylic acid Independent and NPR1 dependent. *Journal of Sugarbeet Research* 44(1):17–34. 10.5274/jsbr.44.1.17.

Beckman, P. A., Brannen, P. M. and Mahaffee, W. F. (1994). Plant response and disease control following seed inoculation with *Bacillus subtills*. In Proceeding of the Third International Workshop on Plant Growth Promoting Rhizobacteria, pp. 3–8. CSIRO, South Australia.

Bélanger, R. R., Labbe, C., Lefebvre, F. and Teichmann, B. (2012). Mode of action of biocontrol agents: all that glitters is not gold. *Canadian Journal of Plant Pathology* 34(4):469–478. 10.1080/07060661.2012.726649.

Beltran-Gracia, E., Macedo-Raygoza, G., Villafaña-Rojas, J., Martinez-Rodriguez, A., Chavez-Castrillon, Y. Y., Espinosa-Escalante, F. M., Di Mascio, P., Ogura, T. and Beltran-Garcia, M. J. (2017). Production of lipopeptides by fermentation processes: endophytic bacteria, fermentation strategies and easy methods for bacterial selection. Fermentation Processes chapter 11. 10.5772/64236.

Bennett, A. J., Leifert, C. and Whipps, J. M. (2003). Survival of biocontrol agents *Coniothyrium minitans* and *bacillus subtills* MBI 600 introduced into pasteurized sterilized and non-sterile soils. *Soil Biology and Biochemistry* 35(12):1565–1573. 10.1016/J.soilbio.2003.08.001.

Bennett, R. K., Steinberg, L. M., Chen, W. and Papoutsakis, E. T. (2018). Engineering the bioconversion of methane and methanol to fuels and chemicals in native and synthetic methylotrophs. *Current Opinion in Biotechnology* 50:81–93. 10.1016/j.copbio.2017.11.010.

Berg, G., Korbel, M., Rybakova, D., Muller, H., Grosch, R. and Smalla, K. (2017). Plant microbial diversity is suggested as the key to future biocontrol and health trends. *FEMS Microbiology Ecology* 93(5). 10.1093/femsec/fix050.

Bernhard, K., Holloway, P. J. and Burges, H. D. (1998). Appendix I: a catalogue of formulation additives: function, nomenclature, proper- ties and suppliers, 333–365. In: Burges, H. D. (Ed) *Formulation of Microbial Biopesticides: Beneficial Microorganisms, Nematodes and Seed Treatments*. Kluwer Academic Publishers, Dordrecht, the Netherlands.

Bevan, B. A. and Malkower, M. (1963). The physiological basis of the killer-character in yeast. Proceeding of the Genetics today, 202–203. In: Geerts, S. J. (Ed). *XIth International Congress of Genetics*, Vol 1. Pergamon Press, Oxford.

Bio-ferm (2017). Botector product user manual, July 2017. Bio-ferm GmbH. Available at: https://bio-ferm.com/en/products/botector, accessed 25 January 2021.

Biotalys (2020). Company website: product pipeline and expected launching dates. Available at: https://www.biotalys.com/our-solutions/product-pipeline, accessed 8 October 2020.

Blacutt, A. A., Mitchell, T. R., Bacon, C. W. and Gold, S. E. (2016). *Bacillus mojavensis* RRC101 lipopeptides provoke physiological and metabolic changes during antagonism against *Fusarium verticillioides*. *Molecular Plant-Microbe Interactions* 29(9):713–723. 10.1094/MPMI-05-16-0093-R.

Blée, E. (2002). Impact of phyto-oxylipins in plant defense. *Trends in Plant Science* 7(7):315–322. 10.1016/s1360-1385(02)02290-2.

Borkar, G. S. and Yumlembam, A. R. (2017). *Bacterial Diseases of Crop Plants* (1st edn.). CRC Press, Boca Raton, FL, 594.

Bozzaro, S. (2019). The past, present and future of *Dictyostelium* as a model system. *International Journal of Developmental Biology* 63(8–9):321–331. 10.1387/ijdb.190128sb.

Brannen, P. M. and Kenney, D. S. (1997). Kodiak–a successful biological control product for suppression of soil borne plant pathogens of cotton. *Journal of Industrial Microbiology and Biotechnology* 19(3):169–171.

Browse, J. (2009). Jasmonate passes muster: a receptor and targets for the defense hormone. *Annual Review of Plant Biology* 60:183–205.

Budge, S. P. and Whipps, J. M. (2001). Potential for integrated control of *Sclerotinia sclerotiorum* in glasshouse lettuce using *Coniothyrium minitans* and reduced fungicide application. *Phytopathology* 91(2):221–227.10.1094/PHYTO.2001.91.2.221.

Bullock, J. and Calvert, D. (2020). Bioformulations 2020, the science, technology and business of bioprotection formulations. Agribusiness. Agrow, IHS Markit, ISBN: 978-1-911627-28-9.

Buttimer, C., McAuliffe, O., Ross, R. P., Hill, C., O'Mahony, J. and Coffey, A. (2017). Bacteriophages and bacterial plant diseases. *Frontiers in Microbiology* 8:34. 10.3389/fmicb.2017.00034.

Carbó, A., Torres, R., Usall, J., Ballesta, J. and Teixidó, N. (2020). Biocontrol potential of *Ampelomyces quisqualis* strain CPA-9 against powdery mildew: conidia production in liquid medium and efficacy on zucchini leaves. *Scientia Horticulturae* 267: 109337, 10.1016/j.scienta.2020.109337.

Carpenter, J. F. and Crowe, J. H. (1988). Modes of stabilization of a protein by organic solutes during desiccation. *Cryobiology* 25(5):459–470. 10.1016/0011-2240(88)90054-5.

Chakravarty, G. and Kalita, M. C. (2011). Comparative evaluation of organic formulations of *Pseudomonas Fluorescents* based bio pesticides and their application in the management of bacterial wilt of Brinjal. *African Journal of Biotechnology* 10(27):7147–7182. 10;5897/AJB11.1209.

Chappell, R. C. R. and Fukami, T. (2018). Nectar yeasts: a natural microcosm for ecology. *Yeast* 35(6):417–423. 10.1002/yea.3311.

Chen, K. N., Chen, C. Y., Lin, Y. C. and Chen, M. J. (2013). Formulation of a Novel Antagonistic Bacterium Based Biopesticide for Fungal Disease Control Using Microencapsulation Techniques. *Journal of Agricultural Science* Vol 5 (3): 5:153–163. doi: 10.5539/jas.v5n3p153

Chet, I. and Sivas, I. (1988). Antifungal composition containing *Trichoderma* active against *Fusarium*, US Patent 4.748,021.

Chin-A-Woeng, T. F. C., Bloemberg, G. V. and Lugtenberg, B. J. J. (2003) Phenazines and their role in biocontrol by Pseudomonas bacteria. *The New Phytologist* 157(3):503-523. 10.1046/j.1469-8137.2003.00686.x.

Chistoserdova, L. and Kalyuzhnaya, M. G. (2018). Current trends in methylotrophy. *Trends in Microbiology* 26(8):703–714. 10.1016/j.tim.2018.01.011.

Chitrampalam, P., Figuli, P. J., Matheron, M. E., Subbarao, K. V. and Pryor, B. M. (2008). Biocontrol of lettuce drop caused by *Sclerotinia sclerotiorum* and *S. minor* in desert agroecosystems. *Plant Disease* 92(12):1625–1634. 10.1094/PDIS-92-12-1625.

Choudhary, D. K., Johri, B. N. and Bhavdish, N. J. (2009). Interactions of *Bacillus* spp. and plants–with special reference to induced systemic resistance (ISR). *Microbiological Research* 164(5):493–513. 10.1016/j.micres.2008.08.007.

Ciotola, M., Di Tommaso, A. and Watson, A. K. (2010). Chlamydospore production, inoculation methods and pathogenicity of *Fusarium oxysporum* M12-4A, a biocontrol for *Striga hermonthica*. *Biocontrol Science and Technology* 10(2):129–145. Accessed on 28 June 2010.

Cochrane, S. A. and Vederas, J. C. (2016). Lipopeptides from bacillus and *Panenibacillus* spp.: a gold mine of antibiotic candidates Msd. *Respiratory Research* 36:4–31. 10.1002/med.21321.

Coley-Smith, J. R. and Cooke, R. C. (1971). Survival and germination of fungal sclerotia. *Annual Review of Phytopathology* 9(1):65–92. 10.1146/annurev.py.09.090171.000433.

Conrath, U. (2009). Chapter 9 priming of induced plant defense responses. *Advances in Botanical Research* 51:361–395 10.1016/S0065-2296(09)51009-9.

Cooke, R. (1983). Morphogenesis of sclerotia, 397–418. In: Smith, J. E. (Ed) *Fungal Differentiation: A Contemporary Synthesis*. Marcel Dekker, New York.

Crowe, J. H., Crowe, L. M., Carpenter, J. F. and Wistrom, C. A. (1987). Stabilization of dry phospholipid bilayers and proteins by sugars. *Biochemical Journal* 242(1):1–10. 10.1042/bj2420001.

Crowe, J. H., Crowe, L. M., Carpenter, J. F., Rudolph, A. S., Wistrom, C. A., Spargo, B. J. and Anchordoguy, T. J. (1988). Interactions of sugars with membranes. *Biochimica et Biophysica Acta* 947(2):367–384. 10.1016/0304-4157(88)90015-9.

Daranas, N., Rosello, G., Caberfiga, J., Donati, I., Francés, J., Badosa, E., Spinelli, F., Montesinos, E. and Bonaterra, A. (2019) Biological control of bacterial plant diseases with *Lactobacillus plantarum* strains selected for their broad-spectrum activity. *The Annals of Applied Biology* 174(1):92–105. PMID 30686827.

De-Coninck, B., Cammue, B. P. A. and Thevissen, K. (2013). Modes of antifungal action and in plants functions of plant defensins and defensin like peptides. *Fungal Biology Reviews* 26(4):109–120.

De-Coninck, B., Verheesen, P., Vos, C. M., Van Daele, I., De Bolle, M. F., Vieira, J. V., Peferoen, M., Cammue, B. P. A. and Thevissen, K. (2017). Fungal glucosylceramide-specific camelid single domain antibodies are characterized by broad spectrum antifungal activity. *Frontiers in Microbiology* 8:1059. 10.3389/fmicb.2017.01059.

De Jonckheere, J. F. (1977). Use of an axenic medium for differentiation between pathogenic and nonpathogenic *Naegleria fowleri* isolates. *Applied and Environment Microbiology* 33:751–757. 10.1128/AEM.33.4.751-757.

De Jonckheere, J., Dive, D., Pussard, M. and Vickerman, K. (1984). *Willaertia magna* gen. nov., sp. nov. (*Vahlkampfiidae*), a thermophilic amoeba found in different habitats. *Protistology* 201:5–13.

De Jonckheere, J. F., Dive, D. G., Pussard, M. and Vickerman, K. (2018). *Willaertia Magna gen. nov., sp. nov. (Vahlkampfiidae)*, a thermophilic amoeba found in different habitats. Available at: https://eurekamag.com/research/001/281/001281223.php, accessed on 12 August 2018.

De los Santos-Villalobos, S., Guzman-Ortiz, D. A., Gomez-Lima, M. A., Delano-Frier, J. P., De-Folter, S., Sanchez, -Garcia, P. and Pena-Cabriaes, J. J. (2013). Potential use of Trichoderma asperellum (Samuels, Licechfeldt et Nirenberg) T8a as a biological control agent against anthracnose in mango (Mangifera Indica L.). *Biological Control* 64:37–44. 10.1016/biocontrol.

De Miccolis Angelini, R. M., Rotolo, C., Gerin, D., Abate, D., Pollastro, S. and Faretra, F. (2019). Global transcriptome analysis and differentially expressed genes in grapevine after application of the yeast derived defense inducer cerevisane. *Pest Management Science* 75(7):2020– 2033. 10.1002/ps.5317.

Del Poeta, M., Nimrichter, L., Rodrigues, M. L. and Luberto, C. (2014). Synthesis and biological properties of fungal glucosylceramide. *PLoS Pathogens* 10(1):e1003832. 10.1371/journal.ppat.1003832.

Demanèche, S., Mirabel, L., Abbe, O., Eberst, J. B. and Souche, J. L. (2020). A new active substance derived from lyzed *Willaertia magna* C2C maky cells to fight grapevine downy mildew. *Plants (Basel)* 9(8):1013. 10.3390/plants9081013.

Dhandhukia, P. C. and Thakkar, V. R. (2007). Standardization of growth and fermentation criteria of 626 *Lasiodiplodia theobromae* for production of jasmonic acid. *African Journal of Biotechnology* 627(6):707–712.

Di Franco, C., Beccari, E., Santini, T., Pisaneschi, G. and Tecce, G. (2002). Colony shape as a genetic trait in the pattern-forming *Bacillus mycoides*. *BMC Microbiology* 2:33.

Dik, A. J., Verhaar, M. A. and Bélanger, R. R. (1998). Comparison of three biological control agents against cucumber powdery mildew (*Sphaerotheca fuliginea*) in semi-commercial-scale glasshouse trials. *European Journal of Plant Pathology* 104(4):413–423. 10.1023/A:1008025416672.

Dömötör, D., Becságh, P., Rákhely, G., Schneider, G. and Kovács, T. (2012). Complete genomic sequence of Erwinia amylovora phage PhiEaH2. *Journal of Virology* 86(19):10899. 10.1128/JVI.01870-12.

Doss, J., Culbertson, K., Hahn, D., Camacho, J. and Barekzi, N. (2017). A review of phage therapy against bacterial pathogens of aquatic and terrestrial organisms. *Viruses* 9(3):50. 10.3390/v9030050.

Dos Santos, A. Z., Araujo Leite, V. L., Mello Innocentini, M. and Lourenco, M. V. (2014a). Evaluation of process parameters in the production of jasmonic acid. *BMC Proceedings* 8(Suppl 4):244. 10.1186/1753-6561-8-S4-P244.

Dos Santos, A. Z., Innocentini, M. D. D. M. and Lourenco, M. V. (2014b). Optimization of jasmonates bioproduction. *BMC Proceedings* 8(S4):208. 10.1186/1753-6561-8-S4-P208.

Droby, S., Cohen, L., Daus, A., Weiss, B., Horev, B., Chalutz, E., Katz, H., Keren, Tzur, M. and Shachnai, A. (1998). Commercial testing of Aspire: a yeast preparation for the biological control of postharvest decay of citrus. *Biological control* 12, 97–101.

Droby, S., Wisniewski, M., Macarisin, D. and Wilson, C. (2009). Twenty years of postharvest biocontrol research: is it time for a new paradigm? *Postharvest Biology and Technology* 52(2):137–145. 10.1016/j.postharvbio.2008.11.009.

Dunlap, C. A., Bowman, M. J. and Schisler, D. A. (2013). Genomic analysis and secondary metabolites production in *Bacillus amyloliqufaciens* AS 43,3: a biocontrol antagonist of *Fusarium* head blight. *Biological Control* 64(2):166–175.

Dunlap, C. A. (2019). Taxonomy of registered *Bacillus* sp. strains used as plant pathogen antagonists. *Biological Control* 134:82–86. 10.1016/j.biocontrol.2019.04.011.

Eiben, U., Karge, M., Lüth, P. and Lortz, B.-M. (2019). Composition essentially free of water and comprising at least one spore forming fungal biological control agent, a polyether-modified trisiloxane and fumed or precipitated silica. USA patent, US10299484B2.

Ekwu, M. A., Tyrrel, S., Harris, J. M., Pawlett, M., Orlova, E. V., Ignatiou, A. and Nocker, A. (2020). Lysis Performance of bacteriophages with different plaque sizes and comparison of lysis kinetics after simultaneous and sequential phage addition. *PHAGE* 1(3):149–157. 10.1089/phage.2020.0005.

Emmert, E. A. B. and Handelsman, J. (1999). Biocontrol of plant disease: a (Gram-) positive perspective. *FEMS Microbiology Letters* 171(1, February):1–9. 10.1111/j.1574-6968.1999.tb13405.x.

Eng, F. (1996). Jasmonic acid production from *Botryodiplodia theobromae* by liquid and solid state 636 fermentation. M.Sc. Thesis. University of Autonomous Metropolitan, Campus Iztapalapa, Mexico. Peer J. Preprints 10.7287/peerj.preprints.26655v1.

Eng, F., Krzysztof Zienkiewicz, K., Gutiérrez-Rojas, M., Favela-Torres, E. and Feussner, I. (2018). Jasmonic acid biosynthesis by microorganisms: derivatives, first evidences on biochemical pathways and culture conditions for production. *PeerJ* 9:e10873. 10.7287/peerj.preprints.26655v1.

EPA (2020). List of approved biopesticide active ingredients by August 2020, United States Environmental Protection Agency Database. Available at: https://www.epa.gov/ingredients-used-pesticide-products/biopesticide-active-ingredients, accessed on 20 September 2020.

Eshel, D., Regev, R., Orenstein, J., Droby, S. and Gan-Mor, S. (2009). Combining physical, chemical and biological methods for synergistic control of postharvest diseases: a case study of black root rot of carrot. *Postharvest Biology and Technology* 54(1, October):48–52. 10.1016/j.postharvbio.2009.04.011.

Estell, D. (2006). Adapting industry practices for the rapid, large-scale manufacture of pharmaceutical proteins. *Bridge* 36:39–44.

European Food Safety Authority (EFSA) (2015). Conclusion on the peer review of the pesticide risk assessment of the active substance *Pseudozyma* flocculosa strain ATTC 64874. *EFSA Journal* 13(10):4250. 10.2903/j.efsa.2015.4250.

European Food Safety Authority (EFSA) (2017a). Peer review of the pesticide risk assessment of the active substance Pseudomonas chlororaphis strain MA 342. *EFSA Journal* 15(1):e04668. 10.2903/j.efsa.2017.4668.

European Food Safety Authority (EFSA) (2017b). Peer review of the pesticide risk assessment of the active substance *Ampelomyces quisqualis* strain AQ10. *EFSA Journal* 15(12):e05078. 10.2903/j.efsa.2017.5078.

European Food Safety Authority (EFSA) (2020). Public consultation on the active substance *Aspergillus flavus*, strain MUCL54911, Rapporteur Assessment Report (Italy), published for consultation on 14 October 2020.

European Food Safety Authority (EFSA) (2021). Peer review of the pesticide risk assessment of the active substance *Bacillus amyloliquefaciens* strain QST 713 (formerly Bacillus subtilis strain QST 713). *EFSA Journal* 19(1):e06381. 10.2903/j.efsa.2021.6381.

Eyal, J., Backer, C. P., Reeder, J. D., DeVane, W. E. and Lumsden, R. D. (1997). Large scale production of Chlamydospores of *Gliocladium virens* strain GL-21 in submerged culture. *Journal of Industrial Microbiology and Biotechnology* 19(3):163–168.

Farbood, M. I., Blocker, R. W., McLean, L. B., Sprecker, M. A., McLean, M. P., Kossiakoff, N., Kim, A. Y. and Hagedorn, M. (2001). Bioprocess for the high-yield production of food flavor-acceptable jasmonic acid and methyl jasmonate, United States Patent 6333180 (International Flavors & 654 Fragrances Inc, IC:C12P 017/02, New York USA).

Ferraz, P., Cassio, F. and Lucas, C. (2019). Potential of yeast as biocontrol agent of the phytopathogen causing cacao witches' broom disease: is microbial warfare a solution?. *Frontiers in Microbiology* 10. 10.3389/fmicb.2019.01766.

Ferrigo, D., Raiola, A., Piccolo, R., Scopel, C. and Causin, R. (2014). *Trichoderma harzanium* T22 induces in maize systemic resistance against *Fusarium verticilloides*. *Journal of Plant Pathology* 96:133–142. 10.4454/JPP. V9611.038.

Fey, P., Kowal, A. S., Gaudet, P., Pilcher, K. E. and Chisholm, R. L. (2007). Protocols for growth and development of *Dictyostelium discoideum*. *Nature Protocols* 2(6):1307–1316. 10.1038/nprot.2007.178. PMID: 17545967.

Fondevilla, S. and Rubiales, D. (2012). Powdery mildew control in pea: a review. *Agronomy for Sustainable Development* 32(2):401–409. 10.1007/s13593-011-0033-1.

Forchetti, G., Masciarelli, O., Alemano, S., Alvarez, D. and Abdala, G. (2007). Endophytic bacteria in sunflowers (*Helianthus annus L.*): isolation, characterization, and production of Jasmonates and abscisic acid in culture medium. *Applied Microbiology and Biotechnology* 76(5):1145–1152.

Frampton, R. A., Pitman, A. R. and Fineran, P. C. (2012). Advances in bacteriophage-mediated control of plant pathogens. *International Journal of Microbiology* 2012:326452. 10.1155/2012/326452.

Fravel, D. R., Connick, W. J. and Lewis, J. A. (1998). Formulation of microorganisms to control plant diseases. In: Burges, H. D. (Ed) *Formulation of Microbial Biopesticides*. Springer, Dordrecht. 10.1007/978-94-011-4926-6_5.

Freimoser, F. M., Reeda-Mejia, M. P., Tillocca, B. and Mighell, Q. (2019). Biocontrol yeasts: mechanisms and applications. *World Journal of Microbiology and Biotechnology* 35(10):154 10.1007/s11274-019-2728-4.

Fuga, C. A. G., Lopes, E. A., Vieira, B. S. and da Cunha, W. V. (2016). Efficiency and compatibility of Trichoderma spp. and Bacillus spp. isolates on the inhibition of Sclerotium cepivorum. *Científica* 44(4):526. 10.15361/1984-5529.2016v44n4p526-531.

Ganeshan, G. and Kumar, M. A. (2005). *Pseudomonas fluorescens*, a potential bacterial antagonist to control plant diseases. *Journal of Plant Interactions* 1(3):123–134.

Garcia, R., Latz, S., Romero, J., Higuera, C., Garcia, K. and Bastias, R. (2019). Bacteriophage production models: an overview. *Frontiers in Microbiology* 4(10):1187. 10.3389/fmicb.2019.01187.

Gasic, S. and Brankica, T. (2013). Biopesticide formulations, possibility of application and future trends. *Pesticidi i Fitomedicina* 28(2):97–102. 10.2298/PIF1302097G.

Gerbore, J., Benhamou, N., Vallance, J., Le Floch, G., Grizard, D., Regnault-Roger, C. and Rey, P. (2014). Biological control of plant pathogens: advantages and limitations seen through the case study of *Pythium oligandrum*. *Environmental Science and Pollution Research International* 21(7):4847–4860. 10.1007/s11356-013-1807-6.

Gerlagh, M., Goossen-van de Geijn, H. M., Fokkema, N. J. and Vereijken, P. F. (1999). Long-term biosanitation by application of *Coniothyrium minitans* on Sclerotinia sclerotiorum-infected crops. *Phytopathology* 89(2):141–147. 10.1094/PHYTO.1999.89.2.141.

Ghasemi, P. A., Sajjadi, S. E. and Parang, K. (2014). A review (research and patents) on 678 jasmonic acid and its derivatives. *Archiv der Pharmazie* 347:229–239. 10.1002/ardp.201300287.

Goettel, M. S., Leger, R. J. S., Bhairi, S., Jung, M. K., Oakley, B. R., Roberts, D. W. and Staples, R. C. (1990). Pathogenicity and growth of *Metarhizium anisopliae* stably transformed to benomyl resistance. *Current Genetics* 17(2):129–132. 10.1007/BF00312857.

Gonçalves, S., Teixeira, A., Abade, J., de Medeiros, L. N., Kurtenbach, E. and Santos, N. C. (2012). Evaluation of the membrane liid selectivity of the pea defenses Psd1. *Biochimica et Biophysica Acta* 1818(5):1420–1426.

Gore-Lloyd, D., Sumann, I., Brachmann, A. O., Schneeberger, K., Ortiz-Merino, R. A., Moreno-Beltrán, M., Schläfli, M., Kirner, P., Santos Kron, A., Rueda-Mejia, M. P., Somerville, V., Wolfe, K. H., Piel, J., Ahrens, C. H., Henk, D. and Freimoser, F. M. (2019). Snf2 controls pulcherriminic acid biosynthesis and antifungal activity of the biocontrol yeast *Metschnikowia pulcherrima*. *Molecular Microbiology* 112(1):317–332. 10.1111/mmi.14272.

Grady, E. N., MacDonald, J., Liu, L., Richman, A. and Yuan, Z. C. (2016). Current knowledge and perspectives of *Paenibacillus*: a review. *Microbial Cell Factories* 15(1):203. 10.1186/s12934-016-0603-7.

Greub, G. and Raoult, D. (2004). Microorganisms resistant to free-living amoebae. *Clinical Microbiology Reviews* 17(2):413–433. 10.1128/CMR.17.2.413-433.2004.

Grose, H. J. (2018). *Fire Quencher an Organic Treatment for Fire Blight IR4, Rutgers Edu*. Brigham Young University. https://ir4.cals.ncsu.edu/biopesticides/workshoppresentations/IR-4FireBlightPresentationSept8.pdf (accessed October 2020).

Hagag, W. M., Saleh, M. A. E., Mostafa, I. and Adel, N. (2013). Mass production, fermentation, formulation of *Pseudomonas putida* for the controlling of die back and Phomopsis disease on grapevine. *Advances in Bioscience and Biotechnology* 4(6):741–750.

Hammer, E. P., Janice, C. J., Michael, G. K. and Uknes, S. J. (2017). Modified biological control agents and their uses. US patent US 9,635,861 B2.

Han, S. H., Lee, S. J., Moon, J. H., Park, K. H., Yang, K. Y., Cho, B. H., Yong Kim, K. Y., Kim, Y. W., Lee, M. C., Anderson, A. J. and Kim, Y. C. (2006). GacS-dependent production of 2R, 3R-butanediol by *Pseudomonas chlororaphis* O6 is a major determinant for eliciting systemic resistance against *Erwinia carotovora* but not against *Pseudomonas syringae* pv. *tabaci* in tobacco. *Molecular Plant-Microbe Interactions* 19(8):924–930. 10.1094/MPMI-19-0924.

Handelsman, J. and Stabb, E. V. (1996). Biocontrol of plant pathogen. *Plant Cell* 8(1855):1869. 10.1105/tpc.8.10.1855.

Hanif, A., Zhang, F., Li, P., Li, C., Xu, Y., Zubair, M., Zhang, M., Jia, D., Zhao, X., Liang, J., Majid, T., Yan, J., Farzand, A., Wu, H. , Gu, Q. and Gao, X. (2019). Fengycin produced by *Bacillus amyloliquefaciens* FZB42 inhibits *Fusarium graminearum* growth and mycotoxins biosynthesis. *Toxins* 11(5):295. 10.3390/toxins11050295.

Harman, G. E., Jin, X., Stasz, T. E., Peruzzotti, G., Leopold, A. C. and Taylor, A. G. (1991), Production of conidial biomass of Trichoderma harzianum for biological control. *Biological Contro* 1(1): 23–28.

Hasni, I., Chelkha, N., Baptiste, E., Mameri, M. R., Lachuer, J., Plasson, F., Colson, P. and La Scola, B. (2019). Investigation of potential pathogenicity of *Willaertia magna* by investigating the transfer of bacteria pathogenicity genes into its genome. *Scientific Reports* 9(1):18318. 10.1038/s41598-019-54580-6.

Hasni, I., Jarry, A., Quelard, B., Carlino, A., Eberst, J. B., Abbe, O. and Demanèche, S. (2020). Intracellular behaviour of three *Legionella pneumophila* Strains within three amoeba strains, including *Willaertia magna* C2C maky. *Pathogens* 9(2):105.

Hause, B., Mrosk, C., Isayenkov, S. and Strack, D. (2007). Jasmonates in arbuscular mycorrhizal interactions. *Phytochemistry* 68(1):101–110.

Hayles, J., Johnson, I., Worthley, C. and Losic, D. (2017). Nanoparticles: a review of current research and prospective, 193–225. In: *New Pesticides and Soil Sensors*. Elsevier, Netherlands.

Heidel, A. J., Lawal, H. M., Felder, M., Schilde, C., Helps, N. R., Tunggal, B., Rivero, F., John, U., Schleicher, M., Eichinger, L., Platzer, M., Noegel, A. A., Schaap, P. and Glöckner, G. (2011). Phylogeny-wide analysis of social amoeba genomes highlights ancient origins for complex intercellular communication. *Genome Research* 21(11):1882–1891. 10.1101/gr.121137.111.

Herman, G. E., Jin, X., Staz, T. E., Peruzzott, G., Leopold, A. C. and Taylor, A. G. (1991). Production of conidial biomass of *Trichoderma harzianum* for biological control. *Biological Control* 1(1):23–28, 10.1016/1049-9644(91)90097-J.

Hernández-Castillo, F. D., Castillo-Reyes, F., Tucuch-Pérez, M. A. and Arredondo-Valdes, R. (2020). Biological efficacy of Trichoderma spp. and Bacillus spp. in the management of plant diseases. Available at: https://www.intechopen.com/online. 10.5772/intechopen.9104.

Hershkovitz, P., et al. (2012). Global changes in gene expression of grapefruit peel tissue in response to the yeast biocontrol agent *Metschnikowia fructicola*. *Molecular Plant Pathology* 13:338–349. 10.1111/j.1364-3703.2011.00750.x.

Heydari, A. and Pessarakli, M. (2010). A review on biological control of fungal plant pathogens using microbial antagonists. *Journal of Biological Sciences* 10(4):273–290.

Hickoop, J. C., Grootenhuis, P. D., Blaauw, M., Veldema, J. S., Van Haastert, P. J. and Linskens, M. H. (2001). Expression of a bioactive, single-chain choriogonadotropin in *Dictyostelium discoideum*. *European Journal of Biochemistry* 256(2):359–363. 10.1046/j.1432-1327.1998.2560359.x. accessed 25 December 2001.

Hilber-Bodmer, M., Schmid, M., Ahrens, C. H. and Freimoser, F. M. (2017). Competition assays and physiological experiments of soil and phyllo-sphere yeasts identify *Candida subhashini* as novel antagonist of filamentous fungi. *BMC Microbiology* 17(1):4. 10.1186/s12866-016-0908-z.

Hinton, B. C. W. (2001). Endophytic and biological control potential of Bacillus mojavensis and related species. *Biological Control* 23(3):274–284. 10.1006/bcon.2001.

Hongzhang, C. (2013). *Modern Solid-State Fermentation: Theory into Practice*. Springer, Dordrecht. 10.1007/978-94-007-6043-1.

Howe, G. A. and Jander, G. (2008). Plant immunity to insect herbivores. *Annual Review of Plant Biology* 59:41–66.

Huang, R., Li, G. Q., Zhang, J., Yang, L., Che, H. J., Jiang, D. H. and Huang, H. C. (2011). Control of postharvest Botrytis fruit rot of strawberry by volatile organic compounds of Candida intermedia. *Phytopathology* 101(7):859–869. 10.1094/PHYTO-09-10-0255.

Huws, S. A., McBain, A. J. and Gilbert, P. (2004). Protozoan grazing and its impact upon population dynamics in biofilm communities. *Journal of Applied Microbiology* 98(1):238–244. 10.1111/j.1365-2672.2004.02449.x.

Jackson, M. A. and Schisler, D. A. (1995). Liquid culture production of microsclerotia of *Colletotrichum truncatum* for use as bioherbicidal propagules. *Mycological Research* 99(7, July):879–884.

Jackson, M. A. and Jaronski, S. T. (2009). Production of microsclerotia of the fungal entomopathogen *Metarhizium anisophliae* and their potential for use as a biocontrol agent for soil inhabiting insects. *Mycological Research* 113(8):842–850.

Jackson, M. A. and Payne, A. R. (2016). Liquid culture production of fungal microsclerotia. *Methods in Molecular Biology* 1477:71–83. 10.1007/978-1-4939-6367-6_7.

Jacques, P. (2011). Surfactin and other lipopeptides from Bacillus spp, 57–91. In: Soberon-Cavez, G. (Ed) *Biosurfactants Microbiology Monographs* (vol. 20). Springer. 10.1007/978-3-642-14490-5_3.

Janahiraman, V., Anandham, R., Kwon, S. W., Sundaram, S., Karthik Pandi, V., Krishnamoorthy, R., Kim, K., Samaddar, S. and Sa, T. (07 November 2016). Control of wilt and rot pathogens of tomato by antagonistic pink pigmented facultative methylotrophic delftia lacustris and bacillus spp. *Frontiers in Plant Science* 7:1626 10.3389/fpls.2016.01626.

Janisiewicz, V. W. and Korsten, L. (2002). Biological control of postharvest diseases of fruits. *Annu. Rev. Phytology* 40:411–441 10.1146/annurev.phyt.40120401.130158.

Jin, X., Taylor, A. G. and Harman, G. E. (1996). Development of media and automated liquid fermentation methods to produce desiccation-tolerant propagules of *Trichoderma harzianum*. *Biological Control* 7(3, December):267–274. 10.1006/bcon.1996.0093.

Jin, X. and Custis, D. (2011). Microencapsulating aerial conidia of trichoderma harzianum through spray drying at elevated temperatures *Biological Control* 56(2), 202–208. 10.1016/j.biovontrol.2010.11.008.

Jiang, C. H., Liao, M. J., Wang, H. K., Zheng, M. Z., Xu, J. J. and Guo, J. H. (2018). *Bacillus velezensis*, a potential and efficient Biocontrol agent in control of pepper gray mold caused by *Botrytis cinera*. *Biological Control* 126:147–157. 10.1016/j.biocontrol.2018.07.017, 126147.

John, S., Wicks, T. J., Hunt, J. S., Lorimer, M. F., Oakey, H. and Scott, E. S. (2005). Protection of grapevine pruning wounds from infection by *Eutypa lata* using *Trichoderma harzianum* and *Fusarium lateritium*. *Australasian Plant Pathology* 34(4):569–575.

Johnsson, L., Hökeberg, M. and Gerhardson, B. (1998). Performance of the *Pseudomonas chlororaphis* biocontrol agent MA 342 against cereal seed-borne diseases in field experiments. *European Journal of Plant Pathology* 104(7):701–711. 10.1023/A:1008632102747.

Jurač, K., Nabergoj, D. and Podgornik, A. (2019). Bacteriophage production processes. *Applied Microbiology and Biotechnology* 103(2):685–694.

Kaneka (2020). Company brochure biosurfactant Kaneka surfactin. Available at: https://www.kaneka.co.jp/en/business/qualityoflife/nbd_002.html, accessed 28 September 2020.

Keswani Cheta, K. B. and Vivek, S. (2016). Formulation technology of biocontrol agents: present status and future prospects. *Bioformulations: For Sustainable Agriculture*.

Kiewnick, S. (2001). Advanced fermentation and formulation technologies for fungal antagonists. *IOBC/WPRS Bulletin* 24(1):77–79.

Kiewnick, S. (2007). Review Practicalities of developing and registering microbial biological control agents. *CAB Reviews* 013:2. 10.1079/PAVSNNR20072013.

Kloepper, J. W., Ryu, C.-M. and Zhang, S. (2004). Induced systemic resistance and promotion of plant growth by Bacillus spp.. *Phytopathology* 94(11):1259–1266.

Kobori, N. N., Mascarin, G. M., Jackson, M. A. and Schisler, D. A. (2014). Liquid culture production of microsclerotia and submerged conidia by *Trichoderma harzianum* active against damping-off disease caused by *Rhizoctonia solani*. *Fungal Biology* 119(4):179–190. 10.1016/j.funbio.2014.12.005.

Köhl, J., Postma, J., Nicot, P., Ruocco, M. and Blum, B. (2011). Stepwise screening of microorganisms for commercial use in biological control of plant-pathogenic fungi and bacteria. *Biological Control* 57(1, April):1–12. 10.1016/j.biocontrol.2010.12.004.

Konstantinidou-Doltsinis, S., Markellou, E., Kasselaki, A.-M., Siranidou, E., Kalamarakis, A., Tzembelikou, K., Schmitt, A., Koumakis, C. and Malathrakis, N. (2007). Control of powdery mildew of grape in Greece using Sporodex® L and Milsana®. *Journal of Plant Diseases and Protection* 114(6):256–262. 10.1007/BF03356226.

Kowalska, J., Drożdżyński, D., Remlein-Starosta, D., Sas-Paszt, L. and Malusá, E. (2012). Use of *Cryptococcus albidus* for controlling grey mold in the production and storage of organically grown strawberries. *Journal of Plant Diseases and Protection* 119(5–6):174–178.

Kumar, A., Verma, H., Singh, V. K., Singh, P. P., Singh, S. K., Ansari, W. A., Yadav, A., Singh, P. K. and Pandey, K. D. (2017). Role of *Pseudomonas* spp. in sustainable agriculture and disease management. 195–215. In: Meena, V., Mishra, P., Bisht, J. and Pattanayak, A. (eds) *Agriculturally Important Microbes for Sustainable Agriculture*. Springer, Singapore. https://doi.org/10.1007/978-981-10-5343-6_7.

Kumar, S. (2013). Trichoderma: a biological weapon for managing plant diseases and promoting sustainability. *International Journal of Agricultural Sciences and Veterinary Medicine* 1(3):1–16.

Kumar, L. R., Ndao, A., Valéro, J. and Tyagi, R. D. (2019). Production of *Bacillus thuringiensis* based biopesticide formulation using starch industry wastewater (SIW) as substrate: a techno-economic evaluation. *Bioresource Technology* 294:122144. 10.1016/j.biortech.2019.122144.

Kumar, V., Ahluwalia, V., Saran, S., Kumar, J., Patel, A. K. and Singhania, R. R. (2020). Recent developments on solid-state fermentation for production of microbial secondary metabolites: challenges and solutions. *Bioresource Technology* 2020:124566, 10.1016/j.biortech.2020.124566.

Kunz, S., Schmitt, A. and Haug, P. (2008). Field testing of strategies for fire blight control in organic fruit growing. *Acta Horticulturae* 896:431–436. 10.17660. Acta. Hortic (2011) 896.62.

Kwok, R. (2010). Five hard truths for synthetic biology. *Nature* 463(7279):288–290. 10.1038/463288a.

Laredo-Alcalá, E., Michelena-Álvarez, G., Iliná, A., Borjas-Ramos, J., Hernández-Castillo, F. and Martínez-Hernández, J. (2016). Uso de *Botryodiplodia theobromae* como productor de ácido 750 jasmónico en dos sistemas de fermentación. *Revista Mexicana de Micología* 44:1–9.

László, M. (2012). A copper resistant, fengycin-producing bacillus mojavensis strain for for controlling vegetable pathogens, its use and compositions containing it. WO EP US ES HU EP2753181A2.

Leathers, T. D., Saunders, L. P., Bowman, M. J., Price, N. P. J., Bischof, K. M., Rich, J. O., Skory, C. D. and Nunally, M. S. (2020). Inhibition of Erwinia amylovora by *Bacillus nakamurai*. *Current Microbiology* 77(5):875–881 10.1007/s00284-019-01845-y.

Lee, G., Lee, S. H., Kim, K. M. and Ryu, C.-M. (2017). Foliar application of the leaf-colonizing yeast *Pseudozyma churashimaensis* elicits systemic defense of pepper against bacterial and viral pathogens. *Scientific Reports* 7:39432. 10.1038/srep39432.

Leggett, M., Leland, J., Kellear, K. and Epp, B. (2011). Formulation of Microbial agents– an industrial perspective. *Canadian Journal of Plant Pathology* 33(2):101–107. 10.1080/07060661.2011.563050.

Levin, B. R., Moineau, S., Bushman, M. and Barrangou, R. (2013). The population and evolutionary dynamics of phage and bacteria with CRISPR-mediated immunity. *PLoS Genetics* 9(3):e1003312.

Levy, S. B. (1978). Emergences of antibiotic-resistant bacteria in the intestinal flora of farm inhabitants. *Journal of Infectious Diseases* 137(5):688–690.

Lewis, J. A. and Papavizas, G. C. (1983). Production of chlamydospores and conidia by *Trichoderma* spp. in liquid and solid growth media. *Soil Biology and Biochemistry* 15(3):351–357.

Li, B. Q. and Tian, S. P. (2006). Effects of trehalose on stress tolerance and biocontrol efficacy of *Cryptococcus laurentii*. *Journal of Applied Microbiology* 100(4):854–861. 10.1111/j.1365-2672.2006.02852.x.

Li, G. Q., Huang, H. C., Miao, H. J., Erickson, R. S., Jiang, D. H. and Xiao, Y. N. (2006). Biological control of Sclerotinia diseases of rapeseed by aerial applications of the mycoparasite *Coniothyrium minitans*. *European Journal of Plant Pathology* 114(4):345–355. 10.1007/s10658-005-2232-6.

Li, B. Q., Zhou, Z. W. and Tian, S. P. (2008). Combined effects of endo- and exogenous trehalose on stress tolerance and biocontrol efficacy of two antagonistic yeasts. *Biological Control* 46(2):187–193. 10.1016/j.biocontrol.2008.04.011.

Li, Y. Q., Song, K., Li, Y.-C. and Chen, J. (2016). Statistical culture-based strategies to enhance chlamydospore production by *Trichoderma harzianum* SH2303 in liquid fermentation. *Journal of Zhejiang University. Science. B* 17(8):619–627. 10.1631/jzus.B1500226.

Li, N., Han, X., Feng, D., Yuan, D. and Huang, L.-J. (2019). Signaling crosstalk between salicylic acid and ethylene/jasmonate in plant defense: do we understand what they are whispering?. *International Journal of Molecular Sciences* 20(3):671. 10.3390/ijms20030671.

Lindow, S. E., McGourthy, G. and Elkins, R. (1996). Interactions of antibiotic with *Pseudomonas fluoresces* A506 in the control of the blight and frost injury of pear. *Phytopathology* 86:618–623.

Lisansky, S. G. and Coombs, J. (1993). *The Worldwide Directory of Agro Technoloy CPL*. Scientific Publishing, Newbury.

Liu, J., Tian, S. P., Li, B. Q. and Qin, G. Z. (2009). Enhancing viability of two biocontrol yeasts in liquid formulation by applying sugar protectant combined with antioxidant. *BioControl* 54(6):817–824. 10.1007/s10526-009-9229-6.

Lu, Y., Beshay, U., Friehs, K. and Flaschel, E. (2004). Mass production of *Dictyostelium discoideum* in homogenous and heterogeneous cultivation systems. *Process Biochemistry* 39(12):1859–1870. 10.1016/j.procbio.2003.09.012.

Lumsden, R. D., Lewis, J. A. and Fravel, D. R. (1995). Formulation and delivery of biocontrol agents for use against soilborne plant pathogens. *ACS Symposium Series* Vol. 595. Biorational Pest Control Agents Chapter 11, 166–182. 10.1021/bk-1995-0595.ch011.

Lutz, M. P., Wenger, S., Maurhofer, M., Défago, G. and Duffy, B. (2004). Signaling between bacterial and fungal biocontrol agents in a strain mixture. *FEMS Microbiology Ecology* 48(3):447–455, 10.1016/j.femsec.2004.03.002.

Malik, D., Sokolov, I., Vinner, G., Mancuso, F., Cinquerrui, S., Vladisavljevic, G., et al. (2017). Formulation, stabilization and encapsulation of bacteriophage for phage therapy. Loughborough University. Journal Contribution. Available at: https://hdl.handle.net/2134/25197.

Mameri, R. M., Bodennec, J., Bezin, L. and Demanèche, S. (2020). Mitigation of expression of virulence genes in *Legionella pneumophila*, internalized in the free-living amoeba *Willaertia magna* C2C maky. *Pathogens* 9(6):447.

Mancera-López, M. E., Izquierdo-Estévez, W. F., Escalante-Sánchez, A., Ibarra, J. E. and Barrera-Cortés, J. (2018). Encapsulation of *Trichoderma harzianum* conidia as a method of conidia preservation at room temperature and propagation in submerged culture. *Biocontrol Science and Technology* 29(2):107–130, 10.1080/09583157.2018.1535053.

Mancuso, F. S., Shi, J. and Malik, D. J. (2018). High throughput manufacturing of bacteriophage using continues stirred tank bioreactor connected in serious to ensure optimum host bacteria physiology for phage production. *Viruses* 10(10):537. 10.3390/v10100537.

Manjula, K. and Podile, A. R. (2001). Chitin-supplemented formulations improve biocontrol and plant growth promoting efficiency of Bacillus subtilis AF 1. *Canadian Journal of Microbiology* 47(7):618–625.

Mascarin, G. M., Kobori, N. N., de Jesus Vital, R. C., Jackson, M. A. and Quintela, E. D. (2014). Production of microsclerotia by Brazilian strains of *Metarhizium* spp. using submerged liquid culture fermentation. *World Journal of Microbiology and Biotechnology (Formerly MIRCEN Journal of Applied Microbiology and Biotechnology)* 30(5):1583–1590. 10.1007/s11274-013-1581-0.

Massart, S. and Jijakli, M. H. (2014). Pichia anomala and candida oleophila in biocontrol of postharvest diseases of fruits: 20 years of fundamental and practical research. In: Prusky, D. and Gullino, M. (Eds) *Post-harvest Pathology: Plant Pathology in the 21st Century* (Contributions to the 9th International Congress), (vol. 7). Springer, Cham. https://doi.org/10.1007/978-3-319-07701-7_10.

Marco, G. M. and Stail, R. E. (1983). Control of bacterial spot of pepper initiated by strains of Xanthomonas campestris *pv, vesicatoria* that differ in sensitive to copper. *Plant Disease* 67:779–781.

Marks, D. (2012). Agricultural compositions containing a jasmonic acid or dihydrojasmonic acid salt. Plant Impact PLC Manchester, European patent application EP 1 740 045 B1.

Martínez-Medina, A., Fernández, I., Sánchez-Guzmán, M. J., Jung, S. C., Pascual, J. A. and Pozo, M. J. (2013). Deciphering the hormonal signaling network behind the systemic resistance induced by *Trichoderma harzianum* in tomato. *Frontiers in Plant Science* 4:206.

Maruyama, C. R., Bilesky-José, N., de Lima, R. and Fraceto, L. F. (2020). Encapsulation of *Trichoderma harzianum* preserves enzymatic activity and enhances the potential for biological control. *Frontiers in Bioengineering and Biotechnology* 8:225. 10.3389/fbioe.2020.00225.

Mauro, A., Garcia-Cela, E., Pietri, A., Cotty, P. J. and Battilani, P. (2018). Biological control products for aflatoxin prevention in Italy: commercial field evaluation of atoxigenic *Aspergillus flavus* active ingredients. *Toxins* 10(1):30. 10.3390/toxins10010030.

Mbili, N. C. (2012). Evaluation of integrated control of postharvest grey mold and blue mold of pome fruit using yeast, potassium silicate and hot water treatments. Dissertation for master degree. University of KwaZulu-Natal. Available at: ukzn-dspace.ukzn.ac.za.

Melin, P., Håkansson, S., Eberhard, T. H. and Schnürer, J. (2006). Survival of the biocontrol yeast *Pichia anomala* after long-term storage in liquid formulations at different temperatures, assessed by flow cytometry. *Journal of Applied Microbiology* 100(2):264–271. 10.1111/j.1365-2672.2005.02778.x.

Melin, P., Håkansson, S. and Schnürer, J. (2007). Optimization and comparison of liquid and dry formulations of the biocontrol yeast *Pichia anomala* J121. *Applied Microbiology and Biotechnology* 73(5):1008–1016. 10.1007/s00253-006-0552-x.

Mercier, J., Jiménez-Santamaría, J. I. and Tamez-Guerra, P. (2007). Development of the volatile-producing fungus *Muscodor albus*. Worapong Strobel and Hess, as a novel antimicrobial biofumigant. *Revista Mexicana de Fitopatología* 25(2):173–9.

Miersch, O., Günther, T., Fritsche, W. and Sembdner, G. (1993). Jasmonates from different fungal species. *Natural Product Letters* 2(4):293–299. 10.1080/10575639308043825.

Miersch, O., Bohlmann, H. and Wasternack, C. (1999a). Jasmonates and related compounds from Fusarium oxysporum. *Phytochemistry* 50(4):517–523.

Miersch, O., Regvar, M. and Wasternack, C. (1999b). Metabolism of jasmonic acid in *Pisolithus tinctorius* cultures. *Phyton–Annales Rei Botanicae* 39:243–247.

Miao, Y., Tenor, J. L., Toffaletti, D. L., Washington, E. J., Liu, J., Shadrick, W. R., Schumacher, M. A., Lee, R. E., Perfect, J. R. and Brennan, R. G. (2016). Structures of trehalose-6-phosphate phosphatase from pathogenic fungi reveal the mechanisms of substrate recognition and catalysis. *Proceedings of the National Academy of Sciences of the United States of America* 113(26):7148–7153. 10.1073/pnas.1601774113.

Mnif, I. and Ghribi, D. (2015). *Review Lipopeptides Biosurfactants: Mean Classes and New Insights for Industrial, Biomedical and Environmental Application*. Wiley. 10.1002/bip.22630.

Moazami, N. (2002). Biopesticide production: encyclopedia of life support systems, Biotechnology vol VI. Availble at: https://www.eolss.net/Sample-Chapters/C17/E6-58-05-08.pdf, accessed 24 September 2020.

Mondello, V., Larignon, P., Armengol, J., Kortekamp, A., Vaczy, K., Prezman, F., Serrano, E., Rego, C., Mugnai, I. and Fontaine, F. (2018). Management of grapevine trunk diseases: knowledge transfer current strategies and innovative strategies adopted in Europe. *Phytopathology, Mediterr.* 57:369–383. 10.14601/Phytopathol.

Montesinos, E. (2003). Development, registration and commercialization of microbial pesticides for plant protection. *International Microbiology* 6(4):245–252. 10.1007/s10123-003-0144-x. Epub 2003 Sep 3.

Moral, J., Garcia-Lopez, M. T., Camiletti, B. X., Jaime, R., Michailides, T. J., Bandyopadhyay, R. and Ortega-Beltran, A. (2020). Present status and perspective on the future use of aflatoxin biocontrol products. *Agronomy* 10(4):491. 10.3390/agronomy10040491.

Mounir, R., Durieux, A., Bodo, E., Allard, C., Simon, J. P., Achbani, E. H., El-Jaafari, S., Douira, A. and Jijakli, M. H. (2007). Production, formulation and antagonistic activity of the biocontrol like-yeast *Aureobasidium pullulans* against *Penicillium expansum*. *Biotechnology Letters* 29(4):553–559. 10.1007/s10529-006-9269-2.

Mutti, M. and Corsini, L. (2019). Robust approaches for the production of active ingredient and drug product for human phage therapy. *Frontiers in Microbiology* 10:2289. 10.3389/fmicb.2019.02289.

Nam, H. S., Anderson, A. J. and Kim, Y. C. (2018). Biocontrol efficacy of formulated *Pseudomonas chlororaphis* O6 against plant diseases and root-knot nematodes. *Plant Pathology Journal* 34(3):241-249.

Neff, R. J. (1957). Purification, Axenic cultivation and description of a soil Amoeba, *Acanthamocha* sp. *Journal of Protozoology* 4:176–182.

Neher, O. T., Johnston, M. R., Zidack, N. K. and Jacobsen, B. J. (2009). Evaluation of *Bacillus mycoides* isolate BmJ and *B. mojavensis* isolate 203-7 for the control of anthracnose of cucurbits caused by *Glomerella cingulata* var. *orbiculare*. *Biological Control* 48(2, February):140–146.

Okman, J. I., Kent, D. J., Wiedmann, M. and Denes, T. (2016). Temperature significantly affects the plaquing and adsorption efficiencies of listeria phages. *Frontiers in Microbiology* 7:631.

Omer, A. D., Thaler, J. S., Granett, J. and Karban, R. (2000). Jasmonic acid induced resistance in grapevines to a root and leaf feeder. *Journal of Economic Entomology* 93(3):840–845. 10.1603/0022-0493-93.3.840.

Ongena, M. and Jacques, P. (2007). Bacillus lipopeptides; versatile weapons for plant diseases biocontrol. *Trends in Microbiology* 16(3):115–125. 10.1016/j.tim.2007.12.009.

Pagadoy, M., Paypoux, F. and Wallach, J. (2005). Solid-phase synthesis of surfactin, a powerful biosurfactant produced by *Bacillus subtills* and of four analogues. *International Journal of Peptide Research and Therapeutics* 11(3):195–202. 10.1007/s10989-005-6790-4.

Papavizas, G. C. (1985). Trichoderma and Gliocladium: biology, ecology, and potential for biocontrol. *Annual Review of Phytopathology* 23(1):23–54.

Patiño-Vera, M., Jiménez, B., Balderas, K., Ortiz, M., Allende, R., Carrillo, A. and Galindo, E. (2005). Pilot-scale production and liquid formulation of *Rhodotorula minuta*, a potential biocontrol agent of mango anthracnose. *Journal of Applied Microbiology* 99(3):540–550. 10.1111/j.1365-2672.2005.02646.x.

Péchy-Tarr, M., Bruck, D. J., Maurhofer, M., Fischer, E. , Vogne, C., Henkels, M. D., Donahuev, K. M., Grunder, J., Loper, J. E. and Keel, C. (2008). Molecular analysis of a novel gene cluster encoding an insect toxin in plant-associated strains of *Pseudomonas fluorescens*. *Environmental Microbiology* 10(9, September):2368–2386.

Pieterse, C. M. and van Loon, L. C. (2004). NPR1: the spider in the web of induced resistance signaling pathways. *Current Opinion in Plant Biology* 7(4):456–464. 10.1016/j.pbi.2004.05.006.

Pieterse, C. M., Leon-Reyes, A., van der Ent, S. and van Wees, S. C. M. (2009). Networking by small molecule hormones in plant immunity. *Nature Chemical Biology* 5(5):308–316. 10.1038/nchembio.164.

Pieterese, M. J., Zamioudis, C., Berendsen, R. L., Weller, D. M., van Wees, S. C. M. and Bakker, P. A. H. M. (2014). Induced systemic resistance by Beneficial microbes. *Annual Review of Phytopathology* 52:347–375. doi: 10.1146/annurev-phyto-082712-102340.

Pinho, E., Grootveld, M., Soares, G. and Henriques, M. (2014). Cyclodextrins as encapsulation agents for plant bioactive compounds. *Carbohydrate Polymers* 101:121–135. 10.1016/j.carbpol.2013.08.078.

Pirnay, J. P., Blasdel, B. G., Bretaudeau, L., Buckling, A., Chanishvili, N., Clark, J. R., Corte-Real, S., Debarbieux, L., Dublanchet, A., De Vos, D., Gabard, J., Garcia, M., Goderdzishvili, M., Górski, A., Hardcastle, J., Huys, I., Kutter, E., Lavigne, R., Merabishvili, M., Olchawa, E., Parikka, K. J., Patey, O., Pouilot, F., Resch, G., Rohde, C., Scheres, J., Skurnik, M., Vaneechoutte, M., Van Parys, L., Verbeken, G., Zizi, M. and Van den Eede, G. (2015). Quality and safety requirements for sustainable phage therapy products. *Pharmaceutical Research* 32(7):2173–2179. 10.1007/s11095-014-1617-7.

Pirzada, T., de Farias, B. V., Mathew, R., Guenther, R. H., Byrd, M. V., Sit, T. L., Pal, L., Opperman, C. H. and Khan, S. A. (2020). Recent advances in biodegradable matrices for active ingredient release in crop protection: towards attaining sustainability in agriculture. *Current Opinion in Colloid and Interface Science* 48:121–136. 10.1016/j.cocis.2020.05.002.

Pisoschi, A. M., Pop, A., Cimpeanu, C., Turcus, V., Predoi, G. and Lordache, F. (2020). Nanoencapsulation techniques for compounds and products with antioxidant and antimicrobial activity–a critical view. *European Journal of Medicinal Chemistry* 157(5):1326–1345. 10.1016/j.ejmech.2018.08.076.

Rahbarizadeh, F., Rassaee, M. J., Forouzandeh-Moghadam, M. and Abdolamir, A. (2005). High expression and purification of the recombinant camelid anti-MUC1 single domain antibodies in Escherichia coli. *Protein Expression and Purification* 44(1):32–38. 10.1016/j.pep.2005.04.008.

Rajaofera, M. J. N., Wang, Y., Jatoi, Z. A., Jin, P., Cui, H., Lin, C. and Miao, W. (2020). *Bacillus atrophaeus* HAB-5 secretion metabolites preventing occurrence of systemic diseases in tobacco plant. *European Journal of Plant Pathology* 156(1):159–172. 10.1007/s10658-019-01873-1.

Rajesh, R. W., Rahul, M. S. and Ambalal, N. S. (2016). *Trichoderma*: a significant fungus for agriculture and environment. *African Journal of Agricultural Research* 11(22):1952–1965. 10.5897/AJAR2015.10584.

Rajnovic, D., Muñoz-Berbel, X. and Mas, J. (2019). Fast phage detection and quantification: an optical density-based approach. *PLoS ONE* 14(5):e0216292. 10.1371/journal.pone.0216292.

Ramezani, M., Ramezani, F. and Gerami, M. (2019). Nanoparticles in pest incidences and plant disease control, 233–272. In: *Nanotechnology for Agriculture: Crop Production & Protection*. 10.1007/978-981-32-9374-8_12.

Ravensberg, J. W. (2011). Mass production and product development of a microbial pest control agent, 59–127. In: *A Roadmap to the Successful Development and Commercialization of Microbial Pest Control Products for Control of Arthropods* (Vol. 10). Springer, Dordrecht. https://doi.org/10.1007/978-94-007-0437-4_3.

Rodrigues, G. A., Shalaev, E., Karami, T. K., Cunningham, J., Slater, N. K. H. and Rivers, H. M. (2019). Pharmaceutical development of AAV-based gene therapy products for the eye. *Pharmaceutical Research* 36(2):29. 10.1007/s11095-018-2554-7.

Rogers, R. S., Abernathy, M., Richardson, D. D., Rouse, J. C., Sperry, J. B., Swann, P., Wypych, J., Yu, C., Zang, L. and Deshpande, R. (2017). A view on the importance of "multi-attribute method" for measuring purity of biopharmaceuticals and improving overall control strategy. *AAPS Journal* 20(1):7. 10.1208/s12248-017-0168-3.

Rotolo, C., De Miccolis Angelini, R. M., Pollastro, S. and Faretra, F. (2015). A Taqman-based qPCR assay for quantitative detection of the biocontrol agents Bacillus subtilis strain QST713 and Bacillus amyloliquefaciens subsp. plantarum strain D747. *BioControl* 61(1):91–101. 10.1007/s10526-015-9701-4.

Russell, K. H. and Boyetchko, M. S. (2006). Research initiatives in the art and science of biopesticide formulations. *Soil Biology and Biochemistry* 38(4):845–849. 10.1016/j.soilbio.2005.07.003.

Ryan, C. A. and Farmer, E. E. (1999). Method of inducing plant defense mechanisms.US Patent 5,935,809.

Ryan, C. A. and Moura, D. S. (2002). Systemic wound signaling in plants: a new perception. *Proceedings of the National Academy of Sciences of the United States of America* 99(10):6519–6520. 10.1073/pnas.112196499.

Said, S. D. (2007). Spore production by biocontrol agent *trichoderma harzianum* in submerged fermentation: effect of agitation and aeration. *Journal of Chemical Engineering and Environments* 6(2):71–6.

Sanches, P. A., Santos, F., Peñaflor, M. F. G. V. and Bento, J. M. S. (2017). Direct and indirect resistance of sugarcane to *Diatraea saccharalis* induced by jasmonic acid. *Bulletin of Entomological Research* 107(6):828–838. 10.1017/S0007485317000372.

Sanders, D., Borys, K. D., Kisa, F., Rakowski, S. A., Lozano, M. and Filutowicz, M. (2017). Multiple dictyostelid species destroy biofilms of Klebsiella oxytoca and other Gram-negative species. *Protist* 168(3):311–325. 10.1016/j.protis.2017.04.001.

Sandra, S., Karabegovic, I. S., Beskosld, V., Nikolic, N., and Lazic, M. (2019). Bacillus based microbial formulation optimization of the production process. Hem, Ind. 73(3):169–182. 10.2298/HEMIND190214014S.

Santos, S. B., Carvalho, C., Azeredo, J. and Ferreira, E. C. (2014). Population dynamics of a Salmonella lytic phage and its host: implications of the host bacterial growth rate in modelling. *PLoS ONE* 9(7):e102507. 10.1371/journal.pone.0102507.

Sarrocco, S., Mauro, A. and Battilani, P. (2019). Use of competitive filamentous fungi as an alternative approach for mycotoxin risk reduction in staple cereals: state of art and future perspectives. *Toxins* 11(12):701; 10.3390/toxins11120701.

Sauvageau, D. and Cooper, D. G. (2010). Two-stage, self-cycling process for the production of bacteriophages. *Microbial Cell Factories* 9:81. 10.1186/1475-2859-9-81.

Schisler, D. A., Jackson, M. A. and Bothast, R. J. (1991). Influence of nutrition during conidiation of *Colletotrichum truncatum* on conidial germination and efficacy in inciting disease in *Sesbania exaltata*. *Phytopathology* 81:567–590.

Schisler, D. A., Slininger, P. J., Behle, R. W. and Jackson, M. A. (2004). Formulation of *Bacillus* spp. for biological control of plant diseases. *Phytopathology* 94(11):1267–1271, 18944465.

Schmidt, C. S., Lorenz, D., Wolf, G. A. and Jager, J. (2001). Biological control of the grapevine dieback fungus Eutypa lata II: influence of formulation additives and transposon mutagenesis on the antagonistic activity of Bacillus subtilis and *Erwinia herbicola*. *Journal of Phytopathology* 149(7–8):437–445.

Schrader, J., Schilling, M., Hollmann, D., Sell, D., Villella Filhn, M., Marx, A. and Vorholt, J. A. (2009). Methanol-based industrial biotechnology current status and future perspective of methylotropic bacteria. *Trends in Biotechnology* 27(2):107–115. 10.1016/j.tibtech.2008.10.009.

Seibold, A., Fried, A., Kunz, S., Moltmann, E., Lange, E. and Jelkmann, W. (2004). Yeasts as antagonists against fireblight. *EPPO Bulletin* 34(3):389–390. 10.1111/j.1365-2338.2004.00766.x.

Serrano, L., Manker, D., Brandi, F. and Cali, T. (2013). The use of Bacillus subtilis QST 713 and Bacillus pumilus QS 2808 as protectant fungicide in conventional application programs for black leaf streak control. *Acta Horticulturae* 986:149–155. 10.17660/ActaHortic.2013.986.

Shafi, J., Tian, H. and Ji, M. (2017). *Bacillus* species as versatile weapons for plant pathogens: a review. *Biotechnology and Biotechnological Equipment* 31(3):446–459. 10.1080/13102818.2017.1286950.

Shearer, J. F. and Jackson, M. A. (2006). Liquid culturing of microsclerotia of *Mycoleptodiscus terrestris*, a potential biological control agent for the management of hydrilla. *Biological Control* 38(3, September):298–306. 10.1016/j.biocontrol.2006.04.012.

Sheridan, L. W., Ruocco, M., Vinale, F., Nigro, M., Marra, R., Lombardi, N., Pascale, A., Lanzuise, S., Manganiello, G. and Lorito, M. (2014). *Trichoderma*-based products and their widespread use in agriculture. *The Open Mycology Journal* 8(1):71–126. 10.2174/1874437001408010071.

Sigler, W. V., Nakatsu, C. H., Reichev, Z. J. and Turco, R. F. (2001). Fate of the biological control agent *Pseudomonas aureofaciens* TX-1 after application to turfgrass. *Applied and Environmental Microbiology* 67(8):3542–3548. 10.1128/AEM.67.8.3542-3548.2001.

Sivasakthi, S., Usharani, G. and Sarana, P. J. (2014). Biocontrol potentiality of plant growth promoting bacteria (PgPr)–*pseudomonas fluorescens* and *Bacillus subtilis*: a review. *African Journal of Agriculture* 9(16):1265–1277. 10.5897/AJAR2013.7914.

Sivasithamparam, C. and Parker, A. (1976). Production of chlamydospores and microsclerotia by *Gaeumannomyces graminis ar. Tritici*. *Soil Biology and Biochemistry* 8(1):71–73.

Slininger, P. J., Côté, G. L., Shea-Andersh, M. A., Dien, B. S. and Skory, C. D. (2020). Application of isomelezitose as an osmoprotectant for biological control agent preservation during drying and storage. *Biocontrol Science and Technology* 31(2):132–152, 10.1080/09583157.2020.1833307.

Spadaro, D. and Gullino, M. L. (2004). State of the art and future prospects of the biological control of postharvest fruit diseases. *International Journal of Food Microbiology* 91(2):185–194. 10.1016/S0168-1605(03)00380-5.

Stein, T. (2005). *Bacillus subtills* antibiotics: structures synthesis and specific function. Mol. *Molecular Microbiology* 56(4):845–857, 10.1111/j.1365-2958.2005.04587.x.

Stockwell, V. O. and Stack, J. P. (2007). Using *Pseudomonas* spp. for integrated biological control. *Phytopathology* 97(2):244–249. 10.1094/PHYTO-97-2-0244.

Storms, Z. J., Teel, M. R., Mercurio, K. and Sauvageau, D. (2019). The virulence index: a metric for quantitative analysis of phage virulence. *Phage* 1(1):27–36

Strobel, G., Bui, V. P., Su, H., Himmel, P., Marrone, P., Xing, L. and Lewis, S. (2013). Muscodor Albus Strain Producing Volatile Organic Compounds and Methods of Use. USA patent US20140086879A1.

Strobel, G., Bui, V. P., Su, H., Himmel, P., Marrone, P., Xing, L. and Lewis, S. (2020). *Muscodor Albus* Strain Producing Volatile Organic Compounds and Methods of Use. U.S. patent number 10,869,482 [Application Number 14/789,742] granted December 2020.

Svircev, A., Roach, D. and Castle, A. (2018). Framing the future with bacteriophages in agriculture. *Viruses* 10(5):218. 10.3390/v10050218.

Sun, M. H., Chen, Y. M., Liu, J. F., Li, S. D. and Ma, G. Z. (2013). Effects of culture conditions on spore types of *Clonostachys rosea*, 67-1 in submerged fermentation. *Letters in Applied Microbiology* 58(4):318–324.

Sun, Z. B., Zhang, J., Sun, M. H. and Li, S. D. (2019). Identification of genes related to chlamydospore formation in *Clonostachys rosea* 67-1.*MicrobiologyOpen* 8(1):e00624. 10.1002/mbo3.624.

Tabachnik, M. (1989). Method of growing Trichoderma. USA patent 4,837,155.

Terras, F. R., Eggermont, K., Kovaleva, V., Raikhel, N. V., Osborn, R. W., Kester, A., Rees, S. B., Torrekens, S., Van Leuven, F. and Vanderleyden, J.. (1995). Small cysteine-rich antifungal proteins from radish their role in host defense. *Plant Cell* 7(5):573–588.

Thevissen, K., de Melo, T., et al. (2012). The Plant defensis RsAFP2 induces cell wall stress septin misallocation and accumulation of ceramide in *candida albicans. Molecular Microbiology* 84:166–180.

Tjamos, S. E., Flemetakis, E., Paplomatas, E. J. and Katinakis, P. (2005). Induction of resistance to *Verticillium dahlia* in *Arabidopsis thaliana* by the biocontrol agent K-165 and pathogenesis-related proteins gene expression. *Molecular Plant-Microbe Interactions* 18(6):555–561.

Tombolini, R., van der Gaag, D. J., Gerhardson, B. and Jansson, J. K. (1999 Aug). Colonization pattern of the biocontrol strain *Pseudomonas chlororaphis* MA 342 on barley seeds visualized by using green fluorescent protein. *Applied and Environmental Microbiology* 65(8):3674–3680. PMC91550.

Torres, R., Usall, J., Teixido, N., Abadias, M. and Vinas, I. (2003). Liquid formulation of the biocontrol agent Candida sake by modifying water activity or adding protectants. *Journal of Applied Microbiology* 94(2):330–339.

Tripathi, D., Raikhy, G. and Kumar, D. (2019). Chemical elicitors of systemic acquired resistance–salicylic acid and its functional analogs. *Current Plant Biology* 17(January):48–59. 10.1016/j.cpb.2019.03.002.

Trotsenko, Y. A. and Torgonskaya, M. L. (2018). Current trends in methylotrophy-based biotech. *Adv Biotech & Micro. Mini Review* 9(3).10.19080/AIBM.2018.09.555763.

Turner, J. T. and Backman, P. A. (1991). Factors related to peanut yield increase after see treatment with *Bacillus subtills. Plant Disease* 75:347–353.

Umashankar, K., Chandralekha, A., Dandavate, T., Tavanandi, H. A. and Raghavarao, K. S. M. S. (2019). A nonconventional method for drying of *Pseudomonas aeruginosa* and its comparison with conventional methods. *Drying Technology* 37(7):839–853. 10.1080/07373937.2018.1469140.

USEPA (2003). Biopesticide registration action document *Aspergillus flavus* AF36. Available at: https://www3.epa.gov/pesticides/chem_search/reg_actions/registration/decision_PC-006456_3-Jul-03.pdf, accessed February 2020.

Van de Mortel, J. E., De Vos, R. C. H., Dekkers, E., Pineda, A., Guillod, L., Bouwmeester, K., van Loon, J. J., Dicke, M. and Raaijmakers, J. M. (2012). Metabolic and transcriptomic changes induced in *Arabidopsis* by the rhizobacterium *Pseudomonas fluorescens* SS101. *Plant Physiology* 160(4):2173–2188.

Vandenberghe, L. P. S., Pandey, A., Carvalho, J. C., Letti, L. A. J., Woiciechowski, A. L., Karp, S. G., Thomaz-Soccol, V., Martínez-Burgos, W. J., Penha, R. O., Herrmann, L. W., Rodrigues, A. O. and Soccol, C. R. (2020). Solid-state fermentation technology and innovation for the production of agricultural and animal feed bioproducts. *Systems Microbiology and Biomanufacturing* 1(2):142–165.

Van der Vaart (2002). Expression of V_{HH} antibody fragments in *Saccharomyces cerevisiae*. In: O'Brien, P. M. and Aitken, R. (Eds) *Antibody Phage Display. Methods in Molecular Biology* (vol. 178). Humana Press. 10.1385/1-59259-240-6:359.

Vemmer, M. and Patel, A. V. (2013). Review of encapsulation methods suitable for microbial biological control agents. *Biological Control* 67(3):380–389. 10.1016/j.biocontrol.2013.09.003.

Vesely, D. (1989). Biological control of damping-off pathogens by treating sugar-beet seed with A powdery preparation of the mycoparasite pythium oligandrum in large-scale field trials. *Developments in Soil Science* 18:445–449. 10.1016/S0166-2481(08)70248-4.

Vidhyasekaran, P. (2002). *Bacterial Disease Resistance in Plants. Molecular Biology and Biotechnological Applications*. 452 pp. The Haworth Press, Binghamton, NY.

Vidhyasekaran, P., Rabindran, R., Muthamilan, M., Nayar, K., Rajappan, K., Subramanian, N. and Vasumathi, K. (1997). Development of a powder formulation of *Pseudomonas fluorescens* for control of rice blast. *Plant Pathology* 46(3):291–297.

Visesato, M., Rella, A., Farnoud, A. M., Singh, A., Munshi, M., Bryan, A., Naseem, S., Konopka, J. B., Ojima, I., Bullesbach, E., Ashbaugh, A., Linke, M. J., Cushion, M., Collins, M., Ananthula, H. K., Sallans, L., Desai, P. B., Wiederhold, N. P., Fothergill, A. W., Kirkpatrick, W. R., Patterson, T., Wong, L. H., Sinha, S., Giaever, G., Nislow, C., Flaherty, P., Pan, X., Cesar, G. V., de Melo Tavares, P., Frases, S., Miranda, K., Rodrigues, M. L., Luberto, C., Nimrichter, L. and Del Poeta, M. (2015). Identification of new class of antifungal targeting the synthesis of fungal Ssphingolipids. *mBio* 6(issue 3):e00647.

Villamizar, L. F., Nelson, T. L., Jones, S. A., Jackson, T. A., Hurst, M. R. H. and Marshall, S. D. G. (2018). Formation of microsclerotia in three species of *Beauveria* and storage stability of a prototype granular formulation. *Biocontrol Science and Technology* 28(12):1097–1113, 10.1080/09583157.2018.1514584.

Villamizar, L. F., Barrera, G., Marshall, S. D. G., Richena, M., Harland, D. and Jackson, T. A. (2020). Three-dimensional cellular aggregates formed by *Beauveria pseudobassiana* in liquid culture with potential for use as a biocontrol agent of the African black beetle (*Heteronychus arator*). *Mycology* 12(2):105–118.

Vriens, K., Cols, T. L., Harvey, P. J., Craik, D. J., Braem, A., Vleugels, J., De Coninck, B., Cammue, B. P. and Thevissen, K. (2016). The radish defensins RsAFP2 act synergistically with caspofungin against *Candida albicans* biofilms. *Peptides* 75:71–79.

Vu, N. T. and Oh, C. S. (2020). Bacteriophage usage for bacterial disease management and diagnosis in plants. *Plant Pathology Journal* 36(3):204–217. 10.5423/PPJ.RW.04.2020.0074.

Waffa, M. H., Malaka, A. E., Inas, M. and Noran, A. (2013). Mass production, fermentation, formulation of Pseudomonas putida for controlling of die back and phomopsis diseases on grapevine. *Advances in Bioscience and Biotechnology* 4: 741–750.

Walters, D. R., Newton, A. C. and Lyon, G. D. (2005). Induced resistance: helping plants to help themselves. *Biologist* 52:28–33.

Wang, H. H., Wang, J. L., Li, Y. P., Liu, X., Wen, J. Z. and Lei, Z. R. (2011). Liquid culturing of microsclerotia of Beauveria bassiana, an entomopathgeonic fungus to control western flower thrip, Frankliniella occidentalis. *Journal of Applied Entomology* 48(3), 588–595.

Wasternack, C. (2006). Oxylipins: biosynthesis, signal traduction and action, 185–222. In: Hedden, P. and Thomas, S. (Eds) *Annual Plant Reviews* (vol. 24). Wiley Online Library.

Wasternack, C. (2014). Jasmonates in plant growth and stress responses phytohormones: a window to metabolism, signaling and biotechnological applications. ISBN: 978-1-4939-0490-7.

Wasternack, C. and Feussner, I. (2018). The oxylipin pathways: biochemistry and function. *Annual Review of Plant Biology* 69:363–386. 10.1146/annurev-arplant-042817-040440.

Wei, Y.-H. and Chu, I.-M. (2002). Mn+2 improves surfactin production by *Bacillus subtills*. *Biotechnology Letters* 24(6):479–482. 10.1023/A:1014534021276.

Weiss, A., Mögel, G. and Kunz, S. (2006). Development of "Boni-Protect" – a yeast preparation for use in the control of post-harvest diseases of apples, 113–117. In: Boos, M. (Ed). *12th International Conference on Cultivation Technique and Phytopathologtical Problems in Organic Fruit-Growing*. Weinsberg, Germany.

Weitere, M., Bergfeld, T., Rice, S. A., Matz, C. and Kjelleberg, S. (2005). Grazing resistance of *Pseudomonas aeruginosa* biofilms depends on type of protective mechanism, developmental stage and protozoan feeding mode. *Environmental Microbiology* 7(10):1593–1601.

Weller, D. M. (2007). *Pseudomonas* biocontrol agents of soilborne pathogens: looking back over 30 years. *Phytopathology* 97(2):250–256.10.1094/PHYTO-97-2-0250.

Whipps, J. M. and Gerlagh, M. (1992). Biology of *Coniothyrium minitans* and its potential for use in disease biocontrol. *Mycological Research* 96(11):897–907.10.1016/S0953-7562(09)80588-1.

Whipps, J. M., Sreenivasaprasad, S., Muthumeenakshi, S., Rogers, C. W. and Challen, M. P. (2008). Use of *Coniothyrium minitans* as a biocontrol agent and some molecular aspects of sclerotial mycoparasitism. *European Journal of Plant Pathology* 121(3):323–330.10.1007/s10658-007-9238-1.

Wicks, A.S. and Pinchuk, G.E. (2015). Porous matrices for culture and formulation of agriculture biopesticide and chemicals. USA patent 2015/03035347.

Wisniewski, M. and Droby, S. (2012). Biopreservation of food and feed by postharvest biocontrol with microorganisms, 57–66. In: Sundh, I., Wilcks, A. and Goettel, M. S. (Eds) *Beneficial Microorganisms in Agriculture, Food and the Environment*. CABI Publishing International, Oxfordshire.

Wisniewski, M., Droby, S., Norelli, J., Liu, J. and Schena, L. (2016). Alternative management technologies for postharvest disease control: the journey from simplicity to complexity. *Postharvest Biology and Technology* 122:3–10. 19.1016/J.postharvbio.2016.05.012.

Wisniewski, M., Wilson, C., Droby, S., Chalutz, E., El Ghaouth, A. and Stevens, C. (2007). Postharvest biocontrol: new concepts and applications. 10.1079/9781845932657.0262.

Won, S. J., Kwon, J. H., Kim, D. H. and Ahn, Y. S. (2019). The effect of *Bacillus licheniformis* MH48 on control of foliar fungal diseases and growth promotion of *Camellia oleifera* seedlings in the coastal reclaimed land of Korea. *Pathogens* 8(1). 10.3390/pathogens8010006.

Woo, S. L., Ruocco, M., Vinale, F., Nigro, M., Marra, R., Lombardi, N., Pascale, A., Lanzuise, S., Manganiello, G. and Lorito, M. (2014). *Trichoderma*-based products and their widespread use in agriculture. *The Open Mycology Journal* 8(1):71–126 10.2174/1874437001408010071.

Worrall, E., Hamid, A., Mody, K. T., Mitter, N. and Pappu, H. R. (2018). Nanotechnology for plant diseases management. *Agronomy* 8(12):285, 10.3390/agronomy8120285.

Wright, S. A. I. and Beer, S. V. (1995). The role of antibiotics in biological control of fire blight by *Erwinia herbicola* strain Eh318. *Acta Horticulturae* 411:309–312.

Wu, J. and Baldwin, I. T. (2010). New insights into plant responses to the attack from insect herbivores. *Annual Review of Genetics* 44:1–24.

Xie, Y., Wahab, L., Gill, J. J., Xie, Y., Wahab, L. and Gill, J. J. (2018). Development and validation of a microtiter plate-based assay for determination of bacteriophage host range and virulence. *Viruses* 10(4):189. 10.3390/v10040189.

Yang, N., Wu, Q. and Xu, Y. (2020). Fe nanoparticles in *Bacillus amyloliquefaciens*. *ACS Omega* 5(12):6321–6329. 10.1021/acsomega.9b03648.

Yanglei, Y., Frenzel, E., Spoelder, J., Elzenga, J. T. M., van Elsas, J. D. and Kuiper, O. P. (2018). Optimized fluorescent proteins for the rhizosphere-associated bacterium

Bacillus mycoides with endophytic and biocontrol agent potential. *Environmental Microbiology Reports* 10(1):57–74. 10.1111/1758-2229.12607.

Ye, Y.-F., Li, Q.-Q., Gang, F., et al. (2012). Identification of antifungal substance (Iturin a2) produced by *Bacillus subtilis* B47 and it effect on southern corn leaf blight. *Journal of Integrative Agriculture* 11(1):58–61.

Yezza, A., Tyagi, R. D., Valèro, J. R., Surampalli, R. Y. and Smith, J. (2004). Scale-up of biopesticide production processes using wastewater sludge as a raw material. *Journal of Industrial Microbiology and Biotechnology* 31(12):545–552. 10.1007/s10295-004-0176-z.

Yuan, S., Wisniewski, M., Droby, S. and Liu, J. (2015). Responses of yeast biocontrol agents to environmental stress. *Applied and Environmental Microbiology* 81(9):2968–2975. 10.1128/AEM.04203-14.

Yuan, Y., Peng, Q., Zhang, S., Liu, T., Yang, S., Yu, Q., Wu, Y. and Gao, M. (2019). Phage reduce stability for regaining infectivity during antagonistic coevolution with host bacterium. *Viruses* 11(2):118.

Yusran, Y., Roemheld, V. and Mueller, T. (2009). Effects of Pseudomonas sp. "Proradix" and Bacillus amyloliquefaciens FZB42 on the Establishment of AMF Infection, Nutrient Acquisition and Growth of Tomato. Proceedings of the International Plant Nutrition Colloquium XVI2009-06-14.

Zajc, J., Gostinčar, C., Černoša, A. and Gunde-Cimerman, N. (2019). Stress-tolerant yeasts: opportunistic pathogenicity versus biocontrol potential. *Genes* 10(1):42. 10.3390/genes10010042.

Zaki, O., Wekers, F., Compere, P., Jacques, P., Thonart, P. and Sabri, A. (2018). Morphological differences between serial and submerged sporidia of bio-fungicide *Pseudozyma flocculosa* CRS: 16788. *PLoS ONE* 13:1–16. 10.1371/journal pone.0201677

Zaki, O., Weekers, F., Thonart, P., Tesch, E., Kuenemann, P. and Jacques, P. (2020). Limiting factors of mycopesticide development. *Biological Control* 144(May): 104220. 10.1016/j.biocontrol.2020.104220.

Zeng, W., Wang, D., Kirk, W. and Hao, J. (2012). Use of *Coniothyrium minitans* and other microorganisms for reducing Sclerotinia sclerotiorum. *Biological Control* 60(2, February):225–232.

Zhang, H. Y., Zheng, X. D., Wang, L., Li, S. S. and Liu, R. F. (2007a). Effect of yeast antagonist in combination with hot water dips on postharvest Rhizopus rot of strawberries. *Journal of Food Engineering* 78(1):281–287. 10.1016/j.jfoodeng.2005.09.027.

Zhang, H. Y., Zheng, X. D. and Yu, T. (2007b). Biological control of postharvest diseases of peach with *Cryptococcus laurentii*. *Food Control* 18(4):287–291. 10.1016/j.foodcont.2005.10.007.

Zhang, J. C. (2013). Method for large scale fermentation production of *Bacillus subtilis* antibacterial lipopeptides. Chinese patent No. 103757075A.

Zhang, M., Zhang, M. W., Zhang, L. and Zhang, L. (2015). Methyl jasmonate and its potential in 901 cancer therapy. *Plant Signaling and Behavior* 10(9):e1062199. 10.1080/15592324.2015.1062199.

Chapter 6

Key issues in the regulation of microbial bioprotectants in the European Union: challenges and solutions to achieve more sustainable crop protection

Rüdiger Hauschild, APIS Applied Insect Science GmbH, Germany; and Willem J. Ravensberg, Koppert Biological Systems, The Netherlands

1 Introduction

Microbial bioprotectants, also known as biopesticides, are subject to registration when used for the control of pests and diseases in crops, forestry and other situations where plants require protection. On a global scale, they are regulated by the same legislation as chemical pesticides. In this chapter, the term bioprotectant is used to describe a plant protection product (PPP) based on an active substance of natural origin, in accordance with the terminology used by the International Biocontrol Manufacturers Association (https://ibma-global.org/wp-content/uploads/2020/12/ibmadefinitionleafletweb.pdf; accessed 4 April 2021) and not the term biopesticide. Bioprotectants have many modes of action but often not a direct '-cidal' effect. In many cases, the term biopesticide is not defined in legislation, except for the USA (https://www.epa.gov/pesticides/biopesticides; accessed 4 April 2021). In European Union (EU) legislation, microorganisms are defined but the term biopesticide does not appear in pesticide regulations.

http://dx.doi.org/10.19103/AS.2021.0093.07

The legislative procedures for pesticides were designed many years ago and since the first bioprotectants were developed they have all been assessed, without exception, using these existing regulations. The resulting key issue is that data requirements for chemicals are not appropriate for living organisms that behave completely differently to chemical molecules. Moreover, the microorganisms currently used are found in the natural world and have co-evolved with all organisms. In contrast, most of the chemical molecules designed as crop protection substances are xenobiotic and unknown to nature.

The registration of pesticides involves two interconnected aspects: the administrative procedures and the data requirements that govern dossier compilation and the subsequent evaluation by means of risk assessment and risk management decisions. Both have an impact on the approval process of the microorganism as an active substance. In this chapter, we focus on data requirements as the key issue, but we will also discuss the procedures and the influence these two aspects have on each other and on microbial protectants entering the market. In the frame of the EU policy support action REBECA (EU Policy Support Action, 2006–2007), Hauschild et al. (2011) compared the data requirements and procedures between the EU and other systems. The data requirements are relatively similar due to the harmonization efforts of the OECD (Organisation for Economic Co-operation and Development), but interpretation differs between regulatory authorities. Procedures are more time-consuming in the EU due to the federal structure of the system. Guest (2018) provides an overview of regulatory procedures for biopesticides in 15 countries spread over the world and recognizes similar shortcomings and flaws in nearly all countries. Many authors have argued that the approval process for bioprotectants does not function efficiently (Villaverde et al., 2014) and that the system severely hampers placing new innovative biological products on the market (Guest, 2018; Grant and Gwynn, 2020), in contrast to the aim of our society to reduce the use of hazardous chemicals and replace them with safer alternatives.

The EU is recognized as having the most stringent approval system for active substances and the most complicated and time-consuming procedures (Balog et al., 2017; Frederiks and Wesseler, 2018; Wesseler, 2019; Guijarro et al., 2020; Sundh and Eilenberg, 2020). From the authors' personal experience, it takes approximately 5 years to obtain approval for a new active substance, and the PPPs based on that active substance to be registered, before it can be put on the market in one or more EU countries and used by farmers. One microorganism (*Metschnikowia fructicola* strain NRRL Y-27328) with almost simultaneous dossier submission in the USA and the EU was first granted product authorization after 21 months in the USA versus 60 months in the EU. Furthermore, in other parts of the world, the approval systems are also a major hurdle for (microbial) bioprotectants (Arora et al., 2016), including Australia (Hauxwell et al., 2010), India (Arjjumend and Koutouki, 2018; Keswani et al.,

2019), Asia (Singh et al., 2016) and Brazil (Bettiol, 2011). The regulatory process for biopesticides in the USA is generally recognized as the most appropriate and many authors recommend the implementation of a similar approach in other jurisdictions. Many authors worldwide argue that we need better systems that have more appropriate risks assessments and faster and cheaper procedures. In this chapter, we will outline the issues and provide suggestions on how an improved procedure can be achieved.

2 Issues of the administrative registration procedure in the European Union

2.1 Procedural and timeline issues

Administrative registration procedures vary substantially between authorities, including timelines, fees, but also dossier formats, methods of (electronic) submission, advice to applicants through pre-submission meetings, helpdesks, guidance documents, etc.; without doubt, the process is the most complicated in the EU. In most instances a national authority governs the process; however, in the EU all 27 (28 until 2020) Member States, the European Food Safety Authority (EFSA) and the European Commission are involved in the whole procedure. Moreover, it is a two-step process with approval of the active substance at EU level and product authorization at national or zonal level, for some applications (greenhouses, indoor treatments, seeds) at European level, making it a very complex, costly and lengthy process. Currently, it takes about 5 years for a product with a new active substance to get to the market; about 3–3.5 years for the approval of the active substance and 1–1.5 years for product registration. This is approximately twice as long as it takes in the USA, as concluded by Frederiks and Wesseler (2018) who compared both regulatory systems for the registration of microbial control agents. In the EU, delays are frequent due to limited resources in the relevant authorities and issues regarding dossier content. Moreover, some Member States have a dedicated authority performing dossier evaluation for the active substance as well as authorization of the PPP in-house, while in most EU countries the evaluation is performed by several institutes, specialized in, e.g. human toxicology or environmental evaluation, etc. The latter case is, as expected, frequently inefficient and contributes to delays and other issues. Another issue for applicants involves the procedural costs, which vary considerably between Member States. For instance, the fees for a Rapporteur Member State (RMS) procedure (writing the Draft Assessment Report, DAR) ranges from some tens of thousands of Euros up to a few hundred thousand Euros. The choice of a Member State as Rapporteur is up to the applicant although the choice is also guided by the availability of a time slot and the experience of an RMS. However, for a renewal procedure, the European

Commission assigns the RMS and an applicant has to accept the decision and pay the expenses of that particular RMS.

Considering the regulatory process of microorganisms in the EU, since 2015, 24 microorganisms have navigated the regulatory process in the EU, 17 as new active substances and 7 as renewals. Just 2 Member States acted as the RMS for the majority of the strains, 14 out of 24, with the other strains were spread over 6 countries. Only four countries invariably commented on the DAR during the peer-review process (Fig. 1). This shows that most EU countries do not take part in the evaluation of microorganisms, which may be due to a lack of resources, expertise or interest. Clearly, the current review process is not functioning efficiently. To improve the system, we recommend the following:

- Appoint specialist countries for groups of microorganisms (bacteria, fungi and other microorganisms, or according to ecological groups as discussed below).
- Build up expertise and, subsequently, risk assessment quality.
- Allow only a few other countries to comment, e.g. three per zone.

In this way, a more efficient review can be performed, which is faster and resource effective. All countries must then accept the evaluation of the

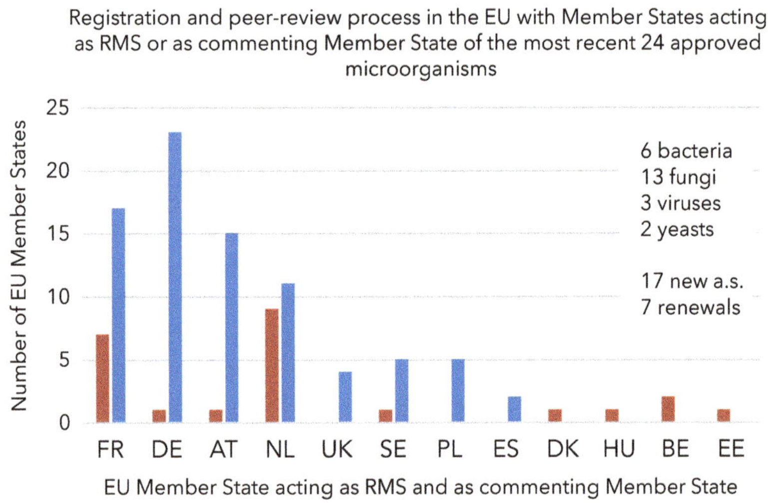

Figure 1 Registration and peer-review process of the most recent 24 approved microorganisms in the EU. Red: Country acted as RMS, blue: country provided comments. Countries: AT = Austria; BE = Belgium; DE = Germany; DK = Denmark; EE = Estonia; ES = Spain; FR = France; HU = Hungary; NL = the Netherlands; PL = Poland; SE = Sweden; UK = United Kingdom.

specialist countries; this approach should work as it is already, more or less, current practice.

The EU has evaluated its regulatory system in a so-called Refit procedure and the report (EC, 2020a) acknowledges many of these issues. The main conclusions, not only for bioprotectants, were that: 'Efficiency stands out as the critical area requiring attention. Due to a lack of resources and capacity in the Member States, most of the procedures set out in the regulations are suffering from severe delays, in particular, those related to the renewal of approval of active substances and the authorization of PPP containing approved substances, which in turn negatively affects their effectiveness'. However, it lacks recommendations for improvement, which will be developed within the Farm to Fork Strategy of the EU (EC, 2020b).

Both bioprotectants and conventional actives are processed by the same people and via the same procedures, whereas fundamental differences between these two types of products warrant a separation. In the USA Environmental Protection Agency (EPA) two separate streams were designed and have functioned well over the decades. Adopting such an approach in the EU would overcome many problems. Details and recommendations to improve the EU system for bioprotectants were detailed by Grant and Gwynn (2020), two crucial improvements include dedicated regulators at all levels with the necessary skills and harmonization and reciprocity of evaluations.

2.2 Low-risk active substances and plant protection products

EU Regulation (EC) 1107/2009 recognizes a low-risk (LR) category for active substances (a.s.) and products if they fulfill specific criteria; these were provided in an annex but designed initially only for chemicals. These criteria were later amended through Regulation (EU) 2017/1432 (EC, 2017) in order to also apply to microorganisms. Incentives to develop LR actives involve a longer initial approval period of 15 years instead of 10 years. However, approval renewal of the active substance is valid for 15 years for both chemicals and microorganisms and an LR advantage is therefore lost. Product authorization by national authorities should be within 120 days, which is ideal but in practice is almost never achieved, with the authorization process still taking approximately 1 year. Another issue is that authorities in an increasing number of cases do not automatically process a product containing a LR active substance as a LR product, which means that the reduced evaluation time of 120 days for product registration is not applied because the LR status is only assigned at the end of the evaluation. The LR status cannot be stated on the product label, which is the first information a farmer encounters before deciding to apply the product; however, it can be mentioned in any advertisement (Art. 66 of Reg. 1107/2009). Data protection for a LR product is 13 years, 3 years more than for a regular product. This means that data used for approval of an active ingredient are exclusive for the first applicant who generated the data for a longer period and cannot be used or referred to by other applicants. This period can be extended up to a maximum of 15 years when it includes extensions for minor uses. This incentive of longer data protection is seen as of little value by the biocontrol industry as much of the information on microorganisms is available in the public domain and other information is strain-specific and cannot be used by competitors anyway. LR products can benefit from reduced requirements for efficacy data: according to the European Plant Protection Organization (EPPO) Guideline PP 1/296 (1) (EPPO, 2017), less efficacy trials are needed for a LR product compared with a conventional product. However, if the LR status of the product is only assigned at the end of the evaluation, the applicant needs to provide the full set of trials anyway.

A specific problem arises from the fact that for precautionary reasons, respiratory protection is requested for products containing microorganisms due to their (theoretical) potential to induce sensitization. Implementing personal protection measures is seen as risk mitigation, which prevents registration as a 'low-risk' PPP. Although the sensitizing potential does not explicitly preclude microorganisms from being low risk, their products are not approved as 'low risk' because a product may not be approved as 'low risk' if specific risk mitigation measures are applied, and although the respiratory protection derives from the precautionary phrase, it is considered a specific risk mitigation measure.

In practice, the 'low-risk' status of the active substance has little effect on the producer or distributor of the product. The advantage of the extended approval period is relatively limited as it only reduces investment and workload at a later stage. Extended data protection is of little value for microorganisms as detailed above. Finally, the most attractive incentive, reduced evaluation times for low-risk products, is rarely achieved because either timelines are not met, or the decision if the product is 'low risk'is only taken after the evaluation for product approval.

All criteria outlined in Commission Regulation (EU) 2017/1432 refer to the hazard of a substance or a microbial strain and not to risk from their use. The term 'low-hazard active substance' would be more appropriate than 'low-risk active substance'. Defining a 'low-risk active substance' is impossible without determining the exposure to this active substance, and 'low risk' cannot be a property of a substance. Defining a risk (for humans, non-target organisms or the environment) requires information on product formulation and information on the use of the product (application rates, frequency and methods, crops, etc.). For this reason, active substance evaluation is coupled to the evaluation of a representative formulation and representative use, and an active substance cannot be approved unless 'safe use' is demonstrated. Consequently, as risk is the multiplication of hazard and the exposure to it, a product containing a 'low-hazard active ingredient' would in most (if not in all) cases automatically qualify as a 'low-risk product'. Finally, the lack of implementation of the 'low-risk' principle results from confusion and inconsistencies in the terminology in Regulation (EC) 1107/2009.

Active substances are only approved as low risk after the full evaluation procedure has taken place. There is no faster procedure for potential low-risk active substances, only for products based on a low-risk active, but not in every case. For many years, the biocontrol industry advocated for faster approval, such as a preliminary approval based on a quick evaluation of the dossier at the point of dossier completeness check. Such an extended and adapted completeness check would also allow the provision of a temporary approval as low risk and a low-risk product (IBMA, 2018). A second option would be to grant this early approval after the DAR has been finished by the Rapporteur Member State; in this case, the low-risk status would be based on a full evaluation. If not confirmed in the peer-review process, this could be corrected after the full evaluation. The advantage is faster market access, which would give the farmer more low-risk tools as promoted in Article 1 of the Sustainable Use of Pesticides Directive 2009/128/EC (EC, 2009b) and has been in force since 2014. Currently, 10 years after the regulation came into force, only 26 active substances, of which 17 are microorganisms, have been approved as low risk according to the EU Pesticide Database (https://ec.europa.eu/food/plant/pesticides/eu-pesticides-database/public/?event=activesubstance.selection&language=EN; accessed 25 March 2020). This has led to a limited number of LR PPPs at national level and due to national differences in interpretation of mitigation measures, some products based on the LR a.s. are not authorized as LR PPP in certain countries. This illustrates the lack of harmonization within the EU Member States, which has prompted clarification. The EU also published a list of potentially LR a.s. that have been approved under former legislation EC 91/414 and are awaiting LR approval after the renewal process that is still ongoing for many of those a.s., of which 30 are microorganisms (EC, 2018).

Not all microorganisms are approved as LR. a.s. as illustrated by strains of *Bacillus*: 17 are approved as an active substance and just 2 as LR a.s. although some were evaluated before the low-risk category was established. Seven are being assessed as potential LR (EC, 2018) and await the status following the renewal decision. Another example is strains of *Beauveria bassiana*: six are approved (in October 2020), two of them under the old legislation and two in 2017 and two in 2019 under Regulation 1107/2009; however, none of them are LR. One of the reasons for not being assessed as low risk could be antibiotic resistance, but it is often data gaps and concerns leading to uncompleted risk assessments and therefore low risk is not granted. Here again, the inappropriate data requirements, often on secondary metabolites, prevent microorganisms from being assessed as low risk. However, the only low-risk criterion for microorganisms is the absence of multiple resistance to antibiotics (or antimycotics) (EC, 2020c). In cases where low risk was not granted because of data gaps for secondary metabolites, the decision was made prior to the low-risk criteria being implemented or by incorrect interpretation of the low-risk criteria. The EU is currently revising the data requirements for microorganisms, which will hopefully provide a better match between questions and answers and lead to less data gaps and concerns. As a consequence, more microorganisms are expected to be approved as low-risk a.s.

The principle of low risk in Regulation 1007/2009 is a positive initiative; however, in practice, very few LR PPPs have become available to farmers in the 10 years since this regulation came into force. The Refit report (EC, 2020a) states: 'The availability of low-risk active substances, including microorganisms, has increased but is still considered by stakeholders as insufficient and procedures are considered lengthy'. The same report mentions: 'The Commission acknowledged the need to find ways to accelerate placing on the market of low-risk active substances and products'. In addition, the new Farm to Fork Strategy of the Commission (EC, 2020b) only discusses a reduction of chemical pesticide use and states: 'There is an urgent need to reduce dependency on pesticides…'. Furthermore, the Farm to Fork Strategy mentions: 'The Commission will also facilitate the placing on the market of pesticides containing biological active substances' and 'It will act to reduce the length of the pesticide authorization process by Member States'. Most likely 'facilitate' refers to revision of the data requirements that will be seen as crucial; however, we doubt this will lead to a faster procedure as the Commission refers to the Member States. The latter are mainly responsible for product authorization, but the a.s. procedure takes considerably longer.

2.3 General Food Law and transparency in risk assessment

In June 2019, the 'Transparency' Regulation EU 2019/1381 (EC, 2019a) on transparency and sustainability of the EU risk assessment in the food chain was published. It establishes new rules under the General Food Law (GFL, Regulation (EC) No 178/2002; EC, 2002) of the procedures for approvals of active substances and publication of information. It amended Regulation (EC) No 1107/2009, and another eight EU regulations and one directive concerning eight sectors of the food chain.

This new law details rules on transparency for products in the food chain regulated by the EFSA, which includes microorganism-based PPPs. It enables more public access to studies and limits protection of proprietary information. The most important topic in the General Food Law is the notification of studies. All studies that are initiated from 27 March 2021 onwards and intended for use in an application dossier must be notified to the EFSA by applicants and study laboratories. If studies have not been notified, or when notified studies are missing in the dossier at the time of submission, this will lead to delays or even rejection of the dossier by authorities, for both renewals as well as a first approval case. The obligation to notify all studies intended for a dossier adds further workload for applicants and requires detailed planning for several years in advance. This is particularly difficult in the case of active substances resulting from public research, where experimental data are not planned with the perspective of a regulatory submission as this may not be predictable during an early research phase.

Further to this, the EU Commission and EFSA decided, without consultation of the industry, to use the International Uniform ChemicaL Information Database (IUCLID) data format for the electronic submission of pesticide dossiers under the GFL. This is considered a step towards further harmonization between the EFSA and European Chemicals Agency (ECHA), which already uses IUCLID for Registration, Evaluation, Authorisation and Restriction of Chemicals under Regulation (EC) 1907/2006 (REACH) and Biocides. REACH is a Regulation of the EU, adopted to improve the protection of human health and the environment from the potential risks posed by chemicals ('except pesticides') (https://echa.europa.eu/regulations/reach/understanding-reach). IUCLID is a software application to record, store, maintain and exchange data on intrinsic and hazardous properties of chemical substances. The format of dossiers is designed for REACH chemicals and biocides and not for PPPs. It needs to be adapted for chemical pesticides, and even more so for microbial active substances. The International Biocontrol Manufacturers Association (IBMA) requested a pilot project, as it was expected the implementation of this format would create difficulties for a dossier involving a microorganism. The pilot with a dossier on an entomopathogenic fungus revealed many difficulties involving how to fill in the studies and statements in the IUCLID format. Furthermore, transferring a dossier from a CADDY ('Computer-Aided Dossier and Data Supply', a special XML format for electronic submission) or EU format takes a couple of hundred hours resulting in costs of many tens of thousands of Euros; consequently, guidance is being prepared by the EFSA. Improvements on the new format were necessary but the implementation date remained March 2021. This is another case where dossiers of microorganisms need to be squeezed into an inappropriate chemical format, accompanied by the burden of considerable extra expense. This will be extremely difficult for small biocontrol companies that develop new microbial PPPs. The new provisions in the GFL originated from issues with chemical PPPs, notably glyphosate, as the EU Parliament and Commission required more transparency in order to prevent issues on, for example, studies included or not for the risk assessments. Through the notification system, this will be prevented and transparency in decision-making improved; however, at the same time, it creates another hurdle for small niche and less harmful products that are required by our society, and will probably deter applicants from submitting their dossier in the EU.

3 Issues and solutions for the data requirements for microbial bioprotectants

3.1 The development of data requirements for microbials

The first bioprotectants were approved around 50 years ago and must have been a difficult process for both the applicant and the evaluating authority with only having data requirements for chemicals. Gradually, certain initiatives emerged that developed more appropriate requirements for microorganisms. The early development of these data requirements started in the 1970s in the USA and later in the EU. Over time, other organizations such as the International Organisation for Biological and Integrated Control (IOBC), OECD, EPPO and the Food and Agriculture Organization of the United Nations (FAO) became involved and developed guidance documents. For a historical overview on the development of data requirements in the USA and Europe see Ravensberg (2011). Despite the many improvements that have been made in this respect, many adaptations are currently still being implemented by authorities and are still requested by the biocontrol industry. Most of the adaptations are in the form of guidance documents, either updates or new documents developed by the OECD, EFSA, EU or FAO. Some of the latest examples are new guidance on secondary metabolites (EC, 2020d) and antimicrobial resistance (EC, 2020c) (published 23 October 2020) and draft guidance on whole genome sequencing and bacteriophages, a new group of biocontrol agents used in PPPs. Although the guidance was introduced, adaptations to the data requirements as laid down in the European Regulations have not been made.

We will use the EU as an example to illustrate why the registration of microbial PPPs remains a major obstacle in getting biopesticides to the market. The need for such improvements is best illustrated by the conclusions of the risk assessments of microorganisms by the EFSA, which generally contain many 'data gaps' and 'areas of concern'. Although the conclusions of the EFSA on risk assessments in general contain these data gaps, the risk management decision, taken by the European Commission and Member States, leads to approval of the active substance, that is, the strain of a microorganism. In our opinion, this shows the fundamental mismatch between the legal data requirements for microorganisms and their products (laid down in Regulation 283/2013 part B and Regulation 284/2013 part B, respectively (EC, 2013a,b)), and the outcome of the registration process, resulting from different interpretations of data requirements. It would be helpful if the legal data requirements in the regulation were more appropriate to the biological characteristics of the microbial active substances, which could lead to less data gaps. However, the guidance documents would still be open to different interpretations. Historically, the questions were developed for chemicals and over time adapted for microorganisms, but that approach has not led to an appropriate and coherent set of relevant questions for a risk assessment of microbes. The European Commission has finally realized this and is currently revising the legal data requirements with the Member States and the EFSA, including the 'Uniform Principles for evaluation and authorization of plant protection products' (Regulation (EU) No. 546/2011 Part II) (EC, 2011). The Uniform Principles describe the harmonized way by which active substances and pesticide formulations need to be evaluated for the national authorization of pesticides. The biocontrol industry, through the IBMA, is also consulted. The IBMA is currently developing a new model based on a decision tree approach, which would basically lead to predictability of which portion of the data requirements can be addressed by public literature on the species' biology and ecology and where strain-specific experimental data are needed to answer only the relevant data requirements for a certain type of microorganism. We will describe this proposal in more detail later in this chapter (Section 3.3).

Another issue in the approval process is the knowledge and experience of the evaluators within authorities (Grant and Gwynn, 2020; Sundh and Eilenberg, 2020). They should be trained in microbial biology or ecology and have scientific experience in microbial interactions related to plants; however, many of them are trained as chemists and lack the scientific background on biology, ecology, etc. related to microorganisms. This is also the case for many evaluators within the EFSA. Living organisms behave completely differently under natural conditions than molecules and therefore this should be strongly considered during a risk assessment. The evaluation of the dossier should be performed by scientists that have a good understanding of the subject. Many chapters in this book illustrate the complex microbial interactions in the microorganism–plant system and between microorganisms. A way to improve the risk assessment is by installing a 'Green TEAM' as evaluators for biopesticides, as done by the Ctgb (Dutch Board for the Authorization of Plant Protection Products and Biocides) in the Netherlands. Scientists with a background in biology are trained to fulfil a risk assessment for biologically active substances. The complexity of microbial interactions is enormous and recognizing which ecological characteristics could pose a risk is a real challenge. A positive step is that the European Commission is setting up a training program for evaluators through the 'Better Training for Safer Food' Initiative on this topic (https://ec.europa.eu/food/safety/btsf_en, accessed on 30 March 2021). The Ctgb will be one of the trainers. Each competent authority, including the EFSA, should ideally have a 'green team' with specialists trained and experienced in the field of (microbial) bioprotectants. For many years, the number of new active substances applied for approval in the EU has been higher for bioprotectants than for traditional chemical active substances, and this trend will surely develop further in the direction of biological substances. In 2020, over 70% of the pending substances are of biological origin; hence, authorities must adapt to this trend. In the USA, the EPA has had a special division for the registration of biopesticides for more than 30 years, which has worked well (Leahy et al., 2014; Arthurs and Dara, 2019). This example should be followed all around the world. Brazil has recently adopted a similar approach and many biopesticides have been approved over the last 2 years, with registration roughly taking only 1 year (Togni et al., 2019; Domingues, 2019).

3.2 Data requirements: issues and solutions

The data requirements currently in place were introduced in June 2001 when Commission Directive 2001/36/EC (EC, 2001) came into force. These data requirements were transferred without change into the new Regulation (EC) No. 1107/2009 (EC, 2009a) with entry into force in April 2013 of 'Commission Regulation (EU) No. 283/2013 (EC, 2013a)' and 'Commission Regulation (EU) No. 284/2013 (EC, 2013b)' referring to the active substance and the PPP, respectively. This means that the experiences gained during the evaluation of microorganisms as 'new' and 'existing' active substances under Directive 91/414 between 2002 and 2013 were not considered for transfer to the new regulation.

Data requirements were derived from those used for chemical active ingredients and, to a certain extent, adapted to evaluate aspects that are specific for living organisms. The crucial difference between a living microorganism and every other plant protection active ingredient is the potential for survival and proliferation after product application. This is addressed with the data requirements on infectivity and pathogenicity to mammals and non-target organisms and persistence and proliferation on crops and in the environment. On the other hand, requirements for long-term toxicity, fate and behaviour in the environment and effects on non-target organisms are reduced compared with chemical active ingredients.

During the use of these data requirements over the last 19 years, interpretation has changed considerably. Data required and evaluated in, e.g. 2004 may either not be requested or would not be sufficient. This became evident during the re-approval process for more than 20 microbial active ingredients in the Renewal Programs AIR 2 to AIR4 (https://ec.europa.eu/food/plant/pesticides/approval_active_substances/approval_renewal_en (accessed February 24, 2021), when data accepted at the time of the first approval as an active ingredient were no longer sufficient for the renewal of approval, although legal data requirements were not changed. Interpretation of data requirements further varies considerably between evaluators from different EU Member States and the EFSA. This has led to a substantial loss of predicting what data would be needed, and to the costs and timelines for applicants.

One possible reason for the differences in interpretation may be due to data requirements being defined very broadly to include all microorganisms. Decisions defining which data are needed for a particular strain are taken on a case-by-case basis, meaning that some data may not be needed for the risk assessment of a particular group due to the biological properties of a microorganism. However, such data can be requested by evaluators, which leads to scientific effort and a regulatory burden for both the applicant and the regulator that is not justified.

Authorities and applicants have gained experience with the current data requirements over the last 19 years during the evaluation of more than 60 microbial strains. One key lesson learned is that new data requirements are necessary.

3.3 Biology as the basis of the risk assessment

Microorganisms should be assigned to groups, depending on their biology and ecology, to facilitate the interpretation of data requirements and the decision of what data are needed for a given microorganism. For the risk assessment, the actual biology and relationships to other organisms are more relevant than the taxonomy alone, and the risk of a biocontrol agent depends on its host range.

A decision tree is being developed by the IBMA based on ecological groups, which allows a more reliable definition of what experimental data are required and where literature information on the species is sufficient.

The decision tree (as well as the current data requirements) starts with the question of identity of the microorganism. Clear molecular identification according to up-to-date techniques and taxonomy is needed to identify the strain, to assign it to a species and to distinguish it from other related strains. Most biocontrol species (and all the species currently approved or under evaluation in the EU) are well known and consequently, their biology, ecology and host range are well described in the literature. Published information on the species and additional strain-specific information on the target organisms and mode(s) of action from the development of the biocontrol agent will allow assigning the species to an ecological group as a first approach for the risk assessment.

If the host range is well defined and limited (few related species of the same order or species from closely related families) the risk assessment is relatively easy because only host species that are exposed to the applied product need to be evaluated. In particular, obligate parasites with a limited host range have a very low risk to organisms outside their host range. Baculoviruses are a good example of this group which was acknowledged with the guidance document (EC, 2008) on the assessment of new isolates of baculovirus species already included in Annex I of Council Directive 91/414/EEC. Such organisms can only proliferate in the presence of their host, in fact, only inside the insect larval stage, which restricts their ability to multiply in the environment. For this group (also including other microorganisms with a comparably restricted host range, such as *Pasteuria* spp. and many bacteriophages), literature information should be sufficient for the risk assessment, and no studies of toxicity, infectivity or pathogenicity to mammals or other non-target organisms should be required. Such an approach is currently proposed for bacteriophages in the draft guidance based on their specificity. Formally, this group can be divided into two subgroups, the first (1.1) with a host range that only includes pest species or pathogens, and the second (1.2) with a broader host range that may also include non-pest species. In the second case, exposure of a non-target host needs to be determined. In fact, most if not all of the examples cited above would fall into the first subgroup.

If the host range is broader than just a few species of the same family, or no defined host exists, the risk assessment needs to be extended to all non-target organisms that are exposed. Such organisms may also be able to proliferate as saprophytes in the absence of a host, which might enable multiplication in the environment. This group 2 can also be subdivided into two subgroups: the first subgroup 2.1 includes specific parasites (e.g. *Ampelomyces quisqualis*, *Coniothyrium minitans*) with a restricted host range and low capacity to proliferate in the absence of their host. As they are not likely to adversely affect any organisms outside their host range, strain-specific data would be limited to survival and proliferation in the environment. The second subgroup, 2.2, includes non-specific parasites with the capacity to survive and proliferate as saprophytes. It includes most of the species commonly used in biocontrol, such as *Trichoderma* spp., *Gliocladium* spp., *Clonostachys* spp., '*Bacillus amyloliquefaciens* functional group', and *Pseudomonas* spp., which are active against plant-pathogenic fungi or bacteria, and insect or nematode pathogens such as *Beauveria* spp., *Isaria* spp., *Metarhizium* spp., *Purpureocillium* spp. or *Bacillus firmus*. For strains of this subgroup, strain-specific data are needed for the effect on non-target organisms (including mammals) and survival and proliferation in the environment.

The third group includes microorganisms that live permanently or occasionally inside a plant as an endophyte. To date, no such microorganisms have been evaluated in the EU. In addition to strain-specific data required for group 2 microorganisms, data on the potential accumulation of metabolites with adverse effects in the plant might be needed.

As described in the following chapter, metabolites produced by biocontrol microorganisms are difficult to evaluate. The decision tree proposes restricting the list of metabolites that need to be assessed to known bacterial or fungal toxins, especially mycotoxins known from plant pathogens or storage pathogens or bacterial toxins that cause problems as food or feed contaminants. These compounds are well known and a list of toxins that need to be addressed for biocontrol agents can be established based on lists for food or feed contaminants. If information detailing the biocontrol microorganism can produce such substances is available, the active ingredient or the formulated product needs to be analyzed for the presence or absence of these compounds. In case the compound is present, a risk assessment similar to the one for chemical active substances can be conducted. If there is no indication from the literature of the production of such compounds by the biocontrol species, and no adverse effects are observed in strain-specific studies, no further data on the production of secondary metabolites are justified.

3.4 Secondary metabolites as the key issue

3.4.1 Regulatory situation

One aspect of the risk assessment that probably triggers most additional requirements during the evaluation is microbial metabolites. Some microorganisms are well known for producing metabolites (primary and secondary) that may have an impact on non-target organisms including humans. Prominent cases linked to adverse effects via toxic metabolites (bacterial toxins or mycotoxins) include *Clostridium botulinum*, *Aspergillus flavus*, plant-pathogenic *Fusarium* spp. and others. The toxic metabolites produced by these microorganisms are well known and tested routinely in food or feed. None of these toxin producers has ever been considered as a biocontrol agent. If strains of these species are used in biocontrol, the absence of production of these metabolites needs to be assured, as in the atoxigenic *Aspergillus flavus* strain (see elsewhere in this book). Obviously, the data requirements for microbial metabolites are intended to prevent a toxin producer being used in plant protection. However, for microorganisms currently used in biocontrol, information is available in published literature or can be generated using genetic or biochemical methods to exclude the occurrence of toxin accumulation. Biocontrol microorganisms should not generally be regarded as toxin producers, and biocontrol companies and research institutes discard such organisms early in their screening processes due to this characteristic (Deising et al., 2017; Koch et al., 2018).

Another issue is a result of the terminology in data requirements for PPPs and their actives. Regulation (EC) 1107/2009 defines 'metabolite' as follows: '"Metabolite" means any metabolite or a degradation product of an active substance, safener or synergist, formed either in organisms or in the environment'. This definition only refers to metabolites of chemical molecules and is not suitable for microbial metabolites. As described above, microbial metabolites are completely different regarding production and biology, but this is not reflected in the regulation and the data requirements for microorganisms. Moreover, EU regulations seem to overemphasize secondary metabolites compared with other countries such as Australia, the USA and Canada (Sundh and Eilenberg, 2020). We therefore suggest discussions take place regarding microbial compounds in order to emphasize their biological origin and avoid the link with chemical metabolites.

3.4.2 Microbial compounds and their biology

For species that are present in the environment such as those currently used in biocontrol (*Bacillus*, *Pseudomonas*, *Trichoderma*, *Clonostachys*, *Beauveria*, *Isaria*, *Metarhizium,* etc.), and their microbial compounds, other organisms have co-evolved and developed mechanisms to degrade or inactivate these compounds in case of adverse effects. Consequently, the effects on non-target organisms are expected to be limited and transient (Köhl et al., 2019a). However, if these compounds are contained in the PPP it must be considered that concentrations of microbial compounds of biocontrol agents can be substantially higher in the environment after product application compared with under 'natural' conditions. In this case, a risk assessment is required and is possible based on the content of the compounds in the product, application rates and data on effects.

The case of *Bacillus thuringiensis* is excluded as the mode of action is at least partly based on crystal proteins (cry proteins), also called endotoxins. These proteins are pro-toxins that are only activated under the alkaline conditions of the insect midgut. The activated toxins are specific to certain insect orders (Lepidoptera, Coleoptera) and no other non-target organisms are affected; therefore, they do not present any risk to the environment or humans.

'Potentially toxic' microbial compounds are also known for some species of biocontrol agents, but to date, no adverse effects on humans or the environment linked to potentially toxic compounds from biocontrol agents have been observed or reported. Many so-called 'toxins' are described in the literature for species of the genera *Bacillus*, *Trichoderma*, *Beauveria*, *Metarhizium* and others frequently used for the selection of biocontrol strains. However, the discovery and description of such toxins are in many cases driven by the search for compounds with activity for medical use, and effects in cell cultures or other *in vitro* systems are often considered sufficient to characterize a molecule as a 'toxin'. Some biocontrol agents (e.g. some *Trichoderma* strains) are able to produce compounds belonging to the same chemical group as known toxins, but the compounds from biocontrol strains are only distantly related to such known toxins and have little chemical similarity. In these cases, the compounds produced by biocontrol agents have far lower toxicity (if any at all) to mammals or other non-target organisms. Although the chemical, biological and toxicological properties of these compounds produced by biocontrol agents are different from such toxins, the fact that these compounds belong to the same chemical group causes concerns, which can trigger extensive data requirements.

In some cases, molecular phylogeny can exclude the production of a group of potentially toxic compounds if such production is limited to a phylogenetic group that does not include the biocontrol agent.

Among microbial compounds, a distinction needs to be made between those involved in the mode of action against the target organisms, and those produced independently of the target organism. For the risk assessment, compounds produced during fermentation and contained in the formulated product need to be distinguished from those produced *in situ*.

If a biocontrol strain belongs to a species known to produce 'potentially relevant metabolites', different approaches are possible to exclude the exposure of humans and the environment to such compounds. If the genes involved in the biosynthesis are known, genome sequence information can be used to verify if these genes are present. However, the presence of such genes does not necessarily mean that these genes are functional, expressed and that they encode functional proteins. Genome data cannot reveal when and under what conditions a gene is expressed. Microbial compounds can further be identified (and possibly quantified) in the fermentation broth, the technical grade active ingredient and the formulated product.

We believe the production of microbial compounds in the environment after application of the biocontrol product ('*in situ* production') does not represent a risk for humans or non-target organisms. Considering microbial ecology, the production and secretion of microbial compounds represent a loss of energy and nutrients for the microorganism. In environments where biocontrol organisms are applied, both are limited and many organisms compete for the same resources. Consequently, microbial compounds are produced and secreted in a highly regulated manner, depending on their function, the presence of a host and the status of the interaction with these (Köhl et al., 2019b). As many of the compounds are not stable under environmental conditions, they will not accumulate. In general, the accumulation of microbial compounds by microorganisms only occurs if substrate and energy are abundant. This is the case for plant pathogens such as *Fusarium* species in cereals that accumulate trichothecenes, *Aspergillus flavus* that accumulates aflatoxins or *Aspergillus* and *Penicillium* species that produce ochratoxins. The production and accumulation of mycotoxins can occur during plant growth or during storage if conditions are permissive for production. In these cases, the fungi have abundant substrates (energy and nutrients) and toxins are also produced to prevent other organisms from colonizing the substrate. The only situation where a biocontrol organism has (sufficient) substrate to accumulate microbial compounds is the case of entomopathogenic fungi. After infesting and killing their host, generally an insect, some strains of fungi such as *Beauveria*, *Metarhizium* or *Isaria* can produce and accumulate microbial compounds while they are growing inside and on the surface of the dead host. However, as these fungi occur naturally, other animals potentially feeding on colonized insects have co-evolved with the fungi and either avoid such feed or have mechanisms to cope with these compounds if they are toxic. Consequently, secondary poisoning in birds or mammals by such compounds produced by biocontrol agents has never been observed although it may be theoretically possible. Taken together, the *in situ* production and accumulation of potentially toxic compounds by biocontrol microorganisms is very unlikely and effects on non-target organisms, including humans, have never been observed. This was confirmed in an extensive review in Annex 1 of the OECD guidance document (OECD, 2018), which collected information on secondary metabolites from different microorganisms and demonstrated that these do not pose a risk to non-target organisms including humans.

3.4.3 Stability of microbial compounds

A technical problem of the current data requirements is the low stability of many microbial compounds. Data requirements include information on the stability and persistence of relevant compounds in the environment. As described above, most of the compounds produced by biocontrol agents are produced in small amounts at discrete sites, and accumulation does not occur (with the exception of entomopathogenic fungi, though only in hosts and thus in a limited quantity). In addition, many of the compounds have a short half-life. If data proving the absence of a certain compound from an environmental matrix (crops, soil) are required, technical difficulties make such proof impossible. Many of the compounds are not commercially available, and the development of methods is very expensive in terms of costs, time and workload. However, even if sensitive methods could be developed, proving absence is mostly impossible because many compounds will be below detection levels. Taken together, low *in situ* production rates and instability make it impossible to demonstrate the absence of a compound, and this often results in a 'data gap' that can prevent the approval of the microorganism as an active substance for plant protection. On the other hand, no reports of adverse effects from compounds produced by biocontrol organisms are known to the authors, and the risk appears to be extremely low. The probability that a biocontrol agent of a well-known species produces a yet unknown toxin can be neglected if no adverse effects are observed in studies on toxicity of the active ingredient or the product with mammals or other non-target organisms. Similarly, for a species that is not well described, studies with the active ingredient or the product should be sufficient to detect potentially toxic compounds. In addition, sufficient knowledge on the biology, ecology and mode of action needs to be provided to be able to assess the possibility of producing and accumulating potentially toxic compounds.

We believe that data requirements for microbial compounds should be limited to toxins that are well described and routinely assessed in food or feed unless data from toxicity studies on mammals or other organisms indicate that toxic compounds are present in the active ingredient or the product.

3.4.4 New EU guidance on secondary metabolites

The EU recently adopted a new guidance document on microbial metabolites (EC, 2020d), with the aim of identifying all compounds potentially produced by a strain, starting with literature information. Data on potential adverse effects and exposure need to be collected. Genome information can be used to exclude certain compounds from the assessment if it can be demonstrated that the gene(s) involved in the biosynthesis are absent. Subsequently, toxicity data are collected and compared with exposure in a risk assessment similar to those for chemical actives. Unless exposure from compounds contained in the product or produced *in situ* after application of the product cannot be excluded by experimental data (which is extremely difficult as detailed above), a risk assessment needs to be conducted. Unfortunately, this guidance starts from the assumption that all compounds are toxins, and that biocontrol microorganisms are toxin producers, similar to some fungi and bacteria cited above. Only unknown compounds produced *in situ* do not need to be considered unless adverse effects are observed in studies with non-target organisms. No practical experience with the guidance document exists to date, but it appears it would considerably increase the workload for applicants to prove the 'innocence' of a biocontrol organism if every compound needs to be assessed as a potential toxin.

3.5 Testing methods

Many data requirements can be addressed by the literature on the species. Others need strain- and/or product-specific experimental data and studies have to be conducted. Testing methods for studies to fulfil regulatory requirements are another aspect where improvements are needed.

3.5.1 Physical, chemical and technical properties of the formulated product

Guidelines for physical-chemical and technical properties are designed for conventional chemical products and the same guidelines are applied to microbial products. This is justified as products containing microorganisms are applied, in most cases, with the same equipment as conventional products, and compatibility and safety for the user need to be assured. However, sometimes methods are difficult to apply, especially if the content of the active substance needs to be determined in a bioassay, which involves activity testing via quantification of the amount of the biocontrol agent needed to kill the host of the biocontrol microorganism, as required for baculoviruses. For example, this is the case for testing suspensibility or spontaneity of dispersion, and many laboratories that can perform the technical test are not able to perform the bioassay, which complicates study conduct.

The determination of microbial contaminants is currently recommended to be performed using traditional plating methods, which were originally developed for food material with a relatively low microbial content. For a microbial biocontrol product consisting of a large proportion of microbial material, sensitivity of the growth media may not be sufficient and potential contaminants may still be overgrown by the biocontrol agent. The detection of contaminants according to the list provided in the Working Document on Microbial Contaminant Limits for Microbial Pest Control Products (EC, 2012) by quantitative polymerase chain reaction (qPCR) has already proven to be more sensitive and reliable in other fields but is still not widely accepted by authorities for PPPs.

Methods used in studies need to be validated, and again the relevant guidance document (EC, 2019b) was developed for chemical actives and not adapted for microorganisms. However, deficiencies in the guidance concerning microorganisms are recognized and adaptations are accepted by authorities.

3.5.2 Toxicity, infectivity and pathogenicity to mammals

Testing methods to assess toxicology, pathogenicity and infectivity to mammals are mostly acceptable. Methods developed in the USA (OCSSP 885.3050, 885.3150 and 885.3200) are widely accepted. These methods enable evaluation of the pathogenicity of the biocontrol microorganism and potential contaminants, infectivity of the biocontrol agent and contaminants, and toxicity of compounds contained in the test material that originates from the biocontrol agent, potential contaminants or impurities from growth media or co-formulants. Technically, clearance assessment for viruses or obligate parasites that do not proliferate in the absence of their host is probably the most difficult part, because their presence in the test animals can only be determined in a bioassay with the target organism. This bioassay might be inhibited by compounds from the test animal tissue matrix. For example, conducting an infectivity test with baculoviruses on their host insect with material from rat tissue may not be possible; however, the clearance assessment can be accomplished by qPCR methods.

Another point of discussion is the necessity to conduct studies for toxicology, pathogenicity and infectivity via three pathways, oral, pulmonary and injection. Pulmonary exposure is mostly achieved by intratracheal instillation instead of inhalation, which is technically difficult because of the particle nature of microbials. Injection application is preferably intravenous for bacteria and viruses and intraperitoneal for fungi and protozoa. Oral exposure is the most likely exposure scenario for humans (consumers, if the microorganism persists long enough) and wild mammals after product application, but on the other hand, both groups are continuously exposed to a plethora of microorganisms and mammalian defence systems are adapted to withstand microorganisms. Inhalation exposure is also relevant, especially for operators or bystanders, and lung infections are probably the most dangerous potential infection by biocontrol organisms, especially in immunocompromised people. Using the intratracheal application, microbial material is placed within the lungs at a site, which microbial particles would probably not access under natural conditions, because they would not stay in the airstream throughout the passage from the upper lung tracts to the trachea. Consequently, the exposure in the study represents more a worst-case scenario than a realistic scenario as the first protection of the upper respiratory tract against infection has been overcome by placing them at a site that would be unreachable under natural conditions. Intraperitoneal or intravenous injection testing represents an unrealistic exposure, and the outcome can, in most cases, be predicted from intratracheal studies. As skin infection through wounds is a more likely scenario of microbial infection than deep body infections, subcutaneous testing may be a preferred method if infectivity testing is required. We recommend a tiered approach: intratracheal testing should be performed first as the 'realistic worst case'. If no adverse effects are observed, no further testing should be required.

In our experience, over the last 20 years, adverse effects have only been observed in one infectivity study out of more than 100 studies involving over 50 microbial strains; all the other studies revealed no adverse effects. This demonstrates that knowledge on the taxonomy and biology of the strain is crucial for the evaluation and that testing can be reduced to a single study on intratracheal toxicity, pathogenicity and infectivity per strain.

3.5.3 Exposure to consumers and the environment

Data requirements do not foresee studies for residues and environmental fate, but information is required for survival, persistence and proliferation of the microorganism on crops, food and feed, and in soil, air and water. Such information on the species can generally be found in the literature. Strain-specific data could be generated along with efficacy trials under realistic application conditions, but they would only be needed for some entomopathogens with a broad host range. However, in general, introduced microorganisms always decline due to the natural biological buffering of the environment to levels that are within common fluctuations and ranges without strongly affecting microbial communities and pose no risk to the environment (Köhl et al., 2019a).

3.5.4 Effects on non-target organisms

Data requirements on the effects on non-target organisms can also be addressed using literature about the species, and knowledge on the taxonomy and biology of the biocontrol agent can exclude a hazard for many non-target organisms. If studies are needed, the guidelines used to date do not always address the critical point appropriately. Similar to human health assessment, studies should be able to address pathogenicity and infectivity by the biocontrol microorganism and potential contaminants, and toxicity of compounds contained in the test material that originates from the biocontrol agent, potential contaminants or impurities from growth media or co-formulants.

Guidelines for birds and fish are widely acceptable, but over recent decades, adverse effects have rarely been observed. *Daphnia* tests are probably the most difficult for microbial products as *Daphnia* are very sensitive to particles in their growth media. As microbial biocontrol products always present such particles, adverse effects are frequently observed in *Daphnia* studies although the particles have more impact than microorganisms *per se*. Determination of infectivity and pathogenicity in daphnids is very difficult due to the size of the test organism and the difficulty of distinguishing between microorganisms in or on the daphnid. Such assessment requires great experience and expertise but is not always possible. The test using *Daphnia* is probably more appropriate for the detection of potentially adverse effects caused by microbial compounds or process impurities than for the assessment of infectivity or pathogenicity. On the other hand, *Daphnia* pathogens are well known. Interestingly, two biocontrol agents currently approved in the EU belong to the same genera as daphnid pathogens (*Metschnikowia* and *Pasteuria*). The biocontrol microorganisms and the *Daphnia* pathogens are only distantly related and differ in their biology, and no adverse effects were observed in the studies. This again highlights that sound information on taxonomy and biology is sufficient to exclude a risk to non-target organisms.

Testing the effects of biocontrol agents on non-target arthropods requires adaptation of methods in many cases. Tests should assess infectivity/pathogenicity rather than toxicity. Finding the right test organism is difficult because some of the established test organisms live for a shorter time than the infection period needed to show an effect. This is the case, for example, for *Aphidius*, where adults only live a few days whereas infection may need up to 2 weeks to establish. Alternative tests would be far more intensive and, in many cases, would need to be developed. On the other hand, effects are expected for entomopathogens with a broad host range and can be excluded for many biocontrol microorganisms based on literature information. Very recently, Reinbacher et al. (2021) proposed a new method to test the effects of entomopathogenic fungi on *Gaeolaelaps* (*Hypoaspis*) *aculeifer*.

No adverse effects from biocontrol microorganisms are expected for soil microorganisms and earthworms and tests are not required. Argumentation based on literature is widely accepted. In fact, these tests are obsolete and do not provide any useful information for a risk assessment; hence, this data requirement should be removed.

A further aspect is the methods determining the effects on sewage organisms or on methods for the determination of drinking water quality. Interestingly, both issues are not included in the data requirements, but in the Uniform Principles. Effects on sewage treatment organisms can be excluded for biocontrol organisms because they are not aquatic organisms. Biocontrol agents may survive in water, but seldom enter in a vegetative phase and very rarely proliferate. Furthermore, they are weak competitors compared with sewage microorganisms and only small populations reach wastewater treatment plants. A risk to sewage microorganisms can therefore be excluded. Methods for drinking water analysis are specific for potential contaminants and designed not to interfere with other organisms. Therefore, any interference is unrealistic. Both data requirements are obsolete.

4 New research, novel products and application methods

Research on the biological control of pests and diseases involves both academia and industry, and regularly delivers innovative products. A search on Google Scholar with the term 'biocontrol agent' presents 30 000 hits over the last 5 years, with even more hits on 'biocontrol products'. How many of these products reach the market is impossible to say, but novel, innovative PPPs have to pass regulations and, as is well known, innovation is often hampered by regulations that were not designed for those novelties. One could even say this for biocontrol products in general that have to pass regulations designed for chemicals. New biocontrol products that we see today or in the near future include bacteriophages, weakened or mild strains of plant viruses, endophytic microorganisms, microbial consortia, amoebae (already developed as a biocide), gene-edited microorganisms, genetically modified microorganisms, etc. Research on the microbiome of plants is a new, exciting field where many microbes are being discovered as well as their functions in relation to the plant. When these relate to plant protection will they fall under the current regulations? We still need to see practical applications originating from this microbiome work, but it is bound to generate new approaches to keep plants healthy. In addition, we see new functions and applications of biocontrol agents, sometimes with multiple modes of action and multiple uses, such as biofertilizers, biostimulants, inoculants, plant defence inducers or elicitors, etc. Novel application methods are developed to deliver the microorganism to their target, for example, using honeybees and bumblebees as vectors to deliver the product at the flower, stem-injection (see elsewhere in this book), applying weak plant viruses by making small leaf surface wounds via a rub-in liquid, various types of seed treatment, and wound treatment via pruning scissors. Furthermore, formulations with two active substances can be developed, such as with two strains, or a microorganism and a chemical or a botanical. With precision agriculture, we expect a different method of controlling pests. Early applications will become more routine based on early observations of pests and diseases in distinct locations, with different application devices, such as robots and drones. What are the consequences of the risk assessment of such technologies? It can be a real challenge to get such innovative products, with their new ways of application, through the registration system. It requires a flexible approach in building a dossier, but also in the manner of risk evaluation by authorities. The product developers understand their new products well, but for evaluators it can be a challenge to reach the same level of understanding and allow a flexible approach; hence, ticking the box on all standard questions is not an appropriate approach. Knowledge of the subject should guide a risk assessment and will be key with such novel PPPs as mentioned above, and the training of evaluators will become even more relevant. The development of a decision tree will help to guide applicants and evaluators through the system and allow only relevant issues to be emphasized. Guidance is crucial to applicants, so consultation with authorities at an early phase is an absolute requirement, such as pre-submission meetings, guidance documents, workshops and training.

5 Conclusion and future trends

5.1 The Pesticide Paradox

Pesticides have been under societal pressure ever since the publication of '*Silent Spring*' by Rachel Carlson in the 1960s. Many organizations advocate for the decreased use of chemical pesticides because they have harmful effects on human health and the environment, even after their approval by authorities. Recently, there have been fierce ongoing debates over chemicals such as glyphosate and neonicotinoids. Switzerland will have a vote to ban synthetic pesticides in early 2021. Another example is the launch in early 2020 of an ambitious roadmap by 24 European research institutes to find alternatives to chemical pesticides (https://www.inrae.fr/sites/default/files/pdf/CP_MoU_VFinal2302_VA_0.pdf; accessed 25 October 2020). Moreover, authorities recognize this and policies are being developed to reduce the use of chemical pesticides, e.g., the Sustainable Use of Pesticides Directive (Dir. 2009/128/EC) (EC, 2009b) and, very recently, the EU Farm to Fork (F2F) Strategy (EC, 2020b) as part of the EU Green Deal. Surprisingly, little or no provision has been developed to allow alternatives to chemical pesticides to enter the market in a timely and appropriate way. Pests and diseases will always be present in agricultural crops and farmers need tools to protect their crops and maintain food security; bioprotectants can be one set of tools that can be used in a sustainable way. The F2F Strategy states: 'The Commission will also facilitate the placing on the market of pesticides containing biologicalactive substances and reinforce the environmental risk assessment of pesticides. It will act to reduce the length of the pesticide authorization process by Member States'. The word 'facilitate' most likely refers to revision of the data requirements for microorganisms. However, to what extent this will make the questions more appropriate remains to be seen. Moreover, the Commission does not appear to be planning to revise the requirements for pheromones and botanicals. Draft guidance, similar to that on secondary metabolites, does not appear to make it easier, but rather more demanding. On the other hand, we see that a risk assessment can be made more appropriate to fit the biology of a bioprotectant, the example we refer to is the draft guidance for bacteriophages. Phages and baculoviruses are the outstanding examples of the direction crop protection products should be taking, to specific solutions that target the intended pests and without harmful effects on other organisms. Many microbial bioprotectants are developed for niche markets as their biology only allows such narrow targeting. Examples are baculoviruses, but also many fungi, yeasts and bacteria that, due to their mode(s) of action, only have effects on a limited number of other organisms. The decision tree discussed above is based on biology and ecology, and a risk assessment should only address the real risks related to the biology of the microorganism. We are convinced that in this way, with specific microorganisms, safe PPPs can be developed for nearly all pests and diseases. With this approach, many small products have the potential to become available for the farmer's toolbox. However, here lies the difficulty, the regulatory system has developed in such a way that the expense and administrative burden has become so enormous that it is impossible for small-scale applicants with niche products to get through the system (Robin and Marchand, 2019); however, these are exactly the type of products that are demanded in our society. The system has become so restrictive that only broad-spectrum products with big markets and developed by multinationals can get through. The lengthy procedure of about 5 years is a compounding factor, as waiting that long for any income is difficult or even impossible to overcome for small companies; it also deters investors in this area. Product development and registration combined take about 10 years and a return on investment from the outset of a project can only be expected after 13–15 years. It is clear that getting niche products to the market is a very complex and extremely risky entrepreneurial exercise. Although our society wants safe and selective PPPs to be used by farmers, the regulatory system is a huge hurdle to achieving that goal. The GFL will only exacerbate this situation although its purpose is to have a more transparent and accepted risk assessment. This is the great paradox of pesticide registration in the EU. Therefore, we advocate for a separate legislative framework for bioprotectants, particularly for products based on microorganisms. An independent system with its own resources and expert evaluators will then be able to achieve timely market access for the innovative and sustainable tools that agriculture needs from now on. We have seen, over a number of years, that a large part or even more than half of the new active substances are based on products of biological origin, so the system should be ready imminently to deal with that. There can be no reversion to traditional chemicals and an ambitious EU agricultural sector cannot wait any longer for such a new approach.

5.2 Crop protection policy

This policy is mainly shaped via efforts to reduce the use of chemical pesticides and the dependency thereon. What we see in practice is that the Sustainable Use Directive (SUD) goal is not reached as Integrated Pest Management (IPM) is hardly used by European farmers and this reduction is not achieved, with dependency on chemicals remaining very high (ECA, 2020). A tool like the comparative assessment of chemicals for candidates for substitution has failed completely. Chemicals are withdrawn from the market when they are re-assessed and this was achieved with the implementation of Regulation 1107/2009 in the process of taking over from Directive 91/414 when approximately 400–500 actives were taken from the market. It is a process that is based solely on the risk assessment of individual chemicals. Market demand and company policies determine whether new actives are submitted for approval in Europe; however, the number of pesticides over the last 10 years has increased (Buckwell et al., 2019). Farmers have complained, and still do, about not having enough tools, but this pesticide policy does not look at whether a farmer has a sufficient set of tools for a certain crop and for an appropriate IPM programme. The European policy, in this respect, is completely insufficient, and to our knowledge, there are no programmes in individual Member States that are performing any better despite several attempts. New attempts are being developed in countries like France and the Netherlands. The Dutch plan aims for resilient agriculture, where resilient crops and minimal use of low-risk pesticides will protect the crops and have less impact on the environment. This approach demands a better and faster registration system for new, selective low-risk pesticides, mainly bioprotectants, and only a new approval regime will be able to deliver that. Farmers will need to be further trained in this field and need advice from IPM experts. The ambitious plans in the EU need to deliver sustainable agriculture within the next 10 years in order to stop the decline in biodiversity, soil fertility, pollinators and other ecosystem services, as well as the debates on the harmful effects of today's pesticides. The future of crop protection is in regenerative agriculture with bioprotectants and biostimulants as the farmers' tools to intervene in cases of abiotic and biotic stresses, in crop-specific IPM programmes. An innovative legislative framework is needed to make that possible. In addition to ambitious plans in the EU Green Deal, efforts are also required to design an appropriate regulatory system to support the transition.

5.3 Recommendations to improve the approval process and accelerate market access of microbial bioprotectants

The registration of a microbial bioprotectant is a huge challenge worldwide and assessing safety for humans and the environment is the key issue. The variety of microorganisms in nature is enormous and new ones are still being discovered, and for the majority of them, scientists do not know what their functions are. With modern technologies, the microbiome of the plant and soil are being studied and both contain a plethora of microorganisms that are crucial to the effective functioning of both plant and soil (see elsewhere in this book). On the other hand, many are known to be beneficial, while others can cause serious problems and diseases in many organisms. These features are most likely the cause of a general fear of 'invisible' microorganisms as we experience today with the coronavirus pandemic.

Is the risk assessment driven by fear of the unknown? Microorganisms make all kinds of compounds to defend themselves from antagonists and predators and these substances can have beneficial as well as harmful effects. Microorganisms can adapt to new situations and even transfer genetic material; these combined factors make them extremely complex and hard to fully understand. However, they are part of nature and have evolved in conjunction with other organisms.

In order to assess their use and safety, their biological origin must be considered in this context. This requires evaluators that understand this aspect of biological sciences and ecological principles. Living microorganisms are totally different from chemical molecules and assessing their safety is a completely different discipline. The first microorganism used as a PPP was evaluated as a chemical product, following the same regime, very similar data requirements and by the same evaluators. This was quite logical and understandable, but of course that system was not suitable for the purpose. Adaptations were made over time, over the course of decades, to improve the process. This was repeated in many countries and today that process is still ongoing. The above sections demonstrate that the current legislative system in the EU has its flaws and weaknesses, and is in fact a complete mismatch, despite all the attempts to repair it with adapted data requirements, guidance, etc. The authors' experience goes back 40 years when it was a struggle in the EU and still is today, and maybe even more so. Recommendations to improve the EU system were formulated in the EU policy support Action REBECA (Ehlers, 2011) and further detailed by Scheepmaker et al. (2019) for microbial metabolites, but very few of these recommendations have been implemented.

Based on the above arguments we conclude the following:

Recommendations for improvements in the procedure:

- Establish specialized teams with well-trained experts;
- Concentrate evaluation of microbials with a few Member States in the EU; and
- Implement conditional approval for low-risk products.

Recommendations for improvements in data requirements:

- Consider the biology and ecology of the biocontrol agent first;
- Restrict evaluation of 'metabolites' to known toxins;
- Conduct studies only if other information is not sufficient; and
- Adapt study protocols.

Remarkably, we are currently witnessing the same situation with biostimulants. This new category of products needs to be regulated due to concerns over their use and safety, and they have been added within legislation that was designed to cover plant nutrients and fertilizers. In particular, biostimulants based on microorganisms do not fit in and the system lacks a procedure and data requirements to assess their safety and approval. They have been placed in a legislative framework that superficially seems acceptable but in fact simply causes problems, in parallel to what happened with microbial PPPs. Sundh and Eilenberg (2020) observed that for many applications of microorganisms in food, feed, agriculture and the environment, regulations and the risk assessments vary greatly and lack proportionality. Similar issues can be expected when microbiome research leads to the use of more microbiological products in various disciplines. These issues may be justified in the early days of such microbial products but when the utilization of microorganisms increases greatly in all these different uses, a new approach is necessary. These microorganisms need to be assessed on safe use and hence, the exposure of humans and the environment to them. Their claimed functional use is hardly relevant as microorganisms are not limited to one claimed area but can have several properties, such as plant protection and biostimulation, which may be expressed depending on the environmental conditions. Their biological characteristics dictate whether or not they can be safely used, and not the administrative box they are put in. Currently, a strain of a microorganism is assessed differently on safety related to which use is claimed and under which legislation with its own data requirements it falls. This makes no sense; therefore, a specially designed legislative framework for microorganisms is required in which safety is assessed independently from claimed functions. Data requirements need to be proportionate to the actual risk, and literature information at species level should be accepted as the basis of the evaluation. Together with solid molecular identification and taxonomy, a microorganism can be assigned to an ecological group, which defines what experimental data are needed in addition to the literature on the species for the risk assessment. The new system should also have scientists educated and experienced with microorganisms, including pathology and ecology. Procedures should be set up and resources allocated specifically for this administrative process. In this way, conflicts with other products under the same regulation can be avoided as we see today with microbials in the chemical system and biostimulants in the fertilizer system. In current systems, we see that attention and resources are mainly used for the biggest groups of products with the biggest concerns. Examples are, respectively, health and environmental concerns with pesticides, and pollution concerns with fertilizers. With a separate system, this can all be avoided, the assessments will be more appropriate and market access can be achieved much faster. Products that are much more sustainable and environmentally friendly will then reach the market, and policy goals on circular economy, less pollution and increased sustainability can be achieved. Therefore, we strongly appeal for a new legislative framework for the approval of microorganisms for use as PPPs, plant biostimulants, feed additives, biocides and maybe more in the future. The EU can lead the way and be an example for other countries around the world. We advise considering the suggestion of the former Ctgb director L van Duijn (2018) who proposed using as a baseline 'biopesticides are low risk unless...'. Many of his other suggestions are in line with many of our recommendations and solutions. His low-risk proposal was submitted to the EU via the Dutch Ministry of Agriculture as a reply to actions needed based on the Refit evaluation of Regulation (EC) 1107/2009.

The daily exposure of humans to millions of microorganisms in our food and our co-evolution can be regarded as the basis for this proposal. With every apple we consume we are exposed to 100 million (10^8) bacteria (Wassermann et al., 2019), the majority of which are unknown and have never been studied. A similar situation is found regarding fungi in strawberries and apples (Abdelfattah et al., 2016a,b). Interestingly, it is broadly recommended to eat fresh fruit every day for our health, which is illustrated by the well-known saying 'an apple a day keeps the doctor away'. The application of a biocontrol agent would result in a population of 1.3×10^6 colony forming units (CFU)/apple, which is about 1% of the overall population on an apple (see Box 1 for the detailed calculation).

Box 1

To estimate the populations of a biocontrol microorganism on apples resulting from the application of a microbial biocontrol product, the following assumptions are made:

- Single application rate for the biocontrol microorganism: 3×10^{12} CFU/ha
- Apple yield: 28 t/ha (average for Germany, in France 38–40 t/ha)

$$3 \times 10^{12} \text{ CFU} / 28 \text{ t} = 1.1 \times 10^5 \text{ CFU/g}$$

Apple weight: 200 g: 2.2×10^7 CFU/apple.

- If 60% of the population end up on plants and 40% on soil, theoretical maximum population levels are 6.6×10^4 CFU/g fruit or 1.3×10^7 CFU/apple. This calculation does not take into account that the majority of cells of the biocontrol agent will end up on leaves and not on fruit because the fruit surface is relatively small when compared with leaves. If we assume that 10% of the biocontrol agent ends up on fruit, the population would be 1.3×10^6 CFU/apple. This is about 1% of the existing bacterial microbiome of the apple.
- In addition, decreased populations of the biocontrol agent between application of the product and harvest through UV radiation or washing off by rain is not considered, but this decrease is substantial and requires repeated applications.

How relevant is the addition of 1% of a well-characterized microbial strain to the mostly uncharacterized microbiome on a fruit and how realistic are risks associated with this additional microorganism? This example justifies, in our opinion, that the evaluation of microorganisms needs to be revised.

6 References

Abdelfattah, A., Wisniewski, M., Li Destri Nicosia, M. G., Cacciola, S. O. and Schena, L. (2016a). Metagenomic analysis of fungal diversity on strawberry plants and the effect of management practices on the fungal community structure of aerial organs. *PLoS ONE* 11(8), e0160470. https://doi.org/10.1371/journal.pone.0160470.

Abdelfattah, A., Wisniewski, M., Droby, S. and Schena, L. (2016b). Spatial and compositional variation in the fungal communities of organic and conventionally grown apple fruit at the consumer point-of-purchase. *Hortic. Res.* 3, 16047. Available at: https://doi.org/10.1038/hortres.2016.47 (accessed 5 October 2016).

Arjjumend, H. and Koutouki, K. (2018). Science of biopesticides and critical analysis of Indian legal frameworks regulating biocontrol agents. *Int. J. Agric. Environ. Biotechnol.* 11(3), 563–571.

Arora, N. K., Verma, M., Prakash, J. and Mishra, J. (2016). Regulation of biopesticides: global concerns and policies. In: *Bioformulations: For Sustainable Agriculture* Arora, N. K. and Mehnaz, S. and Raffaella Balestrini, R. (Eds), Springer, New Delhi, pp. 283–299.

Arthurs, S. and Dara, S. K. (2019). Microbial biopesticides for invertebrate pests and their markets in the United States. *J. Invertebr. Pathol.* 165, 13–21.

Balog, A., Hartel, T., Loxdale, H. D. and Wilson, K. (2017). Differences in the progress of the biopesticide revolution between the EU and other major crop-growing regions. *Pest Manag. Sci.* 73(11), 2203–2208.

Bettiol, W. (2011). Biopesticide use and research in Brazil. *Outlooks Pest Manag.* 22(6), 280–283.

Buckwell, A., de Wachter, E., Nadeu, E. and Williams, A. (2019). *Crop Protection & The EU Food System: Where Are They Going? Report*, Rise Foundation, Brussels.

Deising, H. B., Gase, I. and Kubo, Y. (2017). The unpredictable risk imposed by microbial secondary metabolites: how safe is biological control of plant diseases? *J. Plant Dis. Prot.* 124(5), 413–419. Available at: https://doi.org/10.1007/s41348-017-0109-5.

Domingues, F. (2019). Reforms to the Brazilian regulatory system, the good and the bad areas. *Presentation given at Annual Biocontrol Industry Meeting (ABIM) 2019*. Available at: https://www.abim.ch/index.php?eID=tx_securedownloads&p=124&u=0&g=0&t=1606562638&hash=749d057ae549ca14cc84b323051a91fda7b32909&file=/fileadmin/abim/documents/presentations2019/ABIM_2019_2_02_Fabio_Domingues.pdf (accessed 27 November 2020).

EPPO (2017). Efficacy evaluation of plant protection products PP 1/296 (1) Principles of efficacy evaluation for low-risk plant protection products. *Bulletin OEPP/EPPO Bulletin* 47, 297–304.

EC (2001). Commission directive 2001/36/EC of 16 May 2001 amending council directive 91/414/EEC concerning the placing of plant protection products on the market. Available at: https://eur-lex.europa.eu/legal-ontent/EN/TXT/PDF/?uri=CELEX:32001L0036&from=EN (accessed 5 December 2020).

EC (2002). Regulation (EC) No 178/2002 of the European Parliament and of the Council of 28 January2002 laying down the general principles and requirements of food law, establishing the European Food Safety Authority and laying down procedures in matters of food safety. Available at: https://eur-lex.europa.eu/legalcontent/EN/TXT/PDF/?uri=CELEX:32002R0178&from=EN (accessed 7 December 2020).

EC (2008). Guidance Document on the assessment of new isolates of baculovirus species already included in Annex I of Council Directive 91/414/EEC. SANCO/0253/2008 rev. 2; 22 January 2008. Available at: https://ec.europa.eu/food/sites/food/files/plant/docs/pesticides_aas_guidance_baculovirus.pdf (accessed 23 December 2020).

EC (2009a). Regulation (EC) No 1107/2009 of the European Parliament and of the Council of 21 October 2009 concerning the placing of plant protection products on the market and repealing Council Directives 79/117/EEC and 91/414/EEC. Available at: https://eur-lex.europa.eu/legal-content/EN/TXT/PDF/?uri=CELEX:32009R1107&from=EN (accessed 5 December 2020).

EC (2009b). Directive 2009/128/EC of the European Parliament and of the Council of 21 October 2009 establishing a framework for Community action to achieve the sustainable use of pesticides. Available at: https://eur-lex.europa.eu/legal-content/EN/TXT/PDF/?uri=CELEX:32009L0128&from=EN (accessed 2 December 2020).

EC (2011). Commission Regulation (EU) No 546/2011 of 10 June 2011 implementing Regulation (EC) No 1107/2009 of the European Parliament and of the Council as regards uniform principles for evaluation and authorisation of plant protection products. Available at: https://eur-lex.europa.eu/legalcontent/EN/TXT/PDF/?uri=CELEX:32011R0546&from=EN (accessed 2 December 2020).

EC (2012). Working document on microbial contaminant limits for microbial pest control products SANCO/12116/2012 –rev. 0 September 2012.

EC (2013a). Commission Regulation (EU) No 283/2013 of 1 March 2013 setting out the data requirements for active substances, in accordance with Regulation (EC) No 1107/2009 of the European Parliament and of the Council concerning the placing of plant protection products on the market. Available at: https://eur-lex.europa.eu/legal-content/EN/TXT/PDF/?uri=CELEX:32013R0283&from=EN (accessed 23 December 2020).

EC (2013b). Commission Regulation (EU) No 284/2013 of 1 March 2013 setting out the data requirements for plant protection products, in accordance with Regulation (EC) No 1107/2009 of the European Parliament and of the Council concerning the placing of plant protection products on the market. Available at: https://eur-lex.europa.eu/legal-content/EN/TXT/PDF/?uri=CELEX:32013R0284&from=EN (accessed 23 December 2020).

EC (2017). Commission Regulation (EU) 2017/1432 of 7 August 2017 amending Regulation (EC) No 1107/2009 of the European Parliament and of the Council concerning the placing of plant protection products on the market as regards the criteria for the approval of low-risk active substances. Available at: https://eur-lex.europa.eu/legal-content/EN/TXT/PDF/?uri=CELEX:32017R1432&from=EN (accessed 2 December 2020).

EC (2018). Commission Notice concerning a list of potentially low-risk active substances approved for use in plant protection (2018/C 265/02). Available at: https://eur-lex.europa.eu/legal-content/EN/TXT/PDF/?uri=CELEX:52018XC0727(01)&rid=7 (accessed 2 December 2020).

EC (2019a). Regulation (EU) 2019/1381 of the European Parliament and of the Council of 20 June 2019 on the transparency and sustainability of the EU risk assessment in the food chain and amending Regulations (EC) No 178/2002, (EC) No 1829/2003, (EC) No 1831/2003, (EC) No 2065/2003, (EC) No 1935/2004, (EC) No 1331/2008, (EC) No 1107/2009, (EU) 2015/2283 and Directive 2001/18/EC. Available at: https://eur-lex.europa.eu/legal-content/EN/TXT/PDF/?uri=CELEX:32019R1381&from=EN (accessed 7 December 2020).

EC (2019b). Technical active substance and plant protection products: guidance for generating and reporting methods of analysis in support of pre- and post-registration data requirements for Annex (Section 4) of Regulation (EU) No 283/2013 and Annex (Section 5) of Regulation (EU) No 284/2013. SANCO/3030/99 rev.5 22 March 2019.

Available at: https://ec.europa.eu/food/sites/food/files/plant/docs/pesticides_ppp_app-proc_guide_phys-chem-ana_3030.pdf (accessed 23 December 2020).

EC (2020a). Report from the commission to the European Parliament and the council evaluation of regulation (EC) No 1107/2009 on the placing of plant protection products on the market and of Regulation (EC) No 396/2005 on maximum residue levels of pesticides {COM(2020) 208 final}. Available at: https://eur-lex.europa.eu/legal-content/EN/TXT/PDF/?uri=CELEX:52020SC0087&from=EN (accessed 2 December 2020).

EC (2020b). Farm to fork strategy. For a fair, healthy and environmentally-friendly food system. Available at: https://ec.europa.eu/food/sites/food/files/safety/docs/f2f_action-plan_2020_strategy-info_en.pdf; accessed 2 December 2020).

EC (2020c). Guidance on the approval and low-risk criteria linked to "Antimicrobial Resistance" applicable to microorganisms used for plant protection in accordance with Regulation (EC) No 1107/2009, SANTE/2020/12260, 23 October 2020. Available at: https://ec.europa.eu/food/sites/food/files/plant/docs/pesticides_ppp_app-proc_guide_180652_microorganism-amr_202011.pdf (accessed 23 December 2020).

EC (2020d). Guidance on the risk assessment of metabolites produced by microorganisms used as plant protection active substances in accordance with Article 77 of Regulation (EC) No 1107/2009, SANCO/2020/12258 23 October 2020. Available at: https://ec.europa.eu/food/sites/food/files/plant/docs/pesticides_ppp_app-proc_guide_180653_microorganism-metabolites-concern_202011.pdf (accessed 23 December 2020).

ECA (2020). Sustainable use of plant protection products: limited progress in measuring and reducing risks. Special report 05/2020. Available at: https://www.eca.europa.eu/Lists/ECADocuments/SR20_05/SR_Pesticides_EN.pdf (accessed 7 December 2020).

Ehlers, Ralf-Udo (Ed.) (2011). *Regulation of Biological Control Agents*. Springer Science + Business Media B.V., Dordrecht, Heidelberg, London, New York.

Frederiks, C. and Wesseler, J. H. H. (2018). A comparison of the EU and US regulatory frameworks for the active substance registration of microbial biological control agents. *Pest Manag. Sci.* 75(1), 87–103.

Grant, W. and Gwynn, R. (2020). Improving regulatory approval processes for biopesticides and other new biological technologies in agriculture. In: Biopesticides for *Sustainable Agriculture* Birch, N. and Glare, T. (Eds), Burleigh Dodds Science Publishing Ltd., Cambridge, pp. 45–60.

Guest, P. (2018). Global biopesticide regulations 2018. Report, Agribusiness Intelligence Informa UK, London.

Guijarro, B., Fernández-Getino, A. P. and Alonso-Prados, J. L. (2020). Regulation and requirement challenges for the definition of a successful biocontrol strategy within the EU framework. In: *How Research Can Stimulate the Development of Commercial Biological Control Against Plant Diseases, Progress in Biological Control 21* De Cal, A., De Cal, P., Melgarejo, P. and Magan, N. (Eds), Springer Nature, Switzerland AG, pp. 323–346. Available at: https://doi.org/10.1007/978-3-030-53238-3_19.

Hauschild, R., Speiser, B. and Tamm, L. (2011). Regulation according to EU Directive 91/414: Data requirements and procedure compared with regulation practice in other OECD countries. In: Ehlers, Ralf-Udo (Ed), *Regulation of Biological Control Agents*. Springer Science + Business Media B.V., Dordrecht, Heidelberg, London, New York, Chapter 2, S. 25-77.

Hauxwell, C., Tichon, M., Buerger, P. and Anderson, S. (2010). Registration and use of microbial biopesticide in Australia. In: *The Use and Regulation of Microbial Pesticides in Representative Jurisdictions Worldwide*. Kabaluk, J. T., Svircev, A. M., Goettel, M. S. and Woo, S. G. (Eds). IOBC Global, pp. 80–88. Available at: www.IOC-Global.org

IBMA (2018). New regulatory framework for bioprotection agents, IBMA vision on how to improve regulation in the European Union. IBMA White Paper, Brussels, IBMA.

Keswani, C., Dilnashin, H., Birla, H. and Sigh, S. P. (2019). Regulatory barriers to Agricultural Research commercialization: A case study of biopesticides in India. *Rhizosphere* 11, 10155. Available at: https://doi.org/10.1016/j.rhisph.2019.100155.

Koch, E., Becker, J. O., Berg, G., Hauschild, R., Jehle, J., Köhl, J. and Smalla, K. (2018). Biocontrol of plant diseases is not an unsafe technology! *J. Plant Dis. Prot.* 125(2), 121–125. Available at: https://doi.org/10.1007/s41348-018-0158-4.

Köhl, J., Booij, K., Kolnaar, R. and Ravensberg, W. J. (2019a). Ecological arguments to reconsider data requirements regarding the environmental fate of microbial biocontrol agents in the registration procedure in the European Union. *BioControl* 64(5), 469–487.

Köhl, J., Kolnaar, R. and Ravensberg, W. J. (2019b). Mode of action of microbial biological control agents against plant diseases: relevance beyond efficacy. *Front. Plant Sci.* 10(845), 845. Available at: https://doi.org/10.3389/fpls.2019.00845.

Leahy, J., Mendelsohn, M., Kough, J., Jones, R. and Berckes, N. (2014). Biopesticide oversight and registration at the U.S. Environmental Protection Agency. In: *Biopesticides: State of the Art and Future Opportunities* Gross, A. D., Coats, J. R., Duke, S. O. and Seiber, J. N. (Eds), American Chemical Society, 1172(16), 3–18.

OECD (2018). Working document on the risk assessment of secondary metabolites of microbial biocontrol agents, series on pesticides no. 98, ENV/JM/MONO(2018)33; Annex 1, ENV/JM/MONO(2018)33/ANN1.

Ravensberg W. J. (2011). Critical factors in the successful commercialization of microbial pest control products. In: Ravensberg W. J. (ed) A roadmap to the successful development and commercialization of microbial pest control products for control of arthropods, Vol 10, Progress in biological control. Springer Science Business Media, Dordrecht, pp 295–356

Reinbacher, L., Bacher, S., Praprotnik, E. and Grabenweger, G. (2021). Standard non-target tests for risk assessment of plant protection products are unsuitable for entomopathogenic fungi—a proposal for a new protocol. *J. Soils Sediments* 21(6), 2357–2368. Available at: https://doi.org/10.1007/s11368-021-02919-w.

Robin, D. C. and Marchand, P. A. (2019). Evolution of the biocontrol active substances in the framework of the European pesticide Regulation (EC) No. 1107/2009. *Pest Manag. Sci.* 75(4), 950–958.

Scheepmaker, J. W. A., Busschers, M., Sundh, I., Eilenberg, J. and Butt, T. M. (2019). Sense and nonsense of the secondary metabolites data requirements in the EU for beneficial microbial control agents. *Biol. Control* 136, 104005.

Singh, H. B., Sarma, B. K. and Keswani, C. (2016). *Agriculturally Important Microorganisms. Commercialization and Regulatory Requirements in Asia*, Springer, New Delhi.

Sundh, I. and Eilenberg, J. (2020). Why has the authorization of microbial biological control agents been slower in the EU than in comparable jurisdictions? *Pest Manag. Sci.* 77(5), 2170–2178. Available at: https://onlinelibrary.wiley.com/doi/pdf/10.1002/ps.6177 (accessed 15 December 2020).

Togni, P. H. B., Venzon, M., Lagoa, A. C. G. and Sujii, E. R. (2019). Brazilian legislation leaning Towards fast registration of biological control agents to benefit organic agriculture. *Neotrop. Entomol.* 48(2), 175–185.

van Duijn, L. P. (2018). A member state proposal to improve the regulatory approach for biological PPPs. *Presentation given at Annual Biocontrol Industry Meeting (ABIM) 2018*. Available at: https://www.abim.ch/fileadmin/abim/documents/presentations2018/1_03_Luuk_van_Duijn.pdf (accessed 23 December 2020).

Villaverde, J. J., Sevilla-Morán, B., Sandín-España, P., López-Goti, C. and Alonso-Prados, J. L. (2014). Biopesticides in the framework of the European pesticide Regulation (EC) No. 1107/2009. *Pest Manag. Sci.* 70(1), 2–5.

Wassermann, B., Müller, H. and Berg, G. (2019). An apple a day: which bacteria do we eat With organic and conventional apples? *Front. Microbiol.* 10, 1629. Available at: https://doi.org/10.3389/fmicb.2019.01629 (accessed 24 July 2019).

Wesseler, J. (2019). Perspective: regulation of pest and disease control strategies and why (many) economists are concerned. *Pest Manag. Sci.* 75(3), 578–582.

Chapter 7

Microbial bioprotectants and the marketplace

Mark C. Trimmer, DunhamTrimmer LLC, USA

1 Microbial market history

Microbials have been known and used as bioprotectants for more than two centuries. *Beauveria bassiana* was isolated and described by the Italian entomologist Agostino Bassi in 1815 (Bell 1974). Fungal pathogens dominated the early use of microbials, but production challenges and performance consistency were lacking and interest declined (Lord 2005). *Bacillus thuringiensis* (Bt) was first isolated from silkworms by Ishiwatari Shigetane in 1901 in Japan and subsequently rediscovered by Ernst Berliner who found the same organism infecting flour moths in 1911 (Biological Products Industry Alliance; Sanahuja et al. 2011). Berliner was the first to recognize and suggest the use of Bt as an insect control tool through direct application in the field or greenhouse.

Despite this early knowledge of microbial potential for insect control, their use was extremely limited until the mid-twentieth century. The first commercial Bt product was Sporeine in France in 1938 (Sanahuja et al. 2011). The efficacy of early products was highly variable because there was a lack of understanding of how Bt killed insects, the importance of strain selection, how to optimize production, and how to formulate and apply the product in the field.

http://dx.doi.org/10.19103/AS.2021.0093.08
Published by Burleigh Dodds Science Publishing Limited, 2022.

In 1957, the first significant Bt product in the USA was introduced by Sandoz under the brand name Thuricide (Sanahuja et al. 2011). During the 1960s, researchers began to gain a better understanding of the Bt insecticidal toxin and to identify more effective strains, and standards for determining Bt potency were developed. Molecular biology advances in the 1970s and 1980s greatly advanced the understanding of the genetics of Bt toxin mode of action. Between 1961 and 1995, the US EPA approved 177 products containing Bt (Siegel 2001).

Parallel to the Bt development, baculoviruses were identified in the 1940s as possible biopesticides. It took nearly 50 years for the first commercial baculovirus-based bioprotectants to be launched, but today they are used successfully for control of a range of Lepidopteran pests.

As the understanding of the potential use of various microbials as bioprotectants expanded, the nascent market for baculovirus products began to emerge in the 1970s with the launch of the first commercial product Elcar (*Helicoverpa zea* nuclear polyhedrosis virus) as well as several non-commercial baculovirus-based products launched by the US Forestry Service (Black, et al. 1997). The early products suffered from poor potency, high costs, poor formulation characteristics and inferior efficacy in comparison to chemical alternatives.

The use of microbials for plant disease control also has a long history. *Bacillus subtilis* was first characterized in 1835 by C. G. Ehrenberg who classified it as *Vibrio subtilis* (Ehrenberg 1835). In 1872, it was reclassified and renamed *B. subtilis* by F. Cohn (Cohn 1872). The potential for *B. subtilis* for plant disease prevention was not recognized until the twentieth century, when its ability to trigger induced systemic resistance response in plants as well as the production of compounds that mitigate plant pathogen infections in plants were identified (Shafi, et al. 2017). The company AgraQuest (acquired by Bayer in 2012) was the first to launch a successful *B. subtilis*-based biofungicide in 2000, with the EPA approval of its product Serenade (US Federal Register, August 30, 2000). A closely related species, *Bacillus amyloliquefaciens* was discovered in 1943 in Japan (Fukumoto 1943). For nearly 50 years, bacteriologists debated whether *B. amyloliquefaciens* was a separate species or a subspecies of *B. subtilis*. The question was finally settled in 1987, with the decision that it was a new species (Priest et al. 1987). Since that time, many companies have developed and launched products containing strains of either *B. subtilis* or *B. amyloliquefaciens* as biofungicides, and this group of products is now among the market leaders in the biofungicide segment.

The value of *Trichoderma* spp. as a tool for plant disease prevention was recognized early in the twentieth century. In the 1930s, *Trichoderma lignorum* was identified as a parasite of other soil fungi (Weindling 1932). Other species, including *T. asperellum, T. atroviride, T. harzianum* and *T. virens*, among others, were subsequently identified, characterized and shown to be effective biocontrol tools (Howell 2003). The broad range of species and strains and their value as biofungicides has led to their widespread development and commercialization. A survey published in 2014 estimated there were more than 250 products found in the international market containing *Trichoderma* spp. (Woo et al. 2014).

More recently, *Pseudomonas* spp. have been studied as potential biological control agents. In the 1990s, the first *Pseudomonas* spp.-based products were approved as biopesticides by the US EPA, for control of soil fungal pathogens, the suppression of fire blight (*Erwinia amylovora*) in pome fruit and for the suppression of postharvest diseases (Stockwell and Stack 2007). In the same period, *Streptomyces lydicus* WYEC 108 was isolated from linseed plant roots (Yuan and Crawford 1995). The strain was shown to be effective in preventing soil fungal diseases by excreting lytic enzymes (chitinases, glucanases and peroxidases) which degrade the cell walls of pathogenic fungi. The product Actinovate was developed and commercialized by the company Natural Industries (acquired by Novozymes in 2012). Actinovate today is distributed by Valent BioSciences under a license from Novozymes. These products have found success in some market segments, but they have not seen the broad market adoption observed for *B. subtilis*-, *B. amyloliquefaciens*- or *Trichoderma*-based products.

In the early 1990s, the global bioprotectant market reached its first milestone of US$100 million in revenue (see Fig. 1) (Dunham 2019). At that point, the market was growing at a modest pace, which doubled the market size every eight to nine years. By 2005, this growth rate began accelerating, with the market doubling in size every four to five years. This rate of growth has been maintained over the last 15 years.

The global COVID-19 pandemic will undoubtedly have some impact on the growth trend for the bioprotectant market, potentially slowing the growth predicted in the near term. The economic shutdowns which took place in Europe and the USA in the first half of 2020 resulted in severe short-term reductions in demand, particularly in the cut-flower and perishable fresh fruit and vegetable markets. Despite these negative impacts on market demand, the fundamental drivers of this market remain intact, which will be discussed in more detail.

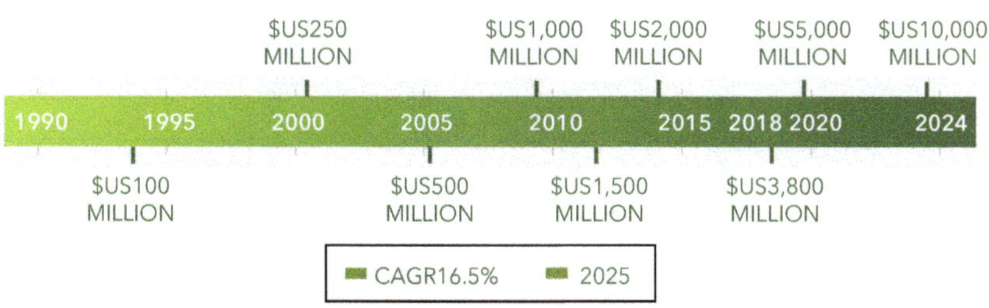

Figure 1 Global bioprotectant market value in US$ from 1990 through 2025.

2 The microbial bioprotectant market

2.1 Microbials' share of global biocontrol

At the end of 2018, DunhamTrimmer estimated the global bioprotectant market value at just over US$3.8 billion (see Fig. 2). Microbials represent over half of this market, at over US$2.0 billion.

The microbial sector of the market is also growing more rapidly than other sectors, primarily due to interest and investment from multinational crop protection companies and significant innovation from a number of smaller companies. At predicted growth rates, the total value of the global microbial market is expected to exceed US$5 billion by 2025.

2.2 Global/regional market for microbials

The North American market is the global leader in microbial bioprotectant sales (see Fig. 3) (Dunham 2019). Microbials have been the primary focus in this region, due to a favorable regulatory environment and a number of companies focused on microbial development. In contrast, the European

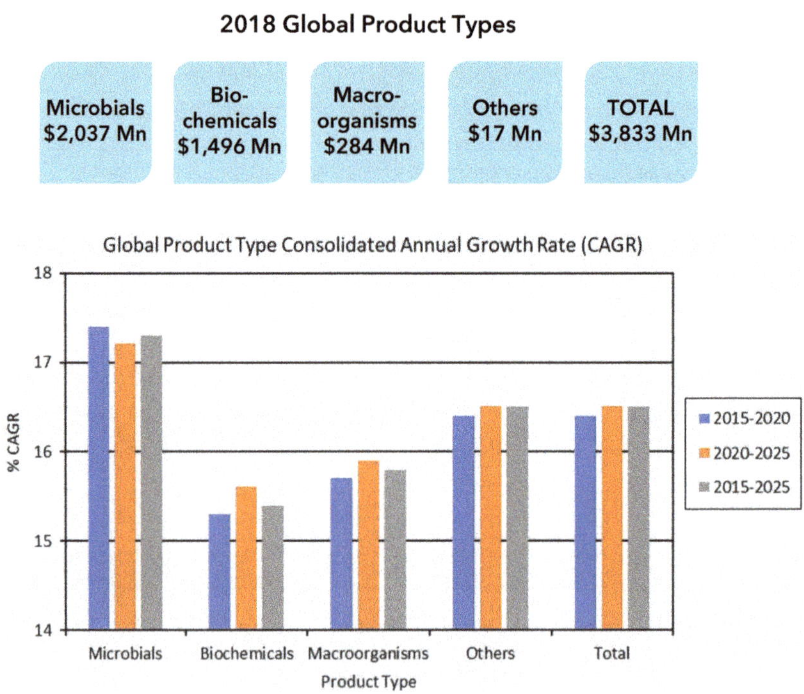

Figure 2 Global bioprotectant market value in 2018 and consolidated annual growth rate by product type.

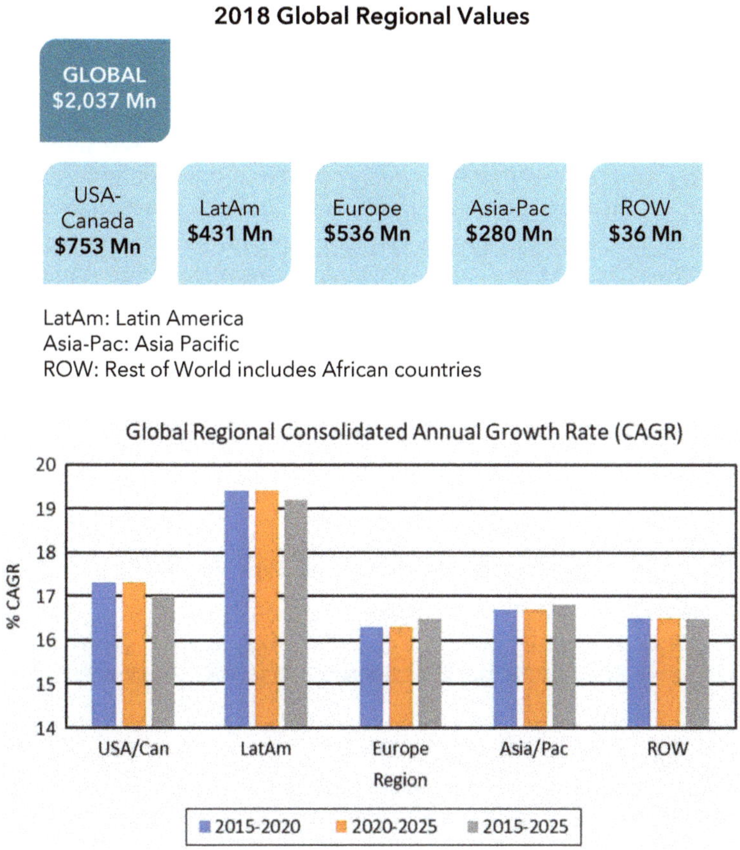

Figure 3 Global microbial bioprotectant market value in 2018 and consolidated annual growth rate by region.

market is much more balanced between microbial and biochemical products. The use of baculovirus products is more prevalent in Europe than in other regions. The microbial market in Latin America is one of the fastest-growing bioprotectant sectors in the world. Growth could be even faster, but it has been restrained by a challenging regulatory environment in the important Brazilian market. This is now improving, with recent revisions to the regulatory guidelines and emphasis on approving biological products leading to approvals in Brazil within 12-18 months after dossier submission. The Asian market is more dominated by biochemicals, although the microbial market is still a significant portion of the total market. Government policies in some Asian countries favor the development of botanical plant extract products over microbials.

2.3 Microbial types

The bacterial segment represents the majority of the microbial market, with nearly 60% of total sales (see Fig. 4) (Dunham 2019). Bt products continue to dominate this market, although many other microbials are now growing at a faster rate. The fit with an established large-scale production model used by conventional chemical pesticide producers has helped establish bacterial products as the segment leaders. Improvements in formulation technology have improved product shelf-life, stability, and consistency of performance.

The fungal segment is nearly half the size of the bacterial segment. It is dominated by products based on *Trichoderma spp., Metarhizium anisopliae,* and *Beauveria bassiana*. The biggest challenges for the fungal products are production cost using a solid-state fermentation system, and the stability of the products. Many fungal products are less stable in sunlight than bacteria and require formulation ingredients to boost stability in UV light.

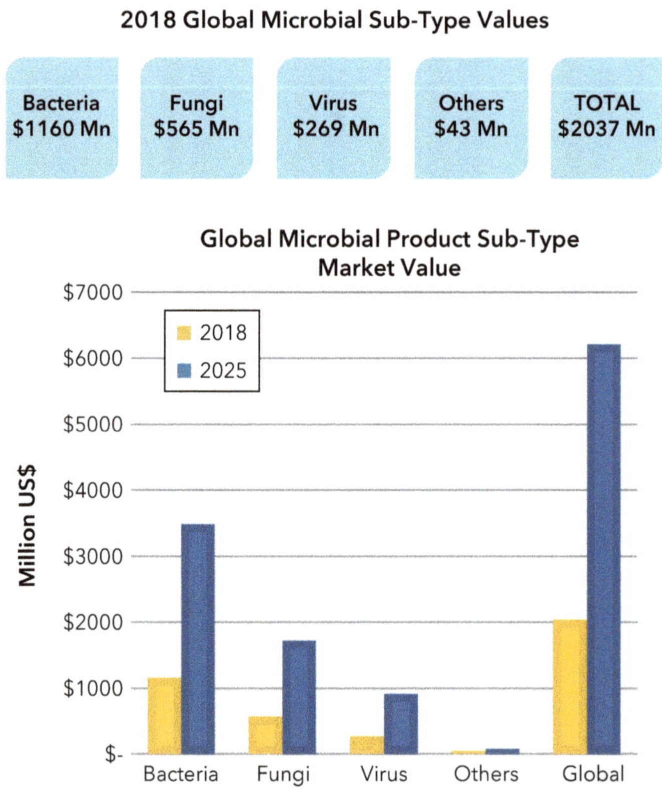

Figure 4 Global microbial bioprotectant market value in 2018 and market share by microbial sub-type.

Published by Burleigh Dodds Science Publishing Limited, 2022.

Viruses are the smallest of the three major segments. Virus bioprotectants have traditionally been focused on tree fruit for the control of Lepidopteran pests such as codling moth (*Cydia pomonella*) in apples, pears, and English walnuts. In recent years, we have seen new virus products under development from companies including AgBiTech (www.AgBiTech.com) and Andermatt Biocontrol (https://www.andermattbiocontrol.com/), for control of pests including *Helicoverpa armigera* and *Spodoptera exigua* in cotton and *Plutella xylostella* in cruciferous vegetables. Successful expansion of virus bioprotectants to markets for these and other important pests will drive increased market share for this group.

2.4 Microbial use segments

Bioinsecticides and biofungicides combined dominate the bioprotectant market. Bioinsecticides represent nearly 50% of the market (see Fig. 5) (Dunham 2019). Biofungicides represent over 40% of the market. Bionematicides have been the fastest growing segment in recent years, mainly due to their use as

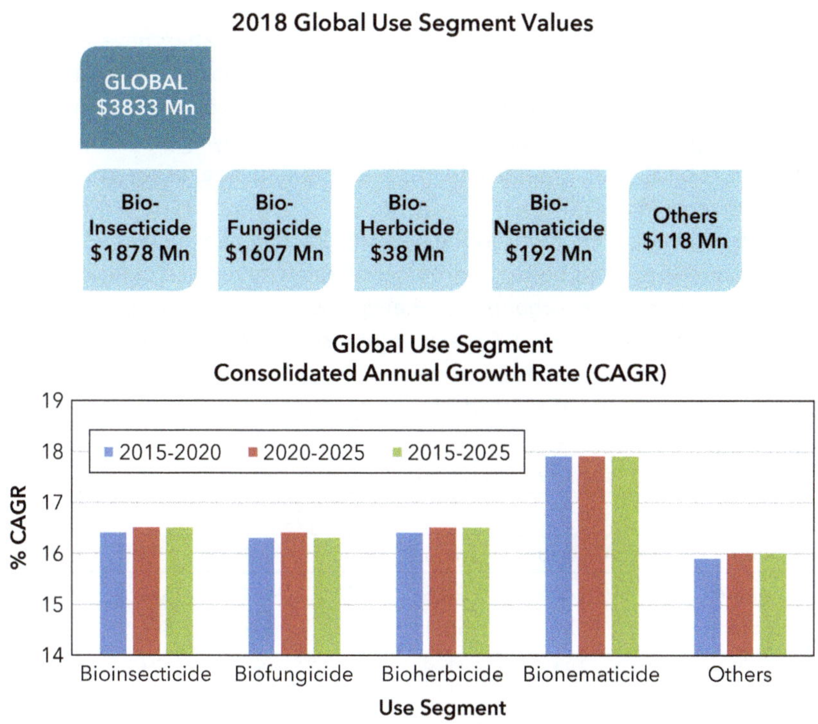

Figure 5 Global bioprotectant market value in 2018 and consolidated annual growth rate by use segment.

seed treatments. While the growth of bionematicides has been rapid, the total value of this segment remains much smaller than the two leading segments. Bioherbicides are not an important segment. There are very few bioherbicides in the market and none of them are widely used. Several companies are seeking to identify and launch new microbial bioherbicides, but none have succeeded yet.

Both North America and Europe lead the bioinsecticide and biofungicide market with nearly identical sales values (see Table 1). Bt-based products continue to represent nearly 50% of the bioinsecticide market. Fungal-based bioinsecticide products are gaining market share, as improved formulations with better protection from solar degradation are developed. The biofungicide market is slightly smaller in value than bioinsecticides.

The fungal biofungicide segment is dominated by *Trichoderma*-based biofungicides. A wide variety of species and strains have been developed, for use as seed and soil treatments to protect against soil pathogens. Bacterial biofungicides based on *B. subtilis* and *B. amyloliquefaciens* have emerged as the fastest growing products in the biofungicide segment over the last 20 years. AgraQuest (acquired by Bayer in 2012) pioneered this segment with the launch of Serenade in 2000. Since that time, many other leading companies, including BASF, Certis USA, Marrone Bio, Novozymes, and others have launched biofungicide products based on these actives. New microbial biofungicides are emerging from the investigation of the plant and soil micriobiome. Companies, including AgBiome, Bioconsortia, and others are identifying and developing novel microbials with potential for use as biofungicide using proprietary screening techniques.

A strong market interest and the need to develop nematicides with lower mammalian toxicity and better environmental safety to replace older chemical nematicides based on organophosphates has created an opportunity for bionematicides (see Table 1). Several large crop protection companies have sought to acquire efficacious bionematicides to fill this need. Bionematicides are predominantly used as seed treatments and are often combined in a

Table 1 Global bioprotectant market value in 2018 by use pattern

	Million US $					
Use Pattern	USA/Canada	Europe	LatAm	Asia-Pac	ROW	Total
Biofungicide	$546	$558	$232	$243	$28	$1,607
Bioinsecticide	$554	$551	$410	$322	$42	$1,879
Bionematicide	$64	$63	$35	$29	$2	$193
Bioherbicide	$12	$12	$7	$6	$1	$38
Global Total	$1,176	$1,184	$684	$600	$73	$3,717

LatAm, Latin America; Asia-Pac, Asia Pacific; ROW, Rest of World.

complete seed treatment package including fungicides, insecticides, starter fertilizer, and possibly biostimulants.

There are few bioherbicides in the market, even though weed control is the largest and most important plant protection use segment for conventional chemical crop protection products (see Fig. 9). The key challenge is the low cost/ha of this segment and the need for bioherbicides which are selective for crops and provide long-term residual control of weeds. Several companies are screening for bioherbicides, with a focus on finding solutions for glyphosate-resistant weeds.

2.5 Bioprotection crop group and application type segments

Bioprotection products are overwhelmingly focused on the fruit and vegetable sector, with over 75% of all global value in these crops (see Fig. 6) (Dunham 2019). While bioprotectant use of both seed treatment and row crops is growing much faster, the small absolute size of these segments today means they will not reach 20% market share until 2025.

Over half of all bioprotection products are applied foliarly, with soil applications representing most remaining applications (see Fig. 7) (Dunham 2019). Foliar and soil applications combined represent 90% of all bioprotection treatments. Seed treatment and post-harvest uses are both growing much faster, but their total market share will remain less than 10% of the market combined through 2025.

3 Microbial market trends and drivers

3.1 Why do microbials dominate?

Over the last decade, the merger and acquisition activity in the bioprotectant area has increased dramatically and a significant portion of the announcements have focused on microbial bioprotectant companies. AgraQuest, Pasteuria Bioscience, LAM International and Prophyta were among the leading microbial companies acquired by others in the decade (2BMonthly). What are the factors leading global companies to make large investments in this space?

- Microbial bioprotectants can be protected from competitors with strong barriers to entry through either intellectual property protection or trade secrets in their fermentation processes and formulation. While the microbe itself in its natural state, as found in nature, cannot be patented, its method of production, formulation, and use can be subject to patent protection. Moreover, if the bioactive molecules produced by a microbial are isolated,

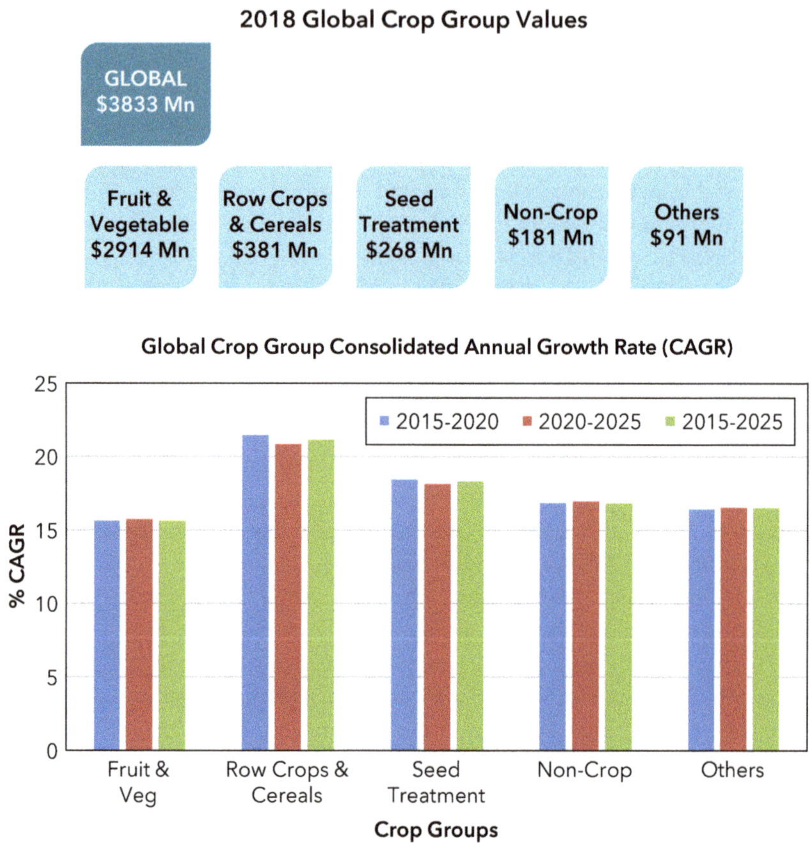

Figure 6 Global bioprotectant market value in 2018 and consolidated annual growth rate by crop group.

purified, identified, or modified in any way, their use as bioprotectants may also be patented in many countries.

- Microbial bioprotectants are produced in fermentation facilities that mirror chemical production, allowing for high degrees of quality control and consistency in the end product. This makes them highly attractive to global crop protection companies, as this production approach is analogous to their practice with chemical plant protection agents. This also differentiates microbials from botanical products which must be obtained and extracted from natural sources.
- The potential for differentiation through the discovery of novel microbial species and strains is high. Many start-up companies are focused on the discovery of new microbial leads in the soil and plant biomes. Molecular biology tools are rapidly advancing these companies' ability to identify and characterize new strains.

Published by Burleigh Dodds Science Publishing Limited, 2022.

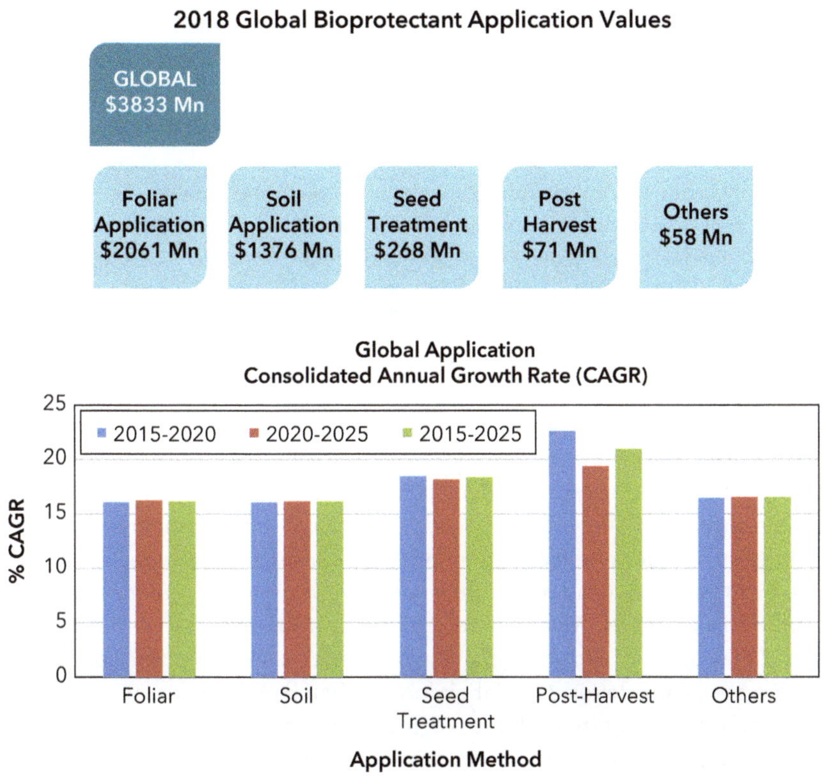

Figure 7 Global bioprotectant market value in 2018 and consolidated annual growth rate by application type.

- The microbial sector of the bioprotectant market is growing much more rapidly than the traditional chemical crop protection market (Dunham 2019). This makes integrating microbials into existing portfolios highly attractive to global crop protection companies.

Companies with novel microbial bioprotectants, with strong barriers to entry and demonstrated cost-effective performance in the field, will continue to attract investor attention as this market continues to grow.

3.2 Market factors driving bioprotectant adoption

Microbial bioprotectants (as well as other types of bioprotectants) offer benefits across the entire value chain (Dunham 2019).

A key benefit for growers is the complex mode of action of bioprotectants that makes them excellent tools to integrate into integrated pest management

(IPM) programs to prevent pests from developing pesticide resistance. In addition, the safety of bioprotectants results in shorter pre-harvest intervals and re-entry intervals on product labels, making them much more flexible for producers of fruit and vegetable crops that are often harvested multiple times. This safety for both the workers and consumers makes bioprotectants extremely useful in late-season pest control, close to the harvest.

Global manufacturers see bioprotectants as highly valuable additions to their portfolio. The pest resistance prevention attribute helps to prevent resistance development in older chemical active ingredients and extends the life cycle of products on which they are based. The higher growth rate of the bioprotectant market is highly attractive as an addition to an existing chemical portfolio. Adding bioprotectants also allows manufacturers to meet consumer demands, as discussed below.

Without doubt, consumer demands are the number-one drivers of bioprotectant market growth. Many consumer surveys have shown that food purchasing decisions are influenced by health considerations (Dunham 2019). The number-one factor mentioned by consumers is the desire to avoid chemical pesticide residues. Popular press reporting on chemical pesticide residues in food has driven higher consumer awareness. While the general public may not fully understand all the implications and risks of residues, they equate any pesticide residue in their food as an unnecessary risk which should be avoided. Recognizing this as a marketing opportunity, many grocery store chains have set their own restrictions on pesticide residues which are more stringent than those required by government regulations. As grocery stores compete based upon who can set the highest restrictions, growers must adapt their pest management systems to minimize chemical pesticide residues, particularly in fresh markets for fruits and vegetables, as these are food items which are often consumed unprocessed for which consumers have the highest concern for pesticide residues. This has created substantial market opportunities for bioprotectants, as they do not contribute to chemical pesticide residues.

Another consumer concern is the use of sustainable production practices for their food. Europe recently announced its Farm to Fork Strategy, which aims to accelerate the transition to a sustainable food system (European Commission). The benefits anticipated from this transition include a neutral or positive environmental impact, mitigation of climate change, a reversal in the loss of biological diversity, assurance of food security and broad consumer access and preservation of food affordability while providing fair economic returns to growers. One of the targets established in this strategy is a substantial reduction in the use of chemical fertilizers and pesticides in crop production. Bioprotectants have a strong fit in supporting many of these goals and have the potential to benefit from this initiative in the EU.

4 Myths about the bioprotectant market

4.1 Bioprotectant use in organic crops

There are several myths from many sources that are reported repeatedly about the bioprotectant market from many sources. The number one among these is: the use of bioprotectants in organic crops drives the market growth. A closer analysis of the data reveals a different story.

FiBL, The Research Institute of Organic Agriculture, (https://www.fibl.org/en.html) reports that in 2018, there were 71.5 million hectares of organic farmland globally, which represents 1.5% of global farmland (see Fig. 8). The number-one country in terms of area is Australia, with 35.6 million hectares or 50% of the global total. Permanent grassland represents over two-thirds of total organic farmland, at 48.2 million hectares. Cropland totals 18.1 million hectares (25%), with 13.3 million hectares in arable crops (19%) and 4.8 million hectares in permanent crops (7%).

Looking more closely at crops, as presented previously in the market data, bioprotectants are predominantly used in fruit and vegetable crops, with minimal use in row crops. Based on FiBL (Research Institute for Organic Agriculture) data for 2018, the data for global area for fruit and vegetable crop groups is summarized in Table 2.

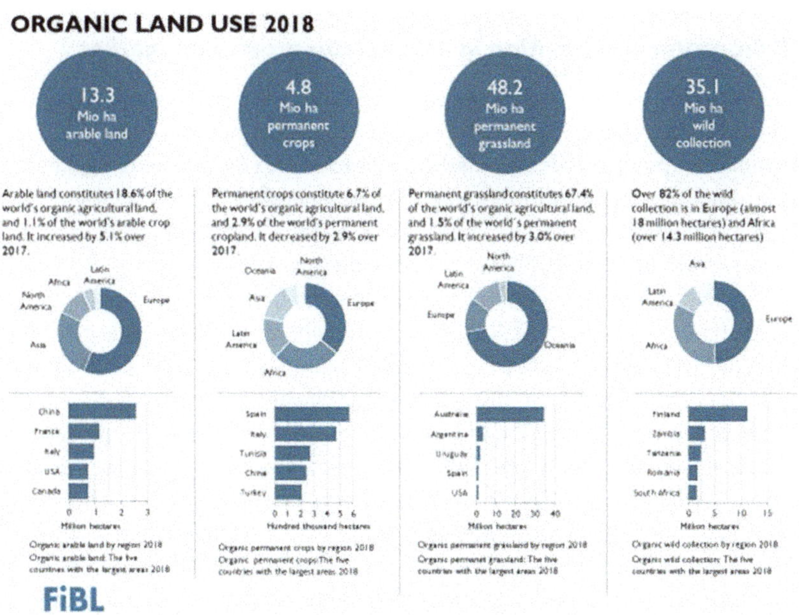

Figure 8 Global organic land use by crop type in 2018.

Table 2 Global area of organic production of fruit and vegetable crops in 2018

Crop Group	Global Total Ha
Berries	61 548
Citrus	90 048
Temperate Fruit	234 760
Tropical & Subtropical Fruit	273 059
Grapes	422 276
Nuts	664 931
Strawberries	7507
Vegetables	387 394
Total	2 141 523

Based on this data, in total, on a global basis, there are 2.1 million hectares of organic fruit and vegetable cultivation. Assuming that bioprotectants are applied on 100% of this area at a cost of US$150/ha, the total value of the bioprotectant market in organic farmland would be just over US$320 million or less than 10% of the global bioprotectant market value in 2018. While organic producers are certainly early adopters of bioprotectants, the increased use of bioprotectants on traditional non-organic hectares is the main driver of growth in this market.

4.2 Bioprotectant position in global crop protection market

Another myth is that bioprotection is a very small part of the total crop protection market and is therefore unimportant. At a quick glance, the global conventional plant protection market, valued at US$64 billion, is significantly larger than the global bioprotectant market, valued at US$3.8 billion in 2018. The bioprotection market is growing at more than 15% annually, versus an annual growth of 5% annually for the conventional market.

However, bioprotection has a much higher presence in the plant protection market than most people realize, even those who are active in the market. To understand this, let us contrast the two markets and examine each to understand their most important market segments and their values.

While bioprotectants are only slightly more than 5% of the global crop protection market overall, the two markets are positioned very differently in terms of uses and crops. While herbicides represent 40% of the conventional market (see Fig. 9), they are nearly absent from the bioprotectant market. Fruits and vegetables dominate the bioprotectant market but are only 20% of the conventional market (see Fig. 10). From a growth perspective, the bioprotectant market is growing at nearly three times the rate of the conventional market.

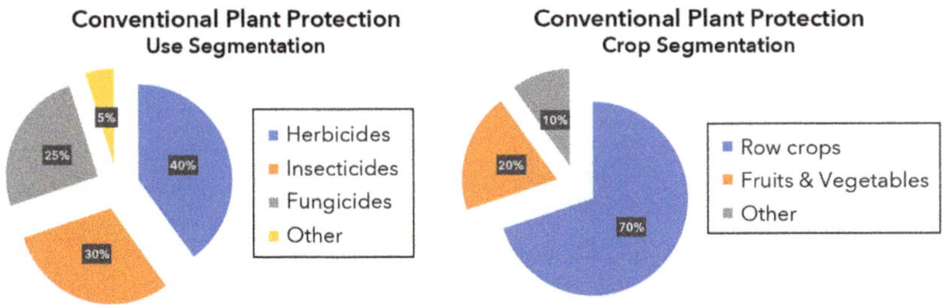

Figure 9 Global conventional plant protection market segmented by use and crop group.

Looking at this a bit closer, it is interesting to look at bioprotectant share in the portion of the market in which it participates. If we do not consider the herbicide market in which bioprotectants are not present, we see the bioprotectant market share increase to 9%. Even more impressively, if we focus on the fruit and vegetable market where bioprotectants are predominantly used, we see a market share for bioprotectants at more than 18% (see Fig. 11).

This analysis shows that in markets where they participate, bioprotectants are an important part of the market.

5 Limitations for microbial bioprotectants

5.1 Market fragmentation

As with bioprotectants as a whole, the microbial market is highly fragmented with many small companies. The proprietary database of bioprotectant companies developed by Dunham Trimmer contains more than 200 bioprotectant companies. The number-one challenge for many smaller companies is market access. Gaining support from the distribution channel for these small companies that often only have a single new bioprotectant product to offer is very difficult. This is one of the major advantages the global crop protection companies have, as they have established relationships and provide the channel with a full portfolio of plant protection products to meet all the needs of the growers.

5.2 Lack of resources

A second limitation for most bioprotectant companies is a lack of resources to fully develop and exploit their technology. Both people and money are often limited for most start-up companies, and the need to prioritize market opportunities is intense. New bioprotectant companies must allocate their precious resources wisely to ensure they deliver the return on investment expected by their investors. This is the leading factor driving merger and acquisition activity in the space, as larger companies can afford to put more

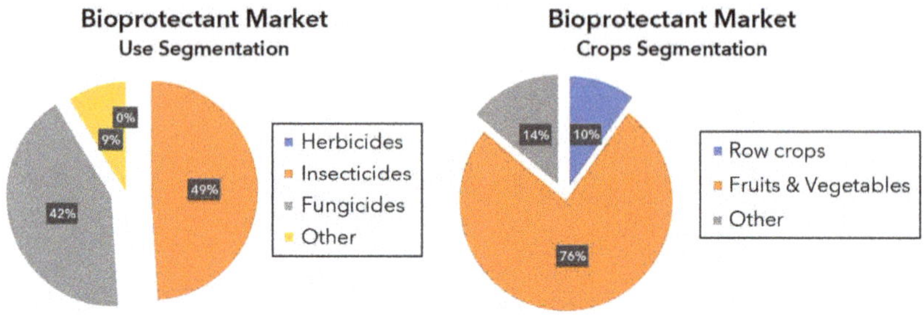

Figure 10 Global bioprotectant market segmented by use and crop group.

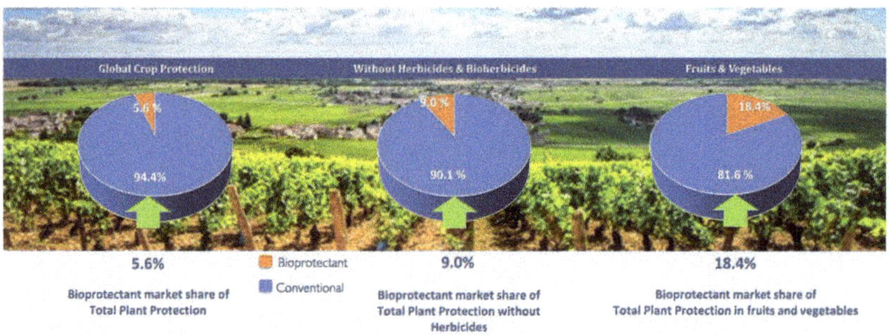

Figure 11 Global bioprotectant share of the global plant protection market in 2018.

resources behind the product development effort. Combining their resources with their market access allows the large crop input companies to solve two of the leading limiting factors for many microbial bioprotectant companies.

6 Future opportunities and threats

6.1 Expanded microbial species

To date, many of the microbial bioprotectants have relied on a limited number of genera, including *Bacillus, Pseudomonas, Streptomyces, Triochoderma* and a few others. While a few companies have brought unique species to the market, this has largely been the exception. The majority of new microbial bioprotectants have been new strains of existing species. In the last decade, a number of start-up companies have been established with the goal of identifying novel microbial species through investigation of the plant/soil microbiome. Some of these have focused more on biostimulant products, but several, such as AgBiome or Bioconsortia, have also sought to identify novel microbial bioprotectants. This trend is building upon similar research approaches in human and animal pharma, and has already identified several novel pipeline microbial bioprotectants that these companies have in development or have recently registered.

6.2 Improved formulations

Formulation development and optimization is an area in which most global crop protection companies invest many millions of dollars with new chemical active ingredients to optimize performance and storage stability of the end-use product. Until recently, microbial bioprotectant manufacturers were not putting the same emphasis on this area. Today, both established microbial companies and the new entry by global companies have led to much more focus on formulations and their importance in achieving optimal performance.

Another opportunity in formulations is the development of combination products with chemicals. Bayer (https://www.cropscience.bayer.com/) pioneered this when they introduced the seed treatment combination of *Bacillus firmus* and the neonicotinoid insecticide clothianidin under the brand name Poncho VOTiVO in 2010. The chemical insecticide provides early-season protection, while the microbial protectant colonizes the maize roots for extended benefits later into the season. This product remains one of the leading seed treatments in the US maize market. Other companies, including Syngenta (https://www.syngenta.com/) and FMC (http://www.fmc.com/), have also launched microbial and chemical combination products in the seed treatment market.

6.3 Expansion to row crops

As discussed in the market review, bioprotectants in general are used primarily in the fruit and vegetable market sectors. One of the key aspects of this sector that facilitated the entry of microbial bioprotectants is that fruit and vegetable crops routinely treated multiple times in a single cropping season. This made it less risky for growers to integrate bioprotectants into the pest management program slowly over time, as they gained experience and confidence with their use. To sustain the current market growth rates of bioprotectants, these products will need to successfully penetrate the large row crop area. Most row crops in North America are not normally treated multiple times. A single treatment is often used for a key pest, meaning the producer must be confident of the product performance. In Latin America, row crops often do receive multiple treatments for a single pest due to differences in climate and growing practices. This may create an opportunity for bioprotectants to more easily find a fit in row crops in Latin America. In North America, the use of bioprotectants in the seed treatment market in row crops may be the entry that leads to future use as foliar or soil treatments.

6.4 Bioherbicide development

As described above, the use of bioprotectants as bioherbicides is extremely limited. The lower cost per hectare of herbicides and the requirement for crop selectivity combined with broad weed-control spectrum make this a challenging market. In addition, the environmental conditions for foliar treatments are often not favorable for microbial bioprotectants. To date, only a few microbial bioherbicides have been developed and none has achieved substantial market success. Several companies are working on pipeline bioherbicide products. Most are either focused on particularly difficult-to-control weed species or a weed species with widespread herbicide resistance, such as glyphosate resistance. A successful microbial bioherbicide that could control one or more important glyphosate-resistant species would be a breakthrough product and would drive future growth of the bioprotectant market.

6.5 Synergy with precision agriculture

One of the keys to success with bioprotectants is close understanding of disease or pest population development in the field. Most bioprotectants perform best when applied early in the disease or pest development cycle. The sensing and monitoring tools used in precision agriculture can be invaluable in recognizing and identifying pest problems early. Once they are identified, other tools like drones for applications can ensure the product placement is accurately delivered to those parts of the field which need treatment. The synergistic development of precision agriculture and bioprotectants is already happening in some regions. Latin America has been among the leaders in this approach. Further expansion of these efforts is expected in other regions.

6.6 Innovative application approaches

As mentioned previously, seed treatment applications have been among the first uses of microbial bioprotectants in row crops. In North America and Latin America, seed treatment packages often include a broad range of products, both conventional and biological, to ensure successful germination and early-season vigor. In North America, there have been some distributors offering complete biological seed treatment packages such as the BIOst biological seed treatment technology platform launched by Albaugh (https://www.albaughllc.com/north-america/) or the seed+ biological seed treatment for maize and soybeans launched by Verdesian Life Sciences (https://www.vlsci.com/) in the USA in 2018.

Other companies have been developing completely novel ways to deliver biofungicides precisely to crops at flowering. Taking advantage of bumble bees being used as pollinators, special hives were developed by both Bee Vectoring Technology (http://www.beevt.com/) (BVT) and Biobest (https://www.biobestgroup.com/) that require the insect to pass through a tray holding a dry powder formulation of a biofungicide. After picking up the biofungicide on their bodies and legs, the insects then deliver the product to flowers during pollination. The biofungicides used are completely harmless to the insect. Using this approach, the product is delivered to the crop at optimal timing to prevent disease infections that target the flower. An additional benefit is the reduced cost for application of the product. Both companies are working to expand to provide a range of products. BVT received US EPA approval for their first active ingredient *Clonostachys rosea*, to be delivered by this system in August 2019. Biobest delivers Prestop 4B based on *Gliocladium catenulatum,* in a system that they have branded Flying Doctors in Belgium and the Netherlands. Both companies are seeking to expand to offer more active ingredients through this system in more countries.

6.7 Regulatory requirements

Depending upon the region, regulatory requirements can be an advantage or can delay the development of microbial bioprotectants. The development of the bioprotectant market in the USA has been facilitated by a regulatory approach that recognizes the inherent safety of these products in comparison to chemical pesticides. In 1994, the Biopesticides and Pollution Prevention Division (BPPD) was established in EPA's Office of Pesticide Programs (OPP) to facilitate the registration of biopesticides (United States EPA website). In other parts of the world, registration approval remains a major hurdle as bioprotectants are forced to comply with regulatory guidelines designed to regulate chemicals, often resulting in longer timelines for approvals as regulators struggle to adapt chemical regulations to microbial products (Ravensberg 2011).

Increasingly, regulators are taking a closer look at the bioactive metabolites produced by microbial bioprotectants. In some countries, regulators now may request isolation and characterization of the metabolites as well as some toxicology or pathogenicity testing for them. These developments may lead to more data requirements and increased time for microbial bioprotectant reviews in some countries.

At the same time, there is increasing regulatory pressure on chemical crop protection products globally. The EU Farm to Fork strategy lists definitive targets including a 50% reduction in the use and risk of chemical pesticides (European Commission Farm to Fork Strategy). Many other countries have increased the regulatory requirements for chemicals related to environmental risk (non-target organisms and environmental fate), user safety and consumer dietary safety. These increasing requirements have led to an increased cost and longer timelines for new chemical pesticide approvals. In their most recent study commissioned by CropLife America (http://www.croplifeamerica.org/), the US trade association of crop protection companies, estimated it takes on average 11 years and costs US$286 million to register a new crop protection chemical (Phillips McDougall 2016). The cost has nearly doubled, and the time required has increased by three years since 1995. These costs and timelines force global manufacturers to focus on only the largest market opportunities, often leaving specialty crop producers with limited tools often based upon older active ingredients for disease and pest control. The lack of products with new and novel modes of action has increased the challenges these growers face with pests developing resistance. This has created substantial market opportunities for bioprotectants to exploit, due to their much lower registration costs, shorter approval timelines and complex modes of action.

7 Summary

Microbial-based products are the leading type of bioprotectant on the market today and represent more than US$2 billion in sales. This sector of the market has attracted significant attention from investors and is growing at higher rates than other types of bioprotectants. Market demand, especially from consumers, is a primary driver. Market fragmentation and limited market access for smaller companies limit growth. Multiple opportunities exist which could drive continued future market growth. Primary among these would be the expansion into row crops and successful development of microbial-based bioherbicides.

8 Where to look for further information

For more information on microbial bioprotectants and their markets, see the following sources.

8.1 Industry Associations

- Biological Products Industry Alliance (https://www.bpia.org/).
- Bioprotection Global (https://www.bioprotectionglobal.org/).
- CropLife Brasil – Produtos Biológicos (https://croplifebrasil.org/produtos-biologicos/).
- International Biocontrol Manufacturers Association (https://ibma-global.org/).
- South African Bioproducts Organisation (https://sabo.org.za/).

8.2 Market Research

- DunhamTrimmer LLC, a market research firm that focuses exclusively on bioprotection, biostimulants, and biofertilizers (http://dunhamtrimmer.com/).

8.3 Publications

- 2BMonthly, a monthly e-newsletter covering the global biostimulants and bioprprotection industries, published by DunhamTrimmer LLC and New Ag International. (http://www.2bmonthly.com/).

8.4 Regulatory Agencies

- OECD Biological Pesticides (https://www.oecd.org/env/ehs/pesticides-biocides/biological-pesticides.htm).
- UK Health and Safety Executive – Biopesticides (https://www.hse.gov.uk/pesticides/pesticides-registration/applicant-guide/biopesticides-home.htm).
- United States Environmental Protection Agency – Biopesticide page (https://www.epa.gov/pesticides/biopesticides).

9 References

2BMonthly e-newsletter. A DunhamTrimmer and NewAg International Publication. Published monthly. Available at: www.2bmonthly.com; accessed on June 19, 2021.

Bell, J. V. M. (1974). *Insect Diseases* (Cantwell, G. E., Ed.). New York: Marcel Dekker Inc. 188 pages.

Biological Products Industry Alliance. History of Biopesticides. Available at: https://www.bpia.org/history-of-biopesticides/; accessed on September 20, 2020.

Black, B. C., Brennan, L. A., Dierks, P. M. and Gard, I. E. (1997). Commercialization of baculoviral insecticides In: Miller, L. K. (Ed.) *The Baculoviruses: The Viruses*. Springer: Boston, MA. pp. 341–387.

Cohn, F. (1872). Untersuchungen uber Bacterien. *Beitrage zur Biologie der Pflanzen* 1: 127–224.

Dunham, W. C. (2019). *Global Biocontrol Market Overview, Trends, Drivers, and Insights*. DunhamTrimmer, LLC.

Ehrenberg, C. G. (1835). *Physikolische Abhandlungen der Koeniglichin Akademie der Wissenschaftern zu Berlin aus dr Jahren 1833–35*. Deutsche Akademie der Wissenschaften zu Berlin, Berlin. pp. 145–336.

European Commission Farm to Fork Strategy. Available at: https://ec.europa.eu/food/farm2fork_en; accessed on October 1, 2020.

FiBL (The Research Institute of Organic Agriculture). Available at: https://www.fibl.org/en.html; accessed on September 21, 2020.

Fukumoto, J. (1943). Studies on the production of bacterial amylase. I. Isolation of bacteria secreting potent amylases and their distribution. *Journal of the Agricultural Chemical Society of Japan* 19(7): 487–503.

Howell, C. R. (2003). Mechanisms employed by Trichoderma species in the biological control of plant diseases: the history and evolution of current concepts. *Plant Disease* 87(1): 4–10.

Lord, J. C. (2005). From Metchnikoff to Monsanto and beyond: the path of microbial control. *Journal of Invertebrate Pathology* 89(1): 19–29.

McDougall, P. (2016). The cost of new agrochemical product discovery, development and registration in 1995, 2000, 2005-8 and 2010 to 2014. R&D expenditure in 2014 and expectations for 2019. A Consultancy Study for CropLife International, CropLife America and the European Crop Protection Association. March 2016.

Priest, F. G., Goodfellow, M., Shute, L. A. and Berkley, R. C. W. (1987). *Bacillus amyloliquefaciens* sp. nov., nom. rev. *International Journal of Systematic Bacteriology* 37(1): 69–71.

Ravensberg, W. J. (2011). *A Roadmap to the Successful Development and Commercialization of Microbial Pest Control Products for Control of Arthropods, Progress in Biological Control 10*. Springer Science+Business Media B.V., Dordrecht, Netherlands. 412 pages.

Sanahuja, G., Banakar, R., Twyman, R. M., Capell, T. and Christou, P. (2011). Bacillus thuringiensis: a century of research, development, and commercial applications. *Plant Biotechnology Journal* 9(3): 283–300.

Shafi, J., Tian, H. and Ji, M. (2017). Bacillus species as a ve4rsatile weapon for plant pathogens: a review. *Biotechnology and Biotechnological Equipment* 31(3): 446–459.

Siegel, J. P. (2001). The mammalian safety of *Bacillus thuringiensis*-based insecticides. *Journal of Invertebrate Pathology* 77(1): 13–21.

Stockwell, V. O. and Stack, J. P. (2007). Using *Pseudomonas* spp. for integrated biological control. *Phytopathology* 97(2): 244–249.

United States Environmental Protection Agency. Biopesticides. Available at: https://www.epa.gov/pesticides/biopesticides#:~:text=Biopesticides%20include%20naturally%20occurring%20substances,%2Dincorporated%20protectants)%20or%20PIPs; accessed September 25, 2020.

United States Federal Register (August 30, 2000). Environmental protection agency notice. 65 Federal Register Pages 52732–33:Document number 00-21921.

Weindling, R. (1932). Trioderma lignorum as a parasite of other soil fungi. *Phytopathology* 22: 837–845.

Woo, S. L., Ruocco, M., Vinale, F., Nigro, M., Marra, R., Lombardi, N., Pascale, A., Lanzuise, S., Manganiello, G. and Lorito, M. (2014). *Trichoderma*-based products and their widespread use in agriculture. *The Open Mycology Journal* 8(1, Suppl-1, M4): 71–126.

Yuan, W. M. and Crawford, D. L. (1995). Characterization of streptomyces lydicus WYEC 108 as a potential biocontrol agent against fungal root and seed rots. *Applied and Environmental Microbiology* 61: 3119–3128.

Part 2

Biological control agents

Chapter 8

The use of *Bacillus* spp. as bacterial biocontrol agents to control plant diseases

Adrien Anckaert, Anthony Arguelles Arias and Grégory Hoff, Gembloux Agro-Bio Tech, ULiege (University of Liège), Belgium; Maryline Calonne-Salmon and Stéphane Declerck, UCLouvain (University of Louvain-la-Neuve), Belgium; and Marc Ongena, Gembloux Agro-Bio Tech, ULiege (University of Liège), Belgium

1 Introduction

Biological control of plant diseases, involving the use of (micro-)organisms or compounds of biological origin, is now considered one of the most promising alternatives to the use of chemical pesticides. Sales of so-called biocontrol

http://dx.doi.org/10.19103/AS.2021.0093.10

agents (BCAs) are expected to reach *circa* 10% of the overall global crop protection market in the near future with an annual growth rate estimated between 15% and 20% (Parnell et al., 2016; Nishimoto, 2019; Zaki et al., 2020). Biocontrol can provide plant protection against many diseases and is environment friendly. Since it often involves several modes of action to restrict the growth of microbial pathogens, biocontrol may limit the development of resistance mechanisms in pathogens but further studies are needed to assess their durability (Geiger et al., 2010; Bardin et al., 2015; Borel, 2017; Kim et al., 2017).

BCAs may act against pathogens via a number of mechanisms: competition for space and nutrients, parasitism, antibiosis and/or by stimulating the host plant defences (Köhl et al., 2019). Biocontrol organisms are natural enemies of plant pathogens including viruses, bacteria, fungi, insects and nematodes. Non-living agents of biological origin include semiochemicals (pheromones, plant volatiles) and biochemical products mainly secreted or extracted from plants and microorganisms (Raymaekers et al., 2020).

Microbial products currently dominate the BCA market, particularly bacteria such as *Bacillus thuringiensis* which is widely used as a bio-insecticide. Selected beneficial bacteria naturally living in close association with plants, and referred to as plant growth-promoting rhizobacteria (PGPR), are also used as BCAs against plant pathogens (Savary et al., 2019). PGPRs constitute a diverse group of bacteria isolated from the rhizosphere and belonging mainly to the genera *Pseudomonas, Streptomyces, Acetobacter, Azospirillum, Paenibacillus, Serratia, Burkholderia, Herbaspirillum*, *Rhodococcus, Rhizobium* and *Bacillus* (Backer et al., 2018).

Like other PGPR, some *Bacillus* spp. can improve plant growth by nitrogen fixation, phosphate solubilization, phytohormone production or by mitigating the impact of some abiotic stress factors (Kumar et al., 2011; Saxena et al., 2020). This genus includes species that are particularly effective in controlling plant diseases, representing some of the most popular BCAs with global sales of US$ 160 million in 2016 (Chen, 2017).

Even though reduction of insect and nematode infection has also been occasionally reported (Mnif and Ghribi, 2015; Engelbrecht et al., 2018; Ruiu, 2020), the biocontrol activity of *Bacillus* species has been mainly established against plant diseases. This chapter illustrates the diversity of pathosystems in which BCA based on *Bacillus* spp. have proved effective. It describes the mechanisms underpinning this biocontrol activity via the production of a wide range of enzymes, proteins and small-size bioactive secondary metabolites (BSMs). As these BSMs are clearly involved in pathogen control, we emphasize the importance of understanding the ecological factors influencing their production. In the last part of the chapter, we highlight the potential interactions between *Bacillus* spp. and other soil microorganisms in developing consortia of biocontrol agents combining species with synergistic activities for plant health improvement.

2 *Bacillus* spp. as biocontrol agents (BCAs)

Most of the *Bacillus* spp. reported as effective BCAs have been isolated from the rhizosphere, occasionally from the phyllosphere, and belong to the so-called *Bacillus subtilis* complex (Cawoy et al., 2011; Fan et al., 2017a). These bacteria may be very competitive in their natural niche. They form populations on plant tissues, as illustrated by *B. velezensis* FZB42, and are able to colonize roots of different monocot and dicot species (Fan et al., 2011, 2012). Root colonization ability contributes to the secretion of antimicrobial compounds and plant resistance elicitors. This metabolite secretion allows to combat a diverse range of bacteria, viruses and fungal pathogens with different lifecycles and modes of virulence (biotrophic, hemibiotrophic or necrotrophic) as illustrated via selected examples in Table 1 (Fira et al., 2018; Miljaković et al., 2020). *Bacillus* spp. biocontrol efficacy has been demonstrated in greenhouse (Fousia et al., 2016; Beris et al., 2018) and field conditions (Matzen et al., 2019; Cucu et al., 2020), as well as at the post-harvest stage for fruit diseases (Punja et al., 2016; Gava et al., 2019).

The ability of *Bacillus* spp. to antagonize Gram-negative bacteria and reduce diseases caused by these pathogens has been established mainly *in vitro* and under controlled conditions (Table 1). A single strain can act against several bacterial pathogens. For example, *B. velezensis* LS69 has been shown to display antibacterial activities against *Erwinia carotovora* and *Ralstonia solanacearum* (Liu et al., 2017). Species can counteract Gram-positive bacterial pathogens such *as Clavibacter michiganensis by B. amyloliquefaciens* S1 (Gautam et al., 2019) and *Streptomyces scabies* by *B. amyloliquefaciens* Ba01 (Lin et al., 2018).

Bacillus spp. have been shown to exhibit antagonistic activity against most economically important fungal plant pathogens (Dean et al., 2012) such as *Botrytis cinerea* (Jiang et al., 2018), *Magnaporthe oryzae* (Rahman et al., 2015), *Fusarium graminearum* (Ntushelo et al., 2019), *Fusarium oxysporum* (Elanchezhiyan et al., 2018), *Blumeria graminis* (Matzen et al., 2019), *Zymoseptoria tritici* (Kildea et al., 2008) or *Colletotrichum acutatum* (Wang et al., 2020b). A single strain can antagonize different fungi, such as *B. velezensis* Y6 and F7 which inhibit both *F. oxysporum* and *Colletrichum gloeosporioides in vitro* (Cao et al., 2018). Antagonism of *Bacillus* strains against oomycetes such as *Pythium aphanidermatum* (Zouari et al., 2016) and *Phytophthora infestans* (Caulier et al., 2018) has also been demonstrated.

Several *Bacillus* spp. have also been reported to control diseases caused by viruses. This has been illustrated on different pathosystems including

Table 1 Selected examples of published studies showing the versatility action of *Bacillus* genus against phytopathogens with different lifecycles or distinct structural features

Pathogen group	Structural feature or lifecycles	Pathogen species	Disease	Biocontrol *Bacillus* species	Crop(s)	Experimental set-up	Reference
Bacteria	**Gram+**	*Streptomyces scabies*	Common scab	*B. amyloliquefaciens*	Potato	*In vitro*, Pot assays, Field	(Lin et al., 2018)
		Clavibacter michiganensis subsp. *michiganensis*	Bacterial canker	*B. amyloliquefaciens*	Tomato	*In vitro*, Pot assays	(Gautam et al., 2019)
	Gram-	*Rhizobium radiobacter*	Crown gall	*B. subtilis/B. amyloliquefaciens*	Tomato	*In vitro, In vivo*	(Frikha-Gargouri et al., 2017)
		Erwinia amylovora	Fire blight	*B. amyloliquefaciens*	Pear/Apple	*In vitro*, Field	(Ait Bahadou et al., 2018)
		Pseudomonas syringae	Leaf spot	*B. amyloliquefaciens/ B. pumilus*	Sugar beet	*In vitro, In vivo*	(Nikolić et al., 2019)
		Xanthomonas axonopodis pv. *vesicatoria*	Bacterial leaf spot	*B. amyloliquefaciens*	Tomato/ Pepper	*In vitro, In vivo*	(Medeot et al., 2020)
Fungi	**Biotrophic**	*Cladosporium fulvum*	Tomato leaf mold	*B. subtilis*	Tomato	*In vitro*, Pot assays	(Wang et al., 2018)
		Blumeria graminis	Powdery mildew	*B. velezensis*	Winter wheat/ Spring barley/ Oats/Triticale	Field, Greenhouse	(Matzen et al., 2019)
		Plasmopara halstedii	Downy mildew	*Bacillus spp.*	Sunflower	Field, Greenhouse	(Nandeeshkumar et al., 2008)
		Plasmopara viticola	Downy mildew	*B. subtilis/B. pumilus*	Grapevine	*In vitro*, Field	(Zhang et al., 2017a)
	Hemibiotrophic	*Mycosphaerella graminicola (Zymoseptoria tritici)*	Leaf blotch	*B. megaterium*	Winter wheat	*In vitro, In vivo*	(Kildea et al., 2008)

		Pyricularia oryzae	Rice blast	*Bacillus* spp.	Rice	*In vitro*, Greenhouse	(Rais et al., 2016)
		Bipolaris sorokiniana	Wheat spot blotch	*B. subtilis*	Wheat	*In vitro*, *In vivo*	(Villa-Rodríguez et al., 2019)
		Colletotrichum acutatum	Anthracnose	*B. amyloliquefaciens*	Loquats	*In vitro*, *In vivo*	(Wang et al., 2020b)
	Necrotrophic	*Botrytis cinerea*	Gray mold	*B. velezensis*	Pepper	*In vitro*, Greenhouse	(Jiang et al., 2018)
		Fusarium oxysporum	Wilt	*B. velezensis*	Tomato	Greenhouse, Field	(Elanchezhiyan et al., 2018)
		Rhizoctonia solani	Damping-off	*B. subtilis*/*B. amyloliquefaciens*	Tomato	*In vitro*, Greenhouse, Field	(Solanki et al., 2015)
		Sclerotinia sclerotiorum	White mold	*B. velezensis*	Tomato	*In vivo*, Greenhouse	(Farzand et al., 2019)
Oomycete		*Pythium aphanidermatum*	Damping-off	*B. amyloliquefaciens*	Tomato	*In vivo*	(Zouari et al., 2016)
		Phytophthora infestans	Late blight	*B. subtilis*	Potato	*In vitro*, Pilot field	(Caulier et al., 2018)
Virus		Tomato yellow leaf curl virus (TYLCV)		*B. velezensis*/*B. amyloliquefaciens*	Tomato	*In vivo*, Pot assays	(Guo et al., 2019)
		Cucumber mosaic virus (CMV)		*B. amyloliquefaciens*	Pepper/*Nicotiana benthamiana*	*In vivo*, Field	(Lee and Ryu, 2016)

tomato plants infected by the tomato yellow leaf curl virus counteracted by *B. amyloliquefaciens* Ba13 (Guo et al., 2019) or pepper and tobacco plants infected by the cucumber mosaic virus and controlled by *B. amyloliquefaciens* 5B6 (Lee and Ryu, 2016).

These bacilli have been commercially used as BCAs partly because of their ability to combat disease and partly because of technological advantages (Table 2). Species belonging to the so-called *B. subtilis* complex are, for example, considered GRAS ('generally regarded as safe') (Schallmey et al., 2004) and can be produced at an industrial scale. These bacteria are aerobic with relatively high growth rates, low nutritional requirements and are able to grow on various nutrient sources. They usually produce a range of enzymes (cellulases, amylases, proteases) degrading various substrates derived from naturally abundant sources such as lignocellulose, starch, proteins, hydrocarbon and biofuels (Chen et al., 2018; Elisashvili et al., 2019). This allows their cultivation in low-cost media such as plant raw materials in industrial bioreactors or in solid-state fermentation (Khardziani et al., 2017; Berikashvili et al., 2018). An important advantage of *Bacillus* spp. is their ability to form endospores which make them particularly resistant to abiotic stresses such as heat and drought (Piggot and Hilbert, 2004; Mutlu et al., 2020). The endospore allows the formulation of *Bacillus*-based products with good long-term storage due to resistance to industrial processing (lyophilization or spray drying of the spore suspensions collected from the fermentation broth) and the ability to mix endospores with appropriate additives, adjuvants or surfactants (Schisler et al., 2004; Stamenkovic-Stojanovic et al., 2019). Products can either be sprayed on aerial parts of the plants, delivered into soil or coated on seeds according to the disease targeted and farmers needs (Rahman, 2016; Toral et al., 2020).

3 The diversity of *Bacillus* spp. metabolites involved in biocontrol

The biocontrol activity of *Bacillus* species has mainly been linked to their ability to produce a wide range of chemically diverse compounds (Fig. 1). Comparative genomics has revealed that species of the *B. subtilis* group, including plant-associated clades, are particularly rich in biosynthetic gene clusters (BGCs) encoding bioactive secondary metabolites (BSMs) (Grubbs et al., 2017; Harwood et al., 2018). Up to 12% of the genome is devoted to the synthesis of those compounds (Chowdhury et al., 2015a; Molinatto et al., 2016; Liu et al., 2017; Pandin et al., 2018). Some BGCs are widespread across species in the *B. subtilis* group while others seem to be more species-specific. *B. velezensis* has the highest number of different genes coding for BSMs (Fan et al., 2018), making this species one of the most efficient and commonly used bacilli in biocontrol (Ye et al., 2018; Rabbee et al., 2019). Other species of the

B. subtilis group produce fewer BSMs in terms of diversity, for example, *B. amyloliquefaciens* and *B. subtilis* (Harwood et al., 2018; Andrić et al., 2020). Many of the commercialized strains have been registered as *B. subtilis* or *B. amyloliquefaciens*. However, according to recent re-classification based on phylogenetic analyses at the whole genome level, most of these strains actually belong to *B. velezensis* species (Dunlap et al., 2016). This is the case for the strains GBO3, MBI600, QST713, FZB42 or D747 mentioned in Table 2 (Fan et al., 2017a; Dunlap, 2019).

BSMs formed by *Bacillus* spp. originate either from classical ribosomal synthesis or from non-ribosomal synthesis involving mega-enzymatic complexes. *Bacillus* species are able to synthesize both ribosomal peptides that can be post-translationally modified (RiPPs) and non-ribosomal metabolites (Fig. 1) (Arguelles Arias et al., 2011).

Among RiPPs, *B. subtilis* and related species produce bacteriocins, including lantibiotics, (Abriouel et al., 2011; Lajis, 2020) such as plantazolicin, subtilin, ericin, mersacidin, amylolysin, subtilosin and amylocyclicin (Abriouel et al., 2011; Scholz et al., 2011, 2014; Arguelles Arias et al., 2013). Some antimicrobial peptides such as LCI peptides are also considered bacteriocin-like inhibitory substances because their structure has not been elucidated yet, or they cannot be classified in an existing group (Abriouel et al., 2011; Salazar et al., 2017).

Unlike RiPPs, non-ribosomally synthesized molecules seem to be much more conserved within the species. These molecules are synthesized by large modular enzymatic complexes (non ribsosomal peptide synthetases and polyketides synthetases) which are classified into two different types: non-ribosomal peptides (NRPs) and polyketides (PKs) (Dutta et al., 2014; Winn et al., 2016; Bozhüyük et al., 2019). These molecules are synthesized using amino acids (for NRPs) and carboxylic acids (for PKs) as building blocks (Chen et al., 2009a; Winn et al., 2016).

There are three main types of PKs produced by species of the *B. subtilis* group: bacillaene, difficidin and macrolactin (Chen et al., 2006; Caulier et al., 2019). Structural variants can often be coproduced by the same strain. For example, transcription of the difficidin operon results in the production of difficidin and its oxidized form oxydifficidin (Caulier et al., 2019). Cyclic lipopeptides (cLPs) of the iturin, surfactin, and fengycin families are the best-studied compounds among NRPs. These share amphiphilic properties due to a similar structure composed of a peptidic moiety (seven amino acids for iturins and surfactin, and ten amino acids in fengycin) linked to a fatty acid tail (ranging from 12 to 19 carbons) (Ongena and Jacques, 2008).

cLPs biosynthesis allows great structural diversity such as the incorporation of non-proteogenic amino acids or *D*-amino acids. Variants exhibiting differentiations in fatty acid chain length, branching type or amino acids

Table 2 Examples of *Bacillus*-based biocontrol agents commercialized for plant disease control (Lakshmanan et al., 2014; Fira et al., 2018; Rabbee et al., 2019; Borriss, 2020)

Product	*Bacillus* strain (species assignment)	Pathogens	Crops	Company
Amylo-X® WG	*B. amyloliquefaciens* D747	*Pseudomonas syringae*, *Monilinia spp*, *Stemphylium* spp., *Botrytis cinerea*, *Sclerotinia* spp, *Pezicula alba*	Vegetable cultivation, arboriculture, grapevine, mushroom, fruit trees	Certis EUROPE
DOUBLE NICKEL 55™	*B. amyloliquefaciens* D747	*Sphaerotheca fuliginea*, *Erysiphe cichoracearum*, *Phytophthora capsici*, *B. cinerea*, *Alternaria solani*, *Sclerotinia minor*, *Sclerotinia sclerotiorum*, *Rhizoctonia solani*	Cucumber, fruiting vegetable, grapevine, lettuce, potatoes	Certis USA, LLC
Eco-Shot	*B. amyloliquefaciens* D747	*Colletotrichum gloeosporioides*, *Podosphaera fuliginea*, *S. sclerotiorum*, *B. cinerea*, *Erysiphe polygoni*, *A. solani*, *Phyllosticta citricarpa*, *Cryptosporiopsis perennans*	Grape, citrus, vegetables, legumes	IHARA
Serifel®	*B. amyloliquefaciens* MBI 600	*B. cinerea*, *Sclerotinia* spp.	Grapes, tomato, lettuce, strawberry, mushroom, pepper, tobacco	BASF, Germany
Integral® Pro (SUBTILEX®)	*B. amyloliquefaciens* MBI 600	*Phoma* spp.	Rapeseed, *Brassicaceae*	BASF, Germany
FZB24®liquide	*B. amyloliquefaciens*/ *velezensis* FZB42	*Rhizoctonia* spp., *Oidium* spp., *B. cinerea*, *Sclerotinia spp*, *Bremia* spp., *Fusarium* spp.	Potatoes, lettuce	Bayer Crop Science/ ABITEP
Taegro®	*B. amyloliquefaciens* FZB42	*Oidium* spp., *B. cinerea*, *Sclerotinia* spp., *Bremia* spp.	Strawberry, chicory, cucumber, lettuce, corn, salad, grapevine	Syngenta

EcoGuard ™ BIOFUNGICIDE	*B. licheniformis* SB3086	*Sclerotinia homoeocarpa, Pyricularia grisea, P. oryzae , Gloeocercospora sorghi, Puccinia graminis, Alternaria* spp., *Typhula* spp., *Microdochium nivale, Drechslera poae, Puccinia* spp., *Uramyces* spp., *Calletatrichum graminicola*	Turf, ornamental plants	Novozymes Biologicals, Inc.
Sonata®	*B. pumilus* QST 2808	*Oidium* spp.	Vegetable cultivation, tobacco, grapevine, fruit bush	Bayer Crop Science, previously AgraQuest
Ballad®	*B. pumilus* QST 2808	*Alternaria brassicae, oidium* spp., *Rhizoctonia oryzae, Xanthomonas* spp., *Puccinia sorghi, Erysiphe graminis, Puccinia* spp., *Xanthomonas* spp., *Sphaerotheca* spp., *Pseudomonas syringe, Cercospora* spp.	Oil seed crops, cereal grains, grass seed, sweet corn	Bayer Crop Science, previously AgraQuest
BSF4®	*B. subtilis* BSF4	*Erwinia amylovora, B. cinerea*	Pear, apple, strawberry, tomato, ornamental plants, lettuce	Agribiotec, Italy
Cease®	*B. subtilis* QST 713	*Peronospora belbahrii, B. cinerea, E. cichoracearum*	Ornamental plant, tomato, lettuce, celery, strawberry, tropical fruit, cannabis	BioWorks®
BIOBAC® WP	*B. subtilis* BSY 1336	*B. cinerea, Sclerotinia* spp., *Leveillula tarurica*	Tomato, grapevine, cherry, strawberry, pepper	Kuanghwa Chemical Co. Ltd., Taiwan
BACSTAR®	*B. subtilis* D747	*E. amylovora, B. cinerea, P. capsici*	Apple, pears, citrus, grapes, kiwifruit, strawberry, lettuce, onions, ornamental plant, turf	Certis USA, LLC

(*Continued*)

Table 2 Examples of *Bacillus*-based biocontrol agents commercialized for plant disease control (Lakshmanan et al., 2014; Fira et al., 2018; Rabbee et al., 2019; Borriss, 2020) (*Continued*)

Product	*Bacillus* strain (species assignment)	Pathogens	Crops	Company
Companion®	*B. subtilis* GB03	*Sclerotinia* spp., *Rhizoctonia* spp., *Fusarium* spp., *Aspergillus* spp.	Pod vegetable, cotton, peanuts, soybean, wheat, barley, corn, strawberry, grapes, tomatoes	Growth Products Ltd., USA
Kodiak ™	*B. subtilis* GB03	*Rhizoctonia* spp., *Fusarium* spp., *Alternaria* spp., *Aspergillus* spp.	Seed treatment, cotton, soybean, wheat, barley,	Bayer Crop Science, previously Gustafsson LLC
Serenade® ASO	*B. subtilis* QST 713	*B. cinerea*, *Oidium* spp., *Golovinomyces cichoracearum*, *Podosphaera* spp.	Fruit bush, vegetable cultivation, tomato, pepper	Bayer Crop Science, previously AgraQuest
Serenade® SOIL	*B. subtilis* QST 713	*Fusarium oxysporum*, *Pythium* spp., *Rhizoctonia* spp.	Root vegetables, bulb vegetables, leafy vegetables, *Brassicaceae*, fruiting vegetable, cucurbits	Bayer Crop Science, previously AgraQuest
Serenade® OPTI	*B. subtilis* QST 713	*Botrytis cinerea*, Sclerotinia spp, *G. cichoracearum*, *Podosphaera* spp., *Albugo candida*, *Oidium* spp., *Phytophthora* spp., *Peronospora* spp., *E. amylovora*	Stone fruits, garden beet, tomatoes, arboriculture, vegetable cultivation, fruit bush	Bayer Crop Science, previously AgraQuest
Serenade® MAX	*B. subtilis* QST 713	*B. cinerea*, *Oidium* spp., *Colletotrichum* spp., *Glomerella* spp.	Avocado, banana, mushroom, cucumber, ornamental plant, arboriculture, lettuce, melon, tomato, grapes	Bayer Crop Science, previously AgraQuest
Rhapsody®	*B. subtilis* QST713	*B. cinerea, Sclerotinia* spp., *Rhizoctonia* spp., *Fusarium* spp., *Pythium* spp., *Oidium* spp., *Monilinia* spp.	Vegetable cultivation, grapevine, fruit bush, ornamental plant, potatoes	Bayer Crop Science, previously AgraQuest

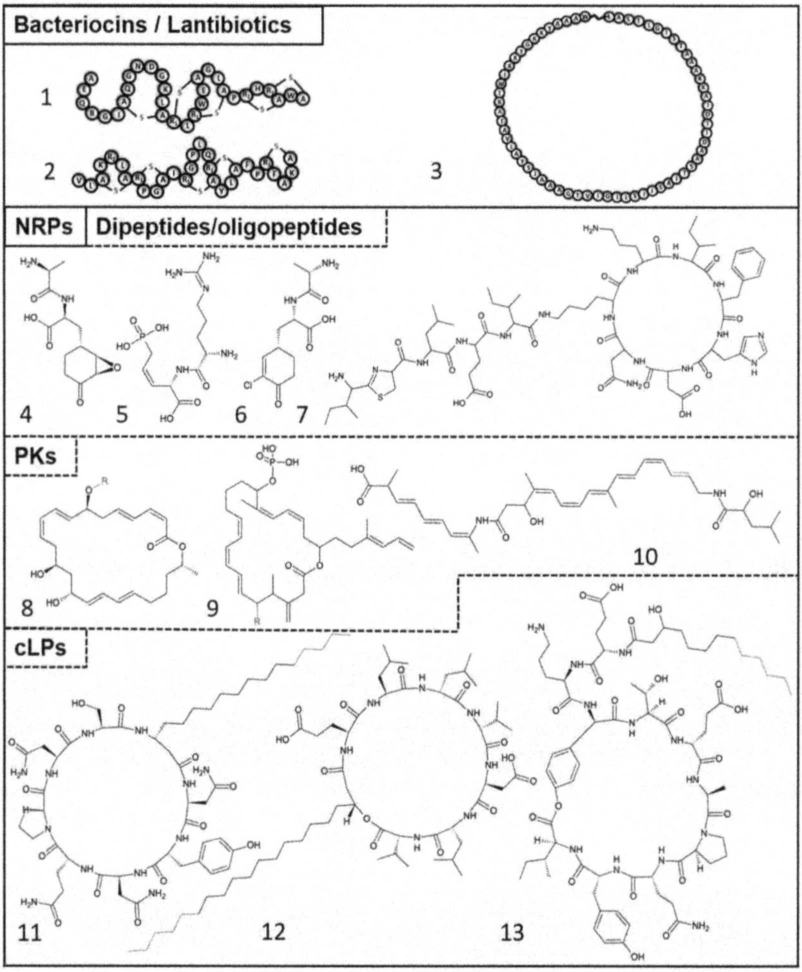

Figure 1 Overview of bioactive secondary metabolites (BSMs) produced by the *Bacillus subtilis* complex. The antimicrobial compounds can be synthesized by the classical ribosomal synthesis, such as the lantibiotics amylolysin (1, R1: aminobutyrate) or ericin A (2, R2: 2,3-didehydroalanine) and the cyclic bacteriocins amylocyclicin (**3**). Besides, some are produced by a non-ribosomal biosynthesis, the non-ribosomal peptides (NRPs), as the unusual peptides (dipeptides and oligopeptides), the cyclic lipopeptides (cLPs), and the polyketides (PKs). The dipeptides and oligopeptides include bacilysin (4), rhizocticin (5), chlorotetain (6), and bacitracin (7). The three main polyketides are synthetized by this genus, macrolactin (8, macrolactin A, R: H or $COCH_2COOH$ or $COCH_2CH_2COOH$), difficidin (9, R: H or OH for oxydifficidin) and bacillaene (10, unsaturated or bacillaene glycosylated or dihydrobacillaene glycosylated). Likewise, the three main families of cLPs produced by *Bacillus* sp. are iturin (11, iturin A, C13 to C17), surfactin (12, surfactin, C12 to C16) and fengycin (13, fengycin A, C14 to C19).

substitution can also be coproduced by the same strain. Surfactin is the only cLP that can be found in all species of the *B. subtilis* group. Some lipopeptides such as lichenysin or pumilacidin are specific to the producing species (*B. licheniformis* and *B. pumilus*, respectively). This structural diversity is also present in the iturin family, regrouping peptidic variants such as bacillomycin, iturin and mycosubtilin, or in the fengycin family that includes fengycin A, fengycin B and plipastatin (Ongena and Jacques, 2008; Raaijmakers et al., 2010; Caulier et al., 2019). Unlike surfactin, these two cLPs families are not produced by all members of the *B. subtilis* group and structural variants do not seem to be linked to the species (Andrić et al., 2020).

Several oligopeptides are known to be produced by non-ribosomal synthesis pathways, including bacilysin, chlorotetain, bacitracin and rhizocticin. These are not synthetized by the NRPS machinery and *de facto* are referred to as unusual peptides (Konz et al., 1997; Ming and Epperson, 2002; Rajavel et al., 2009).

4 *Bacillus* spp. biocontrol mechanisms: root colonization

The effectiveness of *Bacillus* spp. as a BCA relies on three main mechanisms that are (i) establishment on plant tissues, mainly roots, that prevent or reduce colonization of pathogens via competition, (ii) antibiosis towards microbial pathogens, (iii) induction of systemic resistance in the host plant.

Various approaches (comparative studies of strains with very different patterns, specific knock-out mutants, tests of purified molecules) have been used to understand the roles of the molecules involved in these mechanisms (Fig. 2).

4.1 *Chemotaxis*

Plant roots that provide physical support and nutrients via exudates are colonized by a plethora of microbes. As with other members of this root-associated microbiome, a successful *Bacillus* spp.-plant association is based on a fine-tuned molecular 'dialogue'. This is driven by root exudates, the composition which varies according to plant genotype and growth stage (Van Overbeek and Van Elsas, 2008; Zhang et al., 2013; Sasse et al., 2018). The proximity of the root is first sensed by the bacterium which moves towards root tissues via chemotaxis (Yssel et al., 2011; Sourjik and Wingreen, 2012). *B. velezensis* FZB42, for example, is attracted by root exudates of maize which act as a chemoattractant and facilitate motility and biofilm formation (Jin et al., 2019).

The presence of pathogens modifies root exudates, increasing their attractiveness for *Bacillus* spp. In cucumber plants infected by *F. oxysporum*

f. sp. *cucumerinum,* the proportion of chemoattractant compounds (citric acid and fumaric acid) is higher compared to non-infected plants, resulting in enhanced root colonization by *B. velezensis* SQR9 (Liu et al., 2014b, 2020). Better colonization of *B. velezensis* 32 on *Rhizobium radiobacter*-infected tomato roots has also been reported compared to uninfected seedlings, correlated with changes in root exudate composition (Abdallah et al., 2020). *Arabidopsis thaliana* foliar infection by *Pseudomonas syringae* also induced root secretion of L-malic acid, attracting *B. subtilis* FB17 and stimulating biofilm formation (Rudrappa et al., 2008). So far, up to ten specific chemoreceptors/ chemoattractants of root exudate compounds, such as amino acids, organic acids or sugars, have been characterized in *B. subtilis* NCIB 3610 and *B. velezensis* SQR9 (Tan et al., 2013; Zhang et al., 2013; Allard-Massicotte et al., 2016; Feng et al., 2019).

4.2 Motility and biofilm formation

Biofilm formation capacity is widely distributed through the *Bacillus* genus. Biofilms can be defined as a multicellular community encased within an extracellular matrix of self-produced polymers such as exopolysaccharides and proteins (e.g. TasA or Bsla) (Vlamakis et al., 2013; Pandin et al., 2017). Exopolysaccharides promote mobility and nutrient capture. This form of communal life constitutes a divergent lifestyle of planktonic cells in which metabolism remodelling occurs through a complex regulatory network (Kearns et al., 2005; Pisithkul et al., 2019). Biofilms are composed of heterogeneous subpopulations organized through this regulator network. Different communities of cells are present in the same biofilm, including matrix-producing cells, surfactin-producing cells, flagellated motile cells and sporulated cells.

Biofilms provide a strong ecological advantage for *Bacillus* spp., promoting migration (Vlamakis et al., 2013; Flemming et al., 2016). They allow *Bacillus* spp. to colonize and survive on the surface of plant roots. *Bacillus* spp. have been shown to form biofilms on plant roots, for example, *B. velezensis* FZB42 on *Zea mays*, *A. thaliana* and *Lemna minor* plantlets (Fan et al., 2011). The plant stimulates biofilm formation both through root exudates and the presence of polysaccharides derived from the plant cell wall which act as signal molecules. Maize and cucumber root exudates, or individual exudate compounds such as glucose, fructose, citric acid, malic acid and fumaric acid, have been shown to stimulate biofilm formation in *B. velezensis SQR9* and in *B. velezensis* S3-1 (Zhang et al., 2015; Jin et al., 2019; Liu et al., 2020). Biofilm establishment is also triggered in the presence of pectin and arabinogalactan, two important plant polysaccharides, in *B. amyloliquefaciens* S499 and SQY 162, and *B. subtilis* NCIB 3610 (Beauregard et al., 2013; Debois et al., 2015; Wu et al., 2015a).

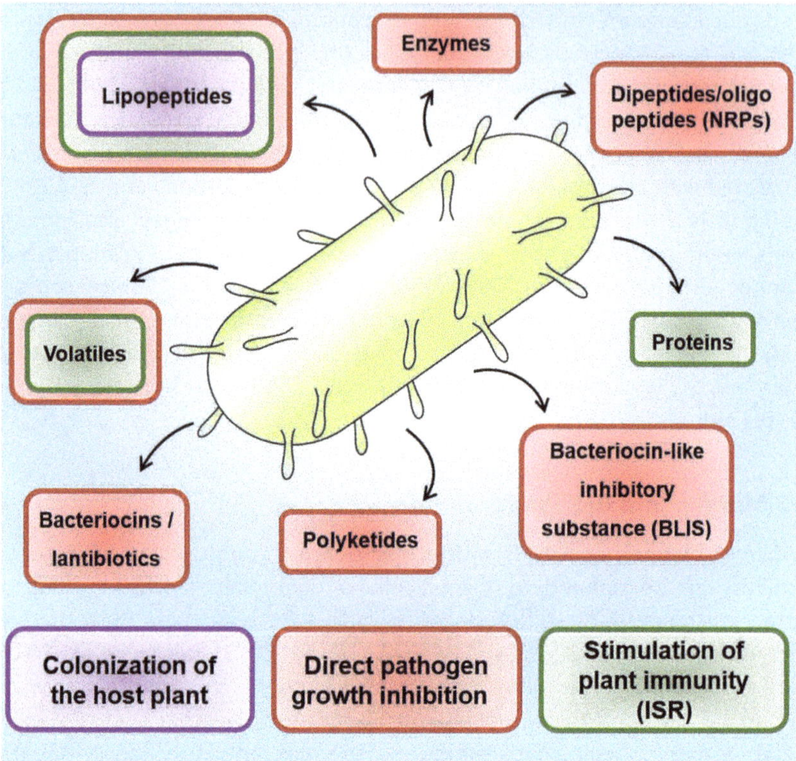

Figure 2 Implication of the different bioactive secondary metabolites (BSMs) in the three main mechanisms of plant protection: by colonization and competition for ecological niche/substrate (in purple), by direct growth inhibition (in red) and/or as an elicitor of induced systemic resistance (ISR) (in green).

Some small-size BSMs, particularly those belonging to the lipopeptide family (especially surfactin), are essential in reducing surface tension and acting as a wetting agent. Surfactin helps to organize cells in bundles to increase biofilm surface area. When deprived of surfactin, for example, mutants of *B. subtilis* 6051 and OKB120 and *B. atrophaeus* ATCC 9372 show impaired biofilm formation (Bais et al., 2004; Aleti et al., 2016). Surfactin plays an important role in the ecological fitness of *Bacillus* spp. for root colonization, both as a surfactant and as a quorum-sensing (QS) molecule (Kearns et al., 2004; Chen et al., 2013; Aleti et al., 2016). Iturin and, to a lesser extent, fengycin have also been reported to facilitate biofilm formation. *In vitro* studies of *B. velezensis* Y6 have shown that these two lipopeptides play important roles in cell motility (Cao et al., 2018). In the case of *B. subtilis* 916, it has been suggested that bacillomycin L (a variant of the iturin family) and surfactin can synergistically contribute to biofilm formation and so to the ecological fitness of *Bacillus* spp.

(Luo et al., 2015). However, a recent study has shown that surfactin production is not required for biofilm formation by *B. subtilis* 3610 (Thérien et al., 2020). The lipopeptide arsenal of *B. subtilis* 3610 is less diverse and contains only surfactin. The involvement of surfactin in biofilm formation may be more complex than expected and could be linked to the lipopeptidome richness of each *Bacillus* species.

Biofilm lifestyle also allows *Bacillus* spp. to resist the presence of other microorganisms. Biofilms act as a fortress to block antimicrobials produced by other competitors (Flemming et al., 2016). The biofilm extracellular matrix has been shown to protect *B. subtilis* 3610 colonies from infiltration by competitors such as *Pseudomonas chlororaphis* (Molina-Santiago et al., 2019). Biofilm formation can also be boosted by the presence of other microorganisms. In *B. subtilis* ATCC6051, the expression of *sinR* and *tasA* genes involved in biofilm formation is modified in contact with the pathogen *Fusarium culmorum*, stimulating matrix multicellular production (Khezri et al., 2016). Biofilm formation also promotes the production of compounds involved in many biocontrol activities (Vlamakis et al., 2013; Pandin et al., 2019; Pisithkul et al., 2019). The biofilm extracellular matrix improves the competitiveness of *Bacillus* spp. in the rhizosphere and thus its biocontrol capability (Pandin et al., 2017). Deficiency in matrix production has been associated with a decrease in biocontrol efficacy by *B. subtilis* 3610 on tomato root against *R. solanacearum* (Chen et al., 2013) and by *B. subtilis* 6051 against *P. syringae* on *A. thaliana* (Bais et al., 2004).

5 *Bacillus* spp. biocontrol mechanisms: antagonistic activity

Bacillus spp. have great potential to control infection through direct inhibition of pathogen growth via antibiosis and/or by interfering with pathogen fitness, notably via quorum quenching (QQ). *Bacillus* spp. antimicrobial potential is related to the synthesis of different classes of metabolites, enzymes, and low-molecular-weight compounds including RiPPs, PKs, NRPs and volatiles (Fig. 2).

5.1 Enzymes

Antagonistic activities of enzymes rely on several modes of action, such as QQ or cell lysis. QQ is an important strategy in plant disease suppression by interfering with QS molecules used by most Gram-negative pathogens. As an example, lactonases, which interfere with the well-known QS molecules N-acyl-L-homoserine lactones (AHL), have been found in numerous *Bacillus* spp. and are considered as QQ enzymes (Dong et al., 2002; Chandra Kalia et al., 2011; Raafat et al., 2019). By blocking cross-talk (via degradation of AHL by QQ) of *P. syringae*, *B. cereus* INT1c reduces pathogen motility, leading to inefficient root colonization (Ananda et al., 2019). QS interference by several *Bacillus* spp. has also been reported to decrease the hypersensitivity response of tomato plants caused by *P. syringae* due to degradation of pathogen AHLs (Jose et al., 2019).

A second enzymatic mode of action involves cell-wall-degrading enzymes. These include chitinases, which degrade chitin (a major component of fungal cell walls) and other enzymes such as chitosanases, glucanases, cellulases and lipases (Caulier et al., 2019; Miljaković et al., 2020). These lytic enzymes have been reported to reduce plant pathogen growth. Chitinase produced by *B. subtilis* ATCC1774 and *B. cereus* CRS7 decreases the growth of *Rhizoctonia solani* (Saber et al., 2015) and *B. cinerea*, respectively (Kishore and Pande, 2007). Alkaline protease purified from *B. amyloliquefaciens* SP1 and heterologously expressed in *Escherichia coli* has been reported to inhibit *F. oxysporum* growth (Guleria et al., 2016).

5.2 Bacteriocins and lantibiotics

Several RiPPs produced by *Bacillus* spp. are effective as antibacterial agents. Bacteriocins exhibit a broad spectrum of antibacterial activity acting by cell lysis, pore formation or inhibition of cell wall biosynthesis (Abriouel et al., 2011; Lajis, 2020). Bacteriocin activity has been occasionally reported against Gram-positive and Gram-negative bacterial plant pathogens. Amylocyclicin, produced by *B. velezensis* FZB42, appears to exhibit activity against *C. michiganensis*, while Bac IH7 and Bac14B inhibit the growth of *R. radiobacter* (formerly called *Agrobacterium tumefaciens*), and *Pseudomonas* spp., *E. carotovora* and *Alternaria solani* (Hammami et al., 2009, 2012; Scholz et al., 2014). LCI peptide, purified from *B. subtilis* A014 and considered as bacteriocin-like inhibitory substances, have antibacterial activities against *Xanthomonas campestris* and *R. solanacearum*. The APC2 protein, a so-called LCI-like peptide produced by *B. amyloliquefaciens FS6*, has been shown to prevent *Fusarium solani* infection (Gong et al., 2011; Fan et al., 2018; Saikia et al., 2019; Wang et al., 2020a).

5.3 Non-ribosomal metabolites

The activity of dipeptides and oligopeptides against plant pathogens has been little studied. Dipeptide bacilysin, a non-ribosomal synthesized molecule produced by *B. velezensis* FZB42, has been reported to inhibit growth of bacteria such as *Erwinia amylovora*, *Xanthomonas oryzae* and oomycete *P. infestans* (Chen et al., 2009b; Wu et al., 2015b; Caulier et al., 2018). PKs are known for their antibacterial activity. A mutant, for example, of *B. velezensis* FZB42, impaired in difficidin, has been found to inhibit *E. amylovora* (Chen et al., 2009b) and *X. oryzae* (Wu et al., 2015b) less than the wild-type strain. Purified macrolactin has also been reported to be active against *R. solanacearum* and *F.*

oxysporum (Yuan et al., 2012a). Bacillaene has been shown to protect *Bacillus* spp. cells against the degrading-enzymatic activity caused by competitors (*Streptomyces* sp. Mg1 and *Myxococcus xanthus*) (Barger et al., 2012; Müller et al., 2014, 2015). *Bacillus* PK type difficidin also shows antimicrobial activity, depending on the structure. The oxydifficidin derivative is approximately three times more active against *R. solanacearum* as compared to difficidin (Im et al., 2020) showing the importance of structural variations among *Bacillus* spp. BSMs.

The antimicrobial activity of lipopeptides is mainly related to members of the fengycin and iturin family. These two families have been reported to damage hyphae and conidia of several fungal pathogens including *F. graminearum* (Gu et al., 2017; Hanif et al., 2019) and *Monilinia fructicola* (Liu et al., 2011). The effects of lipopeptides depend on their amphiphilic nature and chemical structure. Fengycins are active against filamentous fungi such as *Rhizopus stolonifera* (Tao et al., 2011), *Magnaporthe grisea* (Zhang and Sun, 2018) or *R. solani* (Guo et al., 2014). Iturins are efficient not only against filamentous fungal pathogens such as *C. gloeosporioides* (Jin et al., 2020) and *Aspergillus flavus* (Gong et al., 2014) but also against the oomycete *P. infestans* (Wang et al., 2020c). Fengycin can also inhibit damaging mycotoxin synthesis produced by *F. graminearum* (Hanif et al., 2019). Iturins and fengycins are mainly active against fungi. However, the antibacterial activity of iturins and fengycins has also been occasionally reported against *X. campestris* and *Pectobacterium carotovorum* (Zeriouh et al., 2011) and *Xanthomonas axonopodis pv. vesicatoria* (Medeot et al., 2020) or *R. solanacearum* (Villegas-Escobar et al., 2018). Mode of actions of iturins and fengycins are still partially unknown but are most probably related to their capacity to disrupt fungal pathogens plasma membranes by forming pores, leading thus to cell death (Deleu et al., 2008; Wise et al., 2014; Zakharova et al., 2019). Small structural variations can impact bioactivity. Different homologues of bacillomycin D have thus different antimicrobial activities against the same pathogen according to the physiological stage of the fungi (Tanaka et al., 2015; Luna-Bulbarela et al., 2018). Differential activities of these variants can be viewed as a strategy by *Bacillus* spp. to increase the antibiotic spectra of one molecule family.

Surfactin has been found to have antimicrobial activity but in the 50–200 μM range, a higher concentration than in natural conditions (Jourdan et al., 2009; Raaijmakers et al., 2010; Debois et al., 2014; Liu et al., 2014a; Fan et al., 2017b; Sarwar et al., 2018). The inhibitory effect of surfactin is probably not due to direct antagonism but rather the interference with the colonization process of pathogens such as *P. syringae* (Bais et al., 2004) and *R. solanacearum* (Almoneafy et al., 2014). Even if surfactin cannot be considered as an antimicrobial molecule, it still has the ability to interact with biological membranes to induce structural modifications (Deleu et al., 2013). Its antimicrobial activity can be linked to a synergistic effect with other cLPs. The combination of surfactin with bacillomycin D or mycosubtilin leads to more efficient control of *F. oxysporum* f. sp. *iridacearum* (Mihalache et al., 2018) and *B. cinerea* (Tanaka et al., 2015). Combining surfactin and fengycin is effective against *F. oxysporum* f. sp. *iridacearum* (Mihalache et al., 2018) and *P. infestans* (Wang et al., 2020c). Co-production of two antifungal compounds (i.e. iturins and fengycins) by *Bacillus* improves its antifungal properties since these antifungal compounds are active against different pathogens.

5.4 Volatiles

The antimicrobial activities of *Bacillus* spp. are also linked to volatiles. The most relevant group of volatiles are volatile organic compounds (VOCs) including alcohol, aldehydes, ketones, hydrocarbons, acids and terpenes. The majority of VOCs are derived from *Bacillus* spp. metabolic pathways such as 2-nonanone produced by β-oxidation of fatty acid (Korpi et al., 2009; Fincheira et al., 2017). Some of these metabolites have been reported to affect motility and biofilm formation of bacterial pathogens (Raza et al., 2016) and/or decrease fungal spore germination and growth (Yuan et al., 2012b). Antimicrobial activities of VOCs have been extensively studied (Caulier et al., 2019; Kai, 2020), including against fungal pathogens such as *Sclerotinia sclerotiorum* (Lim et al., 2017), *B. cinerea* (Jiang et al., 2018), *A. solani* (Zhang et al., 2020) or *M. fructicola* (Liu et al., 2018a). VOCs synthesized by *Bacillus* spp. also display antimicrobial activity against *R. solanacearum* such as 2-nonanone and 2-undecanone (Raza et al., 2016).

The BSM arsenal of *Bacillus* species such as *B. velezensis* includes many antimicrobials with broad-spectrum activity. However, recent studies strongly suggest no marked and durable effect of these bacteria on the microbiome of the treated crop (Correa et al., 2009; Chowdhury et al., 2013; Kröber et al., 2014; Qiao et al., 2017). Those bacilli thus provide protection to their host plant but have no detrimental effect on the microbiome, which is of prime interest for future application as BCAs.

6 *Bacillus* spp. biocontrol mechanisms: induced-systemic resistance

Bacillus spp. can protect plants by triggering an immune response which is systemically expressed in all organs of the host plant called induced systemic resistance (ISR) (Yu et al., 2017; Ranf, 2018; Schellenberger et al., 2019). ISR involves activation of a latent defence process called priming and characterized by activation of defence responses only after infection. Priming provides an enhanced level of protection and a faster/stronger activation of defences by the whole plant (Pieterse et al., 2014). ISR activated by *Bacillus* spp. protects plants against a broad spectrum of pathogens (Köhl et al., 2019; Miljaković et al., 2020). ISR can, *de facto*, complement resistance induced by pathogen attack such as recognition of pathogen- or microbial-associated molecular patterns (PAMPs or MAMPs, respectively) (Van Wees et al., 2000). *Bacillus* spp. are not generally a source of MAMPs, even though some *Bacillus* spp. can trigger ISR in host plants (Vanthana et al., 2019; Rajamanickam and Nakkeeran, 2020). BSM induction of ISR by *Bacillus* spp. involves different types of metabolites, degradation products due to bacterial activity or secreted molecules and volatiles (Pršić and Ongena, 2020). However, the molecular mechanisms driving recognition of these elicitors of *Bacillus* spp., especially for BSMs, are poorly understood.

Bacillus spp. can produce some protein elicitors of plant defence. The AMEP 412 protein in the *Bacillus* genus triggers ISR in tobacco against *P. syringae* pv. *Tomato* (Shen et al., 2019). The PeBA1 protein from *B. amyloliquefaciens* NC6 induces resistance in tobacco against infection by *B. cinerea* and by the tobacco mosaic virus (Wang et al., 2016).

cLPs are key ISR elicitors in a range of pathosystems. Surfactin and, to a lesser extent, iturin or fengycin, are powerful inducers of plant systemic resistance (Crouzet et al., 2020). Application of surfactin at the root level decreases disease severity due to the necrotrophic fungus *B. cinerea* on bean, tomato and *A. thaliana* leaves (Ongena et al., 2007; Debois et al., 2015). A comparative analysis using different *Bacillus* strains revealed a strong correlation between the amounts of surfactin and its ability to trigger defence immunity in a plant (Cawoy et al., 2014). Studies with *B. velezensis* FZB42 surfactin deficient mutant demonstrated the importance of this lipopeptide in the induction of plant defences in lettuce against *R. solani* (Chowdhury et al., 2015b). ISR activation by surfactin has been reported for several other pathosystems such as melon/*Podosphaera fusca* (García-Gutiérrez et al., 2013) and wheat/*Z. tritici* (Le Mire et al., 2018). Several iturin variants have also been shown to trigger ISR in different pathosystems, such as strawberry/*C. gloeosporioides* (Yamamoto et al., 2015), chili pepper/*Phytophthora capsica* (Park et al., 2016) or wheat/*Z. tritici* (Mejri et al., 2018). The iturin variant mycosubtilin has been reported to induce an immune response in grapevine against *B. cinerea* (Farace et al., 2015). Fengycin has been shown to trigger ISR in tomato against *B. cinerea* (Ongena et al., 2007), and more recently in grapevine and in tomato against *Plasmopara viticola* and *S. sclerotiorum*, respectively (Farzand et al., 2019; Li et al., 2019).

Lipopeptide activity has been shown to be plant and even organ-dependent (Pršić and Ongena, 2020). Surfactin is more efficient on dicots than on monocots (Ongena et al., 2007; Rahman et al., 2015; Mejri et al., 2018). Iturin treatment is more efficient when applied on leaves rather than roots (Han et al., 2015; Yamamoto et al., 2015). Iturin and surfactin have been shown to trigger an immune response by interacting with plant membrane lipids, but the precise mechanism remains unclear (Henry et al., 2011; Deleu et al., 2013; Fiedler and Heerklotz, 2015; Balleza et al., 2019). It appears that surfactin activity is related to plant plasma membrane composition and is also affected by lipopeptide structure. Structural changes in fatty acid chain length (long chains are more efficient), as well as the substitution in the peptidic cycle, modulate their ability to trigger ISR (Jourdan et al., 2009; Kawagoe et al., 2015).

Volatiles produced by *Bacillus* are also involved in the induction of plant immunity. Exposure of roots of *A. thaliana* seedlings to VOC mixtures, produced by *B. subtilis* GBO3 or by *B. amyloliquefaciens* IN937a, resulted in fewer lesions from the pathogen *E. carotovora* subsp. *carotovora* due to immune resistance induction (Ryu et al., 2004). The majority of cases showing VOCs as inducers of systemic resistance relate to acetoin and 2,3-butanediol (Ryu et al., 2004; Rudrappa et al., 2010; Peng et al., 2019). Commercially available compounds such as 3-pentanol are involved in the induced resistance in pepper plants against *X. axonopodis* pv. *vesicatoria* (Choi et al., 2014).

7 Factors influencing the production of bioactive secondary metabolites: cellular regulation

The conditions under which optimal production of BSMs is achieved by *Bacillus* spp. are still poorly understood and studied in a restricted number of metabolites. Evidence for efficient *in situ* production of these metabolites by *Bacillus* spp. inoculants mainly concerns lipopeptides. Surfactin and/or iturin have been recovered in biologically relevant quantities from cucumber roots inoculated with *B. velezensis* strain QST713 (Kinsella et al., 2009) and from the tomato rhizosphere after treatment with *B. subtilis BGS3* (Ongena et al., 2007; Nihorimbere et al., 2009) and strain RB14-C (Asaka and Shoda, 1996). Surfactin synthesis *ad planta* by *B. velezensis* FZB42 has been demonstrated in Lemna plantlets (Fan et al., 2011) and in lettuce together with fengycin and bacillomycin (Chowdhury et al., 2015b). Fengycins and iturins have also been detected on leaves and in fruits, illustrating that these plant tissues may also support cLP biosynthesis (Touré et al., 2004; Romero et al., 2007; Zeriouh et al., 2011, 2014; Debois et al., 2015). This limited research on the production of BSMs under natural conditions reflects the inherent difficulties in detecting and quantifying these small-size compounds in complex environments like soil. These molecules can be chemically unstable, degraded by other (micro)organisms or adsorbed on the surface or within soil particles. Whether BSMs accumulate to biologically efficient concentrations in the rhizosphere is, therefore, still a matter of debate. However, it is crucial to measure such concentrations in order to better understand the real impact of these molecules in eliciting plant defences and/or in the direct biocontrol of pathogens.

Recent studies mainly on *B. subtilis* suggest that regulation of the expression of BGCs at the cellular level is quite complex, at least for NRP/PK products. It involves a range of pleiotropic regulators or transcription factors driving developmental processes such as QS, biofilm formation or sporulation. Natural conditions are also far from *in vitro* cultural conditions in artificial media. Root-associated bacteria feed almost exclusively on plant exudates, cross-communicate with other organisms (with positive or negative outcomes) and have to face multiple abiotic factors in the soil. Production of BSMs may, therefore, also be modulated by multiple ecological parameters.

The regulation of lipopeptides biosynthesis at the cellular level has been quite well studied. Two QS regulatory systems; that is, ComQXPA (Lopez et al., 2009) and Rap-Phr (Auchtung et al., 2006), are key players in surfactin regulation. Phosphorylated ComA is required to initiate *srfA* gene transcription, while the Rap phosphatases dephosphorylate ComA act as repressors. Additional transcription factors such as Spx (Zhang et al., 2006) or codY (Coutte et al., 2015) interfere with the role of ComA (Han et al., 2015) and others, such as ClpX/P, (Nakano et al., 2000) display a chaperone activity to facilitate ComA DNA binding. DegU, another master regulator, has also been shown to have a positive effect on surfactin production (Ogura, 2001). The AbrB/Abh system seems to affect lipopeptide synthesis in a more specific manner depending on the physiology of the cell (Zhi et al., 2017).

Environmental factors such as phosphate limitation or oxidative stress also have an impact on surfactin regulation via transcription factors such as PhoP and PerR (Ogura, 2001; Hayashi et al., 2005). The regulation of iturin and fengycin shares some regulators with surfactin. DegU, ComA, AbrB and GlnR, which are involved in glutamine synthesis (Koumoutsi et al., 2007; Zhang et al., 2017b; Xu et al., 2020), regulate iturin biosynthesis together with YczE, presumably acting at the post-transcriptional level (Koumoutsi et al., 2007; Dang et al., 2019). DegQ, PhoP and AbrB drive fengycin synthesis (Wang et al., 2015; Lu et al., 2016; Guo et al., 2018). Two extra-factors LutR and SinR, both involved in cellular mechanisms, are also positive regulators (İrigül-Sönmez et al., 2014).

The regulation of lantibiotics is determined by cell density. They act as autoinducing peptides via a typical two-component regulatory system composed of a sensor (histidine kinase) and a response regulator (LanRK) present in their operon (Kleerebezem et al., 2004; Schmitz et al., 2006).

The complexity of BSM regulation in *Bacillus* spp. is illustrated in the case of the dipeptide bacilysin. It involves several pleiotropic regulators depending on the physiological state of the cells in culture. In *B. subtilis*, bacilysin biosynthesis is under the control of the ComQXPA and PhrC QS system (Yazgan et al., 2001; Yazgan Karata et al., 2003). ComA binds directly to the promoter region of the *bac* operon and initiates a basal expression of this BGC (Köroğlu et al., 2011). The LutR regulon and the two interconnected regulators (DegU and DegQ) also regulate bacilysin synthesis (Köroğlu et al., 2008; Mariappan et al., 2012). ComA may also act indirectly on bacilysin production by influencing DegU through DegQ activation. AbrB acts as a repressor of *bac* during the exponential growth phase, but its negative regulatory effect is silenced by Spo0A (Köroğlu et al., 2011). The other transcription factors CodY and ScoC also negatively regulate bacilysin production by binding directly to the promoter region.

Less information is available regarding PKS regulation. In *B. subtilis,* expression of the *bae* operon, which is the only PK cluster conserved among species of the *B. subtilis* complex (Fan et al., 2018), is under the control of several transcriptional regulators. Full expression of this operon requires ComA, DegU, the ScoC regulon and CodY, which binds multiple sites in the cluster (Belitsky and Sonenshein, 2013; Vargas-Bautista et al., 2014). The *dfn* and *mln* operons, responsible for the synthesis of the two other PKs (difficidin and macrolactin, respectively) are only present within the *B. velezensis* species, explaining the low number of studies dealing with their regulatory pathways. The antiterminator LoaP, encoded by a gene positioned directly upstream of the *dfn* operon, has been shown to regulate the transcriptional readthrough of termination sites located within the *dfn* and *mln* operons in *B. velezensis* (Goodson et al., 2017). Upon *loaP* deletion, the production of both difficidin and macrolactin is abolished, but bacillaene synthesis is fully conserved (Goodson et al., 2017). Difficidin production is also impaired in mutants not expressing the DegU regulator (Mariappan et al., 2012).

8 Factors influencing the production of bioactive secondary metabolites: biofilm formation

Agitated liquid cultures are commonly used to assess the effect of a given factor on secondary metabolite synthesis. However, the physiology of planktonic cells, when undergoing fast 'latence-exponential-stationary' growth typically observed in batch cultivation, does not reflect the physiology of bacteria developing in the close vicinity of roots. PGPR form biofilm-structured multicellular communities at the surface of roots in nutritionally limited conditions very different from the optimized laboratory conditions. The inoculated strain establishes on plant tissues as biofilm-structured patches. Biofilm formation is associated with the diversification of the cell community into sub-populations with distinct roles and activities (metabolically active vegetative cells, matrix producers, cannibals and spores). This formation results in phenotypic heterogeneity that may influence the synthesis of secondary metabolites (Nihorimbere et al., 2009).

The effect of biofilm formation on the pattern of lipopeptides produced by *B. amyloliquefaciens* has been assessed by growing the bacterium in static liquid cultures conducted in wells of microtiter plates. In these conditions, cells readily aggregate to form pellicles at the liquid-air interface. LC-MS profiling of the secreted lipopeptides revealed clear differences compared to agitated cultures with enhanced production of surfactin production but not iturins and fengycins. The proportions of the three lipopeptide families were similar to those secreted after root colonization and clearly differed from planktonic cells secreting much higher amounts of iturins and fengycins (Nihorimbere et al., 2012; Debois et al., 2014). The formation of biofilm is accompanied by metabolic changes under the control of regulators such as SpoOA. These transcriptional modifications may also influence the production of BSMs like the *bacABCDEF* and *sbo-alb* operons coding, respectively, for bacilysin and subtilosin A that are up-regulated during biofilm formation (Pisithkul et al., 2019). The production of other BSMs under biofilm conditions has not yet been investigated.

The formation, organization and functioning of this complex multi-cellular structure have been extensively studied in *B. subtilis* and related species but almost exclusively *in vitro* by forming macrocolonies on gelified media or as pellicles developing at the air-liquid interface. However, biofilm formation on plant tissues is very different compared to an inert surface, taking into account interactions with the host. In natural conditions, *Bacillus* cells evolve in the middle of other microbes and interactions with other microbial species can modulate motility and biofilm formation (Andrić et al., 2020) (see below). Biofilm formation and functionality have been poorly investigated *in planta*, including all these biotic parameters that may interfere with this process. There is a need to develop new cultivation methods in the lab to mimic biofilm formation by bacterial cells colonizing roots or influenced by plant factors. This initially requires gnotobiotic conditions where the bacterium and the plant interact under controlled physico-chemical settings such as temperature and pH. Population diversification and phenotypic heterogeneity should be studied using fluorescent transcriptional reporters designed to highlight the phenotype of interest using confocal laser-scanning or other microscopy techniques. The use of flow cytometry on microbial cells extracted from biofilm structures may also provide a dynamic view of the microbial population of interest.

9 Factors influencing the production of bioactive secondary metabolites: abiotic factors

The potential of BCAs such as *Bacillus* spp. to produce BSMs has been, in most cases, evaluated *in vitro* using optimal growth conditions for key parameters such as temperature, pH and oxygen availability. However, these conditions fluctuate in soil and may significantly influence BSMs production *in planta*. Research so far has mainly focused on the impact of these abiotic factors on lipopeptide production.

Oxygen depletion is common in the rhizosphere due to consumption by roots and microbes (Dubern and Bloemberg, 2006). Under oxygen-starvation conditions, surfactin production remains unchanged, suggesting that a low oxygen level is not detrimental for surfactin synthesis in the rhizosphere (Nihorimbere et al., 2009, 2012). A positive effect of O_2 limitation has even been reported for surfactin production but under bioreactor conditions (Yeh et al., 2006; Yi et al., 2017). In contrast, expression of iturin and fengycin operons is O_2-dependent (Guez et al., 2008; Nihorimbere et al., 2012; Rangarajan et al., 2015).

Temperature stress is a key factor affecting basal and secondary metabolism production in microbes, significantly affecting the survival and effectiveness of plant-associated microbes used as microbial inoculants. Little is known about the effects of low or fluctuating temperature on the production of BSMs by *Bacillus* spp. Several studies have reported an increase in lipopeptide production with decreasing temperatures (Fickers et al., 2008; Pertot et al., 2013; Pereira Monteiro et al., 2016; Chen et al., 2019). Growth at low temperatures (15°C) led to a marked increase in surfactin production in comparison to high temperatures (>35°C) (Pertot et al., 2013). Low temperatures may affect the expression of ComK and further down-regulate genes coding for the RapF and RapH repressors of surfactin gene expression (Budde et al., 2006; Jacques, 2011). Enhanced production at low temperature may also rely on post-transcriptional events, such as NRPS protein turn-over or modifications in the lipopeptide export process (Fickers et al., 2008). Data from *in planta* assays performed in greenhouses have also suggested that enhanced surfactin production at low temperature may counter-balance the reduced growth of *B. amyloliquefaciens* S499 on roots, maintaining its ISR triggering potential under cold conditions (Pertot et al., 2013). The effect of cold/heat on the production of other *Bacillus* spp. lipopeptides has been little investigated but enhanced iturin synthesis by lowering the temperature from 37°C to 25°C has been observed (Ohno et al., 1995; Jacques et al., 1999; Fickers et al., 2008). The consequences of temperature fluctuations on cLP production seem to vary between strains. Results obtained for surfactin in *B. velezensis* S499 and FZB42 could not be confirmed for *B. velezensis* QST713 (Pertot et al., 2013). It has also been shown that increased temperature can cause a higher production of surfactin in *B. velezensis* FJAT-46737 (Chen et al., 2020). Further investigation covering a wider range of temperatures and evaluating the impact on the whole secreted metabolome is required to better appreciate its impact on the production of BSMs.

The effect of pH on BSMs production is poorly documented. Surfactin production has been reported to be inhibited at pH 5 (Yi et al., 2017). Efficient synthesis of lipopeptides is favoured in mild-acidic to neutral environments, which is often the case for rhizosphere due to multiple processes related notably to plant exudation and root/microflora activity (Cosby et al., 1998; Akpa et al., 2001; Hinsinger et al., 2009; Wang et al., 2020a).

10 Factors influencing the production of bioactive secondary metabolites: biotic factors

10.1 Interactions with the host plant

Root exudates are primarily used as nutrients by plant-associated bacteria to sustain growth but may also stimulate flagella motility, chemotaxis and biofilm formation. They are, therefore, essential for rhizosphere establishment (Kierul et al., 2015; Pandin et al., 2017). The chemical nature of these exudates also influences the production of BSMs.

cLPs production is impacted both qualitatively and quantitatively by changes in exudate components. Carbon (C) sources in root exudates support *B. amyloliquefaciens* S499 growth. Surfactin secretion (but not iturin or fengycin) has been shown to be significantly higher in the presence of organic acids as compared to sugars (Nihorimbere et al., 2012). The relative proportions of different homologues may vary in relation to carbon sources. The synthesis of C_{15} surfactins is promoted in the presence of organic and amino acids compared to sugars.

In response to maize root exudates, the expression of genes involved in the synthesis of bacillaene, difficidin, macrolactin, fengycin and surfactin was up-regulated in *B. velezensis* FZB42 and in *B. velezensis* SQR9 (Fan et al., 2012; Zhang et al., 2015). Root exudates of *Eruca sativa* (rocket salad) have been shown to induce *pks* operon expression, responsible for bacillaene synthesis in *B. subtilis*, allowing the bacterium to attack other root-associated bacteria (Ogran et al., 2019). In *B. velezensis* SQR9, comparative proteomic analysis revealed that PKs, fengycin and surfactin were overproduced upon colonization of cucumber roots (Qiu et al., 2014).

Physical contact with the root surface and, more specifically, perception of plant polymers may initiate processes such as biofilm formation in *Bacillus* species. It has been shown that polymers of the plant cell wall (xylan, pectin) also act as cues for triggering the synthesis of surfactin (Beauregard et al., 2013; Debois et al., 2015; Wu et al., 2015a). The *srfAA* gene is induced when *Bacillus* spp. are placed in contact with rice seedlings (Xie et al., 2015). As surfactin is involved in swarming motility and biofilm formation, improved production may contribute to spread on root tissues and aggregation in biofilm-related multicellular communities.

10.2 Interactions with microbial pathogens

Bacillus spp. have to compete with other microorganisms in the rhizosphere. *Bacillus* spp. thus have to adapt and improve their ecological fitness by modulating BSM production in response to fungal and bacterial competitors. The presence of chitin or S-glucan has been found to increase the production of fungal cell wall degrading enzymes such as chitinase and glucanases by *B. subtilis* JF419701, (Alamri, 2015). It has been shown that co-cultures between plant beneficial bacilli and oomycetes or fungi (including *P. aphanidermatum*, *F. oxysporum* or *S. sclerotiorum*) increase iturin and fengycin production, both known for their antifungal properties (Cawoy et al., 2015; Farzand et al., 2020). Fengycin synthesis by *B. velezensis* S499 has also been reported to be upregulated in the presence of *Rhizomucor variabilis*, suggesting that the bacterium is able to perceive some molecular trigger(s) emitted by the pathogen (Zihalirwa Kulimushi et al., 2017).

In addition to these two antifungal cLPs, surfactin production has also been induced when *Bacillus* spp. are confronted to *Phytophthora parasitica*, *R. solani*, *F. solani*, *S. sclerotiorum*, *R. stolonifera, Fusarium sambucinum and Trichoderma aggressivum* (Chowdhury et al., 2013; DeFilippi et al., 2018; Pandin et al., 2019). Surfactin overproduction could promote competition for nutrients and space (Molina-Santiago et al., 2019; Andrić et al., 2020). Surfactin is involved in QS, stimulates biofilm formation, and contributes to plant root colonization (Kinsinger et al., 2003; Lopez et al., 2009; Raaijmakers et al., 2010). The isoform profile of cLPs changes depending on the competitor pathogen (Cao et al., 2018; DeFilippi et al., 2018). This suggests that bacteria modulate BSM synthesis depending on signals emitted by specific fungal pathogens (Frey-Klett et al., 2011).

Variation in BSM production in response to bacterial competitors has been poorly studied and mostly limited to transcriptional analyses. Expression of genes belonging to iturin, fengycin and surfactin operons has been shown to be up-regulated in the presence of *R. solanacearum* (Almoneafy et al., 2014; Cao et al., 2018). Expression of lipopeptide operons and bacilysin in *B. velezensis* BK7 has been reported to be significantly upregulated in the presence of *Pseudomonas fuscovaginae* (Kakar et al., 2014). Bacterial competitors also stimulate *B. subtilis* motility and biofilm formation, but the causal link between overproduction of biofilm in the presence of some competitors and surfactin production has not yet been demonstrated (Grau et al., 2015; van Gestel et al., 2015; Liu et al., 2018c).

11 Interactions of *Bacillus* spp. with other beneficial microorganisms and their use in biocontrol

Interactions between *Bacillus* spp. and other microorganisms do not always lead to competition and antagonism but may be compatible and result in synergistic effects. The application of *Bacillus* spp. with other beneficial microbes has thus emerged as a promising biocontrol strategy.

Synergistic effects in protecting plants against pathogens and promoting growth have been reported using combinations of *Bacillus* spp. with *Trichoderma* species (Wu et al., 2018; Karuppiah et al., 2019; Izquierdo-García et al., 2020), another BCA acting via competition, mycoparasitism, antibiosis and triggering of ISR (Vinale et al., 2008; Ghorbanpour et al., 2018). Higher production of BSMs, as well as cryptic compounds (depending on specific conditions), has also been reported in liquid co-cultivation of *B. subtilis* 22 and *Trichoderma atroviride* SG3403 as compared with the two microorganisms cultivated individually (Li et al., 2020).

Arbuscular mycorrhizal (AM) fungi are also key microorganisms in soil, forming symbiotic associations with more than 70% of vascular plant species (Brundrett and Tedersoo, 2018). They improve plant nutrition (mainly phosphorus and nitrogen) and water uptake. Plants forming associations with these soil fungi have been found to resist biotic and abiotic stresses better (Bonfante and Genre, 2010; Plouznikoff et al., 2016; Mathieu et al., 2018), partly due to the ability of AM fungi to induce ISR (Gallou et al., 2011, 2012; Fiorilli et al., 2018; Deja-Sikora et al., 2020). Some synergies and improved functionality of PGPR-AM fungi consortia have been reported for plant growth and health (Armada et al., 2018; Zhang et al., 2018; Yadav et al., 2020). Co-inoculation of *B. subtilis* or *B. amyloliquefaciens* with AM fungi did not affect fungal growth but improved resistance to biotic and abiotic stresses due to more efficient nutrient supply in *Artemisia annua* L., *Allium sativum* L. and *Triticum aestivum* L. (Awasthi et al., 2011; Agnolucci et al., 2019; Rashad et al., 2020; Yadav et al., 2020). Metabolism markers such as succinate dehydrogenase and alkaline phosphatase showed a positive impact of *Bacillus* spp. on AM fungi metabolism (Vivas et al., 2003). However, there remains a lack of data demonstrating enhanced biocontrol activity of formulations combining bacilli and AM fungi.

Co-inoculation of *Bacillus* spp. with bacteria belonging to the *Pseudomonas* or *Streptomyces* genera is also promising. These BCA species are competitors, producing antimicrobials affecting *Bacillus* growth. Antagonistic interactions have been reported between *Pseudomonas protegens* PF-5 and *B. subtilis* NCIB3610 (Powers et al., 2015; Molina-Santiago et al., 2019) or between *Streptomyces* sp. MG1 and *B. subtilis* 3610 (Barger et al., 2012; Traxler and Kolter, 2015). However, co-inoculation of *B. licheniformis* B642 and the beneficial *Pseudomonas fluorescens* strain FAP2 may have beneficial effects on the host plant (Ansari and Ahmad, 2019). The combination of these two genera enhanced resistance against *X. campestris* pv. *campestris* compared to their use individually (Mishra and Arora, 2012). The combination of *B. subtilis* GBO3 with *P. fluorescens* CECT 5398 has been shown to improve the control of *F. oxysporum* and *R. solani* in tomato and pepper (Domenech et al., 2006). However, the molecular basis underpinning synergistic effects still remains to be deciphered.

Combining strains belonging to the *Bacillus* genus can also enhance disease suppression. The combination of *B. amyloliquefaciens, B. sphaericus, B. pumilus* strains or of *B. altitudinis and B. velezensis* strains exhibited higher levels of biocontrol against *X. axonopodis* pv. *vesicatoria*, *P. syringae* pv. *tomato* and *R. solanacearum* on tomato but also against *Pythium ultimum* and cucumber mosaic virus on cucumber compared single bacteria (Jetiyanon and Kloepper, 2002; Liu et al., 2018b). Studies in greenhouses have demonstrated that a mixture of *Bacillus* species (*B. pumilus* and *B. amyloliquefaciens*) significantly improved control of *P. capsici* on squash compared to individual strains (Zhang et al., 2010).

Some products combining other PGPR species with *Bacillus* spp. have been commercialized as BCAs. Products include LS213 by Gustafson Inc. (combining *B. subtilis* strain GB03 and *B. amyloliquefaciens* strain IN937a), Bio Protector by Bacto Agro Culture Care Pvt Ltd (2 strains of *Trichoderma*, *Ps. fluorescens* and *B. subtilis*) or BioYield™ by Bayer Crop Science (*B. amyloliquefaciens* GB99 + *B. subtilis GB12*), though most are described as biofertilizers or biostimulants (Domenech et al., 2006; Woo et al., 2014; Le Mire et al., 2016; Borriss, 2020).

Further research is needed to better understand the molecular dialogue existing between *Bacillus* and other BCAs to evaluate factors that inhibit or enhance synergistic interactions. Interaction-mediated variations in colony morphology, motility, biofilm formation, or sporulation illustrate how soil bacilli can protect themselves from antimicrobials emitted by bacterial competitors. The effect on these developmental processes could be coupled with significant modulation in the production of specific BSMs. These BSMs would then act as antimicrobials or in promoting cooperative interspecies communication processes which do not affect growth (Bleich et al., 2015; Liu et al., 2018c). This understanding would help design combinations of *Bacillus* with other microbes for more efficient biocontrol products.

12 Conclusions and future trends

The potential of some *Bacillus* species to control plant diseases caused by a range of pathogens has been amply demonstrated from lab to field, and selected isolates have been successfully commercialized. However, as for other microorganisms, the success of *Bacillus*-based products has been hampered by their highly variable or poor performance across agro-ecological environments and host plant species. There is a need to optimize/adapt their production and formulation (addition of prebiotics) at an industrial scale. It is also crucial to improve our basic knowledge of the processes influencing the expression of biocontrol traits and their persistence in plants following application in the field.

The biocontrol potential of *Bacillus* species mostly relies on their capacity to synthesize a wide range of BSMs involved in the three processes underpinning biocontrol. Some BSMs viewed as 'specialized' metabolites having a single main function may actually play multiple roles. As illustrated for surfactin, many BSMs may be multifunctional and act as signals or antimicrobials depending on the concentration. It is necessary to combine approaches such as loss/gain of functions of mutants and the effect of purified BSMs to clearly identify bioactivities, depending on the concentration and the type of microbial pathogen (for antagonism) or plant genotype (for ISR). Better characterization of the molecular mechanisms of key bioactivities compounds acting individually or in combination is required if we want to understand why they inhibit some pathogens but not others or why they can stimulate immunity in some plants but not others. There is a need to combine carefully designed biotests with other approaches such as experimental and *in silico* biophysics to investigate effects on cellular membranes (depending on lipid composition) and thus on the target organism (Deleu et al., 2014; Balleza et al., 2019).

As they can be affected by abiotic factors and multitrophic interactions, we also need to improve our knowledge on the timing, amount and diversity of BSMs when produced *in planta* under natural conditions. This would help to understand whether these BSMs act as signals at sub-inhibitory concentrations or if they can play antimicrobial functions upon reaching threshold amounts at least locally around the plant tissues. It will also help in determining to what extent effects differ according to plant species, age and physico-chemical conditions. We need to develop new *ex vivo* approaches and exploit technologies like imaging mass spectrometry to spatiotemporally resolve the dynamics of production of those small molecules produced upon colonization or during microbial interspecies interactions (Debois et al., 2014; Boughton et al., 2016; Spraker et al., 2020).

Integrating all those data is necessary to predict the pathosystems against which these bacilli would work as BCAs. This will benefit the practical use of *Bacillus*-based products as sprays to treat aerial parts of the plants or as soil inoculants to protect seedlings from attack by soil-borne microbial pathogens. Even if results are promising (Parnell et al., 2016; Meng and Hao, 2017), prophylactic applications such as soil drenching or seed coating to protect crops in the first stage of their development remain a major challenge for large-scale biocontrol, notably for cereals.

B. velezensis is among the best candidate species for the development of biocontrol agents. Several products with different strains of this species are already on the market (Table 2) but all these strains are genetically close and have a similar arsenal of bioactive metabolites. Soil represents an un-tapped reservoir from which new *Bacillus* species/strains forming unknown but highly active compounds can be isolated, based on rationally designed screening procedures (Köhl et al., 2019).

Another attractive alternative is the development of products with *Bacillus* spp. metabolites as active ingredients (Glare et al., 2012; Heimpel and Mills, 2017). Lipopeptides are naturally formed in high amounts by species such as *B. subtilis* and *B. velezensis* and optimization of their production in bioreactors at pilot scale has been thoroughly investigated (Rangarajan et al., 2015; Motta Dos Santos et al., 2016; Brück et al., 2019). There are methods available for extraction and (semi-)purification, allowing cost-effective processes to obtain bio-sourced products with reasonable purity and good stability (Coutte et al., 2017). If formulated metabolites are developed, their (eco-)toxicological risks need to be carefully assessed in relation to doses used to treat crops and stability in the environment.

13 Funding

Lab research work on *Bacillus* is supported by the EU Interreg V France-Wallonie-Vlaanderen portfolio SmartBiocontrol (Bioprotect and Bioscreen projects, avec le soutien du Fonds européen de développement régional - Met steun van het Europees Fonds voor Regionale Ontwikkeling), by the Microsoilsystem project funded by the Walloon Region (ID D31-1388SPW/DGO3) and by the EOS project ID 30650620 from the FWO/F.R.S.-FNRS (National fund for Scientific Research in Belgium). A. Anckaert is a recipient of the F.R.I.A. fellowship (Formation à la Recherche dans l'Industrie et l'Agriculture) and M. Ongena is senior research associate at the F.R.S.-F.N.R.S.

14 Where to look for further information

The International Organisation for biological and integrated control (IOBC) is a very active society headed by (academic) experts in the field who organizes symposia, workshops, and congresses on all aspects of the biocontrol science. The website https://www.iobc-wprs.org/ also provides valuable information on integrated production and integrated pest management.

For more information about the biocontrol industry and a comprehensive view of biocontrol technologies are used (or being developed) as alternatives to control pests and diseases effectively in an environmental friendly way in agriculture, forestry: https://ibma-global.org/.

15 References

Abdallah, D. B., Krier, F., Jacques, P., Tounsi, S. and Frikha-Gargouri, O. (2020). *Agrobacterium tumefaciens* C58 presence affects Bacillus velezensis 32a ecological fitness in the tomato rhizosphere, *Environ. Sci. Pollut. Res. Int.* 27(22), 28429–28437. doi: 10.1007/s11356-020-09124-1.

Abriouel, H., Franz, C. M. A. P., Omar, N. and Gálvez, A. (2011) Diversity and applications of *Bacillus bacteriocins*, *FEMS Microbiol. Rev.* 35(1), 201–232. doi: 10.1111/j.1574-6976.2010.00244.x.

Agnolucci, M., Avio, L., Pepe, A., Turrini, A., Cristani, C., Bonini, P., Cirino, V., Colosimo, F., Ruzzi, M. and Giovannetti, M. (2019). Bacteria associated with a commercial mycorrhizal inoculum: community composition and multifunctional activity as assessed by illumina sequencing and culture-dependent tools, *Front. Plant Sci.* 9(1956), 1–13. doi: 10.3389/fpls.2018.01956.

Ait Bahadou, S., Ouijja, A., Karfach, A., Tahiri, A. and Lahlali, R. (2018). New potential bacterial antagonists for the biocontrol of fire blight disease (*Erwinia amylovora*) in Morocco, *Microb. Pathog.* 117, 7–15. doi: 10.1016/j.micpath.2018.02.011.

Akpa, E., Jacques, P., Wathelet, B., Paquot, M., Fuchs, R., Budzikiewicz, H. and Thonart, P. (2001). Influence of culture conditions on lipopeptide production by *Bacillus subtilis*, *Appl. Biochem. Biotechnol.* 91-93, 551–561. doi: 10.1385/ABAB:91-93:1-9:551.

Alamri, S. A. (2015). Enhancing the efficiency of the bioagent *Bacillus subtilis* JF419701 against soil-borne phytopathogens by increasing the productivity of fungal cell wall degrading enzymes, *Arch. Phytopathol. Plant Prot.* 48(2), 159–170. doi: 10.1080/03235408.2014.884671.

Aleti, G., Lehner, S., Bacher, M., Compant, S., Nikolic, B., Plesko, M., Schuhmacher, R., Sessitsch, A. and Brader, G. (2016). Surfactin variants mediate species-specific biofilm formation and root colonization in Bacillus, *Environ. Microbiol.* 18(8), 2634–2645. doi: 10.1111/1462-2920.13405.

Allard-Massicotte, R., Tessier, L., Lécuyer, F., Lakshmanan, V., Lucier, J. F., Garneau, D., Caudwell, L., Vlamakis, H., Bais, H. P. and Beauregard, P. B. (2016). *Bacillus subtilis* early colonization of *Arabidopsis thaliana* roots involves multiple chemotaxis receptors, *mBio* 7(6), 1–10. doi: 10.1128/mBio.01664-16.

Almoneafy, A. A., Kakar, K. U., Nawaz, Z., Li, B., Saand, M. A., Chun-lan, Y. and Xie, G.-L. (2014). Tomato plant growth promotion and antibacterial related-mechanisms of four rhizobacterial Bacillus strains against *Ralstonia solanacearum*, *Symbiosis* 63(2), 59–70. doi: 10.1007/s13199-014-0288-9.

Ananda, M., Rusmana, I. and Akhdiya, A. (2019). Quorum quenching of *Bacillus cereus* INT1c against *Pseudomonas syringae*, *J. Phys. Conf. Ser.* 1277(012010), 1–9. doi: 10.1088/1742-6596/1277/1/012010.

Andrić, S., Meyer, T. and Ongena, M. (2020). Bacillus responses to plant-associated fungal and bacterial communities, *Front. Microbiol.* 11(1350), 1350. doi: 10.3389/fmicb.2020.01350.

Ansari, F. A. and Ahmad, I. (2019). Fluorescent Pseudomonas -FAP2 and *Bacillus licheniformis* interact positively in biofilm mode enhancing plant growth and photosynthetic attributes, *Sci. Rep.* 9(1), 4547. doi: 10.1038/s41598-019-40864-4.

Arguelles Arias, A., Craig, M. and Fickers, P. (2011). Gram-positive antibiotic biosynthetic clusters: a review, In: Mendez-Vilas, (ed) *Science Against Microbial Pathogens: Communicating Current Research and Technological Advances*, 977–986.

Arguelles Arias, A., Ongena, M., Devreese, B., Terrak, M., Joris, B. and Fickers, P. (2013). Characterization of amylolysin, a novel lantibiotic from *Bacillus amyloliquefaciens* GA1, *PLoS ONE* 8(12), e83037. doi: 10.1371/journal.pone.0083037.

Armada, E., Leite, M. F. A., Medina, A., Azcón, R. and Kuramae, E. E. (2018). Native bacteria promote plant growth under drought stress condition without impacting the rhizomicrobiome, *FEMS Microbiol. Ecol.* 94(7), 1–13. doi: 10.1093/femsec/fiy092.

Asaka, O. and Shoda, M. (1996). Biocontrol of *Rhizoctonia solani* damping-off of tomato with *Bacillus subtilis* RB14, *Appl. Environ. Microbiol.* 62(11), 4081–4085. doi: 10.1128/AEM.62.11.4081-4085.1996.

Auchtung, J. M., Lee, C. A. and Grossman, A. D. (2006). Modulation of the ComA-dependent quorum response in *Bacillus subtilis* by multiple rap proteins and phr peptides, *J. Bacteriol.* 188(14), 5273–5285. doi: 10.1128/JB.00300-06.

Awasthi, A., Bharti, N., Nair, P., Singh, R., Shukla, A. K., Gupta, M. M., Darokar, M. P. and Kalra, A. (2011). Synergistic effect of Glomus mosseae and nitrogen fixing *Bacillus subtilis* strain Daz26 on artemisinin content in *Artemisia annua* L., *Appl. Soil Ecol.* 49(1), 125–130. doi: 10.1016/j.apsoil.2011.06.005.

Backer, R., Rokem, J. S., Ilangumaran, G., Lamont, J., Praslickova, D., Ricci, E., Subramanian, S. and Smith, D. L. (2018). Plant growth-promoting rhizobacteria: context, mechanisms of action, and roadmap to commercialization of biostimulants for sustainable agriculture, *Front. Plant Sci.* 9(1473), 1473. doi: 10.3389/fpls.2018.01473.

Bais, H. P., Fall, R. and Vivanco, J. M. (2004). Biocontrol of *Bacillus subtilis* against infection of Arabidopsis roots by *Pseudomonas syringae* is facilitated by biofilm formation and surfactin production, *Plant Physiol.* 134(1), 307–319. doi: 10.1104/pp.103.028712.

Balleza, D., Alessandrini, A. and Beltrán García, M. J. (2019). Role of lipid composition, physicochemical interactions, and membrane mechanics in the molecular actions of microbial cyclic lipopeptides, *J. Membr. Biol.* 252(2–3), 131–157. doi: 10.1007/s00232-019-00067-4.

Bardin, M., Ajouz, S., Comby, M., Lopez-Ferber, M., Graillot, B., Siegwart, M. and Nicot, P. C. (2015). Is the efficacy of biological control against plant diseases likely to be more durable than that of chemical pesticides?, *Front. Plant Sci.* 6(566), 566. doi: 10.3389/fpls.2015.00566.

Barger, S. R., Hoefler, B. C., Cubillos-Ruiz, A., Russell, W. K., Russell, D. H. and Straight, P. D. (2012). Imaging secondary metabolism of Streptomyces sp. Mg1 during cellular lysis and colony degradation of competing *Bacillus subtilis*, *Antonie Leeuwenhoek* 102(3), 435–445. doi: 10.1007/s10482-012-9769-0.

Beauregard, P. B., Chai, Y., Vlamakis, H., Losick, R. and Kolter, R. (2013). Bacillus biofilm induction by plant polysaccharides, *Proc. Natl. Acad. Sci. U. S. A.* 110(17), E1621–E1630. doi: 10.1073/pnas.1218984110.

Belitsky, B. R. and Sonenshein, A. L. (2013). Genome-wide identification of *Bacillus subtilis* CodY-binding sites at single-nucleotide resolution, *Proc. Natl. Acad. Sci. U. S. A.* 110(17), 7026–7031. doi: 10.1073/pnas.1300428110.

Berikashvili, V., Sokhadze, K., Kachlishvili, E., Elisashvili, V. and Chikindas, M. L. (2018). *Bacillus amyloliquefaciens* spore production under solid-state fermentation of lignocellulosic residues, *Probiotics Antimicrob. Proteins* 10(4), 755–761. doi: 10.1007/s12602-017-9371-x.

Beris, D., Theologidis, I., Skandalis, N. and Vassilakos, N. (2018). *Bacillus amyloliquefaciens* strain MBI600 induces salicylic acid dependent resistance in tomato plants against Tomato spotted wilt virus and Potato virus Y, *Sci. Rep.* 8(1), 10320. doi: 10.1038/s41598-018-28677-3.

Bleich, R., Watrous, J. D., Dorrestein, P. C., Bowers, A. A. and Shank, E. A. (2015). Thiopeptide antibiotics stimulate biofilm formation in *Bacillus subtilis*, *Proc. Natl. Acad. Sci. U. S. A.* 112(10), 3086–3091. doi: 10.1073/pnas.1414272112.

Bonfante, P. and Genre, A. (2010). Mechanisms underlying beneficial plant-fungus interactions in mycorrhizal symbiosis, *Nat. Commun.* 1(48), 48. doi: 10.1038/ncomms1046.

Borel, B. (2017). CRISPR, microbes and more are joining the war against crop killers, *Nature* 543(7645), 302–304. doi: 10.1038/543302a.

Borriss, R. (2020). Phytostimulation and biocontrol by the plant-associated *Bacillus amyloliquefaciens* FZB42: an update. In: Kumar, M., Kumar, V. and Prasad, R. (eds) *Phyto-Microbiome in Stress Regulation. Environmental and Microbial Biotechnology*. Springer, Singapore, 1–20. doi: 10.1007/978-981-15-2576-6_1.

Boughton, B. A., Thinagaran, D., Sarabia, D., Bacic, A. and Roessner, U. (2016). Mass spectrometry imaging for plant biology: a review, *Phytochem. Rev.* 15(3), 445–488. doi: 10.1007/s11101-015-9440-2.

Bozhüyük, K. A., Micklefield, J. and Wilkinson, B. (2019). Engineering enzymatic assembly lines to produce new antibiotics, *Curr. Opin. Microbiol.* 51, 88–96. doi: 10.1016/j.mib.2019.10.007.

Brück, H. L., Delvigne, F., Dhulster, P., Jacques, P. and Coutte, F. (2019). Molecular strategies for adapting *Bacillus subtilis* 168 biosurfactant production to biofilm cultivation mode, *Bioresour. Technol.* 293, 1–8. doi: 10.1016/j.biortech.2019.122090.

Brundrett, M. C. and Tedersoo, L. (2018). Evolutionary history of mycorrhizal symbioses and global host plant diversity, *New Phytol.* 220(4), 1108–1115. doi: 10.1111/nph.14976.

Budde, I., Steil, L., Scharf, C., Völker, U. and Bremer, E. (2006). Adaptation of *Bacillus subtilis* to growth at low temperature: a combined transcriptomic and proteomic appraisal, *Microbiology (Reading)* 152(3), 831–853. doi: 10.1099/mic.0.28530-0.

Cao, Y., Pi, H., Chandrangsu, P., Li, Y., Wang, Y., Zhou, H., Xiong, H., Helmann, J. D. and Cai, Y. (2018). Antagonism of two plant-growth promoting *Bacillus velezensis* isolates against *Ralstonia solanacearum* and *Fusarium oxysporum*, *Sci. Rep.* 8(1), 4360. doi: 10.1038/s41598-018-22782-z.

Caulier, S., Gillis, A., Colau, G., Licciardi, F., Liépin, M., Desoignies, N., Modrie, P., Legrève, A., Mahillon, J. and Bragard, C. (2018). Versatile antagonistic activities of soil-borne *Bacillus* spp. and *Pseudomonas* spp. against *Phytophthora infestans* and other potato pathogens, *Front. Microbiol.* 9(143), 143. doi: 10.3389/fmicb.2018.00143.

Caulier, S., Nannan, C., Gillis, A., Licciardi, F., Bragard, C. and Mahillon, J. (2019). Overview of the antimicrobial compounds produced by members of the *Bacillus subtilis* group, *Front. Microbiol.* 10(302), 302. doi: 10.3389/fmicb.2019.00302.

Cawoy, H., Bettiol, W., Fickers, P. and Ongena, M. (2011). Bacillus-based biological control of plant diseases. In: *Pesticides in the Modern World–Pesticides Use and Management*. InTechOpen, London. doi: 10.5772/17184.

Cawoy, H., Debois, D., Franzil, L., De Pauw, E., Thonart, P. and Ongena, M. (2015). Lipopeptides as main ingredients for inhibition of fungal phytopathogens by *Bacillus subtilis*/amyloliquefaciens, *Microb. Biotechnol.* 8(2), 281–295. doi: 10.1111/1751-7915.12238.

Cawoy, H., Mariutto, M., Henry, G., Fisher, C., Vasilyeva, N., Thonart, P., Dommes, J. and Ongena, M. (2014). Plant defense stimulation by natural isolates of Bacillus depends on efficient surfactin production, *Mol. Plant Microbe Interact.* 27(2), 87–100. doi: 10.1094/MPMI-09-13-0262-R.

Chandra Kalia, V., Raju, S. C. and Purohit, H. J. (2011). Genomic analysis reveals versatile organisms for quorum quenching enzymes: acyl-homoserine lactone-acylase and -Lactonase, *Open Microbiol. J.* 3, 1–13. doi: 10.2174/1874285801105010001.

Chen, J. (2017). *Global Markets for Biopesticides*. BCC Research, Wellesley.

Chen, L., Gu, W., Xu, H. Y., Yang, G. L., Shan, X. F., Chen, G., Kang, Y. H., Wang, C. F. and Qian, A. D. (2018). Comparative genome analysis of *Bacillus velezensis* reveals a potential for degrading lignocellulosic biomass, *3 Biotech* 8(5), 253. doi: 10.1007/s13205-018-1270-7.

Chen, M., Wang, J., Liu, B., Zhu, Y., Xiao, R., Yang, W., Ge, C. and Chen, Z. (2020). Biocontrol of tomato bacterial wilt by the new strain *Bacillus velezensis* FJAT-46737 and its lipopeptides, *BMC Microbiol.* 20(1), 160. doi: 10.1186/s12866-020-01851-2.

Chen, M. C., Wang, J. P., Zhu, Y. J., Liu, B., Yang, W. J. and Ruan, C. Q. (2019). Antibacterial activity against *Ralstonia solanacearum* of the lipopeptides secreted from the *Bacillus amyloliquefaciens* strain FJAT-2349, *J. Appl. Microbiol.* 126(5), 1519–1529. doi: 10.1111/jam.14213.

Chen, X. H., Koumoutsi, A., Scholz, R., Schneider, K., Vater, J., Süssmuth, R., Piel, J. and Borriss, R. (2009a). Genome analysis of *Bacillus amyloliquefaciens* FZB42 reveals its potential for biocontrol of plant pathogens, *J. Biotechnol.* 140(1–2), 27–37. doi: 10.1016/j.jbiotec.2008.10.011.

Chen, X. H., Scholz, R., Borriss, M., Junge, H., Mögel, G., Kunz, S. and Borriss, R. (2009b). Difficidin and bacilysin produced by plant-associated *Bacillus amyloliquefaciens* are efficient in controlling fire blight disease, *J. Biotechnol.* 140(1–2), 38–44. doi: 10.1016/j.jbiotec.2008.10.015.

Chen, X. H., Vater, J., Piel, J., Franke, P., Scholz, R., Schneider, K., Koumoutsi, A., Hitzeroth, G., Grammel, N., Strittmatter, A. W., Gottschalk, G., Süssmuth, R. D. and Borriss, R. (2006). Structural and functional characterization of three polyketide synthase gene clusters in *Bacillus amyloliquefaciens* FZB 42, *J. Bacteriol.* 188(11), 4024–4036. doi: 10.1128/JB.00052-06.

Chen, Y., Yan, F., Chai, Y., Liu, H., Kolter, R., Losick, R. and Guo, J. H. (2013). Biocontrol of tomato wilt disease by Bacillus subtilis isolates from natural environments depends on conserved genes mediating biofilm formation, *Environ. Microbiol.* 15(3), 848–864. doi: 10.1111/j.1462-2920.2012.02860.x.

Choi, H. K., Song, G. C., Yi, H. S. and Ryu, C. M. (2014). Field evaluation of the bacterial volatile derivative 3-pentanol in priming for induced resistance in pepper, *J. Chem. Ecol.* 40(8), 882–892. doi: 10.1007/s10886-014-0488-z.

Chowdhury, S. P., Dietel, K., Rändler, M., Schmid, M., Junge, H., Borriss, R., Hartmann, A. and Grosch, R. (2013). Effects of *Bacillus amyloliquefaciens* FZB42 on lettuce growth and health under pathogen pressure and its impact on the rhizosphere bacterial community, *Plos One* 8(7), e68818. doi: 10.1371/journal.pone.0068818.

Chowdhury, S. P., Hartmann, A., Gao, X. and Borriss, R. (2015a). Biocontrol mechanism by root-associated *Bacillus amyloliquefaciens* FZB42—a review, *Front. Microbiol.* 6(780), 780. doi: 10.3389/fmicb.2015.00780.

Chowdhury, S. P., Uhl, J., Grosch, R., Alquéres, S., Pittroff, S., Dietel, K., Schmitt-Kopplin, P., Borriss, R. and Hartmann, A. (2015b). Cyclic lipopeptides of *Bacillus amyloliquefaciens* subsp. plantarum colonizing the lettuce rhizosphere enhance plant defense responses toward the bottom rot pathogen *Rhizoctonia solani*, *Mol. Plant Microbe Interact.* 28(9), 984–995. doi: 10.1094/MPMI-03-15-0066-R.

Correa, O. S., Montecchia, M. S., Berti, M. F., Fernández Ferrari, M. C., Pucheu, N. L., Kerber, N. L. and García, A. F. (2009). *Bacillus amyloliquefaciens* BNM122, a potential microbial biocontrol agent applied on soybean seeds, causes a minor impact on rhizosphere and soil microbial communities, *Appl. Soil Ecol.* 41(2), 185–194. doi: 10.1016/j.apsoil.2008.10.007.

Cosby, W. M., Vollenbroich, D., Lee, O. H. and Zuber, P. (1998). Altered srf expression in *Bacillus subtilis* resulting from changes in culture pH Is dependent on the Spo0K oligopeptide permease and the ComQX system of extracellular control, *J. Bacteriol.* 180(6), 1438–1445. doi: 10.1128/JB.180.6.1438-1445.1998.

Coutte, F., Lecouturier, D., Dimitrov, K., Guez, J. S., Delvigne, F., Dhulster, P. and Jacques, P. (2017). Microbial lipopeptide production and purification bioprocesses, current progress and future challenges, *Biotechnol. J.* 12(7), 1–10. doi: 10.1002/biot.201600566(28636078).

Coutte, F., Niehren, J., Dhali, D., John, M., Versari, C. and Jacques, P. (2015). Modeling leucine's metabolic pathway and knockout prediction improving the production of surfactin, a biosurfactant from *Bacillus subtilis*, *Biotechnol. J.* 10(8), 1216–1234. doi: 10.1002/biot.201400541.

Crouzet, J., Arguelles-Arias, A., Dhondt-Cordelier, S., Cordelier, S., Pršić, J., Hoff, G., Mazeyrat-Gourbeyre, F., Baillieul, F., Clément, C., Ongena, M. and Dorey, S. (2020). Biosurfactants in plant protection against diseases: rhamnolipids and lipopeptides case study, *Front. Bioeng. Biotechnol.* 8(1014), 1014. doi: 10.3389/fbioe.2020.01014.

Cucu, M. A., Gilardi, G., Pugliese, M., Gullino, M. L. and Garibaldi, A. (2020). An assessment of the modulation of the population dynamics of pathogenic *Fusarium oxysporum* f. sp. *lycopersici* in the tomato rhizosphere by means of the application of *Bacillus subtilis* QST 713, *Trichoderma* sp. TW2 and two composts, *Biol. Control* 142, 1–10. doi: 10.1016/j.biocontrol.2019.104158.

Dang, Y., Zhao, F., Liu, X., Fan, X., Huang, R., Gao, W., Wang, S. and Yang, C. (2019). Enhanced production of antifungal lipopeptide iturin A by *Bacillus amyloliquefaciens* LL3 through metabolic engineering and culture conditions optimization, *Microb. Cell Fact.* 18(1), 68. doi: 10.1186/s12934-019-1121-1.

Dean, R., Van Kan, J. A. L., Pretorius, Z. A., Hammond-Kosack, K. E., Di Pierto, A., Spanu, P. D., Rudd, J. J., Dickman, M., Kahmann, R., Ellis, J. and Foster, G. D. (2012). The Top 10 fungal pathogens in molecular plant pathology, *Mol. Plant Pathol.* 13(4), 414–430. doi: 10.1111/j.1364-3703.2011.00783.x.

Debois, D., Fernandez, O., Franzil, L., Jourdan, E., de Brogniez, A., Willems, L., Clément, C., Dorey, S., De Pauw, E. and Ongena, M. (2015). Plant polysaccharides initiate underground crosstalk with bacilli by inducing synthesis of the immunogenic lipopeptide surfactin, *Environ. Microbiol. Rep.* 7(3), 570–582. doi: 10.1111/1758-2229.12286.

Debois, D., Jourdan, E., Smargiasso, N., Thonart, P., De Pauw, E. and Ongena, M. (2014). Spatiotemporal monitoring of the antibiome secreted by Bacillus biofilms on plant roots using MALDI mass spectrometry imaging, *Anal. Chem.* 86(9), 4431–4438. doi: 10.1021/ac500290s.

DeFilippi, S., Groulx, E., Megalla, M., Mohamed, R. and Avis, T. J. (2018). Fungal competitors affect production of antimicrobial lipopeptides in *Bacillus subtilis* Strain B9-5, *J. Chem. Ecol.* 44(4), 374–383. doi: 10.1007/s10886-018-0938-0.

Deja-Sikora, E., Kowalczyk, A., Trejgell, A., Szmidt-Jaworska, A., Baum, C., Mercy, L. and Hrynkiewicz, K. (2020). Arbuscular mycorrhiza changes the impact of potato virus Y on growth and stress tolerance of *Solanum tuberosum* L. in vitro, *Front. Microbiol.* 10(2971), 1–12. doi: 10.3389/fmicb.2019.02971.

Deleu, M., Crowet, J. M., Nasir, M. N. and Lins, L. (2014). Complementary biophysical tools to investigate lipid specificity in the interaction between bioactive molecules and the plasma membrane: a review, *Biochim. Biophys. Acta* 1838(12), 3171–3190. doi: 10.1016/j.bbamem.2014.08.023.

Deleu, M., Lorent, J., Lins, L., Brasseur, R., Braun, N., El Kirat, K., Nylander, T., Dufrêne, Y. F. and Mingeot-Leclercq, M. P. (2013). Effects of surfactin on membrane models displaying lipid phase separation, *Biochim. Biophys. Acta* 1828(2), 801–815. doi: 10.1016/j.bbamem.2012.11.007.

Deleu, M., Paquot, M. and Nylander, T. (2008). Effect of fengycin, a lipopeptide produced by *Bacillus subtilis*, on model biomembranes, *Biophys. J.* 94(7), 2667–2679. doi: 10.1529/biophysj.107.114090.

Domenech, J., Reddy, M. S., Kloepper, J. W., Ramos, B. and Gutierrez-Mañero, J. (2006). Combined application of the biological product LS213 with Bacillus, Pseudomonas or Chryseobacterium for growth promotion and biological control of soil-borne diseases in pepper and tomato, *BioControl* 51(2), 245–258. doi: 10.1007/s10526-005-2940-z.

Dong, Y. H., Gusti, A. R., Zhang, Q., Xu, J. L. and Zhang, L. H. (2002). Identification of quorum-quenching N-acyl homoserine lactonases from *Bacillus* species, *Appl. Environ. Microbiol.* 68(4), 1754–1759. doi: 10.1128/AEM.68.4.1754-1759.2002.

Dubern, J. F. and Bloemberg, G. V. (2006). Influence of environmental conditions on putisolvins I and II production in *Pseudomonas putida* strain PCL1445, *FEMS Microbiol. Lett.* 263(2), 169–175. doi: 10.1111/j.1574-6968.2006.00406.x.

Dunlap, C. A. (2019). Taxonomy of registered Bacillus spp. strains used as plant pathogen antagonists, *Biol. Control* 134, 82–86. doi: 10.1016/j.biocontrol.2019.04.011.

Dunlap, C. A., Kim, S. J., Kwon, S. W. and Rooney, A. P. (2016). *Bacillus velezensis* is not a later heterotypic synonym of *Bacillus amyloliquefaciens*; *Bacillus methylotrophicus*, *Bacillus amyloliquefaciens* subsp. *plantarum* and "*Bacillus oryzicola*" are later heterotypic synonym, *Int. J. Syst. Evol. Microbiol.* 66(3), 1212–1217. doi: 10.1099/ijsem.0.000858.

Dutta, S., Whicher, J. R., Hansen, D. A., Hale, W. A., Chemler, J. A., Congdon, G. R., Narayan, A. R. H., Håkansson, K., Sherman, D. H., Smith, J. L. and Skiniotis, G. (2014). Structure of a modular polyketide synthase, *Nature* 510(7506), 512–517. doi: 10.1038/nature13423.

Elanchezhiyan, K., Keerthana, U., Nagendran, K., Prabhukarthikeyan, S. R., Prabakar, K., Raguchander, T. and Karthikeyan, G. (2018). Multifaceted benefits of *Bacillus amyloliquefaciens* strain FBZ24 in the management of wilt disease in tomato caused by *Fusarium oxysporum* f. sp. *lycopersici*, *Physiol. Mol. Plant Pathol.* 103, 92–101. doi: 10.1016/j.pmpp.2018.05.008.

Elisashvili, V., Kachlishvili, E. and Chikindas, M. L. (2019). Recent advances in the physiology of spore formation for Bacillus probiotic production, *Probiotics Antimicrob. Proteins* 11(3), 731–747. doi: 10.1007/s12602-018-9492-x.

Engelbrecht, G., Horak, I., Jansen van Rensburg, P. J. and Claassens, S. (2018). Bacillus -based bionematicides: development, modes of action and commercialisation, *Biocontrol Sci. Technol.* 28(7), 629–653. doi: 10.1080/09583157.2018.1469000.

Fan, B., Blom, J., Klenk, H. P. and Borriss, R. (2017a). *Bacillus amyloliquefaciens*, *Bacillus velezensis*, and *Bacillus siamensis* form an "operational group *B. amyloliquefaciens*" within the *B. subtilis* species complex, *Front. Microbiol.* 8(22), 22. doi: 10.3389/fmicb.2017.00022.

Fan, H., Zhang, Z., Li, Y., Zhang, X., Duan, Y. and Wang, Q. (2017b). Biocontrol of bacterial fruit blotch by *Bacillus subtilis* 9407 via surfactin-mediated antibacterial activity and colonization, *Front. Microbiol.* 8, 1–15. doi: 10.3389/fmicb.2017.01973.

Fan, B., Borriss, R., Bleiss, W. and Wu, X. (2012). Gram-positive rhizobacterium *Bacillus amyloliquefaciens* FZB42 colonizes three types of plants in different patterns, *J. Microbiol.* 50(1), 38–44. doi: 10.1007/s12275-012-1439-4.

Fan, B., Chen, X. H., Budiharjo, A., Bleiss, W., Vater, J. and Borriss, R. (2011). Efficient colonization of plant roots by the plant growth promoting bacterium *Bacillus amyloliquefaciens* FZB42, engineered to express green fluorescent protein, *J. Biotechnol.* 151(4), 303–311. doi: 10.1016/j.jbiotec.2010.12.022.

Fan, B., Wang, C., Song, X., Ding, X., Wu, L., Wu, H., Gao, X. and Borriss, R. (2018). *Bacillus velezensis* FZB42 in 2018: the gram-positive model strain for plant growth promotion and biocontrol, *Front. Microbiol.* 9(2491), 2491. doi: 10.3389/fmicb.2018.02491.

Farace, G., Fernandez, O., Jacquens, L., Coutte, F., Krier, F., Jacques, P., Clément, C., Barka, E. A., Jacquard, C. and Dorey, S. (2015). Cyclic lipopeptides from *Bacillus subtilis* activate distinct patterns of defence responses in grapevine, *Mol. Plant Pathol.* 16(2), 177–187. doi: 10.1111/mpp.12170.

Farzand, A., Moosa, A., Zubair, M., Khan, A. R., Ayaz, M., Massawe, V. C. and Gao, X. (2020). Transcriptional profiling of diffusible lipopeptides and fungal virulence genes during *Bacillus amyloliquefaciens* EZ1509-mediated suppression of *Sclerotinia sclerotiorum*, *Phytopathology®* 110(2), 317–326. doi: 10.1094/PHYTO-05-19-0156-R.

Farzand, A., Moosa, A., Zubair, M., Khan, A. R., Massawe, V. C., Tahir, H. A. S., Sheikh, T. M. M., Ayaz, M. and Gao, X. (2019). Suppression of *Sclerotinia sclerotiorum* by the induction of systemic resistance and regulation of antioxidant pathways in tomato using fengycin produced by *Bacillus amyloliquefaciens* FZB42, *Biomolecules* 9(10), 1–17. doi: 10.3390/biom9100613.

Feng, H., Zhang, N., Fu, R., Liu, Y., Krell, T., Du, W., Shao, J., Shen, Q. and Zhang, R. (2019). Recognition of dominant attractants by key chemoreceptors mediates recruitment of plant growth-promoting rhizobacteria, *Environ. Microbiol.* 21(1), 402–415. doi: 10.1111/1462-2920.14472.

Fickers, P., Leclère, V., Guez, J. S., Béchet, M., Coucheney, F., Joris, B. and Jacques, P. (2008). Temperature dependence of mycosubtilin homologue production in *Bacillus subtilis* ATCC6633, *Res. Microbiol.* 159(6), 449–457. doi: 10.1016/j.resmic.2008.05.004.

Fiedler, S. and Heerklotz, H. (2015). Vesicle leakage reflects the target selectivity of antimicrobial lipopeptides from *Bacillus subtilis*, *Biophys. J.* 109(10), 2079–2089. doi: 10.1016/j.bpj.2015.09.021.

Fincheira, P., Parra, L., Mutis, A., Parada, M. and Quiroz, A. (2017). Volatiles emitted by *Bacillus* sp. BCT9 act as growth modulating agents on Lactuca sativa seedlings, *Microbiol. Res.* 203, 47–56. doi: 10.1016/j.micres.2017.06.007.

Fiorilli, V., Vannini, C., Ortolani, F., Garcia-Seco, D., Chiapello, M., Novero, M., Domingo, G., Terzi, V., Morcia, C., Bagnaresi, P., Moulin, L., Bracale, M. and Bonfante, P. (2018). Omics approaches revealed how arbuscular mycorrhizal symbiosis enhances yield and resistance to leaf pathogen in wheat, *Sci. Rep.* 8(1), 9625. doi: 10.1038/s41598-018-27622-8.

Fira, D., Dimkić, I., Berić, T., Lozo, J. and Stanković, S. (2018). Biological control of plant pathogens by *Bacillus* species, *J. Biotechnol.* 285, 44–55. doi: 10.1016/j.jbiotec.2018.07.044.

Flemming, H. C., Wingender, J., Szewzyk, U., Steinberg, P., Rice, S. A. and Kjelleberg, S. (2016). Biofilms: an emergent form of bacterial life, *Nat. Rev. Microbiol.* 14(9), 563-575. doi: 10.1038/nrmicro.2016.94.

Fousia, S., Paplomatas, E. J. and Tjamos, S. E. (2016). *Bacillus subtilis* QST 713 confers protection to tomato plants against *Pseudomonas syringae* pv tomato and induces plant defence-related genes, *J. Phytopathol.* 164(4), 264–270. doi: 10.1111/jph.12455.

Frey-Klett, P., Burlinson, P., Deveau, A., Barret, M., Tarkka, M. and Sarniguet, A. (2011). Bacterial-fungal interactions: hyphens between agricultural, clinical, environmental, and food microbiologists, *Microbiol. Mol. Biol. Rev.* 75(4), 583–609. doi: 10.1128/MMBR.00020-11.

Frikha-Gargouri, O., Ben Abdallah, D., Bhar, I. and Tounsi, S. (2017). Antibiosis and bmyB gene presence as prevalent traits for the selection of efficient Bacillus biocontrol agents against crown gall disease, *Front. Plant Sci.* 8(1363), 1363. doi: 10.3389/fpls.2017.01363.

Gallou, A., Declerck, S. and Cranenbrouck, S. (2012). Transcriptional regulation of defence genes and involvement of the WRKY transcription factor in arbuscular mycorrhizal potato root colonization, *Funct. Integr. Genomics* 12(1), 183–198. doi: 10.1007/s10142-011-0241-4.

Gallou, A., Lucero Mosquera, H. P., Cranenbrouck, S., Suárez, J. P. and Declerck, S. (2011). Mycorrhiza induced resistance in potato plantlets challenged by *Phytophthora infestans*, *Physiol. Mol. Plant Pathol.* 76(1), 20–26. doi: 10.1016/j.pmpp.2011.06.005.

García-Gutiérrez, L., Zeriouh, H., Romero, D., Cubero, J., de Vicente, A. and Pérez-García, A. (2013). The antagonistic strain *Bacillus subtilis* UMAF6639 also confers protection to melon plants against cucurbit powdery mildew by activation of jasmonate- and salicylic acid-dependent defence responses, *Microb. Biotechnol.* 6(3), 264–274. doi: 10.1111/1751-7915.12028.

Gautam, S., Chauhan, A., Sharma, R., Sehgal, R. and Shirkot, C. K. (2019). Potential of *Bacillus amyloliquefaciens* for biocontrol of bacterial canker of tomato incited by *Clavibacter michiganensis* ssp. *michiganensis*, *Microb. Pathog.* 130, 196–203. doi: 10.1016/j.micpath.2019.03.006.

Gava, C. A. T., Alves, Í. L. S. and Duarte, N. C. (2019). Timing the application of *Bacillus subtilis* QST 713 in the integrated management of the postharvest decay of mango fruits, *Crop Prot.* 121, 51–56. doi: 10.1016/j.cropro.2019.03.013.

Geiger, F., Bengtsson, J., Berendse, F., Weisser, W. W., Emmerson, M., Morales, M. B., Ceryngier, P., Liira, J., Tscharntke, T., Winqvist, C., Eggers, S., Bommarco, R., Pärt, T., Bretagnolle, V., Plantegenest, M., Clement, L. W., Dennis, C., Palmer, C., Oñate, J. J., Guerrero, I., Hawro, V., Aavik, T., Thies, C., Flohre, A., Hänke, S., Fischer, C., Goedhart, P. W. and Inchausti, P. (2010). Persistent negative effects of pesticides on biodiversity and biological control potential on European farmland, *Basic Appl. Ecol.* 11(2), 97-105. doi: 10.1016/j.baae.2009.12.001.

Ghorbanpour, M., Omidvari, M., Abbaszadeh-Dahaji, P., Omidvar, R. and Kariman, K. (2018). Mechanisms underlying the protective effects of beneficial fungi against plant diseases, *Biol. Control* 117, 147–157. doi: 10.1016/j.biocontrol.2017.11.006.

Glare, T., Caradus, J., Gelernter, W., Jackson, T., Keyhani, N., Köhl, J., Marrone, P., Morin, L. and Stewart, A. (2012). Have biopesticides come of age?, *Trends Biotechnol.* 30(5), 250–258. doi: 10.1016/j.tibtech.2012.01.003.

Gong, Q., Zhang, C., Lu, F., Zhao, H., Bie, X. and Lu, Z. (2014). Identification of bacillomycin D from *Bacillus subtilis* fmbJ and its inhibition effects against *Aspergillus flavus*, *Food Control* 36(1), 8–14. doi: 10.1016/j.foodcont.2013.07.034.

Gong, W., Wang, J., Chen, Z., Xia, B. and Lu, G. (2011). Solution structure of LCI, a novel antimicrobial peptide from *Bacillus subtilis*, *Biochemistry* 50(18), 3621-3627. doi: 10.1021/bi200123w.

Goodson, J. R., Klupt, S., Zhang, C., Straight, P. and Winkler, W. C. (2017). LoaP is a broadly conserved antiterminator protein that regulates antibiotic gene clusters in *Bacillus amyloliquefaciens*, *Nat. Microbiol.* 2(17003), 17003. doi: 10.1038/nmicrobiol.2017.3.

Grau, R. R., de Oña, P., Kunert, M., Leñini, C., Gallegos-Monterrosa, R., Mhatre, E., Vileta, D., Donato, V., Hölscher, T., Boland, W., Kuipers, O. P. and Kovács, Á. T. (2015). A duo of potassium-responsive histidine kinases govern the multicellular destiny of *Bacillus subtilis*, *mBio* 6(4), e00581. doi: 10.1128/mBio.00581-15.

Grubbs, K. J., Bleich, R. M., Santa Maria, K. C., Allen, S. E., Farag, S., AgBiome Team, Shank, E. A. and Bowers, A. A. (2017). Large-scale bioinformatics analysis of Bacillus genomes uncovers conserved roles of natural products in bacterial physiology, *mSystems* 2(6), 1–18. doi: 10.1128/mSystems.00040-17.

Gu, Q., Yang, Y., Yuan, Q., Shi, G., Wu, L., Lou, Z., Huo, R., Wu, H., Borriss, R. and Gao, X. (2017). Bacillomycin D produced by *Bacillus amyloliquefaciens* is involved in the antagonistic interaction with the plant-pathogenic fungus *Fusarium graminearum*, *Appl. Environ. Microbiol.* 83(19), 1075-1092. doi: 10.1128/AEM.01075-17.

Guez, J. S., Müller, C. H., Danze, P. M., Büchs, J. and Jacques, P. (2008). Respiration activity monitoring system (RAMOS), an efficient tool to study the influence of the oxygen transfer rate on the synthesis of lipopeptide by *Bacillus subtilis* ATCC6633, *J. Biotechnol.* 134(1–2), 121–126. doi: 10.1016/j.jbiotec.2008.01.003.

Guleria, S., Walia, A., Chauhan, A. and Shirkot, C. K. (2016). Molecular characterization of alkaline protease of *Bacillus amyloliquefaciens* SP1 involved in biocontrol of *Fusarium oxysporum*, *Int. J. Food Microbiol.* 232, 134–143. doi: 10.1016/j.ijfoodmicro.2016.05.030.

Guo, Q., Dong, W., Li, S., Lu, X., Wang, P., Zhang, X., Wang, Y. and Ma, P. (2014). Fengycin produced by *Bacillus subtilis* NCD-2 plays a major role in biocontrol of cotton seedling damping-off disease, *Microbiol. Res.* 169(7–8), 533–540. doi: 10.1016/j.micres.2013.12.001.

Guo, Q., Dong, L., Wang, P., Li, S., Zhao, W., Lu, X., Zhang, X. and Ma, P. (2018). The PhoR/PhoP two-component system regulates fengycin production in *Bacillus subtilis* NCD-2 under low-phosphate conditions, *J. Integr. Agric.* 17(1), 149–157. doi: 10.1016/S2095-3119(17)61669-1.

Guo, Q., Li, Y., Lou, Y., Shi, M., Jiang, Y., Zhou, J., Sun, Y., Xue, Q. and Lai, H. (2019). *Bacillus amyloliquefaciens* Ba13 induces plant systemic resistance and improves rhizosphere microecology against tomato yellow leaf curl virus disease, *Appl. Soil Ecol.* 137, 154–166. doi: 10.1016/j.apsoil.2019.01.015.

Hammami, I., Jaouadi, B., Bacha, A. Ben, Rebai, A., Bejar, S., Nesme, X. and Rhouma, A. (2012). *Bacillus subtilis* bacteriocin Bac 14B with a broad inhibitory spectrum: purification, amino acid sequence analysis, and physicochemical characterization, *Biotechnol. Bioprocess Eng.* 17, 41–49. doi: 10.1007/s12257-010-0401-8.

Hammami, I., Rhouma, A., Jaouadi, B., Rebai, A. and Nesme, X. (2009). Optimization and biochemical characterization of a bacteriocin from a newly isolated *Bacillus subtilis* strain 14B for biocontrol of *Agrobacterium* spp. *strains, Lett. Appl. Microbiol.* 48(2), 253–260. doi: 10.1111/j.1472-765X.2008.02524.x.

Han, Q., Wu, F., Wang, X., Qi, H., Shi, L., Ren, A., Liu, Q., Zhao, M. and Tang, C. (2015). The bacterial lipopeptide iturins induce *Verticillium dahliae* cell death by affecting fungal signalling pathways and mediate plant defence responses involved in pathogen-associated molecular pattern-triggered immunity, *Environ. Microbiol.* 17(4), 1166–1188. doi: 10.1111/1462-2920.12538.

Hanif, A., Zhang, F., Li, P., Li, C., Xu, Y., Zubair, M., Zhang, M., Jia, D., Zhao, X., Liang, J., Majid, T., Yan, J., Farzand, A., Wu, H., Gu, Q. and Gao, X. (2019). Fengycin produced by *Bacillus amyloliquefaciens* FZB42 inhibits *Fusarium graminearum* growth and mycotoxins biosynthesis, *Toxins (Basel)* 11(5), 1–11. doi: 10.3390/toxins11050295.

Harwood, C. R., Mouillon, J. M., Pohl, S. and Arnau, J. (2018). Secondary metabolite production and the safety of industrially important members of the *Bacillus subtilis* group, *FEMS Microbiol. Rev.* 42(6), 721–738. doi: 10.1093/femsre/fuy028.

Hayashi, K., Ohsawa, T., Kobayashi, K., Ogasawara, N. and Ogura, M. (2005). The H_2O_2 stress-responsive regulator PerR positively regulates srfA expression in *Bacillus subtilis*, *J. Bacteriol.* 187(19), 6659–6667. doi: 10.1128/JB.187.19.6659-6667.2005.

Heimpel, G. E. and Mills, N. J. (2017). *Biological Control: Ecology and Applications*. Cambridge University Press, Cambridge. doi: 10.1017/9781139029117.

Henry, G., Deleu, M., Jourdan, E., Thonart, P. and Ongena, M. (2011). The bacterial lipopeptide surfactin targets the lipid fraction of the plant plasma membrane to trigger immune-related defence responses, *Cell. Microbiol.* 13(11), 1824–1837. doi: 10.1111/j.1462-5822.2011.01664.x.

Hinsinger, P., Bengough, A. G., Vetterlein, D. and Young, I. M. (2009). Rhizosphere: biophysics, biogeochemistry and ecological relevance, *Plant Soil* 321(1–2), 117–152. doi: 10.1007/s11104-008-9885-9.

Im, S. M., Yu, N. H., Joen, H. W., Kim, S. O., Park, H. W., Park, A. R. and Kim, J. C. (2020). Biological control of tomato bacterial wilt by oxydifficidin and difficidin-producing *Bacillus methylotrophicus* DR-08, *Pestic. Biochem. Physiol.* 163, 130–137. doi: 10.1016/j.pestbp.2019.11.007.

İrigül-Sönmez, Ö., Köroğlu, T. E., Öztürk, B., Kovács, Á. T., Kuipers, O. P. and Yazgan-Karataş, A. (2014). In *Bacillus subtilis* LutR is part of the global complex regulatory network governing the adaptation to the transition from exponential growth to stationary phase, *Microbiology (Reading)* 160(2), 243–260. doi: 10.1099/mic.0.064675-0.

Izquierdo-García, L. F., González-Almario, A., Cotes, A. M. and Moreno-Velandia, C. A. (2020). *Trichoderma virens* Gl006 and *Bacillus velezensis* Bs006: a compatible interaction controlling *Fusarium wilt* of cape gooseberry, *Sci. Rep.* 10(1), 6857. doi: 10.1038/s41598-020-63689-y.

Jacques, P. (2011). Surfactin and other lipopeptides from *Bacillus* spp. In: Soberón-Chávez, G. (ed) *Biosurfactants. Microbiology Monographs*, vol 20. Springer, Berlin, Heidelberg, 57–91. doi: 10.1007/978-3-642-14490-5_3.

Jacques, P., Hbid, C., Destain, J., Razafindralambo, H., Paquot, M., De Pauw, E. and Thonart, P. (1999). Optimization of biosurfactant lipopeptide production from *Bacillus subtilis* S499 by Plackett-Burman design, *Appl. Biochem. Biotechnol.* 77(1–3), 223–234. doi: 10.1385/ABAB:77:1-3:223.

Jetiyanon, K. and Kloepper, J. W. (2002). Mixtures of plant growth-promoting rhizobacteria for induction of systemic resistance against multiple plant diseases, *Biol. Control* 24(3), 285–291. doi: 10.1016/S1049-9644(02)00022-1.

Jiang, C.-H., Liao, M.-J., Wang, H.-K., Zheng, M.-Z., Xu, J.-J. and Guo, J.-H. (2018). Bacillus velezensis, a potential and efficient biocontrol agent in control of pepper gray mold caused by *Botrytis cinerea*, *Biol. Control* 126, 147–157. doi: 10.1016/j.biocontrol.2018.07.017.

Jin, P., Wang, H., Tan, Z., Xuan, Z., Dahar, G. Y., Li, Q. X., Miao, W. and Liu, W. (2020). Antifungal mechanism of bacillomycin D from *Bacillus velezensis* HN-2 against *Colletotrichum gloeosporioides* Penz, *Pestic. Biochem. Physiol.* 163, 102–107. doi: 10.1016/j.pestbp.2019.11.004.

Jin, Y., Zhu, H., Luo, S., Yang, W., Zhang, L., Li, S., Jin, Q., Cao, Q., Sun, S. and Xiao, M. (2019). Role of maize root exudates in promotion of colonization of *Bacillus velezensis* Strain S3-1 in rhizosphere soil and root tissue, *Curr. Microbiol.* 76(7), 855–862. doi: 10.1007/s00284-019-01699-4.

Jose, P. A., Krishnamoorthy, R., Kwon, S.-W., Janahiraman, V., Senthilkumar, M., Gopal, N. O., Kumutha, K. and Anandham, R. (2019). Interference in quorum sensing and virulence of the phytopathogen *Pseudomonas syringae* pv. *passiflorae* by *Bacillus* and *Variovorax* species, *BioControl* 64(4), 423–433. doi: 10.1007/s10526-019-09932-6.

Jourdan, E., Henry, G., Duby, F., Dommes, J., Barthélemy, J. P., Thonart, P. and Ongena, M. (2009). Insights into the defense-related events occurring in plant cells following perception of surfactin-type lipopeptide from *Bacillus subtilis*, *Mol. Plant Microbe Interact.* 22(4), 456–468. doi: 10.1094/MPMI-22-4-0456.

Kai, M. (2020). Diversity and distribution of volatile secondary metabolites throughout *Bacillus subtilis* isolates, *Front. Microbiol.* 11(559), 559. doi: 10.3389/fmicb.2020.00559.

Kakar, K. U., Duan, Y.-P., Nawaz, Z., Sun, G., Almoneafy, A. A., Hassan, M. A., Elshakh, A., Li, B. and Xie, G. L. (2014). A novel rhizobacterium Bk7 for biological control of brown sheath rot of rice caused by *Pseudomonas fuscovaginae* and its mode of action, *Eur. J. Plant Pathol.* 138(4), 819–834. doi: 10.1007/s10658-013-0356-7.

Karuppiah, V., Sun, J., Li, T., Vallikkannu, M. and Chen, J. (2019). Co-cultivation of *Trichoderma asperellum* GDFS1009 and *Bacillus amyloliquefaciens* 1841 causes differential gene expression and improvement in the wheat growth and biocontrol activity, *Front. Microbiol.* 10(1068), 1068. doi: 10.3389/fmicb.2019.01068.

Kawagoe, Y., Shiraishi, S., Kondo, H., Yamamoto, S., Aoki, Y. and Suzuki, S. (2015). Cyclic lipopeptide iturin A structure-dependently induces defense response in Arabidopsis plants by activating SA and JA signaling pathways, *Biochem. Biophys. Res. Commun.* 460(4), 1015–1020. doi: 10.1016/j.bbrc.2015.03.143.

Kearns, D. B., Chu, F., Branda, S. S., Kolter, R. and Losick, R. (2005). A master regulator for biofilm formation by *Bacillus subtilis*, *Mol. Microbiol.* 55(3), 739–749. doi: 10.1111/j.1365-2958.2004.04440.x.

Kearns, D. B., Chu, F., Rudner, R. and Losick, R. (2004). Genes governing swarming in *Bacillus subtilis* and evidence for a phase variation mechanism controlling surface motility, *Mol. Microbiol.* 52(2), 357–369. doi: 10.1111/j.1365-2958.2004.03996.x.

Khardziani, T., Kachlishvili, E., Sokhadze, K., Elisashvili, V., Weeks, R., Chikindas, M. L. and Chistyakov, V. (2017). Elucidation of *Bacillus subtilis* KATMIRA 1933 potential for spore production in submerged fermentation of plant raw materials, *Probiotics Antimicrob. Proteins* 9(4), 435–443. doi: 10.1007/s12602-017-9303-9.

Khezri, M., Jouzani, G. S. and Ahmadzadeh, M. (2016). *Fusarium culmorum* affects expression of biofilm formation key genes in *Bacillus subtilis*, *Braz. J. Microbiol.* 47(1), 47–54. doi: 10.1016/j.bjm.2015.11.019.

Kierul, K., Voigt, B., Albrecht, D., Chen, X. H., Carvalhais, L. C. and Borriss, R. (2015). Influence of root exudates on the extracellular proteome of the plant growth-promoting bacterium *Bacillus amyloliquefaciens* FZB42, *Microbiology (Reading)* 161(1), 131–147. doi: 10.1099/mic.0.083576-0.

Kildea, S., Ransbotyn, V., Khan, M. R., Fagan, B., Leonard, G., Mullins, E. and Doohan, F. M. (2008). *Bacillus megaterium* shows potential for the biocontrol of *Septoria tritici* blotch of wheat, *Biol. Control* 47(1), 37–45. doi: 10.1016/j.biocontrol.2008.07.001.

Kim, K. H., Kabir, E. and Jahan, S. A. (2017). Exposure to pesticides and the associated human health effects, *Sci. Total Environ.* 575, 525–535. doi: 10.1016/j.scitotenv.2016.09.009.

Kinsella, K., Schulthess, C. P., Morris, T. F. and Stuart, J. D. (2009). Rapid quantification of *Bacillus subtilis* antibiotics in the rhizosphere, *Soil Biol. Biochem.* 41(2), 374–379. doi: 10.1016/j.soilbio.2008.11.019.

Kinsinger, R. F., Shirk, M. C. and Fall, R. (2003). Rapid surface motility in *Bacillus subtilis* is dependent on extracellular surfactin and potassium ion, *J. Bacteriol.* 185(18), 5627–5631. doi: 10.1128/JB.185.18.5627-5631.2003.

Kishore, G. K. and Pande, S. (2007). Chitin-supplemented foliar application of chitinolytic *Bacillus cereus* reduces severity of Botrytis gray mold disease in chickpea under controlled conditions, *Lett. Appl. Microbiol.* 44(1), 98–105. doi: 10.1111/j.1472-765X.2006.02022.x.

Kleerebezem, M., Bongers, R., Rutten, G., Vos, W. M. and Kuipers, O. P. (2004) Autoregulation of subtilin biosynthesis in *Bacillus subtilis*: the role of the spa-box in subtilin-responsive promoters, *Peptides* 25(9), 1415–1424. doi: 10.1016/j.peptides.2003.11.025.

Köhl, J., Kolnaar, R. and Ravensberg, W. J. (2019). Mode of action of microbial biological control agents against plant diseases: relevance beyond efficacy, *Front. Plant Sci.* 10(845), 845. doi: 10.3389/fpls.2019.00845.

Konz, D., Klens, A., Schörgendorfer, K. and Marahiel, M. A. (1997). The bacitracin biosynthesis operon of *Bacillus licheniformis* ATCC 10716: molecular characterization of three multi-modular peptide synthetases, *Chem. Biol.* 4(12), 927-937. doi: 10.1016/S1074-5521(97)90301-X.

Köroğlu, T. E., Kurt-Gür, G., Ünlü, E. C. and Yazgan-Karataş, A. (2008). The novel gene yvfI in *Bacillus subtilis* is essential for bacilysin biosynthesis, *Antonie Leeuwenhoek* 94(3), 471–479. doi: 10.1007/s10482-008-9265-8.

Köroğlu, T. E., Öğülür, I., Mutlu, S., Yazgan-Karataş, A. and Özcengiz, G. (2011). Global regulatory systems operating in bacilysin biosynthesis in *Bacillus subtilis*, *J. Mol. Microbiol. Biotechnol.* 20(3), 144–155. doi: 10.1159/000328639.

Korpi, A., Järnberg, J. and Pasanen, A. L. (2009). Microbial volatile organic compounds, *Crit. Rev. Toxicol.* 39(2), 139–193. doi: 10.1080/10408440802291497.

Koumoutsi, A., Chen, X. H., Vater, J. and Borriss, R. (2007). DegU and YczE positively regulate the synthesis of bacillomycin D by *Bacillus amyloliquefaciens* strain FZB42, *Appl. Environ. Microbiol.* 73(21), 6953–6964. doi: 10.1128/AEM.00565-07.

Kröber, M., Wibberg, D., Grosch, R., Eikmeyer, F., Verwaaijen, B., Chowdhury, S. P., Hartmann, A., Pühler, A. and Schlüter, A. (2014). Effect of the strain *Bacillus amyloliquefaciens* FZB42 on the microbial community in the rhizosphere of lettuce under field conditions analyzed by whole metagenome sequencing, *Front. Microbiol.* 5(252), 252. doi: 10.3389/fmicb.2014.00252.

Kumar, A., Prakash, A. and Johri, B. N. (2011). Bacillus as PGPR in crop ecosystem. In: Maheshwari, D. (ed) *Bacteria in Agrobiology: Crop Ecosystems*. Springer, Berlin, Heidelberg, 37–60. doi: 10.1007/978-3-642-18357-7.

Lajis, A. F. B. (2020). Biomanufacturing process for the production of bacteriocins from Bacillaceae family, *Bioresour. Bioprocess.* 7(1), 1–26. doi: 10.1186/s40643-020-0295-z.

Lakshmanan, V., Selvaraj, G. and Bais, H. P. (2014). Functional soil microbiome: belowground solutions to an aboveground problem, *Plant Physiol.* 166(2), 689–700. doi: 10.1104/pp.114.245811.

Le Mire, G., Nguyen, M. L., Fassotte, B., Du Jardin, P., Verheggen, F., Delaplace, P. and Haissam Jijakli, M. (2016). Implementing plant biostimulants and biocontrol strategies in the agroecological management of cultivated ecosystems: a review, *Biotechnol. Agron. Soc. Environ.* 20 (Suppl.1), 299–313. doi: 10.25518/1780-4507.12717.

Le Mire, G., Siah, A., Brisset, M.-N., Gaucher, M., Deleu, M. and Jijakli, M. (2018). Surfactin protects wheat against *Zymoseptoria tritici* and activates both salicylic acid- and jasmonic acid-dependent defense responses, *Agriculture* 8(1), 1–14. doi: 10.3390/agriculture8010011.

Lee, G. H. and Ryu, C. M. (2016). Spraying of leaf-colonizing *Bacillus amyloliquefaciens* protects pepper from Cucumber mosaic virus, *Plant Dis.* 100(10), 2099–2105. doi: 10.1094/PDIS-03-16-0314-RE.

Li, T., Tang, J., Karuppiah, V., Li, Y., Xu, N. and Chen, J. (2020). Co-culture of *Trichoderma atroviride* SG3403 and *Bacillus subtilis* 22 improves the production of antifungal secondary metabolites, *Biol. Control* 140, 1–8. doi: 10.1016/j.biocontrol.2019.104122.

Li, Y., Héloir, M. C., Zhang, X., Geissler, M., Trouvelot, S., Jacquens, L., Henkel, M., Su, X., Fang, X., Wang, Q. and Adrian, M. (2019). Surfactin and fengycin contribute to the protection of a *Bacillus subtilis* strain against grape downy mildew by both direct effect and defence stimulation, *Mol. Plant Pathol.* 20(8), 1037–1050. doi: 10.1111/mpp.12809.

Lim, S. M., Yoon, M. Y., Choi, G. J., Choi, Y. H., Jang, K. S., Shin, T. S., Park, H. W., Yu, N. H., Kim, Y. H. and Kim, J. C. (2017). Diffusible and volatile antifungal compounds produced by an antagonistic *Bacillus velezensis* G341 against various phytopathogenic fungi, *Plant Pathol. J.* 33(5), 488–498. doi: 10.5423/PPJ.OA.04.2017.0073.

Lin, C., Tsai, C. H., Chen, P. Y., Wu, C. Y., Chang, Y. L., Yang, Y. L. and Chen, Y. L. (2018). Biological control of potato common scab by *Bacillus amyloliquefaciens* Ba01, *PLoS ONE* 13(4), e0196520. doi: 10.1371/journal.pone.0196520.

Liu, C., Yin, X., Wang, Q., Peng, Y., Ma, Y., Liu, P. and Shi, J. (2018a). Antagonistic activities of volatiles produced by two Bacillus strains against *Monilinia fructicola* in peach fruit, *J. Sci. Food Agric.* 98(15), 5756–5763. doi: 10.1002/jsfa.9125.

Liu, K., McInroy, J. A., Hu, C. H. and Kloepper, J. W. (2018b). Mixtures of plant-growth-promoting rhizobacteria enhance biological control of multiple plant diseases and plant-growth promotion in the presence of pathogens, *Plant Dis.* 102(1), 67–72. doi: 10.1094/PDIS-04-17-0478-RE.

Liu, Y., Kyle, S. and Straight, P. D. (2018c). Antibiotic stimulation of a *Bacillus subtilis* migratory response, *mSphere* 3(1), 1–13. doi: 10.1128/mSphere.00586-17.

Liu, G., Kong, Y., Fan, Y., Geng, C., Peng, D. and Sun, M. (2017). Whole-genome sequencing of *Bacillus velezensis* LS69, a strain with a broad inhibitory spectrum against pathogenic bacteria, *J. Biotechnol.* 249, 20–24. doi: 10.1016/j.jbiotec.2017.03.018.

Liu, J., Hagberg, I., Novitsky, L., Hadj-Moussa, H. and Avis, T. J. (2014a). Interaction of antimicrobial cyclic lipopeptides from *Bacillus subtilis* influences their effect on spore germination and membrane permeability in fungal plant pathogens, *Fungal Biol.* 118(11), 855–861. doi: 10.1016/j.funbio.2014.07.004.

Liu, Y., Zhang, N., Qiu, M., Feng, H., Vivanco, J. M., Shen, Q. and Zhang, R. (2014b). Enhanced rhizosphere colonization of beneficial *Bacillus amyloliquefaciens* SQR9 by pathogen infection, *FEMS Microbiol. Lett.* 353(1), 49–56. doi: 10.1111/1574-6968.12406.

Liu, J., Zhou, T., He, D., Li, X. Z., Wu, H., Liu, W. and Gao, X. (2011). Functions of lipopeptides bacillomycin D and fengycin in antagonism of *Bacillus amyloliquefaciens* C06 towards *Monilinia fructicola*, *J. Mol. Microbiol. Biotechnol.* 20(1), 43–52. doi: 10.1159/000323501.

Liu, Y., Feng, H., Fu, R., Zhang, N., Du, W., Shen, Q. and Zhang, R. (2020). Induced root-secreted D-galactose functions as a chemoattractant and enhances the biofilm formation of *Bacillus velezensis* SQR9 in an McpA-dependent manner, *Appl. Microbiol. Biotechnol.* 104(2), 785–797. doi: 10.1007/s00253-019-10265-8.

Lopez, D., Vlamakis, H., Losick, R. and Kolter, R. (2009). Paracrine signaling in a bacterium, *Genes Dev.* 23(14), 1631–1638. doi: 10.1101/gad.1813709.

Lu, H., Qian, S., Muhammad, U., Jiang, X., Han, J. and Lu, Z. (2016). Effect of fructose on promoting fengycin biosynthesis in *Bacillus amyloliquefaciens* fmb-60, *J. Appl. Microbiol.* 121(6), 1653–1664. doi: 10.1111/jam.13291.

Luna-Bulbarela, A., Tinoco-Valencia, R., Corzo, G., Kazuma, K., Konno, K., Galindo, E. and Serrano-Carreón, L. (2018). Effects of bacillomycin D homologues produced by *Bacillus amyloliquefaciens* 83 on growth and viability of *Colletotrichum gloeosporioides* at different physiological stages, *Biol. Control* 127, 145–154. doi: 10.1016/j.biocontrol.2018.08.004.

Luo, C., Zhou, H., Zou, J., Wang, X., Zhang, R., Xiang, Y. and Chen, Z. (2015). *Bacillomycin* L and surfactin contribute synergistically to the phenotypic features of *Bacillus subtilis* 916 and the biocontrol of rice sheath blight induced by *Rhizoctonia solani*, *Appl. Microbiol. Biotechnol.* 99(4), 1897–1910. doi: 10.1007/s00253-014-6195-4.

Mariappan, A., Makarewicz, O., Chen, X. H. and Borriss, R. (2012). Two-component response regulator DegU controls the expression of bacilysin in plant-growth-promoting bacterium *Bacillus amyloliquefaciens* FZB42, *J. Mol. Microbiol. Biotechnol.* 22(2), 114–125. doi: 10.1159/000338804.

Mathieu, S., Cusant, L., Roux, C. and Corradi, N. (2018). Arbuscular mycorrhizal fungi: intraspecific diversity and pangenomes, *New Phytol.* 220(4), 1129–1134. doi: 10.1111/nph.15275.

Matzen, N., Heick, T. M. and Jørgensen, L. N. (2019). Control of powdery mildew (*Blumeria graminis* spp.) in cereals by Serenade®ASO (*Bacillus amyloliquefaciens* (former subtilis) strain QST 713), *Biol. Control* 139, 1–8. doi: 10.1016/j.biocontrol.2019.104067.

Medeot, D. B., Fernandez, M., Morales, G. M. and Jofré, E. (2020). Fengycins from *Bacillus amyloliquefaciens* MEP218 exhibit antibacterial activity by producing alterations on the cell surface of the pathogens *Xanthomonas axonopodis* pv. *vesicatoria* and *Pseudomonas aeruginosa* PA01, *Front. Microbiol.* 10(3107), 1–12. doi: 10.3389/fmicb.2019.03107.

Mejri, S., Siah, A., Coutte, F., Magnin-Robert, M., Randoux, B., Tisserant, B., Krier, F., Jacques, P., Reignault, P. and Halama, P. (2018). Biocontrol of the wheat pathogen

Zymoseptoria tritici using cyclic lipopeptides from *Bacillus subtilis*, *Environ. Sci. Pollut. Res. Int.* 25(30), 29822–29833. doi: 10.1007/s11356-017-9241-9.

Meng, Q. and Hao, J. J. (2017). Optimizing the application of *Bacillus velezensis* BAC03 in controlling the disease caused by *Streptomyces scabies*, *BioControl* 62(4), 535–544. doi: 10.1007/s10526-017-9799-7.

Mihalache, G., Balaes, T., Gostin, I., Stefan, M., Coutte, F. and Krier, F. (2018). Lipopeptides produced by *Bacillus subtilis* as new biocontrol products against fusariosis in ornamental plants, *Environ. Sci. Pollut. Res. Int.* 25(30), 29784–29793. doi: 10.1007/s11356-017-9162-7.

Miljaković, D., Marinković, J. and Balešević-Tubić, S. (2020). The significance of Bacillus spp. in disease suppression and growth promotion of field and vegetable crops, *Microorganisms* 8(7), 1–19. doi: 10.3390/microorganisms8071037.

Ming, L. J. and Epperson, J. D. (2002). Metal binding and structure-activity relationship of the metalloantibiotic peptide bacitracin, *J. Inorg. Biochem.* 91(1), 46–58. doi: 10.1016/S0162-0134(02)00464-6.

Mishra, S. and Arora, N. K. (2012). Evaluation of rhizospheric Pseudomonas and Bacillus as biocontrol tool for *Xanthomonas campestris* pv *campestris*, *World J. Microbiol. Biotechnol.* 28(2), 693–702. doi: 10.1007/s11274-011-0865-5.

Mnif, I. and Ghribi, D. (2015). Potential of bacterial derived biopesticides in pest management, *Crop Prot.* 77, 52–64. doi: 10.1016/j.cropro.2015.07.017.

Molina-Santiago, C., Pearson, J. R., Navarro, Y., Berlanga-Clavero, M. V., Caraballo-Rodriguez, A. M., Petras, D., García-Martín, M. L., Lamon, G., Haberstein, B., Cazorla, F. M., de Vicente, A., Loquet, A., Dorrestein, P. C. and Romero, D. (2019). The extracellular matrix protects *Bacillus subtilis* colonies from Pseudomonas invasion and modulates plant co-colonization, *Nat. Commun.* 10(1), 1919. doi: 10.1038/s41467-019-09944-x.

Molinatto, G., Puopolo, G., Sonego, P., Moretto, M., Engelen, K., Viti, C., Ongena, M. and Pertot, I. (2016). Complete genome sequence of *Bacillus amyloliquefaciens* subsp. plantarum S499, a rhizobacterium that triggers plant defences and inhibits fungal phytopathogens, *J. Biotechnol.* 238, 56–59. doi: 10.1016/j.jbiotec.2016.09.013.

Motta Dos Santos, L. F., Coutte, F., Ravallec, R., Dhulster, P., Tournier-Couturier, L. and Jacques, P. (2016). An improvement of surfactin production by *B. subtilis* BBG131 using design of experiments in microbioreactors and continuous process in bubbleless membrane bioreactor, *Bioresour. Technol.* 218, 944–952. doi: 10.1016/j.biortech.2016.07.053.

Müller, S., Strack, S. N., Hoefler, B. C., Straight, P. D., Kearns, D. B. and Kirby, J. R. (2014). Bacillaene and sporulation protect *Bacillus subtilis* from predation by *Myxococcus xanthus*, *Appl. Environ. Microbiol.* 80(18), 5603–5610. doi: 10.1128/AEM.01621-14.

Müller, S., Strack, S. N., Ryan, S. E., Kearns, D. B. and Kirby, J. R. (2015). Predation by *Myxococcus xanthus* induces *Bacillus subtilis* to form spore-filled megastructures, *Appl. Environ. Microbiol.* 81(1), 203–210. doi: 10.1128/AEM.02448-14.

Mutlu, A., Kaspar, C., Becker, N. and Bischofs, I. B. (2020). A spore quality–quantity tradeoff favors diverse sporulation strategies in *Bacillus subtilis*, *ISME J.* 14(11), 2703–2714. doi: 10.1038/s41396-020-0721-4.

Nakano, M. M., Zhu, Y., Liu, J., Reyes, D. Y., Yoshikawa, H. and Zuber, P. (2000). Mutations conferring amino acid residue substitutions in the carboxy-terminal domain of RNA polymerase α can suppress clpX and clpP with respect to developmentally regulated transcription in *Bacillus subtilis*, *Mol. Microbiol.* 37(4), 869–884. doi: 10.1046/j.1365-2958.2000.02052.x.

Nandeeshkumar, P., Ramachandrakini, K., Prakash, H. S., Niranjana, S. R. and Shekar Shetty, H. (2008). Induction of resistance against downy mildew on sunflower by rhizobacteria, *J. Plant Interact.* 3(4), 255–262. doi: 10.1080/17429140802245697.

Nihorimbere, V., Cawoy, H., Seyer, A., Brunelle, A., Thonart, P. and Ongena, M. (2012). Impact of rhizosphere factors on cyclic lipopeptide signature from the plant beneficial strain *Bacillus amyloliquefaciens* S499, *FEMS Microbiol. Ecol.* 79(1), 176–191. doi: 10.1111/j.1574-6941.2011.01208.x.

Nihorimbere, V., Fickers, P., Thonart, P. and Ongena, M. (2009). Ecological fitness of *Bacillus subtilis* BGS3 regarding production of the surfactin lipopeptide in the rhizosphere, *Environ. Microbiol. Rep.* 1(2), 124–130. doi: 10.1111/j.1758-2229.2009.00017.x.

Nikolić, I., Berić, T., Dimkić, I., Popović, T., Lozo, J., Fira, D. and Stanković, S. (2019). Biological control of *Pseudomonas syringae* pv. *aptataon* sugar beet with *Bacillus pumilus* SS-10.7 and *Bacillus amyloliquefaciens* (SS-12.6 and SS-38.4) strains, *J. Appl. Microbiol.* 126(1), 165–176. doi: 10.1111/jam.14070.

Nishimoto, R. (2019). Global trends in the crop protection industry, *J. Pestic. Sci.* 44(3), 141–147. doi: 10.1584/jpestics.D19-101.

Ntushelo, K., Ledwaba, L. K., Rauwane, M. E., Adebo, O. A. and Njobeh, P. B. (2019). The mode of action of *Bacillus* species against *Fusarium graminearum*, tools for investigation, and future prospects, *Toxins (Basel)* 11(10), 1–14. doi: 10.3390/toxins11100606.

Ogran, A., Yardeni, E. H., Keren-Paz, A., Bucher, T., Jain, R., Gilhar, O. and Kolodkin-Gal, I. (2019). The plant host induces antibiotic production to select the most-beneficial colonizers, *Appl. Environ. Microbiol.* 85(13), 1–15. doi: 10.1128/AEM.00512-19.

Ogura, M. (2001). DNA microarray analysis of *Bacillus subtilis* DegU, ComA and PhoP regulons: an approach to comprehensive analysis of *B.subtilis* two-component regulatory systems, *Nucleic Acids Res.* 29(18), 3804–3813. doi: 10.1093/nar/29.18.3804.

Ohno, A., Ano, T. and Shoda, M. (1995). Effect of temperature on production of lipopeptide antibiotics, iturin A and surfactin by a dual producer, *Bacillus subtilis* RB14, in solid-state fermentation, *J. Ferment. Bioeng.* 80(5), 517–519. doi: 10.1016/0922-338X(96)80930-5.

Ongena, M. and Jacques, P. (2008). *Bacillus lipopeptides*: versatile weapons for plant disease biocontrol, *Trends Microbiol.* 16(3), 115–125. doi: 10.1016/j.tim.2007.12.009.

Ongena, M., Jourdan, E., Adam, A., Paquot, M., Brans, A., Joris, B., Arpigny, J. L. and Thonart, P. (2007). Surfactin and fengycin lipopeptides of *Bacillus subtilis* as elicitors of induced systemic resistance in plants, *Environ. Microbiol.* 9(4), 1084–1090. doi: 10.1111/j.1462-2920.2006.01202.x.

Pandin, C., Le Coq, D., Canette, A., Aymerich, S. and Briandet, R. (2017). Should the biofilm mode of life be taken into consideration for microbial biocontrol agents?, *Microb. Biotechnol.* 10(4), 719–734. doi: 10.1111/1751-7915.12693.

Pandin, C., Le Coq, D., Deschamps, J., Védie, R., Rousseau, T., Aymerich, S. and Briandet, R. (2018). Complete genome sequence of *Bacillus velezensis* QST713: a biocontrol agent that protects *Agaricus bisporus* crops against the green mould disease, *J. Biotechnol.* 278, 10–19. doi: 10.1016/j.jbiotec.2018.04.014.

Pandin, C., Darsonval, M., Mayeur, C., Le Coq, D., Aymerich, S. and Briandet, R. (2019). Biofilm formation and synthesis of antimicrobial compounds by the biocontrol agent *Bacillus velezensis* QST713 in an agaricus bisporus compost micromodel, *Appl. Environ. Microbiol.* 85(12), 1–13. doi: 10.1128/AEM.00327-19.

Park, K., Park, Y.-S., Ahamed, J., Dutta, S., Ryu, H., Lee, S.-H., Balaraju, K., Manir, M. and Moon, S.-S. (2016). Elicitation of induced systemic resistance of chili pepper by iturin A analogs derived from *Bacillus vallismortis* EXTN-1, *Can. J. Plant Sci.* 96(4), 564–570. doi: 10.1139/cjps-2015-0199.

Parnell, J. J., Berka, R., Young, H. A., Sturino, J. M., Kang, Y., Barnhart, D. M. and DiLeo, M. V. (2016). From the lab to the farm: an industrial perspective of plant beneficial microorganisms, *Front. Plant Sci.* 7(1110), 1110. doi: 10.3389/fpls.2016.01110.

Peng, G., Zhao, X., Li, Y., Wang, R., Huang, Y. and Qi, G. (2019). Engineering *Bacillus velezensis* with high production of acetoin primes strong induced systemic resistance in *Arabidopsis thaliana*, *Microbiol. Res.* 227, 1–12. doi: 10.1016/j.micres.2019.126297.

Pereira Monteiro, F., Henrique Vasconcelos de Medeiros, F., Ongena, M., Franzil, L., Estevao de Souza, P. and Teodoro de Souza, J. (2016). Effect of temperature, pH and substrate composition on production of lipopeptides by *Bacillus amyloliquefaciens* 629, *African. J. Microbiol. Res.* 10(36), 1506–1512. doi: 10.5897/AJMR2016.8222.

Pertot, I., Puopolo, G., Hosni, T., Pedrotti, L., Jourdan, E. and Ongena, M. (2013). Limited impact of abiotic stress on surfactin production in planta and on disease resistance induced by *Bacillus amyloliquefaciens* S499 in tomato and bean, *FEMS Microbiol. Ecol.* 86(3), 505–519. doi: 10.1111/1574-6941.12177.

Pieterse, C. M. J., Zamioudis, C., Berendsen, R. L., Weller, D. M., Van Wees, S. C. M. and Bakker, P. A. H. M. (2014). Induced systemic resistance by beneficial microbes, *Annu. Rev. Phytopathol.* 52(1), 347–375. doi: 10.1146/annurev-phyto-082712-102340.

Piggot, P. J. and Hilbert, D. W. (2004). Sporulation of *Bacillus subtilis*, *Curr. Opin. Microbiol.* 7(6), 579–586. doi: 10.1016/j.mib.2004.10.001.

Pisithkul, T., Schroeder, J. W., Trujillo, E. A., Yeesin, P., Stevenson, D. M., Chaiamarit, T., Coon, J. J., Wang, J. D. and Amador-Noguez, D. (2019). Metabolic remodeling during biofilm development of *Bacillus subtilis*, *mBio* 10(3), 1–32. doi: 10.1128/mBio.00623-19.

Plouznikoff, K., Declerck, S. and Calonne-Salmon, M. (2016). Mitigating abiotic stresses in crop plants by arbuscular mycorrhizal fungi. In: Vos, C. and Kazan, K. (eds) *Belowground Defence Strategies in Plants. Signaling and Communication in Plants.* Springer, Cham, 341–400. doi: 10.1007/978-3-319-42319-7_15.

Powers, M. J., Sanabria-Valentín, E., Bowers, A. A. and Shank, E. A. (2015). Inhibition of cell differentiation in *Bacillus subtilis* by *Pseudomonas protegens*, *J. Bacteriol.* 197(13), 2129–2138. doi: 10.1128/JB.02535-14.

Pršić, J. and Ongena, M. (2020). Elicitors of plant immunity triggered by beneficial bacteria, *Front. Plant Sci.* 11, 1–12. doi: 10.3389/fpls.2020.594530.

Punja, Z. K., Rodriguez, G. and Tirajoh, A. (2016). Effects of *Bacillus subtilis* strain QST 713 and storage temperatures on post-harvest disease development on greenhouse tomatoes, *Crop Prot.* 84, 98–104. doi: 10.1016/j.cropro.2016.02.011.

Qiao, J., Yu, X., Liang, X., Liu, Y., Borriss, R. and Liu, Y. (2017). Addition of plant-growth-promoting *Bacillus subtilis* PTS-394 on tomato rhizosphere has no durable impact on composition of root microbiome, *BMC Microbiol.* 17(1), 131. doi: 10.1186/s12866-017-1039-x.

Qiu, M., Xu, Z., Li, X., Li, Q., Zhang, N., Shen, Q. and Zhang, R. (2014). Comparative proteomics analysis of *Bacillus amyloliquefaciens* SQR9 revealed the key proteins involved in in situ root colonization, *J. Proteome Res.* 13(12), 5581–5591. doi: 10.1021/pr500565m.

Raafat, M. M., Ali-Tammam, M. and Ali, A. E. (2019). Quorum quenching activity of *Bacillus cereus* isolate 30b confers antipathogenic effects in *Pseudomonas aeruginosa*, *Infect. Drug Resist.* 12, 1583–1596. doi: 10.2147/IDR.S182889.

Raaijmakers, J. M., De Bruijn, I., Nybroe, O. and Ongena, M. (2010). Natural functions of lipopeptides from Bacillus and Pseudomonas : more than surfactants and antibiotics, *FEMS Microbiol. Rev.* 34(6), 1037–1062. doi: 10.1111/j.1574-6976.2010.00221.x.

Rabbee, M. F., Ali, M. S., Choi, J., Hwang, B. S., Jeong, S. C. and Baek, K. H. (2019). *Bacillus velezensis*: a valuable member of bioactive molecules within plant microbiomes, *Molecules* 24(6), 1–13. doi: 10.3390/molecules24061046.

Rahman, A., Uddin, W. and Wenner, N. G. (2015). Induced systemic resistance responses in perennial ryegrass against *Magnaporthe oryzae* elicited by semi-purified surfactin lipopeptides and live cells of *Bacillus amyloliquefaciens*, *Mol. Plant Pathol.* 16(6), 546–558. doi: 10.1111/mpp.12209.

Rahman, M. (2016). *Bacillus* spp.: a promising biocontrol agent of root, foliar, and postharvest diseases of plants. In: Islam, M., Rahman, M., Pandey, P., Jha, C. and Aeron, A. (eds) *Bacilli and Agrobiotechnology*. Springer, Cham, 113–141. doi: 10.1007/978-3-319-44409-3_6.

Rais, A., Shakeel, M., Hafeez, F. Y. and Hassan, M. N. (2016). Plant growth promoting rhizobacteria suppress blast disease caused by *Pyricularia oryzae* and increase grain yield of rice, *BioControl* 61(6), 769–780. doi: 10.1007/s10526-016-9763-y.

Rajamanickam, S. and Nakkeeran, S. (2020). Flagellin of *Bacillus amyloliquefaciens* works as a resistance inducer against groundnut bud necrosis virus in chilli (*Capsicum annuum* L.), *Arch. Virol.* 165(7), 1585–1597. doi: 10.1007/s00705-020-04645-z.

Rajavel, M., Mitra, A. and Gopal, B. (2009). Role of *Bacillus subtilis* BacB in the synthesis of bacilysin, *J. Biol. Chem.* 284(46), 31882–31892. doi: 10.1074/jbc.M109.014522.

Ranf, S. (2018). Pattern recognition receptors–versatile genetic tools for engineering broad-spectrum disease resistance in crops, *Agronomy* 8(8), 1–13. doi: 10.3390/agronomy8080134.

Rangarajan, V., Dhanarajan, G. and Sen, R. (2015). Bioprocess design for selective enhancement of fengycin production by a marine isolate *Bacillus megaterium*, *Biochem. Eng. J.* 99, 147–155. doi: 10.1016/j.bej.2015.03.016.

Rashad, Y. M., Abbas, M. A., Soliman, H. M., Abdel-Fattah, G. G. and Abdel-Fattah, G. M. (2020). Synergy between endophytic *Bacillus amyloliquefaciens* GGA and arbuscular mycorrhizal fungi induces plant defense responses against white rot of garlic and improves host plant growth, *Phytopathol. Mediterr.* 59(1), 169–186. doi: 10.14601/Phyto-11019.

Raymaekers, K., Ponet, L., Holtappels, D., Berckmans, B. and Cammue, B. P. A. (2020). Screening for novel biocontrol agents applicable in plant disease management–a review, *Biol. Control* 144, 1–18. doi: 10.1016/j.biocontrol.2020.104240.

Raza, W., Ling, N., Yang, L., Huang, Q. and Shen, Q. (2016). Response of tomato wilt pathogen *Ralstonia solanacearum* to the volatile organic compounds produced by a biocontrol strain *Bacillus amyloliquefaciens* SQR-9, *Sci. Rep.* 6, 24856. doi: 10.1038/srep24856.

Romero, D., de Vicente, A., Rakotoaly, R. H., Dufour, S. E., Veening, J. W., Arrebola, E., Cazorla, F. M., Kuipers, O. P., Paquot, M. and Pérez-García, A. (2007). The Iturin and fengycin families of lipopeptides are key factors in antagonism of *Bacillus subtilis* toward *Podosphaera fusca*, *Mol. Plant Microbe Interact.* 20(4), 430–440. doi: 10.1094/MPMI-20-4-0430.

Rudrappa, T., Biedrzycki, M. L., Kunjeti, S. G., Donofrio, N. M., Czymmek, K. J., Paré, P. W. and Bais, H. P. (2010). The rhizobacterial elicitor acetoin induces systemic resistance in *Arabidopsis thaliana*, *Commun. Integr. Biol.* 3(2), 130–138. doi: 10.4161/cib.3.2.10584.

Rudrappa, T., Czymmek, K. J., Paré, P. W. and Bais, H. P. (2008). Root-secreted malic acid recruits beneficial soil bacteria, *Plant Physiol.* 148(3), 1547–1556. doi: 10.1104/pp.108.127613.

Ruiu, L. (2020). Plant-growth-promoting bacteria (PGPB) against insects and other agricultural pests, *Agronomy* 10(6), 1–12. doi: 10.3390/agronomy10060861.

Ryu, C. M., Farag, M. A., Hu, C. H., Reddy, M. S., Kloepper, J. W. and Paré, P. W. (2004). Bacterial volatiles induce systemic resistance in Arabidopsis, *Plant Physiol.* 134(3), 1017–1026. doi: 10.1104/pp.103.026583.

Saber, W. I. A., Ghoneem, K. M., Al-Askar, A. A., Rashad, Y. M., Ali, A. A. and Rashad, E. M. (2015). Chitinase production by *Bacillus subtilis* ATCC 11774 and its effect on biocontrol of Rhizoctonia diseases of potato, *Acta Biol. Hung.* 66(4), 436–448. doi: 10.1556/018.66.2015.4.8.

Saikia, K., Belwal, V. K., Datta, D. and Chaudhary, N. (2019). Aromatic-rich C-terminal region of LCI is a potent antimicrobial peptide in itself, *Biochem. Biophys. Res. Commun.* 519(2), 372–377. doi: 10.1016/j.bbrc.2019.09.013.

Salazar, F., Ortiz, A. and Sansinenea, E. (2017). Characterisation of two novel bacteriocin-like substances produced by *Bacillus amyloliquefaciens* ELI149 with broad-spectrum antimicrobial activity, *J. Glob. Antimicrob. Resist.* 11, 177–182. doi: 10.1016/j.jgar.2017.08.008.

Sarwar, A., Hassan, M. N., Imran, M., Iqbal, M., Majeed, S., Brader, G., Sessitsch, A. and Hafeez, F. Y. (2018). Biocontrol activity of surfactin A purified from Bacillus NH-100 and NH-217 against rice bakanae disease, *Microbiol. Res.* 209, 1–13. doi: 10.1016/j.micres.2018.01.006.

Sasse, J., Martinoia, E. and Northen, T. (2018). Feed your friends: do plant exudates shape the root microbiome?, *Trends Plant Sci.* 23(1), 25–41. doi: 10.1016/j.tplants.2017.09.003.

Savary, S., Willocquet, L., Pethybridge, S. J., Esker, P., McRoberts, N. and Nelson, A. (2019). The global burden of pathogens and pests on major food crops, *Nat. Ecol. Evol.* 3(3), 430–439. doi: 10.1038/s41559-018-0793-y.

Saxena, A. K., Kumar, M., Chakdar, H., Anuroopa, N. and Bagyaraj, D. J. (2020). *Bacillus* species in soil as a natural resource for plant health and nutrition, *J. Appl. Microbiol.* 128(6), 1583–1594. doi: 10.1111/jam.14506.

Schallmey, M., Singh, A. and Ward, O. P. (2004). Developments in the use of *Bacillus* species for industrial production, *Can. J. Microbiol.* 50(1), 1–17. doi: 10.1139/w03-076.

Schellenberger, R., Touchard, M., Clément, C., Baillieul, F., Cordelier, S., Crouzet, J. and Dorey, S. (2019). Apoplastic invasion patterns triggering plant immunity: plasma membrane sensing at the frontline, *Mol. Plant Pathol.* 20(11), 1602–1616. doi: 10.1111/mpp.12857.

Schisler, D. A., Slininger, P. J., Behle, R. W. and Jackson, M. A. (2004). Formulation of *Bacillus* spp. for biological control of plant diseases, *Phytopathology®* 94(11), 1267–1271. doi: 10.1094/PHYTO.2004.94.11.1267.

Schmitz, S., Hoffmann, A., Szekat, C., Rudd, B. and Bierbaum, G. (2006). The lantibiotic mersacidin is an autoinducing peptide, *Appl. Environ. Microbiol.* 72(11), 7270–7277. doi: 10.1128/AEM.00723-06.

Scholz, R., Molohon, K. J., Nachtigall, J., Vater, J., Markley, A. L., Sussmuth, R. D., Mitchell, D. A. and Borriss, R. (2011). Plantazolicin, a novel microcin B17/Streptolysin S-like natural product from *Bacillus amyloliquefaciens* FZB42, *J. Bacteriol.* 193(1), 215–224. doi: 10.1128/JB.00784-10.

Scholz, R., Vater, J., Budiharjo, A., Wang, Z., He, Y., Dietel, K., Schwecke, T., Herfort, S., Lasch, P. and Borriss, R. (2014). Amylocyclicin, a novel circular bacteriocin produced by *Bacillus amyloliquefaciens* FZB42, *J. Bacteriol.* 196(10), 1842–1852. doi: 10.1128/JB.01474-14.

Shen, Y., Li, J., Xiang, J., Wang, J., Yin, K. and Liu, Q. (2019). Isolation and identification of a novel protein elicitor from a *Bacillus subtilis* strain BU412, *AMB Express* 9(1), 117. doi: 10.1186/s13568-019-0822-5.

Solanki, M. K., Singh, R. K., Srivastava, S., Kumar, S., Kashyap, P. L. and Srivastava, A. K. (2015). Characterization of antagonistic-potential of two Bacillus strains and their biocontrol activity against *Rhizoctonia solani* in tomato, *J. Basic Microbiol.* 55(1), 82–90. doi: 10.1002/jobm.201300528.

Sourjik, V. and Wingreen, N. S. (2012). Responding to chemical gradients: bacterial chemotaxis, *Curr. Opin. Cell Biol.* 24(2), 262–268. doi: 10.1016/j.ceb.2011.11.008.

Spraker, J. E., Luu, G. T. and Sanchez, L. M. (2020). Imaging mass spectrometry for natural products discovery: a review of ionization methods, *Nat. Prod. Rep.* 37(2), 150–162. doi: 10.1039/C9NP00038K.

Stamenkovic-Stojanovic, S., Karabegovic, I., Beskoski, V., Nikolic, N. and Lazic, M. (2019). Bacillus based microbial formulations: optimization of the production process, *Hem. Ind.* 73(3), 169–182. doi: 10.2298/HEMIND190214014S.

Tan, S., Yang, C., Mei, X., Shen, S., Raza, W., Shen, Q. and Xu, Y. (2013). The effect of organic acids from tomato root exudates on rhizosphere colonization of *Bacillus amyloliquefaciens* T-5, *Appl. Soil Ecol.* 64, 15–22. doi: 10.1016/j.apsoil.2012.10.011.

Tanaka, K., Amaki, Y., Ishihara, A. and Nakajima, H. (2015). Synergistic effects of [Ile 7] surfactin homologues with bacillomycin D in suppression of gray mold disease by *Bacillus amyloliquefaciens* biocontrol strain SD-32, *J. Agric. Food Chem.* 63(22), 5344–5353. doi: 10.1021/acs.jafc.5b01198.

Tao, Y., Bie, X. M., Lv, F. X., Zhao, H. Z. and Lu, Z. X. (2011). Antifungal activity and mechanism of fengycin in the presence and absence of commercial surfactin against *Rhizopus stolonifer*, *J. Microbiol.* 49(1), 146–150. doi: 10.1007/s12275-011-0171-9.

Thérien, M., Kiesewalter, H. T., Auria, E., Charron-Lamoureux, V., Wibowo, M., Maróti, G., Kovács, Á. T. and Beauregard, P. B. (2020). Surfactin production is not essential for pellicle and root-associated biofilm development of *Bacillus subtilis*, *Biofilm* 2, 1–7. doi: 10.1016/j.bioflm.2020.100021.

Toral, L., Rodríguez, M., Béjar, V. and Sampedro, I. (2020). Crop protection against *Botrytis cinerea* by Rhizhosphere biological control agent *Bacillus velezensis* XT1, *Microorganisms* 8(7), 1–17. doi: 10.3390/microorganisms8070992.

Touré, Y., Ongena, M., Jacques, P., Guiro, A. and Thonart, P. (2004). Role of lipopeptides produced by *Bacillus subtilis* GA1 in the reduction of grey mould disease caused by *Botrytis cinerea* on apple, *J. Appl. Microbiol.* 96(5), 1151–1160. doi: 10.1111/j.1365-2672.2004.02252.x.

Traxler, M. F. and Kolter, R. (2015). Natural products in soil microbe interactions and evolution, *Nat. Prod. Rep.* 32(7), 956–970. doi: 10.1039/C5NP00013K.

van Gestel, J., Vlamakis, H. and Kolter, R. (2015). From cell differentiation to cell collectives: *Bacillus subtilis* uses division of labor to migrate, *PLoS Biol.* 13(4), e1002141. doi: 10.1371/journal.pbio.1002141.

Van Overbeek, L. and Van Elsas, J. D. (2008). Effects of plant genotype and growth stage on the structure of bacterial communities associated with potato (*Solanum tuberosum* L.), *FEMS Microbiol. Ecol.* 64(2), 283–296. doi: 10.1111/j.1574-6941.2008.00469.x.

Vanthana, M., Nakkeeran, S., Malathi, V. G., Renukadevi, P. and Vinodkumar, S. (2019). Induction of in planta resistance by flagellin (Flg) and elongation factor-TU (EF-Tu) of *Bacillus amyloliquefaciens*(VB7) against groundnut bud necrosis virus in tomato, *Microb. Pathog.* 137, 1–11. doi: 10.1016/j.micpath.2019.103757.

Vargas-Bautista, C., Rahlwes, K. and Straight, P. (2014). Bacterial competition reveals differential regulation of the pks genes by *Bacillus subtilis*, *J. Bacteriol.* 196(4), 717–728. doi: 10.1128/JB.01022-13.

Villa-Rodríguez, E., Parra-Cota, F., Castro-Longoria, E., López-Cervantes, J. and de los Santos-Villalobos, S. (2019). Bacillus subtilis TE3: a promising biological control agent against *Bipolaris sorokiniana*, the causal agent of spot blotch in wheat (Triticum turgidum L. subsp. durum), *Biol. Control* 132, 135–143. doi: 10.1016/j.biocontrol.2019.02.012.

Villegas-Escobar, V., González-Jaramillo, L. M., Ramírez, M., Moncada, R. N., Sierra-Zapata, L., Orduz, S. and Romero-Tabarez, M. (2018). Lipopeptides from *Bacillus* sp. EA-CB0959: active metabolites responsible for in vitro and in vivo control of *Ralstonia solanacearum*, *Biol. Control* 125, 20–28. doi: 10.1016/j.biocontrol.2018.06.005.

Vinale, F., Sivasithamparam, K., Ghisalberti, E. L., Marra, R., Woo, S. L. and Lorito, M. (2008). Trichoderma-plant-pathogen interactions, *Soil Biol. Biochem.* 40(1), 1–10. doi: 10.1016/j.soilbio.2007.07.002.

Vivas, A., Marulanda, A., Gómez, M., Barea, J. M. and Azcón, R. (2003). Physiological characteristics (SDH and ALP activities) of arbuscular mycorrhizal colonization as affected by *Bacillus thuringiensis* inoculation under two phosphorus levels, *Soil Biol. Biochem.* 35(7), 987–996. doi: 10.1016/S0038-0717(03)00161-5.

Vlamakis, H., Chai, Y., Beauregard, P., Losick, R. and Kolter, R. (2013). Sticking together: building a biofilm the *Bacillus subtilis* way, *Nat. Rev. Microbiol.* 11(3), 157–168. doi: 10.1038/nrmicro2960.

Wang, H., Shi, Y., Wang, D., Yao, Z., Wang, Y., Liu, J., Zhang, S. and Wang, A. (2018). A biocontrol strain of *Bacillus subtilis* WXCDD105 used to control tomato *Botrytis cinerea* and *Cladosporium fulvum* Cooke and promote the growth of seedlings, *Int. J. Mol. Sci.* 19(5), 1–17. doi: 10.3390/ijms19051371.

Wang, N., Liu, M., Guo, L., Yang, X. and Qiu, D. (2016). A novel protein elicitor (PeBA1) from *Bacillus amyloliquefaciens* NC6 induces systemic resistance in tobacco, *Int. J. Biol. Sci.* 12(6), 757–767. doi: 10.7150/ijbs.14333.

Wang, P., Guo, Q., Ma, Y., Li, S., Lu, X., Zhang, X. and Ma, P. (2015). DegQ regulates the production of fengycins and biofilm formation of the biocontrol agent *Bacillus subtilis* NCD-2, *Microbiol. Res.* 178, 42–50. doi: 10.1016/j.micres.2015.06.006.

Wang, R., Liang, X., Long, Z., Wang, X., Yang, L., Lu, B. and Gao, J. (2020a). An LCI-like protein APC 2 protects ginseng root from *Fusarium solani* infection, *J. Appl. Microbiol.* 130(1), 165–178. doi: 10.1111/jam.14771.

Wang, X., Yuan, Z., Shi, Y., Cai, F., Zhao, J., Wang, J. and Wang, Y. (2020b). *Bacillus amyloliquefaciens* HG01 induces resistance in loquats against anthracnose rot caused by *Colletotrichum acutatum*, *Postharvest Biol. Technol.* 160, 1-7. doi: 10.1016/j.postharvbio.2019.111034.

Wang, Y., Zhang, C., Liang, J., Wang, L., Gao, W., Jiang, J. and Chang, R. (2020c). Surfactin and fengycin B extracted from *Bacillus pumilus* W-7 provide protection against potato late blight via distinct and synergistic mechanisms, *Appl. Microbiol. Biotechnol.* 104(17), 7467–7481. doi: 10.1007/s00253-020-10773-y.

Van Wees, S. C. M., De Swart, E. A. M., Van Pelt, J. A., Van Loon, L. C. and Pieterse, C. M. J. (2000). Enhancement of induced disease resistance by simultaneous activation of salicylate- and jasmonate-dependent defense pathways in *Arabidopsis thaliana*, *Proc. Natl. Acad. Sci. U. S. A.* 97(15), 8711–8716. doi: 10.1073/pnas.130425197.

Winn, M., Fyans, J. K., Zhuo, Y. and Micklefield, J. (2016). Recent advances in engineering nonribosomal peptide assembly lines, *Nat. Prod. Rep.* 33(2), 317–347. doi: 10.1039/C5NP00099H.

Wise, C., Falardeau, J., Hagberg, I. and Avis, T. J. (2014). Cellular lipid composition affects sensitivity of plant pathogens to fengycin, an antifungal compound produced by *Bacillus subtilis* strain CU12, *Phytopathology®* 104(10), 1036–1041. doi: 10.1094/PHYTO-12-13-0336-R.

Woo, S. L., Ruocco, M., Vinale, F., Nigro, M., Marra, R., Lombardi, N., Pascale, A., Lanzuise, S., Manganiello, G. and Lorito, M. (2014). Trichoderma-based products and their widespread use in agriculture, *TOMYCJ* 8(1), 71-126. doi: 10.2174/1874437001408010071.

Wu, K., Fang, Z., Guo, R., Pan, B., Shi, W., Yuan, S., Guan, H., Gong, M., Shen, B. and Shen, Q. (2015a). Pectin enhances bio-control efficacy by inducing colonization and secretion of secondary metabolites by *Bacillus amyloliquefaciens* SQY 162 in the rhizosphere of tobacco, *PLoS ONE* 10(5), e0127418. doi: 10.1371/journal.pone.0127418.

Wu, L., Wu, H., Chen, L., Yu, X., Borriss, R. and Gao, X. (2015b). Difficidin and bacilysin from *Bacillus amyloliquefaciens* FZB42 have antibacterial activity against *Xanthomonas oryzae* rice pathogens, *Sci. Rep.* 5, 12975. doi: 10.1038/srep12975.

Wu, Q., Ni, M., Dou, K., Tang, J., Ren, J., Yu, C. and Chen, J. (2018). Co-culture of *Bacillus amyloliquefaciens* accC11060 and *Trichoderma asperellum* GDFS1009 enhanced pathogen-inhibition and amino acid yield, *Microb. Cell Fact.* 17(1), 155. doi: 10.1186/s12934-018-1004-x.

Xie, S., Wu, H., Chen, L., Zang, H., Xie, Y. and Gao, X. (2015). Transcriptome profiling of *Bacillus subtilis* OKB105 in response to rice seedlings, *BMC Microbiol.* 15(21), 21. doi: 10.1186/s12866-015-0353-4.

Xu, Y., Cai, D., Zhang, H., Gao, L., Yang, Y., Gao, J., Li, Y., Yang, C., Ji, Z., Yu, J. and Chen, S. (2020). Enhanced production of iturin A in *Bacillus amyloliquefaciens* by genetic engineering and medium optimization, *Process Biochem.* 90, 50–57. doi: 10.1016/j.procbio.2019.11.017.

Yadav, R., Ror, P., Rathore, P. and Ramakrishna, W. (2020). Bacteria from native soil in combination with arbuscular mycorrhizal fungi augment wheat yield and biofortification, *Plant Physiol. Biochem.* 150, 222–233. doi: 10.1016/j.plaphy.2020.02.039.

Yamamoto, S., Shiraishi, S. and Suzuki, S. (2015). Are cyclic lipopeptides produced by *Bacillus amyloliquefaciens* S13-3 responsible for the plant defence response in strawberry against *Colletotrichum gloeosporioides*?, *Lett. Appl. Microbiol.* 60(4), 379–386. doi: 10.1111/lam.12382.

Yazgan, A., Özcengiz, G. and Marahiel, M. A. (2001). Tn10 insertional mutations of *Bacillus subtilis* that block the biosynthesis of bacilysin, *Biochim. Biophys. Acta* 1518(1-2), 87–94. doi: 10.1016/S0167-4781(01)00182-8.

Yazgan Karata, A., Çetin, S. and Özcengiz, G. (2003). The effects of insertional mutations in comQ, comP, srfA, spo0H, spo0A and abrB genes on bacilysin biosynthesis in *Bacillus subtilis*, *Biochim. Biophys. Acta Gene Struct. Expr.* 1626(1–3), 51–56. doi: 10.1016/S0167-4781(03)00037-X.

Ye, M., Tang, X., Yang, R., Zhang, H., Li, F., Tao, F., Li, F. and Wang, Z. (2018). Characteristics and application of a novel species of Bacillus: *Bacillus velezensis*, *ACS Chem. Biol.* 13(3), 500–505. doi: 10.1021/acschembio.7b00874.

Yeh, M.-S., Wei, Y.-H. and Chang, J.-S. (2006). Bioreactor design for enhanced carrier-assisted surfactin production with *Bacillus subtilis*, *Process Biochem.* 41(8), 1799-1805. doi: 10.1016/j.procbio.2006.03.027.

Yi, G., Liu, Q., Lin, J., Wang, W., Huang, H. and Li, S. (2017). Repeated batch fermentation for surfactin production with immobilized *Bacillus subtilis* BS-37: two-stage pH control and foam fractionation, *J. Chem. Technol. Biotechnol.* 92(3), 530–535. doi: 10.1002/jctb.5028.

Yssel, A., Reva, O. and Tastan Bishop, O. (2011). Comparative structural bioinformatics analysis of *Bacillus amyloliquefaciens* chemotaxis proteins within *Bacillus subtilis* group, *Appl. Microbiol. Biotechnol.* 92(5), 997–1008. doi: 10.1007/s00253-011-3582-y.

Yu, X., Feng, B., He, P. and Shan, L. (2017). From chaos to harmony: responses and signaling upon microbial pattern recognition, *Annu. Rev. Phytopathol.* 55(1), 109-137. doi: 10.1146/annurev-phyto-080516-035649.

Yuan, J., Li, B., Zhang, N., Waseem, R., Shen, Q. and Huang, Q. (2012a). Production of bacillomycin- and macrolactin-type antibiotics by *Bacillus amyloliquefaciens* NJN-6 for suppressing soilborne plant pathogens, *J. Agric. Food Chem.* 60(12), 2976–2981. doi: 10.1021/jf204868z.

Yuan, J., Raza, W., Shen, Q. and Huang, Q. (2012b). Antifungal activity of *Bacillus amyloliquefaciens* NJN-6 volatile compounds against *Fusarium oxysporum* f. sp. *cubense*, *Appl. Environ. Microbiol.* 78(16), 5942–5944. doi: 10.1128/AEM.01357-12.

Zakharova, A. A., Efimova, S. S., Malev, V. V. and Ostroumova, O. S. (2019). Fengycin induces ion channels in lipid bilayers mimicking target fungal cell membranes, *Sci. Rep.* 9(1), 16034. doi: 10.1038/s41598-019-52551-5.

Zaki, O., Weekers, F., Thonart, P., Tesch, E., Kuenemann, P. and Jacques, P. (2020). Limiting factors of mycopesticide development, *Biol. Control* 144, 1–12. doi: 10.1016/j.biocontrol.2020.104220.

Zeriouh, H., Romero, D., García-Gutiérrez, L., Cazorla, F. M., de Vicente, A. and Pérez-García, A. (2011). The Iturin-like lipopeptides are essential components in the biological control arsenal of *Bacillus subtilis* against bacterial diseases of cucurbits, *Mol. Plant Microbe Interact.* 24(12), 1540–1552. doi: 10.1094/MPMI-06-11-0162.

Zeriouh, H., de Vicente, A., Pérez-García, A. and Romero, D. (2014). Surfactin triggers biofilm formation of *Bacillus subtilis* in melon phylloplane and contributes to the biocontrol activity, *Environ. Microbiol.* 16(7), 2196–2211. doi: 10.1111/1462-2920.12271.

Zhang, D., Yu, S., Yang, Y., Zhang, J., Zhao, D., Pan, Y., Fan, S., Yang, Z. and Zhu, J. (2020). Antifungal effects of volatiles produced by *Bacillus subtilis* against *Alternaria solani* in potato, *Front. Microbiol.* 11(1196), 1196. doi: 10.3389/fmicb.2020.01196.

Zhang, L., Feng, G. and Declerck, S. (2018). Signal beyond nutrient, fructose, exuded by an arbuscular mycorrhizal fungus triggers phytate mineralization by a phosphate solubilizing bacterium, *ISME J.* 12(10), 2339–2351. doi: 10.1038/s41396-018-0171-4.

Zhang, L. and Sun, C. (2018). Fengycins, cyclic lipopeptides from marine *Bacillus subtilis* strains, kill the plant-pathogenic fungus *Magnaporthe grisea* by inducing reactive oxygen species production and chromatin condensation, *Appl. Environ. Microbiol.* 84(18), 1–17. doi: 10.1128/AEM.00445-18.

Zhang, N., Wang, D., Liu, Y., Li, S., Shen, Q. and Zhang, R. (2013). Effects of different plant root exudates and their organic acid components on chemotaxis, biofilm formation and colonization by beneficial rhizosphere-associated bacterial strains, *Plant Soil* 374(1–2), 689–700. doi: 10.1007/s11104-013-1915-6.

Zhang, N., Yang, D., Wang, D., Miao, Y., Shao, J., Zhou, X., Xu, Z., Li, Q., Feng, H., Li, S., Shen, Q. and Zhang, R. (2015). Whole transcriptomic analysis of the plant-beneficial rhizobacterium *Bacillus amyloliquefaciens* SQR9 during enhanced biofilm formation regulated by maize root exudates, *BMC Genomics* 16(685), 685. doi: 10.1186/s12864-015-1825-5.

Zhang, S., White, T. L., Martinez, M. C., McInroy, J. A., Kloepper, J. W. and Klassen, W. (2010). Evaluation of plant growth-promoting rhizobacteria for control of Phytophthora blight on squash under greenhouse conditions, *Biol. Control* 53(1), 129–135. doi: 10.1016/j.biocontrol.2009.10.015.

Zhang, X., Zhou, Y., Li, Y., Fu, X. and Wang, Q. (2017a). Screening and characterization of endophytic Bacillus for biocontrol of grapevine downy mildew, *Crop Prot.* 96, 173–179. doi: 10.1016/j.cropro.2017.02.018.

Zhang, Y., Nakano, S., Choi, S. Y. and Zuber, P. (2006). Mutational analysis of the *Bacillus subtilis* RNA polymerase α C-terminal domain supports the interference model of Spx-dependent repression, *J. Bacteriol.* 188(12), 4300–4311. doi: 10.1128/JB.00220-06.

Zhang, Z., Ding, Z. T., Zhong, J., Zhou, J. Y., Shu, D., Luo, D., Yang, J. and Tan, H. (2017b). Improvement of iturin A production in *Bacillus subtilis* ZK0 by overexpression of the comA and sigA genes, *Lett. Appl. Microbiol.* 64(6), 452–458. doi: 10.1111/lam.12739.

Zhi, Y., Wu, Q. and Xu, Y. (2017). Genome and transcriptome analysis of surfactin biosynthesis in *Bacillus amyloliquefaciens* MT45, *Sci. Rep.* 7, 40976. doi: 10.1038/srep40976.

Zihalirwa Kulimushi, P., Argüelles Arias, A., Franzil, L., Steels, S. and Ongena, M. (2017). Stimulation of fengycin-type antifungal lipopeptides in *Bacillus amyloliquefaciens* in the presence of the maize fungal pathogen *Rhizomucor variabilis*, *Front. Microbiol.* 8(850), 850. doi: 10.3389/fmicb.2017.00850.

Zouari, I., Jlaiel, L., Tounsi, S. and Trigui, M. (2016). Biocontrol activity of the endophytic *Bacillus amyloliquefaciens* strain CEIZ-11 against *Pythium aphanidermatum* and purification of its bioactive compounds, *Biol. Control* 100, 54–62. doi: 10.1016/j.biocontrol.2016.05.012.

Chapter 9

The use of *Pseudomonas* spp. as bacterial biocontrol agents to control plant diseases

Monica Höfte, Ghent University, Belgium

1 Introduction

Bacteria have their origin in marine environments. They split into a group of land-adapted bacteria, the Terrabacteria, and a group that remained in water, the Hydrobacteria, about 3 billion years ago. The genus *Pseudomonas* belongs to the *Gammaproteobacteria*, a class of bacteria that emerged from

http://dx.doi.org/10.19103/AS.2021.0093.11

the Hydrobacteria 1.75 billion years ago (Battistuzzi and Hedges, 2009). The *Pseudomonas* genus diverged well before the colonization of land by plants. The evolutionary history of the *Pseudomonas* genus is founded on hundreds of millions of years spent mostly in aquatic habitats in the absence of higher plants (Morris et al., 2013). *Pseudomonas* is now one of the most ubiquitous genera in the world. *Pseudomonas* bacteria are inhabitants of sea, freshwater and soil-related environments. Many species live in association with plants and animals, mostly as saprophytes, but some are pathogenic for plants or animals. *Pseudomonas* bacteria have useful applications in biotechnology, plant growth promotion, bioremediation and biocontrol (Peix et al., 2009).

This chapter will focus on beneficial plant-associated *Pseudomonas* strains with the capacity to control plant diseases. There is an extensive literature on this topic with thousands of papers. Most *Pseudomonas* biocontrol agents have been isolated from soil, the rhizosphere of plants or from water. They are good root colonizers and well known for their capacity to control soilborne pathogens. Some strains can also protect against leaf pathogens by inducing systemic resistance in plants. They usually do not survive well on above-ground parts of plants, except for a few *Pseudomonas* biocontrol strains from the *P. syringae* group. This review will start with recent advances in *Pseudomonas* taxonomy and a summary of its most important biocontrol traits. An overview will then be given of the most important *Pseudomonas* groups and subgroups harboring biocontrol strains. Examples of well-characterized and representative biocontrol strains will show the links between the phylogeny, ecology and biocontrol traits. The chapter concludes by reviewing commercially available biocontrol strains.

2 *Pseudomonas* taxonomy

The genus *Pseudomonas* is diverse and complex and is currently divided into various phylogenetic groups with more than 220 described species. New species are being described constantly. A good overview of the historical evolution of *Pseudomonas* taxonomy and species described up to 2009 is given by Peix et al. (2009). The same authors have published an update of about 70 new species described from 2009 until 2018 (Peix et al., 2018). Species and phylogenetic groups have been delineated based on multi locus sequence analysis (MLSA) using the genes 16S rDNA, *gyrB, rpoB* and *rpoD* (Gomila et al., 2015; Lalucat et al., 2020; Mulet et al., 2010) and whole-genome sequences (Garrido-Sanz et al., 2016; Hesse et al., 2018). The 16S rRNA gene allows differentiation of the genus *Pseudomonas* but is not at the species level. Particularly useful for species discrimination are the *rpoD, gyrB* and *rpoB* genes. The *rpoD* gene is most discriminative for species delineation, followed by *gyrB*, and by the *rpoB* which is the least discriminative of these markers (Mulet et al., 2010). The taxonomy of *Pseudomonas* is constantly evolving and major breakthroughs have been made by the availability of *rpoD*, MLSA and whole-genome sequences of the species type strains (Girard et al., 2020b; Hesse et al., 2018; Lalucat et al., 2020).

The main phylogenetic groups defined by the 4-gene MLSA are to a large extent consistent with the groupings in genomics analyses. This means that new *Pseudomonas* isolates can easily be taxonomically positioned by sequencing these four genes (Garrido-Sanz et al., 2016) or just the *rpoD* gene (Girard et al., 2020b). Lalucat et al. (2020) have presented a complete MLSA-based phylogenetic tree using 216 *Pseudomonas* species. They distinguish three main lineages represented by the species:

- *P. aeruginosa*;
- *P. fluorescens*; and
- *P. pertucinogena*.

The *P. fluorescens* lineage comprises five phylogenetic groups:

- *P. fluorescens*;
- *P. asplenii*;
- *P. lutea*;
- *P. syringae*; and
- *P. putida*.

The *P. aeruginosa* lineage comprises eight phylogenetic groups:

- *P. straminea*;
- *P. anguilliseptica*;
- *P. oryzihabitans*;
- *P. stutzeri*;
- *P. oleovorans*;
- *P. resinovorans*;
- *P. aeruginosa*;
- *P. linyingensis*; and
- the genus *Azotobacter*.

The *P. fluorescens* group is further divided into eight subgroups:

- *P. fluorescens*;
- *P. gessardii*;
- *P. fragi*;
- *P. mandelii*;

- *P. koreensis*;
- *P. jessenii*;
- *P. corrugata*; and
- *P. chlororaphis*.

Hesse et al. (2018) constructed a phylogeny of 166 type strains within the *Pseudomonas* genus based on protein sequences of 100 single-copy orthologous genes and found 13 groups of *Pseudomonas*. In their protein-based phylogeny, the *P. fluorescens* group is composed of 10 subgroups since they distinguish the *P. protegens* subgroup from the *P. chlororaphis* subgroup and, in their phylogeny, *P. asplenii* is clustered within the *P. fluorescens* group. *P. chlororaphis* and *P. protegens* will be considered as two separate subgroups since their biocontrol traits are clearly different.

3 Plant-beneficial *Pseudomonas* strains

Plant-beneficial effects of *Pseudomonas* bacteria have been extensively studied and the most important findings are summarized in a number of comprehensive reviews. Many *Pseudomonas* strains can promote plant growth directly in the absence of pathogens by increasing the availability and uptake of mineral nutrients via phosphate solubilization, by enhancing root growth via the production or manipulation of phytohormones, or by enhancing tolerance to abiotic stress. Aspects of plant growth promotion have been reviewed elsewhere and will not be further discussed here (Berg, 2009; Dimkpa et al., 2009; Hayat et al., 2010; Lugtenberg and Kamilova, 2009; Ma et al., 2016; Ngumbi and Kloepper, 2016; Rajkumar et al., 2017).

The most important mechanisms of disease suppression by *Pseudomonas* biocontrol agents include:

- competition for nutrients or space (Kamilova et al., 2005);
- antibiosis (Haas and Défago, 2005; Raaijmakers et al., 2002; Raaijmakers and Mazzola, 2012); and
- induced systemic resistance (Bakker et al., 2007; De Vleesschauwer and Höfte, 2009; Pieterse et al., 2014).

A hallmark paper on *Pseudomonas* biocontrol that gives a good introduction to the subject has been published by Haas and Défago (2005). Weller (2007) has given an overview of 30 years of biocontrol research using *Pseudomonas* to control soilborne pathogens. Many good reviews have also appeared in books. A comprehensive overview of the older literature can be found in Thomashow and Weller (1996). More recent reviews have been published by Mercado-Blanco (2015) and Olorunleke et al. (2015b). Reviews about more specific aspects of *Pseudomonas* biocontrol will be cited later in the chapter.

The *P. fluorescens, P. putida, P. syringae* and *P. aeruginosa* groups harbor biocontrol strains. Within the *P. fluorescens* group, *P. chlororaphis, P. protegens* and *P. corrugata* subgroups, the so-called CPC cluster are especially rich in biocontrol agents and biocontrol properties (Vacheron et al., 2016). The *P. fluorescens, P. mandelii, P. jessenii* and *P. koreensis* subgroups, the so-called FMJK cluster, have fewer biocontrol properties but are enriched in phytostimulatory properties such as plant hormone modulation and plant nutrition (Vacheron et al., 2016).

It should be noted that biocontrol agents are not properly taxonomically determined in many published studies. Identification is based only on the 16S rDNA gene which does not have enough resolution to discriminate between subgroups or species. Many species that are described in the literature as *P. fluorescens* actually belong to other or undescribed species. Misidentification is a common problem even for publicly available genomes (Tran et al., 2017). In their analysis of 166 genomes of type strains and 1224 additional genomes of *Pseudomonas* spp. Hesse et al. (2018) identified 350 distinct clusters. They found that 189 clusters potentially represent novel species of *Pseudomonas*. Many of these clusters are singletons consisting of only one strain of *Pseudomonas*. These data show the tremendous diversity in *Pseudomonas* and we can expect the description of many new species in the coming years. Thanks to whole-genome sequencing and major advances made in *Pseudomonas* taxonomy, mainly in the *P. fluorescens* group, it is becoming increasingly clear that there is a strong link between phylogeny and biocontrol phenotypes (Garrido-Sanz et al., 2016). Researchers are urged to identify their *Pseudomonas* biocontrol strains properly by sequencing at least the *rpoD* gene to determine to which *Pseudomonas* group or subgroup they belong.

4 Secondary metabolite production in *Pseudomonas* biocontrol strains

Secondary metabolite production in *Pseudomonas* has been reviewed extensively. A comprehensive overview of the biosynthesis, chemistry and biological significance of *Pseudomonas* secondary metabolites is provided by Gross and Loper (2009). The literature about secondary metabolites involved in biocontrol is summarized in Haas and Défago (2005), Mishra and Arora (2018), Olorunleke et al. (2015b), and Raaijmakers and Mazzola (2012). It has become evident that only a limited number of bioactive compounds play a clear role in biocontrol of plant diseases:

- hydrogen cyanide (HCN);
- 2,4-diacetylphloroglucinol (DAPG);
- phenazines;

- pyrrolnitrin;
- pyoluteorin;
- 2-hexyl-5-propyl-alkylresorcinol;
- siderophores; and
- (cyclic) lipopeptides.

These compounds are usually produced by specific phylogenetic groups or subgroups within the *Pseudomonas* genus (Table 1) and are briefly discussed below.

4.1 Phenazines

Phenazines are redox-active nitrogen-containing tricyclic pigments that are produced by *Pseudomonas* and some other Gram-negative *Proteobacteria* (Mavrodi et al., 2006). More than 100 different phenazine structures have been described but biocontrol strains produce phenazine-1-carboxylic acid (PCA, lemon yellow), phenazine-1-carboxamide (PCN, green), 2-hydroxyphenazine (2-OH-PHZ, brick-red) or 2-hydroxyphenazine-1-carboxylic acid (2-OH-PCA, orange). Pyocyanin (PYO), a blue phenazine pigment, is only found in *P. aeruginosa* where it contributes to infection in mammalians (Price-Whelan et al., 2006).

The core phenazine biosynthesis genes are clustered and highly conserved. A seven-gene locus, named *phzABCDEFG* is responsible for the synthesis of PCA, which starts from chorismic acid, a product from the shikimate pathway. Phenazine diversity results from modification of PCA with dedicated enzymes such as PhzO (modifies PCA into 2-OH-PCA), PhzH (modifies PCA into PCN), PhzS (modifies PCA into 1-OH-PCA) and PhzM (modifies PCA into pyocyanin together with PhzS). In all phenazine-producing *Pseudomonas* strains, phenazine production is regulated by quorum sensing and also requires a functional GacS/GacA two-component signal transduction system. More details about phenazine biosynthesis can be found in Biessy and Filion (2018), Gross and Loper (2009) and Mavrodi et al. (2006).

Phenazines display broad-spectrum activity against fungal, oomycete and bacterial pathogens (Biessy and Filion, 2018), but can also trigger induced systemic resistance (ISR) in various plants (Ma et al., 2016b; De Vleesschauwer and Höfte, 2009). Phenazines have a physiological role in biofilm formation and iron reduction (Mavrodi et al., 2013). Phenazines are typically produced by *Pseudomonas* strains that belong to *P. aeruginosa*, or the *chlororaphis* and *fluorescens* subgroups within the *P. fluorescens* group (Table 1). Extensive information about microbial phenazines can be found in Chincholkar and Thomashow (2013).

Table 1 Examples of secondary metabolites produced by specific taxonomic groups and subgroups within the genus *Pseudomonas* (see text for references)

Taxonomic group/subgroup	Antibiotics	Biosurfactants	Siderophores
P. aeruginosa	HCN, phenazines (PYO, PCA, PCN), PLT	Rhamnolipids	pyoverdine, pyochelin
P. chlororaphis	HCN, phenazines (PCA, 2-OH-PCA, PCN), PRN, HPR	Viscosin group	pyoverdine, achromobactin
P. protegens	HCN, DAPG, PRN, PLT,	Orfamide group, sessilin	pyoverdine, enantio-pyochelin
P. corrugata	HCN, DAPG	Mycin/Peptin group, corrugatin	Pyoverdine, achromobactin, corrugatin
P. koreensis	HCN	Amphisin group, Bananamide group, cocoyamide/gacamide	pyoverdine
P. mandelii	HCN	Mycin/Peptin group	pyoverdine
P. gessardii			pyoverdine, (thio)quinolobactin
P. fluorescens	phenazines (PCA)	Viscosin group, Orfamide group	pyoverdine, pseudomonin
P. putida	HCN	WLIP, xantholysin, entolysin, putisolvin	pyoverdine
P. syringae		Mycin/Peptin group	pyoverdine, achromobactin

Abbreviations: Antibiotics: DAPG: 2,4-diacetylphloroglucinol, HCN: hydrogen cyanide, HPR: 2-hexyl-5-propyl-alkylresorcinol, PCA: phenazines-1-carboxylate, PCN: phenazines-1-carboxamide, 2-OH-PCA: 2-hydroxyphenazine-1-carboxylic acid, PLT: pyoluteorin, PRN: pyrrolnitrin, PYO: pyocyanine, Lipopeptides: Mycin: Syringomycin group lipopeptides, Peptin: Corpeptin, Sclerosin and Syringopeptin group lipopeptides.

4.2 2,4-diacetylphloroglucinol (DAPG)

DAPG is a polyketide antibiotic that is produced by *Pseudomonas* strains that predominantly belong to the *P. protegens* and *P. corrugata* subgroups (Table 1) and a few isolated strains in other taxonomic groups (Almario et al., 2017). It plays a key role in the biocontrol of root and seedling diseases and is active against fungi, oomycetes, nematodes and bacteria. The compound is phytotoxic at high concentrations. DAPG- producing *Pseudomonas* strains have typically been isolated from soils suppressive to take-all, an important wheat root disease caused by *Gaeumannomyces graminis* var. *tritici* (Weller et al., 2007). The DAPG gene cluster is highly conserved and comprises nine genes involved in biosynthesis (*phlACB* and *phlD*), efflux (*phlEI*), degradation (*phlG*) and regulation (*phlH* and *phlF*). The biosynthesis starts from three molecules of malonyl CoA. PhlD is a type III polyketide synthase responsible for the biosynthesis of phloroglucinol. Besides its role in direct antibiosis, the compound can induce systemic resistance in *Arabidopsis* to the downy mildew pathogen *Hyaloperonospora parasitica* (Iavicoli et al., 2003), *Pseudomonas syringae* pv. *tomato* (Weller et al., 2012) and *Botrytis cinerea* (Chae et al., 2020) by triggering jasmonate- and ethylene-mediated defense responses. DAPG can promote amino acid exudation from plant roots by blocking amino acid influx (Phillips et al., 2004) and stimulates root branching in tomato by interacting with the auxin-dependent signaling pathway (Brazelton et al., 2008).

4.3 Pyoluteorin

Pyoluteorin is a hybrid non-ribosomal peptide synthetase/polyketide synthase product consisting of a resorcinol ring attached to a dichlorinated pyrrole moiety. Pyoluteorin is produced by *Pseudomonas* strains belonging to the *P. protegens* subgroup (Ramette et al., 2011) and a few *P. aeruginosa* isolates (Hu et al., 2005). The antibiotic has not been reported in genera other than *Pseudomonas*. Pyoluteorin suppresses seedling damping-off diseases caused by the oomycete pathogen *Pythium ultimum* in cotton and cress (Howell and Stipanovic, 1980; Maurhofer et al., 1994b).

The pyoluteorin biosynthetic gene cluster was first described in *P. protegens* Pf-5. It encompasses 17 genes and contains structural genes (*pltABCDEFGLM*) and genes involved in efflux (*pltIJKNOP*) and regulation (*pltZ* and *pltR*) (Gross and Loper, 2009). The cluster is only present in *Pseudomonas* strains that also have *phlD*, either as part of the DAPG operon, as is the case in *P. protegens* or associated with the pyoluteorin gene cluster in *P. aeruginosa* strains. Nanomolar concentrations of phloroglucinol, the precursor of DAPG that is synthesized by PhlD are required for pyoluteorin production, while micromolar concentrations of phloroglucinol inhibit pyoluteorin production

(Kidarsa et al., 2011). Phloroglucinol is transformed by a halogenase encoded in the pyoluteorin gene cluster by *pltM* into chlorinated derivatives that induce expression of pyoluteorin biosynthetic genes via PltR (Yan et al., 2017).

4.4 Pyrrolnitrin

Pyrrolnitrin is a chlorinated phenylpyrrol derived from tryptophan with a broad-spectrum activity against Ascomycete and Basidiomycete fungi (Ligon et al., 2000). Pyrrolnitrin is produced by various bacterial genera, including *Pseudomonas, Burkholderia, Serratia* and *Myxococcus*. The four-gene pyrrolnitrin biosynthetic gene cluster *prnABCD* was first described in *P. chlororaphis* subsp. *aurantiaca* BL915 (previously identified as *P. fluorescens* and *P. aurantiaca*) (Hill et al., 1994; Kirner et al., 1998) and is highly conserved in pyrrolnitrin-producing strains. The cluster is found in *Pseudomonas* biocontrol strains that belong to the *P. protegens* and *P. chlororaphis* subgroup. *Pseudomonas* strains may have obtained this cluster by horizontal gene transfer (Costa et al., 2009). The compound has been used as a lead structure in the development of a new class of agricultural fungicides, the phenylpyrroles (Nyfeler and Ackermann, 1992). The phenylpyrrole fungicides, Fenpiclonil and Fludioxonil, are registered against multiple fungal crop diseases for seed or foliar treatment (Kilani and Fillinger, 2016).

4.5 2-hexyl-5-propyl-alkylresorcinol (HPR)

2-hexyl-5-propyl-alkylresorcinol (HPR) is structurally related to DAPG but the biosynthesis pathway is completely different. The compound is derived from octanoic acid. The biosynthetic gene cluster for HPR was first elucidated in *P. chlororaphis* subsp. *aurantiaca* BL915 and is composed of three biosynthetic genes *darABC* and two regulatory genes *darS* and *darR* (Nowak-Thompson et al., 2003). The biosynthetic gene cluster is present in various biocontrol strains belonging to the *P. chlororaphis* subgroup (Biessy et al., 2019; Calderón et al., 2013). HPR has antifungal and antibacterial activity.

4.6 Hydrogen cyanide (HCN)

Hydrogen cyanide (HCN) is a respiratory poison that inhibits cytochrome c oxidase, the terminal component of the respiratory chain, in many organisms. In prokaryotes, HCN production seems to be restricted to the *Proteobacteria* and to certain cyanobacteria (Blumer and Haas, 2000a). Within the genus *Pseudomonas* known HCN producers can be found in the *P. aeruginosa* group and in various subgroups of the *P. fluorescens* group. Virtually all DAPG producers in the *P. corrugata* and *P. protegens* subgroup also produce HCN and this trait is also common in the *P. chlororaphis* subgroup (Table 1).

A cluster of three genes, *hcnABC*, is responsible for HCN production from the metabolic precursor glycine (Laville et al., 1998). The *hcnABC* cluster appears to be ancestral in *Pseudomonas* strains associated with roots (Frapolli et al., 2012). Cyanogenesis in *Pseudomonas* is regulated by the GacA/GacS signal transduction system (Sonnleitner et al., 2009) and by the anaerobic regulator ANR and is stimulated by iron (Blumer and Haas, 2000b). HCN production by *P. protegens* CHA0 accounts for the suppression of tobacco black root rot caused by *Thielaviopsis basicola* and iron sufficiency is important for both HCN production and disease suppression (Voisard et al., 1989). HCN production by *Pseudomonas* has also been implicated in the suppression of root-knot nematodes (Siddiqui et al., 2006), aphids (Kang et al., 2019), termites (Devi and Kothamasi, 2009) and other insects (Flury et al., 2017). HCN, together with DAPG is also responsible for the biocontrol activity of *P. brassicacearum* LBUM300 against bacterial canker of tomato caused by *Clavibacter michiganensis* subsp. *michiganensis* (Lanteigne et al., 2012; Paulin et al., 2017).

4.7 Biosurfactants

Pseudomonas strains produce two types of biosurfactants:

- rhamnolipids; and
- nonribosomal lipopeptides.

Rhamnolipids are glycolipids typically produced by *P. aeruginosa* and their role in biocontrol has been summarized in D'aes et al. (2010), Olorunleke et al. (2015b) and Crouzet et al. (2020). Lipopeptides are produced by strains in various *Pseudomonas* groups and subgroups (Table 2). They are comprised of an oligopeptide composed of 8–25 amino acids that are N-terminally acylated with a fatty acid. They are produced by nonribosomal peptide synthetases. The vast majority of *Pseudomonas* lipopeptides is not linear but displays a macrolactone ring that contains 4–9 amino acids.

Pseudomonas CLPs have been classified in at least 15 different groups (Table 2) and their biosynthesis, structures and biological functions have been reviewed recently (Geudens and Martins, 2018; Götze and Stallforth, 2020). A complete overview of all described *Pseudomonas* lipopeptides to date can be found in Götze and Stallforth (2020). Many new structures have been unraveled in recent years and their identification is greatly facilitated by genome mining (Paterson et al., 2017). They play a role in swarming motility, biofilm formation, environmental adaptation, nutrient availability, root colonization and biocontrol (D'aes et al., 2010; Olorunleke et al., 2015b; Raaijmakers et al., 2010).

Table 2 Lipopeptides produced by *Pseudomonas* biocontrol strains

Group	AA[1]	Phylogenetic group/subgroup	CLPs involved in biocontrol	Reference
Bananamides	8:6	*P. koreensis* SG	bananamide D-F	Omoboye et al., 2019a
Corrugatin	8	*P. corrugata SG*		
Viscosin	9:7	*P. fluorescens* SG, *P. chlororaphis SG, P. putida* G	viscosin, viscosinamide, WLIP, massetolide, pseudodesmin	Omoboye et al., 2019b; Oni et al., 2020a; Thrane et al., 2000; Tran et al., 2007
Syringomycin	9:9	*P. syringae* G, *P. corrugata* SG, P. *mandelii* SG	syringomycin, cormycin, thanamycin, nunamycin	Bull et al., 1998; Michelsen et al., 2015a; Van Der Voort et al., 2015
Orfamide	10:8	*P. protegens* SG, *P. fluorescens* SG	orfamide, poaeamide	Ma et al., 2016a; Zachow et al., 2015
Amphisin	11:9	*P. koreensis* SG	lokisin, rhizoamide	Omoboye et al., 2019b, Hultberg et al., 2010a
Cocoyamide	11:5	*P. koreensis* SG	cocoyamide/gacamide	Jahanshah et al., 2019; Oni et al., 2019a
Putisolvin	12:4	*P. putida* G	putisolvin I, II, III, IV	Kuiper et al., 2003
Entolysin	14:5	*P. putida* G	entolysin	Omoboye et al., 2019b; Oni et al., 2019a; Vallet-Gely et al., 2010a
Xantholysin	14:8	*P. putida* G	xantholysin	Omoboye et al., 2019b; Oni et al., 2019a
Tolaasin	18:5	*P. protegens* SG	sessilin	D'aes et al., 2014; Olorunleke et al., 2015a
Fuscopeptin	19:5	*P. asplenii* SG		
Jessenipeptin	19:5	*P. asplenii* SG		
Corpeptin	22:5	*P. corrugata* SG, *P. mandelii* SG	corpeptin, thanapeptin, nunapeptin	Michelsen et al., 2015a; Van Der Voort et al., 2015
Sclerosin	22	*P. corrugata* SG	sclerosin	Berry et al., 2012a
Syringopeptin (SP22)	22:8	*P. syringae* G		
Syringopeptin (SP25)	25:8	*P. syringae* G		

[1]AA: amino acid, the first figure indicates the number of AA in the peptide chain, the second figure the number of AA in the macrolactone ring.

4.8 Strain-specific bio-active molecules

In addition, various strain-specific bio-active metabolites are only produced by a limited number of *Pseudomonas* bacteria. Gene clusters carrying the biosynthetic genes for these metabolites have most likely been obtained by horizontal gene transfer. Examples are:

- rhizoxins;
- promysalin;
- sessilin; and
- L-furanomycin.

Rhizoxins, produced by a few strains from the *P. protegens* subgroup (Loper et al., 2008; Takeuchi et al., 2015) and *P. chlororaphis* MA342 (Ligon et al., 2000) have activity against fungi and oomycetes. Sessilin is an antifungal cyclic lipopeptide (CLP) that appears to be unique to the biocontrol strain *Pseudomonas* sp. CMR12a (D'aes et al., 2014). Promysalin is a salicylic acid-containing antibiotic produced by *P. putida* RW10S1 (Li et al., 2011) with activity against Gram-positive and Gram-negative bacteria (Kaduskar et al., 2017). L-furanomycin is a non-proteinogenic amino acid found in *P. fluorescens* SBW25 that selectively inhibits Gram-negative plant pathogenic bacteria (Trippe et al., 2013).

4.9 Siderophores

Siderophores are implicated in biocontrol activity by competition for iron, direct antagonism or induced systemic resistance. Most *Pseudomonas* strains produce the fluorescent compound pyoverdine (Meyer, 2000) and about 100 distinct pyoverdines have been identified which are structurally divided into four groups. In addition, many *Pseudomonas* strains produce other siderophores in addition to pyoverdine with a lower affinity for ferric iron than pyoverdine. Their chemical structures are diverse and include pyochelin and enantio-pyochelin, achromobactin, pseudomonine, corrugatin and (thio)quinolobactin (Cornelis, 2010; Schalk et al., 2020).

4.10 Insect toxins

Pseudomonas biocontrol strains that belong to the *P. protegens* and *P. chlororaphis* subgroup interact with insects and produce the insecticidal Fit toxin. This is a large protein toxin related to the insect toxin Mcf of *Photorhabdus* and *Xenorhabdus* spp., bacterial symbionts that live in the intestines of entomophagous nematodes. Fit-toxin producers have oral and injectable insecticidal activity against Lepidoptera larvae. The fit locus comprises eight genes, *fitABCDEFG* and the Fit toxin is the product of *fitD*. The other gene products are involved in transport and regulation (Flury et al., 2017, 2016; Péchy-Tarr et al., 2008; Ruffner et al., 2015, 2012).

4.11 Regulation of secondary metabolites

Regulation of secondary metabolites in *Pseudomonas* is complex and will not be discussed in detail in this review. Excellent overviews of antibiotic regulation are provided by Haas and Keel (2003) and Sonnleitner and Haas (2011). Most antibiotics with a role in biocontrol are produced only when the GacS/GacA two-component system is active. This regulatory system is widely conserved in Gram-negative bacteria and regulates biocontrol traits or pathogenicity factors post-transcriptionally. Mutations in *gacS* or *gacA* abolish biocontrol activity in the majority of biocontrol *Pseudomonas* strains. A good introduction can be found in Haas and Défago (2005). On interaction with bacterial signal molecules, which are presumably intermediates of the Krebs cycle (Takeuchi et al., 2009), GacS is autophosphorylated and a phosphate residue is transferred to GacA. This activates the transcription of small RNA genes. These RNAs bind to small proteins that function as post-transcriptional repressors of biocontrol genes. This relieves the translational repression exerted by these proteins at or near the ribosome-binding sites of the target mRNA (Sonnleitner and Haas, 2011).

Spontaneous mutants in *gacA* or *gacS* contribute to the genetic instability of some *Pseudomonas* strains, which is obviously a problem in biocontrol. Particular strains that produce a large number of secondary metabolites accumulate a high proportion of Gac$^-$ mutants in culture (Yan et al., 2018). Colony variants that exhibit a Gac$^-$ phenotype are easy to recognize because they have an increased colony size, are more flat and more fluorescent due to overproduction of pyoverdine (Duffy and Defago, 2000).

5 Secretion systems that play a role in biocontrol

5.1 Type III secretion systems

Many Gram-negative bacteria can interact with diverse eukaryotic hosts via a type III secretion system. This is a protein secretion system that originates from the bacterial flagellar system and spans the two membranes in Gram-negative bacteria. Through this system so-called effector proteins can be secreted directly in the cytoplasm of the host. The type III secretion system plays a role in both parasitic and mutualistic interactions and is found in various *Pseudomonas* biocontrol strains.

T3SS genes are present in *Pseudomonas* biocontrol strains belonging to the *P. fluorescens* and *P. corrugata* subgroups and are enriched in the rhizosphere as compared to corresponding bulk soil. T3SS is involved in interactions with phytopathogens, mycorrhizal fungi and predators (Nazir et al., 2017). Most

Pseudomonas bacteria contain a T3SS that belongs to the Hrp1 family, but some strains also have an additional T3SS that belongs to the SPI-1 family. SPI-1 type T3SS was first identified in *P. fluorescens* F113, a well-studied biocontrol strain of the *P. corrugata* subgroup and enhances resistance to amoeboid grazing (Barret et al., 2013). The Hrp1-type T3SS of *P. fluorescens* strain KD targets the oomycete pathogen *Pythium ultimum,* promoting cucumber protection (Rezzonico et al., 2005). The role of the T3SS in other beneficial *Pseudomonas* bacteria remains elusive, but it was suggested that effectors delivered via this system may suppress host immunity to facilitate root colonization (Yu et al., 2019).

5.2 Type VI secretion systems

Another secretion system that is commonly found in commensal, symbiotic and pathogenic plant-associated bacteria is the type VI secretion system. This secretion system is used to inject toxic effectors in eukaryotic and prokaryotic cells and is involved in interbacterial competition. T6SS clusters are widely distributed in the genus *Pseudomonas* and especially enriched in the *P. putida* group (Bernal et al., 2018). The biocontrol strain *P. putida* KT2440 contains three T6SSs and various T6SS effectors and the systems are used to outcompete various phytopathogenic bacteria in vitro and in planta (Bernal et al., 2017).

6 *Pseudomonas* biocontrol strains: *Pseudomonas protegens* subgroup

Pseudomonas protegens was formally described by Ramette et al. (2011) and comprises isolates that produce the secondary metabolites DAPG and pyoluteorin. Most *P. protegens* strains also produce HCN, pyrrolnitrin, the CLP orfamide and two siderophores: pyoverdine and enantio-pyochelin. The species has insecticidal activity (Flury et al., 2016) and some isolates contain additional gene clusters encoding rhizoxin analogues (Loper et al., 2008) and toxoflavin (Philmus et al., 2015). Isolates belonging to this species have been obtained from the rhizosphere of tobacco, cotton, wheat (Ramette et al., 2011), canola (Zhang et al., 2020b), tomato, cucumber and shepherd's purse (Takeuchi et al., 2014). They have also been found on the phyllosphere of lemon (Michavila et al., 2017) and wheat (Levy et al., 1992), in stored rice grains (Jeong et al., 2018), on lamb's lettuce in hydroponics (Moruzzi et al., 2017), in recycled irrigation water (Yang and Hong, 2020) and on brown macroalga in the Baltic sea (Heiman et al., 2020). Recently it was shown that *P. protegens* can also survive in insects (Flury et al., 2019).

Much of our knowledge about *P. protegens* comes from two of the best-studied biocontrol strains: *P. protegens* Pf-5 and *P. protegens* CHA0. Both strains have become models and many biocontrol traits in *Pseudomonas* were first discovered in these strains. Also our knowledge about the complex regulation of secondary metabolites in *Pseudomonas* biocontrol strains was largely obtained from these two strains (Haas and Défago, 2005; Haas and Keel, 2003).

P. protegens Pf-5 was obtained from cotton roots in Texas and suppresses *Rhizoctonia solani* and *Pythium ultimum* on cotton (Howell and Stipanovic, 1980, 1979). It also controls a variety of other diseases such as fungal diseases on wheat (Pfender et al., 1993) and turfgrass (Rodriguez and Pfender, 1997), Fusarium crown and root rot of tomato (Sharifi-Tehrani et al., 1998) and seed piece decay of potato (Xu and Gross, 1986) (see Table 3). *P. protegens* Pf-5 was the first biocontrol strain of which the whole genome was sequenced (Paulsen et al., 2005). Pf-5 produces the typical *P. protegens* antibiotics HCN, DAPG, pyrrolnitrin and pyoluteorin (Loper et al., 2007). The CLP orfamide was first discovered in *P. protegens* Pf-5 by a combination of genome sequence analysis and isotope-guided fractionation (Gross et al., 2007). Further genome mining revealed the capacity to produce two additional antibiotics: rhizoxin (Loper et al., 2008) and toxoflavin (Philmus et al., 2015).

Rhizoxin is a polyketide macrolide with antifungal, phytotoxic and antitumor properties and binds to ß-tubulin. Rhizoxin-producing endosymbionts have been found in the rice-pathogenic fungus *Rhizopus microspores*. Rhizoxin is the causative agent of rice seedling blight and inhibits mitosis in seedling roots. Rhizoxin biosynthesis genes were first described from the *Rhizopus* endosymbiont *Burkholderia rhizoxina* (Partida-Martinez and Hertweck, 2007) and they are orthologous to the biosynthesis genes found in *P. protegens* Pf-5. The main rhizoxin produced by Pf-5 inhibits the growth of the plant pathogenic fungus *Fusarium oxysporum* (Brendel et al., 2007).

Toxoflavin has antibiotic activity against bacteria and fungi and is phytotoxic against a broad range of monocot and dicot plants. The compound is produced by the plant pathogenic bacteria *Burkholderia glumae* and *B. gladioli* and is a key virulence factor. The toxoflavin gene cluster in Pf-5 shows substantial differences with the *tox* genes from *Burkholderia*. The Pf-5 strain produces trace amounts of toxoflavin with antibiotic activity against various plant pathogenic bacteria (Philmus et al., 2015).

P. protegens CHA0 was isolated from the roots of 7-week-old tobacco plants grown in soil from Morens, Switzerland, with natural suppressiveness to black root rot caused by *Thielaviopsis basicola* (Stutz et al., 1986). Since then, it was shown that strain CHA0 can protect many crop plants including wheat, cucumber, sugar beet, cotton, flax, corn and cress from oomycete and fungal pathogens (Voisard et al., 1994) (Table 3). In addition, *P. protegens* CHA0 can induce systemic resistance to *Hyaloperonospora* in *Arabidopsis* (Iavicoli et al., 2003), root-knot nematodes in tomato (Siddiqui and Shaukat, 2003), *Tobacco Necrosis Virus* in tobacco (Maurhofer et al., 1994a) and *Banana Bunchy Top Virus* in banana (Kavino et al., 2008). Biocontrol traits of *P. protegens* CHA0

Table 3 Selected *P. protegens* group biocontrol strains

Strain	Origin	Activity	Metabolites	Genome sequence	References
P. protegens					
CHA0	black root rot suppressive soil, Switzerland	Tb, TNV, tobacco; Pu, cucumber, wheat and cress; Gg, wheat; Mj and Mp, tomato; BBT, banana; Ha, Arabidopis, insecticidal activity	**DAPG, FIT, HCN, PLT, PRN PCH, PVD, ORF**	Jousset et al., 2014; Smits et al., 2019	Flury et al., 2017, 2016; Kavino et al., 2008; Keel et al., 1992; Maurhofer et al., 1994a; Siddiqui et al., 2006;Stutz et al., 1986; Voisard et al., 1994
Pf-5	cotton rhizosphere, Texas, USA	Rs, cotton; Pu, cotton; Pt, wheat straw; Sh and Dp, turfgrass; Fo, tomato; Pc, potato; insecticidal activity	**DAPG, FIT, HCN, PLT, PRN, PCH, PVD, ORF, RZX, TXF**	Loper et al., 2007; Paulsen et al., 2005	Howell and Stipanovic, 1980, 1979; Pfender et al., 1993; Rangel et al., 2016; Rodriguez and Pfender, 1997; Xu and Gross, 1986
Pf4	roots from hydroponic lamb's lettuce plants, Italy	Rs, lamb's lettuce,	FIT, HCN, ORF, PLT, PVD, PRN, RZX	Polano et al., 2019	Moruzzi et al., 2017; Polano et al., 2019
Cab57	rhizosphere of shepherd's purse, Japan	Pu, cucumber	DAPG, FIT, HCN, PLT, PRN, ORF, PCH, PVD, FIT	Takeuchi et al., 2014	Takeuchi and Someya, 2019
PF	Wheat leaves, Oklahoma, USA	Zt, wheat, insecticidal activity	**DAPG**, HCN, PRN, PLT, RZX, **FIT**	Flury et al., 2016	Levy et al., 1992
Related to *P. protegens*					
CMR12a	rhizosphere of red cocoyam, Cameroon	Pm, cocoyam; Rs, bean and cabbage, Po, rice; Bo, rice; insecticidal activity	**PCA, PCN, SES, ORF, FIT**, HCN	Biessy et al., 2019	D'aes et al., 2014, 2011; Flury et al., 2016; Ma et al., 2017; Z. Ma et al., 2016b; Oni et al., 2019b; Perneel et al., 2007

CMR5c	rhizosphere of red cocoyam, Cameroon	Pm, cocoyam, insecticidal activity	DAPG, **FIT,** HCN, ORF, PCA, PCH, PCN, PLT, PRN,	Biessy et al., 2019	Biessy et al., 2019; Flury et al., 2016; Perneel et al., 2007
OS17	rhizosphere of rice, Japan	Pu, cucumber	DAPG, FIT, HCN, PCH, **RZX**	Takeuchi et al., 2015	Takeuchi et al., 2015

Abbreviations: Pathogens: *BBT: Banana Bunchy Top Virus, Bo: Bipolaris oryzae, Dp: Drechslera poae: Fo: Fusarium oxysporum, Gg: Gaeumannomycis graminis* var. *tritici, Ma: Macrophomina phaseolina, Mj: Meloidogyne javanica, Pm: Pythium myriotylum, Po: Pyricularia oryzae, Pu: Pythium ultimum, Pst: Pseudomonas syringae pv. tomato*; Pt: *Pyrenophora tritici-repentis, Rs: Rhizoctonia solani, Sh: Sclerotinia homeocarpa, Tb: Thielaviopsis baticola*, TNV: *Tobacco Necrosis Virus, Zt: Zymoseptoria tritici*. Metabolites: DAPG: 2,4- diacetylphloroglucinol, HCN: hydrogen cyanide, FIT: insecticidal toxin, ORF: orfamide, PCA: phenazines-1-carboxylate, PCH: pyochelin, PCN: phenazines-1-carboxamide; PLT: pyoluteorin, PRN: pyrrolnitrin, RZX: rhizoxin, TXF: toxoflavin. For metabolites in bold a role in biocontrol has been shown.

have been investigated in great detail. The strain produces the antibiotics HCN, DAPG, pyrrolnitrin, pyoluteorin (Haas and Défago, 2005), orfamide-type CLPs (Ma et al., 2016a), the siderophores pyoverdine and enantio-pyochelin (Youard et al., 2007), and various traits involved in insect toxicity (Flury et al., 2017, 2016). Mutant analysis has revealed that suppression of the different plant diseases involves different mechanisms. HCN and DAPG are involved in black root rot control, DAPG, but not HCN is needed for take-all control, while pyoluteorin is responsible for *Pythium* control (Voisard et al., 1994). Determinants that trigger ISR include DAPG and pyoverdine (De Vleesschauwer and Höfte, 2009).

In nutrient-rich cultures both *P. protegens* Pf-5 and *P. protegens* CHA0 readily accumulate Gac⁻ mutations that completely abolish their biocontrol capacities (Bull et al., 2001; Duffy and Defago, 2000; Yan et al., 2018). Cultures of mutants of *P. protegens* Pf-5 unable to produce pyoluteorin or orfamide accumulated a smaller number of Gac⁻ mutants. Pyoluteorin biosynthesis, but not pyoluteorin itself, contributed significantly to accumulation of Gac⁻ mutants. However, few of these mutants accumulated in the presence of the competitor *Bacillus subtilis* (Yan et al., 2018).

Other *P. protegens* biocontrol strains have been obtained from North America, Europe and Japan and examples are given in Table 3. *P. protegens* CAB 57 originates from the rhizosphere of shepherd's purse in Japan and is active against *P. ultimum* in cucumber (Takeuchi et al., 2014). At least in Japan *P. protegens* strains do not seem to be very common. In a screen of 2800 *Pseudomonas* isolates from the rhizosphere of various plants in Japan, only 5 isolates belonged to the *P. protegens* subgroup (Takeuchi and Someya, 2019). *P. protegens* bacteria were not prevalent either in a collection of 698 *Pseudomonas* isolates from the maize rhizosphere in France and Switzerland (Vacheron et al., 2016). Their high level of effectiveness in short-term greenhouse trials may be counterbalanced by lower rhizosphere competitiveness and survival under field conditions. This suggests a trade-off between rhizosphere prevalence and the ability to produce a large number of plant-beneficial properties (Vacheron et al., 2016).

The *Pseudomonas* strain Os17 isolated from the rhizosphere of rice belongs to the *P. protegens* subgroup but is clearly different from the species *P. protegens*. This strain contains the biosynthetic gene clusters for HCN, DAPG, rhizoxin, pyoverdine and enantio-pyochelin, but gene clusters for pyrrolnitrin, pyoluteorin and orfamide are absent. The strain suppresses damping-off and root rot caused by *P. ultimum* on cucumber. Mutant analysis revealed that rhizoxin is important in *Pythium* biocontrol (Takeuchi et al., 2015).

Pseudomonas sp. CMR12a and CMR5c have been obtained from the rhizosphere of red cocoyam in Cameroon and have antagonistic activity against *Pythium myriotylum*, the causal agent of the destructive cocoyam root rot disease (Perneel et al., 2007). They belong to the *P. protegens* group, but are clearly different from the core *P. protegens* strains (Biessy et al., 2019; Flury et al., 2016). In terms of biocontrol traits they have characteristics of both *P. protegens* and *P. chlororaphis*. Both strains produce the phenazines PCA and PCN, metabolites that are typically produced by *P. chlororaphis* group strains. CMR5c also has the typical *P. protegens* biosynthetic gene clusters for DAPG, PLT and PRN (Biessy et al., 2019; Flury et al., 2016; Perneel et al., 2007).

Both strains produce orfamide-type CLPs (Ma et al., 2016a). In addition, CMR12a produces an unusual CLP, sessilin, which is closely related to tolaasin, a CLP produced by the mushroom pathogen *P. tolaasii* (D'aes et al., 2014). Analysis of their whole genome has revealed that the phenazine biosynthetic gene cluster in both strains is located on a genomic island. In case of CMR12a, this genomic island also contains the sessilin biosynthetic gene cluster which is absent in CMR5c.

Recently, various isolates closely related to CMR12a were found on the root of tissue-cultured derived cocoyam plantlets that had been grown for 2 or 4 weeks in the cocoyam root rot suppressive soil from Boteva, Cameroon. Some of these isolates did not carry the genomic island and did not produce phenazines or sessilin. Intriguingly, these isolates hardly colonized the roots of 8-week-old cocoyam plantlets, indicating that these strains are associated with young cocoyam plantlets only (Omoboye, 2019).

Biocontrol activity of *Pseudomonas* sp. CMR12a has been studied in detail. The strain controls damping-off disease on Chinese cabbage caused by *R. solani* AG 2-1 and root rot disease on bean caused by *R. solani* AG 4-HI. In cabbage, phenazines alone were sufficient to suppress *Rhizoctonia* damping-off, whereas co-production of sessilins and orfamides was required in the absence of phenazines. To suppress *R. solani* on bean, co-production of phenazines, sessilins and orfamides appeared to be important (Olorunleke et al., 2015a). All three metabolites are involved in the suppression of *P. myriotylum* on cocoyam (Oni et al., 2019b). The strain induces resistance to fungal pathogens in rice and bean. Phenazines are solely involved in ISR against the rice blast fungus *P. oryzae*, while phenazines, orfamide and sessilin can trigger ISR in bean against web blight disease caused by *R. solani* AG 2-2 (Ma et al., 2016b). Orfamide at high concentrations is responsible for ISR in rice against the brown spot pathogen *Bipolaris oryzae* (Teleomorph: *Cochliobolus miyabeanus*), but is not effective against the rice blast pathogen *Pyricularia oryzae* (Ma et al., 2017).

The insecticidal activity of *P. protegens* subgroup strains, including *P. protegens* Pf-5, *P. protegens* CHA0, *Pseudomonas* sp. CMR12a and *Pseudomonas* sp. CMR5c, has intensively been studied but will not be further discussed here. Virulence factors that contribute to insect pathogenicity include the Fit toxin, HCN, cyclic lipopeptides, rhizoxins and secreted lytic enzymes (Flury et al., 2017, 2016; Rangel et al., 2016). In addition the T6SS contributes to insect pathogenicity of *P. protegens* in oral infection assays (Vacheron et al., 2019).

7 *Pseudomonas* biocontrol strains: *Pseudomonas chlororaphis* subgroup

P. chlororaphis isolates have been obtained from soil and the rhizosphere of crops such as tomato, potato, maize, radish, beet, alfalfa, soja and clover (Biessy et al., 2019); the stem of sugarcane (Mehnaz et al., 2009); and roots of avocado (Cazorla et al., 2006). The species is divided in four subspecies: *aureofaciens*, *aurantiaca*, *chlororaphis* (Peix et al., 2007) and *piscium*. *P. chlororaphis* subsp. *piscium* was proposed to accommodate two isolates from the intestines of perch and rainbow trout (Burr et al., 2010), but various rhizosphere isolates also cluster in this subspecies (Biessy et al., 2019). Secondary metabolite production in the *P. chlororaphis* subgroup has been studied by Biessy et al. (2019).

All *P. chlororaphis* isolates produce PCA and HCN, but the subspecies differ in the type of phenazines produced and in the production of other secondary metabolites. The subspecies *aureofaciens* and *aurantiaca* produce 2-OH-PCA, the *piscium* subspecies produces PCN, while isolates belonging to the *chlororaphis* subspecies produce either PCN or 2-OH-PCA. All subspecies, except subsp. *piscium* also produce pyrrolnitrin and HPR. NRPS clusters for the production of lipopeptides from the Viscosin group are commonly found in the subspecies *piscium*, *aurantiaca* and *chlororaphis*, but not in the subspecies *aureofaciens*. Many *P. chlororaphis* strains produce two siderophores, pyoverdine and achromobactin. More details can be found in Biessy et al. (2019).

Biocontrol activities of *P. chlororaphis* isolates have been studied extensively and various commercially available biopesticides belong to this species (see Table 6). The biocontrol capacities of *P. chlororaphis* have been reviewed by Anderson and Kim (2020, 2018). *P. chlororaphis* isolates directly control fungi, nematodes, oomycetes and aphids and can induce resistance to fungal and bacterial leaf pathogens. Metabolites involved in biological control activity include phenazines, HCN, pyrrolnitrin, HPR and 2,3-butanediol (Table 4). Well-studied strains include *P. chlororaphis* subsp. *aureofaciens* O6, *P. chlororaphis* subsp. *piscium* PCL1391, *P. chlororaphis* 30-84 and *P. chlororaphis* PCL1606. These and other strains are listed in Table 4.

The role of phenazines in biocontrol has been studied in various *P. chlororaphis* strains. PCA is involved in the suppression of take-all disease of wheat as was shown for *P. chlororaphis* 30-84 (Pierson and Thomashow, 1992). PCN is clearly more effective than PCA in the biocontrol activity against *F. oxysporum* f. sp. *radicis-lycopersici* in tomato. *P. chlororaphis* PCL1391 was active in this pathosystem in contrast to PCA-producing strains such as *P. chlororaphis* 30-84 (Chin-A-Woeng et al., 1998). This has been explained by the fact that PCN is still

Table 4 Selected *P. chlororaphis* group biocontrol strains

Strain	Origin	Activity	Biocontrol metabolites	Genome sequence	References
O6 (subsp. *aurofaciens*)	soil, Utah, USA	Pc and Pstb, tobacco; Cc, cucumber; TMV, tobacco; rkn, tomato, aphids	**BTD, HCN**, PCA, 2-OH-PCA, PRN, HPR	Loper et al., 2012	Han et al., 2006; Kang et al., 2019, 2018; Kim et al., 2004; Lee et al., 2011; Park et al., 2012
PCL1391 (subsp. *piscium*)	rhizosphere tomato, Spain	Fo, tomato; Fo, watermelon; Cl, bean	**PCN**, PCA, HCN, CLP of the Viscosin group	Biessy et al., 2019	Bardas et al., 2009; Biessy et al., 2019; Chin-A-Woeng et al., 1998; Tziros et al., 2007
PCL1606	rhizosphere avocado, Spain	Dn, avocado; Fo, tomato	**HPR,** PRN, HCN	Calderón et al., 2015	Arrebola et al., 2019; Calderón et al., 2015; Cazorla et al., 2006
30-84	wheat rhizosphere, Washington, USA	Gg, wheat	**PCA, 2-OH-PCA,** PRN,HCN, HPR	Loper et al., 2012	Pierson and Thomashow, 1992
Pb-St2	sugarcane stem, Pakistan	Cf, sugarcane	PCA, 2-OH-PCA, PRN, HCN, HPR, WLIP	Mehnaz et al., 2014	Mehnaz et al., 2013, 2009
JD-37	potato rhizosphere, Shanghai, China	Bm, maize	PHZ, PRN, HCN,	Jiang et al., 2014	Fang et al., 2013
R47	potato rhizosphere, Zurich, Switzerland	Pi, potato	PCA, 2-OH-PCA, PRN, HCN, HPR	De Vrieze et al., 2020	De Vrieze et al., 2020; Hunziker et al., 2015
G05	Pepper rhizosphere, China	Fg, wheat; Fo, cucumber	**PRN, PCA**		Chi et al., 2017; Huang et al., 2018

Abbreviations: Pathogens: *Bm: Bipolaris maydis, Cc: Corynespora cassicola, Cf: Colletotrichum falcatum, Cl: Colletotrichum lindemuthianum, Dn: Dematophora necatrix, Fg: Fusarium graminearum, Fo: Fusarium oxysporum, Gg: Gaeumannomycis graminis* var. *tritici, Pi: Phytophthora infestans, Pc: Pectobacterium carotovorum, Pstb: Pseudomonas syringae pv. tobacco*; rkn: root knot nematodes. Metabolites: BTD: 2,3-butanediol, HCN: hydrogen cyanide, PCA: phenazines-1-carboxylate, 2-OH-PCA; PRN: pyrrolnitrin, HPR: 2-hexyl-5-propyl-alkylresorcinol. For metabolites in bold a role in biocontrol has been shown.

active at a pH of 5.7 and higher, while the antagonistic activity of PCA drastically decreases at this pH because the anionic form of PCA will predominate. This means that PCA-producing bacteria are most active at acidic pH.

P. chlororaphis Pcho10 was obtained from *Fusarium* spp.-infested wheat heads in China. This strain strongly reduced *F. graminearum* infections on wheat heads in growth chamber and field conditions and predominantly produces PCN. PCN appeared to be the major compound responsible for activity against *F. graminearum* (Hu et al., 2014). In case of *P. chlororaphis* G05 (Table 4), PCA contributed to the ability to suppress *F. oxysporum* on cucumber (Chi et al., 2017), but pyrrolnitrin rather than PCA was essential to control *Fusarium* head blight on wheat caused by *F. graminearum* (Huang et al., 2018).

P. chlororaphis O6 is known to induce systemic resistance to bacterial and fungal pathogens and stimulates aerial growth. The strain produces the volatile 2, 3-butanediol, a metabolite also known from *Bacillus* spp. The compound is responsible for aerial growth stimulation and ISR against *P. carotovorum* (Han et al., 2006). Genes involved in 2,3-butanediol biosynthesis are also present in the genomes of other *P. chlororaphis* strains (Biessy et al., 2019). In this strain HCN production is involved in nematicidal (Kang et al., 2018; Lee et al., 2011) and aphicidal activity (Kang et al., 2019).

The role of HPR in biocontrol has been investigated in *P. chlororaphis* PCL1606. This isolate was obtained from a screening of rhizosphere bacteria obtained from healthy avocado roots for in vitro antagonism against *Dematophora necatrix* (teleomorph: *Rosellinia necatrix*), the causal agent of white root rot on avocado and various other fruit trees and crops (Cazorla et al., 2006). In contrast to other *P. chlororaphis* biocontrol strains, PCL1606 does not produce phenazines. The *dar* genes, responsible for HPR production in PCL1606, are highly homologous to the *dar* genes present in *P. chlororaphis* subsp. *aurantiaca* BL915. Mutants in HPR are impaired in antagonistic activity in an avocado-*D. necatrix* and a tomato-*F. oxysporum* f. sp. *radicis-lycopersici* pathosystem (Calderón et al., 2013). The HPR biosynthetic gene cluster is present in various other *P. chlororaphis* biocontrol strains including *P. chlororaphis* 30-84 and *P. chlororaphis* O6 (Calderón et al., 2013). Biessy et al. (2019) showed that the HPR biosynthetic cluster is present in the subspecies *aureofaciens*, *aurantiaca* and *chlororaphis*, but not in the subspecies *piscium*. HPR plays a key role in biofilm production in *P. chlororaphis* PCL1606 (Calderón et al., 2019).

8 *Pseudomonas* biocontrol strains: *Pseudomonas corrugata* subgroup

The *Pseudomonas corrugata* subgroup as defined by Garrido-Sanz et al. (2016) and Hesse et al. (2018) comprises the plant pathogenic species *P. corrugata* and *P. mediterranea* and plant beneficial isolates that are classified as *P.*

brassicacearum, *P. kilonenis* or *P. thivervalensis*. This group is called the 'bcm clade' (Melnyk et al., 2019). *P. corrugata* and *P. mediterranea* are closely related and both cause stem pith necrosis on tomato and pepper. *P. corrugata* has also been reported on chrysanthemum and geranium (Catara, 2007). The line between plant-pathogenic and beneficial isolates in this subgroup is very thin (Gislason and de Kievit, 2020).

P. corrugata strains have been isolated from soil and from the rhizosphere of non-diseased plants in various countries (see Catara (2007) for an overview). These strains show activity against a wide range of soilborne fungal, bacterial and oomycete pathogens and postharvest fungal pathogens on fruits. Some *P. corrugata* strains have been patented for biocontrol. *P. corrugata* produces the antimicrobial and phytotoxic CLPs corpeptin A and corpeptin B with 22 amino acid residues, and cormycin, with 9 AA residues (Emanuele et al., 1998; Strano et al., 2015). Pith necrosis symptoms are strongly reduced in mutants unable to produce these compounds. Biocontrol activity in *P. corrugata* and *P. mediterranea* is mediated by these CLPs and volatiles such as HCN (Strano et al., 2017). *P. corrugata and P. mediterranea* strains do not produce the fluorescent siderophore, pyoverdine (Gislason and de Kievit, 2020).

Beneficial isolates within the *P. corrugata* subgroup have been classified as:

- *P. brassicacearium*
- *P. kilonensis*
- *P. thivervalensis*

P. brassicacearum and *P. thivervalensis* were first described by Achouak et al. (2000), while the species *P. kilonensis* was delineated in 2001 (Sikorski et al., 2001). These three species are closely related and it has been suggested that *P. kilonensis* should be considered as a junior synonym of *P. brassicacearum* (Tran et al., 2017). These species generate pyoverdine in contrast to *P. corrugata* and *P. mediterranea*. *P. brassicacearum*-like organisms are frequently found in soils that are disease-suppressive against take-all caused by *G. graminis* var. *tritici* in wheat. Isolates also have been obtained from agricultural soil or rhizosphere of other crops such as canola, potato, strawberry and tomato. They usually show excellent biocontrol capabilities. Many isolates that belong to these species produce the antibiotic DAPG.

Well-studied DAPG-producing strains in the *P. corrugata* subgroup include *Pseudomonas fluorescens* F113 (also called *P. kilonensis* F113), which was originally isolated from the root hairs of a sugar beet plant in Ireland (Shanahan et al., 1992) and *Pseudomonas fluorescens* Q2-87 (originally called *P. aureofaciens* Q2-87) obtained from wheat roots in a suppressive soil in Washington (Harrison et al., 1993). The species name of both isolates is in need of revision since they clearly belong to the *P. corrugata* subgroup (Garrido-Sanz et al., 2016). *Pseudomonas* strain Q2-87 produces DAPG and HCN but only DAPG contributes to its in vitro biocontrol activity against *G. graminis* var. *tritici* (Vincent et al., 1991).

Pseudomonas strain F113 is an excellent rhizosphere colonizer with biocontrol activity against various bacteria, fungi and nematodes including *P. carotovorum* (Cronin et al., 1997a), *F. oxysporum* (Barahona et al., 2011), *P. ultimum* (Fenton et al., 1992) and *Globodera* spp. (Cronin et al., 1997b). Biocontrol activity in this strain is strongly linked to the production of DAPG (Redondo-Nieto et al., 2013). The capacity to produce DAPG is lost in the pathogenic species, *P. corrugata* and *P. mediterranea*, probably because this compound can elicit ISR in plants (Almario et al., 2017).

A subset of *P. brassicacearum* strains does not produce DAPG but contain lipopeptide-encoding NRPS gene clusters located on a genomic island, which has been called the LPQ island by Melnyk et al. (2019). These lipopeptides are very similar to cormycin and corpeptins produced by *P. corrugata*. However, unlike corpeptin, which is a cyclic lipopeptide, sclerosin produced by *P. brassicacearum* DF41 is linear (Berry et al., 2012). *P. brassicacearum* DF41 isolated from canola root tips in Canada (Savchuk and Fernando, 2004) can protect canola against *Sclerotinia sclerotium* which is due to the production of sclerosin (Berry et al., 2010).

Pseudomonas sp. SH-C52 produces lipopeptides that have been characterized as thanapeptin and thanamycin. *Pseudomonas* sp. SH-52 has been found in soils with natural suppressiveness to the fungal pathogen *R. solani* (Mendes et al., 2011). Thanamycin is implicated in the antagonistic activity of SH-52 against *Sclerotium rolfsii* on groundnut (Le et al., 2012) and shows activity against various other fungi in vitro. Thanapeptin is active against oomycetes. Besides thanamycin and thanapeptin, this strain produces a 2-amino acid lipopeptide, brabantamide, with activity against Gram-positive bacteria and oomycetes (Van Der Voort et al., 2015).

SH-52 is closely related to *P. mediterranea* and *P. corrugata* and, like these species, lacks the capacity to produce pyoverdine, but has the biosynthetic gene clusters for the siderophores, achromobactin and corrugatin. Various other *Pseudomonas* strains of the *P. corrugata* group have biosynthetic gene clusters for thanamycin or thanapeptin-like CLPs. More information about lipopeptide production in the *P. corrugata* subgroup can be found in Girard et al. (2020a). Gislason and de Kievit (2020) have mined the genomes of 21 *P. brassicacearum* strains for regions associated with plant beneficial traits and pathogenicity. They found a negative correlation between DAPG and lipopeptide production; strains that produce lipopeptides do not produce DAPG. Selected biocontrol strains belonging to the *P. corrugata* subgroup are listed in Table 5.

Table 5 Selected *P. corrugata* group biocontrol strains

Strain	Origin	Activity	Metabolites	Genome sequence	References
Q2-87	wheat rhizosphere of TAS soil, Washington, USA	Gg, wheat; Pst, *Arabidopsis*	DAPG, HCN, PVD	Loper et al., 2012	Harrison et al., 1993; Weller et al., 2012
F113	root hairs of sugar beet, Ireland	Pc, Fo, Pu, *Globodera* sp.	DAPG, HCN, PVD	Redondo-Nieto et al., 2012	Barahona et al., 2011; Cronin et al., 1997b, 1997a; Fenton et al., 1992; Shanahan et al., 1992
Q8r1-96	wheat rhizosphere of TAS soil, Washington, USA	Gg, wheat	DAPG, HCN, PVD	Loper et al., 2012	
DF41	canola root tips, Canada	Ss, canola	Sclerosin, HCN	Loewen et al., 2014	Berry et al., 2012, 2010; Savchuk and Fernando, 2004
SH-C52	*R. solani* suppressive soil, the Netherlands	Rs, sugar beet	ACH, COR, BRB, HCN, THM, THP	Van Der Voort et al., 2015	Le et al., 2012; Mendes et al., 2011; Van Der Voort et al., 2015

Abbreviations: Pathogens: *Fo: Fusarium oxysporum, Gg: Gaeumannomycis graminis* var. *tritici, Pc: Pectobacterium carotovorum, Pst: Pseudomonas syringae* pv. *tomato, Pu: Pythium ultimum, Rs: Rhizoctonia solani, Ss: Sclerotinia sclerotiorum.* Metabolites: ACH: achromobactin, BRB: brabantamide, COR: corrugatin, DAPG: di-acetyl-phloroglucinol, HCN: hydrogen cyanide, THM: thanamycin, THP: thanapeptin. TAS: take-all suppressive.

9 *Pseudomonas* biocontrol strains: *Pseudomonas fluorescens* subgroup

P. fluorescens forms a large subgroup within the *P. fluorescens* group. At least 20 described type species cluster within this group but many genomes that fall within this subgroup are distinct from the type species. Various well-known and intensively studied biocontrol strains belong to this subgroup. These include two of the so-called Dutch strains that were isolated in the 1980s at the Dutch Phytopathological Laboratory 'Willy Commelin Scholten', WCS374 obtained from the rhizosphere of potato (Geels and Schippers, 1983) and WCS417 from the rhizosphere of wheat grown in a field suppressive to take-all disease (Lamers et al., 1988). These strains have intensively been studied for their plant-growth-promoting and biocontrol abilities. Whole-genome sequencing has clarified their taxonomic position (Berendsen et al., 2015). WCS417 belongs to the species *P. simiae*. WCS374 is closely related to the biocontrol strains *Pseudomonas* sp. A506 isolated from the pear phyllosphere in California, USA (Stockwell and Stack, 2007) and *Pseudomonas* sp. SS101 from the rhizosphere of wheat in the Netherlands (de Souza et al., 2003). It has been proposed to group these strains under the new species name *P. defensor* with WCS374 as the type strain.

P. simiae WCS417 has intensively been studied for its capacity to trigger induced systemic resistance in *Arabidopsis*, grapevine, radish, banana, bean, carnation and tomato. *P. defensor* WCS374 does not trigger ISR in *Arabidopsis* or grapevine, but is able to do so in Eucalypt, radish and rice (reviewed by De Vleesschauwer and Höfte, 2009). Biocontrol activity of these strains is linked to siderophore-mediated competition for iron by pyoverdines. Pyoverdines have also been implicated in the capacity of these strains to elicit ISR, reviewed by Lemanceau et al. (2009). The strains produce bacteriocins, but no antibiotics or CLPs. In addition, the genomes of both strains contain two T6SS loci. Both *P. simiae* WCS417 and *P. defensor* WCS374 possess a type III secretion system and putative effectors have been identified by genome mining (Berendsen et al., 2015). Highly similar T3SS clusters to WCS374 have been found in A506 and SS101 (Stringlis et al., 2019). These T3SS gene clusters are quite distinct from T3SS clusters in phytopathogenic *Pseudomonas* bacteria. Unlike pathogenic bacteria that contain a T3SS cluster, these strains do not trigger a hypersensitive response in tobacco leaves. The presence of T3SS in beneficial rhizosphere bacteria suggests a possible involvement in the suppression of root immune responses, facilitating root colonization.

According to Hesse et al. (2018), another strain that belongs to the same clique group as *P. simiae* WCS417 is *P. fluorescens* PICF7, an endophyte of olive roots and an effective biocontrol agent against *Verticillium* wilt of olive. A cysteine auxotrophic mutant of PICF7 had a reduced root colonization capacity and lost full biocontrol efficacy in olive but was not impaired in the inner colonization of olive root tissues. Swimming motility and pyoverdine production were not involved in endophytic colonization and biocontrol (Maldonado-Gonzalez et al., 2015).

Specific strains within the *P. fluorescens* subgroup are known to produce CLPs that belong to the Viscosin group (viscosin, massetolide, WLIP, viscosinamide) (Oni et al., 2020a). Well-studied strains in this aspect include the massetolide producer *P. fluorescens (lactis)* SS101 and the viscosin-producer *P. fluorescens* SBW25.

P. fluorescens (lactis) SS101 is a plant growth-promoting bacterium and has biocontrol activities against *Pythium intermedium* root rot of flower bulbs (de Souza et al., 2003), tomato late blight caused by *P. infestans* (Tran et al., 2007), *P. syringae* pv. *tomato* and *Spodoptera exigua* on *Arabidopsis*. Massetolide production is involved in the control of *Pythium* root rot on flower bulbs (de Souza et al., 2003) and *P. infestans* on tomato. Massetolide mutants of *P. fluorescens* SS101 were significantly less effective in biocontrol against *P. infestans* on tomato and had an impaired capacity to colonize tomato roots (Tran et al., 2007). Key mechanisms that are linked to growth promotion and ISR against *P. syringae* pv. *tomato* in *Arabidopsis* include modulation of sulfur assimilation, auxin biosynthesis and transport, steroid biosynthesis and carbohydrate metabolism (Cheng et al., 2017).

P. fluorescens SBW25 was originally isolated from the leaf surface of a sugar beet plant grown at the University farm, Wytham in Oxford, UK. *P. fluorescens* SBW25 has extensively been used as a model organism to investigate biofilm formation, experimental evolution and adaptation, plant colonization, plant growth promotion and biocontrol. The strain can colonize the roots and leaves of a wide variety of plants and shows biocontrol activity against *Rhizoctonia* and *Pythium* (Naseby et al., 2001) and triggers ISR in *Arabidopsis thaliana* against *P. syringae* pv. *tomato* (Preston et al., 2003). The strain possesses a rhizosphere-induced type III protein secretion system (Preston et al., 2001). It produces the CLP viscosin which plays a role in swarming motility, biofilm formation and causes lysis of *P. infestans* zoospores (De Bruijn et al., 2007). Another secreted metabolite is L-furanomycin, a non-proteinogenic amino acid with antibacterial activity against several plant pathogens, including *Erwinia amylovora*, *P. syringae* and *Dickeya dadantii* (Trippe et al., 2013).

The endophyte *P. poae* strain RE*1-1-14 was isolated from the internal part of sugar beet roots and is an antagonist of *R. solani*, the causal agent of damping-off and root rot disease of sugar beet. The strain produces the CLP poaeamide, a structurally new member of the Orfamide group. Poaeamide inhibited mycelial growth of *R. solani* and various oomycetes in vitro, and was essential for swarming motility, while a poaemide-deficient mutant colonized the plants at higher densities (Zachow et al., 2015).

Various strains within the *P. fluorescens* subgroup produce the phenazine PCA. Taxonomically, they are associated with the species *P. synxantha*, *P. libanensis*, *P. orientalis* and *P. aridus*. *P. aridus* is a provisionally named species and has not been formally described (Biessy et al., 2019; Parejko et al., 2013). Most of these strains have been obtained from the rhizosphere of dryland wheat in the USA (Mavrodi et al., 2012a,b). All strains produce a T3SS. These strains do not produce other antibiotics or HCN, but some of them have NRPS gene clusters encoding CLPs that based on predictions belong to the Viscosin or Orfamide group. A well-characterized biocontrol strain that belongs to this subgroup is *P. fluorescens* 2–79. The strain was isolated in 1979 from the roots of wheat plants grown in a take-all suppressive soil in Washington state, USA, and is a strong biocontrol agent suppressing *G. graminis* var. tritici in vitro and in planta (Nesemann et al., 2015).

10 *Pseudomonas* biocontrol strains: *Pseudomonas koreensis* subgroup

P. koreensis was first described as a new species by Kwon et al. (2003). The species was obtained from a Korean agricultural soil with low pH and is able to grow at 4°C. Mulet et al. (2010) created the *P. koreensis* subgroup based on partial sequencing of the four housekeeping genes *16S rRNA, gyrB, rpoB* and *rpoD*. Type strains that belong to this subgroup include *P. moraviensis*, *P. granadensis*, *P. helmanticensis* and *P. baetica* (Hesse et al., 2018). *P. fluorescens* Pf0-1, a Gac-deficient model strain of *Pseudomonas* behavior in soil (Seaton et al., 2013), is also a member of the *P. koreensis* subgroup.

Many *P. koreensis*-like organisms are found in tropical regions. Lopes et al. (2018) analyzed the fluorescent *Pseudomonas* population from a low pH Brazilian soil (pH = 5) under sugarcane cultivation. They found that 55 of the 74 isolates belonged to the *P. fluorescens* group and of these 37 (67%) associated to the *P. koreensis* subgroup. In research on the cocoyam rhizosphere in the disease suppressive soil of Boteva in Cameroon (pH = 5.15), it was found that 55% of the *Pseudomonas* isolates belonged to the *P. fluorescens* group. Within this group 84% were members of the *P. koreensis* subgroup (Oni et al., 2020b). Intriguingly, 58% of these isolates produce CLPs that belong to various families.

A great number of isolates produce the newly described CLP cocoyamide, which contains a peptide tail of 11 amino acids. This CLP appears identical to gacamide, a CLP that was recently described in *P. fluorescens* Pf0-1 complemented for GacA (Jahanshah et al., 2019). Other CLPs produced by these isolates are bananamide, lokisin and rhizoamide (Oni et al., 2020b). Lokisin and rhizoamide belong to the amphisin group of CLPs (Table 2). *Pseudomonas* spp. COW3 and COW 65 from the cocoyam rhizosphere produces novel variants of bananamide termed bananamide D, E, F and G (Omoboye et al., 2019a). These compounds display antagonistic activity against the oomycete pathogen *P. myriotylum* and inhibit appressoria formation in the blast fungus *Pyricularia oryzae*. The lokisin producer *Pseudomonas* sp. COR10 and a crude extract containing lokisin induced systemic resistance to *P. oryzae* in rice (Omoboye et al., 2019b).

CLP production appears to be a very common trait within the *P. koreensis* subgroup (Oni et al., 2020b; Stringlis et al., 2018). Virtually all *P. koreensis* isolates from the sugarcane rhizosphere in Brazil are predicted to produce CLPs. *P. granadensis* LMG27940, isolated from a soil sample in Granada, Spain produces the CLP MDN-0066, a compound very closely related to bananamides (Cautain et al., 2015). *Pseudomonas* sp. BW11P2, isolated from the banana rhizoplane in the tropical wetlands of Galgadera, Sri Lanka, produces bananamide A-C (Nguyen et al., 2016). Other putative bananamide producers (predicted by genome mining only) originate from cotton in South Korea and soybean roots in Mississippi, USA (Omoboye et al., 2019a).

P. koreensis 2.74 (CBS 125413) was isolated from the filter skin of a slow filter used in a closed hydroponic system in which tomato was cultivated in Sweden. This strain produces the CLP lokisin and a crude extract of this CLP has a protective effect against *P. ultimum* on tomato (Hultberg et al., 2010a). These authors also showed that *P. koreensis* 2.74 and the CLP could significantly reduce potato late blight caused by *Phytophthora infestans* in a detached-leaf assay (Hultberg et al., 2010b).

P. koreensis RU47 obtained from a take-all suppressive soil in the UK, shows biocontrol effects against *R. solani* AG1-1B in lettuce and *R. solani* AG3 in potato under growth chamber and field conditions (Adesina et al., 2009; Schreiter et al., 2018). The in vitro antagonistic activity of RU47 against *R. solani* AG1-1B is weak, but the strain consistently suppressed the pathogen on lettuce in four independent greenhouse experiments. Other *Pseudomonas* strains with strong in vitro antagonism were inconsistent in the four experiments (Adesina et al., 2009). This example shows that many interesting *Pseudomonas* biocontrol strains are missed if initial screening is based on in vitro antagonism. This is especially true for CLP producers, since they usually do not show clear inhibition zones in vitro. Genome mining revealed that RU47 produces HCN, extracellular enzymes and an amphisin-type CLP that is yet to be characterized (Kuzmanović et al., 2018).

11 *Pseudomonas* biocontrol strains: *Pseudomonas mandelii* subgroup and *Pseudomonas gessardii* subgroup

11.1 Pseudomonas mandelii *subgroup*

P. mandelii originates from mineral water and was first described by Verhille et al. (1999). Other species that belong to this subgroup include

P. frederiksbergensis, *P. lini*, *P. migulae*, *P. proseki* and *P. arsenicoxydans*. These species are typically found in extreme or polluted environments. *P. prosekii* is a recently described species comprising psychrotolerant strains isolated from James Ross Island close to the Antarctic Peninsula (Kosina et al., 2013; Snopkova et al., 2020). *P. migulae* strains have been found in mineral water, oil-contaminated rhizosphere of *Galega orientalis* in Finland (Jussila et al., 2006) and in soil samples in a selenium mining area in southwest China (Li et al., 2015). The type strain of *P. arsenicoxydans* is an arsenite-oxidizing bacterium isolated from sediment samples from the Camarones Valley in the Atacama Desert in Chile (Campos et al., 2010). *P. lini* has been isolated from the rhizosphere of jujube, a drought-tolerant plant perennial fruit tree that is cultivated in arid and semiarid regions (Zhang et al., 2020a). Tellurite-resistant and tellurite-reducing *P. lini* strains were found in Antarctica (Arenas et al., 2014).

Biocontrol strains in the *mandelii* subgroup include *Pseudomonas* sp. PICF141 obtained from the rhizosphere of olive in a commercial nursery in Cordoba, Spain (Gomez-Lama Cabanas et al., 2018) and *Pseudomonas* sp. In5 originating from a potato soil in Greenland suppressive to *R. solani* AG3 (Michelsen et al., 2015b; Michelsen and Stougaard, 2011).

Pseudomonas sp. In5 is closely related to *P. frederiksbergensis* (Girard et al., 2020a) and produces the CLPs nunamycin and nunapeptin (Michelsen et al., 2015b). *P. frederiksbergensis* 39A2 and *P. frederiksbergensis* 38F7, both obtained from Wyoming soil in the USA in 2008 (South et al., 2020) are predicted to also produce the same CLPs (Girard et al., 2020a), reinforcing the close relatedness of these strains. Nunapeptin contains a 22 amino acid peptide similar to corpeptin produced by *P. corrugata*, sclerosin produced by *P. brassicacearum* DF41 and thanapeptin produced by *Pseudomonas* sp. SH-52. All these strains co-produce a second CLP with 9 amino acid peptide, which is nunamycin in case of *Pseudomonas* sp. In5. The presence of nunamycin has also been predicted in the two *P. frederiksbergensis* strains. A crude extract which contains both nunamycin and nunapeptin possessed strong antimicrobial activity against the basidiomycete *R. solani* AG3 and the oomycete *P. aphanidermatum* (Michelsen et al., 2015a). Recently it was shown that fungal-associated molecules, rather than plant signal molecules, positively regulate nunamycin and nunapeptin synthesis in *Pseudomonas* sp. In5 (Christiansen et al., 2020).

Pseudomonas sp. PICF141 is most closely related to *P. lini* in the P. *mandelii* subgroup and shows biocontrol activity against defoliating pathotypes of *Verticillium dahlia* on 5-month-old olive plants. Gene clusters for the production of HCN, 2,3-butanediol, fusaricidin, amphisin and the insecticidal Fit toxin were detected in the genome of PICF141 (Gomez-Lama Cabanas et al., 2018). Fusaricidin is a potent depsipeptide antibiotic produced by *Paenibacillus polymyxa*. It remains to be proven that these compounds are actually produced by *Pseudomonas* sp. PICF141.

11.2 Pseudomonas gessardii *subgroup*

The unusual siderophore, thioquinolobactin, was first described by Matthijs et al. (2007) and Mossialos et al. (2000) in *P. fluorescens* ATCC 17400, a psychrophilic organism isolated from hen's egg (Stanier et al., 1966). This strain shows a strong in vitro antagonism against the oomycete *Pythium debaryanum*. The antagonism is repressed by iron and lost in a mutant impaired in thioquinolobactin production (Matthijs et al., 2007). More recently, the gene cluster encoding this siderophore was found in *Pseudomonas* sp. DTU12.1, a strain obtained from leaf-covered soil in Denmark with bioactivity against *Xanthomonas campestris* and *P. carotovorum*.

Genome mining revealed that the cluster is also present in three other *Pseudomonas* strains, *Pseudomonas* sp. O6C 162, *P. brenneri* BIGb2073 and *P. fluorescens* PS834 (Sazinas et al., 2019). In *Pseudomonas* sp. 06C 126, a strain from Lake Erie (central basin), the quinolobactin gene cluster contributes to the growth inhibition of oomycete plant pathogens (Wagner et al., 2018) and *P. aeruginosa* isolates from cystic fibrosis patients (Chatterjee et al., 2017). Based on *rpoD* sequencing and ANIb values, all these strains clustered closely together and belong to the same species (Sazinas et al., 2019). They are most related to the type strain of *P. proteolytica* (LMG 22170) in the *P. gessardi* subgroup (Hesse et al., 2018). The type strain of *P. proteolytica* was isolated from cyanobacterial mat samples from a pond in Antarctica (Reddy et al., 2004).

12 *Pseudomonas* biocontrol strains: *Pseudomonas putida* group

P. putida is ubiquitous and abundant in temperate soils and waters. Microorganisms belonging to this group are commonly found in polluted soil. *P. putida* is characterized by its physiological robustness, metabolic versatility and high tolerance to stress (Volke et al., 2020). *P. putida* isolates can use natural and man-made toxic organic compounds as a source of carbon and energy and have been studied intensively for their role in the biodegradation of environmental pollutants. *P. putida* is also routinely used as a host in industrial biocatalysis. Readers are referred to the recent review of Kivisaar (2020) for more information about these aspects. The *P. putida* group includes also clinical isolates. They have been detected in human secretions and tissues in hospitals and *P. putida* infections can be fatal in severely ill or immune-compromised patients (Ogura et al., 2019).

The *P. putida* group includes a variety of species and is taxonomically and phylogenetically poorly described. Unlike the *P. fluorescens* group, no subgroups have formally been described and the group is clearly in need of taxonomic revision. Yonezuka et al. (2017) subjected 59 clinical and non-clinical strains of the *P. putida* group to phylogenetic analysis based on ANI analysis. Based on a 95% threshold, they were classified into 26 species and included nine strain clusters and 17 singletons. A MLSA analysis based on nine housekeeping genes agreed with the ANI analysis. Their results indicate a mixed distribution of clinical and non-clinical isolates suggesting that some limited genes are responsible for pathogenicity.

Many *P. putida* strains are excellent colonizers of plant roots and promote their growth. Plant growth promotion is due to their ability to secrete phytohormones such as indole acetic acid that promote root elongation. Many isolates can solubilize phosphates and have the ability to induce systemic resistance or stress tolerance in plants. Inhibition of fungal growth is mainly through the production of siderophores. Isolates of the *P. putida* group do not usually produce the archetypical *Pseudomonas* biocontrol antibiotics such as phenazines, DAPG, pyoluteorin or pyrrolnitrin. But their genomes contain orphan gene clusters with uncharacterized NPRS or hybrid NRPS/PKS orthologs.

Production of CLPs is a common feature and include WLIP (Oni et al., 2020a, 2019b; Rokni-Zadeh et al., 2013), xantholysin (Li et al., 2013a,b; Oni et al., 2019b), entolysin (Oni et al., 2020b; Vallet-Gely et al., 2010) and putisolvin (Kuiper et al., 2003; Oni et al., 2020b, 2019b). WLIP and entolysin can induce systemic resistance to *Pyricularia oryzae* in rice (Omoboye et al., 2019b). Putisolvin-producing *P. putida* strains were enriched in the cocoyam rhizosphere from soils in Nigeria and Cameroon conducive to the cocoyam root rot disease caused by *P. myriotylum* (Oni et al., 2020b) and in the rhizosphere of black pepper in Vietnam (Le et al., 2012). They appear to abound in harsh or contaminated environments where CLP diversity is low, but their role in biocontrol is marginal. Xantholysin can inhibit appressoria formation in *P. oryzae* (Omoboye et al., 2019b) and shows strong insecticidal activity against aphids (Lim et al., 2017).

Various *P. putida* biocontrol strains have been reported in the literature, but often their taxonomic position has only been determined by sequencing the 16S rDNA which is not reliable in differentiating *Pseudomonas* groups, subgroups or species. As an example, *Pseudomonas* CMR12a and CMR5c were initially classified as *P. putida* strains based on their 16S rDNA sequence (Perneel et al., 2007), but turned out to belong to the *P. protegens* group based on MLSA and whole-genome sequencing (Biessy et al., 2019; Flury et al., 2016).

The best-characterized strain in the *P. putida* group is *P. putida* KT2440, a metabolically versatile strain with a remarkable ability to survive in marginal and polluted soil and a workhorse for *Pseudomonas* research and biotechnological applications (Timmis, 2002). The strain is used as a model organism in biodegradation studies. It was the first *Pseudomonas* strain for which the genome was sequenced in 2002 (Belda et al., 2016; Nelson et al., 2002). *P. putida* KT2440 is also an efficient colonizer of the spermosphere and rhizosphere of plants (Molina et al., 2000). The strain induces systemic resistance to *P. syringae* pv. *tomato* in *Arabidopsis* and an extracellular haem-peroxidase was found to be important for both root colonization and induction of ISR (Matilla et al., 2010). *P. putida* KT2240 has three T6SS gene clusters and 10 potential T6SS effectors. *P. putida* KT2440 outcompetes various Gram-negative phytopathogenic bacteria including *P. syringae*, *X. campestris*, *P. carotovorum* and *A. tumefaciens*. This destruction appears to be dependent on the K1-T6SS (Bernal et al., 2017).

Another well-studied plant-growth-promoting and biocontrol strain is *P. putida* WCS358, one of the Dutch strains from the 'Willie Commelin Scholten' collection. WCS358 produces a pyoverdine-type siderophore PVD358 that can be used by very few other *Pseudomonas* strains, while the strain has a large number of pyoverdine receptors. As such it is able to utilize pyoverdines by a wide range of other *Pseudomonas* strains. WCS358 was originally isolated from the rhizosphere of potato (Geels and Schippers, 1983) and can suppress soilborne plant diseases by siderophore-mediated competition for iron (reviewed by Lemanceau et al. (2009)). WCS358 can induce systemic resistance in *Arabidopsis* against *P. syringae* pv. *tomato* and *F. oxysporum* f. sp. *raphani* (Van Wees et al., 1997); in bean against *Colletotrichum lindemuthianum* and *B. cinerea*; and in tomato against *B. cinerea*. Multiple microbe-associated molecular patterns of WCS358 including PVD358, LPS-containing cell walls and flagella induced resistance in *Arabidopsis*, while in bean and tomato, flagella were not effective, whereas PVD358 and LPS-containing cell walls did induce resistance (Meziane et al., 2005). The taxonomic position of this strain within the *P. putida* group has been elucidated (Berendsen et al., 2015) and the species name *P. capeferrum* was proposed because the plant protective effects of this strain are linked to siderophore production.

Recently, a new bioactive metabolite has been discovered in *P. donguensis* SVBP6, an isolate from an agricultural plot in Argentina with broad-spectrum antifungal activity (Agaras et al., 2018; Muzio et al., 2020). The compound was identified as 7-hydroxytropolone, a compound with a seven-membered non-benzenoid aromatic ring. The biosynthetic gene cluster was identified by Tn5 mutagenesis and harbors 15 genes encoding metabolic enzyme functions, a putative efflux transport system and transcriptional regulators. Genome mining revealed that 15 strains among all *Pseudomonas* genomes available at the Pseudomonas Genome Database harbor this gene cluster (Muzio et al., 2020). They are representatives of *P. donghuensis* (8 strains), *P. qingdaonensis* (7 strains) and *P. wadenswilerensis* (1 strain). These species were recently described

(Frasson et al., 2017; Gao et al., 2015; Wang et al., 2019), are closely related and belong to the *P. putida* group. The 7-hydroxytropolone genomic region is associated with antibacterial activity against the black leg and soft rot pathogens *Dickeya solani* and *P. carotovorum* subsp. *brasiliense* in *P. donghuensis* P482 (Krzyzanowska et al., 2016). *P. donguensis* P482 is a tomato rhizosphere isolate from Poland with antibacterial and antifungal activity (Krzyzanowska et al., 2014, 2012). *P. donguensis* HYS, the type strain of the newly established *donguensis* species, was isolated from water of East Lake in Wuhan, China. In this strain 7-hydroxytropolone is used as an iron scavenger (Jiang et al., 2016).

13 *Pseudomonas* biocontrol strains: *Pseudomonas syringae* group and *Pseudomonas aeruginosa* group

13.1 Pseudomonas syringae *group*

The *P. syringae* group is a phylogenetic complex of strains from agricultural and environmental habitats. They are commonly isolated from diseased plants and many organisms that belong to this group are plant pathogens. They can also be found as saprophytes in environments linked to the water cycle such as freshwater, epilithic biofilms, snowpacks, precipitation and wild alpine plants (Morris et al., 2013). The *P. syringae* group contains 15 validly described species, 62 pathovars defined by pathogenic characters, 9 genomospecies defined by DNA-DNA hybridization (Gardan et al., 1999), and 13 phylogroups defined by MLSA (Berge et al., 2014). Based on a combination of methods (3-gene MLSA, amino acid sequences of 139 monocopy genes and average nucleotide identities based on BLAST (ANIb) calculated for 139 genomes), Gomila et al. (2017) distinguished 19 phylogenomic species distributed within 6 phylogenomic branches within the *P. syringae* group.

A few *P. syringae* biocontrol stains have been described. They probably all belong to Genomospecies 1 as defined by Gardan et al. (1999), phylogroup 2 as described by Berge et al. (2014) or phylogenomic branch I as defined by Gomila et al. (2017). The best-studied *P. syringae* biocontrol strains are ESC-10 and ESC-11 (also known as strain L-59-66), isolated from apple fruits (Stockwell and Stack, 2007) and leaves (Janisiewicz and Marchi, 1992), respectively. Both strains have been developed in commercial products Bio-Save 10 and Bio-Save 11 which have been registered in 1995 and are commercially available in the USA (Table 6). The strains are effective biocontrol agents against various postharvest diseases on apple, pear and citrus. Strain ESC-10 produces a hypersensitive response on tobacco and is a weakly virulent pathogen of citrus. Strain ESC-11 is not a pathogen and does not produce a hypersensitive response (Smilanick et al., 1996). Strain ESC-10 is superior for the protection of citrus from green and blue molds, while strain ESC-11 is superior to control postharvest pathogens on apples and pears (Bull et al., 1997). In vitro, both strains produce the CLP syringomycin. Purified syringomycin E inhibited *P. digitatum* spore germination at about 5 μg/mL but about 100 μg syringomycin E per wound was needed to control green mold on lemon to an extent equal to that provided by the biocontrol strains ESC-10 and ESC-11 (Bull et al., 1998). Strain ESC-10 reduced Rhizopus rot on peach caused by *Rhizopus stolonifera* to levels comparable with that of dichloran and iprodione (Northover and Zhou, 2002).

The *P. syringae* strains MA-4 and NSA-6, obtained from the phyllosphere of apple in Canada, also effectively controlled Rhizopus rot on peach. These strains were also effective against brown rot on peach caused by *Monilinia fructicola* and the activity was significantly improved by adding 0.5% $CaCl_2$ to the soak solution (Zhou et al., 1999). According to Zhou et al. (1999) MA-4 and NSA-6 can also control apple scab, caused by *Venturia inaequalis* on apple.

P. *syringae* pv. *syringae* 22d/93 isolated from the phyllosphere of soybean in Jena, Germany (Völksch et al., 1996), is a strong antagonist of *P. syringae* pv. *glycinea*, the causal agent of bacterial blight of soybean. Antagonism has been demonstrated in vitro and in planta under greenhouse and field conditions (May et al., 1997; Völksch and May, 2001). The cyclic lipopeptides, syringomycin and syringopeptin, are produced but have no activity against *P. syringae* pv. *glycinea* (Braun et al., 2010). The rare amino acid 3-methylarginine selectively inhibits *P. syringae* pv. *glycinea* in vitro and the toxin probably acts as an inhibitor of the arginine biosynthesis pathway or an arginine-dependent pathway (Braun et al., 2010). Purified 3-methylarginine suppressed the growth of the pathogen in planta. Mutant analysis revealed that none of these compounds is required for antagonistic activity in planta when the pathogen was applied to artificial wounds (Braun et al., 2010). The strain produces two siderophores, pyoverdine and achromobactin, that play an essential role in leaf colonization but are not directly implicated in biocontrol (Wensing et al., 2010).

13.2 Pseudomonas aeruginosa *group*

Pseudomonas aeruginosa is an ubiquitous organism with minimal requirements for survival that is commonly found in water, soil, animal hosts and associated with plants (Botzenhart and Döring, 1993; Hardalo and Edberg, 1997). *P. aeruginosa* is best known as an opportunistic pathogen of humans. It can cause severe problems in patients hospitalized with cancer, AIDS, cystic fibrosis and severe burns. The bacterium can invade any tissue in which the immune system is compromised (Moradali et al., 2017). *P. aeruginosa* has also been found in association with various other organisms such as nematodes, insects (Mahajan-Miklos et al., 2000) and amoebae (Pukatzki et al., 2002). *P. aeruginosa* has many secondary metabolites in common with biocontrol *Pseudomonas* strains and produces phenazine antibiotics, HCN, the siderophores pyoverdine and

Table 6 Commercially available bioprotectants based on *Pseudomonas* in Europe and the USA

Biocontrol strain	Commercial Name		Usage	Crop/Applications	Pathogen
P. chlororaphis MA342	Cedomon	BioAGri, Sweden; Koppert, The Netherlands	Fungicide	Seed-dressing in field-grown cereals (barley, oats, triticale, wheat)	Seedborne pathogens
	Cerall	BioAGri, Sweden; Koppert, The Netherlands	Fungicide	Seed-dressing in wheat, rye and triticale	Seedborne pathogens
	Cedress	BioAGri, Sweden; Koppert, The Netherlands	Fungicide	Seed-dressing in carrots and pea	*Ascochyta* on pea and *Acrothecium carotae* on carrots
P. aureofaciens strain Tx-1	Spot-Less	Turf Science Laboratories Inc., CA, USA	Fungicide	Lawn and grass management, recreational and ornamental lawns	*Sclerotinia homeocarpa*, *Colletotrichum* spp., *P. aphanidermatum*, *Michrodochium nivale*, *Pyricularia grisea*
P. chlororaphis AFS009	Howler	AFS009 Plant Protection Inc., NC, USA	Fungicide	Fruits, vegetables, ornamentals, hydroponic crops, beans, peas, peanut, trees, turf, home and garden use	*Rhizoctonia*, *Pythium*, *Fusarium*, *Phytophthora*, *Sclerotinia*, *Colletotrichum*, *Botrytis*
	Zio	SePRO	Fungicide	Turfgrass, ornamental plants, vegetable and fruit transplants	
P. fluorescens A506	Blightban, Frostban B	Nufarm Americas, USA	Bactericide, frost prevention	Fruit trees, strawberry, tomato, potato	Frost-forming and russet inducing bacteria, *E. amylovora*
Pseudomonas sp. DSMZ 13134	Proradix	SP Sourcon Padena, Germany	Fungicide, plant growth regulator	Potato	*R. solani*, *Helminthosporium solani*

P. syringae strain ESC-10	Bio-save 10LP, 10NT	Jet Harvest Solutions, FL, USA	ungicide	Postharvest: apples, pears, citrus, potato	*P. digitatum, P. italicum, Geotrichum candidum, B. cinerea, Mucor piriformis, Fusarium sambucinum, H. solani*
P. syringae strain ESC-11	Bio-save 11LP, 11NT	Jet Harvest Solutions, FL, USA	Fungicide	Postharvest: apple, pear, potato, sweet potato,	*P. expansum, B. cinerea, M., Fusarium sambucinum, H. solani, Rhizopus stolonifer*
P. fluorescens CL145A	Zequanox	Marrone Bio Innovations	Molluscicide	Water	Zebra and quagga mussels
P. fluorescens D7	D7	Verdesian Life Sciences, NC, USA	Herbicide	Wheat, barley, triticale, oats, rangeland	downy brome, medusahead, Japanese brome, jointed goatgrass

pyochelin, and rhamnolipid-type biosurfactants. Phenazines produced by *P. aeruginosa* are pyocyanin, a blue pigment, but some strains can also produce 1-OH-PHZ, PCA and PCN (Mavrodi et al., 2006). Hence, it is not a surprise that they are able to inhibit a wide range of plant pathogens by the secretion of antimicrobial compounds or by inducing systemic resistance. *P. aeruginosa* is commonly found in the rhizosphere or as an endophyte of plants, especially in tropical regions. Antagonistic plant-associated *P. aeruginosa* strains have, for instance, been isolated from chickpea (Anjaiah et al., 1998), chilli pepper (Paul et al., 2013), black pepper (Kumar et al., 2013), rice (Yasmin et al., 2017), cotton (Yasmin et al., 2014), ginger (Jasim et al., 2014), tomato (Kumar et al., 2009) and cocoyam (Oni et al., 2020b).

Well-studied *P. aeruginosa* biocontrol strains include PNA1, isolated from chickpea roots in India (Anjaiah et al., 1998), 7NSK2, isolated from the rhizosphere of barley in Belgium (Iswandi et al., 1987) and M18 from the rhizosphere of sweet melon in Shanghai, China (Hu et al., 2005).

P. aeruginosa 7NSK2 produces pyocanin and the siderophores, pyoverdine and pyochelin. Pyoverdine and pyochelin are involved in suppression of *Pythium*-damping-off of tomato (Buysens et al., 1996). ISR against *B. cinerea* in tomato is mediated by pyocyanin and pyochelin (Audenaert et al., 2002), while pyocyanin triggers ISR against *P. oryzae* in rice, but enhances susceptibility to *R. solani* (De Vleesschauwer et al., 2006). *P. aeruginosa* PNA1 produces rhamnolipids, PCA and PCN. Phenazines contributed to the suppression of Fusarium wilt caused by *F. oxysporum* f. sp. *ciceris* in chickpea (Anjaiah et al., 1998) and *Fusarium udum* in pigeonpea (Anjaiah et al., 2003). Phenazines and rhamnolipids acted synergistically in the control of *Pythium splendens* on bean and *P. myriotylum* on cocoyam (Perneel et al., 2008). Ginger is also susceptible to *P. myriotylum* and an endophytic PCA-producing *P. aeruginosa* isolate obtained from the ginger rhizome gave protection against the pathogen (Jasim et al., 2014).

P. aeruginosa M18 can effectively control *Mycosphaerella melonis* on sweat melon plants (Wu et al., 2011) and has strong activity against various fungal, oomycete and bacterial pathogens (Xu and Liu, 2013). The strain predominantly produces PCA rather than pyocyanin especially at 28°C (Huang et al., 2009). In addition it produces pyoluteorin, an antibiotic that is commonly found in *P. protegens* isolates but not in *P. aeruginosa* (Huang et al., 2004). Interestingly, PCA is negatively regulated and pyoluteorin is positively regulated by the global regular GacA (Ge et al., 2004). Expression and regulation of PCA in this strain has been studied in great detail and a high-yielding engineered strain has been constructed (Du et al., 2013) and the fermentation was optimized using surface response methodology (Li et al., 2008). PCA produced by the engineered *P. aeruginosa* M18 strain has been developed in a biopesticide called Shenqinmeisu that was approved in China in 2011 for the control of various fungal and microbial diseases of fruits, vegetables, rice, wheat, potatoes and other crops (Xu and Liu, 2013).

14 Commercial *Pseudomonas*-based bioprotectants

Despite the overwhelming literature on *Pseudomonas* biocontrol, few *Pseudomonas*-based products are commercialized as bioprotectants (Table 6). In Europe, only *P. chlororaphis* MA342 and *Pseudomonas* sp. DSMZ 13134 are registered and sold in various European countries under the commercial names Cedomon, Cerall, Cedress and Proradix.

P. chlororaphis MA342 was isolated from roots of wild crowberry (*Empetrum nigrum* L.) in Sweden (Hökeberg et al., 1997) and included in the EU list of approved active substances in 2004. It is used as a seed treatment on cereals (wheat, rye, triticale), carrot and pea to protect against seedborne pathogens and against foliar and ear pathogens in cereals (EFSA, 2017; Johnsson et al., 1998). It is a good spermosphere, but poor rhizosphere and phyllosphere colonizer. The strain is known to produce 2,3-deepoxy-2,3-didehydro-rhizoxin (DDR). Cerall is a ready-to-use water-based formulation for seed treatment. Cedomon is an oil-based formulation used in barley and oat. Cedress is a flowable suspension for seed treatment of carrots and peas.

It should be noted that the taxonomic identification of MA342 as *P. chlororaphis* is questionable. The strain has not properly been identified and based on 16S rRNA sequence analysis, the strain most probably does not belong to the species *P. chlororaphis*. The production of the metabolite DDR has become an issue for the renewal of the approval of *P. chlororaphis* MA342 as an active substance because of potential genotoxic effects (Hernandez-Jerez et al., 2020).

Field trials in Sweden (Johnsson et al., 1998), Finland (Lötjönen and Torniainen, 2005) and Lithuania (Liatukas et al., 2019) have revealed that *P. chlororaphis* MA 342 can effectively control seedborne cereal pathogens that are present on the seed or near the seed coat such as *Pyrenophora* spp., but does not control seedborne pathogens like the loose smut *Ustilago nuda* that are located in the seed embryo. There is no effect either against soilborne diseases or leaf diseases. Average yield increases in spring barley range between 2% and 5%, but there are strong differences between cultivars. In field trials in Lithuania, seeds of the cultivar 'Alisa DS' were effectively sanitized by the use of Cedomon, leading to yield increases of 5–15%, while this was not observed on the cultivar 'Luoke' (Liatukas et al., 2019). Field experiments in Finland have revealed that Cedomon could control net blotch (*Pyrenophora tera*) in barley nearly as well as chemical products and increased yield with 200–300 kg. Barley life stripe (*Pyrenophora graminae*) was decreased with 40–80%. There was no effect against loose smut (*Ustilago nuda*) of barley or Fusarium.

Seed dressing treatments with Cedomon did not increase the yield of oat compared to untreated controls (Lötjönen and Torniainen, 2005).

Pseudomonas sp. DSMZ 13134 was isolated from the roots of lettuce (EFSA, 2012) in a search for biocontrol agents against *P. ultimum*. The taxonomic position of this strain is uncertain and it is not clear which metabolites it can produce. Its modes of action include competition for space, induced systemic resistance and stimulation of root growth. The strain was included in the EU list of approved active substances in 2013 (EU, 2013). The Proradix product is a wettable powder that is approved for tuber treatment of potato prior to planting against black scurf (*R. solani)* and silver scurf (*Helminthosporium solani)*. The product controls black scurf on potato with an efficacy of 28–40% and silver scurf with an average efficacy of 46% (Moszczynska et al., 2015; CTGB, 2014). In field trials with barley in Germany seed inoculation with Proradix increased yields with 4–19% (Fröhlich et al., 2012).

P. chlororaphis-based products commercially sold in the USA are Spot-Less, Howler and Zio. Howler and Zio are biofungicides containing *P. chlororaphis* strain AFS009 (PC code 006800, registered 2017) to control various oomycete and fungal pathogens on many different crops (EPA, 2017a,b). AgBiome selected the strain after screening a library of more than 60 000 microorganisms. Both products are wettable powders that should be mixed with water and can be applied as a foliar spray, soil drench, in furrow spray, transplant spray or dip, cuttings or bare root dip, or in hydroponic or chemigation application. The products can also be mixed with potting mix or applied dry in furrow. Zio is marketed to the turf and ornamental sector by SePRO. Howler is marketed for use on specialty crops. According to AgBiome, modes of action of *P. chlororaphis* strain AFS009 include production of antibiotics and cell wall degrading enzymes, competition and induced resistance, but further details could not be found.

Spot-less is based on *P. chlororaphis subsp. aureofaciens* Tx-1 (PC code 006473, registered 1999) (Hardebeck et al., 2004; Powell et al., 2000) and is registered for use in turf grass against dollar spot (*Sclerotinia homoeocarpa*), anthracnose (*Colletotrichum* spp.), *Pythium,* pink snow mold (*Microdochium nivale*) and grey leaf spot *(Pyricularia grisea*) (EPA, 2006). The product is a fermented culture of *P. aureofaciens* Tx-1 following 12–14 h of culture in the BioJect® Fermentation and Injection System. The culture is diluted with water and injected into a sprinkler irrigation system. The strain is known to produce PCA. In field trials it was shown that PCA provided significantly better management of dollar spot on creeping bentgrass than untreated plots. Results were comparable to management provided by chlorothalonil (Powell et al., 2000). Field studies carried out at West Lafayette, Indiana in 1998 and 1999 revealed that Tx-1 marginally controls dollar spot, only when disease pressure is low. In combination with fungicides and under low disease pressure, Tx-1 could increase the dollar spot control of fungicides by 37% and increased the duration of control by 2.6 days (Hardebeck et al., 2004). Repeated inoculation with TX-1 resulted in the establishment of the strain in the rhizosphere of turfgrass with a minimal impact on the indigenous bacterial community. The strain also was capable of overwintering. Low activity of TX-1 in the field may be explained by limited PCA production in situ (Sigler et al., 2001).

P. syringae-derived products sold in the USA are the Bio-Save products based on *P. syringae* ESC-10 (PC code 006441, registered 1994) and ESC-11 (PC code 006451, registered 1996) that are used for the control of postharvest pathogens on fruits, potato and sweet potato that have been discussed above. The active ingredients were registered in 1990 and 1996, respectively. Their history has extensively been reviewed by Stockwell and Stack (2007). ESC-10 reduces green and blue mold on lemon with 80–99% effectiveness and on orange with 40–59% effectiveness. ECS-11 gives excellent control of blue mold and grey mold on apple and pear, but host-specific effects are observed for control of postharvest diseases on pear. Both products can be used in combination with fungicides and are very useful to control fungicide-resistant pathogen populations. Their use in the USA is widely adopted (Stockwell and Stack, 2007).

P. fluorescens A506 (PC code 006438, registered 1992) is the active ingredient in Frostban B (Blightban) (EPA, 2017c). The product is used to suppress frost damage on various crops and to suppress fire blight caused by *E. amylovora*. The primary mechanism in fire blight control is competitive exclusion. The commercial product is composed of lyophilized cells of the bacterium that need to be diluted in water. A506 is reviewed by Stockwell and Stack (2007). In non-inoculated orchards in California, *P. fluorescens* A506 could reduce the incidence of fire blight by 50–80%, fruit russet by 30–50% and frost injury by 40–50% (Lindow et al., 1996; Lindow and Suslow, 2003). The strain did not provide significant control in inoculated experimental orchards in Oregon (Stockwell et al., 2010).

Other *Pseudomonas* strains have been registered in the USA (*P. syringae* 724RS, *P. fluorescens* 1629RS, *P. chlororaphis* 63-28) but they are not, or no longer, commercially available. Anderson and Kim (2018) list various bioprotectants based on *P. chlororaphis* O6 that are registered by HyunNong Inc. in South Korea. These include NematoKill against root-knot nematodes, ItaEpi against aphids and three products that are used as microbial fertilizer. However, additional information about these products could not be found.

Bioprotectants or biofertilizers based on various strains of *P. chlororaphis* are commercially available in Russia and listed by Maksimov et al. (2011) with the trade names Pseudobacterin-2, Agat-25K, Helen and Binoram. AGAT-25 is used to control roots rots, mildew, septoriose, brown rust, ear fusariosis, cercosporiosis and pseudoperonosporosis, while Pseudobacterin-2 is used to control root rots (Kabaluk et al., 2010). The active ingredient in Pseudobacterin-2 is *P. aureofaciens* BS1393, a strain obtained from the rhizosphere of barley. Trials in cereals, vegetables and oilseed crops have indicated that Pseudobacterin-2 can reduce disease severity by 65–88%, resulting in a 20–25% yield increase (Thomashow, 2013).

Various *Pseudomonas*-based bioprotectants are registered for commercial use in India. They are all called *P. fluorescens* and have different formulations and accession numbers. A strain with accession number MTCC 5176 is registered to control loose smut on wheat. Strain IPL/PS-01 (accession number MTCC 5727) is available in different formulations to control bacterial leaf blight on rice, and soilborne diseases on tomato. Strain IIHR-PF-2 (accession number ITCCB0034) is approved to control root rot and wilt diseases on tomato, brinjal, carrot and okra. A strain with accession number MTCC 2539 is approved for late leaf spot control on groundnut. Strain TNAU (accession number ITCC BE 0005) is used on groundnut, rice, chili seedlings, tomato and cotton. Strain BIL-331 (accession number MTCC 5866) is approved to control various diseases on rice (Government of India, 2020). The taxonomic status, origin and mode of action of these strains are unclear.

Table 6 lists two other applications of *Pseudomonas* biocontrol. *P.* fluorescens D7 (PC code 16418, registered 2014) is the active ingredient in D7, a product approved in 2014 for use as a herbicide in cereals. D7 is a freeze-dried powder that is dissolved in water and applied as a spray solution to the soil surface. It is used for preemergence applications. Zequanox contains killed cells of *P. fluorescens* CL145A and is registered for control of quagga and zebra mussels in limited open-water environments and in cooling and service water systems for industrial facilities. It was shown that dead *P. fluorescens* CL145A cells achieved the same percentage mussel mortality as live cells and that the mode of action is intoxication (Molloy et al., 2013).

It is not entirely clear why so few *Pseudomonas* strains are commercialized. Possible bottlenecks are inconsistency in field performance, genetic instability, poor shelf life in comparison with spore-forming biocontrol agents, and costs and difficulty of registration (Berg, 2009; Mark et al., 2006; Tabassum et al., 2017). A lot of progress has been made in recent years in the formulation of non-sporulating bacteria (Berninger et al., 2018) and poor shelf life should probably no longer be a major issue. Inconsistency in field performance is, however, a key problem. Good performance in specific trials is often not translated in consistent and effective control in diverse field situations. This can be due to external factors such as soil or climatic conditions, but is also driven by intrinsic traits of the biocontrol agent, such as variable production of secondary metabolites or poor host colonization (Mark et al., 2006). Host-specific responses to bioprotectants are observed, which could in part be driven by quantitative and qualitative differences in plant exudates.

Microbiome studies have revealed that plant species, cultivars and soil type are key drivers of rhizosphere microbiome composition and functioning. A series of elegant studies by the Raaijmakers' group using bean as a model have shown that plant domestication and resistance breeding caused changes in root architecture and altered root exudates. This in turn changed rhizosphere microbiome composition and functioning (Mendes et al., 2019, 2017; Mendes and Mendes, 2018; Pérez- Jaramillo et al., 2016). Specific root exudates may be needed to stimulate *Pseudomonas* colonization and the in situ production of bio-active molecules. The regulation of secondary metabolites in *Pseudomonas* spp. is complex and may require specific nutrients and environmental stimuli or the presence of other microorganisms as was recently shown for *Pseudomonas* sp. In5 (Christiansen et al., 2020). A second major hurdle in the commercialization of microbial biopesticides is the slow, difficult and costly registration process, especially in Europe. These aspects have recently been reviewed (Frederiks and Wesseler, 2018; Scheepmaker et al., 2019).

15 Conclusion

Recently an ever-increasing number of high-quality *Pseudomonas* whole-genome sequences has become available, and insights in the taxonomy of *Pseudomonas* have greatly advanced. Genome mining and the division of the vastly heterogeneous genus *Pseudomonas* in phylogenomic groups and subgroups have clarified the relationship between phenotypic characteristics and phylogeny. Well-known *Pseudomonas* biocontrol strains that have been studied for many years can now be positioned in their correct phylogenetic (sub)group.

Based on their activity, three types of *Pseudomonas* biocontrol strains can be distinguished. The first type comprises the CPC (*chlororaphis, protegens, corrugata* subgroups) cluster defined by Vacheron et al. (2016) and *P. aeruginosa*. CPC isolates produce an arsenal of secondary metabolites, including antibiotics with antagonistic properties such as pyrrolnitrin, DAPG, phenazines and HCN. They have broad-spectrum activity against soilborne pathogens and many strains can induce systemic resistance to foliar pathogens. This cluster has received considerable attention in the past.

CPC strains are readily picked up in in vitro screens for biocontrol strains because of their clear antagonism on plate and broad-spectrum antimicrobial activity. However, several disadvantages come with the production of multiple bioactive compounds, as evidenced in the *P. protegens* group. *P. protegens* strains are prone to *gac* mutations, some of their metabolites are phytotoxic, and they are outcompeted by other bacteria in the rhizosphere of field-grown plants. This may explain why they have not been commercialized.

P. chlororaphis strains are genetically more stable (Yan et al., 2018), and most strains do not produce potential phytotoxic compounds. They are preferred by biopesticide companies. It is striking that a *P. chlororaphis* strain was selected in the 60 000 strain-screen by Agbiome. To my knowledge, there are no commercial biopesticides containing *Pseudomonas* strains from the *P. corrugata* group although Salavida, a product from Sourcon Padena, based on *P. brassicacearum* 3Re-27, is commercialized as a biostimulant in Germany (Berg, 2009). *P. aeruginosa* strains with biocontrol properties also belong to this division. Their spectrum of metabolites is rather similar to *P. chlororaphis* subgroup isolates. The fact that they are opportunistic human pathogens precludes their commercialization, but they are a source of bioactive compounds such as phenazines and rhamnolipids.

A second type is found in the *P. putida* group and the *P. fluorescens*, *P. koreensis*, *P. mandelii* and *P. gessardii* subgroups in the *P. fluorescens* group. Vacheron et al. (2016) has termed *P. fluorescens*, *P. koreensis*, *P. mandelii* and *P. jessenii* the FMJK cluster. It was found that FMJK and *P. putida* are predominantly present in the maize rhizosphere and have fewer biocontrol traits, but more phytostimulatory properties than CPC strains (Vacheron et al., 2016). These (sub)groups harbor many plant growth-promoting *Pseudomonas* strains. An example is *P. jessenii* UW4, an ACC deaminase-producing strain that promotes plant growth in the presence of different environmental stresses, such as flooding, heavy metals, cold, high concentrations of salt and the phytopathogenic bacterium *Agrobacterium* (Duan et al., 2013).

Biocontrol strains from the second type do not produce the archetypical *Pseudomonas* antibiotics. Their biocontrol capacities predominantly depend on siderophores or a diverse arsenal of CLPs. CLPs have interesting antimicrobial and resistance-inducing properties. It has been hypothesized that their large diversity is driven by microbial competition in the rhizosphere (Oni et al., 2020a). An enrichment in CLP production is also seen in a subset of isolates closely related to the plant pathogenic *P. corrugata* and *P. mediterranea* in the *P. corrugata* subgroup. Other traits remain to be investigated such as the targets and roles of the effector proteins secreted by the T3SS and T6SS machinery and the function of strain-specific orphan gene clusters. This second type has significant potential because these strains are often competitive root colonizers and adapted to harsh conditions. The fact that they usually do not produce antibiotics will also facilitate the registration process in case of commercialization.

The third type encompasses biocontrol strains adapted to above-ground plant parts. They predominantly belong to the *P. syringae* group and have antagonistic activities against aerial pathogens. *P. fluorescens* A506 also belongs to this type. Their main mode of action is probably competition. Commercial biofungicides based on *P. syringae* and *P. fluorescens* A506 are already on the market for 30 years to control postharvest and bacterial pathogens.

Despite the overwhelming literature on *Pseudomonas* biocontrol, only few *Pseudomonas* strains are commercialized. This is mainly due to inconsistent field performance, which can be explained by poor adaptation to environmental conditions, suboptimal formulation, and host-specific responses to microorganisms. In addition, the expensive, tedious and time-consuming registration procedure, especially in Europe, hampers the introduction of new microbial bioprotectants.

16 Where to look for further information

16.1 Pseudomonas *biocontrol*

Haas and Défago, 2005; Mercado-Blanco, 2015; Olorunleke et al., 2015b; Müller and Behrendt, 2021.

16.2 Pseudomonas *taxonomy*

Garrido-Sanz et al., 2017; Girard et al., 2020b; Gomila et al., 2015; Hesse et al., 2018; Lalucat et al., 2020; Mulet et al., 2010.

16.3 Secondary metabolites in Pseudomonas

Gross and Loper, 2009.

16.4 Phenazines

Biessy and Filion, 2018; Mavrodi et al., 2013, 2006; Chincholkar and Thomashow, 2013.

16.5 Lipopeptides

Geudens and Martins, 2018; Götze and Stallforth, 2020; Raaijmakers et al., 2010.

16.6 Whole genomes of Pseudomonas

Pseudomonas genome DB https://www.pseudomonas.com/.

16.7 Whole genome comparisons of Pseudomonas *biocontrol strains*

Biessy et al., 2019; Garrido-Sanz et al., 2016; Loper et al., 2012.

16.8 Formulation

Berninger et al., 2018.

17 Acknowledgements

Pseudomonas biocontrol and cyclic lipopeptide research in the laboratory of M. Höfte is supported by grants from the INTERREG France-Wallonie-Vlaanderen Program SmartBiocontrol, Fonds Wetenschappelijk Onderzoek – Vlaanderen (FWO) under Excellence of Science grant RhizoCLiP (EOS Project No. 30650620) and the Concerted Research Action MEMCLiP supported by the Special Research Fund of Ghent University (GOA-028-19).

18 References

Achouak, W., Sutra, L., Heulin, T., Meyer, J. M., Fromin, N., Degraeve, S., Christen, R. and Gardan, L. 2000. *Pseudomonas brassicacearum* sp. nov. and *Pseudomonas thivervalensis* sp. nov., two root-associated bacteria isolated from *Brassica napus* and *Arabidopsis thaliana*. *Int. J. Syst. Evol. Microbiol.* 50(1), 9–18. https://doi.org/10.1099/00207713-50-1-9.

Adesina, M. F., Grosch, R., Lembke, A., Vatchev, T. D. and Smalla, K. 2009. *In vitro* antagonists of *Rhizoctonia solani* tested on lettuce: rhizosphere competence, biocontrol efficiency and rhizosphere microbial community response. *FEMS Microbiol. Ecol.* 69(1), 62–74. https://doi.org/10.1111/j.1574-6941.2009.00685.x.

Agaras, B. C., Iriarte, A. and Valverde, C. F. 2018. Genomic insights into the broad antifungal activity, plant-probiotic properties, and their regulation, in *Pseudomonas donghuensis* strain SVBP6. *PLoS ONE* 13(3), e0194088. https://doi.org/10.1371/journal.pone.0194088.

Almario, J., Bruto, M., Vacheron, J., Prigent-Combaret, C., Moënne-Loccoz, Y. and Muller, D. 2017. Distribution of 2,4-diacetylphloroglucinol biosynthetic genes among the *Pseudomonas* spp. reveals unexpected polyphyletism. *Front. Microbiol.* 8, 1218. https://doi.org/10.3389/fmicb.2017.01218.

Anderson, A. J. and Kim, Y. C. 2020. Insights into plant-beneficial traits of probiotic *Pseudomonas chlororaphis* isolates. *J. Med. Microbiol.* 69(3), 361–371. https://doi.org/10.1099/jmm.0.001157.

Anderson, A. J. and Kim, Y. C. 2018. Biopesticides produced by plant-probiotic *Pseudomonas chlororaphis* isolates. *Crop Prot.* 105, 62–69. https://doi.org/10.1016/j.cropro.2017.11.009.

Anjaiah, V., Cornelis, P. and Koedam, N. 2003. Effect of genotype and root colonization in biological control of fusarium wilts in pigeonpea and chickpea by *Pseudomonas aeruginosa* PNA1. *Can. J. Microbiol.* 49(2), 85–91. https://doi.org/10.1139/w03-011.

Anjaiah, V., Koedam, N., Nowak-Thompson, B., Loper, J. E., Höfte, M., Tambong, J. T. and Cornelis, P. 1998. Involvement of phenazines and anthranilate in the antagonism of *Pseudomonas aeruginosa* PNA1 and Tn5 derivatives toward Fusarium spp. and Pythium spp. *Mol. Plant Microbe Interact.* 11(9), 847–854. https://doi.org/10.1094/MPMI.1998.11.9.847.

Arenas, F. A., Pugin, B., Henriquez, N. A., Arenas-Salinas, M. A., Dias-Vasquez, W. A., Pozo, M., Munoz, C. M., Chasteen, T. G., Perez-Donoso, J. M. and Vasquez, C. C. 2014. Isolation, identification and characterization of highly tellurite-resistant, tellurite-reducing bacteria from Antarctica. *Polar Sci.* 8, 40–52. https://doi.org/10.1016/j.polar.2014.01.001.

Arrebola, E., Tienda, S., Vida, C., De Vicente, A. and Cazorla, F. M. 2019. Fitness features involved in the biocontrol interaction of *Pseudomonas chlororaphis* with host plants: the case study of PcPCL1606. *Front. Microbiol.* 10, 719. https://doi.org/10.3389/fmicb.2019.00719.

Audenaert, K., Pattery, T., Cornelis, P. and Höfte, M. 2002. Induction of systemic resistance to *Botrytis cinerea* in tomato by *Pseudomonas aeruginosa* 7NSK2: role of salicylic acid, pyochelin, and pyocyanin. *Mol. Plant Microbe Interact.* 15(11), 1147–1156.

Bakker, P. A. H. M., Pieterse, C. M. J. and Loon, L. C. Van 2007. Induced systemic resistance by fluorescent *Pseudomonas* spp. *Phytopathology* 97(2), 239–243.

Barahona, E., Navazo, A., Martínez-Granero, F., Zea-Bonilla, T., Pérez-Jiménez, R. M., Martín, M. and Rivilla, R. 2011. *Pseudomonas fluorescens* F113 mutant with enhanced competitive colonization ability and improved biocontrol activity against fungal root pathogens. *Appl. Environ. Microbiol.* 77(15), 5412–5419. https://doi.org/10.1128/AEM.00320-11.

Bardas, G. A., Lagopodi, A. L., Kadoglidou, K. and Tzavella-Klonari, K. 2009. Biological control of three *Colletotrichum lindemuthianum* races using *Pseudomonas chlororaphis* PCL1391 and *Pseudomonas fluorescens* WCS365. *Biol. Control* 49(2), 139–145. https://doi.org/10.1016/j.biocontrol.2009.01.012.

Barret, M., Egan, F., Moynihan, J., Morrissey, J. P., Lesouhaitier, O. and O'Gara, F. 2013. Characterization of the SPI-1 and Rsp type three secretion systems in *Pseudomonas fluorescens* F113. *Environ. Microbiol. Rep.* 5(3), 377–386. https://doi.org/10.1111/1758-2229.12039.

Battistuzzi, F. U. and Hedges, S. B. 2009. A major clade of prokaryotes with ancient adaptations to life on land. *Mol. Biol. Evol.* 26(2), 335–343. https://doi.org/10.1093/molbev/msn247.

Belda, E., van Heck, R. G. A., José Lopez-Sanchez, M., Cruveiller, S., Barbe, V., Fraser, C., Klenk, H. P., Petersen, J., Morgat, A., Nikel, P. I., Vallenet, D., Rouy, Z., Sekowska, A., Martins dos Santos, V. A. P., de Lorenzo, V., Danchin, A. and Médigue, C. 2016. The revisited genome of *Pseudomonas putida* KT2440 enlightens its value as a robust metabolic chassis. *Environ. Microbiol.* 18(10), 3403–3424. https://doi.org/10.1111/1462-2920.13230.

Berendsen, R. L., van Verk, M. C., Stringlis, I. A., Zamioudis, C., Tommassen, J., Pieterse, C. M. J. and Bakker, P. A. H. M. 2015. Unearthing the genomes of plant-beneficial *Pseudomonas* model strains WCS358, WCS374 and WCS417. *BMC Genomics* 16, 539. https://doi.org/10.1186/s12864-015-1632-z.

Berg, G. 2009. Plant–microbe interactions promoting plant growth and health: perspectives for controlled use of microorganisms in agriculture. *Appl. Microbiol. Biotechnol.* 84(1), 11–18. https://doi.org/10.1007/s00253-009-2092-7.

Berge, O., Monteil, C. L., Bartoli, C., Chandeysson, C., Guilbaud, C., Sands, D. C. and Morris, C. E. 2014. A user's guide to a data base of the diversity of *Pseudomonas syringae* and its application to classifying strains in this phylogenetic complex. *PLoS ONE* 9(9), e105547. https://doi.org/10.1371/journal.pone.0105547.

Bernal, P., Allsopp, L. P., Filloux, A. and Llamas, M. A. 2017. The *Pseudomonas putida* T6SS is a plant warden against phytopathogens. *ISME J.* 11(4), 972–987. https://doi.org/10.1038/ismej.2016.169.

Bernal, P., Llamas, M. A. and Filloux, A. 2018. Type VI secretion systems in plant-associated bacteria. *Environ. Microbiol.* 20(1), 1–15. https://doi.org/10.1111/1462-2920.13956.

Berninger, T., González López, Ó, Bejarano, A., Preininger, C. and Sessitsch, A. 2018. Maintenance and assessment of cell viability in formulation of non-sporulating bacterial inoculants. *Microb. Biotechnol.* 11(2), 277–301. https://doi.org/10.1111/1751-7915.12880.

Berry, C., Fernando, W. G. D., Loewen, P. C. and de Kievit, T. R. 2010. Lipopeptides are essential for *Pseudomonas* sp DF41 biocontrol of *Sclerotinia sclerotiorum*. *Biol. Control* 55(3), 211–218. https://doi.org/10.1016/j.biocontrol.2010.09.011.

Berry, C. L., Brassinga, A. K. C., Donald, L. J., Fernando, W. G. D., Loewen, P. C. and de Kievit, T. R. 2012. Chemical and biological characterization of sclerosin, an antifungal lipopeptide. *Can. J. Microbiol.* 58(8), 1027–1034. https://doi.org/10.1139/w2012-079.

Biessy, A. and Filion, M. 2018. Phenazines in plant-beneficial *Pseudomonas* spp.: biosynthesis, regulation, function and genomics. *Environ. Microbiol.* 20(11), 3905–3917. https://doi.org/10.1111/1462-2920.14395.

Biessy, A., Novinscak, A., Blom, J., Léger, G., Thomashow, L. S., Cazorla, F. M., Josic, D. and Filion, M. 2019. Diversity of phytobeneficial traits revealed by whole-genome analysis of worldwide-isolated phenazine-producing *Pseudomonas* spp. *Environ. Microbiol.* 21(1), 437–455. https://doi.org/10.1111/1462-2920.14476.

Blumer, C. and Haas, D. 2000a. Mechanism, regulation, and ecological role of bacterial cyanide biosynthesis. *Arch. Microbiol.* 173(3), 170–177. https://doi.org/10.1007/s002039900127.

Blumer, C. and Haas, D. 2000b. Iron regulation of the *hcnABC* genes encoding hydrogen cyanide synthase depends on the anaerobic regulator ANR rather than on the global activator GacA in *Pseudomonas fluorescens* CHA0. *Microbiology (Reading)* 146(10), 2417–2424. https://doi.org/10.1099/00221287-146-10-2417.

Botzenhart, K. and Döring, G. 1993. Ecology and epidemiology of *Pseudomonas aeruginosa*. In: Campa, M. (Ed), *Pseudomonas Aeruginosa as an Opportunistic Pathogen*. Plenum Press, New York, pp. 1–18. https://doi.org/10.1007/978-1-4615-3036-7_1.

Braun, S. D., Hofmann, J., Wensing, A., Weingart, H., Ullrich, M. S., Spiteller, D. and Völksch, B. 2010. *In vitro* antibiosis by *Pseudomonas syringae* Pss22d, acting against the bacterial blight pathogen of soybean plants, does not influence *in planta* biocontrol. *J. Phytopathol.* 158(4), 288–295. https://doi.org/10.1111/j.1439-0434.2009.01612.x.

Brazelton, J. N., Pfeufer, E. E., Sweat, T. A., Gardener, B. B. M. and Coenen, C. 2008. 2,4-Diacetylphloroglucinol alters plant root development. *Mol. Plant Microbe Interact.* 21(10), 1349–1358. https://doi.org/10.1094/MPMI-21-10-1349.

Brendel, N., Partida-Martinez, L. P., Scherlach, K. and Hertweck, C. 2007. A cryptic PKS–NRPS gene locus in the plant commensal *Pseudomonas fluorescens* Pf-5 codes for the biosynthesis of an antimitotic rhizoxin complex. *Org. Biomol. Chem.* 5(14), 2211-2213. https://doi.org/10.1039/b707762a.

Bull, C. T., Duffy, B., Voisard, C., Défago, G., Keel, C. and Haas, D. 2001. Characterization of spontaneous *gacS* and *gacA* regulatory mutants of *Pseudomonas fluorescens* biocontrol strain C H A0. Antonie van Leeuwenhoek. *Int. J. Gen. Mol. Microbiol.* 79, 327–336. https://doi.org/10.1023/A:1012061014717.

Bull, C. T., Stack, J. P. and Smilanick, J. L. 1997. *Pseudomonas syringae* strains ESC-10 and ESC-11 survive in wounds on citrus and control green and blue molds of citrus. *Biol. Control* 8(1), 81–88. https://doi.org/10.1006/bcon.1996.0476.

Bull, C. T., Wadsworth, M. L., Sorensen, K. N., Takemoto, J. Y., Austin, R. K. and Smilanick, J. L. 1998. Syringomycin E produced by biological control agents controls green mold on lemons. *Biol. Control* 12(2), 89–95. https://doi.org/10.1006/bcon.1998.0622.

Burr, S. E., Gobeli, S., Kuhnert, P., Goldschmidt-Clermont, E. and Frey, J. 2010. *Pseudomonas chlororaphis* subsp. *piscium* subsp. nov., isolated from freshwater fish. *Int. J. Syst. Evol. Microbiol.* 60(12), 2753–2757. https://doi.org/10.1099/ijs.0.011692-0.

Buysens, S., Heungens, K., Poppe, J. and Hofte, M. 1996. Involvement of pyochelin and pyoverdin in suppression of *Pythium*-induced damping-off of tomato by *Pseudomonas aeruginosa* 7NSK2. *Appl. Environ. Microbiol.* 62(3), 865–871.

Calderón, C. E., Pérez-García, A., De Vicente, A. and Cazorla, F. M. 2013. The *dar* genes of *Pseudomonas chlororaphis* PCL1606 are crucial for biocontrol activity via production of the antifungal compound 2-hexyl, 5-propyl resorcinol. *Mol. Plant Microbe Interact.* 26(5), 554–565. https://doi.org/10.1094/MPMI-01-13-0012-R.

Calderón, C. E., Ramos, C., De Vicente, A. and Cazorla, F. M. 2015. Comparative genomic analysis of *Pseudomonas chlororaphis* PCL1606 reveals new insight into antifungal compounds involved in biocontrol. *Mol. Plant Microbe Interact.* 28(3), 249–260. https://doi.org/10.1094/MPMI-10-14-0326-FI.

Calderón, C. E., Tienda, S., Heredia-Ponce, Z., Arrebola, E., Cárcamo-Oyarce, G., Eberl, L. and Cazorla, F. M. 2019. The compound 2-hexyl, 5-propyl resorcinol has a key role in biofilm formation by the biocontrol rhizobacterium *Pseudomonas chlororaphis* PCL1606. *Front. Microbiol.* 10, 396. https://doi.org/10.3389/fmicb.2019.00396.

Campos, V. L., Valenzuela, C., Yarza, P., Kämpfer, P., Vidal, R., Zaror, C., Mondaca, M. A., Lopez-Lopez, A. and Rosselló-Móra, R. 2010. *Pseudomonas arsenicoxydans* sp nov., an arsenite-oxidizing strain isolated from the Atacama Desert. *Syst. Appl. Microbiol.* 33(4), 193–197. https://doi.org/10.1016/j.syapm.2010.02.007.

Catara, V. 2007. *Pseudomonas corrugata*: plant pathogen and/or biological resource?: pathogen. *Mol. Plant Pathol.* 8(3), 233–244. https://doi.org/10.1111/j.1364-3703.2007.00391.x.

Cautain, B., De Pedro, N., Schulz, C., Pascual, J., Sousa, T. D. S., Martin, J., Pérez-Victoria, I., Asensio, F., González, I., Bills, G. F., Reyes, F., Genilloud, O. and Vicente, F. 2015. Identification of the lipodepsipeptide MDN-0066, a novel inhibitor of VHL/HIF pathway produced by a new *Pseudomonas* species. *PLoS ONE* 10, 1–17. https://doi.org/10.1371/journal.pone.0125221.

Cazorla, F. M., Duckett, S. B., Bergström, E. T., Noreen, S., Odijk, R., Lugtenberg, B. J. J., Thomas-Oates, J. E. and Bloemberg, G. V. 2006. Biocontrol of avocado Dematophora root rot by antagonistic *Pseudomonas fluorescens* PCL1606 correlates with the production of 2-hexyl 5-propyl resorcinol. *Mol. Plant Microbe Interact.* 19(4), 418–428. https://doi.org/10.1094/MPMI-19-0418.

Chae, D. H., Kim, D. R., Cheong, M. S., Lee, Y. B. and Kwak, Y. S. 2020. Investigating the induced systemic resistance mechanism of 2,4-diacetylphloroglucinol (DAPG) using DAPG hydrolase-transgenic *Arabidopsis*. *Plant Pathol. J.* 36(3), 255–266. https://doi.org/10.5423/PPJ.OA.02.2020.0031.

Chatterjee, P., Davis, E., Yu, F., James, S., Wildschutte, J. H., Wiegmann, D. D., Sherman, D. H., Mckay, R. M., Lipuma, J. J. and Wildschutte, H. 2017. Environmental pseudomonads inhibit cystic fibrosis patient-derived *pseudomonas aeruginosa*. *Appl. Environ. Microbiol.* 83(2), 1–13. https://doi.org/https://aem.asm.org/content/83/2/e02701-16.

Cheng, X., Etalo, D. W., van de Mortel, J. E., Dekkers, E., Nguyen, L., Medema, M. H. and Raaijmakers, J. M. 2017. Genome-wide analysis of bacterial determinants of plant growth promotion and induced systemic resistance by *Pseudomonas fluorescens*. *Environ. Microbiol.* 19(11), 4638–4656. https://doi.org/10.1111/1462-2920.13927.

Chi, X., Wang, Y., Miao, J., Feng, Z., Zhang, H., Zhai, J., Zhang, H., Tian, L., Xue, W., Yang, T., Huang, R., Hu, X. and Ge, Y. 2017. Development and characterization of a fusion mutant with the truncated *lacZ* to screen regulatory genes for phenazine biosynthesis in *Pseudomonas chlororaphis* G05. *Biol. Control* 108, 70–76. https://doi.org/10.1016/j.biocontrol.2017.02.012.

Chin-A-Woeng, T. F. C., Bloemberg, G. V., Van Der Bij, A. J., Van Der Drift, K. M. G. M., Schripsema, J., Kroon, B., Scheffer, R. J., Keel, C., Bakker, P. A. H. M., Tichy, H. V., De Bruijn, F. J., Thomas-Oates, J. E., Lugtenberg, B. J. J. 1998. Biocontrol by phenazine-1-carboxamide-producing *Pseudomonas chlororaphis* PCL1391 of tomato root rot caused by *Fusarium oxysporum* f. sp. *radicis-lycopersici*. *Mol. Plant-Microbe Interact.* 11, 1069–1077. https://doi.org/10.1094/MPMI.1998.11.11.1069.

Chincholkar, S. and Thomashow, L. S. 2013. *Microbial Phenazines: Biosynthesis, Agriculture and Health*. Springer, Berlin, Heidelberg. https://doi.org/10.1007/978-3-642-40573-0.

Christiansen, L., Alanin, K. S., Phippen, C. B. W., Olsson, S., Stougaard, P. and Hennessy, R. C. 2020. Fungal-associated molecules induce key genes involved in the biosynthesis of the antifungal secondary metabolites nunamycin and nunapeptin in the biocontrol strain *Pseudomonas fluorescens* In5. *Appl. Environ. Microbiol.* 86(21), 1–14. https://doi.org/10.1128/AEM.01284-20.

Cornelis, P. 2010. Iron uptake and metabolism in pseudomonads. *Appl. Microbiol. Biotechnol.* 86(6), 1637–1645. https://doi.org/10.1007/s00253-010-2550-2.

Costa, R., Van Aarle, I. M., Mendes, R. and Van Elsas, J. D. 2009. Genomics of pyrrolnitrin biosynthetic loci: evidence for conservation and whole-operon mobility within Gram-negative bacteria. *Environ. Microbiol.* 11(1), 159–175. https://doi.org/10.1111/j.1462-2920.2008.01750.x.

Cronin, D., Moënne-Loccoz, Y., Fenton, A., Dunne, C., Dowling, D. N. and O'Gara, F. 1997a. Ecological interaction of a biocontrol Pseudomonas fluorescens strain producing 2,4-diacetylphloroglucinol with the soft rot potato pathogen Erwinia carotovora subsp. *atroseptica*. *FEMS Microbiol. Ecol.* 23(2), 95–106.

Cronin, D., Moënne-Loccoz, Y., Fenton, A., Dunne, C., Dowling, D. N. and O'Gara, F. 1997b. Role of 2,4-diacetylphloroglucinol in the interactions of the biocontrol pseudomonad strain F113 with the potato cyst nematode *Globodera rostochiensis*. *Appl. Environ. Microbiol.* 63(4), 1357–1361. https://doi.org/10.1128/AEM.63.4.1357-1361.1997.

Crouzet, J., Arguelles-arias, A., Dhondt-Cordelier, S., Cordelier, S., Pršic, J., Hoff, G., Mazeyrat-Gourbeyre, F., Baillieul, F., Clément, C., Ongena, M. and Dorey, S. 2020. Biosurfactants in plant protection against diseases: rhamnolipids and lipopeptides case study. *Front. Bioeng. Biotechnol.* 8, 1014. https://doi.org/10.3389/fbioe.2020.01014.

CTGB 2014. Proradix Agro. Registration report–Central Zone.

D'aes, J., De Maeyer, K., Pauwelyn, E. and Höfte, M. 2010. Biosurfactants in plant-*Pseudomonas* interactions and their importance to biocontrol. *Environ. Microbiol. Rep.* 2(3), 359-372. https://doi.org/10.1111/j.1758-2229.2009.00104.x.

D'aes, J., Hua, G. K. H., De Maeyer, K., Pannecoucque, J., Forrez, I., Ongena, M., Dietrich, L. E. P., Thomashow, L. S., Mavrodi, D. V. and Höfte, M. 2011. Biological control of *Rhizoctonia* root rot on bean by phenazineand cyclic lipopeptide-producing Pseudomonas CMR12a. *Phytopathology* 101(8), 996–1004. https://doi.org/10.1094/PHYTO-11-10-0315.

D'aes, J., Kieu, N. P., Léclère, V., Tokarski, C., Olorunleke, F. E., De Maeyer, K., Jacques, P., Höfte, M. and Ongena, M. 2014. To settle or to move? The interplay between two classes of cyclic lipopeptides in the biocontrol strain *Pseudomonas* CMR12a. *Environ. Microbiol.* 16(7), 2282–2300. https://doi.org/10.1111/1462-2920.12462.

De Bruijn, I., De Kock, M. J. D., Yang, M., De Waard, P., Van Beek, T. A. and Raaijmakers, J. M. 2007. Genome-based discovery, structure prediction and functional analysis of cyclic lipopeptide antibiotics in *Pseudomonas* species. *Mol. Microbiol.* 63(2), 417-428. https://doi.org/10.1111/j.1365-2958.2006.05525.x.

de Souza, J. T., de Boer, M., de Waard, P., van Beek, T. A. and Raaijmakers, J. M. 2003. Biochemical, genetic, and zoosporicidal properties of cyclic lipopeptide surfactants produced by *Pseudomonas fluorescens*. *Appl. Environ. Microbiol.* 69(12), 7161-7172. https://doi.org/10.1128/aem.69.12.7161-7172.2003.

De Vleesschauwer, D., Cornelis, P. and Höfte, M. 2006. Redox-active pyocyanin secreted by *Pseudomonas aeruginosa* 7NSK2 triggers systemic resistance to *Magnaporthe grisea* but enhances *Rhizoctonia solani* susceptibility in rice. *Mol. Plant Microbe Interact.* 19(12), 1406–1419. https://doi.org/10.1094/MPMI-19-1406.

De Vleesschauwer, D. and Höfte, M. 2009. Chapter 6 Rhizobacteria-induced systemic resistance. *Adv. Bot. Res.* 51, 223–281. https://doi.org/10.1016/S0065-2296(09)51006-3.

De Vrieze, M., Varadarajan, A. R., Schneeberger, K., Bailly, A., Rohr, R. P., Ahrens, C. H. and Weisskopf, L. 2020. Linking comparative genomics of nine potato-associated *pseudomonas isolates* with their differing biocontrol potential against late blight. *Front. Microbiol.* 11, 857. https://doi.org/10.3389/fmicb.2020.00857.

Devi, K. K. and Kothamasi, D. 2009. *Pseudomonas fluorescens* CHA0 can kill subterranean termite *Odontotermes obesus* by inhibiting cytochrome c oxidase of the termite respiratory chain. *FEMS Microbiol. Lett.* 300(2), 195–200. https://doi.org/10.1111/j.1574-6968.2009.01782.x.

Dimkpa, C., Weinand, T. and Asch, F. 2009. Plant–rhizobacteria interactions alleviate abiotic stress conditions. *Plant Cell Environ.* 32(12), 1682–1694. https://doi.org/10.1111/j.1365-3040.2009.02028.x.

Du, X., Li, Y., Zhou, W., Zhou, Q., Liu, H. and Xu, Y. 2013. Phenazine-1-carboxylic acid production in a chromosomally non-scar triple-deleted mutant *pseudomonas aeruginosa* using statistical experimental designs to optimize yield. *Appl. Microbiol. Biotechnol.* 97(17), 7767–7778. https://doi.org/10.1007/s00253-013-4921-y.

Duan, J., Jiang, W., Cheng, Z., Heikkila, J. J. and Glick, B. R. 2013. The complete genome sequence of the plant growth-promoting bacterium *pseudomonas* sp. UW4. *PLoS ONE* 8(3), e58640. https://doi.org/10.1371/journal.pone.0058640.

Duffy, B. K. and Defago, G. 2000. Controlling instability in *gacS-gacA* regulatory genes during inoculant production of *Pseudomonas fluorescens* biocontrol strains. *Appl.*

Environ. Microbiol. 66(8), 3142–3150. https://doi.org/10.1128/AEM.66.8.3142-3150.2000.

EFSA 2012. Conclusion on the peer review of the pesticide risk assessment of the active substance *Pseudomonas* sp. strain DSMZ 13134. *EFSA J.* 10(12), 1–32. https://doi.org/10.2903/j.efsa.2012.2954.

EFSA 2017. Conclusion on the peer review of the pesticide risk assessment of the active substance *Pseudomonas chlororaphis* strain MA. *EFSA J.* 15(1), e04668. https://doi.org/10.2903/j.efsa.2017.4668.

Emanuele, M. C., Scaloni, A., Lavermicocca, P., Jacobellis, N. S., Camoni, L., Di Giorgio, D., Pucci, P., Paci, M., Segre, A. and Ballio, A. 1998. Corpeptins, new bioactive lipodepsipeptides from cultures of *Pseudomonas corrugata*. *FEBS Lett.* 433(3), 317–320. https://doi.org/10.1016/S0014-5793(98)00933-8.

EPA, U. 2017a. US EPA, Pesticide Product Label, Howler T&O.

EPA, U. 2017b. US EPA, Pesticide Product Label, ZIO.

EPA, U. 2017c. Pesticide Product Label, FROSTBAN B.

EPA, U. 2006. Pesticide Product Label, Spot-less.

EU 2013. Commission implementation regulation (EU) No 829. Off. J. Eur. Union L232, 29–32.

Fang, R., Lin, J., Yao, S., Wang, Y., Wang, J., Zhou, C., Wang, H. and Xiao, M. 2013. Promotion of plant growth, biological control and induced systemic resistance in maize by *Pseudomonas aurantiaca* JD37. *Ann. Microbiol.* 63(3), 1177–1185. https://doi.org/10.1007/s13213-012-0576-7.

Fenton, A. M., Stephens, P. M., Crowley, J., O'Callaghan, M. and O'Gara, F. 1992. Exploitation of gene(s) involved in 2,4-diacetylphloroglucinol biosynthesis to confer a new biocontrol capability to a *Pseudomonas* strain. *Appl. Environ. Microbiol.* 58(12), 3873–3878. https://doi.org/10.1128/AEM.58.12.3873-3878.1992.

Flury, P., Aellen, N., Ruffner, B., Péchy-Tarr, M., Fataar, S., Metla, Z., Dominguez-Ferreras, A., Bloemberg, G., Frey, J., Goesmann, A., Raaijmakers, J. M., Duffy, B., Höfte, M., Blom, J., Smits, T. H. M., Keel, C. and Maurhofer, M. 2016. Insect pathogenicity in plant-beneficial pseudomonads: phylogenetic distribution and comparative genomics. *ISME J.* 10(10), 2527–2542. https://doi.org/10.1038/ismej.2016.5.

Flury, P., Vesga, P., Dominguez-Ferreras, A., Tinguely, C., Ullrich, C. I., Kleespies, R. G., Keel, C. and Maurhofer, M. 2019. Persistence of root-colonizing *Pseudomonas protegens* in herbivorous insects throughout different developmental stages and dispersal to new host plants. *ISME J.* 13(4), 860–872. https://doi.org/10.1038/s41396-018-0317-4.

Flury, P., Vesga, P., Péchy-Tarr, M., Aellen, N., Dennert, F., Hofer, N., Kupferschmied, K. P., Kupferschmied, P., Metla, Z., Ma, Z., Siegfried, S., de Weert, S., Bloemberg, G., Höfte, M., Keel, C. J. and Maurhofer, M. 2017. Antimicrobial and insecticidal: cyclic lipopeptides and hydrogen cyanide produced by plant-beneficial *Pseudomonas* strains CHA0, CMR12a, and PCL1391 contribute to insect killing. *Front. Microbiol.* 8, 100. https://doi.org/10.3389/fmicb.2017.00100.

Frapolli, M., Pothier, J. F., Défago, G. and Moënne-Loccoz, Y. 2012. Evolutionary history of synthesis pathway genes for phloroglucinol and cyanide antimicrobials in plant-associated fluorescent pseudomonads. *Mol. Phylogenet. Evol.* 63(3), 877–890. https://doi.org/10.1016/j.ympev.2012.02.030.

Frasson, D., Opoku, M., Picozzi, T., Torossi, T., Balada, S., Smits, T. H. M. and Hilber, U. 2017. *Pseudomonas wadenswilerensis* sp. nov. and *Pseudomonas reidholzensis* sp.

nov., two novel species within the *Pseudomonas putida* group isolated from forest soil. *Int. J. Syst. Evol. Microbiol.* 67(8), 2853–2861. https://doi.org/10.1099/ijsem.0.002035.

Frederiks, C. and Wesseler, J. H. 2018. A comparison of the EU and US regulatory frameworks for the active substance registration of microbial biological control agents. *Pest Manag. Sci.* 75(1), 87–103. https://doi.org/10.1002/ps.5133.

Fröhlich, A., Buddrus-Schiemann, K., Durner, J., Hartmann, A. and von Rad, U. 2012. Response of barley to root colonization by *Pseudomonas* sp. DSMZ 13134 under laboratory, greenhouse, and field conditions. *J. Plant Interact.* 7(1), 1–9. https://doi.org/10.1080/17429145.2011.597002.

Gao, J., Xie, G., Peng, F. and Xie, Z. 2015. *Pseudomonas donghuensis* sp. nov., exhibiting high-yields of siderophore. *Antonie Leeuwenhoek Int. J. Gen. Mol. Microbiol.* 107(1), 83–94. https://doi.org/10.1007/s10482-014-0306-1.

Gardan, L., Shafik, H., Belouin, S., Broch, R., Grimont, F. and Grimont, P. A. D. 1999. DNA relatedness among the pathovars of *Pseudomonas syringae* and description of *Pseudomonas tremae* sp. nov. and *Pseudomonas cannabina* sp. nov. (ex Sutic and Dowson 1959). *Int. J. Syst. Bacteriol.* 49(2), 469–478. https://doi.org/10.1099/00207713-49-2-469.

Garrido-Sanz, D., Arrebola, E., Martínez-Granero, F., García-Méndez, S., Muriel, C., Blanco-Romero, E., Martín, M., Rivilla, R. and Redondo-Nieto, M. 2017. Classification of isolates from the *Pseudomonas fluorescens* complex into phylogenomic groups based in group-specific markers. *Front. Microbiol.* 8, 413. https://doi.org/10.3389/fmicb.2017.00413.

Garrido-Sanz, D., Meier-Kolthoff, J. P., Göker, M., Martín, M., Rivilla, R. and Redondo-Nieto, M. 2016. Genomic and genetic diversity within the *Pseudomonas fluorescens* complex. *PLoS ONE* 11(2), e0150183. https://doi.org/10.1371/journal.pone.0150183.

Ge, Y., Huang, X., Wang, S., Zhang, X. and Xu, Y. 2004. Phenazine-1-carboxylic acid is negatively regulated and pyoluteorin positively regulated by gacA in *Pseudomonas* sp. M18. *FEMS Microbiol. Lett.* 237, 41–47. https://doi.org/10.1016/j.femsle.2004.06.028.

Geels, F. P. and Schippers, B. 1983. Selection of antagonistic fluorescent *Pseudomonas* spp. and their root colonization and persistence following treatment of seed potatoes. *J. Phytopathol.* 108(3–4), 193–206. https://doi.org/10.1111/j.1439-0434.1983.tb00579.x.

Geudens, N. and Martins, J. C. 2018. Cyclic lipodepsipeptides from *Pseudomonas* spp.– biological Swiss-army knives. *Front. Microbiol.* 9, 1867. https://doi.org/10.3389/fmicb.2018.01867.

Girard, L., Höfte, M. and De Mot, R. 2020a. Lipopeptide families at the interface between pathogenic and beneficial *Pseudomonas*-plant interactions. *Crit. Rev. Microbiol.* 46(4), 397–419. https://doi.org/10.1080/1040841X.2020.1794790.

Girard, L., Lood, C., Rokni-zadeh, H., Noort, V. Van, Lavigne, R. and De Mot, R. 2020b. Reliable identification of environmental *Pseudomonas* isolates using the *rpoD* gene. *Microorganisms* 8(8), 1–12. https://doi.org/10.3390/microorganisms8081166.

Gislason, A. S. and de Kievit, T. R. 2020. Friend or foe? Exploring the fine line between *Pseudomonas brassicacearum* and phytopathogens. *J. Med. Microbiol.* 69(3), 347–360. https://doi.org/10.1099/jmm.0.001145.

Gomez-Lama Cabanas, C., Legarda, G., Ruano-Rosa, D., Pizarro-Tobias, P., Valverde-Corredor, A., Niqui, J. L., Trivino, J. C., Roca, A. and Mercado-Blanco, J. 2018. Indigenous *Pseudomonas* spp. strains from the olive (*Olea europaea* L.) rhizosphere as effective biocontrol agents against *Verticillium dahliae*: from the host roots to the bacterial genomes. *Front. Microbiol.* 9, 277. https://doi.org/10.3389/fmicb.2018.00277.

Gomila, M., Busquets, A., Mulet, M., García-Valdés, E. and Lalucat, J. 2017. Clarification of taxonomic status within the *Pseudomonas syringae* species group based on a phylogenomic analysis. *Front. Microbiol.* 8, 2422. https://doi.org/10.3389/fmicb.2017.02422.

Gomila, M., Peña, A., Mulet, M., Lalucat, J. and García-Valdés, E. 2015. Phylogenomics and systematics in *Pseudomonas*. *Front. Microbiol.* 6, 214. https://doi.org/10.3389/fmicb.2015.00214.

Götze, S. and Stallforth, P. 2020. Structure, properties, and biological functions of nonribosomal lipopeptides from pseudomonads. *Nat. Prod. Rep.* 37(1), 29–54. https://doi.org/10.1039/c9np00022d.

Government of India 2020. Major uses of Biopesticides (Registered under the insecticides act, 1968) [WWW Document]. Available at: http://ppqs.gov.in/divisions/cib-rc/major-uses-of-pesticides.

Gross, H. and Loper, J. E. 2009. Genomics of secondary metabolite production by *Pseudomonas* spp. *Nat. Prod. Rep.* 26(11), 1408–1446. https://doi.org/10.1039/b817075b.

Gross, H., Stockwell, V. O., Henkels, M. D., Nowak-Thompson, B., Loper, J. E. and Gerwick, W. H. 2007. The genomisotopic approach: a systematic method to isolate products of orphan biosynthetic gene clusters. *Chem. Biol.* 14(1), 53–63. https://doi.org/10.1016/j.chembiol.2006.11.007.

Haas, D. and Défago, G. 2005. Biological control of soil-borne pathogens by fluorescent pseudomonads. *Nat. Rev. Microbiol.* 3(4), 307–319. https://doi.org/10.1038/nrmicro1129.

Haas, D. and Keel, C. 2003. Regulation of antibiotic production in root-colonizing *Pseudomonas* spp. and relevance for biological control of plant disease. *Annu. Rev. Phytopathol.* 41, 117–153. https://doi.org/10.1146/annurev.phyto.41.052002.095656.

Han, S. H., Lee, S. J., Moon, J. H., Park, K. H., Yang, K. Y., Cho, B. H., Kim, K. Y., Kim, Y. W., Lee, M. C., Anderson, A. J. and Kim, Y. C. 2006. GacS-dependent production of 2R, 3R-butanediol by *Pseudomonas chlororaphis* O6 is a major determinant for eliciting systemic resistance against *Erwinia carotovora* but not against *Pseudomonas syringae* pv. *tabaci* in tobacco. *Mol. Plant Microbe Interact.* 19(8), 924–930. https://doi.org/10.1094/MPMI-19-0924.

Hardalo, C. and Edberg, S. C. 1997. *Pseudomonas aeruginosa*: assessment of risk from drinking water. *Crit. Rev. Microbiol.* 23(1), 47–75. https://doi.org/10.3109/10408419709115130.

Hardebeck, G. A., Turco, R. F., Latin, R. and Reicher, Z. J. 2004. Application of *Pseudomonas aureofaciens* Tx-1 through irrigation for control of dollar spot and brown patch on fairway-height turf. *HortSci.* 39(7), 1750–1753. https://doi.org/10.21273/HORTSCI.39.7.1750.

Harrison, L. A., Letendre, L. E. O., Kovacevich, P., Pierson, E. and Weller, D. 1993. Purification of an antibiotic effective against *Gaeumannomyces graminis* var. *tritici* produced by a biocontrol agent, *Pseudomonas aureofaciens*. *Soil Biol. Biochem.* 25(2), 215–221.

Hayat, R., Ali, S., Amara, U., Khalid, R. and Ahmed, I. 2010. Soil beneficial bacteria and their role in plant growth promotion: a review. *Ann. Microbiol.* 60(4), 579–598. https://doi.org/10.1007/s13213-010-0117-1.

Heiman, C. M., Wiese, J., Kupferschmied, P., Maurhofer, M., Keel, C. and Vacheron, J. 2020. Draft genome sequence of *Pseudomonas* sp. strain LD120, isolated from the marine alga *Saccharina latissima*. *Microbiol. Resour. Announc.* 9(8), e01305-19. https://doi.org/10.1128/MRA.01305-19.

Hernandez-Jerez, A. F., Adriaanse, P., Aldrich, A., Berny, P., Coja, T., Duquesne, S., Marinovich, M., Millet, M., Pelkonen, O., Pieper, S., Tiktak, A., Topping, C. J., Wolterink, G., Herman, L. and Chiusolo, A. 2020. Statement on the translocation potential by *Pseudomonas chlororaphis* MA342 in plants after seed treatment of cereals and peas and assessment of the risk to humans. *EFSA J.* 18, 6276. https://doi.org/10.2903/j.efsa.2020.6276.

Hesse, C., Schulz, F., Bull, C. T., Shaffer, B. T., Yan, Q., Shapiro, N., Hassan, K. A., Varghese, N., Elbourne, L. D. H., Paulsen, I. T., Kyrpides, N., Woyke, T. and Loper, J. E. 2018. Genome-based evolutionary history of *Pseudomonas* spp. *Environ. Microbiol.* 20(6), 2142–2159. https://doi.org/10.1111/1462-2920.14130.

Hill, D. S., Stein, J. I., Torkewitz, N. R., Morse, A. M., Howell, C. R., Pachlatko, J. P., Becker, J. O. and Ligon, J. M. 1994. Cloning of genes involved in the synthesis of pyrrolnitrin from *Pseudomonas fluorescens* and role of pyrrolnitrin synthesis in biological control of plant disease. *Appl. Environ. Microbiol.* 60(1), 78–85. https://doi.org/10.1128/AEM.60.1.78-85.1994.

Hökeberg, M., Gerhardson, B. and Johnsson, L. 1997. Biological control of cereal seed-borne diseases by seed bacterization with greenhouse-selected bacteria. *Eur. J. Plant Pathol.* 103(1), 25–33. https://doi.org/10.1023/A:1008681608400.

Howell, C. R. and Stipanovic, R. D. 1980. Suppression of *Pythium ultimum*-induced damping-off of cotton seedlings by *Pseudomonas fluorescens* and its antibiotic, pyoluteorin. *Phytopathology* 70, 712–715.

Howell, C. R. and Stipanovic, R. D. 1979. Control of *Rhizoctonia solani* on cotton seedlings with *Pseudomonas fluorescens* and with an antibiotic produced by the bacterium. *Phytopathology* 69(5), 480–482.

Hu, H. B., Xu, Y.-Q., Chen, F., Zhang, X. H. and Hur, B. K. 2005. Isolation and characterization of a new fluorescent *Pseudomonas* strain that produces both phenazine-1-carboxylic acid and pyoluteorin. *J. Microbiol. Biotechnol.* 15, 86–90.

Hu, W., Gao, Q., Hamada, M. S., Dawood, D. H., Zheng, J., Chen, Y. and Ma, Z. 2014. Potential of *Pseudomonas chlororaphis* subsp. *aurantiaca* strain Pcho10 as a biocontrol agent against *Fusarium graminearum*. *Phytopathology* 104(12), 1289–1297. https://doi.org/10.1094/PHYTO-02-14-0049-R.

Huang, J., Xu, Y., Zhang, H., Li, Y., Huang, X., Ren, B. and Zhang, X. 2009. Temperature-dependent expression of *phzM* and its regulatory genes *lasI* and *ptsP* in rhizosphere isolate *Pseudomonas* sp. strain M18. *Appl. Environ. Microbiol.* 75(20), 6568–6580. https://doi.org/10.1128/AEM.01148-09.

Huang, R., Feng, Z., Chi, X., Sun, X., Lu, Y., Zhang, B., Lu, R., Luo, W., Wang, Y., Miao, J. and Ge, Y. 2018. Pyrrolnitrin is more essential than phenazines for *Pseudomonas chlororaphis* G05 in its suppression of *Fusarium graminearum*. *Microbiol. Res.* 215, 55–64. https://doi.org/10.1016/j.micres.2018.06.008.

Huang, X., Zhu, D., Ge, Y., Hu, H., Zhang, X. and Xu, Y. 2004. Identification and characterization of *pltZ*, a gene involved in the repression of pyoluteorin biosynthesis in *Pseudomonas* sp. M18. *FEMS Microbiol. Lett.* 232(2), 197–202. https://doi.org/10.1016/S0378-1097(04)00074-6.

Hultberg, M., Alsberg, T., Khalil, S. and Alsanius, B. 2010a. Suppression of disease in tomato infected by *Pythium ultimum* with a biosurfactant produced by *Pseudomonas koreensis*. *BioControl* 55(3), 435–444. https://doi.org/10.1007/s10526-009-9261-6.

Hultberg, M., Bengtsson, T. and Liljeroth, E. 2010b. Late blight on potato is suppressed by the biosurfactant-producing strain *Pseudomonas koreensis* 2.74 and its biosurfactant. *BioControl* 55(4), 543–550. https://doi.org/10.1007/s10526-010-9289-7.

Hunziker, L., Bönisch, D., Groenhagen, U., Bailly, A., Schulz, S. and Weisskopf, L. 2015. *Pseudomonas* strains naturally associated with potato plants produce volatiles with high potential for inhibition of *Phytophthora infestans*. *Appl. Environ. Microbiol.* 81(3), 821–830. https://doi.org/10.1128/AEM.02999-14.

Iavicoli, A., Boutet, E., Buchala, A. and Métraux, J. P. 2003. Induced systemic resistance in *Arabidopsis thaliana* in response to root inoculation with *Pseudomonas fluorescens* C H A0. *Mol. Plant. Microbe Interact.* 16(10), 851–858. https://doi.org/10.1094/MPMI.2003.16.10.851.

Iswandi, A., Bossier, P., Vandenabeele, J. and Verstraete, W. 1987. Relation between soil microbial activity and the effect of seed inoculation with the rhizopseudomonas strain 7NSK2 on plant growth. *Biol. Fertil. Soils* 3(3), 147–151.

Jahanshah, G., Yan, Q., Gerhardt, H., Pataj, Z., Lämmerhofer, M., Pianet, I., Josten, M., Sahl, H. G., Silby, M. W., Loper, J. E. and Gross, H. 2019. Discovery of the cyclic lipopeptide gacamide A by genome mining and repair of the defective GacA regulator in *Pseudomonas fluorescens* Pf0-1. *J. Nat. Prod.* 82(2), 301–308. https://doi.org/10.1021/acs.jnatprod.8b00747.

Janisiewicz, W. J. and Marchi, A. 1992. Control of storage rots on various pear cultivars with a saprophytic strain of *Pseudomonas syringae*. *Plant Dis.* 76, 555–560.

Jasim, B., Anisha, C., Rohini, S., Kurian, J. M., Jyothis, M. and Radhakrishnan, E. K. 2014. Phenazine carboxylic acid production and rhizome protective effect of endophytic *Pseudomonas aeruginosa* isolated from *Zingiber officinale*. *World J. Microbiol. Biotechnol.* 30(5), 1649–1654. https://doi.org/10.1007/s11274-013-1582-z.

Jeong, J. J., Moon, H. J., Pathiraja, D., Park, B., Choi, I. G. and Kim, K. D. 2018. Draft genome sequences of *Bacillus megaterium* KU143, *Microbacterium testaceum* KU313, and *Pseudomonas protegens* AS15, isolated from stored rice grains. *Genome Announc.* 6(22), e00468-18. https://doi.org/10.1128/genomeA.00468-18.

Jiang, Q., Xiao, J., Zhou, C., Mu, Y., Xu, B., He, Q. and Xiao, M. 2014. Complete genome sequence of the plant growth-promoting rhizobacterium *Pseudomonas aurantiaca* strain JD37. *J. Biotechnol.* 192(A), 85–86. https://doi.org/10.1016/j.jbiotec.2014.10.021.

Jiang, Z., Chen, M., Yu, X. and Xie, Z. 2016. 7-Hydroxytropolone produced and utilized as an iron-scavenger by *Pseudomonas donghuensis*. *BioMetals* 29(5), 817–826. https://doi.org/10.1007/s10534-016-9954-0.

Johnsson, L., Hökeberg, M. and Gerhardson, B. 1998. Performance of the *Pseudomonas chlororaphis* biocontrol agent MA 342 against cereal seed-borne diseases in field experiments. *Eur. J. Plant Pathol.* 104(7), 701–711. https://doi.org/10.1023/A:1008632102747.

Jousset, A., Schuldes, J., Keel, C., Maurhofer, M., Daniel, R., Scheu, S. and Thuermer, A. 2014. Full-genome sequence of the plant growth-promoting bacterium *pseudomonas protegens* CHA0. *Genome Announc.* 2(2), e00322-14. https://doi.org/10.1128/genomeA.00322-14.

Jussila, M. M., Jurgens, G., Lindström, K. and Suominen, L., 2006. Genetic diversity of culturable bacteria in oil-contaminated rhizosphere of *Galega orientalis*. *Environ. Pollut.* 139, 244–257. https://doi.org/10.1016/j.envpol.2005.05.013.

Kabaluk, J. T., Svircev, A. M., Goettel, M. S. and Woo, S. G. 2010. *The Use and Regulation of Microbial Pesticides in Representative Jurisdictions Worldwide*. IOBC Global 99. Available at: www.IOBC-Global.org

Kaduskar, R. D., Scala, G. D., Al Jabri, Z. J. H., Arioli, S., Musso, L., Oggioni, M. R., Dallavalle, S. and Mora, D. 2017. Promysalin is a salicylate-containing antimicrobial with a cell-membrane-disrupting mechanism of action on Gram-positive bacteria. *Sci. Rep.* 7(1), 8861. https://doi.org/10.1038/s41598-017-07567-0.

Kamilova, F., Validov, S., Azarova, T., Mulders, I. and Lugtenberg, B. 2005. Enrichment for enhanced competitive plant root tip colonizers selects for a new class of biocontrol bacteria. *Environ. Microbiol.* 7(11), 1809–1817. https://doi.org/10.1111/j.1462-2920.2005.00889.x.

Kang, B. R., Anderson, A. J. and Kim, Y. C. 2019. Hydrogen cyanide produced by *Pseudomonas chlororaphis* O6 is an aphicidal metabolite. *Can. J. Microbiol.* 65(3), 185–190. https://doi.org/10.1139/cjm-2018-0372.

Kang, B. R., Anderson, A. J. and Kim, Y. C. 2018. Hydrogen cyanide produced by *Pseudomonas chlororaphis* O6 exhibits nematicidal activity against *Meloidogyne hapla*. *Plant Pathol. J.* 34(1), 35–43. https://doi.org/10.5423/PPJ.OA.06.2017.0115.

Kavino, M., Harish, S., Kumar, N., Saravanakumar, D. and Samiyappan, R. 2008. Induction of systemic resistance in banana (*Musa* spp.) against Banana bunchy top virus (BBTV) by combining chitin with root-colonizing Pseudomonas fluorescens strain CHA0. *Eur. J. Plant Pathol.* 120(4), 353–362. https://doi.org/10.1007/s10658-007-9223-8.

Keel, C., Schnider, U., Maurhofer, M., Voisard, C., Laville, J., Burger, U., Wirthner, P., Haas, D. and Défago, G. 1992. Suppression of root diseases by *Pseudomonas fluorescens* CHA0: importance of the bacterial secondary metabolite 2,4-diacetylphloroglucinol. *Mol. Plant Microbe Interact.* 5(1), 4–13.

Kidarsa, T. A., Goebel, N. C., Zabriskie, T. M. and Loper, J. E. 2011. Phloroglucinol mediates cross-talk between the pyoluteorin and 2,4-diacetylphloroglucinol biosynthetic pathways in *Pseudomonas fluorescens* Pf-5. *Mol. Microbiol.* 81(2), 395–414. https://doi.org/10.1111/j.1365-2958.2011.07697.x.

Kilani, J. and Fillinger, S. 2016. Phenylpyrroles: 30 years, two molecules and (nearly) no resistance. *Front. Microbiol.* 7, 2014. https://doi.org/10.3389/fmicb.2016.02014.

Kim, M. S., Kim, Y. C. and Cho, B. H. 2004. Gene expression analysis in cucumber leaves primed by root colonization with *Pseudomonas chlororaphis* O6 upon challenge-inoculation with *Corynespora cassiicola*. *Plant Biol. (Stuttg)* 6(2), 105–108. https://doi.org/10.1055/s-2004-817803.

Kirner, S., Hammer, P. E., Hill, D. S., Altmann, A., Fischer, I., Weislo, L. J., Lanahan, M., van Pée, K. H. and Ligon, J. M. 1998. Functions encoded by pyrrolnitrin biosynthetic genes from *Pseudomonas fluorescens*. *J. Bacteriol.* 180(7), 1939–1943.

Kivisaar, M. 2020. Narrative of a versatile and adept species *Pseudomonas putida*. *J. Med. Microbiol.* 69, 324–338. https://doi.org/10.1099/JMM.0.001137.

Kosina, M., Bartak, M., Maslanova, I., Vavrova Pascutti, A., Sedo, O., Lexa, M. and Sedlacek, I. 2013. *Pseudomonas prosekii* sp. nov., a novel psychrotrophic bacterium from Antarctica. *Curr. Microbiol.* 67, 637–646. https://doi.org/10.1007/s00284-013-0406-6.

Krzyzanowska, D. M., Ossowicki, A. and Jafra, S. 2014. Genome sequence of *Pseudomonas* sp. strain P482, a tomato rhizosphere isolate with broad-spectrum antimicrobial activity. *Genome Announc.* 2(3), e00394-14. https://doi.org/10.1128/genomeA.00394-14.

Krzyzanowska, D. M., Ossowicki, A., Rajewska, M., Maciag, T., Jablonska, M., Obuchowski, M., Heeb, S. and Jafra, S. 2016. When genome-based approach meets the "Old but Good": revealing genes involved in the antibacterial activity of *Pseudomonas* sp. P482 against soft rot pathogens. *Front. Microbiol.* 7, 782. https://doi.org/10.3389/fmicb.2016.00782.

Krzyzanowska, D. M., Potrykus, M., Golanowska, M., Polonis, K., Gwizdek-Wisniewska, A., Lojkowska, E. and Jafra, S. 2012. Rhizosphere bacteria as potential biocontrol agents against soft rot caused by various *Pectobacterium* and *Dickeya* spp. strains. *J. Plant Pathol.* 94, 367–378. https://doi.org/10.4454/JPP.FA.2012.042.

Kuiper, I., Lagendijk, E. L., Pickford, R., Derrick, J. P., Lamers, G. E. M., Thomas-Oates, J. E., Lugtenberg, B. J. J. and Bloemberg, G. V. 2003. Characterization of two *Pseudomonas putida* lipopeptide biosurfactants, putisolvin I and II, which inhibit biofilm formation and break down existing biofilms. *Mol. Microbiol.* 51(1), 97–113. https://doi.org/10.1046/j.1365-2958.2003.03751.x.

Kumar, A., Munder, A., Aravind, R., Eapen, S. J., Tümmler, B. and Raaijmakers, J. M. 2013. Friend or foe: genetic and functional characterization of plant endophytic *Pseudomonas aeruginosa*. *Environ. Microbiol.* 15(3), 764–779. https://doi.org/10.1111/1462-2920.12031.

Kumar, T., Wahla, V., Pandey, P., Dubey, R. C. and Maheshwari, D. K. 2009. Rhizosphere competent *Pseudomonas aeruginosa* in the management of *Heterodera cajani* on sesame. *World J. Microbiol. Biotechnol.* 25(2), 277–285. https://doi.org/10.1007/s11274-008-9890-4.

Kuzmanović, N., Eltlbany, N., Ding, G., Baklawa, M., Min, L., Wei, L. and Smalla, K. 2018. Analysis of the genome sequence of plant beneficial strain *Pseudomonas* sp. RU47. *J. Biotechnol.* 281, 183–192. https://doi.org/10.1016/j.jbiotec.2018.07.023.

Kwon, S. W., Kim, J. S., Park, I. C., Yoon, S. H., Park, D. H., Lim, C. K. and Go, S. J. 2003. *Pseudomonas koreensis* sp. nov., *Pseudomonas umsongensis* sp. nov. and *Pseudomonas jinjuensis* sp. nov., novel species from farm soils in Korea. *Int. J. Syst. Evol. Microbiol.* 53(1), 21–27. https://doi.org/10.1099/ijs.0.02326-0.

Lalucat, J., Mulet, M., Gomila, M. and García-Valdés, E. 2020. Genomics in bacterial taxonomy: impact on the genus Pseudomonas. *Genes (Basel)* 11(2). https://doi.org/10.3390/genes11020139.

Lamers, J. G., Schippers, B. and Geels, F. P. 1988. Soil-borne diseases of wheat in the Netherlands and results of seed bacterization with pseudomonads against *Gaeumannomyces graminis* var. *tritici*, associated with disease resistance. In: Jorna, M. L. and Slootmaker, L. A. J. (Eds), *Cereal Breeding Related to Integrated Cereal Production. Proceedings of the Conference of the Cereal Section of EUCARPIA*. Wageningen, pp. 134–139.

Lanteigne, C., Gadkar, V. J., Wallon, T., Novinscak, A. and Filion, M. 2012. Production of DAPG and HCN by *Pseudomonas* sp. LBUM300 contributes to the biological control of bacterial canker of tomato. *Phytopathology* 102(10), 967–973. https://doi.org/10.1094/PHYTO-11-11-0312.

Laville, J., Blumer, C., Von Schroetter, C., De Gaia, V., Défago, G., Keel, C. and Haas, D. 1998 Characterization of the hcnABC gene cluster encoding hydrogen cyanide synthase and anaerobic regulation by ANR in the strictly aerobic biocontrol agent Pseudomonas fluorescens CHA0. *J. Bacteriol.* 180(12), 3187–3196. https://doi.org/10.1128/JB.180.12.3187-3196.1998.

Le, C. N., Kruijt, M. and Raaijmakers, J. M. 2012. Involvement of phenazines and lipopeptides in interactions between *Pseudomonas* species and *Sclerotium rolfsii*, causal agent of stem rot disease on groundnut. *J. Appl. Microbiol.* 112(2), 390–403. https://doi.org/10.1111/j.1365-2672.2011.05205.x.

Lee, J. H., Ma, K. C., Ko, S. J., Kang, B. R., Kim, I. S. and Kim, Y. C. 2011. Nematicidal activity of a nonpathogenic biocontrol bacterium, *Pseudomonas chlororaphis* O6. *Curr. Microbiol.* 62(3), 746–751. https://doi.org/10.1007/s00284-010-9779-y.

Lemanceau, P., Expert, D., Gaymard, F., Bakker, P. A. H. M. and Briat, J. -F. 2009. Role of iron in plant-microbe interactions. *Adv. Bot. Res.* 51, 491–549. https://doi.org/10.1016/S0065-2296(09)51012-9.

Levy, E., Gough, F. J., Berlin, K. D., Guiana, P. W. and Smith, J. T. 1992 Inhibition of Septoria tritici and other phytopathogenic fungi and bacteria by Pseudomonas fluorescens and its antibiotics. *Plant Pathol.* 41(3), 335–341. https://doi.org/10.1111/j.1365-3059.1992.tb02355.x.

Li, W., De Vleeschouwer, M., Ghequire, M. G. K., Sinnaeve, D., Rokni-Zadeh, H., Madder, A., Rozenski, J., Martins, J. and De Mot, R. 2013a. Xantholysin and promysalin, novel antimicrobial secondary metabolites from Pseudomonas putida rhizosphere isolates. In: *Pseudomonas, 14th International Conference, Abstracts*. Federation of European Microbiological Societies (FEMS).

Li, W., Estrada de los Santos, P., Matthijs, S., Xie, G. L., Busson, R., Cornelis, P., Rozenski, J. and De Mot, R. 2011. Promysalin, a salicylate-containing *Pseudomonas putida* antibiotic, promotes surface colonization and selectively targets other *Pseudomonas*. *Chem. Biol.* 18(10), 1320–1330. https://doi.org/10.1016/j.chembiol.2011.08.006.

Li, W., Rokni-Zadeh, H., De Vleeschouwer, M., Ghequire, M. G. K. K., Sinnaeve, D., Xie, G.-L. L., Rozenski, J., Madder, A., Martins, J. C. and De Mot, R. 2013b. The antimicrobial compound xantholysin defines a new group of *Pseudomonas* cyclic lipopeptides. *PLoS ONE* 8(5), e62946. https://doi.org/10.1371/journal.pone.0062946.

Li, X., Kot, W., Wang, D., Zheng, S., Wang, G. and Hansen, L. H. 2015. Draft genome sequence of Se (IV)-reducing bacterium *Pseudomonas migulae* ES3-33. *Genome Announc.* 3, e00406-15. https://doi.org/10.1128/genomeA.00406-15.Copyright.

Li, Y., Jiang, H., Xu, Y. and Zhang, X. 2008. Optimization of nutrient components for enhanced phenazine-1-carboxylic acid production by *gacA*-inactivated *Pseudomonas* sp. M18G using response surface method. *Appl. Microbiol. Biotechnol.* 77(6), 1207–1217. https://doi.org/10.1007/s00253-007-1213-4.

Liatukas, Ž., Supronienė, S., Ruzgas, V. and Leistrumaitė, A. 2019. Effects of organic seed treatment methods on spring barley seed quality, crop, productivity and disease incidence. *Zemdirbyste Agric.* 106(3), 241–248. https://doi.org/10.13080/z-a.2019.106.031.

Ligon, J. M., Hill, D. S., Hammer, P. E., Torkewitz, N. R., Hofmann, D., Kempf, H. J. and Van Pée, K. H. 2000. Natural products with antifungal activity from *Pseudomonas* biocontrol bacteria. *Pest Manag. Sci.* 56, 688–695. https://doi.org/10.1002/1526-4998(200008)56:8.

Lim, D. J., Yang, S. Y., Noh, M. Y., Lee, C. W., Kim, J. C. and Kim, I. S. 2017. Identification of lipopeptide xantholysins from *Pseudomonas* sp. DJ15 and their insecticidal activity against *Myzus persicae*. *Entomol. Res.* 47(6), 337–343. https://doi.org/10.1111/1748-5967.12241.

Lindow, S. E., McGourty, G. and Elkins, R. 1996. Interactions of antibiotics with *Pseudomonas fluorescens* strain A506 in the control of fire blight and frost injury to pear. *Phytopathology* 86(8), 841–848.

Lindow, S. E. and Suslow, T. V. 2003. Temporal dynamics of the biocontrol agent *Pseudomonas fluorescens* strain A506 in flowers in inoculated pear trees. *Phytopathology* 93(6), 727–737.

Loewen, P. C., Switala, J., Fernando, W. G. D. and de Kievit, T. 2014. Genome sequence of *Pseudomonas brassicacearum* DF41. *Genome Announc.* 2(3), e00390-14. https://doi.org/10.1128/genomeA.00390-14.

Loper, J. E., Hassan, K. A., Mavrodi, D. V., Davis, E. W., Lim, C. K., Shaffer, B. T., Elbourne, L. D. H., Stockwell, V. O., Hartney, S. L., Breakwell, K., Henkels, M. D., Tetu, S. G., Rangel, L. I., Kidarsa, T. A., Wilson, N. L., van de Mortel, J. E., Song, C., Blumhagen, R., Radune, D., Hostetler, J. B., Brinkac, L. M., Durkin, A. S., Kluepfel, D. A., Wechter, W. P., Anderson, A. J., Kim, Y. C., Pierson, L. S., Pierson, E. A., Lindow, S. E., Kobayashi, D. Y., Raaijmakers, J. M., Weller, D. M., Thomashow, L. S., Allen, A. E. and Paulsen, I. T. 2012. Comparative genomics of plant-associated *pseudomonas* spp.: insights into diversity and inheritance of traits involved in multitrophic interactions. *PLoS Genet.* 8(7), e1002784. https://doi.org/10.1371/journal.pgen.1002784.

Loper, J. E., Henkels, M. D., Shaffer, B. T., Valeriote, F. A. and Gross, H. 2008. Isolation and identification of rhizoxin analogs from *Pseudomonas fluorescens* Pf-5 by using a genomic mining strategy. *Appl. Environ. Microbiol.* 74(10), 3085–3093. https://doi.org/10.1128/AEM.02848-07.

Loper, J. E., Kobayashi, D. Y. and Paulsen, I. T. 2007. The genomic sequence of *Pseudomonas fluorescens* Pf-5: insights into biological control. *Phytopathology* 97(2), 233–238. https://doi.org/10.1094/PHYTO-97-2-0233.

Lopes, L. D., Davis, E. W., Pereira E Silva, M. C., Weisberg, A. J., Bresciani, L., Chang, J. H., Loper, J. E. and Andreote, F. D. 2018. Tropical soils are a reservoir for fluorescent *Pseudomonas* spp. biodiversity. *Environ. Microbiol.* 20(1), 62–74. https://doi.org/10.1111/1462-2920.13957.

Lötjönen, T. and Torniainen, M. 2005. Does Cedomon work? *Forskningsnytt* 1, 26–27.

Lugtenberg, B. and Kamilova, F. 2009. Plant-growth-promoting rhizobacteria. *Annu. Rev. Microbiol.* 63, 541–556. https://doi.org/10.1146/annurev.micro.62.081307.162918.

Ma, Y., Rajkumar, M., Zhang, C. and Freitas, H. 2016. Beneficial role of bacterial endophytes in heavy metal phytoremediation. *J. Environ. Manage.* 174, 14–25. https://doi.org/10.1016/j.jenvman.2016.02.047.

Ma, Z., Geudens, N., Kieu, N. P., Sinnaeve, D., Ongena, M., Martins, J. C. and Höfte, M. 2016a. Biosynthesis, chemical structure, and structure-activity relationship of orfamide lipopeptides produced by *Pseudomonas protegens* and related species. *Front. Microbiol.* 7, 382. https://doi.org/10.3389/fmicb.2016.00382.

Ma, Z., Hua, G. K. H., Ongena, M. and Höfte, M. 2016b. Role of phenazines and cyclic lipopeptides produced by *Pseudomonas* sp. CMR12a in induced systemic resistance on rice and bean. *Environ. Microbiol. Rep.* 8(5), 896–904. https://doi.org/10.1111/1758-2229.12454.

Ma, Z., Ongena, M. and Höfte, M. 2017. The cyclic lipopeptide orfamide induces systemic resistance in rice to *Cochliobolus miyabeanus* but not to *Magnaporthe oryzae*. *Plant Cell Rep.* 36(11), 1731–1746. https://doi.org/10.1007/s00299-017-2187-z.

Mahajan-Miklos, S., Rahme, L. G. and Ausubel, F. M. 2000. Elucidating the molecular mechanisms of bacterial virulence using non-mammalian hosts. *Mol. Microbiol.* 37(5), 981–988. https://doi.org/10.1046/j.1365-2958.2000.02056.x.

Maksimov, I. V., Abizgil, R. R. and Pusenkova, L. I. 2011. Plant growth promoting rhizobacteria as alternative to chemical crop protectors from pathogens (review). *Appl. Biochem. Microbiol.* 47, 373–385. https://doi.org/10.1134/S0003683811040090.

Maldonado-González, M. M., Schilirò, E., Prieto, P. and Mercado-Blanco, J. 2015. Endophytic colonization and biocontrol performance of Pseudomonas fluorescens PICF7 in olive (*Olea europaea* L.) are determined neither by pyoverdine production nor swimming motility. *Environ. Microbiol.* 17, 3139–3153. https://doi.org/10.1111/1462-2920.12725.

Mark, G. L., Morrissey, J. P., Higgins, P. and O'Gara, F. 2006. Molecular-based strategies to exploit *Pseudomonas* biocontrol strains for environmental biotechnology applications. *FEMS Microbiol. Ecol.* 56(2), 167–177. https://doi.org/10.1111/j.1574-6941.2006.00056.x.

Matilla, M. A., Ramos, J. L., Bakker, P. A. H. M., Doornbos, R., Badri, D. V., Vivanco, J. M. and Ramos-González, M. I. 2010. *Pseudomonas putida* KT2440 causes induced systemic resistance and changes in *Arabidopsis* root exudation. *Environ. Microbiol. Rep.* 2(3), 381–388. https://doi.org/10.1111/j.1758-2229.2009.00091.x.

Matthijs, S., Tehrani, K. A., Laus, G., Jackson, R. W., Cooper, R. M. and Cornelis, P. 2007. Thioquinolobactin, a *Pseudomonas* siderophore with antifungal and anti-*Pythium* activity. *Environ. Microbiol.* 9(2), 425–434. https://doi.org/10.1111/j.1462-2920.2006.01154.x.

Maurhofer, M., Hase, C., Meuwly, P., Metraux, J. P. and Defago, G. 1994a. Induction of systemic resistance of tobacco to tobacco necrosis virus by the root-colonizing *Pseudomonas fluorescens* strain CHA0: influence of the *gacA* gene and of pyoverdine production. *Phytopathology* 84(2). https://doi.org/10.1094/Phyto-84-139.

Maurhofer, M., Keel, C., Haas, D. and Défago, G. 1994b. Pyoluteorin production by *Pseudomonas fluorescens* strain CHA0 is involved in the suppression of *Pythium* damping-off of cress but not of cucumber. *Eur. J. Plant Pathol.* 100(3–4), 221–232. https://doi.org/10.1007/BF01876237.

Mavrodi, D. V., Blankenfeldt, W. and Thomashow, L. S. 2006. Phenazine compounds in fluorescent *Pseudomonas* spp. biosynthesis and regulation. *Annu. Rev. Phytopathol.* 44, 417–445. https://doi.org/10.1146/annurev.phyto.44.013106.145710.

Mavrodi, D. V., Mavrodi, O. V., Parejko, J. A., Bonsall, R. F., Kwak, Y. S., Paulitz, T. C., Thomashow, L. S. and Weller, D. M. 2012a. Accumulation of the antibiotic phenazine-1-carboxylic acid in the rhizosphere of dryland cereals. *Appl. Environ. Microbiol.* 78(3), 804–812. https://doi.org/10.1128/AEM.06784-11.

Mavrodi, O. V., Mavrodi, D. V., Parejko, J. A., Thomashow, L. S. and Weller, D. M. 2012b. Irrigation differentially impacts populations of indigenous antibiotic-producing *pseudomonas* spp. In the rhizosphere of wheat. *Appl. Environ. Microbiol.* 78(9), 3214–3220. https://doi.org/10.1128/AEM.07968-11.

Mavrodi, D. V., Parejko, J. A., Mavrodi, O. V., Kwak, Y. S., Weller, D. M., Blankenfeldt, W. and Thomashow, L. S. 2013. Recent insights into the diversity, frequency and ecological roles of phenazines in fluorescent *Pseudomonas* spp. *Environ. Microbiol.* 15(3), 675-686. https://doi.org/10.1111/j.1462-2920.2012.02846.x.

May, R., Völksch, B. and Kampmann, G. (1997). Antagonistic activities of epiphythic bacteria from soybean leaves against *Pseudomonas syringae pv. glycinea in vitro* and in planta. *Microb Ecol* 43:118–124.

Mehnaz, S., Baig, D. N., Jamil, F., Weselowski, B. and Lazarovits, G. 2009. Characterization of a phenazine and hexanoyl homoserine lactone producing *Pseudomonas aurantiaca* strain PB-St2, isolated from sugarcane stem. *J. Microbiol. Biotechnol.* 19(12), 1688–1694. https://doi.org/10.4014/jmb.0904.04022.

Mehnaz, S., Bauer, J. S. and Gross, H. 2014. Complete genome sequence of the sugar cane endophyte *Pseudomonas aurantiaca* PB-St2, a disease-suppressive bacterium with antifungal activity toward the plant pathogen *Colletotrichum falcatum*. *Genome Announc.* 2(1), e01108-13. https://doi.org/10.1128/genomeA.01108-13.

Mehnaz, S., Saleem, R. S. Z., Yameen, B., Pianet, I., Schnakenburg, G., Pietraszkiewicz, H., Valeriote, F., Josten, M., Sahl, H. G., Franzblau, S. G. and Gross, H. 2013. Lahorenoic acids A-C, Ortho-dialkyl-substituted aromatic acids from the biocontrol strain *Pseudomonas aurantiaca* PB-St2. *J. Nat. Prod.* 76(2), 135–141. https://doi.org/10.1021/np3005166.

Melnyk, R. A., Hossain, S. S. and Haney, C. H. 2019. Convergent gain and loss of genomic islands drive lifestyle changes in plant-associated *pseudomonas*. *ISME J.* 13(6), 1575–1588. https://doi.org/10.1038/s41396-019-0372-5.

Mendes, L. W., de Chaves, M. G., Fonseca, M. C., Mendes, R., Raaijmakers, J. M. and Tsai, S. M. 2019. Resistance breeding of common bean shapes the physiology of the rhizosphere microbiome. *Front. Microbiol.* 10, 2252. https://doi.org/10.3389/fmicb.2019.02252.

Mendes, L. W. and Mendes, R. 2018. Breeding for soil-borne pathogen resistance impacts active rhizosphere microbiome of common bean. *ISME J.* 12(12), 3038–3042. https://doi.org/10.1038/s41396-018-0234-6.

Mendes, L. W., Raaijmakers, J. M., Hollander, M. De, Mendes, R. and Tsai, S. M. 2017. Influence of resistance breeding in common bean on rhizosphere microbiome composition and function. *ISME J.* 12(1), 212–224. https://doi.org/10.1038/ismej.2017.158.

Mendes, R., Kruijt, M., De Bruijn, I., Dekkers, E., Van Der Voort, M., Schneider, J. H. M., Piceno, Y. M., DeSantis, T. Z., Andersen, G. L., Bakker, P. A. H. M. and Raaijmakers, J. M. 2011. Deciphering the rhizosphere microbiome for disease-suppressive bacteria. *Science* 332(6033), 1097–1100. https://doi.org/10.1126/science.1203980.

Mercado-Blanco, J. 2015. *Pseudomonas* strains that exert biocontrol of plant pathogens. In: Ramos, J.-L., Goldberg, J. B. and Filloux, A. (Eds), *Pseudomonas: New Aspects of Pseudomonas Biology* (vol. 7). Springer, Dordrecht, pp. 121–172. https://doi.org/10.1007/978-94-017-9555-5.

Meyer, J. M. 2000. Pyoverdines: pigments, siderophores and potential taxonomic markers of fluorescent *Pseudomonas* species. *Arch. Microbiol.* 174(3), 135–142. https://doi.org/10.1007/s002030000188.

Meziane, H., Van Der Sluis, I., van Loon, L. C., Höfte, M. and Bakker, P. A. H. M. 2005. Determinants of *Pseudomonas putida* WCS358 involved in inducing systemic resistance in plants. *Mol. Plant Pathol.* 6(2), 177–185. https://doi.org/10.1111/j.1364-3703.2005.00276.x.

Michavila, G., Adler, C., De Gregorio, P. R., Lami, M. J., Caram Di Santo, M. C., Zenoff, A. M., de Cristobal, R. E. and Vincent, P. A. 2017. *Pseudomonas protegens* CS1 from the lemon phyllosphere as a candidate for citrus canker biocontrol agent. *Plant Biol. (Stuttg)* 19(4), 608–617. https://doi.org/10.1111/plb.12556.

Michelsen, C. F., Jensen, H., Venditto, V. J., Hennessy, R. C. and Stougaard, P. 2015a. Bioactivities by a crude extract from the Greenlandic *Pseudomonas* sp. In5 involves the nonribosomal peptides, nunamycin and nunapeptin. *PeerJ* 3, e1476. https://doi.org/10.7717/peerj.1476.

Michelsen, C. F., Watrous, J., Glaring, M. A., Kersten, R., Koyama, N. and Dorrestein, P. C. 2015b. Nonribosomal peptides, key biocontrol components for *Pseudomonas fluorescens* In5 , isolated from a Greenlandic suppressive soil. *mBio* 6, e00079-15. https://doi.org/10.1128/mBio.00079-15.Editor.

Michelsen, C. F. and Stougaard, P. 2011. A novel antifungal *Pseudomonas fluorescens* isolated from potato soils in Greenland. *Curr. Microbiol.* 62(4), 1185–1192. https://doi.org/10.1007/s00284-010-9846-4.

Mishra, J. and Arora, N. K. 2018. Secondary metabolites of fluorescent pseudomonads in biocontrol of phytopathogens for sustainable agriculture. *Appl. Soil Ecol.* 125, 35–45. https://doi.org/10.1016/j.apsoil.2017.12.004.

Molina, L. Á., Ramos, C., Duque, E., Ronchel, M. C., García, J. M., Wyke, L. and Ramos, J. L. 2000. Survival of *Pseudomonas putida* KT2440 in soil and in the rhizosphere of plants under greenhouse and environmental conditions. *Soil Biol. Biochem.* 32(3), 315–321. https://doi.org/10.1016/S0038-0717(99)00156-X.

Molloy, D. P., Mayer, D. A., Giamberini, L. and Gaylo, M. J. 2013. Mode of action of *Pseudomonas fluorescens* strain CL145A, a lethal control agent of dreissenid mussels (Bivalvia: Dreissenidae). *J. Invertebr. Pathol.* 113(1), 115–121. https://doi.org/10.1016/j.jip.2012.12.013.

Moradali, M. F., Ghods, S. and Rehm, B. H. A. 2017. *Pseudomonas aeruginosa* lifestyle: a paradigm for adaptation, survival, and persistence. *Front. Cell. Infect. Microbiol.* 7, 39. https://doi.org/10.3389/fcimb.2017.00039.

Morris, C. E., Monteil, C. L. and Berge, O. 2013. The life history of *Pseudomonas syringae*: linking agriculture to earth system processes. *Annu. Rev. Phytopathol.* 51, 85–104. https://doi.org/10.1146/annurev-phyto-082712-102402.

Moruzzi, S., Firrao, G., Polano, C., Borselli, S., Loschi, A., Ermacora, P., Loi, N. and Martini, M. 2017. Genomic-assisted characterisation of *Pseudomonas* sp. strain Pf4, a potential biocontrol agent in hydroponics. *Biocontrol Sci. Technol.* 27(8), 969–991. https://doi.org/10.1080/09583157.2017.1368454.

Mossialos, D., Meyer, J. M., Budzikiewicz, H., Wolff, U., Koedam, N., Baysse, C., Anjaiah, V. and Cornelis, P. 2000. Quinolobactin, a new siderophore of *Pseudomonas fluorescens* ATCC 17400, the production of which is repressed by the cognate pyoverdine. *Appl. Environ. Microbiol.* 66(2), 487–492. https://doi.org/10.1128/AEM.66.2.487-492.2000.

Moszczynska, E., Pytlarz-Kozicka, M. and Grzeszczuk, J. 2015. The impact of applying biological treatment on the infection of potato tubers by the fungus *Rhizoctonia solani* and the bacterium *Streptomyces scabiei*. *J. Res. Appl. Agric. Eng.* 60(4), 46–50.

Mulet, M., Lalucat, J. and García-Valdés, E. 2010. DNA sequence-based analysis of the *Pseudomonas* species. *Environ. Microbiol.* 12(6), 1513–1530. https://doi.org/10.1111/j.1462-2920.2010.02181.x.

Müller, T. and Behrendt, U. 2021. Exploiting the biocontrol potential of plant-associated pseudomonads – a step towards pesticide-free agriculture? *Biol. Control* 155, 104538. https://doi.org/10.1016/j.biocontrol.2021.104538.

Muzio, F. M., Agaras, B. C., Masi, M., Tuzi, A., Evidente, A. and Valverde, C. 2020. 7-hydroxytropolone is the main metabolite responsible for the fungal antagonism of *Pseudomonas donghuensis* strain SVBP6. *Environ. Microbiol.* 22(7), 2550–2563. https://doi.org/10.1111/1462-2920.14925.

Naseby, D. C., Way, J. A., Bainton, N. J. and Lynch, J. M. 2001. Biocontrol of *Pythium* in the pea rhizosphere by antifungal metabolite producing and non-producing Pseudomonas strains. *J. Appl. Microbiol.* 90, 421–429.

Nazir, R., Mazurier, S., Yang, P., Lemanceau, P. and van Elsas, J. D. 2017. The ecological role of type three secretion systems in the interaction of bacteria with fungi in soil and related habitats is diverse and context-dependent. *Front. Microbiol.* 8, 38. https://doi.org/10.3389/fmicb.2017.00038.

Nelson, K. E., Weinel, C., Paulsen, I. T., Dodson, R. J., Hilbert, H., Martins Dos Santos, V. A. P., Fouts, D. E., Gill, S. R., Pop, M., Holmes, M., Brinkac, L., Beanan, M., DeBoy, R. T., Daugherty, S., Kolonay, J., Madupu, R., Nelson, W., White, O., Peterson, J., Khouri, H., Hance, I., Lee, P. C., Holtzapple, E., Scanlan, D., Tran, K., Moazzez, A., Utterback, T., Rizzo, M., Lee, K., Kosack, D., Moestl, D., Wedler, H., Lauber, J., Stjepandic, D., Hoheisel, J., Straetz, M., Heim, S., Kiewitz, C., Eisen, J. A., Timmis, K. N., Düsterhöft, A., Tümmler, B. and Fraser, C. M. 2002. Complete genome sequence and comparative analysis of the metabolically versatile *Pseudomonas putida* KT2440. *Environ. Microbiol.* 4(12), 799–808.

Nesemann, K., Braus-stromeyer, S. A., Thuermer, A., Daniel, R., Mavrodi, D. V. Thomashow, L. S., Weller, D. M. and Braus, H. 2015. Draft genome sequence of the phenazine-producing *Pseudomonas fluorescens* strain 2-79. *Genome Announc.* 3, 5–6. https://doi.org/10.1128/genomeA.00130-15.Copyright.

Ngumbi, E. and Kloepper, J. 2016. Bacterial-mediated drought tolerance : current and future prospects. *Appl. Soil Ecol.* 105, 109–125. https://doi.org/10.1016/j.apsoil.2016.04.009.

Nguyen, D. D., Melnik, A. V., Koyama, N., Lu, X., Schorn, M., Fang, J., Aguinaldo, K., Lincecum, T. L., Ghequire, M. G. K., Carrion, V. J., Cheng, T. L., Duggan, B. M., Malone, J. G., Mauchline, T. H., Sanchez, L. M., Kilpatrick, A. M., Raaijmakers, J. M., De Mot, R., Moore, B. S., Medema, M. H. and Dorrestein, P. C. 2016. Indexing the pseudomonas specialized metabolome enabled the discovery of poaeamide B and the bananamides. *Nat. Microbiol.* 2, 16197. https://doi.org/10.1038/nmicrobiol.2016.197.

Northover, J. and Zhou, T. 2002. Control of rhizopus rot of peaches with postharvest treatments of tebuconazole, fludioxonil, and *Pseudomonas syringae*. *Can. J. Plant Pathol.* 24(2), 144–153. https://doi.org/10.1080/07060660309506989.

Nowak-Thompson, B., Hammer, P. E., Hill, D. S., Stafford, J., Torkewitz, N., Gaffney, T. D., Lam, S. T. and Ligon, J. M. 2003. 2, 5-Dialkylresorcinol biosynthesis in Pseudomonas aurantiaca : novel head-to-head condensation of two fatty acid-derived precursors. *J. Bacteriol.* 185, 860–869. https://doi.org/10.1128/JB.185.3.860.

Nyfeler, R. and Ackermann, P. 1992. Phenylpyrroles, a new class of agricultural fungicides related to the natural antibiotic pyrrolnitrin III. *ACS Symposium Series*. American Chemical Society, 395-404. https://doi.org/10.1021/bk-1992-0504.ch036.

Ogura, K., Shimada, K. and Miyoshi-Akiyama, T. 2019. A multilocus sequence typing scheme of *Pseudomonas putida* for clinical and environmental isolates. *Sci. Rep.* 9(1), 13980. https://doi.org/10.1038/s41598-019-50299-6.

Olorunleke, F. E., Hua, G. K. H., Kieu, N. P., Ma, Z. and Höfte, M. 2015a. Interplay between orfamides, sessilins and phenazines in the control of *Rhizoctonia* diseases by *Pseudomonas* sp. CMR12a. *Environ. Microbiol. Rep.* 7(5), 774–781. https://doi.org/10.1111/1758-2229.12310.

Olorunleke, F. E., Kieu, N. P., Höfte, M., Kieu Phuong, N. and Höfte, M. 2015b. Recent advances in *Pseudomonas* biocontrol. In: Murillo, J., Vinatzer, B. A., Jackson, R. W. and Arnold, D. L. (Eds), *Bacteria-Plant Interactions: Advanced Research and Future Trends*. Caister Academic Press, UK, pp. 167–198. https://doi.org/10.21775/9781908230584.07.

Omoboye, O. O. 2019. Cyclic lipopeptide diversity and biocontrol versatility of Pseudomonas species associated with the cocoyam rhizospere. PhD thesis. Ghent University, Belgium.

Omoboye, O. O., Geudens, N., Duban, M., Chevalier, M., Flahaut, C., Martins, J. C., Leclère, V., Oni, F. E. and Höfte, M. 2019a. *Pseudomonas* sp. COW3 produces new bananamide-type cyclic lipopeptides with antimicrobial activity against *Pythium myriotylum* and *Pyricularia oryzae*. *Molecules* 24(22), 4170. https://doi.org/10.3390/molecules24224170.

Omoboye, O. O., Oni, F. E., Batool, H., Yimer, H. Z., De Mot, R. and Höfte, M. 2019b. *Pseudomonas* cyclic lipopeptides suppress the rice blast fungus *Magnaporthe oryzae* by induced resistance and direct antagonism. *Front. Plant Sci.* 10, 901. https://doi.org/10.3389/fpls.2019.00901.

Oni, F. E., Geudens, N., Adiobo, A., Omoboye, O. O., Enow, E. A., Onyeka, J. T., Salami, A. E., De Mot, R., Martins, J. C. and Höfte, M. 2020a. Biosynthesis and antimicrobial activity of pseudodesmin and viscosinamide cyclic lipopeptides produced by pseudomonads associated with the cocoyam rhizosphere. *Microorganisms* 8(7), 1-26. https://doi.org/10.3390/microorganisms8071079.

Oni, F. E., Geudens, N., Onyeka, J. T., Olorunleke, O. F., Salami, A. E., Omoboye, O. O., Arias, A. A., Adiobo, A., De Neve, S., Ongena, M., Martins, J. C. and Höfte, M. 2020b. Cyclic lipopeptide-producing *Pseudomonas koreensis* group strains dominate the cocoyam rhizosphere of a *Pythium* root rot suppressive soil contrasting with *P. putida* prominence in conducive soils. *Environ. Microbiol.* 22(12), 5137–5155. https://doi.org/10.1111/1462-2920.15127.

Oni, F. E., Geudens, N., Omoboye, O. O., Bertier, L., Hua, H. G. K., Adiobo, A., Sinnaeve, D., Martins, J. C. and Höfte, M. 2019a. Fluorescent *Pseudomonas* and cyclic lipopeptide diversity in the rhizosphere of cocoyam (Xanthosoma sagittifolium). *Environ. Microbiol.* 21(3), 1019–1034. https://doi.org/10.1111/1462-2920.14520.

Oni, F. E., Olorunleke, O. F. and Höfte, M. 2019b. Phenazines and cyclic lipopeptides produced by *Pseudomonas* sp. CMR12a are involved in the biological control of *Pythium myriotylum* on cocoyam (*Xanthosoma sagittifolium*). *Biol. Control* 129, 109-114. https://doi.org/10.1016/j.biocontrol.2018.10.005.

Parejko, J. A., Mavrodi, D. V., Mavrodi, O. V., Weller, D. M. and Thomashow, L. S. 2013. Taxonomy and distribution of phenazine-producing *Pseudomonas* spp. in the dryland agroecosystem of the Inland Pacific Northwest, United States. *Appl. Environ. Microbiol.* 79(12), 3887–3891. https://doi.org/10.1128/AEM.03945-12.

Park, J. Y., Yang, S. Y., Kim, Y. C., Kim, J. C., Dang, Q. L. Le, Kim, J. J. and Kim, I. S. 2012. Antiviral peptide from *Pseudomonas chlororaphis* O6 against tobacco mosaic virus (TMV). *J. Korean Soc. Appl. Biol. Chem.* 55(1), 89–94. https://doi.org/10.1007/s13765-012-0015-2.

Partida-Martinez, L. P. and Hertweck, C. 2007. A gene cluster encoding rhizoxin biosynthesis in *Burkholderia rhizoxina*, the bacterial endosymbiont of the fungus *Rhizopus microsporus*. ChemBioChem. *Chembiochem* 8(1), 41–45. https://doi.org/10.1002/cbic.200600393.

Paterson, J., Jahanshah, G., Li, Y., Wang, Q., Mehnaz, S. and Gross, H. 2017. The contribution of genome mining strategies to the understanding of active principles of PGPR strains. *FEMS Microbiol. Ecol.* 93(3), 1–12. https://doi.org/10.1093/femsec/fiw249.

Paul, N. C., Ji, S. H., Deng, J. X. and Yu, S. H. 2013. Assemblages of endophytic bacteria in chili pepper (*Capsicum annuum* L.) and their antifungal activity against phytopathogens *in vitro*. *Plant Omi. J.* 41, 441–448.

Paulin, M. M., Novinscak, A., Lanteigne, C., Gadkar, V. J. and Filion, M. 2017. Interaction between 2,4-diacetylphloroglucinol- and hydrogen cyanide-producing *Pseudomonas brassicacearum* LBUM300 and *Clavibacter michiganensis* subsp. *michiganensis* in the tomato rhizosphere. *Appl. Environ. Microbiol.* 83(13), 1–13. https://doi.org/10.1128/AEM.00073-17.

Paulsen, I. T., Press, C. M., Ravel, J., Kobayashi, D. Y., Myers, G. S. A., Mavrodi, D. V., DeBoy, R. T., Seshadri, R., Ren, Q., Madupu, R., Dodson, R. J., Durkin, A. S., Brinkac, L. M., Daugherty, S. C., Sullivan, S. A., Rosovitz, M. J., Gwinn, M. L., Zhou, L., Schneider, D. J., Cartinhour, S. W., Nelson, W. C., Weidman, J., Watkins, K., Tran, K., Khouri, H., Pierson, E. A., Pierson, L. S., Thomashow, L. S. and Loper, J. E. 2005. Complete genome sequence of the plant commensal *Pseudomonas fluorescens* Pf-5. *Nat. Biotechnol.* 23(7), 873–878. https://doi.org/10.1038/nbt1110.

Péchy-Tarr, M., Bruck, D. J., Maurhofer, M., Fischer, E., Vogne, C., Henkels, M. D., Donahue, K. M., Grunder, J., Loper, J. E. and Keel, C. 2008. Molecular analysis of a novel gene cluster encoding an insect toxin in plant-associated strains of *Pseudomonas fluorescens*. *Environ. Microbiol.* 10(9), 2368–2386. https://doi.org/10.1111/j.1462-2920.2008.01662.x.

Peix, A., Ramírez-Bahena, M. H. and Velázquez, E. 2018. The current status on the taxonomy of*Pseudomonas* revisited: an update. *Infect. Genet. Evol.* 57, 106–116. https://doi.org/10.1016/j.meegid.2017.10.026.

Peix, A., Ramírez-Bahena, M. H. and Velázquez, E. 2009. Historical evolution and current status of the taxonomy of genus *Pseudomonas*. *Infect. Genet. Evol.* 9(6), 1132–1147. https://doi.org/10.1016/j.meegid.2009.08.001.

Peix, A., Valverde, A., Rivas, R., Igual, J. M., Ramírez-Bahena, M. H., Mateos, P. F., Santa-Regina, I., Rodríguez-Barrueco, C., Martínez-Molina, E. and Velázquez, E. 2007. Reclassification of *Pseudomonas aurantiaca* as a synonym of *Pseudomonas chlororaphis* and proposal of three subspecies, *P. chlororaphis* subsp. *chlororaphis* subsp. nov. *P. chlororaphis* subsp. *aureofaciens* subsp. nov. *Int. J. Syst. Evol. Microbiol.* 57(6), 1286–1290. https://doi.org/10.1099/ijs.0.64621-0.

Pérez-Jaramillo, J. E., Mendes, R. and Raaijmakers, J. M. 2016. Impact of plant domestication on rhizosphere microbiome assembly and functions. *Plant Mol. Biol.* 90(6), 635–644. https://doi.org/10.1007/s11103-015-0337-7.

Perneel, M., D'Hondt, L., De Maeyer, K., Adiobo, A., Rabaey, K. and Höfte, M. 2008. Phenazines and biosurfactants interact in the biological control of soil-borne diseases caused by *Pythium* spp. *Environ. Microbiol.* 10(3), 778–788. https://doi.org/10.1111/j.1462-2920.2007.01501.x.

Perneel, M., Heyrman, J., Adiobo, A., De Maeyer, K., Raaijmakers, J. M., De Vos, P. and Höfte, M. 2007. Characterization of CMR5c and CMR12a, novel fluorescent *Pseudomonas* strains from the cocoyam rhizosphere with biocontrol activity. *J. Appl. Microbiol.* 103(4), 1007–1020. https://doi.org/10.1111/j.1365-2672.2007.03345.x.

Pfender, W. F., Kraus, J. and Loper, J. E. 1993. A genomic region from *Pseudomonas fluorescens* Pf-5 required for pyrrolnitrin production and inhibition of *Pyrenophora tritici-repentis* in wheat straw. *Phytopathology* 83(11). https://doi.org/10.1094/Phyto-83-1223.

Phillips, D. A., Fox, T. C., King, M. D., Bhuvaneswari, T. V. and Teuber, L. R. 2004. Microbial products trigger amino acid exudation from plant roots. *Plant Physiol.* 136(1), 2887–2894. https://doi.org/10.1104/pp.104.044222.

Philmus, B., Shaffer, B. T., Kidarsa, T. A., Yan, Q., Raaijmakers, J. M., Begley, T. P. and Loper, J. E. 2015. Investigations into the biosynthesis, regulation, and self-resistance of toxoflavin in *Pseudomonas protegens* Pf-5. *Chembiochem* 16(12), 1782–1790. https://doi.org/10.1002/cbic.201500247.

Pierson, L. S. and Thomashow, L. S. 1992. Cloning and heterologous expression of the phenazine biosynthetic locus from *Pseudomonas aureofaciens* 30-84. *Mol. Plant Microbe Interact.* 5(4), 330–339 https://doi.org/10.1094/mpmi-5-330.

Pieterse, C. M. J., Zamioudis, C., Berendsen, R. L., Weller, D. M., Van Wees, S. C. M. and Bakker, P. A. H. M. 2014. Induced systemic resistance by beneficial microbes. *Annu. Rev. Phytopathol.* 52, 347–375. https://doi.org/10.1146/annurev-phyto-082712-102340.

Polano, C., Martini, M., Savian, F., Moruzzi, S., Ermacora, P. and Firrao, G. 2019. Genome sequence and antifungal activity of two niche-sharing *pseudomonas protegens* related strains isolated from hydroponics. *Microb. Ecol.* 77(4), 1025–1035. https://doi.org/10.1007/s00248-018-1238-5.

Powell, J. F., Vargas, J. M., Nair, M. G., Detweiler, A. R. and Chandra, A. 2000. Management of dollar spot on creeping bentgrass with metabolites of *Pseudomonas aureofaciens* (TX-1). *Plant Dis.* 84(1), 19–24. https://doi.org/10.1094/PDIS.2000.84.1.19.

Preston, G., Spiers, A., Jackson, R., Gal, M., Gehrig, S., Malone, J., Moon, C., Godfrey, S., Bertrand, N., Field, D. and Rainey, P. 2003. Pseudomonas in the underworld: the secret life of Pseudomonas fluorescens SBW25. In: Iacobellis, N. S., Collmer, A., Hutcheson, S., Mansfield, J., Morris, C. E., Murillo, J., Schaad, N., Stead, D., Surico, G. and Ulrich, M. (Eds), *Pseudomonas Syringae and Related Pathogens. Biology and Genetics*. Kluwer Academic Publishers, pp. 347–353. Amsterdam, the Netherlands.

Preston, G. M., Bertrand, N. and Rainey, P. B. 2001. Type III secretion in plant growth-promoting *pseudomonas fluorescens* SBW25. *Mol. Microbiol.* 41(5), 999–1014. https://doi.org/10.1046/j.1365-2958.2001.02560.x.

Price-Whelan, A., Dietrich, L. E. P. and Newman, D. K. 2006. Rethinking "secondary" metabolism: physiological roles for phenazine antibiotics. *Nat. Chem. Biol.* 2(2), 71–78. https://doi.org/10.1038/nchembio764.

Pukatzki, S., Kessin, R. H. and Mekalanos, J. J. 2002. The human pathogen *Pseudomonas aeruginosa* utilizes conserved virulence pathways to infect the social amoeba

Dictyostelium discoideum. *Proc. Natl. Acad. Sci. U. S. A.* 99(5), 3159–3164. https://doi.org/10.1073/pnas.052704399.

Raaijmakers, J. M., de Bruijn, I., Nybroe, O. and Ongena, M. 2010. Natural functions of lipopeptides from *Bacillus* and *Pseudomonas*: more than surfactants and antibiotics. *FEMS Microbiol. Rev.* 34(6), 1037–1062. https://doi.org/10.1111/j.1574-6976.2010.00221.x.

Raaijmakers, J. M. and Mazzola, M. 2012. Diversity and natural functions of antibiotics produced by beneficial and plant pathogenic bacteria. *Annu. Rev. Phytopathol.* 50, 403–424. https://doi.org/10.1146/annurev-phyto-081211-172908.

Raaijmakers, J. M., Vlami, M. and de Souza, J. T. 2002. Antibiotic production by bacterial biocontrol agents. *Antonie Leeuwenhoek Int. J. Gen. Mol. Microbiol.* 81(1–4), 537–547. https://doi.org/10.1023/A:1020501420831.

Rajkumar, M., Bruno, L. B., Banu, J. R., Bruno, L. B. and Banu, J. R. 2017. Alleviation of environmental stress in plants: the role of beneficial *Pseudomonas* spp. *Crit. Rev. Environ. Sci. Technol.* 47, 372–407. https://doi.org/10.1080/10643389.2017.1318619.

Ramette, A., Frapolli, M., Saux, M. F. Le, Gruffaz, C., Meyer, J. M., Défago, G., Sutra, L. and Moënne-Loccoz, Y. 2011. *Pseudomonas protegens* sp. nov., widespread plant-protecting bacteria producing the biocontrol compounds 2,4-diacetylphloroglucinol and pyoluteorin. *Syst. Appl. Microbiol.* 34(3), 180–188. https://doi.org/10.1016/j.syapm.2010.10.005.

Rangel, L. I., Henkels, M. D., Shaffer, B. T., Walker, F. L., Davis, E. W., Stockwell, V. O., Bruck, D., Taylor, B. J. and Loper, J. E. 2016. Characterization of toxin complex gene clusters and insect toxicity of bacteria representing four subgroups of *Pseudomonas fluorescens*. *PLoS ONE* 11(8), e0161120. https://doi.org/10.1371/journal.pone.0161120.

Reddy, G. S. N., Matsumoto, G. I., Schumann, P., Stackerbrandt, E. and Shivaji, S. 2004. Psychrophilic pseudomonads from Antarctica: *Pseudomonas antartica* sp. nov., *Pseudomonas meridiana* sp. nov. and *Pseudomonas proteolytica* sp. nov. *Int. J. Syst. Evol. Microbiol.* 54(3), 713–719. https://doi.org/10.1099/ijs.0.02827-0.

Redondo-Nieto, M., Barret, M., Morrisey, J. P., Germaine, K., Martínez-Granero, F., Barahona, E., Navazo, A., Sánchez-Contreras, M., Moynihan, J. A., Giddens, S. R., Coppoolse, E. R., Muriel, C., Stiekema, W. J., Rainey, P. B., Dowling, D., O'Gara, F., Martín, M. and Rivilla, R. 2012. Genome sequence of the biocontrol strain *Pseudomonas fluorescens* F113. *J. Bacteriol.* 194(5), 1273–1274. https://doi.org/10.1128/JB.06601-11.

Redondo-Nieto, M., Barret, M., Morrissey, J., Germaine, K., Martínez-Granero, F., Barahona, E., Navazo, A., Sánchez-Contreras, M., Moynihan, J. A., Muriel, C., Dowling, D., O'Gara, F., Martín, M. and Rivilla, R. 2013. Genome sequence reveals that *Pseudomonas fluorescens* F113 possesses a large and diverse array of systems for rhizosphere function and host interaction. *BMC Genomics* 14, 54. https://doi.org/10.1186/1471-2164-14-54.

Rezzonico, F., Binder, C., Défago, G. and Moënne-loccoz, Y. 2005. The Type III secretion system of biocontrol *Pseudomonas fluorescens* KD targets the phytopathogenic Chromista *Pythium ultimum* and promotes cucumber protection. *Mol. Plant Microbe Interact.* 18(9), 991–1001. https://doi.org/10.1094/MPMI-18-0991 -, pp. 18–0991.

Rodriguez, F. and Pfender, W. F. 1997. Antibiosis and antagonism of *Sclerotinia homoeocarpa* and *Drechslera poae* by *Pseudomonas fluorescens* Pf-5 in vitro and in planta. *Phytopathology* 87(6), 614–621. https://doi.org/10.1094/PHYTO.1997.87.6.614.

Rokni-Zadeh, H., Li, W., Yilma, E., Sanchez-Rodriguez, A. and De Mot, R. 2013. Distinct lipopeptide production systems for WLIP (white line-inducing principle) in *Pseudomonas fluorescens* and *Pseudomonas putida*. *Environ. Microbiol. Rep.* 5(1), 160–169. https://doi.org/10.1111/1758-2229.12015.

Ruffner, B., Péchy-Tarr, M., Höfte, M., Bloemberg, G., Grunder, J., Keel, C. and Maurhofer, M. 2015. Evolutionary patchwork of an insecticidal toxin shared between plant-associated pseudomonads and the insect pathogens *Photorhabdus* and *Xenorhabdus*. *BMC Genomics* 16, 609. https://doi.org/10.1186/s12864-015-1763-2.

Ruffner, B., Péchy-Tarr, M., Ryffel, F., Hoegger, P., Obrist, C., Rindlisbacher, A., Keel, C. and Maurhofer, M. 2012. Oral insecticidal activity of plant-associated pseudomonads. *Environ. Microbiol.* 15(3), 751–763. https://doi.org/10.1111/j.1462-2920.2012.02884.x.

Savchuk, S. and Fernando, W. G. D. 2004. Effect of timing of application and population dynamics on the degree of biological control of *Sclerotinia sclerotiorum* by bacterial antagonists. *FEMS Microbiol. Ecol.* 49(3), 379–388. https://doi.org/10.1016/j.femsec.2004.04.014.

Sazinas, P., Hansen, M. L., Aune, M. I., Fischer, M. H. and Jelsbak, L. 2019. A rare thioquinolobactin siderophore present in a bioactive *Pseudomonas* sp. DTU12.1. *Genome Biol. Evol.* 11(12), 3529–3533. https://doi.org/10.1093/gbe/evz267.

Schalk, I. J., Rigouin, C. and Godet, J. 2020. An overview of siderophore biosynthesis among fluorescent pseudomonads and new insights into their complex cellular organization. *Environ. Microbiol.* 22(4), 1447–1466. https://doi.org/10.1111/1462-2920.14937.

Scheepmaker, J. W. A., Busschers, M., Sundh, I., Eilenberg, J. and Butt, T. M. 2019. Sense and nonsense of the secondary metabolites data requirements in the EU for beneficial microbial control agents. *Biol. Control* 136. https://doi.org/10.1016/j.biocontrol.2019.104005, 104005.

Schreiter, S., Babin, D., Smalla, K. and Grosch, R. 2018. Rhizosphere competence and biocontrol effect of *Pseudomonas* sp. RU47 independent from plant species and soil type at the field scale. *Front. Microbiol.* 9, 97. https://doi.org/10.3389/fmicb.2018.00097.

Seaton, S. C., Silby, M. W. and Levy, S. B. 2013. Pleiotropic effects of GacA on *Pseudomonas fluorescens* Pf0-1 *in vitro* and in soil. *Appl. Environ. Microbiol.* 79(17), 5405–5410. https://doi.org/10.1128/AEM.00819-13.

Shanahan, P., O'Sullivan, D. J., Simpson, P., Glennon, J. D. and O'Gara, F. 1992. Isolation of 2,4-diacetylphloroglucinol from a fluorescent pseudomonad and investigation of physiological parameters influencing its production. *Appl. Environ. Microbiol.* 58(1), 353–358. https://doi.org/10.1128/AEM.58.1.353-358.1992.

Sharifi-Tehrani, A., Zala, M., Natsch, A., Moënne-Loccoz, Y. and Défago, G. 1998. Biocontrol of soil-borne fungal plant diseases by 2,4-diacetylphloroglucinol-producing fluorescent pseudomonads with different restriction profiles of amplified 16S rDNA. *Eur. J. Plant Pathol.* 104(7), 631–643. https://doi.org/10.1023/A:1008672104562.

Siddiqui, I. A. and Shaukat, S. S. 2003. Suppression of root-knot disease by *Pseudomonas fluorescens* CHA0 in tomato: importance of bacterial secondary metabolite, 2,4-diacetylpholoroglucinol. *Soil Biol. Biochem.* 35(12), 1615–1623. https://doi.org/10.1016/j.soilbio.2003.08.006.

Siddiqui, I. A., Shaukat, S. S., Sheikh, I. H. and Khan, A. 2006. Role of cyanide production by *Pseudomonas fluorescens* CHA0 in the suppression of root-knot nematode, *Meloidogyne javanica* in tomato. *World J. Microbiol. Biotechnol.* 22(6), 641–650. https://doi.org/10.1007/s11274-005-9084-2.

Sigler, W. V., Nakatsu, C. H., Reicher, Z. J. and Turco, R. F. 2001. Fate of the biological control agent *Pseudomonas aureofaciens* TX-1 after application to turfgrass. *Appl. Environ. Microbiol.* 67(8), 3542–3548. https://doi.org/10.1128/AEM.67.8.3542-3548.2001.

Sikorski, J., Stackebrandt, E. and Wackernagel, W. 2001. *Pseudomonas kilonensis* sp. nov., a bacterium isolated from agricultural soil. *Int. J. Syst. Evol. Microbiol.* 51(4), 1549–1555. https://doi.org/10.1099/00207713-51-4-1549.

Smilanick, J. L., Gouin-Behe, C. C., Margosan, E. A., Bull, C. T. and Mackey, B. E. 1996. Virulence on citrus of *Pseudomonas syringae* strains that control postharvest green mould of citrus fruit. *Plant Dis.* 80(10), 1123–1128.

Smits, T. H. M., Rezzonico, F., Frasson, D., Vesga, P., Vacheron, J., Blom, J., Pothier, J. F., Keel, C., Maurhofer, M. and Sievers, M. 2019. Updated genome sequence and annotation for the full genome of *Pseudomonas protegens* CHA0. *Microbiol. Resour. Announc.* 8(39), e01002-19. https://doi.org/10.1128/MRA.01002-19.

Snopková, K., Čejková, D., Dufková, K., Sedláček, I. and Šmajs, D. 2020. Genome sequences of two Antarctic strains of *Pseudomonas prosekii*: insights into adaptation to extreme conditions. *Arch. Microbiol.* 202, 447–454. https://doi.org/10.1007/s00203-019-01755-4.

Sonnleitner, E., Abdou, L. and Haas, D. 2009. Small RNA as global regulator of carbon catabolite repression in *Pseudomonas aeruginosa*. *Proc. Natl. Acad. Sci. U. S. A.* 106(51), 21866–21871. https://doi.org/10.1073/pnas.pnas.0910308106.

Sonnleitner, E. and Haas, D. 2011. Small RNAs as regulators of primary and secondary metabolism in *Pseudomonas* species. *Appl. Microbiol. Biotechnol.* 91(1), 63–79. https://doi.org/10.1007/s00253-011-3332-1.

South, K. A., Peduto Hand, F. and Jones, M. L. 2020. Beneficial bacteria identified for the control of *Botrytis cinerea* in Petunia greenhouse production. *Plant Dis.* 104(6), 1801–1810. https://doi.org/10.1094/PDIS-10-19-2276-RE.

Stanier, R. Y., Palleroni, N. J. and Doudoroff, M. 1966. The aerobic pseudomonads: a taxonomic study. *J. Gen. Microbiol.* 43(2), 159–271. https://doi.org/10.1099/00221287-43-2-159.

Stockwell, V. O., Johnson, K. B., Sugar, D. and Loper, J. E. 2010. Control of fire blight by *Pseudomonas fluorescens* A506 and *Pantoea vagans* C9-1 applied as single strains and mixed inocula. *Phytopathology* 100(12), 1330–1339. https://doi.org/10.1094/PHYTO-03-10-0097.

Stockwell, V. O. and Stack, J. P. 2007. Using *Pseudomonas* spp. for integrated biological control. *Phytopathology* 97(2), 244–249. https://doi.org/10.1094/PHYTO-97-2-0244.

Strano, C. P., Bella, P., Licciardello, G., Caruso, A. and Catara, V. 2017. Role of secondary metabolites in the biocontrol activity of *Pseudomonas corrugata* and *Pseudomonas mediterranea*. *Eur. J. Plant Pathol.* 149(1), 103–115. https://doi.org/10.1007/s10658-017-1169-x.

Strano, C. P., Bella, P., Licciardello, G., Fiore, A., Lo Piero, A. R., Fogliano, V., Venturi, V. and Catara, V. 2015. *Pseudomonas corrugata* crpCDE is part of the cyclic lipopeptide corpeptin biosynthetic gene cluster and is involved in bacterial virulence in tomato and in hypersensitive response in Nicotiana benthamiana. *Mol. Plant Pathol.* 16(5), 495–506. https://doi.org/10.1111/mpp.12207.

Stringlis, I. A., Zamioudis, C., Berendsen, R. L., Bakker, P. A. H. M. and Pieterse, C. M. J. 2019. Type III secretion system of beneficial rhizobacteria *Pseudomonas simiae* WCS417 and *Pseudomonas defensor* WCS374. *Front. Microbiol.* 10, 1631. https://doi.org/10.3389/fmicb.2019.01631.

Stringlis, I. A., Zhang, H., Pieterse, C. M. J., Bolton, M. D. and De Jonge, R. 2018. Microbial small molecules-weapons of plant subversion. *Nat. Prod. Rep.* 35(5), 410–433. https://doi.org/10.1039/c7np00062f.

Stutz, E. W., Défago, G. and Kern, H. 1986. Naturally occurring fluorescent pseudomonads involved in suppression of black root rot of tobacco. *Phytopathology* 76(2). https://doi.org/10.1094/Phyto-76-181.

Tabassum, B., Khan, A., Tariq, M., Ramzan, M., Iqbal Khan, M. S., Shahid, N. and Aaliya, K. 2017. Bottlenecks in commercialisation and future prospects of PGPR. *Appl. Soil Ecol.* 121, 102–117. https://doi.org/10.1016/j.apsoil.2017.09.030.

Takeuchi, K., Kiefer, P., Reimmann, C., Keel, C., Dubuis, C., Rolli, J., Vorholt, J. A. and Haas, D. 2009. Small RNA-dependent expression of secondary metabolism is controlled by Krebs cycle function in *Pseudomonas fluorescens*. *J. Biol. Chem.* 284(50), 34976–34985. https://doi.org/10.1074/jbc.M109.052571.

Takeuchi, K., Noda, N., Katayose, Y., Mukai, Y., Numa, H., Yamada, K. and Someya, N. 2015. Rhizoxin analogs contribute to the biocontrol activity of a newly isolated *Pseudomonas* strain. *Mol. Plant Microbe Interact.* 28(3), 333–342. https://doi.org/10.1094/MPMI-09-14-0294-FI.

Takeuchi, K., Noda, N. and Someya, N. 2014. Complete genome sequence of the biocontrol strain *Pseudomonas protegens* Cab57 discovered in Japan reveals strain-specific diversity of this species. *PLoS ONE* 9(4), e93683. https://doi.org/10.1371/journal.pone.0093683.

Takeuchi, K. and Someya, N. 2019. An overview of recent genomic research on biocontrol pseudomonad strains isolated from the field in Japan. *Jpn. Agric. Res. Q.* 53(2), 87–91. https://doi.org/10.6090/jarq.53.87.

Thomashow, L. S. 2013. Phenazines in the environment: microbes, habitats, and ecological relevance. In: Chincholkar, S. and Thomashow, L. (Eds), *Microbial Phenazines*. Springer, Berlin, Heidelberg, pp. 199–216. https://doi.org/10.1007/978-3-642-40573-0_10.

Thomashow, L. S. and Weller, D. M. 1996. Current concepts in the use of introduced bacteria for biological disease control: mechanisms and antifungal metabolites. In: Stacey, G. and Keen, N. T. (Eds), *Plant-Microbe Interactions* (vol. 1). Chapman & Hall, New York, pp. 187–236.

Thrane, C., Harder Nielsen, T., Neiendam Nielsen, M., Sørensen, J. and Olsson, S. 2000. Viscosinamide-producing *Pseudomonas fluorescens* DR54 exerts a biocontrol effect on *Pythium ultimum* in sugar beet rhizosphere. *FEMS Microbiol. Ecol.* 33(2), 139–146. https://doi.org/10.1111/j.1574-6941.2000.tb00736.x.

Timmis, K. N. 2002. *Pseudomonas putida*: a cosmopolitan. *Environ. Microbiol.* 4(12), 779–781.

Tran, H., Ficke, A., Asiimwe, T., Höfte, M. and Raaijmakers, J. M. 2007. Role of the cyclic lipopeptide massetolide a in biological control of *Phytophthora infestans* and in colonization of tomato plants by *Pseudomonas fluorescens*. *New Phytol.* 175(4), 731–742. https://doi.org/10.1111/j.1469-8137.2007.02138.x.

Tran, P. N., Savka, M. A. and Gan, H. M. 2017. In-silico taxonomic classification of 373 genomes reveals species misidentification and new genospecies within the genus *Pseudomonas*. *Front. Microbiol.* 8, 1296. https://doi.org/10.3389/fmicb.2017.01296.

Trippe, K., McPhail, K., Armstrong, D., Azevedo, M. and Banowetz, G. 2013. *Pseudomonas fluorescens* SBW25 produces furanomycin, a non-proteinogenic amino acid with selective antimicrobial properties. *BMC Microbiol.* 13, 111. https://doi.org/10.1186/1471-2180-13-111.

Tziros, G. T., Lagopodi, A. L. and Tzavella-Klonari, K. 2007. Reduction of Fusarium wilt in watermelon by *Pseudomonas chlororaphis* PCL1391 and *P. fluorescens* WCS365. *Phytopathol. Mediterr.* 46, 320–323. https://doi.org/10.14601/Phytopathol_Mediterr-2245.

Vacheron, J., Moënne-Loccoz, Y., Dubost, A., Gonçalves-Martins, M., Muller, D. and Prigent-Combaret, C. 2016. Fluorescent *Pseudomonas* strains with only few plant-beneficial properties are favored in the maize rhizosphere. *Front. Plant Sci.* 7, 1212. https://doi.org/10.3389/fpls.2016.01212.

Vacheron, J., Péchy-Tarr, M., Brochet, S., Heiman, C. M., Stojiljkovic, M., Maurhofer, M. and Keel, C. 2019. T6SS contributes to gut microbiome invasion and killing of an herbivorous pest insect by plant-beneficial *Pseudomonas protegens*. *ISME J.* 13(5), 1318–1329. https://doi.org/10.1038/s41396-019-0353-8.

Vallet-Gely, I., Novikov, A., Augusto, L., Liehl, P., Bolbach, G., Péchy-Tarr, M., Cosson, P., Keel, C., Caroff, M. and Lemaitre, B. 2010. Association of hemolytic activity of *Pseudomonas entomophila*, a versatile soil bacterium, with cyclic lipopeptide production. *Appl. Environ. Microbiol.* 76(3), 910–921. https://doi.org/10.1128/AEM.02112-09.

Van Der Voort, M., Meijer, H. J. G., Schmidt, Y., Watrous, J., Dekkers, E., Mendes, R., Dorrestein, P. C., Gross, H. and Raaijmakers, J. M. 2015. Genome mining and metabolic profiling of the rhizosphere bacterium *Pseudomonas* sp. SH-C52 for antimicrobial compounds. *Front. Microbiol.* 6, 693. https://doi.org/10.3389/fmicb.2015.00693.

Van Wees, S. C. M., Pieterse, C. M. J., Trijssenaar, A., Van't Westende, Y. A. M., Hartog, F. and Van Loon, L. C. 1997. Differential induction of systemic resistance in *Arabidopsis* by biocontrol bacteria. *Mol. Plant Microbe Interact.* 10(6), 716–724. https://doi.org/10.1094/MPMI.1997.10.6.716.

Verhille, S., Baida, N., Dabboussi, F., Izard, D. and Leclerc, H. 1999. Taxonomic study of bacteria isolated from natural mineral waters: proposal of *Pseudomonas jessenii* sp. nov. and *Pseudomonas mandelii* sp. nov. *Syst. Appl. Microbiol.* 22(1), 45–58. https://doi.org/10.1016/S0723-2020(99)80027-7.

Vincent, M. N., Harrison, L. A., Brackin, J. M., Kovacevich, P. A., Mukerji, P., Weller, D. M. and Pierson, E. A. 1991. Genetic analysis of the antifungal activity of a soilborne *Pseudomonas aureofaciens* strain. *Appl. Environ. Microbiol.* 57, 2928–2934.

Voisard, C., Bull, C. T., Keel, C., Laville, J., Maurhofer, M., Schnider, U., Défago, G. and Haas, D. 1994. Biocontrol of root diseases by *Pseudomonas fluorescens* CHA0: current concepts and eperimental approaches. In: O'Gara, F., Dowling, D. N. and Boesten, B. (Eds), *Molecular Ecology of Rhizosphere Microorganisms: Biotechnology and the Release of GMOs*, pp. 67–89. https://doi.org/10.1002/9783527615810.ch6.

Voisard, C., Keel, C., Haas, D. and Dèfago, G. 1989. Cyanide production by *Pseudomonas fluorescens* helps suppress black root rot of tobacco under gnotobiotic conditions. *EMBO J.* 8(2), 351–358. https://doi.org/10.1002/j.1460-2075.1989.tb03384.x.

Volke, D. C., Calero, P. and Nikel, P. I. 2020. *Pseudomonas putida*. *Trends Microbiol.* 28, 512–513. https://doi.org/10.1016/j.tim.2020.02.015.

Völksch, B. and May, R. 2001. Biological control of *Pseudomonas syringae* pv. *glycinea* by epiphytic bacteria under field conditions. *Microb. Ecol.* 41(2), 132–139. https://doi.org/10.1007/s002480000078.

Völksch, B., Nüske, J. and May, R. 1996. Characterization of two epiphytic bacteria from soybean leaves with antagonistic activities against *Pseudomonas syringae* pv. *glycinea*. *J. Basic Microbiol.* 36(6), 453–462. https://doi.org/10.1002/jobm.3620360611.

Wagner, A., Norris, S., Chatterjee, P., Morris, P. F. and Wildschutte, H. 2018. Aquatic pseudomonads inhibit oomycete plant pathogens of *Glycine max*. *Front. Microbiol.* 9, 1007. https://doi.org/10.3389/fmicb.2018.01007.

Wang, M. Q., Wang, Z., Yu, L. N., Zhang, C. S., Bi, J. and Sun, J. 2019. *Pseudomonas qingdaonensis* sp. nov., an aflatoxin-degrading bacterium, isolated from peanut rhizospheric soil. *Arch. Microbiol.* 201(5), 673–678. https://doi.org/10.1007/s00203-019-01636-w.

Weller, D. M. 2007. *Pseudomonas* biocontrol agents of soilborne pathogens: looking back over 30 years. *Phytopathology* 97(2), 250–256. https://doi.org/10.1094/PHYTO-97-2-0250.

Weller, D. M., Landa, B. B., Mavrodi, O. V., Schroeder, K. L., De La Fuente, L., Blouin Bankhead, S., Allende Molar, R., Bonsall, R. F., Mavrodi, D. V. and Thomashow, L. S. 2007. Role of 2,4-diacetylphloroglucinol-producing fluorescent *Pseudomonas* spp. in the defense of plant roots. *Plant Biol. (Stuttg)* 9(1), 4–20. https://doi.org/10.1055/s-2006-924473.

Weller, D. M., Mavrodi, D. V., van Pelt, J. A., Pieterse, C. M. J., van Loon, L. C. and Bakker, P. A. H. M. 2012. Induced systemic resistance in *Arabidopsis thaliana* against *Pseudomonas syringae* pv. tomato by 2,4-diacetylphloroglucinol-producing *Pseudomonas fluorescens*. *Phytopathology* 102(4), 403–412. https://doi.org/10.1094/PHYTO-08-11-0222.

Wensing, A., Braun, S. D., Buettner, P., Expert, D., Voelksch, B., Ullrich, M. S. and Weingart, H. 2010. Impact of siderophore production by *Pseudomonas syringae* pv. syringae 22d/93 on epiphytic fitness and biocontrol activity against *Pseudomonas syringae* pv. glycinea 1a/96. *Appl. Environ. Microbiol.* 76(9), 2704–2711. https://doi.org/10.1128/AEM.02979-09.

Wu, D. Q., Ye, J., Ou, H. Y., Wei, X., Huang, X., He, Y. W. and Xu, Y. 2011. Genomic analysis and temperature-dependent transcriptome profiles of the rhizosphere originating strain *Pseudomonas aeruginosa* M18. *BMC Genom.* 12, 1–17. https://doi.org/10.1186/1471-2164-12-438.

Xu, G.-W. and Gross, D. C. 1986. Selection of fluorescent pseudomonads antagonistic to *Erwinia carotovora* and suppressive of potato seed piece decay. *Phytopathology* 76(4), 414–422. https://doi.org/10.1094/Phyto-76-414.

Xu, Y. and Liu, H. 2013. The expression and regulation of two *pHz* gene clusters in PGPR strain of *Pseudomonas aeruginosa*. In: *3rd Asian Conference on Plant Growth-Promoting Rhizobacteria (PGPR) and Other Microbials*. Manila, Philippines (vol. M18), pp. 54–55.

Yan, Q., Lopes, L. D., Shaffer, B. T., Kidarsa, T. A., Vining, O., Philmus, B., Song, C., Stockwell, V. O., Raaijmakers, J. M., McPhail, K. L., Andreote, F. D., Chang, J. H. and Loper, J. E. 2018. Secondary metabolism and interspecific competition affect accumulation of spontaneous mutants in the GacS-GacA regulatory system in *Pseudomonas protegens*. *mBio* 9(1), e01845-17. https://doi.org/10.1128/mBio.01845-17.

Yan, Q., Philmus, B., Chang, J. H. and Loper, J. E. 2017. Novel mechanism of metabolic co-regulation coordinates the biosynthesis of secondary metabolites in *Pseudomonas protegens*. *eLife* 6, e22835. https://doi.org/10.7554/eLife.22835.

Yang, X. and Hong, C. 2020. Biological control of Phytophthora blight by *Pseudomonas protegens* strain 14D5. *Eur. J. Plant Pathol.* 156(2), 591–601. https://doi.org/10.1007/s10658-019-01909-6.

Yasmin, S., Hafeez, F. Y., Mirza, M. S., Rasul, M., Arshad, H. M. I., Zubair, M. and Iqbal, M. 2017. Biocontrol of bacterial leaf blight of rice and profiling of secondary metabolites produced by rhizospheric *Pseudomonas aeruginosa* BRp3. *Front. Microbiol.* 8, 1895. https://doi.org/10.3389/fmicb.2017.01895.

Yasmin, S., Hafeez, F. Y. and Rasul, G. 2014. Evaluation of *Pseudomonas aeruginosa* Z5 for biocontrol of cotton seedling disease caused by *Fusarium oxysporum*. *Biocontrol Sci. Technol.* 24(11), 1227–1242. https://doi.org/10.1080/09583157.2014.932754.

Yonezuka, K., Shimodaira, J., Tabata, M., Ohji, S., Hosoyama, A., Kasai, D., Yamazoe, A., Fujita, N., Ezaki, T. and Fukuda, M. 2017. Phylogenetic analysis reveals the taxonomically diverse distribution of the *Pseudomonas putida* group. *J. Gen. Appl. Microbiol.* 63(1), 1–10. https://doi.org/10.2323/jgam.2016.06.003.

Youard, Z. A., Mislin, G. L. A., Majcherczyk, P. A., Schalk, I. J. and Reimmann, C. 2007. *Pseudomonas fluorescens* CHA0 produces enantio-pyochelin, the optical antipode of the *Pseudomonas aeruginosa* siderophore pyochelin. *J. Biol. Chem.* 282(49), 35546–35553. https://doi.org/10.1074/jbc.M707039200.

Yu, K., Pieterse, C. M. J., Bakker, P. A. H. M. and Berendsen, R. L. 2019. Beneficial microbes going underground of root immunity. *Plant Cell Environ.* 42(10), 2860–2870. https://doi.org/10.1111/pce.13632.

Zachow, C., Jahanshah, G., De Bruijn, I., Song, C., Ianni, F., Pataj, Z., Gerhardt, H., Pianet, I., Lämmerhofer, M., Berg, G., Gross, H. and Raaijmakers, J. M. 2015. The novel lipopeptide poaeamide of the endophyte *Pseudomonas poae* RE∗1-1-14 is involved in pathogen suppression and root colonization. *Mol. Plant Microbe Interact.* 28(7), 800–810. https://doi.org/10.1094/MPMI-12-14-0406-R.

Zhang, M., Yang, L., Hao, R., Bai, X., Wang, Y. and Yu, X. 2020a. Drought-tolerant plant growth-promoting rhizobacteria isolated from jujube (*Ziziphus jujuba*) and their potential to enhance drought tolerance. *Plant Soil* 452(1–2), 423–440. https://doi.org/10.1007/s11104-020-04582-5.

Zhang, Q. X., Kong, X. W., Li, S. Y., Chen, X. J. and Chen, X. J. 2020b. Antibiotics of *Pseudomonas protegens* FD6 are essential for biocontrol activity. *Australas. Plant Pathol.* 49, 307–317. https://doi.org/10.1007/s13313-020-00696-7.

Zhou, T., Northover, J. and Schneider, K. E. 1999. Biological control of postharvest diseases of peach with phyllosphere isolates of *Pseudomonas syringae*. *Can. J. Plant Pathol.* 21(4), 375–381. https://doi.org/10.1080/07060669909501174.

Chapter 10

Are there bacterial bioprotectants besides *Bacillus* and *Pseudomonas* species?

Emilio Montesinos and Anna Bonaterra, Institute of Food and Agricultural Technology, University of Girona, Spain

1 Introduction

Many strains of bacteria are active in the control of plant pathogens, such as bacteria and fungi causing aerial and root diseases, or are effective against postharvest diseases (Montesinos and Bonaterra, 2017). Most studies of the biological control of plant pathogens have focused on the use of strains belonging to the genus *Bacillus* spp. or *Pseudomonas* spp. However, other groups also have the potential as bioprotectants. There are important species/strains of non-*Bacillus* and *Pseudomonas* (NBP) which can be used to develop microbial biopesticides. Some are distributed among the Gram-negative bacteria of the families Rhizobiaceae, Enterobacteriaceae, and Xanthomonadaceae. Others can be found among Gram-positive bacteria such as Lactobacillaceae, Leuconostocaceae, and Streptomycetaceae (for detailed references, see Table 1).

Different bacterial species have received increasing attention as plant bioprotectants. These include species of *Rhizobium*, found to inhibit the

http://dx.doi.org/10.19103/AS.2021.0093.12

Table 1 Groups of bacteria other than *Bacillus* and *Pseudomonas* that have strains of bioprotectants

Family/Phylum	Genus	References
Cellulomonadaceae	*Cellulomonas*	Lourenço Junior et al., 2006
Comamonadaceae	*Delftia*	Janahiraman et al., 2016
Corynebacteriaceae	*Rhodococcus*	Latour et al., 2013
Enterobacteriaceae	*Citrobacter*	Janisiewicz et al., 2013
	Enterobacter	Xue et al., 2009
	Pantoea (Erwinia)	Walterson and Stavrinides, 2015
	Rahnella	Daayf et al., 2003
	Serratia	Kalbe et al., 1996
Flavobacteriaceae	*Chryseobacterium*	Huang et al., 2013
	Flavobacterium	"
Lactobacillaceae	*Lactobacillus*	Roselló et al., 2013
Leuconostocaceae	*Leuconostoc*	Trias et al., 2008
Methylobacteriaceae	*Methylobacterium*	Dourado et al., 2015
Pasteuriaceae	*Pasteuria*	Chen and Dickson, 1998
Pseudomonadaceae	*Burkholderia*	Compant et al., 2005
Rhizobiaceae	*Bradyrhizobium*	Das et al., 2017
	Rhizobium (Agrobacterium)	Volpiano et al., 2018
	Sinorhizobium	Kumar et al., 2010
Sphingomonadaceae	*Sphingomonas*	Innerebner et al., 2011
Streptomycetaceae	*Streptomyces*	Viaene et al., 2016
Xanthomonadaceae	*Lysobacter*	McSpadden Gardener et al., 2014
	Stenotrophomonas	Messiha et al., 2007

growth of various soil-borne bacterial and fungal pathogens including *Agrobacterium tumefaciens*, *Macrophomina phaseolina*, *Rhizoctonia solani*, *Sclerotium* spp. and *Fusarium* spp. in both leguminous and non-leguminous plants. In addition, species of Enterobacteriaceae, such as *Serratia* spp. or *Pantoea* spp., have shown the ability to control fungal soilborne diseases and leaf pathogens as well as suppress postharvest diseases. Bacterial strains belonging to the Actinomycetales have been described as potent antagonists of fungal pathogens, notably *Streptomyces* spp. isolates which produce antibiotics with biocontrol activity against plant pathogens. In addition, other Gram-positive bacteria such as *Lactobacillus* spp. have potential as plant bioprotectants, since they are antagonistic to plant pathogenic bacteria and fungi and are well known as probiotics in other environments. Several less-well-known or unexplored bacteria have also been described as bioprotectants. These include *Acinetobacter* spp., *Chryseobacterium* spp., *Clostridium* spp., *Delftia* spp., *Enterobacter* spp.,

Flavobacterium spp., *Methylobacterium* spp., *Sphingomonas* spp. and *Stenotrophomonas* spp.

2 Modes of action

The major biocontrol mechanisms displayed by NBP antagonists include: antibiosis through antibiotic production, mycoparasitism, production of cell wall lytic enzymes, competition for nutrients and space and induction of host resistance. There are examples of strains that use a single mechanism and those that combine several mechanisms. Examples of these modes of action by bacterial bioprotectants are listed in Table 2 (which includes detailed references).

Many bacterial NBP strains produce one or more antibiotics with antimicrobial activities that supress plant pathogens. *Streptomyces* spp. has been the most studied, since several strains of this genus produce bioactive compounds that have been shown to inhibit pathogens *in vitro* and have been reported to be effective in the control of various bacterial and fungal plant pathogens (Viaene et al., 2016). Examples of such metabolites include macrolide benzoquinones, aminoglycosides, polyenes and nucleosides. *Streptomyces* strains are also known for their ability to produce extracellular enzymes active in fungal cell wall degradation, such as β-1,3-glucanases and chitinases. These hydrolases may be responsible for the mycoparasitic potential of some strains and the partial suppression of plant diseases. Other bioactive metabolites are produced, including volatile organic compounds (VOCs), as signalling molecules to regulate plant growth and immunity in response to biotic and abiotic stresses. In addition, some strains can suppress plant diseases through the induction of systemic resistance (ISR) in plants. ISR elicited by *Streptomyces* strains occurs via the activation of the jasmonic acid/ethylene and salicylic acid pathways.

Various strains of *Pantoea* spp. have been shown to demonstrate biocontrol activity through various mechanisms, including competitive colonisation, production of antibiotics and/or induction of host systemic defence. Some strains of *Pantoea* species have been shown to target a wide spectrum of plant pathogens including bacteria, fungi, oomycetes and parasitic nematodes via secretion of antimicrobial compounds such as pantocins, herbicolins, microcins and phenazines (Pusey et al., 2011; Walterson and Stavrinides, 2015). Other strains such as *Pantoea agglomerans* EPS125 require direct cell-to-cell interaction to combat the fungal pathogen *Monilinia laxa*, without relying on the production of antibiotic substances or nutrient competition (Bonaterra et al., 2003). *Pantoea* species can also produce N-acyl-homoserine lactone (AHL), affecting quorum sensing in pathogens which, coupled with promoting environmental fitness in plants, may contribute to pathogen suppression (Smits et al., 2010).

Table 2 Mechanisms of action in selected biocontrol strains other than *Bacillus* and *Pseudomonas*

Microorganism and strain	Target pathogen or disease	Mechanism involved	References
Pantoea agglomerans EPS125	Postharvest fungi	CE	Bonaterra et al., 2003
P. agglomerans CPA-2	Postharvest fungi	CE	Poppe et al., 2003
Pantoea vagans C9-1	*Erwinia amylovora*	Ab-peptides, CE	Kamber et al., 2012
Serratia marcescens ETR17	Root rot disease in tea	Ab-Prn, Pro, HCN, P-Chit	Dhar Purkayastha et al., 2018
Serratia plymuthica HRO-C48	Soil-borne pathogens	Ab-Prn, P-Chit	Liu et al., 2007
S. plymuthica R1GC4	*Pythium ultimum*	IR	Benhamou et al., 2000
Streptomyces globisporus JK-1	*Penicillium italicum*	Ab-VOC	Li et al., 2010
Streptomyces griseoviridis K61	Soil-borne pathogens	CE, P, Ab-polyenes	Kortemaa et al., 1994
Streptomyces sp. EN27	*Fusarium oxysporum*	IR	Conn et al., 2008
Streptomyces AcH 505	Oak powdery mildew	IR	Kurth et al., 2014
Streptomyces S4-7	*Fusarium*	Ab-thiopeptide	Cha et al., 2016
Rahnella aquatilis BNM 0523	*Botrytis cinerea*	P-Ent	Sansone et al., 2011
Rhizobium etli G12	*Globodera pallida*	IR-LPS	Reitz et al., 2000
Rhizobium rhizogenes K84/K1026	*Agrobacterium tumefaciens*	Ab-opines	Penyalver et al., 2000
Rhodococcus erythropolis R138	*Pectobacterium atrosepticum*	QQ	Latour et al., 2013

Ab, antibiosis; CE, competitive exclusion; P, parasitism involving the production of lytic enzymes; IR, induced resistance; NC, nutrient competition; Prn, pyrrolnitrin; Pro, prodigiosin; Chit, chitinases; Ent, enterochelin; VOC, volatile organic compounds; LPS, Lipopolysaccharides; QQ, quorum quenching; HCN, hydrogen cyanide.

Several *Serratia* spp. strains produce lytic enzymes with antagonistic activity against pathogens. *Serratia marcescens* produces antifungal factors including lytic enzymes and antibiotics such as prodigiosin or pyrrolnitrin, as well as hydrogen cyanide. *S. plymuthica* reduces disease incidence by stimulating plant defence mechanisms. Mechanisms include the accumulation of callose and pectin to strengthen host cell walls and the activation of secondary responses (oxidation and polymerisation of pre-existing phenols and the synthesis of new phenolic compounds) with direct effects on pathogens (Kalbe et al., 1996).

Lactic acid bacteria (LAB) can also produce one or more antimicrobial metabolites. These include organic acids, carbon dioxide, diacetyl hydroxide peroxide and proteinaceous compounds like bacteriocins and antifungal peptides. LAB may also exclude pathogens by pre-emptively colonising plant tissues susceptible to infection, by competition for nutrients and space, or by inducing defence responses in plants.

Antagonistic activity in rhizobia is mainly attributed to production of antibiotics such as opines and bacteriocins (trifolitoxin and rhisobitoxine), hydrogen cyanide, mycolytic enzymes, and siderophores under iron-limiting conditions. Rhizobia are also reported to induce systemic resistance and enhance the expression of plant-defence-related genes, which effectively immunize the plants against pathogens (Das et al., 2017).

Methylobacterium spp. can protect host plants by synthesising antimicrobial compounds, nutrient competition with pathogens or by inducing systemic resistance through VOCs or plant cell wall degradation enzymes such as glycosidases, cellulases, hemicellulases and pectidases (Dourado et al., 2015). Some strains induce increased enzyme activity in the host plant associated with induced systemic resistance.

Sphingomonas spp. protect plants against pathogens through substrate competition since they are able to metabolize a wide range of carbon sources in plant environments (Delmotte et al., 2009). Bacteria can also degrade the chemical signal messengers necessary for quorum sensing, used by many plant-pathogenic bacteria to start the infection process in the host. *Rhodococcus erythropolis*, for example, prevents disease by disrupting quorum sensing-based communication of Gram-negative soft-rot bacteria, primarily by degrading N-acyl-homoserine lactone signalling molecules (Latour et al., 2013).

3 Building a strain collection

3.1 Sources of isolates and methods of isolation

Bacterial NBP antagonists that prevent or limit disease development are naturally present in the plant environment (phyllosphere, rhizosphere and endosphere) and other sources such as soil. These bacteria colonise plant surfaces such as roots, leaves, flowers and fruits as epiphytes, or inhabit the apoplast, conductive vessels (phloem and xylem) and intercellular spaces of plants as endophytes. Sampling these areas increases the probability of obtaining NBP candidates as bacterial bioprotectants (Montesinos and Bonaterra, 2017). Samples may, for example, be taken from plant materials where there is evidence of the presence of beneficial microorganisms, such as healthy plants that have survived disease in epidemic areas (Montesinos, 2003). They can also be obtained near pathogen infection sites or from crop residues of fields where the disease has been present, in order to obtain ecologically adapted microorganisms (Köhl et al., 2011). NBP antagonists may also be isolated from samples obtained from the same natural environment where they will be introduced for disease management to ensure their ecological adaptation.

The isolation of hundreds of candidates is recommended since the presence of microorganisms with high antagonistic activity is relatively rare and antagonistic activity is often strain-specific (Montesinos, 2003). A careful choice of the isolation technique using enrichment and selective media allows successful isolation of specific microbial groups of interest. De Man, Rogosa and Sharpe agar media allow the isolation of LAB. Yeast extract mannitol agar is used to isolate *Rhizobium* spp. Chitin–vitamin agar medium, supplemented with cycloheximide and nalidixic acid to suppress the growth of fungi and Gram-negative bacteria respectively, is used to isolate *Streptomyces* spp. (Table 3).

Since relatively few bacterial genera have been systematically evaluated for their biocontrol activities, a greater understanding of the plant microbiome would lead to the discovery of more potential antagonists. Next-generation sequencing (NGS) technologies, such as characterisation of 16S rRNA gene sequences with NGS, have become a standard method in microbial ecology, allowing detailed study of the diversity of microbial communities in the soil, rhizosphere, phyllosphere and endosphere of plants. This approach allows the discovery of new NBP with interesting traits as bacterial bioprotectants. An example is a study based on the metaproteogenomic analysis of microbial communities from *Arabidopsis thaliana* (grown in natural conditions) and clover and soybean plants (grown as agricultural crops). This showed that *Sphingomonas* spp. are abundant in the phyllosphere of these plants and have a plant-protective effect by suppressing disease symptoms and diminishing pathogen growth (Innerebner et al., 2011). However, an important problem is that only a small fraction of the bacteria detected can currently be cultured. Isolation methods need to be improved in order to obtain new NBP candidates as bioprotectants.

3.2 Screening for antagonistic bacteria

After the completion of a bacterial collection, the isolates must be identified, characterised and tested for control efficacy in the presence of the target pathogen. Trials must then be performed under field or greenhouse conditions to assess control efficacy in commercial practice. Screening for bacteria with antagonistic activity against plant pathogens is a critical step because the type of bacterial bioprotectants selected depends on the screening method used. Screening procedures are usually specific to a given mechanism of action, such as antibiosis, competition or induction of plant defence (Köhl et al., 2011;

Montesinos and Bonaterra, 2017). This means that only a portion of antagonistic microbiota will be selected by a particular method.

Primary screening of beneficial bacteria for antagonism to phytopathogens is typically performed using a rapid-throughput approach (Table 3). *In vitro* dual culture assays on agar or liquid media are used extensively. In these assays, the target pathogen and candidate biocontrol agents are cultured together to assay for direct pathogen growth suppression (Pliego et al., 2011). These assays are fast, efficient and reproducible, and produce clear results. However, they mainly identify microorganisms producing antimicrobial compounds as the mode of action, and may not identify microbial agents that control plant pathogens through other mechanisms such as ISR or niche competition (Köhl et al., 2019).

An alternative screening procedure is based on whole-plant bioassays performed in controlled environments such as greenhouses or growth chambers. In these experiments, pathogen and antagonists interact with the host in controlled conditions conducive to pathogen infection. This allows different mechanisms of action to be assessed (Parikh et al., 2018). However, these assays require significant resources and are time-consuming. There is a need to develop *ex vivo* bioassays in seeds, detached leaves, flowers and fruits that permit faster, reliable and efficient screening (Montesinos et al., 1996; Pusey, 1997). These types of assays allow interaction between the host, the pathogen and the antagonist with the benefit of better correlation with biocontrol efficacy in the field. These assays also reduce the disease incubation period and plant material size and can be used under different biotic and abiotic conditions. A multi-pathogen approach is recommended to select strains with a broad spectrum of activity (Wang et al., 2015). Besides antagonistic efficacy, other characteristics related to suitability for commercial production and registration should be taken into account in selecting attractive candidate antagonists. These characteristics include good field performance, biosafety, suitability for mass production, the ability to establish legal property rights and marketing.

4 Screening techniques

4.1 Molecular and biochemical marker-assisted screening

The use of molecular markers to identify antagonists by detecting genes involved in bioprotection activity can increase the efficiency of screening procedures (Joshi and McSpadden Gardener, 2006). An understanding of the mechanisms of action of many bacterial bioprotectants has made it possible to identify key target genes that can be used as tools for molecular marker-assisted screening of an isolate collection. Advances in genome sequencing, and annotation of many biological control agents or representative strains of a given species, have provided an enormous database in the search for markers (Nelkner et al., 2019).

Table 3 Sources of non-*Bacillus* and *Pseudomonas* isolates and methods of isolation and screening

Microorganism and strain	Origin	Isolation media	Target pathogen or disease	Screening method	Reference
Lactobacillus plantarum SLG17,	Silage	MRS	Fusarium species	*In vitro*	Baffoni et al., 2015
L. plantarum PM411 and TC92	Pear and tomato fruits	MRS	*Erwinia amylovora*, *Pseudomonas syringae* pv *actinidiae*, *Xanthomonas arboricola* pv. *pruni*, *X. fragariae*	*In vitro* and *in planta* assays in a multiple pathosystem approach	Roselló et al., 2013 Daranas et al., 2019
Methylobacterium spp.	Branches of sweet-orange and tangerine plants	TSA	*Xylella fastidiosa*	-	Araújo et al., 2002
Pantoea agglomerans Pl6T4-S6l7	Plum fruits	NYDA	*Monilinia fructicola*	*In vivo* in fruits	Janisiewicz et al., 2013
P. agglomerans EPS125	Pear fruit	KB	*Monilinia laxa* and *Rhizopus stolonifer*	*In vivo* in fruits	Bonaterra et al., 2003
Rhizobium sp. NBRI9513	Roots of chickpea plants	NA, TGY, YEM	*Fusarium oxysporum* sp. *ciceri*, *Rhizoctonia bataticola* and *Pythium* sp.	*In vitro*	Nautiyal, 1997
Rhizobium SEMIA 439 and 4088	Common bean nodules	YEM	*Sclerotium rolfsii*	*In vitro*	Volpiano et al., 2018
Serratia plymuthica HRO C48	Roots of oilseed rape	NA	*Verticillium dahliae* *Rhizoctonia solani* and *Sclerotinia sclerotiorum*	*In vitro*	Kalbe et al., 1996
Sinorhizobium fredii KCC5	Nodules of pigeon pea	YEM	*Fusarium udum*	*In vitro*	Kumar et al., 2010
Streptomyces	Rhizosphere-associated soils	WYE, YCED	*Pythium ultimum*	*In vitro* and *in vivo*	Crawford et al., 1993

Streptomyces globisporus JK-1	Indoor air spora	PDA	*Magnaporthe oryzae*	*In vitro* and *in vivo*	Li et al., 2010
Streptomyces hydrogenans DH16	Soil	SCNA	*Alternaria brassicicola*	*In vitro* and *in vivo*	Manhas and Kaur, 2016
Streptomyces CA-2 and AA-2	Roots of native plants of Algerian Sahara	Chitin-vitamin agar supplemented with cycloheximide and nalidixic acid	*R. solani*	*In vitro* and *in vivo*	Goudjal et al., 2014

MRS, De Man, Rogosa and Sharpe agar; TSA, tryptic soy agar; NYDA, nutrient yeast-dextrose agar; NA, nutrient agar; TGY, tryptone-glucose-yeast extract agar; YEM, yeast extract-mannitol agar with Congo red; WYE, water-yeast extract agar; YCED, casamino acids-yeast extract-glucose agar; PDA, potato dextrose agar; SCNA, starch casein nitrate agar.

An efficient, high-yielding procedure to build up a collection of bacterial bioprotectant candidate strains is shown in Fig. 1. Samples are taken from habitats of potential biocontrol agents. They can then be extracted with the appropriate buffer to obtain a suspension for DNA extraction. A batch of primers is prepared with polymerase chain reaction (PCR) detection of a candidate species/strain, using a genome sequence database targeting genes relating to mechanisms of action (e.g. lipopeptides, chitinases, bacteriocins, etc.) or other characteristics. Positive samples from the PCR analysis are retained for further processing, whilst negative samples are discarded. Sample extracts are used to isolate culturable bacteria in the appropriate culture media. Candidate colonies of bacterial bioprotectants are then identified by ribotyping or partial sequence of housekeeping genes, and selected strains stored for characterisation and assessment of performance. The advantage of this method is that it significantly increases the yield of antagonistic strains during the isolation process compared to non-marker-assisted procedures, as has been reported in bacterial biocontrol agents such as *Bacillus* strains (Mora et al., 2015).

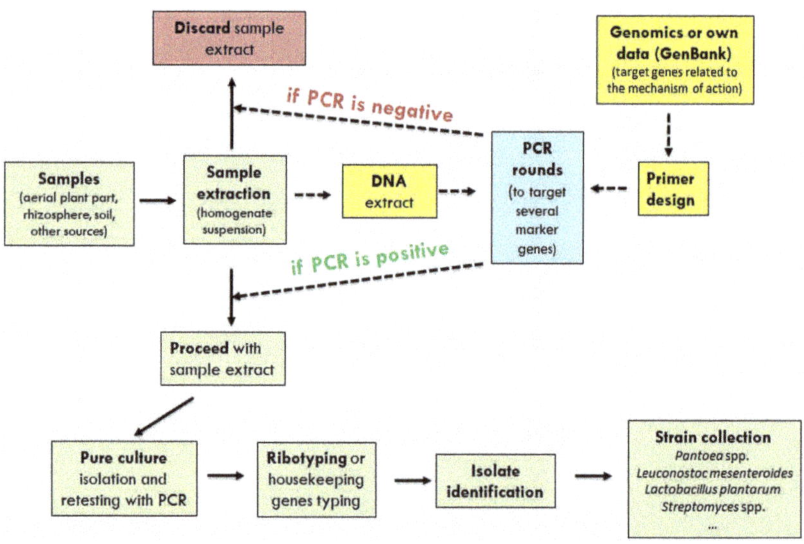

Figure 1 A procedure to build up a strain collection of bacterial bioprotectant strains based on mechanism-oriented molecular marker-assisted screening. The sample is processed to obtain a homogenate suspension, which is submitted for DNA extraction. The DNA extracts are screened by PCR rounds using primers designed against target genes. If the PCR of the extract is positive, the sample is used for isolation in pure culture of the culturable microbiota components; otherwise, it is discarded. Pure cultures of the isolated bacteria are identified by sequencing of 16S rDNA or housekeeping genes. From the selected isolates, a strain collection is built up and stored for further studies.

4.2 The example of LAB

Strains of *Lactobacillus plantarum* suitable for the development of biocontrol agents have been isolated and selected using the approach described in Section 4.1 (Roselló et al., 2013). Many LAB strains produce bacteriocins, including plantaricin, mesentericin or nisin, which have antimicrobial activity in target cells. Since gene sequences of bacteriocins produced by LAB are already available, PCR tools targeted to these genes were designed and used for molecular marker-assisted screening. A large collection of LAB isolated from plant environments in the Mediterranean area was screened using this method and several strains of *L. plantarum* (producing the bacteriocin plantaricin) and *Leuconostoc mesenteroides* (producing mesentericin) were obtained (Roselló et al., 2013).

A set of selected strains of *L. plantarum* was identified and characterised using DNA typing techniques such as multi-locus sequence typing (MLST), able to distinguish strains by their sequence profile. Molecular marker-assisted screening identified strains with *in vitro* antagonistic activity against several bacterial plant pathogens, the ability to colonise flowers as well as other plant organs such as leaf surfaces and a high degree of efficiency in fire blight biocontrol in greenhouse and field tests, as compared to other bacterial biopesticides and reference antibiotics (Daranas et al., 2019) (Fig. 2). These selected strains have the additional advantage that *L. plantarum* species are considered as biosafe by the Environmental Protection Agency (USA) and European Food Safety Authority (European Union) and are thus promising candidates as active ingredients in future microbial pesticides for bacterial disease control.

5 Biosafety issues

In most cases, species/strains of NBP candidates as potential biopesticides are considered safe. Some species of the genera are associated with infections in humans. There are clinical reports of nosocomial or opportunistic infections in immunocompromised patients, although the causal agent was not accurately identified in some cases. Clinical isolates cannot generally be discriminated from strains of other habitats such as water, soil or plants using phylogenetic methods. Biosafety assessment therefore needs to be undertaken on a case-by-case basis.

Most bacterial bioprotectant species are common microbiota of environments such as water, soil, food, animals or plants, although populations in these habitats are lower than after application as a biocontrol agent to plant crops. Some NBP bacteria may produce biofilms that can exist on a wide range of surfaces. Such biofilms allow microorganisms to survive adverse conditions

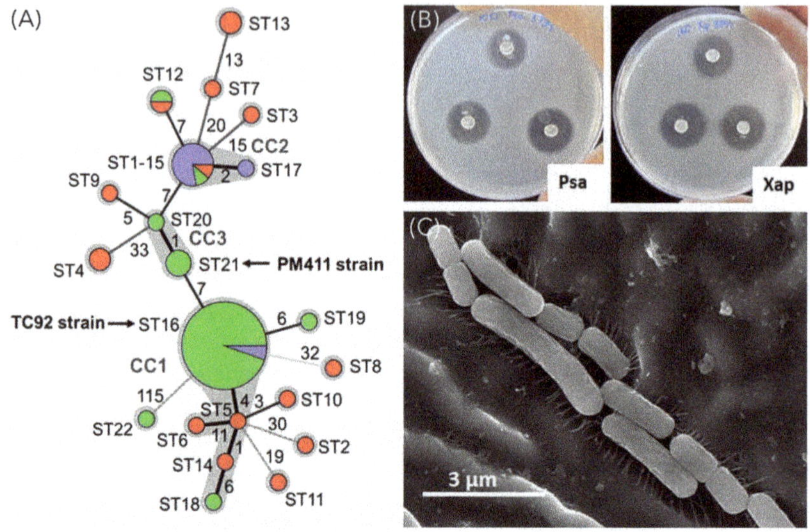

Figure 2 Minimum spanning tree of 71 *Lactobacillus plantarum* isolates based on allelic profiles of the genes *pgm*, *ddl*, *gyrB*, *purK1*, *gdh* and *mutS* (A). Each circle corresponds to a sequence type (ST) and the size of the circle is proportional to the number of isolates within any given ST. Colour codes (CC) of isolates: Green, INTEA-CIDSAV collection; red and blue, strains of other collections. The type of line between isolates indicates the strength of the genetic relationship between them (black, strong relationship; grey, intermediate relationship; and dotted line, weak relationship). The number of mutations between STs is also indicated for each relationship. STs that belong to the same CC are shown as circles grouped in a grey area. Black arrows indicate that the PM411 strain with ST21 belongs to CC3, and the TC92 strain with ST16 belongs to CC1; Antimicrobial activity obtained by the disc test against Psa, *Pseudomonas syringae* pv. *actinidiae* and Xap, *Xanthomonas arboricola* pv. *pruni* (B); Scanning electron micrographs of PM411 *L. plantarum* on GF-677 *Prunus* leaf surface. Micrographs were taken from recently sprayed plants (C).

such as dryness, disinfection chemicals and use of antibiotics. The presence of biofilms on medical devices has been associated with several clinical outbreaks.

A good example of biosafety issues affecting a bacterial biocontrol agent is *Burkholderia cepacia* (named *B. cepacia* complex). This was developed in the USA for a range of commercial uses, including protection against damping-off disease. These strains were subsequently identified as the species *B. ambifaria* and were associated with cystic fibrosis in immunocompromised patients. After assessing the risks of *Burkholderia* biopesticides, a moratorium was placed on their registration in 1999, until they could be proven safe (Hauser and Orsini, 2015).

Species of *Sphingomonas* are isolated from soil and plant root habitats as well as from the phyllosphere. In rare cases they are pathogenic and associated with nosocomial and septic shock infections in immunocompromised individuals, as in the case of *S. paucimobilis* (Hsueh et al., 1998). *Methylobacterium*, a genus of the Rhizobiales, has been found to contaminate medical devices such as endoscopes (Kovaleva et al., 2014). Enterobacteria such as the genera *Pantoea* (*P. agglomerans*/*P. vagans*/*P. ananatis*) and *Serratia* cause wound infection, catheter contamination and sepsis, mainly in immunocompromised patients (Rezzonico et al., 2009). *Agrobacterium radiobacter* has been linked to contaminated catheters, peritonitis, urinary infections and endocarditis, particularly in immunocompromised patients (Edmond et al., 1993). *Streptomyces* infections have also been reported in immunocompromised patients (Kapadia et al., 2007). In contrast, LAB, such as species in the genera *Lactobacillus*, have been granted the status of 'generally recognized as safe' (GRAS) (by USA regulation 21 CFR 170.30) or qualified presumption of safety (QPS) (EFSA, 2020). This is partly due to the long experience of safe use of LAB in food and as probiotics. Another biosafety issue is the use of genetically modified (GM) biocontrol agents to improve performance. The regulation of GM microorganisms and active ingredients in microbial biopesticides differs widely by country (Scheepmaker et al., 2020).

6 Strain improvement

6.1 Physiological improvement

One of the main limitations of bacterial bioprotectants is that, upon application in a commercial environment, survival can be limited and population levels decreased. The establishment of a sufficient suppressive population of bacterial strain on plants is critical for effective biological control of plant diseases. Colonisation and survival depend on environmental conditions. Whilst the sporulating nature of bacterial bioprotectants such as *Streptomyces* spp. allows them to survive unfavourable environmental conditions, this is not the case for non-spore-forming bacteria. Aerial plant surfaces are inhospitable for bacterial growth, because of nutrient limitations, exposure to ultraviolet radiation and changes in water availability and temperature (Lindow and Brandl, 2003). The development of commercial products also requires mass production and dehydration to get a stable product, which reduces viability and subsequent performance. Improving tolerance to environmental stresses can therefore contribute significantly to the survival and effectiveness of bacterial bioprotectants (Hagen et al., 2009).

The physiological adaptation of bacteria to stress by osmoadaptation has been reported as an effective strategy to improve performance (Bonaterra et al., 2005; Pusey and Wend, 2012). Osmoadaptation consists of inoculum preparation under osmotic stress conditions that provokes the intracellular accumulation of compatible solutes (osmolytes) to counterbalance the osmotic pressure of the environment and maintain cell turgor. These osmolytes include sugars, amino acids and their derivatives synthesised or taken from the growth medium. Cultivation under these artificial stress conditions can improve overall environmental stress tolerance. An example of osmoadaptation of the biocontrol agent *P. agglomerans* EPS125 involves combining saline osmotic stress and osmolyte amendment to the growth medium. Osmoadapted cells of EPS125 accumulated trehalose and glycine betaine (GB) intracellularly and showed a higher tolerance to desiccation than non-osmoadapted cells. Osmoadaptation significantly improved mould control compared to treatments with non-osmoadapted cells (Bonaterra et al., 2005).

Another approach has been used to increase the tolerance of *L. plantarum* PM411 to water stress, using hyperosmotic and acidic conditions during bacterial growth. This combined-stress adaptation treatment in PM411 provides protection against water stress and inhospitable conditions on aerial plant surfaces, resulting in more consistent disease suppression (Daranas et al., 2018).

6.2 Genetic improvement

Knowledge of the mechanisms of action and factors affecting performance of the bacterial bioprotectants permits the design of strategies for genetic improvement. However, despite the major advances in genomics, there are few examples of genetic improvement of NBP bioprotectants. One example is the genetic modification of the Ti plasmid in the *Rhizobium rhizogenes* (formely *Agrobacterium radiobacter*) strain K1026 to prevent its transmission to *A. tumefaciens* pathogens, associated with the acquisition of resistance by *A. tumefaciens* to the biocontrol agent *A. radiobacter* K84. Strain K1026 was obtained by transposon mutagenesis of the *tra* genes in the plasmid (Shim et al., 1987; Stockwell et al., 1996; Penyalver et al., 2000; Scheepmaker et al., 2020).

Most advances in genetic improvement have been achieved with members of the enterobacteria. The endochitinase gene (*ChiA*) from *S. marcescens* B2 has been introduced in *Erwinia ananas*, colonising rice leaves (Someya and Akutsu, 2005).

There is significant research on improving secondary-metabolite production in actinomycetes such as *Streptomyces* in antibiotic production for pharmaceutical use (Baltz, 2010).

7 Commercial exploitation

7.1 Patents

Patenting the use of strains for plant disease protection is critical for their commercial use. Approval can be complex because it must demonstrate that a strain is novel and unique. Tables 4 and 5 list 62 patents from the European Patent Office (EPO) database (https://worldwide.espacenet.com/) for NBP strains

Table 4 Relevant patents of use of Gram-negative strains other than *Bacillus* and *Pseudomonas*, as biological control agents of plant diseases

Species/strain/patent reference	Type[a]	Target
Agrobacterium radiobacter H3501/ TBR27 (JPH0940515)	B	Soil bacterial blight tobacco
Agrobacterium radiobacter K84 mutants (US5354684)	B	Crown gall tumours
Agrobacterium vitis F2/5 (US2002090354)	B	Crown gall disease
Methylobacterium oryzae (M. fujisavaensis) CBMB20 (KR20130101310)	F, B	Various fungi and bacterial spot
Methylobacterium spp. *(M. extorquens)* NLS0066, and more (EP 3098303)	F	Various fungi
Pantoea agglomerans 59-4 (KR20080007038)	F	Various fungi
P. agglomerans EPS125 (ES2188375)	F	Postharvest rot of fruits
P. agglomerans JH21 (CN102191198)	F, B	Rice fungal and bacterial diseases
P. agglomerans CECT-4920 (ES2149131)	F	Postharvest fruit rot
Pantoea ananatis R31/R32 JP20150559090)	B	Rice bacterial diseases
P. ananatis H2 (CN109679881)	F	Antracnosis and fusarium wilt
Rhizobium leguminosarum bv. *viciae* IDAC200704 (US2006029576)	F	*Pythium*
Rhizobium vitis ARK-1/ARK-2 (WO20120671227)	B	Crown gall disease
Serratia marcescens CK-B (JPH041109)	F	Fusarium wilt
Serratia plymuthica A21-4 (KR20070080775)	F	Phytophthora diseases
Serratia plymuthica A30 (PL2663659)	B	Several bacterial diseases in potato
Serratia plymuthica CCGG2742 (CN102884173)	F	Grey mould
Serratia plymuthica KCTC101873BP (KR20030070755)	F	*Sclerotium cepivorum*
Serratia proteamaculans 3095 (KR20010019369)	F	Fusarium wilt
Serratia sp. IM2 (KR20180117322)	F	Various fungi
Serratia sp. KUDC3025 (KR20190052847)	B	*Pectobacterium* and *Xanthomonas* in tobacco
Sphingomonas paucimobilis A260 (JPH09194316)	B	Soil bacterial blight tobacco
Sphingomonas TF1401, TF1402 (JP2007332110)	F	Clubroot disease of cabbage

[a]Bactericidal, B; fungicidal, F.

Table 5 Relevant patents of use of Gram-positive strains other than *Bacillus* and *Pseudomonas*, as biological control agents of plant diseases

Species/strain/patent reference	Type[a]	Target
Lactobacillus fermentum CGMCC8735 (CN104745510)	F	Several fungi
Lactobacillus plantarum CM-3 (CN110551649)	F	Grey mould of strawberry
L. plantarum CECT8297 (US2016113286) EU	B	Fireblight of apple and pear
Lactobacillus spp. Lb23 and more strains (NZ630117)	F, B	Fusarium wilt
Streptomyces acidiscabies JA(II)10 (KR20120035534)	F	Grey mould rot
Streptomyces albidoflavus SY-FX-14 (CN103992967)	F	Several fungi
Streptomyces antibioticus subsp. *Xian* (CN101463332)	F, B	NS
Streptomyces aureus A01 (CN107988120A)	F	Several fungi
Streptomyces badius SF7B6 (KR20170000514)	F	Strawberry fungal diseases
Streptomyces badius SP6C4 (US2018146680)	F	Several fungi
Streptomyces blastomyceticus KBC K-1 (KR20100132257)	F	Several fungi
Streptomyces carpaticus T33 (CN110373344)	F	Antracnose
Streptomyces chungwhensis LA-5 (CN107699526)	F	Several fungi
Streptomyces corchorussi NF0919 (CN102899356)	F	Several fungi
Streptomyces diastatochromogenes D (CN101225368)	F	Several fungal diseases
Streptomyces eurocidicus HU2014 (CN105087438)	F	Several fungi
Streptomyces graminearus St3 (CN106434500)	F	*Phytophthora* in soybean
Streptomyces griseofuscus VSG-16 (KR100726864)	F	Several fungi
Streptomyces griseoviridis TA54/TA5411 (FI19880001761)	F	Several fungi
Streptomyces griseus S4-7 (KR20150015309)	F	Fusarium wilt and more
Streptomyces lavendulae DTJ-24 (CN104560826)	F	Anthracnose of strawberry
Streptomyces longissimus SL01 (CN104673723)	F	Valsa canker of apple
Streptomyces luteogriseus ECO0001 (CN20061048662)	F	Grey mould
Streptomyces lydicus CCTCC M2016038 (CN105754891)	F	Several fungi
Streptomyces melanosporofaciens EF-76 (WO2010/115802A1)	F	Several fungi
Streptomyces melanosporofaciens AGL225 (EP3467097A1)	B, F, N	Several diseases
Streptomyces misionesis A2 (CN108977385)	F	Soybean grey mould
Streptomyces roseoverticillatus Sr-63 (CN110373358)	B	Rice bacterial blight

(Continued)

Table 5 (*Continued*)

Species/strain/patent reference	Type[a]	Target
Streptomyces sampsonii SY-FX-31 (CN104388356)	F	Various pathogenic fungi
Streptomyces sampsonii KJ40 (CN102703343)	F	Poplar violet root rot
Streptomyces scabies avirulent strain (US6074638)	B	Potato scab
Streptomyces shandongensis (CN105601358)	F	Strawberry red core disease
Streptomyces sioyaensis F58/107 (CN108795826)	F	Several fungi
Streptomyces sp. AG-P (KR20020050672)	F	Oomycetes
Streptomyces sp. CGMCC9034 (CN1039661133)	F	Soybean sclerotinia and poplar canker
Streptomyces sp. CK-A. (JPH041109)	F	Fusarium diseases
Streptomyces sp. OG-12/OG-13. (JP2003277213)	F	Soil-borne fungal diseases
Streptomyces sp. R-5 (US2001031258)	NS	NS
Streptomyces sp. VSV-12 (KR20080008600)	F	Potato grey mould and root rot

[a]Bactericidal, B; fungicidal, F.

involved in the biological control of plant diseases. Unfortunately, many patents have inadequate data on strain taxonomy and provide insufficient evidence of efficacy in biological control. The Budapest Treaty requires that all strains are deposited in an official microorganism culture collection as part of the patent application, which means they can be verified with the owner's permission.

The majority of strains are described as fungicides and are clustered within the *Streptomyces* genera (35 strains and 26 species), followed by *Serratia* (8 strains and 3 species), *Pantoea* (6 strains and 2 species), *Lactobacillus*, *Agrobacterium*, *Methylobacterium*, *Sphingomonas* and *Rhizobium*. The predominance of *Streptomyces* is due to their diversity and number of antimicrobial compounds produced by actinomycetes. In addition to the fungicidal activity, some strains are described as bactericidal or as sharing fungicidal or bactericidal activities with plant growth promotion.

Most patents give intellectual property protection in the country of the first application (country of priority date) but are not extended to other countries. China and South Korea are very prolific in patenting these types of biological control agents. However, only a few have been licensed or are in commercial use.

7.2 Commercial products

Numerous commercial products based on bacterial bioprotectants are available as registered plant protection products produced by biocontrol companies. They include living microorganisms, antimicrobial metabolites produced by selected microbial organisms or a combination of both. A total of 101 microbial biopesticides were registered in 2017 for disease control in Australia, Brazil, Canada, Europe, Japan, New Zealand and the United States (Van Lenteren et al., 2018). There are fewer products registered in Europe due to strict registration criteria and the complicated regulatory system.

There are still only a few commercial products based on bacteria of the NBP group (Table 6). An increase in the biocontrol market for these products is expected due to consumer demand to reduce synthetic chemical pesticides to protect human health and the environment. There is great potential for the use of microbial inoculants in developing countries.

There are only a few commercial plant protection products based on *Streptomyces* spp. This may be in part because many strains are commercialised primarily as biostimulants to promote plant growth. These products are based on *S. griseoviridis* K61 (Mycostop®, Finland), *S. lydicus* WYEC 108 (Actinovate®, Micro108®, Action Iron®, USA), *S. saraceticus* KH400 (YAN TEN, Taiwan), *S.*

Table 6 Commercial microbial pesticides including strains of genera other than *Bacillus* and *Pseudomonas*

Active ingredient	Commercial name	Target pathogen/disease	Host
Rhizobium rhizogenes K84 and K1026 (*Agrobacterium radiobacter)*	Nogall, Galltrol-A	Crown gall	Fruit and nut trees, ornamentals
Pantoea spp.	Bloomtime, Blossom bless, Blightban C9-1	Fire blight	Pome fruit trees, ornamentals
Serratia plymuthica HRO-C48	RhizoStar*	*Verticillium dahliae*, *Phytophthora cactorum*	Strawberry, oilseed rape
Streptomyces colombiensis	Mycocide KIBC	Powdery mildews, grey mould, brown patch	n.r.
Streptomyces griseoviridis K61	Mycostop	Root rot, grey mould and foot decay	Ornamental and vegetable crops
Streptomyces kasugaensis	Safegrow KIBC	Sheath blight, large patch	Rice, turf grass
Streptomyces lydicus WYEC 108	Actinovate, Micro108, Action Iron	Root rot, grey mould and damping-off	Greenhouse and nursery crops, turf grass
Streptomyces saraceticus KH400	YAN TEN	Nematodes, phytophthora blight, rice damping off, tomato late blight, rice blast, root rot	Fruit trees and vegetables

*Not currently marketed.
n.r., not reported.

colombiensis (Mycocide, KIBC, Korea) and *S. kasugaensis* (Safegrow, KIBC, Korea). They are registered for use as seed treatment, soil drench and foliar spray that can be applied in greenhouse, nursery and field crops, and target a wide range of pathogens. In the case of *Pantoea* spp., commercially formulated biopesticides are currently available such as those based on *P. vagans* C9-1 (BlightBan, Nufarm, Canada), *P. agglomerans* E325 (Bloomtime Biological, Northwest Agricultural Products, USA) and *P. agglomerans* P10c (Blossombless, Grochem, New Zealand), which are used for combating fire blight in apple and pear (Johnson and Stockwell, 1998; Pusey, 2002). *Rhizobia* spp. strains are extensively used as biofertilisers for plant growth. Only a few strains of *R. rhizogenes* are commercialised as plant bioprotectants, such as the strains K84 and K1026 that are the active ingredient of the products Galltrol-A (AgBioChem, USA) and Nogall (Evergreengrowers, Australia), respectively, both for treating crown gall disease. *Serratia* spp. include strains such as *S. plymuthica* HRO-C48, the active ingredient of Rhizostar (E-nema GmbH, Germany) used in strawberries and oilseed rape to control fungal soil pathogens such as *Verticillium dahliae* or *Phytophthora cactorum*.

8 Summary and future trends

There have been significant recent advances in research on bioprotective bacterial agents, based mainly on *Bacillus* and *Pseudomonas* strains. However, species and strains other than NBP, including enterobacteria, actinomycetes, *Sphingomonas*, *Methylobacterium*, *Agrobacterium–Rhizobium* and *Lactobacillus,* offer a broader range of strains, mechanisms of action and commercial opportunities. This potential is underpinned by growing knowledge of the genomics, physiology and ecology of NBP bacterial groups.

Future trends include developing strain-specific monitoring methods for bacterial bioprotectants to study the ecology of strains in the environment. Thanks to the establishment of the pangenome in several species and of new tools of genetic modification, their fitness and efficacy can be improved, but more knowledge is needed on the impact of GMO microorganisms in crop ecosystems.

9 Where to look for further information

The following sources are recommended:

- Baltz, R. H. (2008), 'Renaissance in antibacterial discovery from actinomycetes', *Curr. Opin. Pharmacol.*, 8(5), 557–563.
- Bardin, M., Ajouz, S., Comby, M., Lopez-Ferber, M., Graillot, B., Siegwart, M. and Nicot, P. C. (2015), 'Is the efficacy of biological control against plant

diseases likely to be more durable than that of chemical pesticides?', *Front. Plant Sci.*, 6, 1-14.

- Berg, G. (2009), 'Plant-microbe interactions promoting plant growth and health: perspectives for controlled use of microorganisms in agriculture', *Appl. Microbiol. Biotechnol.*, 84, 11-18.
- De Silva, N. I., Brooks, S., Lumyong, S. and Hyde, K. D. (2019), 'Use of endophytes as biocontrol agents', *Fungal Biol. Rev.*, 33, 133-148.
- Dukare, A. S., Sangeeta, P., Nambi, V. E., Gupta, R. K., Singh, R., Sharma, K. and Vishwakarma, R. K. (2018), 'Exploitation of microbial antagonists for the control of postharvest diseases of fruits: a review', *Crit. Rev. Food Sci. Nutr.*, 59(9), 1498-1513.
- Ellis, J. G. (2017), 'Can plant microbiome studies lead to effective biocontrol of plant diseases?', *Mol Plant Microbe Interact.*, 30(3), 190-193.
- Finkel, O. M., Castrillo, G., Herrera Paredes, S., Salas González, I. and Dangl, J. L. (2017), 'Understanding and exploiting plant beneficial microbes', *Curr. Opin. Plant Biol.*, 38, 155-163.
- Glare, T., Caradus, J., Gelernter, W., Jackson, T., Keyhani, N., Köhl, J., Marrone, P., Morin, L. and Stewart, A. (2012), 'Have biopesticides come of age?', *Trends Biotechnol.*, 30, 250-258.
- Keswani C., Singh, H. B., García-Estrada C., Caradus, J., He, Y. W., Mezaache-Aichour, S., Glare, T. R., Borriss R. and Sansinenea, E. (2020), 'Antimicrobial secondary metabolites from agriculturally important bacteria as next-generation pesticides', *Appl. Microbiol. Biotechnol.*, 104, 1013-1034.
- Lamichhane, J. R., Dachbrodt-Saaydeh, S., Kudsk, P. and Messéan, A. (2016), 'Toward a reduced reliance on conventional pesticides in European agriculture', *Plant Dis.*, 100, 10-24.
- Legein, M., Smets, W., Vandenheuvel, D., Eilers, T., Muyshondt, B., Prinsen, E., Samson, R. and Lebeer, S. (2020), 'Modes of action of microbial biocontrol in the phyllosphere', *Front. Microbiol.* 11, 1619.
- Park, J.-K., Lee, S.-H., Lee, J.-H., Han, S., Kang, H., Kim, J.-C., Kim, Y. C. and McSpadden Gardener, B. B. (2013), 'Sampling and selection factors that enhance the diversity of microbial collections: application to biopesticide development', *Plant Pathol. J.*, 29: 144-153.
- Pertot, I., Caffi, T., Rossi, V., Mugnai, L., Hoffmann, C., Grando, M.S., Gary, C., Lafond, D., Duso, C., Thiery, D., Mazzoni, V. and Anfora, G. (2017), 'A critical review of plant protection tools for reducing pesticide use on grapevine and new perspectives for the implementation of IPM in viticulture', *Crop Prot.*, 97, 70-84.
- Raymaekers, K., Ponet, L., Holtappels, D., Berckmans, B. and Cammue, B. P. A. (2020), 'Screening for novel biocontrol agents applicable in plant disease management - a review', *Biol. Control*, 144, 104240.

- Robin, D. C. and Marchand, P. A. (2019), 'Evolution of the biocontrol active substances in the framework of the European Pesticide Regulation (EC) No. 1107/2009', *Pest Manage. Sci.*, 75, 950–958.
- Syed, A. B., Rahman, S. F., Singh, E., Pieterse, C. M. J. and Schenk, P. M. (2018), 'Emerging microbial biocontrol strategies for plant pathogens', *Plant Sci.*, 267, 102–111.

10 References

Araujo, W. L., Marcon, J., Maccheroni, W. Jr., van Elsas, J. D., van Vuurde, J. W. L. and Azevedo, J. L. (2002). Diversity of endophytic bacterial populations and their interaction with *Xylella fastidiosa* in citrus plants, *Appl. Environ. Microbiol.* 68(10), 4906–4914.

Baffoni, L., Gaggia, F., Dalanaj, N., Prodi, A., Nipoti, P., Psi, A., Biavati, B. and Di Gioia, D. (2015). Microbial inoculants for the biocontrol of *Fusarium* spp. in durum wheat, *BMC Microbiol.* 15, 242.

Baltz, R. H. (2010). *Streptomyces* and *Saccharopolyspora* hosts for heterologous expression of secondary metabolite gene clusters, *J. Ind. Microbiol. Biotechnol.* 37(8), 759–772.

Benhamou, N., Gagné, S., Le Quéré, D. and Dehbi, L. (2000). Bacterial mediated induced resistance in cucumber: beneficial effect of the endophytic bacterium *Serratia plymuthica* on the protection against infection by *Pythium ultimum*, *Phytopathology* 90(1), 45–56.

Bonaterra, A., Camps, J. and Montesinos, E. (2005). Osmotically induced trehalose and glycine betaine accumulation improves tolerance to desiccation, survival and efficacy of the postharvest biocontrol agent *Pantoea agglomerans* EPS125, *FEMS Microbiol. Lett.* 250(1), 1–8.

Bonaterra, A., Mari, M., Casalini, L. and Montesinos, E. (2003). Biological control of *Monilinia laxa* and *Rhizopus stolonifer* in postharvest of stone fruit by *Pantoea agglomerans* EPS125 and putative mechanisms of antagonism, *Int. J. Food Microbiol.* 84(1), 93–104.

Cha, J. Y., Han, S., Hong, H. J., Cho, H., Kim, D., Kwon, Y., Kwon, S. K., Crüsemann, M., Bok Lee, Y., Kim, J. F., Giaever, G., Nislow, C., Moore, B. S., Thomashow, L. S., Weller, D. M. and Kwak, Y. S. (2016). Microbial and biochemical basis of a Fusarium wilt-suppressive soil, *ISME J.* 10(1), 119–129.

Chen, Z. X. and Dickson, D. W. (1998). Review of *Pasteuria penetrans*: biology, ecology, and biological control potential, *J. Nematol.* 30(3), 313–340.

Compant, S., Reiter, B., Sessitsch, A., Nowak, J., Clément, C. and Barka, E. A. (2005). Endophytic colonization of *Vitis vinifera* L. by plant growth-promoting bacterium *Burkholderia* sp. strain PsJN, *Appl. Environ. Microbiol.* 71(4), 1685–1693.

Conn, V. M., Walker, A. R. and Franco, C. M. (2008). Endophytic Actinobacteria induce defense pathways in *Arabidopsis thaliana*, *Mol. Plant Microbe Interact.* 21(2), 208–218.

Crawford, D. L., Lynch, J. M., Whipps, J. M. and Ousley, M. A. (1993). Isolation and characterization of actinomycete antagonists of a fungal root pathogen, *Appl. Environ. Microbiol.* 59(11), 3899–3905.

Daayf, F., Adam, L. and Fernando, W. G. D. (2003). Comparative screening of bacteria for biological control of potato late blight (strain US-8), using in-vitro, detached-leaves, and whole-plant testing systems, *Can. J. Plant Pathol.* 25(3), 276–284.

Daranas, N., Badosa, E., Francés, J., Montesinos, E. and Bonaterra, A. (2018). Enhancing water stress tolerance improves fitness in biological control strains of *Lactobacillus plantarum* in plant environments, *PLoS ONE* 13(1), e0190931.

Daranas, N., Roselló, G., Cabrefiga, J., Donati, I., Francés, J., Badosa, E., Spinelli, F., Montesinos, E. and Bonaterra, A. (2019). Biological control of bacterial plant diseases with *Lactobacillus plantarum* strains selected for their broad-spectrum activity, *Ann. Appl. Biol.* 174(1), 92–105.

Das, K., Prasanna, R. and Saxena, A. K. (2017). Rhizobia: a potential biocontrol agent for soilborne fungal pathogens, *Folia Microbiol.* 62(5), 425–435.

Delmotte, N., Knief, C., Chaffron, S., Innerebner, G., Roschitzki, B., Schlapbach, R., von Mering, C. and Vorholt, J. A. (2009). Community proteogenomics reveals insights into the physiology of phyllosphere bacteria, *Proc. Natl. Acad. Sci. U. S. A.* 106(38), 16428–16433.

Dhar Purkayastha, G., Mangar, P., Saha, A. and Saha, D. (2018). Evaluation of the biocontrol efficacy of a *Serratia marcescens* strain indigenous to tea rhizosphere for the management of root rot disease in tea, *PLoS ONE* 13(2), e0191761.

Dourado, M. N., Camargo Neves, A. A., Santos, D. S. and Araújo, W. L. (2015). Biotechnological and agronomic potential of endophytic pink-pigmented methylotrophic *Methylobacterium* spp., *BioMed Res. Int.* 2015, 909016.

Edmond, M. B., Riddler, S. A., Baxter, C. M., Wicklund, B. M. and Pasculle, A. W. (1993). *Agrobacterium radiobacter*: a recently recognized opportunistic pathogen, *Clin. Infect. Dis.* 16(3), 388–391.

EFSA (2020). Update of the list of QPS-recommended biological agents intentionally added to food or feed as notified to EFSA 12: suitability of taxonomic units notified to EFSA until March 2020, *EFSA J.* 18(7), e06174.

Goudjal, Y., Toumatia, O., Yekkour, A., Sabaou, N., Mathieu, F. and Zitouni, A. (2014). Biocontrol of *Rhizoctonia solani* damping-off and promotion of tomato plant growth by endophytic actinomycetes isolated from native plants of Algerian Sahara, *Microbiol. Res.* 169(1), 59–65.

Hagen, M. J., Stockwell, V. O., Whistler, C. A., Johnson, K. B. and Loper, J. E. (2009). Stress tolerance and environmental fitness of *Pseudomonas fluorescens* A506, which has a mutation in *rpoS*, *Phytopathology* 99(6), 679–688.

Hauser, N. and Orsini, J. (2015). Cepacia syndrome in a non-cystic fibrosis patient, *Case Rep. Infect. Dis.* 2015, 537627.

Hsueh, P. R., Teng, L. J., Yang, P. C., Chen, Y. C., Pan, H. J., Ho, S. W. and Luh, K. T. (1998) Nosocomial infections caused by Sphingomonas paucimobilis: clinical features and microbiological characteristics, *Clin. Infect. Dis.* 26(3), 676–681.

Huang, J., Wei, Z., Tan, S., Mei, X., Yin, S., Shen, Q. and Xu, Y. (2013). The rhizosphere soil of diseased tomato plants as a source for novel microorganisms to control bacterial wilt, *Appl. Soil Ecol.* 72, 79–84.

Innerebner, G., Knief, C. and Vorholt, J. A. (2011). Protection of *Arabidopsis thaliana* against leaf-pathogenic *Pseudomonas syringae* by *Sphingomonas* strains in a controlled model system, *Appl. Environ. Microbiol.* 77(10), 3202–3210.

Janahiraman, V., Anandham, R., Kwon, S. W., Sundaram, S., Karthik Pandi, V., Krishnamoorthy, R., Kim, K., Samaddar, S. and Sa, T. (2016). Control of wilt and rot pathogens of tomato by antagonistic pink pigmented facultative methylotrophic *Delftia lacustris* and *Bacillus* spp., *Front. Plant Sci.* 7, 1626.

Janisiewicz, W. J., Jurick, W. M., Vico, I., Peter, K. A. and Buyer, J. S. (2013). Culturable bacteria from plum fruit surfaces and their potential for controlling brown rot after harvest, *Postharvest Biol. Technol.* 76, 145–151.

Johnson, K. B. and Stockwell, V. O. (1998). Management of fire blight: a case study in microbial ecology, *Annu. Rev. Phytopathol.* 36, 227–248.

Joshi, R. and McSpadden Gardener, B. B. (2006). Identification and characterization of novel genetic markers associated with biological control activities in *Bacillus subtilis*, *Phytopathology* 96(2), 145–154.

Kalbe, C., Marten, P. and Berg, G. (1996). Strains of the genus *Serratia* as beneficial rhizobacteria of oilseed rape with antifungal properties, *Microbiol. Res.* 151(4), 433–439.

Kamber, T., Lansdell, T. A., Stockwell, V. O., Ishimaru, C. A., Smits, T. H. and Duffy, B. (2012). Characterization of the biosynthetic operon for the antibacterial peptide herbicolin in *Pantoea vagans* biocontrol strain C9-1 and incidence in *Pantoea* species, *Appl. Environ. Microbiol.* 78(12), 4412–4419.

Kapadia, M., Rolston, K. V. and Han, X. Y. (2007). Invasive *Streptomyces* infections: six cases and literature review, *Am. J. Clin. Pathol.* 127(4), 619–624.

Köhl, J., Kolnaar, R. and Ravensberg, W. J. (2019). Mode of action of microbial biological control agents against plant diseases: relevance beyond efficacy, *Front. Plant Sci.* 10, 845.

Köhl, J., Postma, J., Nicot, P., Ruocco, M. and Blum, B. (2011). Stepwise screening of microorganisms for commercial use in biological control of plant-pathogenic fungi and bacteria, *Biol. Control* 57(1), 1–12.

Kortemaa, H., Rita, H., Haahtela, K. and Smolander, A. (1994). Root colonization ability of antagonistic *Streptomyces griseoviridis*, *Plant Soil* 163(1), 77–83.

Kovaleva, J., Degener, J. E. and van der Mei, H. C. (2014). *Methylobacterium* and its role in health care-associated infection, *J. Clin. Microbiol.* 52(5), 1317–1321.

Kumar, H., Bajpai, V. K., Dubey, R. C., Maheshwari, D. K. and Kang, S. C. (2010). Wilt disease management and enhancement of growth and yield of *Cajanus cajan* (L) var. Manak by bacterial combinations amended with chemical fertilizer, *Crop Prot.* 29(6), 591–598.

Kurth, F., Mailänder, S., Bönn, M., Feldhahn, L., Herrmann, S., Große, I., Buscot, F., Schrey, S. D. and Tarkka, M. T. (2014). *Streptomyces*-induced resistance against oak powdery mildew involves host plant responses in defense, photosynthesis, and secondary metabolism pathways, *Mol. Plant Microbe Interact.* 27(9), 891–900.

Latour, X., Barbey, C., Chane, A., Groboillot, A. and Burini, J. F. (2013). *Rhodococcus erythropolis* and its γ-lactone catabolic pathway: an unusual biocontrol system that disrupts pathogen quorum sensing communication, *Agronomy* 3(4), 816–838.

Li, Q. L., Ning, P., Zheng, L., Huang, J. B., Li, G. Q. and Hsiang, T. (2010). Fumigant activity of volatiles of *Streptomyces globisporus* JK-1 against *Penicillium italicum* on *Citrus microcarpa*, *Postharvest Biol. Technol.* 58(2), 157–165.

Lindow, S. E. and Brandl, M. T. (2003). Microbiology of the phyllosphere, *Appl. Environ. Microbiol.* 69(4), 1875–1883.

Liu, X., Bimerew, M., Ma, Y., Müller, H., Ovadis, M., Eberl, L., Berg, G. and Chernin, L. (2007). Quorum-sensing signaling is required for production of the antibiotic pyrrolnitrin in a rhizospheric biocontrol strain of *Serratia plymuthica*, *FEMS Microbiol. Lett.* 270(2), 299–305.

Lourenço Júnior, V., Maffia, L. A., Romeiro, R. d. S. and Mizubuti, E. S. G. (2006). Biocontrol of tomato late blight with the combination of epiphytic antagonists and rhizobacteria, *Biol. Control* 38(3), 331-340.

Manhas, R. K. and Kaur, T. (2016). Biocontrol potential of *Streptomyces hydrogenans* strain DH16 toward *Alternaria brassicicola* to control damping off and black leaf spot of *Raphanus sativus*, *Front. Plant Sci.* 7, 1869.

McSpadden Gardener, B. B. M., Kim, I. S., Kim, K. Y. and Kim, Y. C. (2014). Draft genome sequence of a chitinase-producing biocontrol bacterium, *Lysobacter antibioticus* HS124, *Res. Plant Dis.* 20(3), 216-218.

Messiha, N. A. S., van Diepeningen, A. D., Farag, N. S., Abdallah, S. A., Janse, J. D. and van Bruggen, A. H. C. (2007). *Stenotrophomonas malthophilia*: a new potential biocontrol agent of *Ralstonia solanacearum*, causal agent of potato brown rot, *Eur. J. Plant Pathol.* 118(3), 211-225.

Montesinos, E. (2003). Development, registration and commercialization of pesticides for plant protection, *Int. Microbiol.* 6(4), 245-252.

Montesinos, E. and Bonaterra, A. (2017). Pesticides, microbial. In: Schmidt, T. M (Ed.), *Reference Module in Life Sciences*, Amsterdam, Netherlands, Elsevier. Encyclopedia of Microbiology (4th. Ed.), Academic Press, 473-484.

Montesinos, E., Bonaterra, A., Ophir, Y. and Beer, S. V. (1996). Antagonism of selected bacterial strains to *Stemphylium vesicarium* and biological control of brown spot of pear under controlled environmental conditions, *Phytopathology* 86, 856-863.

Mora, I., Cabrefiga, J. and Montesinos, E. (2015). Cyclic lipopeptide biosynthetic genes and products, and inhibitory activity of plant-associated *Bacillus* against phytopathogenic bacteria, *PLoS ONE* 10(5), e0127738.

Nautiyal, C. S. (1997). Rhizosphere competence of *Pseudomonas* sp. NBRI9926 and *Rhizobium* sp. NBRI9513 involved in the suppression of chickpea (*Cicer arietinum* L.) pathogenic fungi, *FEMS Microbiol. Ecol.* 23(2), 145-158.

Nelkner, J., Tejerizo, G. T., Hassa, J., Lin, T. W., Witte, J., Verwaaijen, B., Winkler, A., Bunk, B., Spröer, C., Overmann, J., Grosch, R., Pühler, A. and Schlüter, aA. (2019). Genetic potential of the biocontrol agent *Pseudomonas brassicacearum* (formerly *P. trivialis*) 3Re2-7 unravelled by genome sequencing and mining, comparative genomics and transcriptomics, *Genes* 10(8), 601.

Parikh, L., Eskelson, M. J. and Adesemoye, A. O. (2018). Relationship of *in vitro* and *in planta* screening: improving the selection process for biological control agents against Fusarium root rot in row crops, *Arch. Phytopathol. Plant Prot.* 51(3-4), 156-169.

Penyalver, R., Vicedo, B. and López, M. M. (2000). Use of the genetically engineered *Agrobacterium* strain K1026 for biological control of crown gall, *Eur. J. Plant Pathol.* 106(9), 801-810.

Pliego, C., Ramos, C., de Vicente, A. and Cazorla, F. M. (2011). Screening for candidate bacterial biocontrol agents against soilborne fungal plant pathogens, *Plant Soil* 340(1-2), 505-520.

Poppe, L., Vanhoutte, S. and Höfte, M. (2003). Modes of action of *Pantoea agglomerans* CPA-2, an antagonist of postharvest pathogens on fruits, *Eur. J. Plant Pathol.* 109(9), 963-973.

Pusey, P. L. (1997). Crab apple blossoms as a model for research on biological control of fire blight, *Phytopathology* 87(11), 1096-1102.

Pusey, P. L. (2002). Biological control agents for fire blight of apple compared under conditions limiting natural dispersal, *Plant Dis.* 86(6), 639-644.

Pusey, P. L., Stockwell, V. O., Reardon, C. L., Smits, T. H. M. and Duffy, B. (2011). Antibiosis activity of *Pantoea agglomerans* biocontrol strain E325 against *Erwinia amylovora* on apple flower stigmas, *Phytopathology* 101(10), 1234–1241.

Pusey, P. L. and Wend, C. (2012). Potential of osmoadaptation for improving *Pantoea agglomerans* E325 as biocontrol agent for fire blight of apple and pear, *Biol. Control* 62(1), 29-37.

Reitz, M., Rudolph, K., Schröder, I., Hoffmann-Hergarten, S., Hallmann, J. and Sikora, R. A. (2000). Lipopolysaccharides of *Rhizobium etli* strain G12 act in potato roots as an inducing agent of systemic resistance to infection by the cyst nematode *Globodera pallida*, *Appl. Environ. Microbiol.* 66(8), 3515–3518.

Rezzonico, F., Smits, T. H. M., Montesinos, E., Frey, J. E. and Duffy, B. (2009). Genotypic comparison of *Pantoea agglomerans* plant and clinical strains, *BMC Microbiol.* 9, 204.

Roselló, G., Bonaterra, A., Francés, J., Montesinos, L., Badosa, E. and Montesinos, E. (2013). Biological control of fire blight of apple and pear with antagonistic *Lactobacillus plantarum*, *Eur. J. Plant Pathol.* 137(3), 621–633.

Sansone, G., Rezza, I., Fernandez, G., Calvente, V., Benuzzi, D. and Sanz, M. I. (2011). Inhibitors of polygalacturonase and laccase of *Botrytis cinerea* and their application to the control of this fungus, *Int. Biodeterior. Biodegrad.* 65(1), 243-247.

Scheepmaker, J. W. A., Hogervost, P. A. M. and Glandorf, D. C. M. (2020). Future introductions of genetically modified microbial biocontrol agents in the EU. Are current EU legislation and risk assessment fit for purpose?, *RIVM Lett. Rep.* 2016-0057.

Shim, J.-S., Farrand, S. K. and Kerr, A. (1987). Biological control of crown gall: construction and testing of new biocontrol agents, *Phytopathology* 77(3), 463–466.

Smits, T. H. M., Rezzonico, F., Pelludat, C., Goesmann, A., Frey, J. E. and Duffy, B. (2010). Genomic and phenotypic characterization of a non-pigmented variant of *Pantoea vagans* biocontrol strain C9-1 lacking the 530 kb megaplasmid pPag3, *FEMS Microbiol. Lett.* 308(1), 48–54.

Someya, N. and Akutsu, K. (2005). Biocontrol of plant diseases by genetically modified microorganisms: current status and future prospects. In: *PGPR: Biocontrol and Biofertilization* Siddiqui, Z. A. (Ed), Dordrecht, Springer, 297–312.

Stockwell, V. O., Kawalek, M. D., Moore, L. W. and Loper, J. E. (1996). Transfer of pAgK84 from the biocontrol agent *Agrobacterium radiobacter* K84 to *A. tumefaciens* under field conditions, *Phytopathology* 86(1), 31–37.

Trias, R., Bañeras, L., Montesinos, E. and Badosa, E. (2008). Lactic acid bacteria from fresh fruit and vegetables as biocontrol agents of phytopathogenic bacteria and fungi, *Int. Microbiol.* 11(4), 231- 236.

Van Lenteren, J. C., Bolckmans, K., Köhl, J., Ravensberg, W. J. and Urbaneja, A. (2018). Biological control using invertebrates and microorganism: plenty of new opportunities, *BioControl* 63(1), 39-59.

Viaene, T., Langendries, S., Beirinckx, S., Maes, M. and Goormachtig, S. (2016). *Streptomyces* as a plant's best friend?, *FEMS Microbiol. Ecol.* 92(8), fiw119.

Volpiano, C. G., Lisboa, B. B., São José, J. F. B., de Oliveira, A. M. R., Beneduzi, A., Passaglia, L. M. P. and Vargas, L. K. (2018), Rhizobium strains in the biological control of the phytopathogenic fungi Sclerotium (Athelia) rolfsii on the common bean, *Plant Soil* 432, 229-243.

Walterson, A. M. and Stavrinides, J. (2015). *Pantoea*: insights into a highly versatile and diverse genus within the Enterobacteriaceae, *FEMS Microbiol. Rev.* 39(6), 968–984.

Wang, L. Y., Xie, Y. S., Cui, Y. Y., Xu, J., He, W., Chen, H. G. and Guo, J. H. (2015). Conjunctively screening of biocontrol agents (BCAs) against fusarium root rot and fusarium head blight caused by *Fusarium graminearum*, *Microbiol. Res.* 177, 34–42.

Xue, Q. Y., Chen, Y., Li, S. M., Chen, L. F., Ding, G. C., Guo, D. W. and Guo, J. H. (2009). Evaluation of the strains of *Acinetobacter* and *Enterobacter* as potential biocontrol agents against Ralstonia wilt of tomato, *Biol. Control* 48(3), 252–258.

Chapter 11

The use of *Trichoderma* spp. to control plant diseases

Enrique Monte and Rosa Hermosa, Spanish-Portuguese Institute for Agricultural Research (CIALE) – University of Salamanca, Spain

1 Introduction

Trichoderma (teleomorph: *Hypocrea*) is a genus of anamorphic fungi belonging to the Phylum Ascomycota. It is characterized by the production of high quantities of hydrolytic enzymes. Some species of this genus harbor the highest number of chitinolytic enzymes ever described within the kingdom Fungi - and a strong expansion of genes for solute transport (Kubicek et al., 2011). These fungi also demonstrate an amazing environmental opportunism, with a very efficient cell damage repair system, that allows them to colonize and live on practically any substrate and resist different types of environmental stresses (Druzhinina et al., 2011). These features confer survival skills on *Trichoderma* spp. to produce a great variety of expression patterns, which allow these fungi to overcome challenging situations, enabling them to be present in many different environments (soil, water, dead tissues, rhizosphere, inside the plant, etc.) (Lorito et al., 2010). They are therefore distributed throughout the world in all continents, from the Caribbean tropical rainforest to the dry shores of the Sahara Desert or the frozen tundra of Antarctica (Hermosa et al., 2004; Sadfi-Zouaoui et al., 2009).

http://dx.doi.org/10.19103/AS.2021.0093.13

However, the most outstanding feature of *Trichoderma* is the ability to antagonize, parasitize or kill other fungi, oomycetes and even nematodes, working as biocontrol agents in plant pathology by means of different mechanisms of action. Such mechanisms (Monte, 2001) are not mutually exclusive, since they may be displayed jointly and simultaneously. They can be summarized as follows:

- Feeding directly on or in a pathogen to destroy it, a process called parasitism.
- Releasing molecules that slow down or kill the pathogen in the vicinity of such molecules, a process known as antibiosis.
- Growing faster or using its food source more efficiently than the pathogen, a process known as competition.
- Causing a plant to modify its metabolism and make products that discourage or kill the pathogen, a process defined as induced resistance.

Direct *Trichoderma*-based plant disease biocontrol relies mainly on the production of fungal cell wall degrading enzymes (CWDEs) and/or antibiotics as well as competition for nutrients (Lorito et al., 2010), while the triggering of plant defenses can be considered an indirect form of biocontrol.

Mycoparasitic *Trichoderma* spp. are able to degrade and grow within the sclerotia produced as resting structures by other fungi. Comparative genomics studies suggest that the ancestors of *Trichoderma* shared a common ancestor with the ant-mutualistic *Escovopsis* and entomoparasitic fungi (Druzhinina et al., 2018). They also turned into mycoparasites on wood-degrading basidiomycetes that acquired saprotrophic characteristics to follow their prey into substrates, exploring wood and other plant structures as novel ecological niches (Kubicek et al., 2011). It has been proposed that, in such niches, nearly half of the *Trichoderma* genes encoding plant CWDEs were obtained, via a process of lateral gene transfer from plant-associated filamentous fungi belonging to different phylogenetically close classes of Ascomycota hosts (Druzhinina et al., 2018). We have recently observed that the mycoparasitic activity of *Trichoderma* on ascomycetes not only helps them acquire many types of genes but also provides them with the capacity to capture and transfer mycoviruses related to hypovirulence in other species of ascomycetous fungi, such as *Cryphonectria parasitica* (Romón-Ochoa, personal communication).

Trichoderma spp. also produce an astonishing number of secondary metabolites (SM), many of them with human applications such as pharmaceutical and agri-biotechnological uses (Hermosa et al., 2014). SM have a role as signaling molecules that allow *Trichoderma* to communicate with the environment. SM can thus act as microbe-associated molecular patterns (MAMPs) activating plant immune responses and as auxin-like compounds at low concentrations (Vinale et al., 2008). SM produced by *Trichoderma* can also exert an antimicrobial effect at high doses. However, in natural conditions, volatile metabolites are released but do not accumulate to effective antimicrobial concentrations.

Information about the biosynthetic pathways of *Trichoderma* SM, fungal metabolism and ecological interactions have emerged from genetic and genomic studies (Hermosa et al., 2014). It is important to emphasize that the chemical structures of many *Trichoderma* SM were discovered before molecular tools helped to clarify the taxonomy of this genus. It is, for example, incorrect to assert that *T. harzianum* produces trichothecenes when the production of these SM is restricted to the clade Brevicompactum (Nielsen et al., 2005). Similarly, *T. harzianum* T22, probably the best-known *Trichoderma* commercial strain, has been relocated within *T. afroharzianum* (Chaverri et al., 2015). This is particularly important for those who want to register active ingredients with commercial applications.

As rhizosphere colonizers, *Trichoderma* spp. have developed opportunistic mechanisms for adaptation to abiotic stresses as well as for nutrient uptake and solute transport. In the plant, these processes are facilitated by the induction of cell wall extension and expansion, secondary root development, lateral root hair production and a higher photosynthetic rate (Hermosa et al., 2013). The presence of potential fungal preys and plant root-derived nutrients in the rhizosphere has helped *Trichoderma* spp. to colonize plant roots, boosting the evolution to endophytism and further positive interactions between *Trichoderma* and plants (Druzhinina et al., 2011; Hermosa et al., 2012). The fact that *Trichoderma* spp. are beneficial to plants makes them an ideal microbial model to study plant-beneficial (non-pathogenic) fungus mechanisms. A direct benefit of advances in *Trichoderma* knowledge is the appearance on the market of a new generation of products based on strains selected both for their antagonist ability and other positive growth effects on crops (Lorito et al., 2010; Woo and Pepe, 2018).

2 Biocontrol mechanisms: mycoparasitism

A fundamental part of the *Trichoderma* antifungal system relies on the recognition, binding and enzymatic disruption of the host-fungus cell wall. Chitinases and glucanases produced by *Trichoderma* degrade the immature cell wall at hyphal apices. They also degrade the strong chitin-glucan complexes of mature cell walls, as well as structures such as sclerotia and chlamydospores, reducing both disease symptoms and pathogen spread (Monte, 2001). A major part of the antifungal system of *Trichoderma* consists of a series of genes coding for a variety of lytic enzymes, including endochitinases, N-acetyl-β-glucosaminidases, chitin 1,4-β-chitobiosidases, proteases, endo- and exoglucan β-1,3-glucosidases, endoglucan β-1,6-glucosidases, xylanases, mannanases, pectinases, pectin lyases, amylases, esterases/lipases, phospholipases, RNases and DNases (Lorito, 1998).

Thirty years ago, the most common biological control agents of the genus *Trichoderma* were *T. harzianum*, *T. virens*, *T. viride* and *T. koningii* strains (Table 1). However, studies based on CWDEs, β-1,3-glucanase, β-1,6-glucanase, cellulase, chitinase and protease, isoenzyme patterns (Sanz et al., 2004) served to demonstrate that:

- strains of the same species show different antifungal behavior.
- in addition to *T. harzianum* as the most recognized biocontrol agent in this genus, novel strains of the emerging species *T. asperellum* and *T. atroviride* have different CWDE profiles that allow them to efficiently disrupt the cell wall of plant pathogenic fungi and oomycetes by hydrolyzing biopolymers not present in plant tissues.

Tests also showed that *Trichoderma* chitinases and glucanases had no additional harmful effects on the plant even when relatively large quantities of CWDE were injected, or the corresponding genes were expressed, into the plants (Monte, 2001; Nicolás et al., 2014). Moreover, purified CWEDs (e.g., chitinases CHIT42, CHIT36 or CHIT33; N-acetyl-β-D-glucosaminidase NAG1; α-1,3-glucanases AGN13.1 or AGN13.2; β-1,3-glucanase GLUC78, β-1,6-glucanases BGN16.1 or BGN16.3; and proteases PRB1, PR1 or P6281) from different *Trichoderma* spp. have demonstrated to be highly effective in inhibiting mycelial growth and spore germination in a broad range of pathogens from important genera such as *Alternaria, Colletotrichum, Fusarium, Rhizoctonia, Ustilago, Phytophthora, Pythium* and especially *Botrytis*.

Trichoderma mycoparasitic development, and gene expression and signal transduction during mycoparasitism have been reviewed by Viterbo and Horwitz (2010). The observation that *Trichoderma* CWDEs-encoding genes are differentially regulated, depending on culture conditions, suggests specific functional roles for the resulting enzymes (Suárez et al., 2007; Morán-Diez et al., 2019). Although a large number of genes are inducible by the host, there are also those that can be induced upon release of catabolite repression. This means that a mycoparasite, growing initially as a saprophyte in a nutrient-scarce environment, may express genes that will enable it to attack a host whenever it becomes available (Viterbo and Horwitz, 2010).

Genomic approaches have highlighted that among the abundance of chitinolytic enzymes present in *T. atroviride* and *T. virens*, there is an elevated number of chitinases with carbohydrate-binding domains – particularly enriched in mycoparasitic *Trichoderma* spp. Other chitinases have LysM modules that indicate other/additional roles for these proteins in fungal biology and as effectors in *Trichoderma*-plant communication (Kubicek et al., 2011).

Table 1 Some biocontrol success examples, achieved against fungi, oomycetes and nematodes, after application of *Trichoderma* strains at the greenhouse (G) and field (F) level

Species/strain	Disease/Pathogen	Crop	G/F	Reference
T. harzianum TH	Damping-off *Rhizoctonia solani* *Sclerotinia rolfsii*	Bean Cotton Tomato	F	Elad et al. (1980)
T. harzianum T315	Damping-off *Pythium aphanidermatum*	Cucumber Pea, pepper Tomato	G	Sivan et al. (1984)
T. koningii T8 *T. harzianum* T12	Seed rot *Pythium* spp.	Pea Beans	F	Hadar et al. (1984)
T. harzianum T35	Fusarium wilt *F. oxysporum* f.sp. *vasinfectum* *F. oxysporum* f.sp. *melonis* *F. culmorum*	Cotton Muskmelon Wheat	G	Sivan and Chet (1986)
T. harzianum 017 *T. harzianum* 043 *T. virens* 041	Root and crown rot *Phytophthora cactorum*	Apple seedlings	G	Smith et al. (1990)
T. afroharzianum T22	Several examples Damping-off, root rot *Pythium ultimum* Anthracnose *Colletotrichum graminicola*	Various Maize	G/F G	Harman (2000) Harman et al. (2004)
T. harzianum Th5 *T. koningii* Tk6	Tan spot *Pyrenophora tritici-repentis*	Wheat	G	Perelló et al. (2003)
T. harzianum T39	Anthracnose *Colletotrichum acutatum* Grey mould *Botrytis cinerea*	Strawberry	G	Freeman et al. (2004)
T. viride T112 *T. viride* MO	Root rot *Bipolaris sorokiniana*	Wheat seedlings	G	Salehpour et al. (2005)
T. hamatum LU593	Lettuce drop *S. minor*	Lettuce	F	Rabeendran et al. (2006)
T. martiale ALF247	Black pod rot *Phytophthora palmivora*	Cocoa	F	Hanada et al (2009)
T. harzianum T2	Septoria blotch *Zymoseptoria tritici*	Wheat	F	Perelló et al. (2009)
T. harzianum Th3013	Purple blotch *Alternaria porri*	Onion	G	Albo-Elyousr et al. (2014)
T. hamatum	Vascular wilt *F. oxysporum* f.sp. *lentis*	Lentil	G	El-Hassan et al. (2013)

(*Continued*)

Table 1 (*Continued*)

Species/strain	Disease/Pathogen	Crop	G/F	Reference
T. koningiopsis T51	Grey mould *B. cinerea*	Tomato seedlings	G	You et al. (2016)
T. longibrachiatum T6	Cereal cyst nematode *Heterodera avenae*	Wheat	G	Zhang et al. (2017)
T. atroviride T11	Root-knot nematode *Meloidogyne javanica*	Tomato	G	Medeiros et al. (2017)
T. asperelloides T25	White mold *S. sclerotiorum*	Soybean	F	Sumida et al. (2018)
T. ghanense T8	Crown and root rot *F. oxysporum* f.sp. *radicis lycopersici*	Tomato	G	Debbi et al. (2018)
T. gamsii T6085	Fusarium head blight *Fusarium graminearum*	Wheat	F	Sarrocco et al. (2019)
T. asperellum BCC1	White rot *Sclerotinia cepivorum*	Onion	F	Rivera-Méndez et al. (2020)

Trichoderma chitinases with LysM-binding modules are likely to be involved in the scavenging of the chitin fragments released as a result of plant chitinase activity on fungal cell walls during infection. This prevents them from acting as MAMPs and thereby dampening plant defenses while facilitating root colonization (Hermosa et al., 2013).

There are still major gaps in our understanding of the molecular determinants responsible for the onset of antifungal behavior and gene regulation in the course of host recognition, attachment and mycoparasitism. It is known that fungal prey signals can act at a distance without the existence of physical contact and that most CWDEs are inducible enzymes whose concentration increases dramatically when their substrates are present. β-1,3-glucanase, chitinase and protease activities of *Trichoderma* in particular may lead to the release of host signals that, acting at a distance without physical contact, can attract *Trichoderma* towards its prey. *Trichoderma* spp. are able to sense other fungi before contact (Seidl et al., 2009; Morán-Diez et al., 2019). Electron microscopy and cytochemistry studies have shown that *Trichoderma* has receptors with galactose-binding affinity at the cell surface which can recognize the fine glucan-enriched extracellular matrices on the host cell surface (Benhamou and Chet, 1993).

Fungal-host recognition and adhesion during the pathogenesis of other fungi, insects and plants have been reported to be mediated by hydrophobins, which are very well represented in *Trichoderma* genomes. Hydrophobins were overexpressed in different *Trichoderma* spp. cDNA libraries obtained under simulated mycoparasitism and plant polymer interaction conditions (Vizcaíno et al., 2006). *Trichoderma* has the highest number of hydrophobins among the ascomycetes. These hydrophobins are cysteine-rich active proteins located on the outer surfaces of cell walls of hyphae and conidia that enable self-assembly at the interface between the hydrophilic cell wall and a hydrophobic environment. Upon contact with the host, a signaling cascade comprising G-proteins and MAPKs is activated, modulating the *Trichoderma* protein expression patterns involved in mycoparasitism (Reithner et al., 2007) and SM biosynthesis (Beier et al., 2020). *Trichoderma* coils around or grows along the host hyphae and forms hook-like structures that aid in penetrating the host's cell wall. Although coiling around the host is visually striking, it is not clear to what extent coiling and attachment are fully required for antagonism (Viterbo and Horwitz, 2010). Depending on the *Trichoderma* strain, *Verticillium dahliae* is inhibited but not always overgrown by *T. atroviride* and *T. asperellum* (Carrero-Carrón et al., 2016). This observation fits within the high complexity existing in a biocontrol process, which can be influenced by the nature of the pathogen as well as the *Trichoderma* strains involved (Druzhinina et al., 2011). The availability of specific genes rather than the regulation of gene expression seems to be the key element for successful antagonism (Atanasova et al., 2013).

Mycoparasitic abilities are not uniform throughout the genus and most biocontrol studies have been performed on *T. harzianum* (and *T. afroharzianum*), *T. virens*, *T. viride*, *T. koningii*, *T. hamatum*, *T. polysporum*, *T. atroviride* and *T. asperellum* (and *T. asperelloides*) strains, as well as those from the most recent *T. gamsii*, *T. guizhouense*, *T. parareesei*, *T. martiale*, *T. koningiopsis* or *T. ghanense* species, among others (Table 1). A comparative transcriptomics study using *Rhizoctonia solani* as a target confirmed the increased mycoparasitic potential of *T. atroviride* and *T. virens* compared to that of *T. reesei* and also highlighted the different mycoparasitic strategies in *Trichoderma* (Atanasova et al., 2013). While *T. atroviride* and *T. virens* express a number of genes for which a function in antagonism can be interpreted, the saprotroph *T. reesei* mainly expresses genes for nutrient acquisition. In this same study, the two strong mycoparasitic species also differed in the tools that they used for combating and killing *R. solani*.

Antibiosis as well as the action of CWDEs is used by *T. atroviride*, indicating a parasitic interaction that does not aim to kill the host. However, *T. virens* acts by poisoning *R. solani* with toxins, showing clear necrotrophic behavior (Atanasova et al., 2013). Both examples illustrate the need for carefully choosing the right biocontrol agent to be applied to a given pathosystem. In this way, *Trichoderma* strains producing the greatest number of chitinases or β-1,3-glucanases, and the highest expression of endochitinase genes, are not necessarily the most appropriate antagonists. Proteases are often forgotten biocontrol agents. These enzymes may fulfill important roles in host cell lysis by attacking proteins and lipoproteins, which are also part of the host cell wall, and in the inactivation of enzymes secreted by phytopathogens during the infection process in plants

(Elad and Kapat, 1999), as well as in protein recycling (Ramada et al., 2016). Studies based on transcriptomic analyses have indicated that proteolysis is a major biological process involved in mycoparasitism by *Trichoderma* when overgrowing its host (Reithner et al., 2011; Atanasova et al., 2013; Steindorff et al., 2014). Serine, aspartic and metallopeptidases have been found to be related to overgrowing biocontrol (Morán-Diez et al., 2019).

Different *Trichoderma* functional genomics and proteomics studies have consistently shown that the hexagonal peroxisome protein 1 (HEX1) is one of the most expressed transcripts in *Trichoderma* strains during mycoparasitism (Lorito et al., 2010). The HEX1 protein self-assembles to form the typical Woronin body associated with the hyphal septum in ascomycetes. It maintains cellular integrity by sealing the septal pore after cellular damage, and also acts as a host factor that enhances the accumulation of RNA, including the accumulation of dsRNA mycovirus in an infected filamentous fungus (Son et al., 2013). HEX1 could have a dual function in *Trichoderma* mycoparasitism since:

- this protein is related to RNA uptake and replication.
- considering that HEX1 is formed from peroxisomes, it could be involved in the generation of oxidative stress by *Trichoderma* strains necessary to successfully compete with other fungi and oomycetes in soil and the rhizosphere.

Oxidative stress increases CWDEs activity and therefore the biocontrol efficiency, as demonstrated in the direct confrontation of *T. harzianum* with *Pythium ultimum* (Montero-Barrientos et al., 2011).

The antifungal activity of *Trichoderma* CWDEs can be enhanced synergistically by combining enzymes with different hydrolytic activities. As with biocontrol of *Botrytis cinerea*, the combined application of purified chitinases and glucanases from *T. harzianum* can be as effective as most of the chemical fungicides are used against this pathogen (Monte, 2001). Small doses of one or two CWDEs can be also combined with chemical fungicides. They can be synergistic with SM and plant pathogenesis-related proteins such as peptaibols and thaumatin-like proteins, respectively. However, the practical implementation of purified CWDEs is far from being cost-effective in conventional agriculture. *Trichoderma* protein extracts with high CWDEs activities obtained as by-products can be produced from culture supernatants of the liquid fermentations made to produce *Trichoderma* biomass.

3 Biocontrol mechanisms: antibiosis

Trichoderma spp. produce more than 120 SM structures in a strain-dependent manner (Yu and Keller, 2005). Some of them exhibit antibiotic properties against fungi, oomycetes, bacteria and even insects, and can act synergistically with CWDEs to inhibit the growth of fungal pathogens and pests (Schirmböck et al., 1994; Lorito et al., 1994; Coppola et al., 2017). In comparison with other filamentous fungi, gene clusters involved in the biosynthesis of polyketides (PK) and small non-ribosomal peptides (NRP) are higher in biocontrol species such as *T. atroviride* and *T. virens* (Kubicek et al., 2011).

Many *Trichoderma* SM have demonstrated antimicrobial activity since they can inhibit the growth of fungi and oomycetes such as *Botrytis*, *Colletotrichum*, *Fusarium*, *Gaeumannomyces*, *Penicillium*, *Phytophthora, Pythium*, *Rhizoctonia*, *Sclerotinia* or *Stachybotrys* (Hermosa et al., 2014). Biocontrol properties of SM isolated from *Trichoderma* spp. have been reviewed by different authors (Reino et al., 2008; Hermosa et al., 2014; Contreras-Cornejo et al., 2016; Keswani et al., 2019; Khan et al., 2020). The principal SM are 6-penthyl pyrone (6-PP), pirrolidindiones (harzianic acid), furanosteroids (viridin), diketopiperazines (gliotoxin and gliovirin), NRP (the peptaibols trichokonins, trichorzianines and alamethicins), terpenes (the trichothecenes trichodermin and harzianum A) and PK (aspinolides).

The volatile organic compound (VOC) 6PP is the representative metabolite common to *Trichoderma* and responsible for the yellow pigment observed in agar plates and the coconut aroma associated with this genus. Several studies have shown the inhibitory properties of 6PP and its derivatives against *R. solani*, *B. cinerea*, *P. ultimum*, *Fusarium oxysporum* and *Sclerotium rolfsii* (Hermosa et al., 2014). At low concentrations, 6PP induces plant defense mechanisms and may also act as an auxin-like compound and/or auxin inducer (Vinale et al., 2008). This behavior suggests the existence of shared common components of plant defense and developmental responses of plants to *Trichoderma*. 6PP released by *T. atroviride* from the belowground rhizosphere of maize plants worked as airborne signaling VOC that increased the parasitism rate of wasps on the fall armyworm, *Spodoptera frugiperda* (Contreras-Cornejo et al., 2018), highlighting the relevance of *Trichoderma* SM in pest control management.

Harzianic acid is a siderophore that significantly alters iron availability in the soil environment for other microorganisms (Vinale et al., 2013). It has shown antibiotic activity against *R. solani*, *Sclerotinia sclerotiorum* and *Pythium irregulare* (Vinale et al., 2009). Harzianic acid also boosts plant immune responses in a salicylic acid (SA)- and ethylene (ET)/jasmonic acid (JA)-dependent way (Manganiello et al., 2018). Viridins are potent irreversible inhibitors of phosphoinositide-3-kinases that are intracellular signal transducer enzymes involved in cell growth and proliferation. Viridins from *Trichoderma* cultures reduce spore germination of *Botrytis allii*, *Colletotrichum lini*, *Fusarium caeruleum*, *Penicillium expansum*, *Aspergillus niger* and *Stachybotrys atra* (Singh et al., 2005).

Gliotoxin and gliovirin, initially described in *T. viride*, are diketopiperazines with a sulfur bridge responsible for the generation of reactive oxygen species (ROS) that can cause an imbalance in redox homeostasis and severe oxidative stress (Scharf et al., 2016). The so-called 'Q' strains of *T. virens* produce gliotoxin (but not gliovirin) and are more effective in the antagonism of *R. solani*. The 'P' strains produce gliovirin (but not gliotoxin) and are not able to antagonize *R. solani*, but inhibit the oomycete *P. ultimum* (Howell and Stipanovic, 1983; Howell et al., 1993). Gliotoxin exhibits bioactivity against the pathogenic fungus *Aspergillus fumigatus* and plays an important role in the biocontrol activity of *Trichoderma* against phytopathogens such as *S. rolfsii*, *Macrophomina phaseolina*, *Pythium debaryanum* and *Pythium aphanidermatum* (Khan et al., 2020). Gliotoxin synergism with endochitinase enzymes inhibits *B. cinerea* (Lorito et al., 1994) and increases the entomopathogenic capacity of this SM (Vargas et al., 2014).

Peptaibols are short non-ribosomal peptide chains, with an unusual amino acid content, that can exhibit antifungal, antibacterial and anticancer properties. These properties are driven primarily by their membrane insertion and pore-forming abilities due to their linear and amphipathic nature (Rebuffat et al., 1992). Trichokonins induce apoptotic cell death in *F. oxysporum*, *B. cinerea*, *V. dahliae*, *Ascochyta citrullina* and *Phytophthora parasitica* (Shi et al., 2012). Interaction with plants also elicits systemic resistance against viral attack (Luo et al., 2010), increases auxin content and disrupts auxin response gradients in root tips (Shi et al., 2016). Trichorzianines inhibit the β-glucan synthase activity of *B. cinerea* and, in combination with CWDEs, can synergistically inhibit spore germination and hyphal growth of this fungus (Schirmböck et al., 1994; Lorito et al., 1996). Alamethicins permeabilize membrane vesicles and have the capacity to induce plant defenses (Kredics et al., 2013). Although alamethicins have been reported to be toxic for plants at certain concentrations (Rippa et al., 2010), the cellulase activity of *T. viride* induces plant resistance to these peptaibols (Aidemark et al., 2010).

Trichothecenes are sesquiterpenes that prevent peptide bond formation at the peptidyl transferase center of the 60S ribosomal subunit during protein synthesis. Trichodermin produced by *T. brevicompactum* has shown anti-yeast and anti-filamentous fungi activity, such activity generally being higher than conventional amphotericin B and hygromycin antifungal antibiotics (Tijerino et al., 2011a). It has been shown in greenhouse assays that tomato seed treatments with *T. brevicompactum* strains increased leaf necrotic lesions produced by *B. cinerea*, mainly when the tomato seeds were coated with trichodermin-overproducing transformants, suggesting these strains of *T. brevicompactum* are not appropriate as agricultural biocontrol agents (Tijerino et al., 2011b). In contrast, *T. arundinaceum* produces the 'beneficial' trichothecene harzianum A, able to inhibit the germination of *B. cinerea* spores, reduce the growth of *R. solani* and prime SA- and JA-responsive defense pathways in tomato plants (Malmierca et al., 2012).

T. arundinaceum also synthesizes aspinolides that have a significant antibiotic effect against *B. cinerea* (Malmierca et al., 2015). However, these polyketidic metabolites are particularly important when it comes to understanding the molecular crosstalk between a biocontrol agent and its prey. There is a bidirectional transcriptional regulation in which *T. arundinaceum* triggers the synthesis of high levels of several aspinolides when the production of harzianum A is blocked. The sesquiterpene botrydial produced by *B. cinerea* disrupts the synthesis of harzianum A, but the aspinolides of *T. arundinaceum* overcome the pathogen response by downregulating the expression of genes involved in the production of botrydial (Malmierca et al., 2016). Aspinolides can also induce the expression of tomato SA-related genes, which indicates that they may play a role in the sensitization of tomato defense-related genes (Malmierca et al., 2015). *Trichoderma* trichothecenes and PK can be considered as key molecules in the fungal metabolomics network and they play a role in the environmental interactions established by these fungi with other organisms present in the rhizosphere.

Despite the high efficiency of *Trichoderma* SM in inhibiting phytopathogens, few studies have assessed the efficacy of these SM against a particular disease under field conditions (Keswani et al., 2014). Antibiotic production is often correlated with biocontrol ability. 6-PP, aspinolides, harzianolide, gliotoxin, harzianic acid and harzianum A are some examples of *Trichoderma* SM involved in plant defense and/or growth stimulation. It has been proven that the application of pure SM has been found to show effects on the host and pathogen similar to those obtained by using the corresponding living microbe. However, as with the direct application of CWDEs, it is not currently cost-effective to obtain purified SM for direct use in crops. Novel formulation and application strategies, such as microcapsules and nanoparticles (Fraceto et al., 2018), will be able to exploit the synergism of SM with plant beneficial microbes, contributing to the development of new generations of biocontrol agents and biostimulants with increased efficacy in crop production (Marra et al., 2019).

4 Biocontrol mechanisms: competition

Soil microbiota and plants compete against one another by recruiting limiting resources, such as carbon and nitrogen, into their biomass. Competition for nutrients, oxygen, physical space or light can occur at varying rates. The ubiquity of *Trichoderma* spp. in natural and agricultural soils (Hermosa et al., 2004) indicates they are an excellent competitor for space and nutritional resources. After overcoming the early defenses of the plant, *Trichoderma* is able to compete with pathogens for apoplast space (Carrero-Carrón et al., 2018). It has the ability to modify the root architecture to promote the plant's uptake of nutrients, having a greater inner and outer surface area for colonization and availability of root exudates (Samolski et al., 2012). The dynamics of nutrients in the rhizosphere are often complex and *Trichoderma* has developed competitive skills to overcome the shortage of resources, adapt to the environment and gain access to roots. Gene expression analysis during the interaction of *T. atroviride* with *B. cinerea*/*R. solani* allowed the identification of 66 genes that were overexpressed during the onset of mycoparasitism (Seidl et al., 2009). The most abundant genes detected in this study were not CWDEs, but those involved in post-translational processing and amino acid metabolism. They included components of the stress response reaction to nitrogen shortage, signal transduction and lipid catabolism.

The expression of peptide transporter activity was increased when *T. harzianum* interacted with *B. cinerea* (Vizcaíno et al., 2006). Genes encoding heat-shock proteins (HSP) and detoxification responses were upregulated when *T. atroviride* overgrew the colonies of *V. dahliae* (Morán-Diez et al., 2019). *hsp* genes were also induced during the initial contact of *T. asperellum* hyphae with those of *S. sclerotiorum* and *F. oxysporum*, as well as during thermal stress (Mota et al., 2019). These signs all seem to indicate that *Trichoderma* strains prepare themselves to withstand stress conditions, enabling them to gain competitive advantage. Having an array of HSP allows *Trichoderma* to tolerate cold, heat, oxidative, osmotic or saline stresses (Montero-Barrientos et al., 2008). The upregulation of an ATP-binding cassette (ABC) transporter gene in the presence of pathogen-secreted metabolites and specific mycotoxins helped *T. atroviride* to overcome the stress associated with mycoparasitism, while their deletion caused the loss of protection of tomato plants from *P. ultimum* and *R. solani* attack (Ruocco et al., 2009).

Imbalance between ROS production in plants and their detoxification by enzymatic and non-enzymatic reactions causes oxidative stress. ROS act as signaling molecules involved in growth and developmental processes, pathogen defense responses such as hypersensitive reaction and systemic acquired resistance (SAR). *Trichoderma* spp. are good competitors in the rhizosphere since they tolerate high levels of ROS (Morán-Diez et al., 2010). They induce ROS production which, in turn, facilitates CWDE activity against pathogens, most of them with a lower range of oxidative stress tolerance (Montero-Barrientos et al., 2011).

In the first stages of root adherence and colonization, *Trichoderma* low molecular mass hydrophobic cell wall proteins (hydrophobins) and expansin-like proteins are able to recognize cellulose and modify root architecture (Mendoza-Mendoza et al., 2018). The *T. harzianum qid74* gene encodes a cysteine-rich cell wall protein that is highly upregulated in interaction with tomato roots (Samolski et al., 2012). QID74 is involved in lateral root growth and hair formation and elongation. This results in increasing total absorptive surface, facilitating nutrient uptake and translocation of nutrients in the shoots, resulting in increased plant biomass through an efficient use of NPK and micronutrients. Intracellular invertase identified in *T. virens* hydrolyzed plant-derived sucrose in the rhizosphere, which conferred a competitive advantage in colonizing the roots (Vargas et al., 2014). The release by *Trichoderma* spp. of Fe^{3+}-chelating molecules, called siderophores, can suppress the growth of pathogenic microorganisms by depriving the pathogens of iron (Vinale et al., 2012). The unavailability of iron determines the control of Fusarium wilt disease by *T. asperellum* in tomato plants (Segarra et al., 2010). The dual role of harzianic acid as both antibiotic and siderophore serves to increase the biocontrol potential of strains producing this SM (Vinale et al., 2013).

Applying *Trichoderma* to crops affects soil bacterial and fungal communities in a pH- and N supply-dependent manner, respectively, but also regulates available nutrients in the soil and chemical compounds in the rhizosphere (Zhang et al., 2019). It has been recently observed that the application of *T. harzianum* triggers wheat bacterial diversity under high N fertilization, although such application was ineffective against non-target organisms (Illescas et al., 2020). At this point, a key question is what is going on with mycorrhizal fungi. *Trichoderma* strains are potential competitors with mycorrhizae, although their combined application increases crop yields (Buysens et al., 2016). Mycorrhizae facilitate *Trichoderma* entry into the roots (De Jaeger et al., 2010). *Trichoderma* strains also promote access of mycorrhizal fungi even to non-host canola or *Arabidopsis* roots (Poveda et al., 2019). Since it is not possible to generalize, it is best to consider the application of these two types of beneficial fungi on a case-by-case basis.

5 Plant beneficial effects: systemic defense and growth promotion

The interplay of signals/hormones involved in *Trichoderma*-induced plant defense has been previously reviewed (Hermosa et al. 2012, 2013; Mendoza-Mendoza et al., 2018). Plants have cell-surface pattern recognition receptors (PRRs) that recognize microbe-derived MAMPs (e.g. chitin or β-glucans) and plant cell wall oligomers released by microbial hydrolytic activity (Morán-Diez et al., 2009; Saravanakumar et al., 2016) which are known as damage-associated molecular patterns (DAMPs). In plant-*Trichoderma*-pathogen systems, DAMPs are also derived from the activity of *Trichoderma* CWDEs on pathogens' cell walls (Hermosa et al., 2013). As a result, an innate immune response referred to as MAMP-triggered immunity (MTI) is activated. Effective *Trichoderma* strains are able to induce a stronger MTI response in the plant compared to pathogen-triggered immunity (Lorito et al., 2010). This defense response causes polysaccharide accumulation to strengthen plant cell wall, synthesis of defensive molecules such as phytoalexins and pathogenesis-related (PR) proteins, and accumulation of defensive phytohormones (e.g. SA, JA, ET) which send alarm signals that are systemically transmitted through the cells that have not been attacked.

Upon *Trichoderma* root colonization, the plant reacts with an increase in SA levels. SA is the key phytohormone controlling early root colonization of *Trichoderma* since it is necessary for the plant to limit *Trichoderma* to the apoplastic space of the epidermis and cortex, preventing access to the vascular tissues (Alonso-Ramírez et al., 2014). In plant-*Trichoderma* compatible interactions, SA levels decrease as the fungus colonizes the apoplast pathway of the roots. After the onset of *Trichoderma* establishment, JA biosynthesis and JA-responsive genes are expressed in the root cells. As plants accumulate JA, *Trichoderma* is able to negatively modulate the expression of positive regulators of SA-dependent defense (Brotman et al., 2013) and produce its own SA (Pérez et al., 2015) to fine-tune the process. A pioneer plant-*Trichoderma* interaction study (Shoresh et al., 2005) described the involvement of JA/ET signaling in induced systemic resistance (ISR) triggered by *T. asperellum* in cucumber plants at 24-h post-inoculation. The composition and concentration of the phytohormone blend produced by plant immune responses to coordinate the defense in response to *Trichoderma* in different parts of the plant not only vary over time but also with the fungal strain, its concentration and how long the interaction lasts.

Successful invaders (as pathogens often do) have evolved to suppress MTI by delivering a plethora of small molecules, referred to as effectors, at the interface of the host plant (apoplastic effectors) or inside the cell (cytoplasmic effectors). *Trichoderma* spp. are able to increase the level of the first MTI response (initiated by MAMPs and DAMPs) by producing a variety of apoplastic effectors (Hermosa et al., 2013; Contreras-Cornejo et al., 2016; Mendoza-Mendoza et al., 2018; Ramírez-Valdespino et al., 2019). These are likely to be detected by cell-surface PRR receptors, limit the loss of resistance and therefore keep the plant response to a level above or just below the threshold for effective resistance (Lorito et al., 2010). This behavior allows *Trichoderma* to establish an avirulent relationship with the plant and to elicit a durable defense (priming) that allows the plant to deploy a more rapid and stronger defense response compared with a non-primed plant.

Although SA accumulation is typically found in SAR induced by biotrophic pathogens, and JA and ET have been described as signal transduction molecules for ISR due to the effect of beneficial microbes (and also herbivores and necrotrophic pathogens), the resistance promoted by *Trichoderma* is modulated by both SA and JA/ET pathways (Salas-Marina et al., 2011; Malmierca et al., 2012; Martínez-Medina et al., 2013). It is often stated that

Trichoderma induces SAR or ISR at a given time point. This can be misleading. Plant-*Trichoderma* crosstalk is dynamic and the expression of defense-related genes of the SA and/or JA/ET pathways may overlap in an undulating manner, apparently in keeping with the ability of *Trichoderma* to grow both on and in the roots, priming defense responses against biotic and abiotic stresses (Rubio et al., 2014; Rivera-Méndez et al., 2020). *Trichoderma* induces minor changes in the plant transcriptome over time, and several weeks after contact with the plant, the waves of the undulating responses are likely to disappear (Domínguez et al., 2016). Plants remember previous stresses and, depending on the type of primary stimulus and the pathosystem involved, primed plants can deploy a diverse set of defense mechanisms (Balmer et al., 2015).

In summary, *Trichoderma* ingress into root tissues and modulate the defense signaling network to circumvent the plant immune system, while enabling plants to make use of their priming memory to cope with different pathogens. This trait allows effective *Trichoderma* strains to exploit their biocontrol potential across the plant and, for that purpose they produce a vast array of effector molecules (Table 2). Nevertheless, signaling events that follow PRR activation - phosphorylation, signal transduction, transcriptional reprograming mediated by transcription factors (TF) and phytohormone feedback - are numerous because, in many instances, negatively-acting TFs are activated to repress transcription of defense-related genes in a negative feedback loop that fine-tunes signaling (Couto and Zipfel, 2016). Moreover, several growth-promoting phytohormones have been associated with plant immunity. We have proposed a *Trichoderma*-plant crosstalk model (Hermosa et al., 2012) in which *Trichoderma* auxin and gibberellin production (Contreras-Cornejo et al., 2009; Jaroszuk-Sciseł et al., 2019) and ET availability (Viterbo et al., 2010) interconnect with plant development and defense responses. *Trichoderma* modulate the expression of TF genes (e.g. MYC2) that act on the nodes (e.g. DELLA-JAZ binding), regulating trade-offs needed to achieve a balance between growth and defense (Hermosa et al., 2013). In the absence of disease, *Trichoderma* can enhance plant biomass production and promote lateral root growth through an auxin-dependent mechanism (Contreras-Cornejo et al., 2009). Selected *Trichoderma* strains positively affect the development of maize (Shoresh and Harman, 2008) and tomato shoots (Fiorini et al., 2016), increasing leaf pigment content and improving photosynthetic activity, and also facilitating tomato growth under abiotic stress conditions (Rubio et al., 2017). The ability of *Trichoderma* to promote growth is not restricted to herbaceous plants but has also been described in woody plants such as olive trees (Carrero-Carrón et al., 2016).

Continuing with the priming memory of plants inoculated with *Trichoderma*, a recent study has demonstrated that, in addition to a direct biocontrol activity against the nematode *Meloidogyne javanica*, *T. atroviride* is able to induce

Table 2 *Trichoderma* effectors. Secreted proteins and secondary metabolites released by *Trichoderma* into the root apoplast that are able to trigger plant defense responses. For more details, see Hermosa et al. (2013), Contreras-Cornejo et al. (2016) and Ramírez-Valdespino et al. (2019)

Effector proteins	Species of *Trichoderma*
Cerato-platanins Sm1/2/3 and Epl1/2/3	*T. virens* and *T. atroviride*
Chitin-binding LysM proteins	*T. virens*, *T. atroviride*, *Trichoderma* spp.
Chitin-binding Avr4 homologue	*T. afroharzianum*, *T. atroviride*
Carboxipeptidase Avr9 homologue	*T. afroharzianum*, *T. atroviride*
Swollenin	*T. asperellum*, *T. reesei*
Hydrophobins	*T. asperellum*, *T. virens*, *T. longibrachiatum*
Cellulases	*T. longibrachiatum*, *T. harzianum*, *T. viride*
Endopoligalacturonase ThPG1	*T. harzianum*
Xylanase Eix	*T. viride*
Effector metabolites	
6-pentyl pyrones	*T. viride*, *T. harzianum*, *T. atroviride*
Harzianic acid	*T. harzianum*
Aspinolides	*T. arundinaceum*
Terpenes	
Harzianum A	*T. arundinaceum*
Trichodiene (*tri5* gene)	*T. harzianum*
β-ocimene	*T. virens*
Peptaibols	
Alamethicins	*T. viride*
Trichokonins	*T. psuedokoningii*
Trichovirin II	*T. viride*, *T. harzianum*
18mer TvBI and TvBII	*T. virens*
Phytohormones	
Salicylic acid	*T. parareesei*
Auxin	*T. virens*, *T. atroviride*
Gibberellins	*T. koningiopsis*

systemic resistance without establishing any contact with the pathogen. However, tomato progeny inherited resistance to the nematode as well as growth promotion without compromising the level of resistance of the plant's offspring to the pathogen (Medeiros et al., 2017). This means that primed plants, in fact, inherit the *Trichoderma*-regulated fitness costs associated with induction of defenses and growth.

6 Registration and commercialization

There is abundant evidence that *Trichoderma* strains are excellent biocontrol agents, resulting in many plant protection products (PPP) on the market. It is important to recognize early work to register pioneering *Trichoderma*-based active ingredients in the EU (Directive 91/414, Annex I list of positive substances, approved 01/05/2009) for use in biological control (Woo et al., 2014): Remedier (*T. asperellum* ICC012 and *T. gamsii* ICC080), Xedavir (*T. asperellum* TV1), Binab (*T. atroviride* IMI206040 and *T. polysporum* IMI206039), Trianum (*T. afroharzianum* T22) and TUSAL (*T. atroviride* T11 and *T. asperellum* T25). These as well as other commercial biocontrol agents are approved in different countries – Rootshield (*T. afroharzianum* T22), Trichodex (*T. harzianum* T39), SoilGard (*T. virens* GL21), Floragard (*T. hamatum* TH382), Trichopel (*T. atroviride* LC52), and so on – (Monte, 2001). They have made possible subsequent registration of *Trichoderma* formulations. Dossiers completed in those early years have contributed to a more rational approach today to the process of authorization.

Although single *Trichoderma* strains are used as PPP, many are also known to have properties that result in plant growth promotion and other beneficial effects, typically not indicated in the registered product details (Woo and Pepe, 2018). Conversely, there are officially approved plant biostimulants (e.g., mycorrhizae) that also possess biocontrol properties by inducing systemic defenses (Cameron et al., 2013). Registration depends on the microbial trait that the manufacturer is targeting. Since the PPP registration process is cumbersome, many *Trichoderma* formulations for commercial use against plant pathogens are not registered as PPP. The biostimulant abilities of a given strain are highlighted rather than the potential biocontrol properties. *Trichoderma* strains are marketed as plant inoculants or plant-strengthening agents, also called biostimulants, circumventing the time and expense necessary for the registration of biocontrol products (Harman et al., 2010). Fertilizer regulations only cover biostimulants, and pesticide regulations only cover PPP. Policymakers are still investigating whether multiple use, as both biostimulant and PPP, is possible and how this can be regulated. Setting aside the direct mycoparasitic activity of many strains, the capacity of many *Trichoderma* strains to induce systemic defenses (officially considered a PPP trait) and to promote plant growth (biostimulation) are determined by the host, depending on which one the plant wants to tap. The current legal regulatory framework does not reflect the range of benefits of *Trichoderma*. Since the role of *Trichoderma* is to regulate plant fitness costs associated with balancing defense and growth, adequate registration procedures are urgently needed that are based on the scientific understanding of this ecologically important and fascinating fungal genus.

7 Conclusions

Some key conclusions are:

- *Trichoderma* are amongst the most efficient and wider biofungicides existing in the market. However, biocontrol properties must be identified at strain level.

- Different *Trichoderma* biocontrol mechanisms may be displayed jointly and simultaneously.
- The genetic background of *T. asperellum* offers the highest chance to find isolates with strong biocontrol abilities.
- The practical implementation of purified CWDEs and SM is far from being cost-effective in conventional agriculture.
- In addition to the high efficiency of *Trichoderma* SM in inhibiting phytopathogens they can act as signaling molecules at lower concentrations.
- Effective *Trichoderma* strains are able to induce a strong MTI response in the plant since they can activate the PRR receptors twice, as MAPMs and as secreted apoplastic effectors.
- *Trichoderma* interconnect plant development and defense responses, acting on the nodes that regulate the trade-offs needed to achieve a balance between growth and defense.
- Plant progeny inherit the *Trichoderma*-regulated fitness costs associated with induction of defenses and growth.
- *Trichoderma* capacities as biocontrol agents or biostimulants are determined by the host, depending on which capacities the plant wants to tap.
- The current registration frameworks do not justice to biology of *Trichoderma* and interactions with plants and plant pathogens, and an appropriate legal framework needs to be designed.

8 Where to look for further information

The selection and practical application of *Trichoderma* strains as biocontrol agents in agriculture require proper identification. A multilocus identification system for *Trichoderma* can be found in Dou et al. (2020) (doi: https://doi.org/10.1128/AEM.01532-20). A *Trichoderma* taxonomy browser can be found on the NCBI website (https://www.ncbi.nlm.nih.gov/Taxonomy/Browser/wwwtax.cgi?id=5543). The main updates on the taxonomy of *Trichoderma*, list of species and identification tools can be found on the website of the International Commission on *Trichoderma* Taxonomy (ICTT) (https://trichoderma.info/)

More information on the research conducted on the use of *Trichoderma* as a biocontrol agent and its beneficial interaction with plants can be found on the website of the Spanish-Portuguese Institute for Agricultural Research (CIALE) (https://ciale.usal.es/grupos-de-investigacion/grupo.aspx?id=9). The site includes a list of publications and projects on this subject. The following articles are highlighted on:

- Molecular ecology and genomics of *Trichoderma*: Druzhinina et al. (2011) (doi: 10.1038/nrmicro2637).

- Beneficial effects of *Trichoderma* to plants: Hermosa et al. (2012) (10.1099/mic.0.052274-0).
- Effect of *Trichoderma* on plant growth and defense balance: Hermosa et al. (2013) (doi: 10.2436/20.1501.01.181).
- Transgenic plants expressing *Trichoderma* genes: Nicolás et al. (2014) (10.1016/j.plantsci.2014.03.005).
- *Trichoderma*-induced heritable plant responses: Medeiros et al. (2017) (10.1038/srep40216.), Moran-Diez et al. (2021) (https://doi.org/10.3390/jof7040318).
- Molecular dialogues between *Trichoderma* and roots: Mendoza-Mendoza et al. (2018) (doi: https://doi.org/10.1016/j.fbr.2017.12.001).

Information regarding the *Trichoderma* genomics released can be found on the JGI MycoCOSM portal:

(https://mycocosm.jgi.doe.gov/mycocosm/home/releases?flt=trichoderma&seq=all&pub=all&grp=fungi&srt=released&ord=desc).

9 Acknowledgements

We thank the European Regional Development Fund (FEDER) under the Regional Government of Castile and Leon support (Projects SA094P20 and Escalera de Excelencia CLU-2018-04), the Spanish Government (Project RTI2018-099986-B-I00) and the Diputación de Salamanca (Projects-2018).

10 References

Aidemark, M., Tjellström, H., Sandelius, A. S., Stålbrand, H., Andersson, E., Rasmusson, A. G. and Widell, S. (2010). *Trichoderma viride* cellulase induces resistance to the antibiotic pore-forming peptide alamethicin associated with changes in the plasma membrane lipid composition of tobacco BY-2 cells, *BMC Plant Biol.* 10, 1-13.

Albo-Elyousr, K. A. M., Abdel-Hafez, S. I. I. and Abdel-Rahim, I. R. (2014). Isolation of *Trichoderma* and evaluation of their antagonistic potential against *Alternaria porri*, *J. Phytopathol.* 162(9), 567-574.

Alonso-Ramírez, A., Poveda, J., Martín, I., Hermosa, R., Monte, E. and Nicolás, C. (2014). Salicylic acid prevents *Trichoderma harzianum* from entering the vascular system of roots, *Mol. Plant Pathol.* 15(8), 823-831.

Atanasova, L., Le Crom, S., Gruber, S., Coulpier, F., Seidl-Seiboth, V., Kubicek, C. P. and Druzhinina, I. S. (2013). Comparative transcriptomics reveals different strategies of *Trichoderma* mycoparasitism, *BMC Genomic* 14, 121.

Balmer, A., Pastor, V., Gamir, J., Flors, V. and Mauch-Mani, B. (2015). The prime-ome: towards a holistic approach to priming, *Trends Plant Sci.* 20(7), 443-452.

Beier, S., Hinterdobler, W., Monroy, A. A., Bazafkan, H. and Schmoll, M. (2020). The kinase USK1 regulates cellulase gene expression and secondary metabolite biosynthesis in *Trichoderma reesei*, *Front. Microbiol.* 11, 974.

Benhamou, N. and Chet, I. (1993). Hyphal interactions between *Trichoderma harzianum* and *Rhizoctonia solani*: ultrastructure and gold cytochemistry of the mycoparasitic process, *Phytopathology* 83(10), 1062–1071.

Brotman, Y., Landau, U., Cuadros-Inostroza, Á, Takayuki, T., Fernie, A. R., Chet, I., Viterbo, A. and Willmitzer, L. (2013). *Trichoderma*-plant root colonization: escaping early plant defense responses and activation of the antioxidant machinery for saline stress tolerance, *PLoS Pathog.* 9(3), e1003221.

Buysens, C., César, V., Ferrais, F., Dupré de Boulois, H. and Declerck, S. (2016). Inoculation of *Medicago sativa* cover crop with *Rhizophagus irregularis* and *Trichoderma harzianum* increases the yield of subsequently- grown potato under low nutrient conditions, *Appl. Soil Ecol.* 105, 137–143.

Cameron, D. D., Neal, A. L., van Wees, S. C. and Ton, J. (2013). Mycorrhiza-induced resistance: more than the sum of its parts?, *Trends Plant Sci.* 18(10), 539–545.

Carrero-Carrón, I., Rubio, M. B., Niño-Sánchez, J., Navas-Cortés, J. A., Jiménez-Díaz, R. M., Monte, E. and Hermosa, R. (2018). Interactions between *Trichoderma harzianum* and defoliating *Verticillium dahliae* in resistant and susceptible wild olive clones, *Plant Pathol.* 67(8), 1758–1767.

Carrero-Carrón, I., Trapero-Casas, J. L., Olivares-García, C., Monte, E., Hermosa, R. and Jiménez-Díaz, R. M. (2016). *Trichoderma asperellum* is effective for biocontrol of Verticillium wilt in olive caused by the defoliating pathotype of *Verticillium dahliae*, *Crop Prot.* 88(10), 45–52.

Chaverri, P., Branco-Rocha, F., Jaklitsch, W., Gazis, R., Degenkolb, T. and Samuels, G. J. (2015). Systematics of the *Trichoderma harzianum* species complex and the re-identification of commercial biocontrol strains, *Mycologia* 107(3), 558–590.

Contreras-Cornejo, H. A., del-Val, E., Macías-Rodríguez, L., Alarcón, A., González-Esquivel, C. E. and Larsen, J. (2018). *Trichoderma atroviride*, a maize root associated fungus, increases the parasitism rate of the fall armyworm *Spodoptera frugiperda* by its natural enemy *Campoletis sonorensis*, *Soil Biol. Biochem.* 122, 196–202.

Contreras-Cornejo, H. A., Macías-Rodríguez, L., Cortés-Penagos, C. and López-Bucio, J. (2009). *Trichoderma virens*, a plant beneficial fungus, enhances biomass production and promotes lateral root growth through an auxin-dependent mechanism in *Arabidopsis*, *Plant Physiol.* 149(3), 1579–1592.

Contreras-Cornejo, H. A., Macías-Rodríguez, L., del-Val, E. and Larsen, J. (2016). Ecological functions of *Trichoderma* spp. and their secondary metabolites in the rhizosphere: interactions with plants, *FEMS Microbiol. Ecol.* 92(4), fiw036.

Coppola, M., Cascone, P., Chiusano, M. L., Colantuono, C., Lorito, M., Pennacchio, F., Rao, R., Woo, S. L., Guerrieri, E. and Digilio, M. C. (2017). *Trichoderma harzianum* enhances tomato indirect defense against aphids, *Insect Sci.* 24(6), 1025–1033.

Couto, D. and Zipfel, C. (2016). Regulation of pattern recognition receptor signalling in plants, *Nat. Rev. Immunol.* 16(9), 537–552.

De Jaeger, N., Declerck, S. and De la Providencia, I. E. (2010). Mycoparasitism of arbuscular mycorrhizal fungi: a pathway for the entry of saprotrophic fungi in roots, *FEMS Microbiol. Ecol.* 73(2), 312–322.

Debbi, A., Boureghda, H., Monte, E. and Hermosa, R. (2018). Distribution and genetic variability of *Fusarium oxysporum* associated with tomato diseases in Algeria and a biocontrol strategy with indigenous *Trichoderma* spp., *Front. Microbiol.* 9, 282.

Domínguez, S., Rubio, M. B., Cardoza, R. E., Gutiérrez, S., Nicolás, C., Bettiol, W., Hermosa, R. and Monte, E. (2016). Nitrogen metabolism and growth enhancement in tomato plants challenged with *Trichoderma harzianum* expressing the *Aspergillus nidulans* acetamidase *amdS* gene, *Front. Microbiol.* 7, 1182.

Druzhinina, I. S., Chenthamara, K., Zhang, J., Atanasova, L., Yang, D., Miao, Y., Rahimi, M. J., Grujic, M., Cai, F., Pourmehdi, S., Salim, K. A., Pretzer, C., Kopchinskiy, A. G., Henrissat, B., Kuo, A., Hundley, H., Wang, M., Aerts, A., Salamov, A., Lipzen, A., LaButti, K., Barry, K., Grigoriev, I. V., Shen, Q. and Kubicek, C. P. (2018). Massive lateral transfer of genes encoding plant cell wall-degrading enzymes to the mycoparasitic fungus *Trichoderma* from its plant-associated hosts, *PLoS Genet.* 14(4), e1007322.

Druzhinina, I. S., Seidl-Seiboth, V., Herrera-Estrella, A., Horwitz, B. A., Kenerley, C. M., Monte, E., Mukherjee, P. K., Zeilinger, S., Grigoriev, I. V. and Kubicek, C. P. (2011). *Trichoderma*: the genomics of opportunistic success, *Nat. Rev. Microbiol.* 9(10), 749–759.

El-Hassan, S. A., Gowen, S. R. and Pembroke, B. (2013). Use of *Trichoderma hamatum* for biocontrol of lentil vascular wilt disease: efficacy, mechanisms of interaction and future prospects, *J. Plant Prot. Res.* 53(1), 12–26.

Elad, Y., Chet, I. and Katan, J. (1980). *Trichoderma harzianum*: a biocontrol agent effective against *Sclerotium rolfsii* and *Rhizocionia solani*, *Phytopathology* 70(2), 119–121.

Elad, Y. and Kapat, A. (1999). The role of *Trichoderma harzianum* protease in the biocontrol of *Botrytis cinerea*, *Eur. J. Plant Pathol.* 105(2), 177–189.

Fiorini, L., Guglielminetti, L., Mariotti, L., Curadi, M., Picciarelli, P., Scartazza, A., Sarrocco, S. and Vannacci, G. (2016). *Trichoderma harzianum* T6776 modulates a complex matabolic network to stimulate tomato *cv.* Micro-Tom growth, *Plant Soil* 400(1), 351–366.

Fraceto, L. F., de Lima, R., Oliveira, H. C., Ávila, D. S. and Chen, B. (2018). Future trends in nanotechnology aiming environmental applications, *Energ. Ecol. Environ.* 3(2), 69–71.

Freeman, S., Minz, D., Kolenski, I., Barbul, O., Zvebil, A., Maymon, M., Nitzani, Y., Kirshner, B., Rav-David, D., Bilu, A., Dag, A., Shafir, S. and Elad, Y. (2004). *Trichoderma* biocontrol of *Colletotrichum acutatum* and *Botrytis cinerea* and survival in strawberry, *Eur. J. Plant Pathol.* 110(4), 361–370.

Hadar, Y., Harman, G. E. and Tavlor, A. G. (1984). Evaluation of *Trichoderma koningii* and *T. harzianum* from New York soils for biological control of seed rot caused by *Pythium* spp., *Phytopathology* 74(1), 106–110.

Hanada, R. E., Pomella, A. W. V., Soberanis, W., Loguercio, L. L. and Pereira, J. O. (2009). Biocontrol potential of *Trichoderma martiale* against the black-pod disease (*Phytophthora palmivora*) of cacao, *Biol. Control* 50(2), 143–149.

Harman, G. E. (2000). Myths and dogmas of biocontrol changes in perceptions derived from research on *Trichoderma harzianum* T-22, *Plant Dis.* 84(4), 377–393.

Harman, G. E., Obregón, M. A., Samuels, G. J. and Lorito, M. (2010). Changing models for commercialization and implementation of biocontrol in the developing and the developed world, *Plant Dis.* 94(8), 928–939.

Harman, G. E., Petzoldt, R., Comis, A. and Chen, J. (2004). Interactions between *Trichoderma harzianum* strain T22 and maize inbred line Mo17 and effects of these interactions on diseases caused by *Pythium ultimum* and *Colletotrichum graminicola*, *Phytopathology* 94(2), 147–153.

Hermosa, M. R., Keck, E., Chamorro, I., Rubio, B., Sanz, L., Vizcaíno, J. A., Grondona, I. and Monte, E. (2004). Genetic diversity shown in *Trichoderma* biocontrol isolates, *Mycol. Res.* 108(8), 897–906.

Hermosa, R., Cardoza, R. E., Rubio, M. B., Gutiérrez, S. and Monte, E. (2014). Secondary metabolism and antimicrobial metabolites of *Trichoderma*. In: Gupta, V. K., Schmoll, M. and Herrera-Estrella, A. (Eds), *Biotechnology and Biology of Trichoderma*, Amsterdam: Elsevier, pp. 125–137.

Hermosa, R., Rubio, M. B., Cardoza, R. E., Nicolás, C., Monte, E. and Gutiérrez, S. (2013). The contribution of *Trichoderma* to balancing the costs of plant growth and defense, *Int. Microbiol.* 16(2), 69–80.

Hermosa, R., Viterbo, A., Chet, I. and Monte, E. (2012). Plant-beneficial effects of *Trichoderma* and of its genes, *Microbiology (Reading)* 158(1), 17–25.

Howell, C. R. and Stipanovic, R. D. (1983). Gliovirin, a new antibiotic from *Gliocladium virens*, and its role in the biological control of *Pythium ultimum*, *Can. J. Microbiol.* 29(3), 321–324.

Howell, C. R., Stipanovic, R. D. and Lumsden, R. D. (1993). Antibiotic production by strains of *Gliocladium virens* and its relation to biocontrol of cotton seedling diseases, *Biocontrol Sci. Technol.* 3(4), 435–441.

Illescas, M., Rubio, M. B., Hernández-Ruiz, V., Morán-Diez, M. E., Martínez de Alba, A. E., Nicolás, C., Monte, E. and Hermosa, R. (2020). Effect of inorganic N top dressing and *Trichoderma harzianum* seed-inoculation on crop yield and the shaping of root microbial communities of wheat plants cultivated under high basal N fertilization, *Front. Plant Sci.* 11, 575861.

Jaroszuk-Scisel, J., Tyskiewicz, R., Nowak, A., Ozimek, E., Majewska, M., Hanaka, A., Tyśkiewicz, K., Pawlik, A. and Janusz, G. (2019). Phytohormones (auxin, gibberellin) and ACC deaminase in vitro synthesized by the mycoparasitic *Trichoderma* DEMTkZ3A0 strain and changes in the level of auxin and plant resistance markers in wheat seedlings inoculated with this strain conidia, *Int. J. Mol. Sci.* 20(19), 4923.

Keswani, C., Mishra, S., Sarma, B. K., Singh, S. P. and Singh, H. B. (2014). Unraveling the efficient applications of secondary metabolites of various *Trichoderma* spp., *Appl. Microbiol. Biotechnol.* 98(2), 533–544.

Keswani, C., Singh, H. B., Hermosa, R., García-Estrada, C., Caradus, J., He, Y. W., Mezaache-Aichour, S., Glare, T. R., Borriss, R., Vinale, F. and Sansinenea, E. (2019). Antimicrobial secondary metabolites from agriculturally important fungi as next biocontrol agents, *Appl. Microbiol. Biotechnol.* 103(23–24), 9287–9303.

Khan, R. A. A., Najeeb, S., Hussain, S., Xie, B. and Li, Y. (2020). Bioactive secondary metabolites from *Trichoderma* spp. against phytopathogenic fungi, *Microorganisms* 8(6), 817.

Kredics, L., Szekeres, A., Czifra, D., Vágvölgyi, C. and Leitgeb, B. (2013). Recent results in alamethicin research, *Chem. Biodivers.* 10(5), 744–771.

Kubicek, C. P., Herrera-Estrella, A., Seidl-Seiboth, V., Martinez, D. A., Druzhinina, I. S., Thon, M., Zeilinger, S., Casas-Flores, S., Horwitz, B. A., Mukherjee, P. K., Mukherjee, M., Kredics, L., Alcaraz, L. D., Aerts, A., Antal, Z., Atanasova, L., Cervantes-Badillo, M. G., Challacombe, J., Chertkov, O., McCluskey, K., Coulpier, F., Deshpande, N., von Döhren, H., Ebbole, D. J., Esquivel-Naranjo, E. U., Fekete, E., Flipphi, M., Glaser, F., Gómez-Rodríguez, E. Y., Gruber, S., Han, C., Henrissat, B., Hermosa, R., Hernández-Oñate, M., Karaffa, L., Kosti, I., Le Crom, S., Lindquist, E., Lucas, S., Lübeck, M., Lübeck, P. S., Margeot, A., Metz, B., Misra, M., Nevalainen, H., Omann, M., Packer, N., Perrone, G., Uresti-Rivera, E. E., Salamov, A., Schmoll, M., Seiboth, B., Shapiro, H., Sukno, S., Tamayo-Ramos, J. A., Tisch, D., Wiest, A., Wilkinson, H. H., Zhang, M., Coutinho, P. M., Kenerley, C. M., Monte, E., Baker, S. E. and Grigoriev, I. V. (2011). Comparative genome sequence analysis underscores mycoparasitism as the ancestral life style of *Trichoderma*, *Genome Biol.* 12(4), R40.

Lorito, M. (1998). Chitinolytic enzymes and their genes. In: Kubicek, C. P. and Harman, G. E. (Eds), *Trichoderma and Gliocladium vol. 2*, London: Taylor Francis, pp. 73–99.

Lorito, M., Farkas, V., Rebuffat, S., Bodo, B. and Kubicek, C. P. (1996). Cell wall synthesis is a major target of mycoparasitic antagonism by *Trichoderma harzianum*, *J. Bacteriol.* 178(21), 6382–6385.

Lorito, M., Hayes, C. K., Zoina, A., Scala, F., Del Sorbo, G., Woo, S. L. and Harman, G. E. (1994). Potential of genes and gene products from *Trichoderma* sp. and *Gliocladium* sp. for the development of biological pesticides, *Mol. Biotechnol.* 2(3), 209–217.

Lorito, M., Woo, S. L., Harman, G. E. and Monte, E. (2010). Translational research on *Trichoderma*: from 'omics to the field', *Annu. Rev. Phytopathol.* 48, 395–417.

Luo, Y., Zhang, D. D., Dong, X. W., Zhao, P. B., Chen, L. L., Song, X. Y., Wang, X. J., Chen, X. L., Shi, M. and Zhang, Y. Z. (2010). Antimicrobial peptaibols induce defense responses and systemic resistance in tobacco against tobacco mosaic virus, *FEMS Microbiol. Lett.* 313(2), 120–126.

Malmierca, M. G., Barúa, J., McCormick, S. P., Izquierdo-Bueno, I., Cardoza, R. E., Alexander, N. J., Hermosa, R., Collado, I. G., Monte, E. and Gutiérrez, S. (2015). Novel aspinolide production by *Trichoderma arundinaceum* with a potential role in *Botrytis cinerea* antagonistic activity and plant defence priming, *Environ. Microbiol.* 17(4), 1103–1118.

Malmierca, M. G., Cardoza, R. E., Alexander, N. J., McCormick, S. P., Hermosa, R., Monte, E. and Gutiérrez, S. (2012). Involvement of *Trichoderma* trichothecenes in the biocontrol activity and induction of plant defense-related genes, *Appl. Environ. Microbiol.* 78(14), 4856–4868.

Malmierca, M. G., Izquierdo-Bueno, I., McCormick, S. P., Cardoza, R. E., Alexander, N. J., Moraga, J., Gomes, E. V., Proctor, R. H., Collado, I. G., Monte, E. and Gutiérrez, S. (2016). Botrydial and botcinins produced by *Botrytis cinerea* regulate the expression of *Trichoderma arundinaceum* genes involved in trichothecene biosynthesis, *Mol. Plant Pathol.* 17(7), 1017–1031.

Manganiello, G., Sacco, A., Ercolano, M. R., Vinale, F., Lanzuise, S., Pascale, A., Napolitano, M., Lombardi, N., Lorito, M. and Woo, S. L. (2018). Modulation of tomato response to *Rhizoctonia solani* by *Trichoderma harzianum* and its secondary metabolite harzianic acid, *Front. Microbiol.* 9, 1966.

Marra, R., Lombardi, N., D'Errico, G., Troisi, J., Scala, G., Vinale, F., Woo, S. L., Bonanomi, G. and Lorito, M. (2019). Application of *Trichoderma* strains and metabolites enhances soybean productivity and nutrient content, *J. Agric. Food Chem.* 67(7), 1814–1822.

Martínez-Medina, A., Fernández, I., Sánchez-Guzmán, M. J., Jung, S. C., Pascual, J. A. and Pozo, M. J. (2013). Deciphering the hormonal signalling network behind the systemic resistance induced by *Trichoderma harzianum* in tomato, *Front. Plant Sci.* 4, 206.

Medeiros, H. A., Araújo Filho, J. V., Freitas, L. G., Castillo, P., Rubio, M. B., Hermosa, R. and Monte, E. (2017). Tomato progeny inherit resistance to the nematode *Meloidogyne javanica* linked to plant growth induced by the biocontrol fungus *Trichoderma atroviride*, *Sci. Rep.* 7, 40216.

Mendoza-Mendoza, A., Zaid, R., Lawry, R., Hermosa, R., Monte, E., Horwitz, B. A. and Mukherjee, P. K. (2018). Molecular dialogues between *Trichoderma* and roots: role of the fungal secretome, *Fungal Biol. Rev.* 32(2), 62–85.

Monte, E. (2001). Understanding *Trichoderma*: between agricultural biotechnology and microbial ecology, *Int. Microbiol.* 4(1), 1–4.

Montero-Barrientos, M., Hermosa, R., Cardoza, R. E., Gutiérrez, S. and Monte, E. (2011). Functional analysis of the *Trichoderma harzianum nox1* gene, encoding an NADPH oxidase, relates production of reactive oxygen species to specific biocontrol activity against *Pythium ultimum*, *Appl. Environ. Microbiol.* 77(9), 3009–3016.

Montero-Barrientos, M., Hermosa, R., Nicolás, C., Cardoza, R. E., Gutiérrez, S. and Monte, E. (2008). Overexpression of a *Trichoderma HSP70* gene increases fungal resistance to heat and other abiotic stresses, *Fungal Genet. Biol.* 45(11), 1506–1513.

Morán-Diez, E., Hermosa, R., Ambrosino, P., Cardoza, R. E., Gutiérrez, S., Lorito, M. and Monte, E. (2009). The ThPG1 endopolygalacturonase is required for the *Trichoderma harzianum*-plant beneficial interaction, *Mol Plant Microbe Int.* 22(8), 1021–1031.

Morán-Diez, M. E., Cardoza, R. E., Gutiérrez, S., Monte, E. and Hermosa, R. (2010). *TvDim1* of *Trichoderma virens* is involved in redox-processes and confers resistance to oxidative stresses, *Curr. Genet.* 56(1), 63–73.

Morán-Diez, M. E., Carrero-Carrón, I., Rubio, M. B., Jiménez-Díaz, R. M., Monte, E. and Hermosa, R. (2019). Transcriptomic analysis of *Trichoderma atroviride* overgrowing plant-wilting *Verticillium dahliae* reveals the role of a new M14 metallocarboxypeptidase CPA1 in biocontrol, *Front. Microbiol.* 10, 1120.

Mota, T. M., Oshiquiri, L. H., Lopes, ÉC. V., Barbosa Filho, J. R., Ulhoa, C. J. and Georg, R. C. (2019). Hsp genes are differentially expressed during *Trichoderma asperellum* self-recognition, mycoparasitism and thermal stress, *Microbiol. Res.* 227, 126296.

Nicolás, C., Hermosa, R., Rubio, B., Mukherjee, P. K. and Monte, E. (2014). *Trichoderma* genes in plants for stress tolerance-status and prospects, *Plant Sci.* 228, 71–78.

Nielsen, K. F., Gräfenhan, T., Zafari, D. and Thrane, U. (2005). Trichothecene production by *Trichoderma brevicompactum*, *J. Agric. Food Chem.* 53(21), 8190–8196.

Perelló, A., Mónaco, C., Simón, M. R., Sisterna, M. and Dal Bello, G. (2003). Biocontrol efficacy of *Trichoderma* isolates for tan spot of wheat in Argentina, *Crop Prot.* 22(9), 1099–1106.

Perelló, A. E., Moreno, M. V., Mónaco, C., Simón, M. R. and Cordo, C. (2009). Biological control of *Septoria tritici* blotch on wheat by *Trichoderma* spp. under field conditions in Argentina, *BioControl* 54(1), 113–122.

Pérez, E., Rubio, M. B., Cardoza, R. E., Gutiérrez, S., Bettiol, W., Monte, E. and Hermosa, R. (2015). The importance of chorismate mutase in the biocontrol potential of *Trichoderma parareesei*, *Front. Microbiol.* 6, 1181.

Poveda, J., Hermosa, R., Monte, E. and Nicolás, C. (2019). *Trichoderma harzianum* favours the access of arbuscular mycorrhizal fungi to non-host Brassicaceae roots and increases plant productivity, *Sci. Rep.* 9(1), 11650.

Rabeendran, N., Jones, E. E., Moot, D. J. and Stewart, A. (2006). Biocontrol of Sclerotinia lettuce drop by *Coniothyrium minitans* and *Trichoderma hamatum*, *Biol. Control* 39(3), 352–362.

Ramada, M. H. S., Steindorff, A. S., Bloch Jr., C. and Ulhoa, C. J. (2016). Secretome analysis of the mycoparasitic fungus *Trichoderma harzianum* ALL 42 cultivated in different media supplemented with *Fusarium solani* cell wall or glucose, *Proteomics* 16(3), 477–490.

Ramírez-Valdespino, C. A., Casas-Flores, S. and Olmedo-Monfil, V. (2019). *Trichoderma* as a model to study effector-like molecules, *Front. Microbiol.* 10, 1030.

Rebuffat, S., Duclohier, H., Auvin-Guette, C., Molle, G., Spach, G. and Bodo, B. (1992). Membrane-modifying properties of the pore-forming peptaibols saturisporin SA IV and harzianin HA V, *FEMS Immunol. Microbiol.* 105(1-3), 151-160.

Reino, J. L., Guerrero, R. F., Hernández-Galán, R. and Collado, I. G. (2008). Secondary metabolites from species of the biocontrol agent *Trichoderma*, *Phytochem. Rev.* 7(1), 89-123.

Reithner, B., Ibarra-Lachette, E., Mach, R. L. and Herrera-Estrella, A. (2011). Identification of mycoparasitism-related genes in *Trichoderma atroviride*, *Appl. Environ. Microbiol.* 77(13), 4361-4370.

Reithner, B., Schuhmacher, R., Stoppacher, N., Pucher, M., Brunner, K. and Zeilinger, S. (2007). Signaling via the *Trichoderma atroviride* mitogen-activated protein kinase Tmk1 differentially affects mycoparasitism and plant protection, *Fungal Genet. Biol.* 44(11), 1123-1133.

Rippa, S., Eid, M., Formaggio, F., Toniolo, C. and Béven, L. (2010). Hypersensitive-like response to the pore-former peptaibol alamethicin in *Arabidopsis thaliana*, *ChemBioChem* 11(14), 2042-2049.

Rivera-Méndez, W., Obregón, M., Morán-Diez, M. E., Hermosa, R. and Monte, E. (2020). *Trichoderma asperellum* biocontrol activity and induction of systemic defenses against *Sclerotium cepivorum* in onion plants under tropical climate conditions, *Biol. Control* 141, 10415.

Rubio, M. B., Hermosa, R., Vicente, R., Gómez-Acosta, F. A., Morcuende, R., Monte, E. and Bettiol, W. (2017). The combination of *Trichoderma harzianum* and chemical fertilization leads to the deregulation of phytohormone networking, preventing the adaptive responses of tomato plants to salt stress, *Front. Plant Sci.* 8, 294.

Rubio, M. B., Quijada, N. M., Pérez, E., Domínguez, S., Monte, E. and Hermosa, R. (2014). Identifying *Trichoderma parareesei* beneficial qualities for plants, *Appl. Environ. Microbiol.* 80(6), 1864-1873.

Ruocco, M., Lanzuise, S., Vinale, F., Marra, R., Turra, D., Woo, S. L. and Lorito, M. (2009). Identification of a new biocontrol gene in *Trichoderma atroviride*: the role of an ABC transporter membrane pump in the interaction with different plant-pathogenic fungi, *Mol. Plant Microbe Int.* 22(3), 291-301.

Sadfi-Zouaoui, N., Hannachi, I., Rouaissi, M., Hajlaoui, M. R., Rubio, M. B., Monte, E., Boudabous, A. and Hermosa, M. R. (2009). Biodiversity of *Trichoderma* strains in Tunisia, *Can. J. Microbiol.* 55(2), 154-162.

Salas-Marina, M. A., Silva-Flores, M. A., Uresti-Rivera, E. E., Castro-Longoria, E., Herrera-Estrella, A. and Casas-Flores, S. (2011). Colonization of *Arabidopsis* roots by *Trichoderma atroviride* promotes growth and enhances systemic disease resistance through jasmonic acid/ethylene and salicylic acid pathways, *Eur. J. Plant Pathol.* 131(1), 15-26.

Salehpour, M., Etebarian, H. R., Roustaei, A., Khodakaramian, G. and Aminian, H. (2005). Biological control of common root rot of wheat (*Bipolaris sorokiniana*) by *Trichoderma* isolates, *Plant Pathol. J.* 4(1), 85-90.

Samolski, I., Rincón, A. M., Pinzón, L. M., Viterbo, A. and Monte, E. (2012). The *qid74* gene from *Trichoderma harzianum* has a role in root architecture and plant biofertilization, *Microbiology (Reading)* 158(1), 129-138.

Sanz, L., Montero, M., Grondona, I., Vizcaíno, J. A., Llobell, A., Hermosa, R. and Monte, E. (2004). Cell wall-degrading isoenzyme profiles of *Trichoderma* biocontrol strains show correlation with rDNA taxonomic species, *Curr. Genet.* 46(5), 277-286.

Saravanakumar, K., Fan, L., Fu, K., Yu, C., Wang, M., Xia, H., Sun, J., Li, Y. and Chen, J. (2016). Cellulase from trichoderma harzianum interacts with roots and triggers induced systemic resistence to foliar disease in maize, *Sci. Rep.*, 10, 35543.

Sarrocco, S., Valenti, F., Manfredini, S., Esteban, P., Bernardi, R., Puntoni, G., Baroncelli, R., Haidukowski, M., Moretti, A. and Vannacci, G. (2019). Is exploitation competition involved in a multitrophic strategy for the biocontrol of Fusarium head blight?, *Phytopathology* 109(4), 560–570.

Scharf, D. H., Brakhage, A. A. and Mukherjee, P. K. (2016). Gliotoxin- bane or boon?, *Environ. Microbiol.* 18(4), 1096–1109.

Schirmböck, M., Lorito, M., Wang, Y. L., Hayes, C. K., Arisan-Atac, I., Scala, F., Harman, G. E. and Kubicek, C. P. (1994). Parallel formation and synergism of hydrolytic enzymes and peptaibol antibiotics, molecular mechanisms involved in the antagonistic action of *Trichoderma harzianum* against phytopathogenic fungi, *Appl. Environ. Microbiol.* 60(12), 4364–4370.

Segarra, G., Casanova, E., Avilés, M. and Trillas, I. (2010). *Trichoderma asperellum* strain T34 controls Fusarium wilt disease in tomato plants in soilless culture through competition for iron, *Microb. Ecol.* 59(1), 141–149.

Seidl, V., Song, L., Lindquist, E., Gruber, S., Koptchinskiy, A., Zeilinger, S., Schmoll, M., Martínez, P., Sun, J., Grigoriev, I., Herrera-Estrella, A., Baker, S. E. and Kubicek, C. P. (2009). Transcriptomic response of the mycoparasitic fungus *Trichoderma atroviride* to the presence of a fungal prey, *BMC Genom.* 10, 567.

Shi, M., Chen, L., Wang, X. W., Zhang, T., Zhao, P. B., Song, X. Y., Sun, C. Y., Chen, X. L., Zhou, B. C. and Zhang, Y. Z. (2012). Antimicrobial peptaibols from *Trichoderma pseudokoningii* induce programmed cell death in plant fungal pathogens, *Microbiology (Reading)* 158(1), 166–175.

Shi, W. L., Chen, X. L., Wang, L. X., Gong, Z. T., Li, S., Li, C. L., Xie, B. B., Zhang, W., Shi, M., Li, C., Zhang, Y. Z. and Song, X. Y. (2016). Cellular and molecular insight into the inhibition of primary root growth of *Arabidopsis* induced by peptaibols, a class of linear peptide antibiotics mainly produced by *Trichoderma* spp., *J. Exp. Bot.* 67(8), 2191–2205.

Shoresh, M. and Harman, G. E. (2008). The molecular basis of shoot responses of maize seedlings to *Trichoderma harzianum* T22 inoculation of the roots: a proteomic approach, *Plant Physiol.* 147(4), 2147–2163.

Shoresh, M., Yedidia, I. and Chet, I. (2005). Involvement of jasmonic acid/ethylene signaling pathway in the systemic resistance induced in cucumber by *Trichoderma asperellum* T203, *Phytopathology* 95(1), 76–84.

Singh, S., Dureja, P., Tanwar, R. S. and Singh, A. (2005). Production and antifungal activity of secondary metabolites of *Trichoderma virens*, *Pestic. Res. J.* 17(2), 26–29.

Sivan, A. and Chet, I. (1986). Biological control of *Fusarium* spp. in cotton, wheat and muskmelon by *Trichoderma harzianum*, *J. Phytopathol.* 116(1), 39–47.

Sivan, A., Elad, Y. and Chet, I. (1984). Biological control effects of a new isolate of *Trichoderma harzianum* on *Pythium aphanidermatum*, *Phytopathology* 74(4), 498–501.

Smith, V. L., Wilcox, W. F. and Harman, G. E. (1990). Potential for biological control of *Phytophthora* root and crown rots of apple by *Trichoderma* and *Gliocladium* spp., *Phytopathology* 80(9), 880–885.

Son, M., Lee, K. M., Yu, J., Kang, M., Park, J. M., Kwon, S. J. and Kim, K. H. (2013). The *HEX1* gene of *Fusarium graminearum* is required for fungal asexual reproduction and pathogenesis and for efficient viral RNA accumulation of *Fusarium graminearum* virus 1, *J. Virol.* 87(18), 10356–10367.

Steindorff, A. S., Ramada, M. H. S., Coelho, A. S. G., Miller, R. N. G., Pappas Jr., G. J. P., Ulhoa, C. J. and Noronha, E. F. (2014). Identification of mycoparasitism-related genes against the phytopathogen *Sclerotinia sclerotiorum* through transcriptome and expression profile analysis in *Trichoderma harzianum*, *BMC Genom.* 15, 204.

Suárez, M. B., Vizcaíno, J. A., Llobell, A. and Monte, E. (2007). Characterization of genes encoding novel peptidases in the biocontrol fungus *Trichoderma harzianum* CECT 2413 using the TrichoEST functional genomics approach, *Curr. Genet.* 51(5), 331–342.

Sumida, C. H., Daniel, J. F. S., Araujod, A. P. C. S., Peitl, D. C., Abreu, L. M., Dekker, R. F. H. and Canteri, M. G. (2018). *Trichoderma asperelloides* antagonism to nine *Sclerotinia sclerotiorum* strains and biological control of white mold disease in soybean plants, *Biocontrol Sci. Technol.* 28(2), 142–156.

Tijerino, A., Cardoza, R. E., Moraga, J., Malmierca, M. G., Vicente, F., Aleu, J., Collado, I. G., Gutiérrez, S., Monte, E. and Hermosa, R. (2011a). Overexpression of the trichodiene synthase gene *tri5* increases trichodermin production and antimicrobial activity in *Trichoderma brevicompactum*, *Fungal Genet. Biol.* 48(3), 285–296.

Tijerino, A., Hermosa, R., Cardoza, R. E., Moraga, J., Malmierca, M. G., Aleu, J., Collado, I. G., Monte, E. and Gutierrez, S. (2011b). Overexpression of the *Trichoderma brevicompactum tri5* gene: effect on the expression of the trichodermin biosynthetic genes and on tomato seedlings, *Toxins* 3(9), 1220–1232.

Vargas, W. A., Mukherjee, P. K., Laughlin, D., Wiest, A., Morán-Diez, M. E. and Kenerley, C. M. (2014). Role of gliotoxin in the symbiotic and pathogenic interactions of *Trichoderma virens*, *Microbiology (Reading)* 160(10), 2319-2330.

Vinale, F., Flematti, G., Sivasithamparam, K., Lorito, M., Marra, R., Skelton, B. W. and Ghisalberti, E. L. (2009). Harzianic acid, an antifungal and plant growth promoting metabolite from *Trichoderma harzianum*, *J. Nat. Prod.* 72(11), 2032-2035.

Vinale, F., Nigro, M., Sivasithamparam, K., Flematti, G., Ghisalberti, E. L., Ruocco, M., Varlese, R., Marra, R., Lanzuise, S., Eid, A., Woo, S. L. and Lorito, M. (2013). Harzianic acid: a novel siderophore from *Trichoderma harzianum*, *FEMS Microbiol. Lett.* 347(2), 123–129.

Vinale, F., Sivasithanparam, K., Ghisalberti, E. L., Marra, R., Barbetti, M. J., Li, H., Woo, S. L. and Lorito, M. (2008). A novel role for *Trichoderma* secondary metabolites in the interactions with plants, *Physiol. Mol. Plant Pathol.* 72(1–3), 80–86.

Vinale, F., Sivasithamparam, K., Ghisalberti, E. L., Ruocco, M., Woo, S. L. and Lorito, M. (2012). *Trichoderma* secondary metabolites that affect plant metabolism, *Nat. Prod. Commun.* 7(11), 1545-1550.

Viterbo, A. and Horwitz, B. A. (2010). Mycoparasitism. In: Borkovich, K. A. and Ebbole, D. J. (Eds), *Cellular and Molecular Biology of Filamentous Fungi*, Washington DC: ASM Press, pp. 676–693.

Viterbo, A., Landau, U., Kim, S., Chernin, L. and Chet, I. (2010). Characterization of ACC deaminase from the biocontrol and plant growth-promoting agent *Trichoderma asperellum* T203, *FEMS Microbiol. Lett.* 305(1), 42–48.

Vizcaíno, J. A., González, F. J., Suárez, M. B., Redondo, J., Heinrich, J., Delgado-Jarana, J., Hermosa, R., Gutiérrez, S., Monte, E., Llobell, A. and Rey, M. (2006). Generation, annotation and analysis of ESTs from *Trichoderma harzianum* CECT 2413, *BMC Genom.* 7(1), 193.

Woo, S. L. and Pepe, O. (2018). Microbial consortia: promising probiotics as plant biostimulants for sustainable agriculture, *Front. Plant Sci.* 9, 1801.

Woo, S. L., Ruocco, M., Vinale, F., Nigro, M., Marra, R., Lombardi, N., Pascale, A., Lanzuise, S., Manganiello, G. and Lorito, M. (2014). *Trichoderma*-based products and their widespread use in agriculture, *TOMYCJ* 8(1), 71–126.

You, J., Zhang, J., Wu, M., Yang, L., Chen, W. and Li, G. (2016). Multiple criteria-based screening of *Trichoderma* isolates for biological control of *Botrytis cinerea* on tomato, *Biol. Control* 101, 31–38.

Yu, J. H. and Keller, N. P. (2005). Regulation of secondary metabolism in filamentous fungi, *Annu. Rev. Phytopathol.* 43(1), 437–458.

Zhang, F., Xu, X., Huo, Y. and Xiao, Y. (2019). *Trichoderma*-inoculation and mowing synergistically altered soil available nutrients, rhizosphere chemical compounds and soil microbial community, potentially driving alfalfa growth, *Front. Microbiol.* 9, 3241.

Zhang, S., Gan, Y., Ji, W., Xu, B., Hou, B. and Liu, J. (2017). Mechanisms and characterization of *Trichoderma longibrachiatum* T6 in suppressing nematodes (*Heterodera avenae*) in wheat, *Front. Plant Sci.* 8, 1491.

Chapter 12

Clonostachys rosea to control plant diseases

Dan Funck Jensen and Mukesh Dubey, Swedish University of Agricultural Sciences, Sweden; Birgit Jensen, University of Copenhagen, Denmark; and Magnus Karlsson, Swedish University of Agricultural Sciences, Sweden

1 Introduction

The ascomycete fungus *Clonostachys rosea* was reported as an aggressive mycoparasite in the late 1950s (Barnett and Lilly, 1962), and initial attempts to use it for biological control of plant diseases soon followed (Shigo, 1958). Since then, there has been a wealth of new knowledge emerging concerning the ecology, physiology and genetics of *C. rosea*, as well as concerning its applied use as a biological control agent (BCA) including formulation, application strategy, efficiency and safety. In this chapter, we use the definition of biological control as the use of living organisms for the control of plant pathogens/diseases in line with the recent update on the terminology, where biological control falls under the umbrella 'bioprotection', with the term BCAs being used only for living organisms, whereas products based on non-living, nature-based substances are another separate part of bioprotection (Stenberg et al., 2021). Due to the extensive literature available on *C. rosea*, this chapter does

http://dx.doi.org/10.19103/AS.2021.0093.14

not represent a comprehensive review but rather aims to highlight selected aspects of *C. rosea* with respect to ecology, mechanisms of action, targeted crops and diseases and product development.

2 Taxonomy and sources

Based on morphology, *C. rosea* (Link) Schroers, Samuels, Seifert & W. Gams was identified as the anamorph of the teleomorph *Bionectria ochroleuca* (Schwein.) Schroers & Samuels (Schroers et al., 1999). This was later confirmed based on DNA sequence data, including internal transcribed spacer (ITS) ribosomal DNA and β-tubulin (*tub*) gene sequences (Schroers, 2001). Following the one-fungus, one-name principle, the use of *C. rosea* as the preferred species label was proposed due to its established use in the scientific literature (Rossman et al., 2013). Until 1999, strains of *C. rosea* were referred to as *Gliocladium roseum* Bainier, now considered a synonym that is sometimes still in use, especially in a more applied, biocontrol context. Two variants of *C. rosea* can be found in the literature, *C. rosea* forma (f.) *rosea* (*G. roseum*) and *C. rosea* f. *catenulata* (*G. catenulatum*), primarily distinguished by the colour of the conidia (white/yellow/salmon and green, respectively). However, a recent study using genealogical concordance phylogenetic species recognition indicates that the two variants constitute a single species (Moreira et al., 2016). Although the vast majority of reports of biological control of plant diseases involves the species *C. rosea*, there is evidence to suggest that certain strains from other, closely related species, also possess biocontrol properties (Table 1, Broberg et al., 2021; Sun et al., 2017; Krauss et al., 2006; García et al., 2003).

Strains of *C. rosea* have been isolated from all continents except Antarctica and from a wide range of habitats (Sun et al., 2020a; Sutton et al., 1997), indicating a cosmopolitan distribution. Strains are typically isolated from soil, fungi, plant debris and from plant parts including roots, leaves and flowers (Walker and Maude, 1975; Nobre et al., 2005; Mueller and Sinclair, 1986; García et al., 2003), but isolations from nematodes and insects are also reported (Verdejo-Lucas et al., 2002; Haarith et al., 2020). Strains of *C. rosea* are even present as endophytes in several halophyte plant species in coastal areas (You et al., 2017). This habitat distribution should be viewed in light of the ecological generalist lifestyle of *C. rosea*, which includes plant endophytism, rhizosphere competence, polyphagous ability and mycoparasitism (Shigo, 1958; Li et al., 2002; Chatterton and Punja, 2012; Saraiva et al., 2015; Maillard et al., 2020). The traits that form the basis of the nutritional versatility that characterises generalist behaviour in *C. rosea* is tightly connected with its ability to control plant diseases and its use as a BCA.

Table 1 Species of *Clonostachys* with reported biocontrol properties.

Species	Authors	Mycobank ID	Mycobank link
Clonostachys byssicola	Schroers	#485119	http://www.mycobank.org/MB/485119
Clonostachys chloroleuca	G. M. Moreira, L. M. Abreu, L. H. Pfenning & H. J. Schroers	#816994	http://www.mycobank.org/MB/816994
Clonostachys rhizophaga	Schroers	#485120	http://www.mycobank.org/MB/485120
Clonostachys rosea	(Link) Schroers, Samuels, K. A. Seifert & W. Gams	#461067	http://www.mycobank.org/MB/461067
Clonostachys solani	(Harting) Schroers & W. Gams	#456098	http://www.mycobank.org/MB/456098
Clonostachys sp.	Not formally described		

3 Mechanisms of action

The mycoparasitic behaviour of *C. rosea* has attracted a lot of attention since its first description (Barnett and Lilly, 1962) and has been considered an important biocontrol trait for combatting plant pathogens (Karlsson et al., 2018). Biocontrol interactions leading to efficient biocontrol of plant diseases, however, can rely on a range of mechanisms of action beyond parasitism (Baker and Cook, 1974; Jensen et al., 2017; Köhl et al., 2019) and several of these may work in concert (Köhl et al., 2019). As mentioned above, the generalist lifestyle of *C. rosea* has equipped it with traits enabling competition for resources and space, and interference competition through antibiosis, in addition to its mycoparasitic ability (Sutton et al., 1997; Fatema et al., 2018). Its endophytic ability allows for establishment in plant organs close to potential pathogen entry points (Saraiva et al., 2015), and will in some cases result in activation of inducible plant defence responses (Kamou et al., 2020; Wang et al., 2019) leading to induced systemic resistance (ISR) (Lahoz et al., 2004). In the following sections, we will explore the contribution of these mechanisms to *C. rosea* biocontrol, and how it varies depending on the host plant and the pathogen which causes the disease.

3.1 Competition for space and nutrients

Competition for space and resources through priority colonisation ahead of the pathogen (Jensen et al., 2017) is reported to be important for the ability of *C. rosea* to control grey mould, caused by *Botrytis. cinerea*, in strawberry and raspberry flowers (Sutton et al., 1997). Reduced germination of *B. cinerea* conidia on raspberry and rose leaves, and subsequent control of grey mould, was shown to depend on the competition for scarce nutrients (Yu and Sutton, 1997b; Morandi et al., 2000). It was also shown that competition with indigenous *Penicillium* and *Alternaria* spp. on rose leaves reduced control of *B. cinerea* by *C. rosea* (Morandi et al., 2000).

3.2 Mycoparasitism

The initial reports of mycoparasitic behaviour in *C. rosea* were based on agar plate interaction studies where *C. rosea* was able to overgrow and destroy established cultures of a range of fungi (Shigo, 1958; Barnett and Lilly, 1962). The attack was characterised by collapse of the surface mycelium of the fungal prey and the destruction of the dark pigment produced by some species. The attack involves attachment to the hyphae of the fungal prey and production of an appressorium, followed by penetration (Makkonen and Pohjakallio, 1960; Walker and Maude, 1975). Confocal fluorescence microscopy studies of the interaction between *C. rosea* expressing the green fluorescent protein and *F. oxysporum* forma specialis (f. sp.) *radicis lycopersici* expressing the red fluorescent protein confirmed production of an appressorium during the penetration (Karlsson et al., 2015). However, scanning electron microscopy analysis of the interaction between *C. rosea* and *B. cinerea* showed examples of direct penetration of *B. cinerea* conidia and germ tubes without the formation of appressoria, resulting in cytoplasmic disintegration (Li et al., 2002). Mycoparasitism of oomycete plant pathogens such as *Pythium aphanidermatum* and *P. ultimum* by *C. rosea* was also reported (Chatterton and Punja, 2009; Mamarabadi et al., 2009). Mycoparasitism was reported as an important mode of action for controlling *B. cinerea* on raspberry stems (Yu and Sutton, 1997b). Significant biocontrol of *Zymoseptoria tritici* causing septoria tritici blotch (STB) has been obtained in field experiments over several years using *C. rosea* (Jensen et al., 2019). As *C. rosea* was sprayed on the wheat crop after the initial pathogen infection of the leaves, mycoparasitism seems to contribute to biocontrol of STB.

3.3 Secretion of fungal cell wall-degrading enzymes

Secretion of fungal cell wall-degrading enzymes, such as chitinases, glucanases and proteases, is a component of the mycoparasitic attack (Pachenari and Dix, 1980). Chitinases and glucanases produced by *C. rosea* were confirmed to degrade the cell walls of taxonomically diverse plant pathogens from the oomycete genus *Pythium* and from the fungal genus *Fusarium* (Inglis and Kawchuk, 2002; Chatterton and Punja, 2009). However, the exact contribution of chitinases to biocontrol in *C. rosea* is difficult to assess; deletion of the *chiC2*, *ech37*, *ech42* and *ech58* chitinase genes resulted in mutants being impaired in their antagonistic ability towards other fungi but there was no reduction of their biocontrol ability (Table 2, Tzelepis et al., 2015; Mamarabadi et al., 2008b). Overexpression of the *ech37* ortholog (*chi67-1*) in the closely related species *C. chloroleuca* resulted in a mutant with higher chitinase activity in liquid cultures, higher rates of parasitism of *Sclerotinia sclerotiorum* sclerotia and higher efficiency to control *S. sclerotiorum* on soybean (Sun et al., 2017), thereby establishing a link between chitinase activity and biocontrol in *Clonostachys*. Nematodes are also a target for enzymes secreted from *C. rosea*; the extracellular serine protease PrC was shown to exhibit nematicidal activity against *Panagrellus redivivus* (Li et al., 2006). Nematode cuticle degradation products released by the proteolytic activity of PrC were also shown to protect *C. rosea* against oxidative stress by scavenging reactive oxygen species (Zou et al., 2010), a novel mechanism for alleviating environmental stress.

3.4 Secretion of antibiotic compounds

Production of secreted compounds with antifungal activity is another component of the mycoparasitic attack. Furthermore, competition for space and resources is intimately connected with the ability of *C. rosea* to defend occupied resources against other fungi through antibiosis. Antibiosis is therefore considered an important trait of *C. rosea* in biocontrol interactions, due to its ability to produce various secondary metabolites with antagonistic effects towards plant pathogens (Han et al., 2020; Saraiva et al., 2020). For example, the polyketide compounds Clonorosein A and B were shown to inhibit germ tube growth in both *B. cinerea* and *F. graminearum* (Fatema et al., 2018). Furthermore, deletion of the polyketide synthase gene *pks29* in *C. rosea* resulted in mutants with an impaired ability to control fusarium foot rot on barley (Table 2, Fatema et al., 2018). Non-ribosomal peptides are another important group of secondary metabolites in fungi, and *C. rosea* was shown to produce a mix of peptaibol compounds that inhibited growth of *S. sclerotiorum* (Rodriguez et al., 2011). Deletion of the non-ribosomal peptide synthetase genes *nps1*, *nps4* and *nps5* in *C. rosea* compromised the ability of the mutants to protect wheat seedlings against fusarium foot rot and nematode root disease (Table 2, Iqbal et al., 2019; Iqbal et al., 2020). *Clonostachys rosea* was also reported to produce glisoprenin compounds that specifically inhibited appressorium formation by the rice blast fungus *Magnaporthe oryzae*, but without any observable antifungal or antibacterial activities (Thines et al., 1998). Production of compounds with antibiotic effect towards bacteria (Zhai et al., 2016) and nematodes (Dong et al., 2005) by *C. rosea* illustrate the need to defend resources against other microorganisms inhabiting the same ecological niche and may be an important ability for establishment of *C. rosea* in the rhizosphere and phyllosphere. For example, germination of spores of the pathogen *Bipolaris solani* was inhibited on barley leaf surfaces by *C. rosea*, clearly indicating the involvement of antibiosis in competition (Jensen et al., 2016a).

Table 2 *Clonostachys* genes functionally validated for their role in antagonism, xenobiotic tolerance or biocontrol of plant diseases.

Family	Gene[1]	*Clonostachys* strain	Validated function	Reference
Chitinase	*ech58* *ech42* *ech37*	*C. rosea* IK726	Involved in antagonism against *Fusarium culmorum*	Mamarabadi et al., 2008b
	chi67-1	*C. chloroleuca* 67-1	Mycoparasitism of *Sclerotinia sclerotiorum* sclerotia Biocontrol of sclerotinia stem rot of soybean	Sun et al., 2017
	chiC2	*C. rosea* IK726	Antagonism against *Botrytis cinerea* and *Rhizoctonia solani*	Tzelepis et al., 2015
Transaldolase	*tal67*	*C. chloroleuca* 67-1	Antagonism against *B. cinerea* Mycoparasitism against *S. sclerotiorum* sclerotia Biocontrol of sclerotinia stem rot of soybean	Liu et al., 2016b
Phosphatase	*ssd1*	*C. chloroleuca* 67-1	Antagonism against *B. cinerea* Mycoparasitism against *S. sclerotiorum* sclerotia Biocontrol of sclerotinia stem rot of soybean	Lv et al., 2020
Hydrolase	*zhd101*	*C. rosea* IK726	Antagonism against *F. graminearum* Biocontrol of fusarium foot rot on wheat Tolerance to zearalenone mycotoxin	Kosawang et al., 2014b
Protein kinase	*mapk*	*C. chloroleuca* 67-1	Mycoparasitism of *S. sclerotiorum* sclerotia Biocontrol of sclerotinia stem rot of soybean	Sun et al., 2020b
Heat shock protein	*hsp*	*C. chloroleuca* 67-1	Mycoparasitism of *S. sclerotiorum* sclerotia	Sun et al., 2019
Perilipin	*per3*	*C. rosea HL-1-1*	Mycoparasitism of *S. sclerotiorum* sclerotia	Sun et al., 2015c
Hydrophobin	*hyd1* *hyd2* *hyd3*	*C. rosea IK726*	Antagonism against *B. cinerea*, *F. graminearum* and *R. solani* Plant root colonization	Dubey et al., 2014b
LysM protein	*lysm1* *lysm2*	*C. rosea IK726*	Antagonism against *B. cinerea* Biocontrol of grey mold on *Arabidopsis* and fusarium foot rot on wheat Root colonization	Dubey et al., 2020

ABC transporter	*abcG5*	*C. rosea IK726*	Antagonism against *F. graminarum* Biocontrol of fusarium foot rot on barley Tolerance to zearalenone mycotoxin and certain fungicides	Dubey et al., 2014a
	abcG29	*C. rosea IK726*	Biocontrol of grey mould on *Arabidopsis* and fusarium foot rot on barley	Dubey et al., 2016
	abcG6	*C. rosea IK726*	Tolerance to certain fungicides	Broberg et al., 2021
MFS transporter	*mfs464*	*C. rosea IK726*	Antagonism against *F. graminarum*	Nygren et al., 2018
Polyketide synthase	*pks22* *pks 29*	*C. rosea IK726*	Antagonism against *B. cinerea* and *F. graminearum* Biocontrol of fusarium foot rot on barley[2]	Fatema et al., 2018
Non-ribosomal peptide synthetase	*nps1* *nps4* *nps5*	*C. rosea IK726*	Antagonism against plant parasitic nematodes, *B. cinerea* and *F. graminearum* Biocontrol of nematode root disease[3] and fusarium foot rot on wheat	Iqbal et al., 2019, 2020

[1] There is currently no agreement on the use of gene name nomenclature in *Clonostachys*. We propose to use the gene nomenclature currently used for *Neurospora*, *Aspergillus* and *Trichoderma*, where a gene name consists of three small letters and a number (all italicised) while the corresponding protein is denoted by the same letters and number written in capital letters (non-italicized); e.g. *acl1* is the gene encoding the ATP citrate lyase protein ACL1. Abbreviations of species names should not be part of the gene name.

[2] Only deletion of *pks29*, not *pks22*, attenuated biocontrol of fusarium foot rot on barley.

[3] Measured as reduced numbers of plant pathogenic nematodes in wheat roots.

3.5 Tolerance towards antifungal compounds

The strong ability of *C. rosea* for interference competition through antibiosis as well as the production of toxic secondary metabolites from the fungal prey during mycoparasitism emphasises the need for toxin tolerance/detoxification mechanisms in *C. rosea*. *Clonostachys rosea* has indeed been shown to be highly tolerant towards the *Fusarium* mycotoxin zearalenone, with strong antifungal activity (Utermark and Karlovsky, 2007). This ability was shown to depend partly on direct detoxification of zearalenone to less toxic compounds by the ZHD101 lactone hydrolase (Takahashi-Ando et al., 2002; Kosawang et al., 2014b), and partly on active efflux from the cell with the ABCG5 ATP-binding cassette (ABC) transporter (Table 2, Dubey et al., 2014a). Both detoxification and efflux contribute to the ability of *C. rosea* to control fusarium foot rot disease on cereals (Dubey et al., 2014a; Kosawang et al., 2014b). Growth of *C. rosea* was not inhibited by high concentrations of the mycotoxin fumonisin B1, which suggests involvement of efflux transporters in the tolerance as the fumonisin B1 was not degraded (Chatterjee et al., 2016). In contrast, tolerance towards deoxynivalenol-type mycotoxins in *C. rosea* is likely to involve glycosylation followed by efflux (Demissie et al., 2020). Furthermore, *C. rosea* is also shown to be relatively tolerant towards phenazine produced by *Pseudomonas chlororaphis* (Karlsson et al., 2015) and certain xenobiotic substances including fungicides (Roberti et al., 2006; Dubey et al., 2014a).

3.6 Induction of plant disease resistance

There is clear evidence that many *Clonostachys* spp. strains, including *C. rosea*, can live as endophytes in plants (Maillard et al., 2020; Saraiva et al., 2015; Sutton et al., 2002; Chatterton and Punja, 2010; Mueller and Sinclair, 1986). This fact alone is significant for biocontrol as the BCA can be present at the sites of infection of plant pathogens. There is also accumulating evidence that this intimate interaction between *C. rosea* and plants can trigger defence gene expression in plants, as shown in tomato and wheat (Kamou et al., 2020; Mouekouba et al., 2014; Wang et al., 2019). This can be interpreted as the recognition of microbe-associated molecular patterns (MAMPs) from *C. rosea* by the plant and subsequent induction of pattern-triggered immunity (PTI) (Jones and Dangl, 2006; Köhl et al., 2019). However, to what extent this induction of plant defence gene expression by *C. rosea* translates into induced local or systemic resistance (ISR) is less clear. Colonisation of wheat seedlings by *C. rosea* resulted in induction of pathogenesis-related proteins that in turn resulted in significant growth inhibition of the pathogen *F. culmorum* (Roberti et al., 2008). Similarly, *C. rosea* inoculated on roots of tobacco plants triggered ISR in leaves against the biotrophic powdery mildew pathogen *Erysiphe orontii* mediated by increased activity of 1,3-β-glucanases, 1,4-β-glycosidases, chitinases and peroxidases in the plant leaves (Lahoz et al., 2004). Recent data also show that *C. rosea* strains applied in soil resulted in reduced stem lesion length caused by the pitch canker pathogen *F. circinatum* in Monterey pine, indicating induced disease resistance in forest tree seedlings (Moraga-Suazo et al., 2016). Induced resistance triggered by *C. rosea* was also suggested as a possible biocontrol mechanism in canola against club root disease caused by *Plasmodiophora brassicae* (Lahlali and Peng, 2014) and in tomatoes against grey mould (Mouekouba et al., 2014; Wang et al., 2019), although it was difficult to clearly separate induced resistance from other mechanisms in these studies.

In addition to these mechanisms of biocontrol, *C. rosea* can also trigger an increased plant growth response (Fig. 1, Ravnskov et al., 2006; Johansen et al., 2005). Although plant growth promotion is not considered a biocontrol mechanism *per se*, it can for example result in avoidance of seedling damping-off caused by *Pythium* spp. if the plant seedlings establish faster in the field due to the microbial treatment.

4 Lessons from genomics and transcriptomics

Application of comparative genomics approaches can be very useful in research and application of biocontrol solutions. For example, it allows for accurate identification of species, populations and strains, it can be used for understanding modes of action and for identification of genetic markers associated with biocontrol traits. Genomic information from *Clonostachys* species has increased rapidly during the recent past. Genome sequence data are currently available for 56 different strains of *C. rosea* (Demissie et al., 2021; Wang et al., 2021; Broberg et al., 2018; Karlsson et al., 2015), 4 strains of *C. byssicola* (Broberg et al., 2021), 4 strains of *C. chloroleuca* (Broberg et al., 2021; Sun et al., 2015a), 1 strain of *C. solani* (Broberg et al., 2021), 3 strains of *C. rhizophaga* (Broberg et al., 2021; Liu et al., 2016a) and 1 strain representing an undescribed *Clonostachys* species (Broberg et al., 2021). The gene content in a species is partly shaped by selection and therefore reflects adaptations towards the ecological niche of the species. Hence, comparing the gene content in *Clonostachys* with other closely related species can provide important clues to the mechanistic basis of traits that are important for their use in biological control.

4.1 Genes encoding proteins involved in secondary metabolite biosynthesis and efflux

One feature that stands out when comparing gene content in *Clonostachys* with plant pathogenic *Fusarium* and mycoparasitic *Trichoderma* species, is the high number of genes involved in the biosynthesis of secondary metabolites,

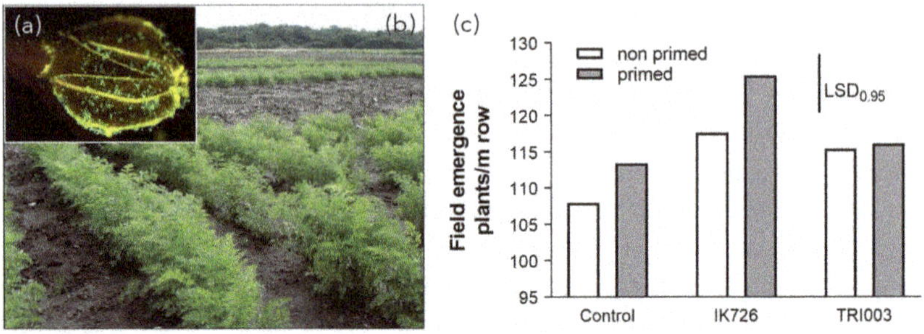

Figure 1 Seed coating and biopriming of carrot seed with *C. rosea* strain IK726 and *Trichoderma harzianum* strain TRI003 to improve field emergence of carrot. (a) Germination of an IK726 bioprimed seed showing hyphal regrowth of the *gfp*-marked IK726 mutant *d11*, (b) Establishment of the carrot plants in the field following seed coating and biopriming, and (c) Effect of IK726 and TRI003 seed coating and biopriming on plant establishment (Jensen, B., unpublished data).

including polyketide synthases, non-ribosomal peptide synthetases and cytochrome P450s (Karlsson et al., 2015; Broberg et al., 2021). In fungi, secondary metabolites perform a variety of functions including protection from biotic and abiotic stresses and interaction with other organisms (Keller et al., 2005; Osbourn, 2010). In *C. rosea*, 75% of the 32 predicted polyketide synthase genes are located in secondary metabolite biosynthetic clusters (Fatema et al., 2018), which is a higher proportion compared with the mentioned mycoparasitic *Trichoderma* species (50%). Gene expression analyses also show induced expression of 17 polyketide synthase genes in *C. rosea* during interactions with *B. cinerea* and *F. graminearum* (Fatema et al., 2018; Demissie et al., 2020; Nygren et al., 2018). Induced expression of 20 polyketide synthase genes was also correlated with pigmentation in *C. rosea* (Fatema et al., 2018). There are 17 predicted non-ribosomal peptide synthetase genes in the *C. rosea* genome (Karlsson et al., 2015; Broberg et al., 2021), from which *nps1*, *nps4*, *nps5* and *nps13* are shown to be induced during mycoparasitic interactions (Nygren et al., 2018; Iqbal et al., 2019; Iqbal et al., 2020). These data fit well with the idea of mycoparasitism and interference competition through antibiosis being an important component of the biocontrol ability of *C. rosea* (Karlsson et al., 2018).

The high numbers of genes associated with secondary metabolite biosynthesis are, not surprisingly, accompanied by equally high numbers of membrane transporter genes predicted to be involved in drug efflux (Karlsson et al., 2015; Nygren et al., 2018; Broberg et al., 2021). More specifically, this relates to the ABC transporter families (Kovalchuk and Driessen, 2010) ABC-B (multidrug resistance proteins), ABC-C (multidrug resistance-associated proteins) and ABC-G (pleiotropic drug resistance proteins), and the major facilitator superfamily (MFS) Drug:H+ Antiporter-2 family. Transcriptomic analyses show that several members of these groups are induced in *C. rosea* and *C. chloroleuca* during interactions with other fungi and during exposure to fungal metabolites and mycotoxins (Kosawang et al., 2014a; Lysøe et al., 2017; Demissie et al., 2018; Nygren et al., 2018; Demissie et al., 2020; Sun et al., 2015b), but also during exposure to bacterial metabolites (Karlsson et al., 2015; Kamou et al., 2016). The ability to neutralise compounds with antifungal activity produced by other microorganisms or defence molecules produced by plants by efflux mechanisms may be an important trait contributing to the biocontrol property of *Clonostachys*. In addition, several sugar and small organic compound MFS transporter gene families contained high gene numbers in *C. rosea* (Nygren et al., 2018), perhaps involved in nutrient uptake.

4.2 Genes encoding fungal cell wall-degrading enzymes

As mentioned in the previous section, secretion of fungal cell wall-degrading enzymes such as chitinases and proteases is one of the suggested mechanisms involved in *C. rosea* biocontrol (Pachenari and Dix, 1980). However, this view is only partially corroborated from a genomics point of view. High numbers of serine protease genes in *C. rosea*, as well as in mycoparasitic *Trichoderma* species, suggests an involvement of these proteases in biotic interactions (Iqbal et al., 2018a). In contrast, *C. rosea* only possesses a moderate number of chitinases (14 genes) compared with certain *Trichoderma* species (Tzelepis et al., 2015). However, both protease and chitinase genes are induced during mycoparasitic interactions in *C. rosea* (Tzelepis et al., 2015; Iqbal et al., 2018a; Lysøe et al., 2017; Mamarabadi et al., 2008a) and *C. chloroleuca* (Sun et al., 2015b). High gene numbers of carbohydrate-active enzymes targeting components of plant cell walls, in particular xylan and rhamnose/pectin (Broberg et al., 2021; Karlsson et al., 2015; Atanasova et al., 2018), may provide the basis of the saprophytic capability of *Clonostachys* and be important for its establishment in soil and the rhizosphere.

4.3 Genes encoding small secreted proteins

Another notable difference between *Clonostachys* and *Trichoderma* species is the low numbers of hydrophobin and LysM protein genes in *Clonostachys*, compared with *Trichoderma*. Hydrophobins are small, cysteine-rich secreted proteins found only in fungi (Wösten, 2001). These proteins aggregate on the outer surface of fungal cell walls and develop amphipathic layers that perform a variety of biological functions in the life cycle of filamentous fungi, including a role during interactions between the fungus and the environment (Wösten, 2001). The *C. rosea* genome contains three class II hydrophobin (*hyd*) genes, which is in strong contrast with the *T. atroviride* and *T. virens* mycoparasites that contain 10 and 9 hydrophobin-encoding genes, respectively (Dubey et al., 2014b). Gene deletion strains of *hyd1* and *hyd3* displayed more aggressive behaviour towards *B. cinerea*, *F. graminearum* and *Rhizoctonia solani*, which also translated into an increased ability to control *B. cinerea* infection of *Arabidopsis thaliana* leaves (Table 2, Dubey et al., 2014b). *Hyd1* and *hyd2* double deletion strains displayed enhanced root colonisation compared with the *C. rosea* wild type strain, while the Δ*hyd3* strain showed reduced root colonisation ability (Dubey et al., 2014b). Taken together, these data show that hydrophobins have an important role in mediating biotic interactions in *C. rosea*.

Lysin motif (LysM) domains are approximately 50 amino acids long carbohydrate-binding modules, reported in proteins from all kingdoms of life including fungi (Kombrink and Thomma, 2013). In fungi, LysM modules can be found with varying numbers of LysM modules either together with catalytic protein modules (referred to as LysM-containing proteins) or without any known catalytic module (referred to as LysM effectors) (de Jonge and Thomma, 2009). LysM effectors act as a virulence factor in plant pathogenic, entomopathogenic and mycoparasitic fungi, either by scavenging chitin oligomers, a well-known MAMP molecule, or by protecting the fungal cell wall against hydrolytic enzymes (Kombrink and Thomma, 2013; Cen et al., 2017; Romero-Contreras et al., 2019). *Clonostachys rosea* only contains three *lysm* genes, compared with 12 and 18 genes in *T. atroviride* and *T. virens*, respectively (Dubey et al., 2020). Gene deletion mutants of the two LysM effector genes *lysm1* and *lysm2* were reduced in their ability to control plant diseases caused by *F. graminearum* and *B. cinerea* (Fig. 2, Table 2, Dubey et al., 2020). Furthermore, a *lysm1* and *lysm2* double deletion strain displayed reduced ability to colonise wheat roots (Dubey et al., 2020).

5 Product development and commercialisation

5.1 Selecting the right strain

A strategy for developing a commercial BCA product embraces a whole range of criteria to be fulfilled such as selection, production and formulation of a microorganism into a storable product that can be easily applied for disease control in greenhouses and in fields (Köhl et al., 2011). A crucial part is the screening step for selection of strains efficient in plant disease control. An example of a screening procedure for selection of efficient biocontrol strains of *C. rosea* is from a Nordic research program 1990–1993. This screening system included the plant in question and simulated the natural conditions where the BCA is to be used (Knudsen et al., 1997; Teperi et al., 1998). This turned out to be a very successful strategy leading to selection of several *C. rosea* strains in Denmark and Finland. Among those the strain J1446 used in the commercial products LALSTOP G46 WG®, Prestop® and Gliomix® (Table 3) and the strain *C. rosea* IK726 (Table 4). Both strains have since been studied intensively in

Figure 2 LysM effector proteins LYSM1 and LYSM2 contribute to the biocontrol of fusarium foot rot disease on wheat. Wheat seeds were coated with *C. rosea* conidia and planted in moist sand together with a *F. graminearum* agar plug. *Clonostachys* strains include wild type (WT), *lysm* gene deletion mutants (Δ*lysm1/2*) and deletion mutants complemented with a functional *lysm1/2* gene (Δ*lysm1/2*+). Seedlings were harvested 3 weeks post-inoculation and disease symptoms were scored on 0–4 scale. The figure is adapted from Dubey et al., 2020.

Denmark and Sweden for their efficacy and biocontrol traits (Jensen et al., 2007; Karlsson et al., 2015).

Reliance on *in vitro* screening procedures, such as dual agar plate tests, is not encouraged due to the complexity of the biocontrol trait, involving multiple and complex mechanisms depending on the BCA, pathogen, host plant and environment (Harman et al., 2004; Jensen et al., 2017; Köhl et al., 2019; Rojas et al., 2020). For example, the *Clonostachys* sp. strain CBS 192.96 efficiently controlled fusarium foot rot disease on wheat seedlings, caused by *F. graminearum*, despite its inability to suppress the growth of *F. graminearum in vitro* (Broberg et al., 2021). Another example is *C. rosea* strain IK726 that failed to suppress *in vitro* growth of *F. culmorum* but efficiently controlled foot rot disease on wheat seedlings caused by the same pathogen (Knudsen et al., 1997). We will not address all relevant steps outlined in Köhl et al. (2011) but in the following give some selected examples of aspects addressed for developing *C. rosea* strains into marketable BCAs.

5.2 Production and formulation: effects on viability and shelf life

Economical mass production of storable high-quality propagules, e.g. spores and chlamydospores, is a prerequisite for the successful development of fungal BCAs. The view that BCAs based on filamentous fungi should be produced by solid-state fermentation (i.e. (Köhl et al., 2011)) probably has its origin in the work with production of *Trichoderma* spp., where it was shown that solid fermentation resulted in the best spore quality and shelf life (Agosin et al., 1997). Solid media are also often applied for the production of *C. rosea* and a high spore production can be obtained, e.g. on wheat grains (James and Sutton, 1996; Sutton et al., 1997), a mixture of sphagnum peat and wheat bran (Fig. 3a and b, Jensen et al., 2000) or on mixtures of wheat bran and cornmeal (Zhang et al.,

2013; Zhang et al., 2015). Maximum spore concentration on a solid substrate is typically reached after 10–14 days (Jensen et al., 2002; Zhang et al., 2015). Nevertheless, the quality of the spores can in some cases be improved if the production period is extended from, e.g., 12 to 18 days (Jensen et al., 2002). This suggests that extending the production period a few days after maximum spore concentration is reached could be relevant in order to maintain viability and biocontrol efficacy for longer periods. High concentrations of *C. rosea* spores can also be produced during submerged liquid fermentation (de Andrade Carvalho et al., 2018). However, direct drying of propagules strongly decreases their viability (Jensen, 1999). Nevertheless, mixing *C. rosea* spores produced in liquid with, e.g. diatomic clay before drying can improve survival, shelf life and biocontrol efficacy significantly (Fig. 3c, Jensen, 1999; Jensen et al., 2002).

Likewise, the difficulties in stabilising spore viability are demonstrated for *T. harzianum* too (Muñoz et al., 1995; Agosin et al., 1997; Harman et al., 1991). The authors showed that spores from solid fermentation developed a thicker outer cell wall layer as compared to spores from liquid culture and suggested that this is important for the desiccation tolerance and enhanced shelf life of *Trichoderma* spores. This trait might also be relevant to study further for improving shelf life of *C. rosea* spores.

Another often overlooked trait of BCA spores is their ability to germinate fast as this should enhance their capability to control pathogens such as *Pythium* sp. that can infect a seed in less than 4 h after sowing (Taylor et al., 1991). Hence, *C. rosea* spores start germinating within 4–6 h (Fig. 4a, Jensen, 1999) which was considerably faster than for, e.g. *T. harzianum* spores (Fig. 4b). In fact, after 12 h >70% *C. rosea* spores had germinated as compared to <1% for *T. harzianum* (Fig. 4b). Similarly, Sutton and Peng (1993) showed faster germination of *C. rosea* than for *Trichoderma* and *Penicillium* on strawberry leaf disks incubated at 10–20°C. However, it should be noted that drying spores can strongly reduce both germinability and speed of germination (Fig. 4a and b). This should be taken into consideration when testing the efficacy of BCAs towards commercial application as tests only with dosages of freshly harvested spores might give misleading results (Jensen et al., 2000).

There is limited information available in the public domain concerning large-scale solid-state production of *C. rosea* at an industrial level. However, a two-step submerged/solid-state scale-up production process has been demonstrated to enhance *C. rosea* spore production (Krauss et al., 2002). Furthermore, a novel solid-state fermenter type based on enhanced growth area for *C. rosea* spores in the medium using optimised response surface methodology has been developed (Zhang et al., 2013; Zhang et al., 2015). Production of *Clonostachys* chlamydospores, i.e. thick-walled highly desiccation-tolerant resting structures, is an alternative approach that seems to work for the species *C. chloroleuca* (Sun et al., 2014). This method might also be relevant for *C. rosea* production.

Table 3 Examples of commercial products based on *Clonostachys rosea*.

Product	Strain	Strain(s) registered in	Production company	Distributor(s)	Region
Baikekong 2®	Strain info not available	China	Harbin Baikekong Biotech. Ltd		China
Gliogen®	VKPM-F1324	Russia	Ecogen LLC	Ecogen LLC	Russia
Guangenbao®, Guanjunling®, Zhongbaofenzuan®	Strain info not available (Product mixed with *Bacillus subtilis*)	China	Chinese Acad. Agric. Sci., Beijing Qigao Biotechnol. Co., Ltd. and Beijing Green Agric. Sci. & Technol. Group Co. Ltd		China
Kamoi®	CPQBA 040-11 DRM 07	Brazil	Agrivalle Brasil Indústria		Brazil
Lalstop46WG®	JI446	Canada, USA	Lallemand	Lallemand, Jetharvest	USA, Canada
Prestop®	J1446	EU, USA, Canada	[1]Danstar Ferment AG	[1]Verdera Danstar Ferment AG	EU (some countries), USA, Canada
Prestop 4B® (bee-vector delivery)	J1446	EU, USA, Canada,	[1]Danstar Ferment AG	Biobest (Flying doctors)	
Vectorite with CR-7®	CR-7	USA	Bee-Vectoring-Technology BVT	BVT	USA

[1]Danstar Ferment AG and Verdera are Lallemand Inc's subsidiary companies.

Table 4 Selected examples of biological control of diseases by application of *Clonostachys rosea*.

Plant/Pathogen	Disease	Isolate origin	Isolate ID	Experimental conditions	Application method	Effects on disease development	Other remarks	Reference
Strawberry								
B. cinerea	Grey mould	Soil, Finland	Verdana B[4] PreStop	Greenhouse	Bee delivery	Reduced berry infection and improve shelf-life of berries	*C. rosea* vectored by bumblebees	Van Delm et al., 2015
B. cinerea	Grey mould	Collection of microorganisms Embrapa Meio Ambiente, Brazil	LQC 62	Field trial	Spraying	Reduced grey mould incidence of fruits	UV-B tolerant *C. rosea* isolate. High UV-B radiation had no influence on efficacy	Nechet et al., 2017
B. cinerea	Grey mould	Strawberry fruit, Canada	PG-A-Fr-88-710	Field trial	Spraying and bee delivery	Reduced grey mould incidence of flowers and fruits	Weekly spraying treatment compared to bee delivery	Peng et al., 1992
B. cinerea	Grey mould	Brazilian ecosystems	Mixture of four isolates	Field trial	Spraying	Reduced flower and fruit infection. Increased yield	Integration with fungicide application	Cota et al., 2009
Raspberry								
B. cinerea	Grey mould	Strawberry fruit, Canada	PG-A-Fr-88-710	Field trial	Spraying and bee delivery	Reduced fruit rot	Comparing honeybee and bumblebee vectoring	Yu and Sutton, 1997a
Rose								
B. cinerea	Grey mould	Strawberry fruit, Canada	PG-A-Fr-88-710	Growth chamber	Drop inoculation	Reduction of *B. cinerea* sporulation	Interactions with indigenous fungi (*Alternaria*, *Aspergillus*, *Penicillium*)	Morandi et al., 2000

Tomato								
B. cinerea	Grey mould	Strawberry fruit, Canada	PG-A-Fr-88-710	Greenhouse, hydroponic	Spraying to deleafed stem wounds	Reduction of *B. cinerea* sporulation in wounds	*C. rosea* established endophytically in stems	Sutton et al., 2002
B. cinerea	Grey mould	Brazil	NCR19/F, NCR60/F, NCR61/F, NCR62/F	Growth chamber, 18°C	Spraying	Reduced incidence and severity of stem symptoms	Wounded stems of whole plants. Inoculation 1 day or at the same time as Bc	Borges et al., 2015
B. cinerea	Grey mould	China	Unknown	Growth chamber	Spraying detached tomato fruits	Reduced incidence and severity on fruits	*C. rosea* either 12 h before or 12 h after *B. cinerea*. Incubated at 25°C at, high humidity	Gong et al., 2017
B. cinerea	Grey mould	Turfy soil, China	Unknown	Growth chamber	Spraying detached leaves	Reduced severity	*B. cinerea* and *C. rosea* applied at the same time	Mouekouba et al., 2014
Alternaria solani, A. alternata	Early blight	Soil, Finland	Prestop, J1446	Greenhouse and field trial	Spraying	Lowered early blight disease severity	*C. rosea* was as effective as copper	Egel et al., 2019
Fusarium oxysporum f. sp. *radices lycopersici*	Root rot	Barley root, Denmark	IK726	Growth chamber	Seedling dip	Reduced disease severity	Microscopy showing *C. rosea* appresoria on *F. oxysporum* hyphae	Karlsson et al., 2015
F. oxysporum	Vascular wilt	Barley root, Denmark	IK726	Greenhouse	Root dip and drenching	Reduced wilt severity	Endophytic colonisation of root and stem	Højer, 2014
Cereals (wheat, barley)								
F. culmorum	Seedling blight, root rot	Barley root, Denmark	IK726	Field trials and growth chamber	Seed treatment	Reduced seedling blight, increased yield in field trial	Significant efficacy at soil temperatures ranging from 6-12°C	Jensen et al., 2000; Knudsen et al., 1995

(*Continued*)

Table 4 (*Continued*).

Plant/Pathogen	Disease	Isolate origin	Isolate ID	Experimental conditions	Application method	Effects on disease development	Other remarks	Reference
F. culmorum	Seedling blight	Wheat crown infected with *F. culmorum*	CR47	Growth chamber	Seedling treatment	Reduced seedling blight		Roberti et al., 2000
F. graminearum	Fusarium head blight	Pea plant, Canada	ACM941	Field trial	Spraying	Reduced FHB and DON content in grains, increased yield	Effects compared to the fungicide tebuconazole	Xue et al., 2009
F. graminearum	Fusarium head blight	Pea plant, Canada	CLO-1 (isolate ACM941)	Field trial	Spraying	Reduced FHB and DON content in grains, increased yield	3 cultivars, highest efficacy in resistant cultivar, dose-response relationship	Xue et al., 2014
Zymoseptoria tritici	Septoria tritici blotch	Barley root, Denmark	IK726	Field trial	Spraying	Reduced STB severity		Jensen et al., 2019
Puccinia triticina, Pu. hordei, Pu. coronata f. sp. *avenacea*	Rusts	Isolated from rust pustules on wheat and oat plants, Australia	H2	Growth chamber	Spraying	Reduced pustle number	Detached leaves on water agar	Wilson et al., 2020
Bipolaris sorokiniana	Leaf blotch	Barley roots, Denmark	IK726	Growth chamber	Spraying	Reduced severity and sporulation	Stored clay formulation. Timing and dosis decisive for control efficacy	Jensen et al., 2016a, b
Potato								
Rhizoctonia solani	Black scurf	*R. solani* hyphae	MpA	Greenhouse	Stem inoculation by agar plug	Reduced black scurf incidence and severity, increased yield of healthy tuber	*C. rosea* and *R. solani* was co-cultivated on stems	Salamone et al., 2018

F. avenaceum, F. coeruleum	Dry rot	Barley roots, Denmark	IK726	Growth chamber	Drop inoculation in wounds	Reduced dry rot severity		Samils et al., 2016
Helminthosporium solani	Silver scurf	Barley roots, Denmark	IK726	Growth chamber	Tuber treatment	Reduced silver scurf severity		Lysøe et al., 2017
Carrot								
A. dauci, A. radicina	Seedling blight	Barley root, Denmark	IK726	Growth chamber	Bio-priming	Eradication of seed-borne *Alternaria* spp., increased seedling stand	Colonisation visualized using *gfp*-tagged IIK726 mutant	Jensen et al., 2004
A. dauci, A. radicina	Seedling blight	Barley root, Denmark	IK726	Growth chamber and field trial	Seed treatment	Increased number of healthy seedlings		Koch et al., 2010
Pulses (pea, soy bean)								
Fusarium, Alternaria, Aphanomyces, Pythium	Pea root rot complex	Pea plant, Canada	ACM941 (ATTCC74447)	Greenhouse, field trial	Seed treatment	Enhanced seedling emergence and yield	*C. rosea* colonize seed, hypocotyl and root	Xue, 2003
Sclerotinia sclerotiorum	Stem rot	Vegetable crop, China	67-1 (*C. chloroleuca*)	Greenhouse	Leaf inoculation	Reduced stem rot severity		Sun et al., 2020b
S. sclerotiorum	Root rot	Suppressive soil, Argentina	BAFC3874	Greenhouse	Mixing into soil	Increased survival		Rodriguez et al., 2011
Pythium ultimum	Damping-off	Unknown	Unknown	Greenhouse	Bark inoculum mixed into the substrate	Reduced damping-off severity		Steinmetz and Schönbeck, 1994

(*Continued*)

Table 4 (*Continued*).

Plant/Pathogen	Disease	Isolate origin	Isolate ID	Experimental conditions	Application method	Effects on disease development	Other remarks	Reference
Oilseed rape								
Plasmodiophora brassicae	Clubroot	Barley roots, Denmark	IK726	Greenhouse	Seed treatment	Reduced disease severity and incidence, reduced *P. brassicae* biomass in roots	Highest control efficacy in resistant cultivar	Andersen et al., 2018
Pl. brassicae	Clubroot	Soil, Finland	Prestop (J1446)	Greenhouse	Soil drench at seeding	Reduced incidence and disease severity, reduced *P. brassicae* biomass in roots	The Prestop product more effective than a conidial suspension	Lahlali and Peng, 2014
Pl. brassicae	Clubroot	Soil, Finland	Prestop (J1446)	Greenhouse, field trials	Soil drench at seeding and seed treatment	Reduced club root severity	Soil drench is more efficient than seed treatment. In field only biocontrol effect in the resistant cultivar	Peng et al., 2011
Chinese cabbage								
Pl. brassicae	Clubroot	Soil, Finland	Prestop (J1446)	Field trial	Soil drench at seeding	Reduced clubroot disease severity	Biocontrol in susceptible cultivar	Peng et al., 2011

Py. tracheiphilum	Chinese cabbage	Leaf and head rot	IK726	Field trial	Spray application to soil surface under leaves	Reduced percentage of attacked plants and increased number of marketable heads	Sprayed to soil and leaf before head setting	Møller et al., 2003
Pine								
F. circinatum	Pitch canker	Pine tissues and soil	Cr7, Cr8	Growth chamber	Drenching	Reduced lesion length	Effective only with a resistant variety	Moraga-Suazo et al., 2016
Cocoa								
Moniliophthora roreri, *Phytophthora* spp., *Theobroma cacao*	Frosty pod rot, black pod	Panama	Several strains	Field trial	Spraying	Reduced disease severity, increased percentage healthy pods		Krauss et al., 2006
Rosellinia spp.	Root rot	Soil, Costa Rica	Several strains (*C. byssicola*, *C. rhizophaga*)	Greenhouse	Mixing into soil	Reduced disease severity		García et al., 2003

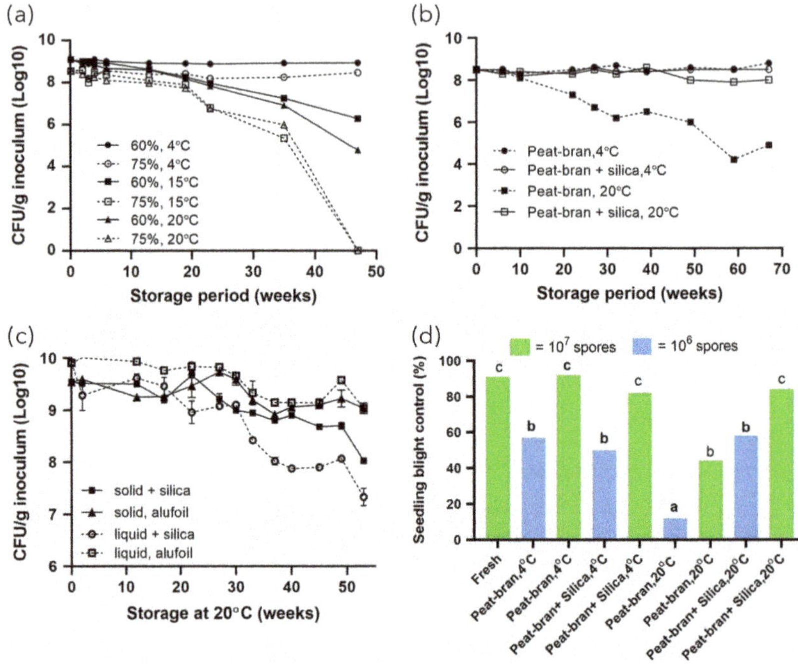

Figure 3 Influence of production, formulation and packaging of *C. rosea* isolate IK726 inoculum on shelf life (CFU/g inoculum) at 4°C and 20°C and on biocontrol efficacy. (a) Effect of water content in the solid peat-bran substrate, (b) Effect of packaging peat-bran inoculum with and without a silica desiccant, (c) Effect of formulating liquid or solid-state produced spores in a clay carrier and packaging the inoculum with silica or in sealed alufoil bags, and (d) Control of fusarium seedling blight using peat-bran inoculum packaged with and without a silica desiccant (Fig 2b) and stored for 29 weeks at 4°C and 20°C. Seeds infested with *F. culmorum* were coated with 10^6 or 10^7 spores/ml of *C. rosea* recovered from the stored formulations. The figures are modified from Jensen (1999). The *C. rosea* peat-bran inoculum was produced in a mixture of sphagnum peat, wheat bran and water (15:26:59, w/w/w) incubated for 2 weeks at 21°C. Subsequently, the inoculum was air-dried, milled and stored (See Jensen et al., 2000 and Jensen et al., 2002 for more details).

Shelf life and vitality of produced propagules can be optimised using various formulation and packing methods and by proper handling after production. Temperature and moisture content are key factors that influence the storage shelf life of *C. rosea* propagules produced both by solid and liquid fermentation (Fig. 3a–c). Significantly longer storage times of spores (months) with preserved biocontrol efficacy (Jensen et al., 2000; Jensen et al., 2002), or even several years at 4°C (Jensen, D. F., unpublished data), was obtained by drying the spores down and keeping them stored at low relative humidity. Raising the storage temperature above 15°C, however, can drastically reduce viability of the inoculum within a few months if the moisture content is not kept low. By optimising the packaging of the inoculum, e.g. with the desiccant silica (Fig. 3b) or formulating the propagules in clay followed by airtight packaging (Fig. 3c) can increase shelf-life for more than half a year at 20°C and the biocontrol efficacy can be maintained (Fig. 3d, Jensen et al., 2002). Treatments of seed with *Clonostachys* spores several months before sowing is also a possibility but might give problems with shelf life depending on how dry the seed is and how the humidity and temperature are controlled during storage of treated seeds (Jensen et al., 2002).

5.3 Commercial products based on C. rosea

On a global scale, at least 10 different commercial products based on *C. rosea* strains are available (Table 3). Some products are available in several countries whereas others are only available in certain regions or countries. So, BCA products are registered in Brazil, China, the EU, Russia and the USA. According to product information, they can be used in a wide range of crops. These include grain crops, cabbage, legumes, vegetables, beans, tomato, cucumber, pepper, tomato, strawberry, raspberry, black current and other *Ribes* spp., blueberry, pome fruit (pear, apple, quince, various *Crataegus* spp.), stone fruit (apricot, peach, plums), herbs and aromatic plants, ornamental plants, potted plants, cut flowers, forest nursery tree seedlings and tree nuts (butternut, chestnut, macadamia, pecan, pistachio). Furthermore, these products are listed to target a wide range of disease types, including damping-off (caused by *Pythium* spp.

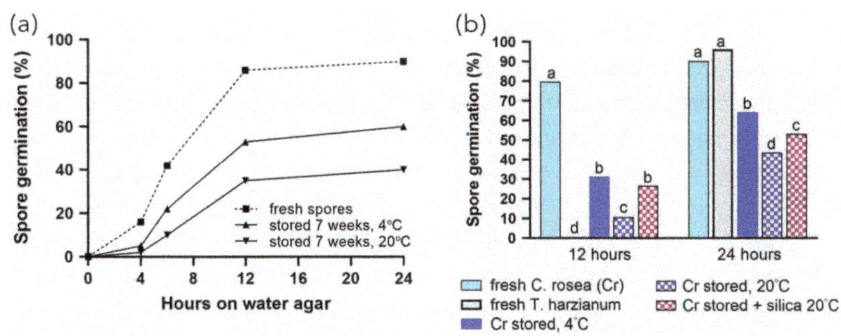

Figure 4 Time-dependent germination of *C. rosea* isolate IK726 on water agar. (a) Spores freshly harvested and undried or recovered from dried peat-bran inoculum stored at 4°C and 20°C, respectively and (b) Freshly harvested and undried spores of *C. rosea* and *T. harzianum* and spores of *C. rosea* recovered from dried peat-bran inoculum stored for 18 weeks with and without silica desiccant at 4°C and 20°C. The figures are modified from Jensen (1999). The *C. rosea* peat-bran inoculum was produced in a mixture of sphagnum peat, wheat bran and water (15:26:59, w/w/w) incubated for 2 weeks at 21°C. Subsequently, the inoculum was air-dried, milled and stored (See Jensen et al., 2000 for more details).

and *R. solani*), grey mould (*B. cinerea*), root- and stem base rot (*Phytophthora* spp. and *Sclerotinia* spp.), leaf spot diseases and fusarium head blight (*Fusarium* spp.).

6 Delivery and action of *C. rosea* as a biological control agent

Application of *C. rosea* has shown potential for biological control of diseases in many important crops. In Table 4, we have listed selected examples of experiments where significant biocontrol efficacy of various *C. rosea* isolates was demonstrated in field trials, greenhouse or in a growth chamber. The majority of examples involve the isolates J1446 (now commercialised in the products LALSTOP G46 WG®, Prestop® and GlioMix®) and IK726, both isolated in a Nordic research programme. Both soil-borne and seed-borne pathogens as well as pathogens attacking leaf, stem, flower and fruit can be controlled by *C. rosea*. It is important to note that the outcome of interactions between host plant genotype and the target pathogen is strongly influenced by the surrounding biotic and abiotic environment which decide the severity of the disease. Realising that an introduced BCA has to perform its action in this highly complex setting makes it difficult to predict if the BCA is going to be successful in a given niche in soil or on the plant. Therefore, successful biological disease control relies on finding the most efficient method for delivery of the BCA in an active state, at the right place, at the right time with the right dose.

6.1 *Using pollinators for* C. rosea *delivery*

The idea of using pollinators to deliver BCAs to flowers goes back to the work of Peng et al. (1992) and Sutton et al. (1997). They used honeybees or bumblebees to deliver spores of *C. rosea* to strawberry and raspberry flowers to control grey mould caused by *B. cinerea* (Table 4). Infection by the pathogen mainly occurs in newly opened flowers. In fact, honeybee vectoring (Peng et al., 1992) and bumblebee vectoring (Van Delm et al., 2015) during flowering resulted in a more efficient spore delivery and thereby better grey mould control in strawberry flowers and fruits as compared to weekly spraying with *C. rosea*. In raspberry, bumblebee delivery of spores controlled grey mould in flowers more efficiently than one *C. rosea* spray application at the beginning of flowering (Yu and Sutton, 1997a). The success of bee delivery is probably achieved because they visit the newly opened flowers delivering the spores with more precise timing, prior to the natural colonisation of flowers by *B. cinerea* (Peng et al., 1992). Using bumblebees instead of honeybees gave more stable results as honeybees were more prone to attraction to other crops flowering at the same time and therefore delivering the BCA to the wrong crop – a problem not seen with bumblebees (Sutton et al., 1997). This method with bumblebees and a special hive construction to facilitate vectoring of *C. rosea* is described in more detail in Yu and Sutton (1997a) and is now used commercially both in Europe and in the USA (Table 3).

6.2 Seed coating for C. rosea delivery

Delivering the BCA with the seed is a common strategy used to control seed-borne, damping-off and seedling diseases (Table 4). Especially for seed-borne pathogens, it is an obvious approach (Knudsen et al., 1995; Jensen et al., 2000; Jensen et al., 2004; Bennett et al., 2009; Koch et al., 2010). For example, using *C. rosea* seed treatment for control of seedling diseases in cereals caused by *F. culmorum* and *B. sorokiniana* have consistently reduced the diseases in several field trials (Knudsen et al., 1995; Jensen et al., 2000). Root diseases can also be controlled by seed delivery as demonstrated for *P. brassicae*, the cause of clubroot in *Brassicae* species (Andersen et al., 2018; Peng et al., 2011), in pea against several soil-borne pathogens (Xue, 2003) and enhance field establishment of carrot plants (Fig. 1, Jensen, B. unpublished data). Efficient seed and root colonisation by *C. rosea* is probably required to obtain effects against soil-borne diseases (Fig. 1a, Xue, 2003; Jensen et al., 2004).

A special case of seed delivery is biopriming - a method first reported for *T. harzianum* (Harman et al., 1989) where the BCA was applied during the seed priming process. Priming is basically done by imbibition or controlled hydration of seed followed by a priming period at a reduced moisture content allowing seeds to go through the first reversible stage of germination but do not allow radical protrusion through the seed coat. The priming process can be completed after 12 to 14 days and after drying back the seeds, they can be stored until sowing. In general, priming results in more rapid germination and seedling emergence in the field, which is important to vegetables, like carrot, where seeds often are sown at low soil temperature and other unfavourable conditions for seedling establishment, e.g. pathogens causing seedling damping-off. Integration of *C. rosea* into the priming process has shown promising potential. Hence, biopriming of carrot seed with *C. rosea* resulted in a significant enhancement of the carrot plant stand in the field as compared to both primed and unprimed seed (Fig. 1, Jensen, B., unpublished data). Likewise, Bennett et al. (2009) showed that drum priming with and without different BCAs consistently improved the emergence of carrot seed in glasshouse trials and that *C. rosea* further shortened emergence time by two days as compared to unprimed seeds. However, in field experiments, no consistent effects on emergence and yield were seen for BCA primed seed (Bennett et al., 2009). The positive effects of biopriming on seedling establishment are probably related to the ability of *C. rosea* to colonise the seed during priming and to colonise root and rhizosphere after planting (Jensen et al., 2004; Bennett and

Whipps, 2008a,b). In some cases, the expected positive effects of seed priming can disappear or even result in drastically reduced seed quality if the seed lot harbours pathogens that are activated by the priming hydration (Jensen et al., 2004). However, the use of biopriming can minimise the risk of such adverse effects. For example, it was shown that the priming of carrot seeds naturally infected by *Alternaria* spp. lead to a lower healthy seedling stand than for nonprimed seed, mainly due to a high degree of post-emergence seedling death. In contrast, *C. rosea* biopriming resulted in a seedling stand that was significantly better than that of both nonprimed and seed primed without the BCA (Jensen et al., 2004).

6.3 Delivering C. rosea to soil or plant growth substrates

In Chinese cabbage, bottom rot caused by the soil-borne pathogen *P. tracheiphilum* can be a devastating problem. *Clonostachys rosea* spray application to the soil surface below the plants resulted in significant disease control and increased yield under commercial field production of Chinese cabbage (Table 4, Møller et al., 2003). Incorporation of the BCA into soil or growth substrate is another approach to control soil-borne pathogens in various crops. For practical use, the focus has mainly been on protecting crops in screen- or glasshouse production or in nurseries, but biocontrol effects have also been shown in field trials (Lahlali and Peng, 2014; Peng et al., 2011). Control of plant pathogenic nematodes by soil treatment with *C. rosea* has been demonstrated on small scale (Iqbal et al., 2018b) but controlling nematodes with *C. rosea* on larger field-scale need further testing. Summing up the available information from companies marketing *C. rosea* BCAs shows that different methods are in use for incorporation into the soil or plant growth substrates. Examples are watering or drip irrigation with the BCA that is used in protected high-value crops as well as dipping roots of small plants or cuttings in a spore suspension of *C. rosea* before planting. *C. rosea* could also have the potential for controlling several fungal plant diseases by watering or incorporating the BCA into golf greens.

6.4 Spray application of C. rosea

Spray application of *C. rosea* is relevant for controlling diseases in the phyllosphere such as the cereal diseases spot blotch in barley (Jensen et al., 2016a), FHB (Xue et al., 2009, 2014) and STB in wheat (Fig. 5, Jensen et al., 2019). Recently, Egel et al. (2019) demonstrated a significant reduction of early blight caused by *Alternaria solani* in tomatoes by spraying of Prestop® in field trials. This shows that spray application can be an important option for diseases in large agriculture field crops. Spray application is also used for the control of grey

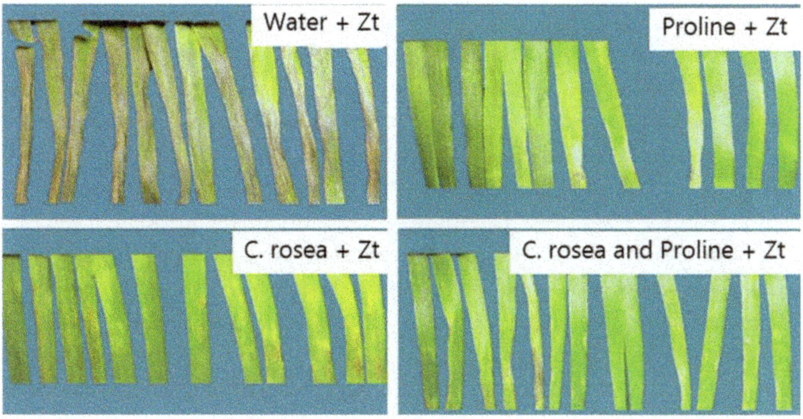

Figure 5 Control of septoria tritici blotch on wheat plants by spray application with *C. rosea* strain IK726 and the fungicide Proline® (a.i. Prothioconazole) in a growth chamber experiment. Leaves were sprayed with either *C. rosea*, Proline® or with both *C. rosea* and Proline® before inoculation with the pathogen *Z. tritici* (Zt) 24 h later. (Photos by Henrik Bendix Aas).

mould in strawberries and tomatoes (Sutton et al., 2002; Cota et al., 2008; Gong et al., 2017; Nechet et al., 2017). The Prestop® product is recommended for spray application on stems and wounds in vegetables against several diseases.

6.5 Delivery of C. rosea in consortia

Consortia, i.e. where two or more different BCA strains are combined, has often been suggested aiming at either additive or synergistic biocontrol effects against one disease or an approach for controlling different diseases by exploiting BCAs targeting different pathogens (e.g. Hoopen et al., 2010; Xu et al., 2011a,b; Krauss et al., 2013; Jensen et al., 2016b). Another strategy is to combine BCA(s) with other microorganisms for controlling plant diseases and at the same time alleviate other biotic or abiotic constraints to crop production. The complex plant microbiome can have an effect trait leading to healthy plants. On the other hand, microbiome function might also have an impact that makes it difficult to successfully establish BCAs or BCA consortia in the crop. However, research is at present mainly descriptive without much evidence for how to regulate the complex microbial communities and thereby facilitate their functions, e.g. biocontrol effect traits or how their functions can be compatible with implementing efficient BCA consortia. As reviewed by Xu et al. (2011a), it has generally been difficult to find published work demonstrating statistically significant improved consortia effects even if only two BCAs have been combined. There is also the issue that consortia members can be antagonistic towards each other leading to unsuccessful control, which is seen in several cases (Xu et al., 2011a,b). Consortia formulations with *C. rosea* show, however, promising results in some cases. A combination of *C. rosea* and the arbuscular mycorrhiza *Glomus intraradices* was delivered to the rhizosphere of tomato plants. The two fungi showed mutual inhibition in the rhizosphere, but nevertheless resulted in synergistic plant growth promotion when combined (Ravnskov et al., 2006). Crops are often suffering from both insect pests and diseases, which necessitate multiple control measures. Therefore, a dual treatment approach involving BCAs targeting the different organisms without compromising the biocontrol traits of each other would be ideal. In a study of wheat, it was demonstrated that entomopathogenic fungi from the genus *Metarhizium* and *C. rosea* could be used in concert to control a root-feeding insect and a seed-borne disease in a single seed treatment (Keyser et al., 2016). Furthermore, an additive effect on biocontrol of tomato foot and root rot disease caused by *F. oxysporum* f. sp. *radicis lycopersici* was achieved by combining *C. rosea* with the phenazine-producing bacterium *P. chlororaphis* (Karlsson et al., 2015).

6.6 Role in integrated pest management (IPM)

Biological control should have a central role in IPM strategies aiming at reduced use of chemical pesticides as discussed elsewhere (Jensen et al., 2016b). *Clonostachys rosea* has shown tolerance to several chemical pesticides (Dubey et al., 2014a; Roberti et al., 2006), and therefore the BCA can be used in combination with various chemical pesticide treatments either in full recommended or in reduced doses of, e.g. fungicides, or in application schemes where BCAs and fungicides are alternated (Cota et al., 2009). Depending on the sensitivity to a pesticide, *C. rosea* might be applied together with the pesticide or with a time distance of a few days between the pesticide and the biocontrol treatments as outlined in the Prestop® info letter: https://verdera.fi/index.php/download_file/view/470/174/, from Lallemand/Verdera. The BCA could also be delivered at other time points in the cropping season to target other pathogens or even between seasons. Applying the BCA in the 'pre-harvest period' in which chemical control measures are not allowed is another option both in IPM and organic production (Jensen et al., 2016b). In addition to reducing the input of chemical pesticides, ongoing research also investigates if this strategy can prevent the build-up of pesticide resistance in pathogen populations when one or two pesticide applications are substituted with *C. rosea* treatments.

Plant disease resistance is a key factor in IPM strategies. Interestingly, it has been shown that plant cultivars harbouring resistance, or are less susceptible towards a disease, facilitate more efficient biocontrol traits as compared to the application of the BCA to a more susceptible cultivar (Yu and Sutton, 1997a; Andersen et al., 2018; Xue et al., 2014; Moraga-Suazo et al., 2016). Thus, when breeding for disease resistance the possibilities to exploit plant genotypes that also facilitate biocontrol effects in the crop should be in focus.

6.7 Is C. rosea pathogenic on plants?

As mentioned in the previous section, there is an extensive amount of literature that reports on the biocontrol properties of *C. rosea* (Table 4, Sun et al., 2020a), without any negative effects on plant growth. However, over the years there

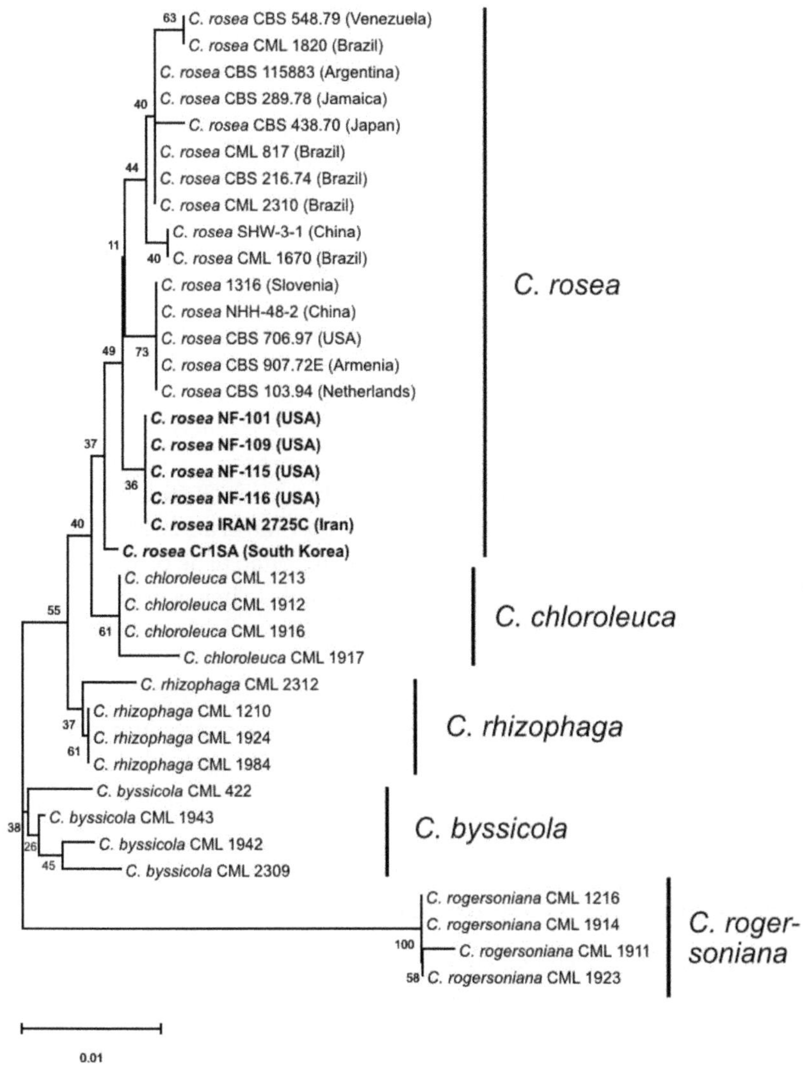

Figure 6 Phylogenetic analysis of *C.* strains. The tree is rooted with *C. rogersoniana*, and based on the internal transcribed spacer (ITS) and neighbour-joining analysis. Bootstrap support values are associated with branches. Sequence identifiers include species, strain ID and geographic origin. The bar marker indicates an average number of substitutions per site. Strains indicated in bold are reported to be pathogenic on plants.

Published by Burleigh Dodds Science Publishing Limited, 2022.

have been a few reports on *C. rosea* being pathogenic to plants. For example, strains identified as *C. rosea* were reported to cause dry rot on potatoes (Theron and Holz, 1991), root rot on soybean (Bienapfl et al., 2012), wilt and crown rot on faba bean (Afshari and Hemmati, 2017) and root rot on orchids (Lee et al., 2020). This discrepancy between plant-beneficial and plant-detrimental properties of different strains of *C. rosea* is intriguing but not easily explained. The intimate association between *C. rosea* and plants, sometimes even involving systemic, asymptomatic colonisation (Saraiva et al., 2015; Mueller and Sinclair, 1986), indicate a delicate balance between the colonisation of *C. rosea* and the immune responses by the plant host. It is plausible that poor physiological status of the plant, a high inoculum of *C. rosea*, as well as certain genotype-by-genotype (*C. rosea* vs. plant) combinations may distort this balance and result in disease symptoms.

In some cases, sequencing of the ITS region was used together with morphology for species identification of plant pathogenic strains (Bienapfl et al., 2012; Afshari and Hemmati, 2017; Lee et al., 2020). A phylogenetic analysis of the ITS sequences of these strains, together with selected *C. rosea* strains representing a worldwide distribution (Broberg et al., 2018) and strains representing closely related *Clonostachys* species (Moreira et al., 2016) is presented in Fig. 6. First, this analysis confirms that the ITS region does not provide enough resolution for distinguishing between different *Clonostachys* species, as reported previously (Schroers, 2001; Abreu et al., 2014). However, it is interesting to note that all the *C. rosea* strains reported to be plant pathogenic clusters in a basal position within the *C. rosea* clade (albeit with low bootstrap support), distinct from the strains representing the worldwide collection. This may suggest that plant pathogenic *C. rosea* strains indeed form a genetically distinct group. Whether they form a separate, as yet undescribed, cryptic species remains to be investigated using other genetic markers such as ATP citrate lyase (*acl1*) and RNA polymerase II large subunit (*rpb1*) that are reported to be more suitable to resolve species boundaries in *Clonostachys* (Moreira et al., 2016).

7 Conclusion and future trends

A key to high efficacy and consistency in biocontrol is to identify and target vulnerable stages in the pathogen lifecycle and plant development. This can include the targeting of pathogen resting structures, temporarily protecting plant wounds and other pathogen entry points, or a more continuous interference with the plant tissue colonisation and dissemination of the pathogen. Basic studies of pathogen biology combined with advanced methods for tracing the presence and activity of the BCA is, therefore, an important aspect for successful implementation of biocontrol solutions. For *C. rosea*, the availability of strains expressing the green fluorescent protein (Lübeck et al., 2002), and more recently, accurate and validated tools for DNA quantification (Gimeno et al., 2019) is therefore promising.

Different strains of *C. rosea* can display a considerable variation in biocontrol-related traits (Iqbal et al., 2020), which emphasises the importance of choosing the correct strain with a high biocontrol ability for the particular pathosystem in question. This is typically done in large screening experiments, involving the plant and the pathogen, in greenhouse or field settings resembling the conditions where the BCA will act (Köhl et al., 2011; Jensen et al., 2016b). The availability of high numbers of whole-genome sequenced *C. rosea* strains (Broberg et al., 2018) opens up new possibilities for including genetic markers coupled with specific traits as a decision-support in the screening procedure. Such genetic markers can also find applications in future attempts to breed for new *C. rosea* strains with specific traits, either through protoplast fusion approaches, which are used in *Trichoderma* (Stasz et al., 1988), or through sexual crosses followed by progeny selection. Given the intimate association between *C. rosea* and plants, genotype-by-genotype interaction effects are likely to have a considerable effect on biocontrol efficacy. This may, in fact, be exploited in plant breeding, where plant compatibility with beneficial microorganisms, including *C. rosea*, can be included as a breeding target alongside yield, quality and disease resistance.

Genetic improvement of BCA strains using CRISPR-Cas9 (Clustered Regularly Interspaced Short Palindromic Repeats – CRISPR Associated protein 9) genome-editing technology is a promising approach for increasing the speed for developing biocontrol solutions, with increased genetic precision (Muñoz et al., 2019). In *Clonostachys*, there are already now several examples of genetic modifications that increase the biocontrol efficacy that may be a future target for CRISPR-Cas9 technology. For example, deletion of the *hyd1* and *hyd3* hydrophobin genes in *C. rosea* resulted in increased biocontrol ability towards *B. cinerea* on leaves (Dubey et al., 2014b). Furthermore, overexpression of the *chi67-1* endochitinase gene in *C. chloroleuca* increased chitinase production and subsequently the ability to control sclerotinia stem rot on soybean (Sun et al., 2017).

Microbiome research is leading to an increasing amount of detailed information not only on what microbial communities are to be found in plant/soil microbiomes but also on how microbiome function is emerging. How environment, plant cultivar and crop management affect microbiome functions will be important in forming strategies for sustainable healthy crop production in the future. Especially for the use of augmentative biocontrol, new detailed information on microbiome changes and their related functions will be important for establishing the correct timing and place to deliver the BCA(s) to the crop plant. In focus are also the pathobiomes where several pathogen species are interacting in a way that modifies the outcome of infection. This is probably in most cases through direct interference with the plant defence responses, but can also relate to biocontrol interactions. An example of this is in FHB on wheat (Tan et al., 2021) where two *Fusarium* spp. interact, leading to a reduced effect of a bacterial biocontrol strain or chemical pesticide treatment in FHB control. Thus, having a focus on how successful *C. rosea* can be in controlling several pathogens found in complex natural pathobiomes should be important for future research. How the whole plant microbiome affects augmentative biocontrol effect traits is an important topic for research. Based on the information brought in this book chapter we believe that *C. rosea* will be an important model organism for such future studies. *Clonostachys rosea* is already today an important factor in sustainable plant protection strategies, and the recent developments in our understanding of its ecology, genetics and application promise an even more significant role in the future.

8 Where to look for further information

The following book chapter provides a good introduction to *C. rosea* and biological control:

- Jensen, D. F., Karlsson, M., Sarrocco, S. and Vannacci, G. (2016). Biological control using microorganisms as an alternative to disease resistance. In: Collinge, D. B. (Ed) *Plant Pathogen Resistance Biotechnology*. Wiley, New York and London, 341–363.

A key scientific conference involving *C. rosea* is the International Workshop on *Trichoderma* and *Gliocladium* (TG), held every second year. Other relevant conferences are the International Congress of Plant Pathology (ICPP) and the IOBC/WPRS Working Group meeting, Biological control of fungal and bacterial plant pathogens.

9 Acknowledgements

This work was financially supported by the SLU Centre for Biological Control, SLU Grogrund and the Swedish Research Council for Environment, Agricultural Sciences and Spatial Planning (FORMAS) (grant number 2018-01420). We acknowledge Lucas Abreu, Andrei Golubev, Ekaterina Sinyaeva, Willem Ravensberg, Jürgen Köhl, Eirian Jones and Manhong Sun for providing information concerning commercial products based on *C. rosea*. We thank Dean Malvick and John C. Bienapfl for providing ITS sequences from *Clonostachys* strains.

10 References

Abreu, L. M., Moreira, G. M., Ferreira, D., Rodrigues-Filho, E. and Pfenning, L. H. (2014). Diversity of *Clonostachys* species assessed by molecular phylogenetics and MALDI-TOF mass spectrometry. *Fungal Biol.* 118(12): 1004–1012.

Afshari, N. and Hemmati, R. (2017). First report of the occurrence and pathogenicity of *Clonostachys rosea* on faba bean. *Australas. Plant Pathol.* 46(3): 231–234.

Agosin, E., Cotoras, M., Munoz, G., San Martin, R. and Volpe, D. (1997). Comparative properties of *Trichoderma harzianum* spore produced under solid state and submerged culture conditions. In: Roussos, S., et al. (Eds) *Advances in Solid State Fermentation*. Dordrecht: Springer Sciences and Business Media, pp. 463–473.

Andersen, C., Jørgensen, H., Manzotti, A. and Jensen, B. (2018). Seed coating with the fungal biocontrol agent *Clonostachys rosea* controls clubroot in oilseed rape. *IOBC WPRS Bull.* 136: 157–163.

Atanasova, L., Dubey, M., Grujic, M., Gudmundsson, M., Lorenz, C., Sandgren, M., Kubicek, C. P., Jensen, D. F. and Karlsson, M. (2018). Evolution and functional characterization of pectate lyase PEL12, a member of a highly expanded *Clonostachys rosea* polysaccharide lyase 1 family. *BMC Microbiol.* 18(1): 178.

Baker, K. F. and Cook, R. J. (1974). *Biological Control of Plant Pathogens*. San Francisco: Freeman.

Barnett, H. L. and Lilly, V. G. (1962). A destructive mycoparasite, *Gliocladium roseum*. *Mycologia* 54(1): 72–77.

Bennett, A. J., Mead, A. and Whipps, J. M. (2009). Performance of carrot and onion seed primed with beneficial microorganisms in glasshouse and field trials. *Biol. Control* 51(3): 417–426.

Bennett, A. J. and Whipps, J. M. (2008a). Beneficial microorganism survival on seed, roots and in rhizosphere soil following application to seed during drum priming. *Biol. Control* 44(3): 349–361.

Bennett, A. J. and Whipps, J. M. (2008b). Dual application of beneficial microorganisms to seed during drum priming. *Appl. Soil Ecol.* 38(1): 83–89.

Bienapfl, J. C., Floyd, C. M., Percich, J. A. and Malvick, D. K. (2012). First report of *Clonostachys rosea* causing root rot of soybean in the United States. *Plant Dis.* 96(11): 1700–1700.

Borges, ÁV., Saraiva, R. M. and Maffia, L. A. (2015). Biocontrol of gray mold in tomato plants by *Clonostachys rosea*. *Trop. Plant Pathol.* 40(2): 71–76.

Broberg, M., Dubey, M., Iqbal, M., Gudmundssson, M., Ihrmark, K., Schroers, H. J., Funck Jensen, D., Brandström Durling, M. and Karlsson, M. (2021). Comparative genomics highlights the importance of drug efflux transporters during evolution of mycoparasitism in *Clonostachys* subgenus *Bionectria* (fungi, Ascomycota, Hypocreales). *Evol. Appl.* 14(2): 476–497.

Broberg, M., Dubey, M., Sun, M. H., Ihrmark, K., Schroers, H. J., Li, S. D., Jensen, D. F., Brandström Durling, M. and Karlsson, M. (2018). Out in the cold: identification of genomic regions associated with cold tolerance in the biocontrol fungus *Clonostachys rosea* through genome-wide association mapping. *Front. Microbiol.* 9: 2844.

Cen, K., Li, B., Lu, Y. Z., Zhang, S. and Wang, C. (2017). Divergent LysM effectors contribute to the virulence of *Beauveria bassiana* by evasion of insect immune defenses. *PLoS Pathog.* 13(9): e1006604.

Chatterjee, S., Kuang, Y., Splivallo, R., Chatterjee, P. and Karlovsky, P. (2016). Interactions among filamentous fungi *Aspergillus niger*, *Fusarium verticillioides* and *Clonostachys rosea*: fungal biomass, diversity of secreted metabolites and fumonisin production. *BMC Microbiol.* 16: 83.

Chatterton, S. and Punja, Z. K. (2009). Chitinase and beta-1,3-glucanase enzyme production by the mycoparasite *Clonostachys rosea* f. catenulata against fungal plant pathogens. *Can. J. Microbiol.* 55(4): 356–367.

Chatterton, S. and Punja, Z. K. (2010). Factors influencing colonization of cucumber roots by Clonostachys rosea f. catenulata, a biological disease control agent. *Biocont. Sci. Technol.* 20: 37–55.

Chatterton, S. and Punja, Z. K. (2012). Colonization of geranium foliage by *Clonostachys rosea* f. catenulata , a biological control agent of botrytis grey mould. *Botany* 90(1): 1–10.

Cota, L. V., Maffia, L. A., Mizubuti, E. S. G. and Macedo, P. E. F. (2009). Biological control by *Clonostachys rosea* as a key component in the integrated management of strawberry gray mold. *Biol. Control* 50(3): 222–230.

Cota, L. V., Maffia, L. A., Mizubuti, E. S. G., Macedo, P. E. F. and Antunes, R. F. (2008). Biological control of strawberry gray mold by *Clonostachys rosea* under field conditions. *Biol. Control* 46(3): 515–522.

de Andrade Carvalho, A. L., de Rezende, L. C., Bertoldo Costa, L., Halfeld-Vieira, BdA., Vegette Pinto, Z., Boechat Morandi, M. A., de Medeiros, F. H. V. and Bettiol, W. (2018). Optimizing the mass production of *Clonostachys rosea* by liquid-state fermentation. *Biol. Control* 118: 16–25.

de Jonge, R. and Thomma, B. P. H. J. (2009). Fungal LysM effectors: extinguishers of host immunity? *Trends Microbiol.* 17(4): 151–157.

Demissie, Z. A., Foote, S. J., Tan, Y. F. and Loewen, M. C. (2018). Profiling of the transcriptomic responses of *Clonostachys rosea* upon treatment with *Fusarium graminearum* secretome. *Front. Microbiol.* 9: 1061.

Demissie, Z. A., Robinson, K. A. and Loewen, M. C. (2021). Draft genome resources for the plant-beneficial fungi *Clonostachys rosea* strains ACM941 and 88–710. *Mol. Plant. Microbe Interact.* 34(4): 453–456 doi: 10.1094/MPMI-10-20-0294-A.

Demissie, Z. A., Witte, T., Robinson, K. A., Sproule, A., Foote, S. J., Johnston, A., Harris, L. J., Overy, D. P. and Loewen, M. C. (2020). Transcriptomic and exometabolomic profiling reveals antagonistic and defensive modes of *Clonostachys rosea* action against *Fusarium graminearum*. *Mol. Plant. Microbe Interact.* 33(6): 842–858.

Dong, J. Y., He, H. P., Shen, Y. M. and Zhang, K. Q. (2005). Nematicidal epipolysulfanyldi-oxopiperazines from *Gliocladium roseum*. *J. Nat. Prod.* 68(10): 1510–1513.

Dubey, M., Jensen, D. F. and Karlsson, M. (2016). The ABC transporter ABCG29 is involved in H_2O_2 tolerance and biocontrol traits in the fungus *Clonostachys rosea*. *Mol. Genet. Genomics* 291(2): 677–686.

Dubey, M., Velez, H., Broberg, M., Jensen, D. F. and Karlsson, M. (2020). LysM proteins regulate fungal development and contribute to hyphal protection and biocontrol traits in *Clonostachys rosea*. *Front. Microbiol.* 11: 679.

Dubey, M. K., Jensen, D. F. and Karlsson, M. (2014a). An ATP-binding cassette pleiotropic drug transporter protein is required for xenobiotic tolerance and antagonism in the fungal biocontrol agent *Clonostachys rosea*. *Mol. Plant. Microbe Interact.* 27(7): 725–732.

Dubey, M. K., Jensen, D. F. and Karlsson, M. (2014b). Hydrophobins are required for conidial hydrophobicity and plant root colonization in the fungal biocontrol agent *Clonostachys rosea*. *BMC Microbiol.* 14: 18.

Egel, D. S., Hoagland, L., Davis, J., Marchino, C. and Bloomquist, M. (2019). Efficacy of organic disease control products on common foliar diseases of tomato in field and greenhouse trials. *Crop Prot.* 122: 90–97.

Fatema, U., Broberg, A., Jensen, D. F., Karlsson, M. and Dubey, M. (2018). Functional analysis of polyketide synthase genes in the biocontrol fungus *Clonostachys rosea*. *Sci. Rep.* 8(1): 15009.

García, R. A. M., Martijn ten Hoopen, G., Kass, D. C. J., Sánchez Garita, V. A. and Krauss, U. (2003). Evaluation of mycoparasites as biocontrol agents of *Rosellinia* root rot in cocoa. *Biol. Control* 27(2): 210–227.

Gimeno, A., Sohlberg, E., Pakula, T., Limnell, J., Keller, B., Laitila, A. and Vogelgsang, S. (2019). Taqman qPCR for quantification of *Clonostachys rosea* used as a biological control agent against *Fusarium graminearum*. *Front. Microbiol.* 10: 1627.

Gong, C., Liu, Y., Liu, S. Y., Cheng, M., Zhang, Y., Wang, R., Chen, H., Li, J., Chen, X. and Wang, A. (2017). Analysis of *Clonostachys rosea*-induced resistance to grey mould disease and identification of the key proteins induced in tomato fruit. *Postharvest Biol. Technol.* 123: 83–93.

Haarith, D., Bushley, K. E. and Chen, S. Y. (2020). Fungal communities associated with *Heterodera glycines* and their potential in biological control: a current update. *J. Nematol.* 52: 1–17.

Han, P. P., Zhang, X. P., Xu, D., Zhang, B., Lai, D. and Zhou, L. (2020). Metabolites from *Clonostachys* fungi and their biological activities. *J. Fungi (Basel)* 6(4): 229.

Harman, G. E., Howell, C. R., Viterbo, A., Chet, I. and Lorito, M. (2004). *Trichoderma* species - Opportunistic, avirulent plant symbionts. *Nat. Rev. Microbiol.* 2(1): 43–56.

Harman, G. E., Jin, X., Stasz, T. E., Peruzzotti, G., Leopold, A. C. and Taylor, A. G. (1991). Production of conidial biomass of *Trichoderma harzianum* for biological control. *Biol. Control* 1(1): 23–28.

Harman, G. E., Taylor, A. G. and Stasz, T. E. (1989). Combining effective strains of *Trichoderma harzianum* and solid matrix priming to improve biological seed treatments. *Plant Dis.* 73(8): 631–637.

Højer, A. (2014). *C. rosea* as an endophyte and a biocontrol agent of *Fusarium* wilt in tomato. MSc thesis. Denmark: University of Copenhagen, pp. 1–82.

Hoopen, G. M., George, A., Martinez, A., Stirrup, T., Flood, J. and Krauss, U. (2010). Compatibility between *Clonostachys* isolates with a view to mixed inocula for biocontrol. *Mycologia* 102(5): 1204–1215.

Inglis, G. D. and Kawchuk, L. M. (2002). Comparative degradation of oomycete, ascomycete, and basidiomycete cell walls by mycoparasitic and biocontrol fungi. *Can. J. Microbiol.* 48(1): 60–70.

Iqbal, M., Broberg, M., Haarith, D., Broberg, A., Bushley, K. E., Brandström Durling, M., Viketoft, M., Funck Jensen, D., Dubey, M. and Karlsson, M. (2020). Natural variation of root lesion nematode antagonism in the biocontrol fungus *Clonostachys rosea* and identification of biocontrol factors through genome-wide association mapping. *Evol. Appl.* 13(9): 2264–2283.

Iqbal, M., Dubey, M., Broberg, A., Viketoft, M., Jensen, D. F. and Karlsson, M. (2019). Deletion of the nonribosomal peptide synthetase gene *nps1* in the fungus

Clonostachys rosea attenuates antagonism and biocontrol of plant pathogenic *Fusarium* and nematodes. *Phytopathology* 109(10): 1698–1709.

Iqbal, M., Dubey, M., Gudmundsson, M., Viketoft, M., Jensen, D. F. and Karlsson, M. (2018a). Comparative evolutionary histories of fungal proteases reveal gene gains in the mycoparasitic and nematode-parasitic fungus *Clonostachys rosea*. *BMC Evol. Biol.* 18(1): 171.

Iqbal, M., Dubey, M., McEwan, K., Menzel, U., Franko, M. A., Viketoft, M., Jensen, D. F. and Karlsson, M. (2018b). Evaluation of *Clonostachys rosea* for control of plant-parasitic nematodes in soil and in roots of carrot and wheat. *Phytopathology* 108(1): 52–59.

James, T. D. W. and Sutton, J. C. (1996). Biological control of botrytis leaf blight of onion by *Gliocladium roseum* applied as sprays and with fabric applicators. *Eur. J. Plant Pathol.* 102(3): 265–275.

Jensen, B. (1999). *Udvikling af et biologisk bejdsemiddel til frø. Miljøprojekt* (vol. 463). København, Denmark: Frontlinien, Miljøministeriet.

Jensen, B., Knudsen, I. M. B. and Jensen, D. F. (2000). Biological seed treatment of cereals with fresh and long-term stored formulations of *Clonostachys rosea*: biocontrol efficacy against *Fusarium culmorum*. *Eur. J. Plant Pathol.* 106(3): 233–242.

Jensen, B., Knudsen, I. M. B. and Jensen, D. F. (2002). Survival of conidia of *Clonostachys rosea* on stored barley seeds and their biocontrol efficacy against seed-borne *Bipolaris sorokiniana*. *Sci. Technol.* 12(4): 427–441.

Jensen, B., Knudsen, I. M. B., Madsen, M. and Jensen, D. F. (2004). Biopriming of infected carrot seed with an antagonist, *Clonostachys rosea*, selected for control of seedborne *Alternaria* spp. *Phytopathology* 94(6): 551–560.

Jensen, B., Lübeck, P. S. and Jorgensen, H. J. L. (2016a). *Clonostachys rosea* reduces spot blotch in barley by inhibiting prepenetration growth and sporulation of *Bipolaris sorokiniana* without inducing resistance. *Pest Manag. Sci.* 72(12): 2231–2239.

Jensen, D. F., Karlsson, M., Sarrocco, S. and Vannacci, G. (2016b). Biological control using microorganisms as an alternative to disease resistance. In: Collinge, D. B. (Ed) *Plant Pathogen Resistance Biotechnology*. New York, London: Wiley, pp. 341–363.

Jensen, D. F., Karlsson, M. and Lindahl, B. (2017). Fungal-fungal interactions: From natural ecosystems to managed plant production with emphasis on biological control of plant diseases. In: Dighton, J. and White, J. (Eds) *The Fungal Community: Its Organization and Role in the Ecosystem*. Boca Raton: CRC Press Press, pp. 549–562.

Jensen, D. F., Knudsen, I. M. B., Lubeck, M., Mamarabadi, M., Hockenhull, J. and Jensen, B. (2007). Development of a biocontrol agent for plant disease control with special emphasis on the near commercial fungal antagonist *Clonostachys rosea* strain 'IK726'. *Australas. Plant Pathol.* 36(2): 95–101.

Jensen, D. F., Mikkelsen, B., Karlsson, M. and Hökeberg, M. (2019). *BCA Control of STB. Patent Cooperation Treaty, WO 2019/125294 A1*. Stockholm: Swedish Intellectual Property Office.

Johansen, A., Knudsen, I. M. B., Binnerup, S. J., Winding, A., Johansen, J., Jensen, L., Andersen, K., Svenning, M. and Bonde, T. (2005). Non-target effects of the microbial control agents *Pseudomonas fluorescens* DR54 and *Clonostachys rosea* IK726 in soils cropped with barley followed by sugar beet: a greenhouse assessment. *Soil Biol. Biochem.* 37(12): 2225–2239.

Jones, J. D. G. and Dangl, J. L. (2006). The plant immune system. *Nature* 444(7117): 323–329.

Kamou, N. N., Cazorla, F., Kandylas, G. and Lagopodi, A. L. (2020). Induction of defense-related genes in tomato plants after treatments with the biocontrol agents *Pseudomonas chlororaphis* ToZa7 and *Clonostachys rosea* IK726. *Arch. Microbiol.* 202(2): 257–267.

Kamou, N. N., Dubey, M., Tzelepis, G., Menexes, G., Papadakis, E. N., Karlsson, M., Lagopodi, A. L. and Jensen, D. F. (2016). Investigating the compatibility of the biocontrol agent *Clonostachys rosea* IK726 with prodigiosin-producing *Serratia rubidaea* S55 and phenazine-producing *Pseudomonas chlororaphis* ToZa7. *Arch. Microbiol.* 198(4): 369–377.

Karlsson, M., Atanasova, L., Jensen, D. F. and Zeilinger, S. (2018). Necrotrophic mycoparasites and their genomes. In: Heitman, J., Howlett, B., Crous, P., Stukenbrock, E., James, T., Gow, N. (Eds) *The Fungal Kingdom*. Washington, DC: ASM Press, pp. 1005–1026.

Karlsson, M., Durling, M. B., Choi, J., Kosawang, C., Lackner, G., Tzelepis, G. D., Nygren, K., Dubey, M. K., Kamou, N., Levasseur, A., Zapparata, A., Wang, J., Amby, D. B., Jensen, B., Sarrocco, S., Panteris, E., Lagopodi, A. L., Pöggeler, S., Vannacci, G., Collinge, D. B., Hoffmeister, D., Henrissat, B., Lee, Y. H. and Jensen, D. F. (2015). Insights on the evolution of mycoparasitism from the genome of *Clonostachys rosea*. *Genome Biol. Evol.* 7(2): 465–480.

Keller, N. P., Turner, G. and Bennett, J. W. (2005). Fungal secondary metabolism - From biochemistry to genomics. *Nat. Rev. Microbiol.* 3(12): 937–947.

Keyser, C. A., Jensen, B. and Meyling, N. V. (2016). Dual effects of *Metarhizium* spp. and *Clonostachys rosea* against an insect and a seed-borne pathogen in wheat. *Pest Manag. Sci.* 72(3): 517–526.

Knudsen, I. M. B., Hockenhull, J. and Jensen, D. F. (1995). Biocontrol of seedling diseases of barley and wheat caused by *Fusarium culmorum* and *Bipolaris sorokiniana*: Effects of selected fungal antagonists on growth and yield components. *Plant Pathol.* 44: 467–477.

Knudsen, I. M. B., Hockenhull, J., Jensen, D. F., Gerhardson, B., Hökeberg, M., Tahvonen, R., Teperi, E., Sundheim, L. and Henriksen, B. (1997). Selection of biological control agents for controlling soil and seed-borne diseases in the field. *Eur. J. Plant Pathol.* 103(9): 775–784.

Koch, E., Schmitt, A., Stephan, D., Kromphardt, C., Jahn, M., Krauthausen, H., Forsberg, G., Werner, S., Amein, T., Wright, S. A. I., Tinivella, F., Gullino, M. L., Roberts, S. J., van der Wolf, J. and Groot, S. P. C. (2010). Evaluation of non-chemical seed treatment methods for the control of *Alternaria dauci* and *A. radicina* on carrot seeds. *Eur. J. Plant Pathol.* 127(1): 99–112.

Köhl, J., Kolnaar, R. and Ravensberg, W. J. (2019). Mode of action of microbial biological control agents against plant diseases: relevance beyond efficacy. *Front. Plant Sci.* 10: 845.

Köhl, J., Postma, J., Nicot, P., Ruocco, M. and Blum, B. (2011). Stepwise screening of microorganisms for commercial use in biological control of plant-pathogenic fungi and bacteria. *Biol. Control* 57(1): 1–12.

Kombrink, A. and Thomma, B. P. H. J. (2013). LysM effectors: secreted proteins supporting fungal life. *PLOS Pathog.* 9(12): e1003769.

Kosawang, C., Karlsson, M., Jensen, D. F., Dilokpimol, A. and Collinge, D. B. (2014a). Transcriptomic profiling to identify genes involved in *Fusarium* mycotoxin deoxynivalenol and zearalenone tolerance in the mycoparasitic fungus *Clonostachys rosea*. *BMC Genomics* 15: 55.

Kosawang, C., Karlsson, M., Velez, H., Rasmussen, P. H., Collinge, D. B., Jensen, B. and Jensen, D. F. (2014b). Zearalenone detoxification by zearalenone hydrolase is important for the antagonistic ability of *Clonostachys rosea* against mycotoxigenic *Fusarium graminearum*. *Fungal Biol.* 118(4): 364–373.

Kovalchuk, A. and Driessen, A. J. M. (2010). Phylogenetic analysis of fungal ABC transporters. *BMC Genomics* 11: 177.

Krauss, U., Hoopen, G. M., Hidalgo, E., Martínez, A., Stirrup, T., Arroyo, C., García, J. and Palacios, M. (2006). The effect of cane molasses amendment on biocontrol of frosty pod rot (*Moniliophthora roreri*) and black pod (*Phytophthora* spp.) of cocoa (*Theobroma cacao*) in Panama. *Biol. Control* 39(2): 232–239.

Krauss, U., Hoopen, M., Rees, R., Stirrup, T., Argyle, T., George, A., Arroyo, C., Corrales, E. and Casanoves, F. (2013). Mycoparasitism by *Clonostachys byssicola* and *Clonostachys rosea* on *Trichoderma* spp. from cocoa (*Theobroma cacao*) and implication for the design of mixed biocontrol agents. *Biol. Control* 67(3): 317–327.

Krauss, U., Martinez, A., Hidalgo, E., ten Hoopen, M. and Arroyo, C. (2002). Two-step liquid/solid state scaled-up production of *Clonostachys rosea*. *Mycol. Res.* 106(12): 1449–1454.

Lahlali, R. and Peng, G. (2014). Suppression of clubroot by *Clonostachys rosea* via antibiosis and induced host resistance. *Plant Pathol.* 63(2): 447–455.

Lahoz, E., Contillo, R. and Porrone, F. (2004). Induction of systemic resistance to *Erysiphe orontii* cast in tobacco by application on roots of an isolate of *Gliocladium roseum* bainier. *J. Phytopathol.* 152(8–9): 465–470.

Lee, S. A., Kang, M. J., Kim, T. D. and Park, E. J. (2020). First report of *Clonostachys rosea* causing root rot of *Gastrodia elata* in Korea. *Plant Dis.* 104(11): 3069.

Li, G. Q., Huang, H. C., Kokko, E. G. and Acharya, S. N. (2002). Ultrastructural study of mycoparasitism of *Gliocladium roseum* on *Botrytis cinerea*. *Bot. Bull. Acad. Sin.* 43: 211–218.

Li, J., Yang, J. K., Huang, X. W. and Zhang, K. (2006). Purification and characterization of an extracellular serine protease from *Clonostachys rosea* and its potential as a pathogenic factor. *Process Biochem.* 41(4): 925–929.

Liu, S., Chang, Y. W., Hu, X. J., Gong, X., Di, Y., Dong, J. and Hao, X. (2016a). Draft genome sequence of fungus *Clonostachys rosea* strain YKD0085. *Genome Announc.* 4(3): e00538-00516.

Liu, J. Y., Li, S. D. and Sun, M. H. (2016b). Transaldolase gene *Tal67* enhances the biocontrol activity of *Clonostachys rosea* 67-1 against *Sclerotinia sclerotiorum*. *Biochem. Biophys. Res. Commun.* 474(3): 503–508.

Lübeck, M., Knudsen, I. M. B., Jensen, B., Thrane, U., Janvier, C. and Jensen, D. F. (2002). GUS and GFP transformation of the biocontrol strain *Clonostachys rosea* IK726 and the use of these marker genes in ecological studies. *Mycol. Res.* 106(7): 815–826.

Lv, B., Jiang, N., Hasan, R., Chen, Y., Sun, M. and Li, S. (2020). Cell wall biogenesis protein phosphatase CrSsd1 is required for conidiation, cell wall integrity, and mycoparasitism in *Clonostachys rosea*. *Front. Microbiol.* 11: 1640.

Lysøe, E., Dees, M. W. and Brurberg, M. B. (2017). A three-way transcriptomic interaction study of a biocontrol agent (*Clonostachys rosea*), a fungal pathogen (*Helminthosporium solani*), and a potato host (*Solanum tuberosum*). *Mol. Plant. Microbe Interact.* 30(8): 646–655.

Maillard, F., Andrews, E., Moran, M., Kennedy, P. G., Van Bloem, S. J. and Schilling, J. S. (2020). Stem-inhabiting fungal communities differ between intact and snapped trees after hurricane Maria in a Puerto Rican tropical dry forest. *For. Ecol. Manag.* 475: 118350.

Makkonen, R. and Pohjakallio, O. (1960). On the parasites attacking the sclerotia of some fungi pathogenic to higher plants and on the resistance of these sclerotia to their parasites. *Acta Agri. Scand.* 10(2–3): 105–126.

Mamarabadi, M., Jensen, B., Jensen, D. F. and Lübeck, M. (2008a). Real-time RT-PCR expression analysis of chitinase and endoglucanase genes in the three-way interaction between the biocontrol strain *Clonostachys rosea* IK726, *Botrytis cinerea* and strawberry. *FEMS Microbiol. Lett.* 285(1): 101–110.

Mamarabadi, M., Jensen, B. and Lübeck, M. (2008b). Three endochitinase-encoding genes identified in the biocontrol fungus *Clonostachys rosea* are differentially expressed. *Curr. Genet.* 54(2): 57–70.

Mamarabadi, M., Jensen, D. F. and Lübeck, M. (2009). An N-acetyl-beta-D-glucosaminidase gene, *cr-nag1*, from the biocontrol agent *Clonostachys rosea* is up-regulated in antagonistic interactions with *Fusarium culmorum*. *Mycol. Res.* 113(1): 33–43.

Møller, K., Jensen, B., Andersen, H. P., Stryhn, H., Hockenhull, J. (2003). Biocontrol of *Pythium tracheiphilum* in Chinese cabbage by *Clonostachys rosea* under field conditions. *Biocont. Sci. Technol.* 13: 171–182.

Moraga-Suazo, P., Sanfuentes, E. and Le-Feuvre, R. (2016). Induced systemic resistance triggered by *Clonostachys rosea* against *Fusarium circinatum* in *Pinus radiata*. *Forest Res.* 5: 174.

Morandi, M. A. B., Sutton, J. C. and Maffia, L. A. (2000). Effects of host and microbial factors on development of *Clonostachys rosea* and control of *Botrytis cinerea* in rose. *Eur. J. Plant Pathol.* 106(5): 439–448.

Moreira, G. M., Abreu, L. M., Carvalho, V. G., Schroers, H. and Pfenning, L. H. (2016). Multilocus phylogeny of *Clonostachys* subgenus *Bionectria* from Brazil and description of *Clonostachys chloroleuca* sp nov. *Mycolog. Prog.* 15(10–11): 1031–1039.

Mouekouba, L. D. O., Zhang, L. L., Guan, X., Chen, X., Chen, H., Zhang, J., Zhang, J., Li, J., Yang, Y. and Wang, A. (2014). Analysis of *Clonostachys rosea*-induced resistance to tomato gray mold disease in tomato leaves. *PLoS ONE* 9(7): e102690.

Mueller, J. D. and Sinclair, J. B. (1986). Occurrence and role of Gliocladium roseum in field-grown soybeans in Illinois. Trans. Brit. Mycol. Soc. 86(4): 677–680.

Muñoz, G., Agosin, E., Cotoras, M., San Martin, R. and Volpe, D. (1995). Comparison of aerial and submerged spore properties for *Trichoderma harzianum*. *FEMS Microbiol. Lett.* 125: 63–69.

Muñoz, I. V., Sarrocco, S., Malfatti, L., Baroncelli, R. and Vannacci, G. (2019). CRISPR-Cas for fungal genome editing: a new tool for the management of plant diseases. *Front. Plant Sci.* 10: 135.

Nechet, K. L., Vilela, E. S. D., Heck, D. W., Terao, D. and Halfeld-Vieira, B. A. (2017). Effect of increased UV-B radiation on biological control of the gray mold by *Clonostachys rosea* and on the expression of strawberry defense-related enzymes. *Australas. Plant Pathol.* 46(2): 107–113.

Nobre, S. A. M., Maffia, L. A., Mizubuti, E. S. G., Cota, L. V. and Dias, A. P. S. (2005). Selection of *Clonostachys rosea* isolates from Brazilian ecosystems effective in controlling *Botrytis cinerea*. *Biol. Control* 34(2): 132–143.

Nygren, K., Dubey, M., Zapparata, A., Iqbal, M., Tzelepis, G. D., Durling, M. B., Jensen, D. F. and Karlsson, M. (2018). The mycoparasitic fungus *Clonostachys rosea* responds with both common and specific gene expression during interspecific interactions with fungal prey. *Evol. Appl.* 11(6): 931–949.

Osbourn, A. (2010). Secondary metabolic gene clusters: evolutionary toolkits for chemical innovation. *Trends Genet.* 26(10): 449–457.

Pachenari, A. and Dix, N. J. (1980). Production of toxins and wall degrading enzymes by *Gliocladium roseum*. *Trans. Brit. Mycol. Soc.* 74(3): 561–566.

Peng, G., McGregor, L., Lahlali, R., Gossen, B. D., Hwang, S. F., Adhikari, K. K., Strelkov, S. E. and McDonald, M. R. (2011). Potential biological control of clubroot on canola and crucifer vegetable crops. *Plant Pathol.* 60(3): 566–574.

Peng, G., Sutton, J. C. and Kevan, P. G. (1992). Effectiveness of honey-bees for applying the biocontrol agent *Gliocladium roseum* to strawberry flowers to suppress *Botrytis cinerea*. *Can. J. Plant Pathol.* 14(2): 117–129.

Ravnskov, S., Jensen, B., Knudsen, I. M. B., Bødker, L., Funck Jensen, D., Karliński, L. and Larsen, J. (2006). Soil inoculation with the biocontrol agent *Clonostachys rosea* and the mycorrhizal fungus *Glomus intraradices* results in mutual inhibition, plant growth promotion and alteration of soil microbial communities. *Soil Biol. Biochem.* 38(12): 3453–3462.

Roberti, R., Badiali, F., Pisi, A., Veronesi, A., Pancaldi, D. and Cesari, A. (2006). Sensitivity of *Clonostachys rosea* and *Trichoderma* spp. as potential biocontrol agents to pesticides. *J. Phytopathol.* 154(2): 100–109.

Roberti, R., Flori, P., Pisi, A., et al. (2000). Evaluation of biological seed treatment of wheat for the control of seed-borne *Fusarium culmorum*. *J. Plant Dis. Protect.* 107: 484–493.

Roberti, R., Veronesi, A., Cesari, A., Cascone, A., Di Berardino, I., Bertini, L. and Caruso, C. (2008). Induction of PR proteins and resistance by the biocontrol agent *Clonostachys rosea* in wheat plants infected with *Fusarium culmorum*. *Plant Sci.* 175(3): 339–347.

Rodriguez, M. A., Cabrera, G., Gozzo, F. C., Eberlin, M. N. and Godeas, A. (2011). *Clonostachys rosea* BAFC3874 as a *Sclerotinia sclerotiorum* antagonist: mechanisms involved and potential as a biocontrol agent. *J. Appl. Microbiol.* 110(5): 1177–1186.

Rojas, E. C., Jensen, B., Jorgensen, H. J. L., Latz, M. A. C., Esteban, P., Ding, Y. and Collinge, D. B. (2020). Selection of fungal endophytes with biocontrol potential against Fusarium head blight in wheat. *Biol. Control* 144: 104222.

Romero-Contreras, Y. J., Ramirez-Valdespino, C. A., Guzman-Guzman, P., Macías-Segoviano, J. I., Villagómez-Castro, J. C. and Olmedo-Monfil, V. (2019). Tal6 from *Trichoderma atroviride* is a LysM effector involved in mycoparasitism and plant association. *Front. Microbiol.* 10: 2231.

Rossman, A. Y., Seifert, K. A., Samuels, G. J., Minnis, A. M., Schroers, H. J., Lombard, L., Crous, P. W., Põldmaa, K., Cannon, P. F., Summerbell, R. C., Geiser, D. M., Zhuang, W. Y., Hirooka, Y., Herrera, C., Salgado-Salazar, C. and Chaverri, P. (2013). Genera in *Bionectriaceae*, *Hypocreaceae*, and *Nectriaceae* (*Hypocreales*) proposed for acceptance or rejection. *IMA Fungus* 4(1): 41–51.

Salamone, A., Gundersen, B. and Inglis, D. (2018). *Clonostachys rosea*, a potential biological control agent for *Rhizoctonia solani* AG-3 causing black scurf on potato. *Biocont. Sci. Technol.* 28: 895–900.

Samils, N., Karlsson, M., Assefa, T. and Jensen, D. F. (2016). Biological control against postharvest diseases on potato tubers. *IOBC WPRS Bull.* 115: 241–242.

Saraiva, R. M., Borges, A. V., Borel, F. C., et al. (2020). Compounds produced by *Clonostachys rosea* deleterious to *Botrytis cinerea*. *Braz. J. Agr.* 95: 34–47.

Saraiva, R. M., Czymmek, K. J., Borges, A. V., Caires, N. P. and Maffia, L. A. (2015). Confocal microscopy study to understand *Clonostachys rosea* and *Botrytis cinerea* interactions in tomato plants. *Biocont. Sci. Technol.* 25: 56–71.

Schroers, H. J. (2001). A monograph of *Bionectria* (Ascomycota, Hypocreales, Bionectriaceae) and its *Clonostachys* anamorphs. *Stud. Mycol.* 46: 1–214.

Schroers, H. J., Samuels, G. J., Seifert, K. A. and Gams, W. (1999). Classification of the mycoparasite *Gliocladium roseum* in *Clonostachys* as *C. rosea*, its relationship to *Bionectria ochroleuca*, and notes on other *Gliocladium*-like fungi. *Mycologia* 91: 365–385.

Shigo, A. L. (1958). Fungi isolated from oak-wilt trees and their effects on *Ceratocystis fagacearum*. *Mycologia* 50(5): 757–769.

Stasz, T. E., Harman, G. E. and Weeden, N. F. (1988). Protoplast preparation and fusion in two biocontrol strains of *Trichoderma harzianum*. *Mycologia* 80(2): 141–150.

Steinmetz, J. and Schönbeck, F. (1994). Conifer bark as growth medium and carrier for *Trichoderma harzianum* and *Gliocladium roseum* to control *Pythium ultimum* on pea. *J. Plant Dis. Protect.* 101: 200–211.

Stenberg, J., Sundh, I., Becher, P., Björkman, C., Dubey, M., Egan, P., Friberg, H., Gil, J., Jensen, D. F., Jonsson, M., Karlsson, M., Khalil, S., Ninkovic, V., Rehermann del Rio, G., Vetukuri, R. and Viketoft, M. (2021). When is it biological control? A framework of definitions, mechanisms, and classifications. *J. Pest Sci.* 94: 665–676.

Sun, M. H., Chen, Y. M., Liu, J. F., Li, S. D. and Ma, G. Z. (2014). Effects of culture conditions on spore types of *Clonostachys rosea* 67-1 in submerged fermentation. *Lett. Appl. Microbiol.* 58(4): 318–324.

Sun, Z. B., Li, S. D., Ren, Q., Xu, J. L., Lu, X. and Sun, M. H. (2020a). Biology and applications of *Clonostachys rosea*. *J. Appl. Microbiol.* 129(3): 486–495.

Sun, Z. B., Wang, Q., Sun, M. H. and Li, S. D. (2020b). The mitogen-activated protein kinase gene *Crmapk* is involved in *Clonostachys chloroleuca* mycoparasitism. *Mol. Plant. Microbe Interact.* 33(7): 902–910.

Sun, Z. B., Sun, M. H. and Li, S. D. (2015a). Draft genome sequence of mycoparasite *Clonostachys rosea* strain 67-1. *Genome Announc.* 3(3): e00546-00515.

Sun, Z. B., Sun, M. H. and Li, S. D. (2015b). Identification of mycoparasitism-related genes in *Clonostachys rosea* 67-1 active against *Sclerotinia sclerotiorum*. *Sci. Rep.* 5: 18169.

Sun, Z. B., Li, S. D., Zhong, Z. M. and Sun, M. H. (2015c). A perilipin gene from *Clonostachys rosea* f. Catenulata HL-1-1 is related to sclerotial parasitism. *Int. J. Mol. Sci.* 16(3): 5347–5362.

Sun, Z. B., Sun, M. H., Zhou, M. and Li, S. D. (2017). Transformation of the endochitinase gene *Chi67-1* in *Clonostachys rosea* 67-1 increases its biocontrol activity against *Sclerotinia sclerotiorum*. *AMB Express* 7(1): 1.

Sun, Z. B., Wang, Q., Sun, M. H. and Li, S. D. (2019). The heat shock protein 70 gene is involved for colony morphology, sporulation and mycoparasitism of *Clonostachys rosea*. *FEMS Microbiol. Lett.* 366(15): fnz188.

Sutton, J. C. and Peng, G. (1993). Biocontrol of *Botrytis cinerea* in strawberry leaves. *Phytopathology* 83(6): 615–621.

Sutton, J. C., Li, D. W., Peng, G., Yu, H., Zhang, P. and Valdebenito-Sanhueza, R. M. (1997). *Gliocladium roseum* - A versatile adversary of *Botrytis cinerea* in crops. *Plant Dis.* 81(4): 316–328.

Sutton, J. C., Liu, W., Huang, R. and Owen-Going, N. (2002). Ability of *Clonostachys rosea* to establish and suppress sporulation potential of *Botrytis cinerea* in deleafed stems of hydroponic greenhouse tomatoes. *Sci. Technol.* 12(4): 413–425.

Takahashi-Ando, N., Kimura, M., Kakeya, H., Osada, H. and Yamaguchi, I. (2002). A novel lactonohydrolase responsible for the detoxification of zearalenone: enzyme purification and gene cloning. *Biochem. J.* 365(1): 1–6.

Tan, J., De Zutter, N., De Saeger, S., De Boevre, M., Tran, T. M., van der Lee, T., Waalwijk, C., Willems, A., Vandamme, P., Ameye, M. and Audenaert, K. (2021). Presence of the weakly pathogenic *Fusarium poae* in the fusarium head blight disease complex hampers biocontrol and chemical control of the virulent *Fusarium graminearum* pathogen. *Front. Plant Sci.* 12: 641890.

Taylor, A. G., Min, T.-G., Harman, G. E. and Jin, X. (1991). Liquid coating for the application of biological seed treatments of *Trichoderma harzianum*. *Biol. Control* 1(1): 16–22.

Teperi, E., Keskinen, M., Ketoja, E. and Tahvonen, R. (1998). Screening for fungal antagonists of seedborne *Fusarium culmorum* on wheat using *in vivo* tests. *Eur. J. Plant Pathol.* 104(3): 243–251.

Theron, D. J. and Holz, G. (1991). Dry rot of potatoes caused by *Gliocladium roseum*. *Plant Pathol.* 40(2): 302–305.

Thines, E., Eilbert, F., Anke, H. and Sterner, O. (1998). Glisoprenins C, D and E, new inhibitors of appressorium formation in *Magnaporthe grisea*, from cultures of *Gliocladium roseum*. 1. Production and Biological Activities. *J. Antibiot.* 51(2): 117–122.

Tzelepis, G., Dubey, M., Jensen, D. F. and Karlsson, M. (2015). Identifying glycoside hydrolase family 18 genes in the mycoparasitic fungal species *Clonostachys rosea*. *Microbiology (Reading)* 161(7): 1407–1419.

Utermark, J. and Karlovsky, P. (2007). Role of zearalenone lactonase in protection of *Gliocladium roseum* from fungitoxic effects of the mycotoxin zearalenone. *Appl. Environ. Microbiol.* 73(2): 637–642.

Van Delm, T., Van Beneden, S., Mommaerts, V., Melis, P., Stoffels, K., Wäckers, F. and Baets, W. (2015). Control of *Botrytis cinerea* in strawberries with *Gliocladium catenulatum* vectored by bumblebees. *J. Berry Res.* 5(1): 23–28.

Verdejo-Lucas, S., Ornat, C., Sorribas, F. J. and Stchiegel, A. (2002). Species of root-knot nematodes and fungal egg parasites recovered from vegetables in Almeria and Barcelona, Spain. *J. Nematol.* 34(4): 405–408.

Walker, J. A. and Maude, R. B. (1975). Natural occurrence and growth of *Gliocladium roseum* on mycelium and sclerotia of *Botrytis allii*. *Trans. Brit. Mycol. Soc.* 65: 335–338.

Wang, Q. Y., Chen, X. L., Chai, X. F., Xue, D., Zheng, W., Shi, Y. and Wang, A. (2019). The involvement of jasmonic acid, ethylene, and salicylic acid in the signaling pathway of *Clonostachys rosea*-induced resistance to gray mold disease in tomato. *Phytopathology* 109(7): 1102–1114.

Wang, H., Dong, Y., Liao, W., Zhang, X., Wang, Q., Li, G., Xu, J. R. and Liu, H. (2021). High-quality genome resource of *Clonostachys rosea* strain CanS41 by Oxford nanopore long-read sequencing. *Plant Dis.* doi: 10.1094/PDIS-12-20-2615-A.

Wilson, A., Cuddy, W. S., Park, R. F., Harm, G. F. S., Priest, M. J., Bailey, J. and Moffitt, M. C. (2020). Investigating hyperparasites as potential biological control agents of rust pathogens on cereal crops. *Australas. Plant Pathol.* 49(3): 231–238.

Wösten, H. A. (2001). Hydrophobins: multipurpose proteins. *Annu. Rev. Microbiol.* 55: 625–646.

Xu, X. M., Jeffries, P., Pautasso, M. and Jeger, M. J. (2011a). Combined use of biocontrol agents to manage plant diseases in theory and practice. *Phytopathology* 101(9): 1024–1031.

Xu, X. M., Jeffries, P., Pautasso, M. and Jeger, M. J. (2011b). A numerical study of combined use of two biocontrol agents with different biocontrol mechanisms in controlling foliar pathogens. *Phytopathology* 101(9): 1032–1044.

Xue, A. G. (2003). Biological control of pathogens causing root rot complex in field pea using *Clonostachys rosea* strain ACM941. *Phytopathology* 93(3): 329–335.

Xue, A. G., Chen, Y. H., Voldeng, H. D., Fedak, G., Savard, M. E., Längle, T., Zhang, J. and Harman, G. E. (2014). Concentration and cultivar effects on efficacy of CLO-1 biofungicide in controlling fusarium head blight of wheat. *Biol. Control* 73: 2–7.

Xue, A. G., Voldeng, H. D., Savard, M. E., Fedak, G., Tian, X. and Hsiang, T. (2009). Biological control of fusarium head blight of wheat with *Clonostachys rosea* strain ACM941. *Can. J. Plant Pathol.* 31(2): 169–179.

You, Y. H., Park, J. M., Seo, Y. G., Lee, W., Kang, M. S. and Kim, J. G. (2017). Distribution, characterization, and diversity of the endophytic fungal communities on Korean seacoasts showing contrasting geographic conditions. *Mycobiology* 45(3): 150–159.

Yu, H. and Sutton, J. C. (1997a). Effectiveness of bumblebees and honeybees for delivering inoculum of *Gliocladium roseum* to raspberry flowers to control *Botrytis cinerea*. *Biol. Control* 10(2): 113–122.

Yu, H. and Sutton, J. C. (1997b). Morphological development and interactions of *Gliocladium roseum* and *Botrytis cinerea* in raspberry. *Can. J. Plant Pathol.* 19(3): 237–246.

Zhai, M. M., Qi, F. M., Li, J., Jiang, C. X., Hou, Y., Shi, Y. P., Di, D. L., Zhang, J. W. and Wu, Q. X. (2016). Isolation of secondary metabolites from the soil-derived fungus *Clonostachys rosea* YRS-06, a biological control agent, and evaluation of antibacterial activity. *J. Agric. Food Chem.* 64(11): 2298–2306.

Zhang, Y. Y., Gao, X., Liu, J. H. and Ge, Y. (2015). Pilot production of *Clonostachys rosea* conidia in a solid-state fermentor optimized using response surface methodology. *Eng. Life Sci.* 15(8): 772–778.

Zhang, Y. Y., Liu, J. H., Zhou, Ym, Zhang, Y. Y., Liu, Y., Gong, T. Y. and Wang, J. (2013). A new two-phase kinetic model of sporulation of *Clonostachys rosea* in a new solid-state fermentation reactor. *Process Biochem.* 48(8): 1119–1125.

Zou, C. G., Xu, Y. F., Liu, W. J., Zhou, W., Tao, N., Tu, H. H., Huang, X. W., Yang, J. K. and Zhang, K. Q. (2010). Expression of a serine protease gene *prC* is up-regulated by oxidative stress in the fungus *Clonostachys rosea*: implications for fungal survival. *PLoS ONE* 5(10): e13386.

Chapter 13

Bacteriophages to control plant diseases

Manoj Choudhary and Mathews Paret, University of Florida and North Florida Research and Education Center, IFAS, University of Florida, USA; Aleksa Obradović, University of Belgrade, Serbia; Katarina Gašić, Institute for Plant Protection and Environment, Serbia; and Jeffrey B. Jones, University of Florida, USA

1 Introduction

Diseases of agricultural crops are major constraints in increasing crop production. Throughout early human civilization, epidemics, mainly due to plant pathogens, were common and plant disease management approaches were extremely limited. Even today, globally 20–30% of food production is lost due to disease annually (He et al., 2016). Plant diseases are caused by fungi, bacteria, viruses, and virus-like organisms. Among these, more than 200 are caused by plant pathogenic bacteria (phytobacteria) (Kering et al., 2019). The main challenges for management of plant pathogenic bacteria result from the high degree of bacterial diversity, the lack of durable plant resistance, the ability of bacteria to multiply exponentially to reach high populations in a relatively

http://dx.doi.org/10.19103/AS.2021.0093.18

short period in a conducive environment, and the lack of effective chemical control (Jones et al., 2012).

Biological control is the use of natural antagonists of phytopathogens to control plant diseases. Among these agents, bacteriophages (phages), also referred to as phages (from the Greek word 'phagein', which means 'to devour'), are viruses with the ability to infect and kill bacteria, and appear to be a promising and re-emerging approach for the management of plant bacterial diseases (Jones et al., 2007; Sieiro et al., 2020; Kering et al., 2019).

2 Advantages of bacteriophages for biocontrol over other strategies

Unlike chemical bactericides, bacteriophages occur naturally in the environment (diseased plants, water, soil etc.). The use of lytic phages or their products for the treatment of bacterial diseases, known as 'phage therapy', has several potential advantages over other biological and chemical control strategies. 1) Phages are self-replicating and grow rapidly and exponentially in the presence of host bacteria; thus, a small quantity is required to control disease (Vandamme and Mortelmans, 2019). 2) Unlike copper-based bactericides and synthetic chemicals, phages do not have a toxic effect on the environment. 3) Phages are highly specific and attack only targeted bacteria, without affecting nontarget organisms. Phages generally have very narrow host ranges, only infecting a particular bacterial species or strain. Phages can thus be used against a specific bacterial species without affecting beneficial microbiota in the environment (Liu et al., 2020). Thus, the application of phages helps in maintaining higher beneficial or desired bacterial populations by suppressing the targeted undesired or harmful bacteria; for example, the *Bradyrhizobium japonicum* strain USDA 469-specific phage was used to suppress indigenous inferior root-nodulating bacteria (rhizobia) in the rhizosphere, in order to promote growth of superior rhizobia to enhance nitrogen fixation in soybean (Basit et al., 1992). 4) Phages can be used to target copper- and antibiotic-resistant bacteria, thus eliminating resistant bacterial strains (Hathaway et al., 2017; Abrahamian et al., 2019). 5) Phages can be used as a tank mix with the majority of chemical bactericides and other chemical or biological crop protection products applied for control of pest and diseases, without significant effects on their efficacy (Lang et al., 2007; Gašić et al., 2011). 6) Phages do not target eukaryotic cells (Greer, 2005), thus they can be used to treat fresh-cut vegetables and fruit against post-harvest bacterial diseases (Zaccardelli et al., 1992) and human pathogens (Leverentz et al., 2001). The demand for organic food (free from chemicals and preservatives) is increasing day by day and phages are well suited as biopesticides that can be used in the field for eliminating bacterial pathogens.

3 Early research on bacteriophages

Hankin (1896) described a substance in the waters of the Ganges and Yamuna rivers in India that could pass through a very fine porcelain filter and had antibacterial properties against cholera. The first scientific confirmation was in 1915, when the science of virology was in its infancy. Frederick Twort observed mysterious 'glassy and transparent' cleared areas in bacterial cultures, formed by the lysing of bacteria, and hypothesized that the causal agent was either an ultramicroscopic virus or a bacterial enzyme (Twort, 1915). Two years later in 1917, Felix D'Herelle, working independently, observed a similar finding and described it as a new type of virus that infects bacteria, which he dubbed a bacteriophage (phage) (D'Herelle, 1917).

The idea of using bacteriophages to control plant diseases was inspired from work done in the medical field. In the early 1920s, Mallmann and Hemstreet (1924) reported that the use of filtrate collected from decomposing cabbage inhibits growth of the 'cabbage-rot organism' *Xanthomonas campestris* pv. *campestris*. A year later, Kotila and Coons (1925) isolated phages from various sources, such as rotting carrots, soil, and river water that were specific to *Pectobacterium carotovorum* or *Agrobacterium tumefaciens*. In bioassays where *P. atrosepticum* and *P. carotovorum* ssp. *carotovorum* were applied alone or in combination with bacterial-specific phages, respectively, to potato tuber and carrot root slices, soft rot was only observed where bacteria alone were applied (Coons and Kotila, 1925; Kotila and Coons, 1925).

Following the success in controlling bacteria under controlled conditions, the first field trial was done by Thomas (1935) to test the efficacy of phages for controlling Stewart's wilt disease caused by *Pantoea stewartii*. Treatment of seed with phages was quite effective and resulted in a reduction in disease incidence from 18% (untreated) to 1.4% (phage). The first three decades of phage research for disease control were considered the golden period. Soon after the discovery of antibiotics, they were determined to be easier to use than phages and were also more effective for disease control. As a result, interest in phages waned in the mid- to the late twentieth century.

Antibiotic use in agriculture began in the 1950s, but due to injudicious and widespread use, bacteria quickly developed resistance to streptomycin, tetracycline, and kasugamycin (Thayer, 1961; Sundin and Wang, 2018). Similarly, copper-based bactericides, which have been used extensively since 1885, resulted in many plant pathogenic bacteria acquiring resistance to copper (Lamichhane et al., 2018). As a result of bacteria acquiring resistance to these bactericides, the public concerns about the use of synthetic chemicals in plant protection, and the demand for non-chemical, sustainable, and eco-friendly alternatives to control bacterial diseases, recent interest has increased significantly in the use of bacteriophages for

Table 1 Application of bacteriophages for management of bacterial diseases since 2000

Pathogen	Disease name with host	Source of phage isolation	Percent disease reduction compared to control	Application type	Number of phage isolates/ strains	Phage conc. (PFU/ ml)* used for green house/ field study	Research findings	References
Acidovorax citrulli	Bacterial fruit blotch of melon	Soil, leaves, and stems of watermelon	66	Soil application	Single	10^9	Phage application after symptom development resulted in 27% disease severity, compared to 80% for the non-treated control. Phage detected in foliar tissue 8 h after addition to soil and leaf tip after 24 h.	Rahimi-Midani and Choi, 2020
Dickeya solani	Potato soft rot	Diseased potato plants and tubers	5	Seed tuber treatment	Single	10^{10}	Phage treatment of infected potato tubers led to a 13% yield increase and 5% reduction in disease incidence.	Adriaenssens et al., 2012
Erwinia amylovora	Fire blight of pear and apple	Soil	54-96	Pre spray with carrier bacteria	Single	10^8 -10^9	Phages reduced infection of detached pear tree blossoms by 84 to 96%, with *Pantoea agglomerans* as a carrier. Infection of potted apple tree blossoms could be reduced by 54% with phage Ea1337-26 and *P. agglomerans.*	Boulé et al., 2011
Pectobacterium atrosepticum	Soft rot of potato	Waste water, Organic waste, and potato skin	64	Seed treatment	Cocktail (ready to use mix of 6)	2.5×10^8	Use of the phage cocktail reduced both disease incidence and disease severity by 61% and 64% respectively.	Carstens et al., 2019

Pectobacterium carotovorum ssp. *carotovorum*, *Pectobacterium wasabia and Dickeya solani*	Soft rot of potato	Potato tuber	80–95	Tuber spray inoculation	Single and mix of 2	10^5	Bioassays with individual single phage reduced severity of soft rot of tubers by 80% on potato slices and by 95% with whole tubers.	Czajkowski et al., 2015
Pectobacterium spp. and *Pantoea* spp.	Soft rot of onion	Potato wastewater	100	Spray and immersion	Cocktail of 4	10^6–10^8	Immersion and spray reduced disease severity, result was uniform over all 4 years.	Zaczek-Moczydłowska et al., 2020
Pseudomonas syringae pv . *actinidiae*	Bacterial blight of kiwifruit	Soil	36	Leaf spray	Cocktail of 4	10^8	Bacteria load over kiwifruit leaves 24 h post-infection reduced by more than 75% in comparison with the untreated plants. Despite this bacterial reduction, no significant differences were observed within one or two phage applications on the leaf surfaces. Phages provide better disease control than copper bactericides.	Flores et al., 2020
Pseudomonas syringae pv. *porri*	Bacterial blight of leek	Soil	84	Immersion and spray	Cocktail of 6	10^7–10^8	Phages decreased symptom development but were not able to completely stop infection. Higher phage concentration and more frequent applications can possibly lead to better disease control. Disease control is not significant over control in different varieties.	Rombouts et al., 2016

(*Continued*)

Table 1 (*Continued*)

Pathogen	Disease name with host	Source of phage isolation	Percent disease reduction compared to control	Application type	Number of phage isolates/ strains	Phage conc. (PFU/ ml)* used for green house/ field study	Research findings	References
Pseudomonas syringae pv. *actinidiae*	Bacterial canker of kiwifruit	Soil, water and leaves of kiwifruit	50	-	Cocktail of 5	10^6	Phages were stable and infective up to 50°C temperature and 4–11 pH. Phages had *in vitro* lytic activity for up to 80 h.	Yu et al., 2015
Pseudomonas tolaasii	Brown blotch of mushroom	Sewage	40% by 7 h	Spray on mushroom buds	Single	10^6	Blotches were completely blocked by co-incubation of phages with pathogen. Lysis of bacterial cells was prominent after 4-5 h of inoculation.	Nguyen et al., 2012
Ralstonia solanacearum	Bacterial wilt of tomato	Surface water	-	Soil drench	Cocktail of 6	6×10^9	Phage cocktail could kill 98% of live bacteria spiked in the sterilized soil at one week after spraying. Treatment efficiencies depended on the timing of application of the phages. Phage therapy should be used soon after early signs of the bacterial wilt.	Wei et al., 2017
Ralstonia solanacearum	Bacterial wilt of tomato	-	Delayed wilting	Pre-treatment soaking	Single	10^6	Tomato plants treated with phage showed no symptoms of bacterial wilt during the experimental period; whereas all untreated plants showed wilting symptom and wilted in 14 days after first symptom. Four months post infection, active phage particles recovered from soil infection.	Fujiwara et al., 2011

Ralstonia solanacearum	Bacterial wilt of tomato	Soil and plants	80	Soil drenching	Cocktail of 2 to 4	10^6 (PFU/g)	Phages decreased the incidence of disease by up to 80% in field experiments. Increase in number of phages in the cocktail reduced disease severity.	Wang et al., 2019
Ralstonia solanacearum	Moko wilt of banana	Soils cultivated with banana and plantain	100% in field	Soil mixture	Cocktail of 2	3.3×10^7 (PFU/g)	Phage cocktail provided 100% protection from Moko disease, while plants treated with single-phage treatment displayed symptoms of the disease.	Ramírez et al., 2020
Ralstonia. solanacearum	Tomato bacterial wilt	Soil	100	Soil drenching	Single	2×10^{11}	Phage suspension added one day before bacterial inoculation controlled disease.	Elhalag et al., 2018
Streptomyces scabies	Common scab of potato	Soil	94.7	24-h immersion	Single	10^9	Seed tubers treated with phage resulted in tuber progeny with reduced levels of surface lesion of scab (1.2%) compared with tubers harvested from non-treated seed tubers (23%).	McKenna et al., 2001
Xanthomonas axonopodis pv. *citrumelo*	Citrus bacterial spot	Citrus nursery	35–48	Spray	Cocktail of 3	3-7x10^8	Phage treatments reduced citrus spot occurrence by 35 and 48% in 2 trials in a commercial citrus nursery. The Control by phages was equal or less effective than copper-mancozeb. Combination of phage with copper did not increase disease control over use of only copper bactericide.	Balogh et al., 2008

(*Continued*)

Table 1 (*Continued*)

Pathogen	Disease name with host	Source of phage isolation	Percent disease reduction compared to control	Application type	Number of phage isolates/ strains	Phage conc. (PFU/ ml)* used for green house/ field study	Research findings	References
Xanthomonas axonopodis pv. *allii*	Xanthomonas leaf blight of onion	Commercial	26– 50	Spray	Cocktail and Mix with ASM	10^5–10^8	Field trial with weekly and biweekly applications of phage reduced disease severity, a result which was similar to treatments of weekly applications of copper-mancozeb.	Lang et al., 2007
Xanthomonas axonopodis pv. *citri*	Asiatic citrus canker of grapefruit	Citrus nursery	87	Spray	Cocktail of 4 phages	10^8–10^{10}	Greenhouse experiments utilizing phage treatment could reduce disease severity by 59% compared to control. However, using a skim milk formulation of phage did not have an effect on disease control.	Balogh et al., 2008
Xanthomonas campestris pv. *vesicatoria*	Bacterial leaf spot of tomato	Tomato/ pepper plant parts (leaves, stems) or soil, field runoff, river and stream waters	12–43	Spray	Cocktail of 6	10^{10}	Greenhouse experiments with formulated phage cocktails could reduce disease severity with formulated phage cocktails providing better protection in comparison to unformulated phage.	Balogh et al., 2003
Xanthomonas campestris pv. *campestris*	Black rot of broccoli	Broccoli plant	16.7–55	Spray	Single	10^7	Only nonpathogenic *Xanthomonas* sp. strain mixed with bacteriophage reduced disease. Application of phage alone had no effect on disease.	Nagai et al., 2017

Xanthomonas citri subsp. *citri*	Citrus Canker	Leaves	86.1	Spoil drench + spray	Phage+ ASM	$5x10^9$	Non formulated phage with ASM showed 42.4–56.9% reduction in disease incidence, compared with 82.1–86.1% reduction incidence by formulated phage combined with ASM.	Ibrahim et al.,2017
Xanthomonas euvesicatoria	Pepper Bacterial Spot	Rhizosphere	50–67	Spray	Single and mixed	10^8	Twice, (before and after bacterial inoculation) phage application was most effective among three only-phage treatments for disease control. However, integrated application of phages (2 h before inoculation) and copper-hydroxide (24 h before inoculation) proved the most effective treatments.	Gašić et al., 2018
Xanthomonas oryzae pv. *oryzae*	Bacterial leaf blight of rice	Floodwater in paddy fields	79	-	Mixture of 3	$5x10^8$	Phage formulation with skim milk significantly reduced the occurrence of BLB to 18.1% compared to 87% in untreated control.	Chae et al., 2014
Xanthomonas oryzae pv. *oryzae*	Bacterial blight of rice	Diseased leaves	95	Post-inoculation by dipping ; Pre-inoculation seed treatment	Single	10^8 -10^9	Disease severity was reduced by 73.9%, 49.6% and 28.9%, respectively, when rice seedlings were sprayed with phage two, four, and six days after pathogen inoculation. Disease severity was reduced by 83.1% if the phage was sprayed before inoculation of pathogen while seed treatment reduced disease severity by 95.4%.	Ogunyemi et al., 2019

(*Continued*)

Table 1 (*Continued*)

Pathogen	Disease name with host	Source of phage isolation	Percent disease reduction compared to control	Application type	Number of phage isolates/ strains	Phage conc. (PFU/ ml)* used for green house/ field study	Research findings	References
Xylella fastidiosa	Pierce's disease of grapevine	Plant and sewage	100	Needle inoculation	Cocktail of 4	10^{10}	Pierce's disease symptoms could be stopped using phage treatment post-infection as well as applying phage prophylactically to grapevines. Post- and pre-treatment with phage significantly reduced bacterial population (10–1000 fold,) compared to non-treated control.	Das et al., 2015

*Plaque forming unit (PFU)/ml is the number of bacteriophage particles present in one ml of suspension that is capable of forming plaque.

disease control. A list of recent uses of bacteriophages for disease control is presented in Table 1.

4 Sources of bacteriophages

Bacteriophages are ubiquitous, being present in all terrestrial and aquatic habitats where their host bacteria are present, controlling those bacterial populations. There are an estimated 10^{31} phage particles in the biosphere (Keen, 2015), an extraordinarily large number that translates to approximately a trillion phages for every grain of sand in the world. Most well-characterized phages belong to the *Myoviridae, Siphoviridae, Podoviridae*, and *Microviridae* families of the *Caudovirales* order (Sieiro et al., 2020; Czajkowski, 2015). Phage particles are produced only from the lysed cells of their host bacteria. A general rule of thumb for isolating bacteriophages is to look where the host bacterium is present. Therefore, phages for soilborne pathogens can be isolated from soil, water sources, and infected plant tissue, and foliar pathogen phages from infected plant tissue. However, Gašić et al. (2011) isolated *X. euvesicatoria*-specific phages from pepper rhizosphere soil, seed, and irrigation water but not from pepper stems, leaves, and fruits. Although diseased plant tissue is often a good source of specific phages, besides tomato/pepper plants, *X. campestris* pv. *vesicatoria*-specific phage was isolated from soil, field run-off, and river and stream waters (Balogh et al., 2003). Most phages were have been isolated from the plant phyllosphere and rhizosphere, and soil near or in the crop production fields, as they have higher phage populations due to continuous availability of the host bacteria (Table 1). However, it is not always the case. Carstens et al. (2019) isolated phages from wastewater and organic waste. In another study, Chae et al. (2014) isolated *X. oryzae* pv. *oryzae* phage from floodwater surrounding rice fields.

Isolation of phages is usually simple and does not require sophisticated protocols. It is recommended to use an overnight enrichment procedure by mixing an environmental sample suspension or washing with the target bacterium suspension in order to increase the number of phages present at low titer. This procedure can be skipped if a high-titer phage population is expected, and one of the protocols described in Table 2 may be followed (Carstens et al., 2019; Balogh, 2006; Iriarte et al., 2012; Katznelson, 1937; Hyman, 2019). If the isolation results in plaque formation, the procedure continues with the phage purification and multiplication steps.

5 Bacteriophage-host interactions

Bacteriophages are completely dependent on host cells for their multiplication and utilize host cell components for replication. Bacteriophages mainly have

Table 2 Method for isolating bacteriophages

Method	Description	Advantage	Disadvantage
Spot testing	Small drop is plated on bacterial lawn on solid media. Zone of lysis suggests presence of phage.	Simple and easy to use. More than one phage can be tested in a single plate.	Solid media affect growth of host bacteria. Phage binding on solid media may result in smaller lysis zone.
Plaque testing	Mixture of phage filtrate with host bacteria is spread on plate and overlaid by soft agar. Plaque formations indicate presence of phage.	Lytic vs. temperate life cycle can be determined by plaque appearance. Plaque size may suggest phage size due to diffusion effects (larger phages diffuse more slowly, etc.).	Only one phage can be tested in a single plate. Diffusion of phage in agar influences plaque size and phage efficacy.
Culture lysis	Phage filtrate is added to broth culture of host bacteria and incubated. Bacterial cell lysis by phage is indicated by loss of turbidity. Live cells can be monitored by using of metabolic dye like tetrazolium.	Useful with bacteria that grow slower on solid media. Phage efficacy is monitored continually over time. Live cell monitoring provides more accurate results.	Cell debris from cells lysed during early infections may bind to and inactivate free phages, interfering with later infections. Fast-evolving bacterial mutant to phage can lead to false negative result as mutants maintain population. Purification of phage needs to be done by spot or plaque method.

three types of life cycles: lytic, temperate (lysogenic), and a third life cycle termed 'chronic cycle' (Howard-Varona et al., 2017), all of which are depicted in Fig. 1. The lytic cycle of phages is responsible for the control of lysis of bacterial cells; therefore, these are also called virulent phages (Vu and Oh, 2020). The infection cycle of bacteriophages begins with recognition of bacterial cells, followed by adsorption of bacteriophages with specific receptors located on the cell surface of susceptible bacteria, followed by phage nucleic acid transfer to host cell. Phages multiply in the host cell and make a number of new phage virion particles (Stone et al., 2019). During the lytic stage, phages encode the protein holin and the enzyme endolysin to lyse bacterial cells for release of the phage progenies (Dy et al., 2018). In general, high titers of lytic phages are required against specific bacterial pathogens to improve efficacy and avoid the development of resistance in the bacterial host.

In the lysogenic cycle, the injected phage genomic DNA integrates as a prophage in the bacterial chromosome or plasmid. Prophages multiply along with the bacterial chromosome and are passed on to the daughter cells.

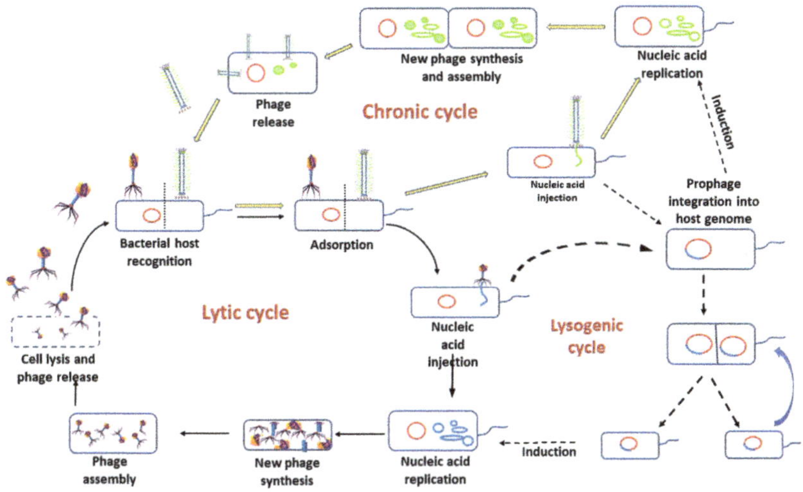

Figure 1 Bacteriophage-host interactions: a generalized life cycle of bacteriophage.

Prophages can switch to a lytic cycle and kill their host spontaneously when bacterial cells are exposed to environmental stresses including antibiotics and certain metabolic conditions of host bacteria (Nanda et al., 2015; Davies et al., 2016).

A third recently defined mechanism, the chronic cycle, of phage-host interaction resulting from chronic infection, has been observed in filamentous phages which are members of the *Inoviridae* family (Yamada, 2013). The chronic cycle of phages does not cause host cell lysis, but establishes a persistent association between the host and phage, producing and releasing phage particles from the growing and dividing host cells (Howard-Varona et al., 2017; Horiuk et al., 2020). In this case, phages form a non-lethal chronic infection with continuous production of phages (Yamada, 2013; Sieiro et al., 2020). *Ralstonia solanacearum* infection with filamentous φRSS1 phage led to abnormal behavior, including reduced turbidity and frequent aggregation in the liquid culture, but more interestingly, φRSS1-infected cells showed enhanced virulence on tobacco (Yamada et al., 2007) and tomato plants (Addy et al., 2012; Yamada, 2013). However, another filamentous phage φRSS0 reduced virulence of *R. solanacearum* (Yamada, 2013).

Ideally, a phage for biocontrol applications should be exclusively lytic and possess a host range which allows productive infection in all strains of the pathogen genus/species to provide complete protection. An effective phage should produce a high number of progeny and diffuse easily in the environment following application (Buttimer et al., 2017). Although highly lytic phages are considered more effective as biological control agents, in a study comparing three lytic phages for bacterial wilt control, phage (φRSL1) of *R. solanacearum*, which was less lytic than the other two phages, provided good disease control (Fujiwara et al., 2011). Furthermore, this study showed that coexistence of bacteria and phages resulted in excellent disease control.

Phages with greater lytic potential under *in vitro* conditions are not necessarily effective in controlling phytopathogens in greenhouse or field conditions. Balogh et al. (2008) reported that three phages of *X. citri* pv. *citri* exhibited lytic activity in overlay plate assays, but two of these phages were unable to lyse their host bacterium on grapefruit leaves and were ineffective in suppressing citrus canker in greenhouse trials. There was no correlation between tomato bacterial leaf spot control efficacy and *in vitro* phage multiplication, while a low correlation was observed between phage persistence and disease control (Balogh et al., 2018).

6 Concerns in using bacteriophages as biocontrol agents

Although phage therapy was shown to be effective in controlling bacterial diseases during the early years, it did not live up to expectations. Phage effectiveness in the field depends on many factors: initial phage concentration, method of application, timing of application on host plant surface, location of host bacterium, and most importantly phage persistence. We can ideate phage replication as equivalent to a 'wild fire' in a forest that expands rapidly till no plants are available. Once the forest is gutted, it stops immediately. Thus, phage applications need to be made many times in the cropping season to reduce phytopathogen population buildup. Disease control is only achieved with a certain density of phage in the environment (Gill and Abedon, 2003). Okabe and Goto (1963) described the use of phage as an ineffective disease control strategy, indicating the limitations of this strategy. According to Kahn et al. (2019), most of the applied phage research has been conducted in academic settings on a few diseases, without input from private companies, given that private companies are wary of getting involved because of profit uncertainties and the fact that the market is too small to offset the cost of development.

Phage commercialization, like for other biocontrol agents, mandates regulatory requirements. Phage registration requires risk assessment pertaining to human health (how toxic is the substance and at what concentration) and environmental effects on terrestrial and aquatic life (EPA, no date (n.d.)). Phage products are exempt from tolerance for residues in the United States of America (USA) (*Code of Federal Regulations*, 2014). However, there are no guidelines in most parts of the world separately for bacteriophages, as they follow the same rules as microorganisms. In the European Union (EU), guidance is in preparation for the registration of bacteriophages, as the available guidance does not address bacteriophages. Phage formulations need to be continuously updated to incorporate new phages against emerging bacterial strains for effective control. However, EU regulations (1107/2009 EC) require a new registration for each change in either number or type of phage population, which requires significant time and added expenses (European Union, 2009; Buttimer et al., 2017). These regulatory requirements lead to a significant delay in the registration and successful application of phages for disease control. The EU guidance in preparation may have provisions to make such changes easier to register.

7 Factors affecting the efficacy of bacteriophages for plant disease control

7.1 Initial concentration of bacteriophages

A minimum number of phages is required to effectively control bacteria. According to Kasman et al. (2002), the effective concentration of phages to kill bacteria depends upon host bacterial density and rate of adsorption. Balogh (2002) demonstrated that a threshold bacteriophage concentration must be present on the tomato phyllosphere to control bacterial spot of tomato disease pathogen *X. perforans*. They found that a minimum concentration of 10^6 plaque-forming units (PFU)/ml was required to control bacterial leaf spot in the greenhouse, and a higher concentration (10^8 PFU/ml) provided statistically similar results, while a concentration of 10^4 PFU/ml was ineffective.

Cabbage black rot bacteria (*X. campestris* pv. *campestris*) biofilm formation is greatly reduced at 10^8 PFU/ml bacteriophage application, in comparison to 10^6 PFU/ml, due to disaggregation of the biofilm by galactose present in phage capsids (Papaianni et al., 2020). Adachi et al. (2012) found that rice seedling blight (*Burkholderia plantarii*) incidence was lower at 10^7 PFU/ml in comparison to 10^6 PFU/ml in stagnant water conditions, but phage concentrations of 10^5-10^8 PFU/ml in circulating water had similar disease control. This indicates that constant agitation of water increases phage interaction with bacteria. Lang et al. (2007) also found no difference in Xanthomonas leaf blight of onion severity at 10^5–10^8 PFU/ml bacteriophage cocktail application.

7.2 Development of resistance to bacteriophages

Bacteria have a high probability of becoming resistant to phages, a problem since the early days of phage therapy. Katznelson (1937) expressed concern about bacteria developing resistance to phages. Bacteria develop resistance by mutation of an essentially required protein or modification of a phage adsorption receptor (Qiao et al., 2010) and/or an altruistic abortive infection (Abi), a complex mechanism in which phage-infected cells commit suicide in order to prevent phage reproduction and provide bacterial host population immunity, similar to a hypersensitive response in plants (Buttimer et al., 2017; Fineran et al., 2009). Another mechanism of resistance is based on bacteria utilizing the CRISPR/CAS system against foreign DNA. The latter system is present in 40% of bacteria (Burstein et al., 2016). This system is also found in the plant pathogen *P. atrosepticum* (Przybilski et al., 2011), *Erwinia amylovora* (Rezzonico et al., 2011), and *X. oryzae* (Semenova et al., 2009). Bacteria developing resistance to phages may also have a fitness cost due to a reduction in virulence, resulting in reduced disease severity. Mutation in type IV pili in *Xanthomonas* spp. and *Xylella fastidiosa* impart resistance to phages, but reduce the virulence of bacteria due to loss of twitching motility (Ahern et al., 2014).

7.3 Strain variation in target bacteria

Bacterial strains that cause plant diseases of various hosts have significant variation, which poses a serious concern for use of phages as a biocontrol agent. This problem is magnified by narrow host-range bacteriophages that only infect a particular strain of a bacterial species. For example, bacteria causing the bacterial leaf spot of tomato and pepper, formerly considered one bacterial species, *X. campestris* pv. *vesicatoria,* is now classified as four different species with each bacterial species consisting of highly diverse strains (Schwartz et al., 2015). For example, in a study by Bouzar et al. (1999), 120 *X. euvesicatoria* strains isolated from various countries in the Caribbean region were typed using 26 bacteriophages, and at least 26 different phage lysis patterns were identified. In another study, soft rot bacterial phages LIMEstone1 and LIMEstone2 were specific to *Dickeya solani* (Adriaenssens et al., 2012), while φD3 and φD5 were broad host-range phages to *D. solani, D. dianthicola, D. zeae, D. dadantii,* and *D. chrysanthemi* (Czajkowski et al., 2015).

7.4 Persistence of bacteriophages in rhizospheres and phyllospheres

Maintaining high populations on the host surface for a sufficient time is an essential requirement for adsorption on bacteria for effective control of the phytopathogen population. This is one of the factors limiting successful phage-based biocontrol. Phytopathogen density reduction is directly proportional to the number of different types of bacteriophages in the phyllosphere or in the rhizosphere that are available for infecting individual bacterial cells (Wang et al., 2019).

Bacteriophage decay in the phyllosphere is mainly due to ultraviolet (UV) radiation, and in varying degrees due to high temperature, high and low pH, and desiccation on host plant surface, and also results from dislodging of phage particles due to rain (Iriarte et al., 2007; Jones et al., 2007). Short-term persistence on the plant leaf surface is a major limiting factor of phage phyllosphere application. UV A & B of sunlight (spectrum 320–400 nm), especially UVB, cause DNA damage (Pfeifer, 1997). UV light is the most detrimental to bacteriophages in the phyllosphere as directly exposed to sunlight on the plant canopy. *X. campestris* pv. *vesicatoria* phage sprayed on tomato leaves at mid-morning (10 AM) when UV light intensity is maximum, declined from 6 to 0 log units in 4 hours. A UV dose of 20 mJ/cm^2 completely eliminated phage on leaf surface (Iriarte et al., 2007). In *in vitro* experiments with *E. amylovora* phage Y2, a five-minute exposure to UV light reduced infective phage population to 0.21% of the initial level (Born et al., 2015), while *D. solani* phages were completely eliminated after ten minutes of UV exposure (Czajkowski et al., 2014).

Phage survival is also affected by pH. Gašić et al. (2018) observed that the *X. euvesicatoria* phage Kφ1 phage virions were completely inactivated at pH values below 2 and above 12, within 24 hours. A similar result was observed by Czajkowski et al. (2014) in *D. solani* phages, which decayed at pH below 2. Considering the effect of temperature on phage survival, Iriarte et al. (2007) observed that 28°C was the optimal temperature for survival in the phyllosphere, while a low temperature supported phage survival and a high temperature was detrimental. Although phage persistence was not affected by many chemicals, copper-mancozeb applied with phage had detrimental effects (Iriarte et al., 2007). Therefore, phage virion decay in the phyllosphere mainly depended on plant canopy and environmental factors.

The rhizosphere is a more hospitable niche compared to the phyllosphere, but it also possesses different types of obstacles for the phage to persist and provide effective disease control. Gill and Abedon (2003) mentioned constraints like pH, moisture levels, the presence of organic matter, and soil type. Different soil types have different properties that affect phage survival. Soils with low moisture content have low phage-diffusion rates (Kleczkowska, 1957; Sykes et al., 1981). Clay soil has better phage survival at higher temperature and low moisture (Kimura et al., 2008). However, clay soil minerals possess positively and negatively charged surfaces to which phage can be adsorbed and trapped on clay particles (Chattopadhyay and Puls, 1999). Phage sorption is also influenced by pH in soil. Sykes et al. (1981) discovered that *Streptomyces* spp. phages survive best in neutral soil The pH and persistance are negligible in acidic soil. Microbiota biofilms also can trap phages in the soil (Storey and Ashbolt, 2001). Nevertheless, many phages are isolated from soil (Vu and Oh, 2020) and can persist several weeks in the soil and rhizosphere (Fujiwara et al., 2011), in spite of these obstacles.

8 Approaches for optimum efficacy of bacteriophage for plant disease control

8.1 Application strategies

Bacteriophages can be applied in various ways in greenhouse or field crops, based upon the nature of the pathogen and the type of plants, in order to enhance direct interaction with bacteria for improved efficacy of disease management. For foliar pathogens, a suspension of bacteriophage is commonly applied by spraying onto the leaf surface (Iriarte et al., 2007; Balogh et al., 2003), whereas soil drenching is done for soil-inhabitant bacterial pathogens (Iriarte et al., 2012). For vascular pathogens, application is via stem injection or infiltration (Das et al., 2015), while seed/tubers are immersed in phage suspension for seed/tuber pathogens (Carstens et al., 2019; Zaczek-Moczydłowska et al., 2020) (Fig. 2).

Phages generally have a narrow host range and can infect only one or a few strains of one pathogenic bacterial species. Therefore, it often requires the use of a phage mixture or cocktail to target diverse bacteria for managing plant diseases. Phage cocktails also provide unique advantages for reducing the chance of development of phage resistance in bacterial pathogens, because it is more difficult for a bacterial pathogen to develop resistance simultaneously to multiple phages. Phage cocktails function by one of these mechanisms: (a) each phage directly kills a group of bacteria; (b) diminishes competitive fitness; or (c) increases beneficial microbes due to reduced pathogen populations (Liu et al., 2020). Taken together, all of these mechanisms of phages either directly

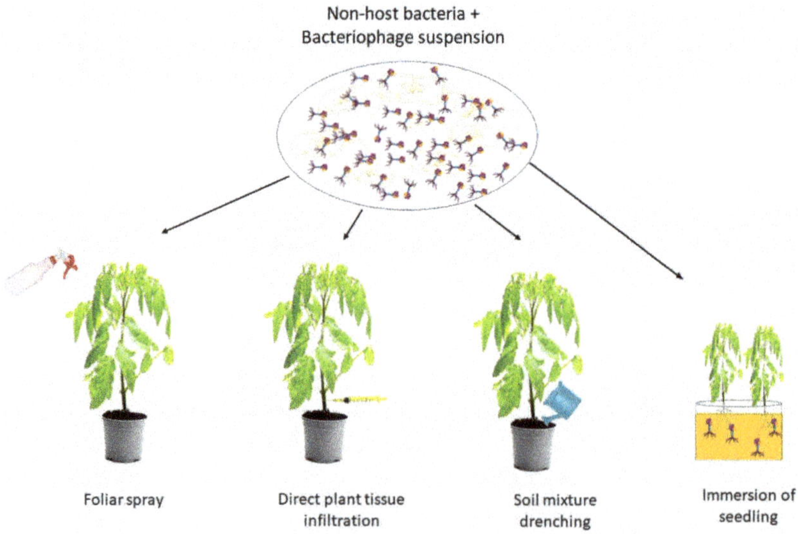

Figure 2 Methods of bacteriophage application in greenhouse and field.

or indirectly dramatically reduce the target bacterial population. Wang et al. (2019) demonstrated that *Ralstonia solanacearum* phage cocktails provided higher efficacy for controlling the bacterial wilt of tomato. They indicated that increasing the types of phage strains in the cocktail from two to four proportionally reduced the disease severity. Use of either two- or three-strain phage cocktails, applied with irrigation water, had higher efficiency in tomato bacterial wilt control as compared to single-phage strain application (Álvarez et al., 2019). Information on a number of research studies relating to the use of phage mixtures to manage diseases is given in Table 1.

Development of resistance to phages can be managed by using strategies developed by Jackson (1989), in which a mixture of host-range wild-type phages is used to lyse phage-resistant bacteria. Flaherty et al. (2000) used a mixture of four phages to reduce disease severity of bacterial leaf spot of tomato to 0.9%, compared to 10.5% in control. In bacterial blight of geranium (*X. campestris* pv. *pelargonii*), disease severity was reduced by 50% or more compared to the control by using a cocktail of five phages daily in foliar sprays (Flaherty et al., 2001). Monitoring the resistance of bacterial strains is essential for the success of phages for long-term disease control; it can be done periodically by isolating bacteria present in the field and testing for sensitivity to phages (Jones et al., 2007; OmniLytics, 2016).

Phages and several compounds have been reported to inhibit biofilm formation in bacteria. This results in enhanced virulent phage efficacy. *E. amylovora* phage L1 capsids produce the enzyme depolymerase that inhibits biofilm formation by degrading extracellular polysaccharide (EPS). Biofilm degradation allows the broad host-range *E. amylovora* phage Y2 to adsorb on bacterial cells for infection (Born et al., 2014). Papaianni et al. (2021) observed a synergistic effect of long fatty acid (LFA) and phage in biofilm eradication of *X. campestris* pv. *campestris*. LFA mimics the diffusible signaling factor (DSF) to induce biofilm dispersal, which makes bacteria more susceptible to phage lysis.

8.2 Strategies for improved persistence in the phyllosphere and rhizosphere

Protection of phages in the phyllosphere from environmental factors (especially UV) for an extended period can be accomplished by applying phages at a suitable time. Most research supports phage application at low intensity of UV light (late evening) to extend survival of viable phages in association with host bacteria, to improve the chance of disease control (Adachi et al., 2012; Balogh et al., 2003, 2018; Iriarte et al., 2007; Jones et al., 2012). Phages survived longer on tomato leaf surfaces during applications in late evening, while with early morning and daytime applications, they lasted only a few hours (Balogh et al., 2003; Iriarte et al., 2007). Evening application of phages was used to control bacterial spot of pepper (Gašić et al., 2018), citrus canker (Ibrahim et al., 2017), and bacterial blight of leek (Rombouts et al., 2016). Iriarte et al. (2012) investigated the effect of different application regimes of *Ralstonia solanacearum* strain RS5 phage mixture of 10^8 PFU/ml on bacterial wilt of tomato. Phages were applied in soil immediately after inoculation (IAI); three days before inoculation(3DBI); 3DBI +IAI; three days after inoculation (3DAI); 3DBI+IAI+3 DAI and IAI+3 DAI. The phage mixture was effective in all combinations except 3 DAI, while single-phage treatment was most effective when applied near the time of inoculation and less effective in 3 DBI, while ineffective at 3 DAI. In contrast to this, Das et al. (2015) reported that Pierce's Disease of grapevines, caused by the vascular pathogen *X. fastidiosa*, was significantly reduced both therapeutically (three weeks after inoculation) and prophylactically (three weeks before inoculation) following stem infiltration with a mixture of four phages.

There is little agreement among researchers on the frequency of application or the interval between repeated applications of phages for optimum disease control. Different levels of control have been reported from different frequencies (single, daily, biweekly) of phage application (Kering et al., 2019), indicating that phage application frequency must be determined for every particular pathosystem. Ideally, phages should be applied with free moisture available on the leaf surface to extend the time period of host bacterium and phage interaction.

Bacteriophage survivial and persistance can be improved by the use of phage-sensitive but non-virulent or antagonistic bacteria as a carrier. These carrier bacteria help in the early establishment of phages in the phyllosphere, and thus, survival when the host bacteria are not present. Bacteriophages applied in a mixture with an avirulent but phage-hosting *R. solanacearum* strain effectively controlled the phage-sensitive strain causing tobacco bacterial wilt (Tanaka et al., 1990). The epiphytic bacteria, *Pantoea agglomerans* strain Eh21-5, were used as a carrier for bacteriophages that target the bacterial pathogen *E. amylovora*, causal agent of fire blight of apple and pear (Boulé et al., 2011), and demonstrated that infection was reduced by 56% compared to the phosphate buffer control. Lehman (2007) applied a phage cocktail with carrier avirulent antagonistic bacteria *P. agglomerans* strains (active component in Bloomtime™) at 40–50% and 80–90% apple bloom. Disease incidence was significantly lower in phage + carrier bacteria in comparison with only carrier antagonistic bacteria. In a study on tomato bacterial leaf spot, Iriarte et al. (2012) applied attenuated mutant *X. perforans* 91-118:ΔOPGH strain prior to phage application, to improve phage persistence on tomato leaves. After seven days, phage population was below the detectable level with only phage application, while phages could be recovered from leaves for up to ten days with prior application with attenuated bacteria.

Another way to improve phage phyllosphere persistence has resulted from the use of a protective formulation to reduce the detrimental effects of UV light. Various compounds have been identified as useful for protecting viruses and microbes from sunlight UV or rain wash-off, including antioxidants and oxidative enzymes, aromatic/heterocyclic amino acids, dyes that are light-adsorbing or reflecting, activated charcoal, starch, flour, alkaline gluten, casein, and lignin-based formulations (Jones et al., 2007). Some of these were tested to improve phage persistence for disease management. In a study by Balogh et al. (2003) to enhance control of bacterial spot of tomato with different formulations of bacteriophages, a bacteriophage mix (Agriphage, OmniLytics Inc., Salt Lake City, UT) was formulated using either (i) PCF : 0.5% pregelatinized cornflour + 0.5% sucrose; (ii) Casecrete: 0.5% Casecrete NH-400 + 0.5% sucrose + 0.25% PCF; or (iii) skim milk: 0.75% powdered skim milk + 0.5% sucrose. These formulations increased phage survival on tomato leaf surface from a few hours to one to two days. Phage population was considerably higher, by approximately log 3, 4 and 5 for PCF, Casecrete, and skim milk formulations, respectively. Although these formulations do not have much effect on disease reduction in the greenhouse, in the field, the Casecrete-formulation reduced disease in three of three trials, while the PCF formulation was effective in one of three trials. Successful use of skim milk and sucrose to improve phage survival and persistence was reported by Obradovic et al. (2004, 2005). The positive effect of skim milk phage formulation was corroborated by Iriarte et al. (2007) for protection of phages under intense UV light, even though the phage population was decreased to much lower levels, but still higher than non-formulated phage. Gašić et al. (2018) also found higher survival of *X. euvesicatoria* phage Kφ 1 mixed with skim milk compared to non-formulated. Chae et al. (2014) had similar results with a skim milk formulation for bacterial leaf blight control in rice. Asiatic citrus canker disease incidence was reduced with formulated phage (0.75% skim milk powder + 0.25% sucrose) in combination with acibenzolar-S-methyl (ASM), which was significantly more effective in reducing disease incidence compared to non-formulation phage (Ibrahim et al., 2017).

Born et al. (2015) used natural extracts from carrot, red pepper, beetroot, casein, and soy peptone in suspension, and purified substances such as astaxanthin, aromatic amino acids, and Tween 80 to minimize the effect of UV on *E. amylovora* phage. All treatments significantly improved phage survival. After a five-minute phage exposure to UV light, 10% carrot or 5% red pepper juice had approximately 2 log higher phage population than the control. Poly-γ-glutamic acid (γ-PGA), a naturally occurring biodegradable biopolymer, improved T2 phage survival under detrimental temperature, pH, and UV conditions *in vitro* (Khalil et al., 2016). Tom et al. (2018) exposed phage to 30 cycles of 302 nm UV light (120 s each) under *in vitro* conditions. This led to increased survival of phage following exposure to UV light. Phage population after 30 UV cycles had a ~50-fold improvement in survival over 10^{-4} killing of the ancestral phage.

8.3 Integration of bacteriophages in disease management

Plant disease management is a holistic approach to manage disease. Bacterial disease control with a single method does not always fetch good results. Bacterial disease management is difficult with the limited number of bactericides available, and is becoming more complicated with development of resistance to antibiotics and copper compounds. Phages can be used with other chemicals and biological control agents to improve efficacy. Bacteriophages can be easily used either as a tank mix or in combination with other chemicals including systemic acquired resistance (SAR) inducers. Obradovic et al. (2005) found that the combination of phage with the SAR inducer ASM not only provided significant tomato bacterial leaf spot control, but also reduced the intensity of hypersensitive response as the SAR reaction was induced by ASM treatment. Using a combination of phage with ASM, citrus canker infection on leaves was reduced from 70.2% in the control to 12.5% (Ibrahim et al., 2017). The severity of bacterial leaf blight of onion was reduced to 50% with ASM+ phage treatment compared to 31% in copper-mancozeb treatment (Lang et al., 2007). Šević et al. (2019) studied the efficacy of pepper bacterial spot phage KΦ1 with biocontrol agents *Bacillus subtilis* strain AAac and QST 713, SAR inducer ASM, a commercial microbial fertilizer (Slavol), copper-based compounds (copper hydroxide, copper oxychloride and mancozeb) and antibiotics (streptomycin sulfate and kasugamycin). In that study, phage alone was not as effective as copper compounds and antibiotics; however, phage combination with copper hydroxide and ASM was most effective and reduced disease severity by 96–98% compared to control. All phage combinations with ASM and copper compounds improved yield and reduced disease severity compared to control. Recently, Abrahamian et al. (2019) reported that a combination of cymoxanil + famoxadone + phosphoric acid + phage significantly improved tomato bacterial leaf spot (*X. perforans*) management compared to the copper-based grower-standard treatment. Phage treatment integrated with other disease management practices is widely used in greenhouses and production fields in Florida as part of a standard integrated management program for tomato bacterial spot (Jones et al., 2007; Obradovic et al., 2008)

9 Bacteriophage therapy challenges that need to be addressed

Bacteriophage application still has several challenges that need to be addressed in order to achieve effective and reliable disease control. Every plant environment is unique and complex; therefore, a more in-depth understanding of the different bacteria-phage systems is required. Although phages may be effective in one crop, they may not be as effective in other pathosystems (Balogh et al., 2018). This requires a variety of experiments, as *in vitro* and greenhouse trials may not fully reflect the harsh conditions of the real environment. To improve survival of phage during long-term storage under ambient temperatures, more superior formulations are needed. There is much variation in how to produce phage for field-level application. The production process needs to be optimized for stability, purity, and quality of phage.

Although available research results demonstrate that various substances can be used to protect phages from UV-induced decay of phages, better protective formulations are still needed. The optimum concentrations and protective effects of these formulations need to be evaluated under field conditions on different crops in large-scale trials. These protective formulations should reduce bacterial population without impeding the lytic cycles of phages.

The method of phage application, timing, phage mixture, and association with protective compounds or amendment with non-host avirulent bacteria as carriers, should be optimized to prevent the emergence of resistant bacteria. The number and type of phages that can be used to make a phage mixture that will be effective against all strains of a pathogenic bacterium need to be ascertained. A large number of phage collections (phage banks) must be developed that are readily available. Although phage integration with other plant protection management practices has been studied in a few crops with encouraging results, more research needs to be done for the prophylactic efficacy and protective use of phages. Genetically engineered phages, with improved UV resistance, extended survival, and increasing host range, may provide a solution for environmental factors limiting phage performance in the future.

In the past two decades, a few phage-based biopesticides have been registered for control of bacterial plant pathogens. A list of registered products with phage concentration is provided in Table 3. Omnilytic, USA, is a leading company for research and commercalization of phages. In recent years, there has been increased interest in the commercialization and use of phages. 'Phagefire' is a joint collaborative project funded by the European Union, for the development of the first-phage-based biopestices to control fire blight in Europe (PhageFire, n.d.). Bayer started a joint venture with Israeli company Ecophage for phage-based biopesticide development (Ecophage, n.d.). There is a need to develop collaborative efforts between public institutions and private companies for the successful use of phage for disease control in farmers' fields.

Table 3 List of phage products available for for bacterial disease control

Product Name	Country	Active ingredient (Bacteriophage against pathogen)	PFU/ml	Use for diseases and crop	Company	Reference
Agriphage	USA	*Xanthomonas campestris* pv. *vesicatoria* and *Pseudomonas syringae* pv. *tomato* phage	1.55×10^{13}	Bacterial leaf spot of tomato and pepper	OmniLytics, Inc	OmniLytics, n.d. b
AgriPhage-CMM	USA and Canada	*Clavibacter michiganensis* subsp. *michiganensis* phage	3.8×10^{12}	Canker of tomato	OmniLytics, Inc	OmniLytics, n.d. a
Agriphage-Fire blight	USA	*Erwinia amylovora* phage	5×10^{12}	Fire blight of apple and pear	OmniLytics, Inc	OmniLytics, 2018
Agriphage-Citrus canker	USA	*Xanthomonas citri* subsp. *citri* phage	5×10^{12}	Canker of citrus	OmniLytics, Inc	OmniLytics, n.d. c
Erwiphage Plus	Hungary	*Erwinia amylovora* phage	2×10^{5}	Fire blight of apple, pear, quinces and loquat	Enviroinvest	Enviroinvest, n.d.
Xylphi-PD	USA	*Xylella fastidiosa* phage	5×10^{9}	Pierce's disease of grape	A&P Inphatec, n.d	A&P Inphatec, n.d
BioLyse-PB	United Kingdom	Soft rot bacteria of potato (*Enterobacteriacea*) phage		Soft rot of potato	APS Biocontrol	APS Biocontrol,n.d.

10 Conclusions

Since the discovery of phages more than a century ago, phage research has been up and down. After being eclipsed by the antibiotics era, phage therapy once again came to the limelight in the late twentieth and the twenty-first centuries, when more and more bacteria were reported to have developed resistance and the popularity of organic food products increased. Bacteriophages may provide environmentally safe and sustainable ways to manage economically important plant bacterial diseases. Today, phages are available to lyse many plant pathogenic bacteria, but their effectiveness is compromised by many factors among which the most limiting one is phage survival, particularly on plant surfaces, due to UV light. Phage application during the evening time enhances phage survival in the phyllosphere. Different protective formulations also help in prolonging phage survival. The use of epiphytic, non-host avirulent bacteria has been implemented for maintaining phage on the host plant and extending phage availability and multiplication to lyse bacteria for effective disease control. Development of resistance by bacteria can be minimized by using phage mixtures instead of single phages. Phages proved to be more effective in combination with copper compounds and SAR inducers, as phage treatments can be easily integrated with already existing disease management practices. With global warming and more awareness by consumers regarding their food choices, the coming years are expected to be boom years for research on phages as biocontrol agents for bacterial diseases.

11 Where to look for further information

The United States Environmental Protection Agency (EPA) provide detailed guidelines for biopesticides including phages registration:

- https://www.epa.gov/pesticide-registration/biopesticide-registration.

A list of all registered bacteriophage in the United States and their technical information can be accessed from the EPA:

- https://www.epa.gov/ingredients-used-pesticide-products/biopesticide-active-ingredients.

12 References

A&P Inphatec Xylphi-PD: Bactericide for use in grapevines, Biologically-based reduction of Pierce's Disease in grapevines. Available at: https://inphatec.com/wp-content/themes/inphatec/pdf/Labels_Label-EPA_approved.pdf. Accessed: 2 August 2021.

Abrahamian, P., Jones, J. B. and Vallad, G. E. (2019). Efficacy of copper and copper alternatives for management of bacterial spot on tomato under transplant and field production, *Crop Protection* 126. Elsevier Ltd, 104919. doi: 10.1016/j.cropro.2019.104919.

Adachi, N., Tsukamoto, S., Inoue, Y. and Azegami, K. (2012). Control of bacterial seedling rot and seedling blight of rice by bacteriophage, *Plant Disease* 96(7), 1033–1036. doi: 10.1094/PDIS-03-11-0232-RE.

Addy, H. S., Askora, A., Kawasaki, T., Fujie, M. and Yamada, T. (2012). The filamentous phage φRSS1 enhances virulence of phytopathogenic *Ralstonia solanacearum* on tomato, *Phytopathology* 102(3). The American Phytopathological Society, 244–251. doi: 10.1094/PHYTO-10-11-0277.

Adriaenssens, E. M., Van Vaerenbergh, J., Vandenheuvel, D., Dunon, V., Ceyssens, P. J., De Proft, M., Kropinski, A. M., Noben, J. P., Maes, M. and Lavigne, R. (2012). T4-related bacteriophage LIMEstone isolates for the control of soft rot on potato caused by "Dickeya solani", *PLoS ONE* Johnson, E. A. (Ed) 7(3). Public Library of Science, e33227. doi: 10.1371/journal.pone.0033227.

Ahern, S. J., Das, M., Bhowmick, T. S., Young, R. and Gonzalez, C. F. (2014). Characterization of novel virulent broad-host-range phages of Xylella fastidiosa and Xanthomonas, *Journal of Bacteriology* 196(2), 459–471. doi: 10.1128/JB.01080-13.

Álvarez, B., López, M. M. and Biosca, E. G. (2019). Biocontrol of the major plant pathogen ralstonia solanacearum in irrigation water and host plants by novel waterborne lytic bacteriophages, *Frontiers in Microbiology* 10, 2813. doi: 10.3389/fmicb.2019.02813.

Balogh, B. (2002). Strategies for improving the efficacy of bacteriophages for controlling bacterial spot of tomato. MS thesis. University of Florida.

Balogh, B., Jones, J. B., Momol, M. T., Olson, S. M., Obradovic, A., King, P. and Jackson, L. E. (2003). Improved efficacy of newly formulated bacteriophages for management of bacterial spot on tomato, *Plant Disease* 87(8), 949–954. doi: 10.1094/PDIS.2003.87.8.949.

Balogh, B. (2006). Characterization and use of bacteriophages associated With citrus bacterial pathogens for disease control. Ph.D. thesis. University of Florida.

Balogh, B., Canteros, B. I., Stall, R. E. and Jones, J. B. (2008). Control of citrus canker and citrus bacterial spot with bacteriophages, *Plant Disease* 92(7), 1048–1052. doi: 10.1094/PDIS-92-7-1048.

Balogh, B., Nga, N. T. T. and Jones, J. B. (2018). Relative level of bacteriophage multiplication in vitro or in phyllosphere may not predict in planta efficacy for controlling bacterial leaf spot on tomato caused by xanthomonas perforans, *Frontiers in Microbiology* 9(Sep), 2176. doi: 10.3389/fmicb.2018.02176.

Basit, H. A., Angle, J. S., Salem, S. and Gewaily, E. M. (1992). Phage coating of soybean seed reduces nodulation by indigenous soil bradyrhizobia, *Canadian Journal of Microbiology* 38(12). NRC Research Press, Ottawa, Canada, 1264–1269. doi: 10.1139/m92-208.

Born, Y., Bosshard, L., Duffy, B., Loessner, M. J. and Fieseler, L. (2015). Protection of Erwinia amylovora bacteriophage Y2 from UV-induced damage by natural compounds, *Bacteriophage* 5(4), e1074330. doi: 10.1080/21597081.2015.1074330.

Born, Y., Fieseler, L., Klumpp, J., Eugster, M. R., Zurfluh, K., Duffy, B. and Loessner, M. J. (2014). The tail-associated depolymerase of erwinia amylovora phage L1 mediates host cell adsorption and enzymatic capsule removal, which can enhance infection by other phage, *Environmental Microbiology* 16(7), 2168–2180. doi: 10.1111/1462-2920.12212.

Boulé, J., Sholberg, P. L., Lehman, S. M., O'Gorman, D. T. and Svircev, A. M. (2011). Isolation and characterization of eight bacteriophages infecting *Erwinia amylovora* and their potential as biological control agents in British Columbia, Canada, *Canadian Journal of Plant Pathology* 33(3). Taylor & Francis, 308–317. doi: 10.1080/07060661.2011.588250.

Bouzar, H., Jones, J. B., Stall, R. E., Louws, F. J., Schneider, M., Rademaker, J. L., de Bruijn, F. J. and Jackson, L. E. (1999). Multiphasic analysis of xanthomonads causing bacterial spot disease on tomato and pepper in the Caribbean and Central America: evidence for common lineages within and between countries, *Phytopathology* 89(4). American Phytopathological Society, 328–335. doi: 10.1094/PHYTO.1999.89.4.328.

Burstein, D., Sun, C. L., Brown, C. T., Sharon, I., Anantharaman, K., Probst, A. J., Thomas, B. C. and Banfield, J. F. (2016). Major bacterial lineages are essentially devoid of CRISPR-Cas viral defence systems, *Nature Communications* 7. Nature Publishing Group, 10613. doi: 10.1038/ncomms10613.

Buttimer, C., McAuliffe, O., Ross, R. P., Hill, C., O'Mahony, J. and Coffey, A. (2017). Bacteriophages and bacterial plant diseases, *Frontiers in Microbiology* 8, 34. doi: 10.3389/fmicb.2017.00034.

Carstens, A. B., Djurhuus, A. M., Kot, W. and Hansen, L. H. (2019). A novel six-phage cocktail reduces Pectobacterium atrosepticum soft rot infection in potato tubers under simulated storage conditions, *FEMS Microbiology Letters* 366(9). Oxford University Press. doi: 10.1093/femsle/fnz101.

Chae, J. C., Hung, N. B., Yu, S. M., Lee, H. K. and Lee, Y. H. (2014). Diversity of bacteriophages infecting Xanthomonas oryzae pv. oryzae in paddy fields and its potential to control bacterial leaf blight of rice, *Journal of Microbiology and Biotechnology* 24(6), 740–747. doi: 10.4014/jmb.1402.02013.

Chattopadhyay, S. and Puls, R. W. (1999). Adsorption of bacteriophages on clay minerals, *Environmental Science and Technology* 33(20), 3609–3614. doi: 10.1021/es9811492.

Code of Federal Regulations (2014). Tolerances and exemptions for pesticide chemical r. Available at: https://www.govinfo.gov/content/pkg/CFR-2014-title40-vol24/xml/CFR-2014-title40-vol24-part180.xml (Accessed: 14 February 2021).

Coons, G. and Kotila, J. (1925). The transmissible lytic principle (bacteriophage) in relation to plant pathogens, *Phytopathology* 15, 357–370.

Czajkowski, R. (2015). Bacteriophages of soft rot Enterobacteriaceae- a minireview, *FEMS Microbiology Letters* 363(2), 230. doi: 10.1093/femsle/fnv230.

Czajkowski, R., Ozymko, Z. and Lojkowska, E. (2014). Isolation and characterization of novel soilborne lytic bacteriophages infecting Dickeya spp. biovar 3 ('D. solani'), *Plant Pathology* 63(4). Blackwell Publishing Ltd, 758–772. doi: 10.1111/ppa.12157.

Czajkowski, R., Ozymko, Z., de Jager, V., Siwinska, J., Smolarska, A., Ossowicki, A., Narajczyk, M. and Lojkowska, E. (2015). Genomic, proteomic and morphological characterization of two novel broad host lytic bacteriophages ΦPD10.3 and ΦPD23.1 infecting pectinolytic Pectobacterium spp. and Dickeya spp, *PLoS ONE* 10(3), e0119812. doi: 10.1371/journal.pone.0119812.

D'Herelle, F. (1917). Sur un microbe invisible antagonist des bacilles disenterique, *comptes rendus de l'Academie de Science de Paris* 165, 373–375.

Das, M., Bhowmick, T. S., Ahern, S. J., Young, R. and Gonzalez, C. F. (2015). Control of pierce's disease by phage, *PLoS ONE* 10(6), e0128902. doi: 10.1371/journal.pone.0128902.

Davies, E. V., Winstanley, C., Fothergill, J. L. and James, C. E. (2016). The role of temperate bacteriophages in bacterial infection, *FEMS Microbiology Letters* 363(5). Oxford University Press, fnw015. doi: 10.1093/femsle/fnw015.

Dy, R. L., Rigano, L. A. and Fineran, P. C. (2018). Phage-based biocontrol strategies and their application in agriculture and aquaculture, *Biochemical Society Transactions* 46(6). Portland Press Ltd, 1605–1613. doi: 10.1042/BST20180178.

Ecophage (n.d.). Bayer launches ecophage to develop viruses against crop disease. Available at: https://www.labiotech.eu/more-news/bayer-ecophage-bacteriophage-agriculture/ (Accessed: 14 February 2021).

Elhalag, K., Nasr-Eldin, M., Hussien, A. and Ahmad, A. (2018). Potential use of soilborne lytic Podoviridae phage as a biocontrol agent against Ralstonia solanacearum, *Journal of Basic Microbiology* 58(8). Wiley-VCH Verlag, 658–669. doi: 10.1002/jobm.201800039.

Enviroinvest (n.d.). Usage information Erwiphage PLUS. Available at: http://www.erwiphage.com/felhasznalasi-informaciok/ (Accessed: 14 February 2021).

EPA (n.d. a). Biopesticide registration. Available at: https://www.epa.gov/pesticide-registration/biopesticide-registration (Accessed: 15 February 2021).

European Union (2009). Regulation (Ec) No 1107/2009 of the European Parliament and of the Council of 21 October 2009, Official Journal of the European Union.

Fineran, P. C., Blower, T. R., Foulds, I. J., Humphreys, D. P., Lilley, K. S. and Salmond, G. P. (2009). The phage abortive infection system, ToxIN, functions as a protein-RNA toxin-antitoxin pair, *Proceedings of the National Academy of Sciences of the United States of America* 106(3). National Academy of Sciences, 894–899. doi: 10.1073/pnas.0808832106.

Flaherty, J. E., Somodi, G. C., Jones, J. B., Harbaugh, B. K. and Jackson, L. E. (2000). Control of bacterial spot on tomato in the greenhouse and field with H-mutant bacteriophages, *Hortscience* 35(5), 882–884. doi: 10.21273/HORTSCI.35.5.882.

Flaherty, J. E., et al. (2001). H-mutant bacteriophages as a potential biocontrol of bacterial blight of geranium, *American Society for Horticultural Science* 36(1), 98–100. doi: 10.21273/hortsci.36.1.98.

Flores, O., Retamales, J., Núñez, M., León, M., Salinas, P., Besoain, X., Yañez, C. and Bastías, R. (2020). Characterization of bacteriophages against pseudomonas syringae pv. Actinidiae with potential use as natural antimicrobials in kiwifruit plants, *Microorganisms* 8(7). MDPI AG, 1–17. doi: 10.3390/microorganisms8070974.

Fujiwara, A., Fujisawa, M., Hamasaki, R., Kawasaki, T., Fujie, M. and Yamada, T. (2011). Biocontrol of Ralstonia solanacearum by treatment with lytic bacteriophages, *Applied and Environmental Microbiology* 77(12), 4155–4162. doi: 10.1128/AEM.02847-10.

Gašić, K., et al. (2011). Isolation and characterization of Xanthomonas euvesicatoria bacteriophages, *Journal of Plant Pathology* 93(2), 415–423. doi: 10.4454/jpp.v93i2.1197.

Gašić, K., Kuzmanović, N., Ivanović, M., Prokić, A., Šević, M. and Obradović, A. (2018). Complete genome of the Xanthomonas euvesicatoria specific bacteriophage KΦ1, its survival and potential in control of pepper bacterial spot, *Frontiers in Microbiology* 9(Aug), 2021. doi: 10.3389/fmicb.2018.02021.

Gill, J. and Abedon, S. T. (2003). Bacteriophage ecology and plants, *APSnet Feature Articles*. doi: 10.1094/APSnetFeature-2003-1103.

Greer, G. G. (2005). Bacteriophage control of foodborne bacteriat, *Journal of Food Protection* 68(5). IAMFES, 1102–1111. doi: 10.4315/0362-028X-68.5.1102.

Hankin, E. H. (1896). 'L'action bactericide des eaux de la Jumna et du Gange sur le vibrion du cholera. In: *Annales de l'Institut Pasteur*, 790. Available at: https://archive.org/stream/annalesdelinstit10inst#page/511/mode/1up.

Hathaway, H., Ajuebor, J., Stephens, L., Coffey, A., Potter, U., Sutton, J. M. and Jenkins, A. T. (2017). Thermally triggered release of the bacteriophage endolysin CHAPK and the bacteriocin lysostaphin for the control of methicillin resistant Staphylococcus aureus (MRSA), *Journal of Controlled Release* 245. Elsevier B.V., 108–115. doi: 10.1016/j.jconrel.2016.11.030.

He, D. C., Zhan, J. S. and Xie, L. H. (2016). Problems, challenges and future of plant disease management: from an ecological point of view, *Journal of Integrative Agriculture* 15(4). Chinese Academy of Agricultural Sciences, 705–715. doi: 10.1016/S2095-3119(15)61300-4.

Horiuk, Y., Horiuk, V., Kukhtyn, M., Tsvihun, A. and Kernychnyi, S. (2020). Characterization of lytic activity of phage SAvB14 on Staphylococcus aureus variant bovis, *Journal of Advanced Veterinary and Animal Research* 7(3), 509–513. doi: 10.5455/javar.2020.g447.

Howard-Varona, C., Hargreaves, K. R., Abedon, S. T. and Sullivan, M. B. (2017). Lysogeny in nature: mechanisms, impact and ecology of temperate phages, *ISME Journal* 11(7). Nature Publishing Group, 1511–1520. doi: 10.1038/ismej.2017.16.

Hyman, P. (2019). Phages for phage therapy: isolation, characterization, and host range breadth, *Pharmaceuticals* 12(1). MDPI AG. doi: 10.3390/ph12010035.

Ibrahim, Y. E., Saleh, A. A. and Al-Saleh, M. A. (2017). Management of Asiatic citrus canker under field conditions in Saudi Arabia using bacteriophages and acibenzolar-s-methyl, *Plant Disease* 101(5). American Phytopathological Society, 761–765. doi: 10.1094/PDIS-08-16-1213-RE.

Iriarte, F. B., Balogh, B., Momol, M. T., Smith, L. M., Wilson, M. and Jones, J. B. (2007). Factors affecting survival of bacteriophage on tomato leaf surfaces, *Applied and Environmental Microbiology* 73(6), 1704–1711. doi: 10.1128/AEM.02118-06.

Iriarte, F. B., Obradović, A., Wernsing, M. H., Jackson, L. E., Balogh, B., Hong, J. A., Momol, M. T., Jones, J. B. and Vallad, G. E. (2012). Soil-based systemic delivery and phyllosphere in vivo propagation of bacteriophages: two possible strategies for improving bacteriophage persistence for plant disease control, *Bacteriophage* 2(4), 215–224. doi: 10.4161/bact.23530.

Jackson, L. E. (1989). Bacteriophage prevention and control of harmful plant bacteria, *International Journal of Biochemistry* 21(10), iv–v. doi: 10.1016/0020-711X(89)90090-6.

Jones, J. B., Jackson, L. E., Balogh, B., Obradovic, A., Iriarte, F. B. and Momol, M. T. (2007). Bacteriophages for plant disease control, *Annual Review of Phytopathology* 45. Annual Reviews Inc., 245–262. doi: 10.1146/annurev.phyto.45.062806.094411.

Jones, J. B., Vallad, G. E., Iriarte, F. B., Obradović, A., Wernsing, M. H., Jackson, L. E., Balogh, B., Hong, J. C. and Momol, M. T. (2012). Considerations for using bacteriophages for plant disease control, *Bacteriophage* 2(4), 208–214. doi: 10.4161/bact.23857.

Kahn, L. H., Bergeron, G., Bourassa, M. W., De Vegt, B., Gill, J., Gomes, F., Malouin, F., Opengart, K., Ritter, G. D., Singer, R. S., Storrs, C. and Topp, E. (2019). From farm management to bacteriophage therapy: strategies to reduce antibiotic use in animal agriculture, *Annals of the New York Academy of Sciences* 1441(1). Blackwell Publishing Inc., 31–39. doi: 10.1111/nyas.14034.

Kasman, L. M., Kasman, A., Westwater, C., Dolan, J., Schmidt, M. G. and Norris, J. S. (2002). Overcoming the phage replication threshold: a mathematical model with implications for phage therapy, *Journal of Virology* 76(11). American Society for Microbiology, 5557–5564. doi: 10.1128/jvi.76.11.5557-5564.2002.

Katznelson, H. (1937). Bacteriophage in relation to plant diseases, *The Botanical Review* 3(10), 499–521. doi: 10.1007/BF02870486.

Keen, E. C. (2015). A century of phage research: bacteriophages and the shaping of modern biology, *BioEssays: News and Reviews in Molecular, Cellular and Developmental Biology* 37(1). John Wiley and Sons Inc., 6–9. doi: 10.1002/bies.201400152.

Kering, K. K., Kibii, B. J. and Wei, H. (2019). Biocontrol of phytobacteria with bacteriophage cocktails, *Pest Management Science* 75(7). John Wiley and Sons Ltd, 1775–1781. doi: 10.1002/ps.5324.

Khalil, I. R., Irorere, V. U., Radecka, I., Burns, A. T. H., Kowalczuk, M., Mason, J. L. and Khechara, M. P. (2016). Poly-γ-glutamic acid: biodegradable polymer for potential protection of beneficial viruses, *Materials* 9(1). MDPI AG, 28. doi: 10.3390/ma9010028.

Kimura, M., Jia, Z., Nakayama, N. and Asakawa, S. (2008). Ecology of viruses in soils: past, present and future perspectives, *Soil Science and Plant Nutrition* 54(1). Taylor & Francis Group, 1–32. doi: 10.1111/j.1747-0765.2007.00197.x.

Kleczkowska, J. (1957). A study of the distribution and the effects of bacteriophage of root nodule bacteria in the soil, *Canadian Journal of Microbiology* 3(2). Canadian Science Publishing, 171–180. doi: 10.1139/m57-021.

Kotila, J. and Coons, G. (1925). Investigations on the blackleg disease of potato, *Michigan Agric. Exp. Stn. Tech. Bull.* 67, 3–29.

Lamichhane, J. R., Osdaghi, E., Behlau, F., Köhl, J., Jones, J. B. and Aubertot, J. (2018). Thirteen decades of antimicrobial copper compounds applied in agriculture: a review, *Agronomy for Sustainable Development* 38(3), 28. doi: 10.1007/s13593-018-0503-9.

Lang, J. M., Gent, D. H. and Schwartz, H. F. (2007). Management of Xanthomonas leaf blight of onion with bacteriophages and a plant activator, *Plant Disease* 91(7), 871–878. doi: 10.1094/PDIS-91-7-0871.

Lehman, S. (2007). Development of a bacteriophage-based biopesticide for fire blight. Ph.D. thesis. Brock University.

Leverentz, B., Conway, W. S., Alavidze, Z., Janisiewicz, W. J., Fuchs, Y., Camp, M. J., Chighladze, E. and Sulakvelidze, A. (2001). Examination of bacteriophage as a biocontrol method for Salmonella on fresh-cut fruit: a model study, *Journal of Food Protection* 64(8). IAMFES, 1116–1121. doi: 10.4315/0362-028X-64.8.1116.

Liu, N., Lewis, C., Zheng, W. and Fu, Z. Q. (2020). Phage cocktail therapy: multiple ways to suppress pathogenicity, *Trends in Plant Science* 25(4). Elsevier Ltd, 315–317. doi: 10.1016/j.tplants.2020.01.013.

Mallmann, W. L. and Hemstreet, C. (1924). Isolation of an inhibitory substance FROM plants 1, *Agricultural Research* 28, 599–602.

McKenna, F., El-Tarabily, K. A., Hardy, G. E. S. J. and Dell, B. (2001). Novel in vivo use of a polyvalent Streptomyces phage to disinfest Streptomyces scabies-infected seed potatoes, *Plant Pathology* 50(6). John Wiley & Sons, Ltd, 666–675. doi: 10.1046/j.1365-3059.2001.00648.x.

Nagai, H., Miyake, N., Kato, S., Maekawa, D., Inoue, Y. and Takikawa, Y. (2017). Improved control of black rot of broccoli caused by Xanthomonas campestris pv. campestris using a bacteriophage and a nonpathogenic Xanthomonas sp. strain, *Journal of General Plant Pathology* 83(6). Springer Tokyo, 373–381. doi: 10.1007/s10327-017-0745-4.

Nanda, A. M., Thormann, K. and Frunzke, J. (2015). Impact of spontaneous prophage induction on the fitness of bacterial populations and host-microbe interactions, *Journal of Bacteriology* 197(3). American Society for Microbiology, 410–419. doi: 10.1128/JB.02230-14.

Nguyen, H. T. D., Yoon, S., Kim, M. H., Kim, Y. K., Yoon, M. Y., Cho, Y. H., Lim, Y., Shin, S. H. and Kim, D. E. (2012). Characterization of bacteriophage φPto-bp6g, a novel phage that lyses Pseudomonas tolaasii causing brown blotch disease in mushrooms, *Journal of Microbiological Methods* 91(3). Elsevier, 514–519. doi: 10.1016/j.mimet.2012.09.032.

Obradovic, A., Jones, J. B., Momol, M. T., Balogh, B. and Olson, S. M. (2004). Management of tomato bacterial spot in the field by foliar applications of bacteriophages and SAR inducers, *Plant Disease* 88(7), 736–740. doi: 10.1094/PDIS.2004.88.7.736.

Obradovic, A., Jones, J. B., Momol, M. T., Olson, S. M., Jackson, L. E., Balogh, B., Guven, K. and Iriarte, F. B. (2005). Integration of biological control agents and systemic acquired resistance inducers against bacterial spot on tomato, *Plant Disease* 89(7), 712–716. doi: 10.1094/PD-89-0712.

Obradovic, A., et al. (2008). Integrated management of tomato bacterial spot. In: *Integrated Management of Diseases Caused by Fungi, Phytoplasma and Bacteria*. Springer Netherlands, pp. 211–223. doi: 10.1007/978-1-4020-8571-0_11.

Ogunyemi, S. O., Chen, J., Zhang, M., Wang, L., Masum, M. M. I., Yan, C., An, Q., Li, B. and Chen, J. (2019). Identification and characterization of five new OP2-related Myoviridae bacteriophages infecting different strains of Xanthomonas oryzae pv. oryzae, *Journal of Plant Pathology* 101(2), 263–273. doi: 10.1007/s42161-018-0188-6.

Okabe, N. and Goto, M. (1963). Bacteriophages of plant pathogens, *Annual Review of Phytopathology* 1(1), 397–418.

OmniLytics (2016). Application instructions & recommendations. Available at: https://www.agriphage.com/wp-content/uploads/2018/02/AgriPhage-Application-Instructions-2016.pdf (Accessed: 14 February 2021).

OmniLytics (2018). Notice of pesticides: X registration reregistration agriphage-fire blight. Available at: www.omnilytics.com (Accessed: 14 February 2021).

OmniLytics (n.d.a). Agriphage-CMM for organic production, bactericide for use on tomatoes [biological control for bacterial canker on tomatoes]. Available at: www.omnilytics.com (Accessed: 14 February 2021).

OmniLytics (n.d.b). Agriphage for organic production: bactericide for use on tomatoes and peppers biological control for bacterial spot and bacterial speck on tomatoes and peppers. Available at: www.omnilytics.com (Accessed: 14 February 2021).

OmniLytics (n.d.c). Notice of pesticides: X registration reregistration agriphage-*Citrus canker*. Available at: www.omnilytics.com (Accessed: 14 February 2021).

Papaianni, M., Paris, D., Woo, S. L., Fulgione, A., Rigano, M. M., Parrilli, E., Tutino, M. L., Marra, R., Manganiello, G., Casillo, A., Limone, A., Zoina, A., Motta, A., Lorito, M. and Capparelli, R. (2020). Plant dynamic metabolic response to bacteriophage treatment after xanthomonas campestris pv. campestris Infection, *Frontiers in Microbiology* 11, 732. doi: 10.3389/fmicb.2020.00732.

Papaianni, M., Ricciardelli, A., Casillo, A., Corsaro, M. M., Borbone, F., della Ventura, B., Velotta, R., Fulgione, A., Woo, S. L., Tutino, M. L., Parrilli, E. and Capparelli, R. (2021). The union is strength: the synergic action of long fatty acids and a bacteriophage against *Xanthomonas campestris* biofilm, *Microorganisms* 9(1), 1–18. doi: 10.3390/microorganisms9010060.

Pfeifer, G. P. (1997). Formation and processing of UV photoproducts: effetcs of DNA sequence and chromatin environment, *Photochemistry and Photobiology* 65(2), 270–283. doi: 10.1111/j.1751-1097.1997.tb08560.x.

PhageFire (n.d.). Project background–PhageFire. Available at: https://www.phagefire.eu/project-background/ (Accessed: 15 February 2021).

Przybilski, R., Richter, C., Gristwood, T., Clulow, J. S., Vercoe, R. B. and Fineran, P. C. (2011). Csy4 is responsible for CRISPR RNA processing in pectobacterium atrosepticum, *RNA Biology* 8(3). Taylor and Francis Inc., 517–528. doi: 10.4161/rna.8.3.15190.

Qiao, J., Qiao, X., Sun, Y. and Mindich, L. (2010). Role of host protein glutaredoxin 3 in the control of transcription during bacteriophage Φ2954 infection, *Proceedings of the National Academy of Sciences of the United States of America* 107(13). National Academy of Sciences, 6000–6004. doi: 10.1073/pnas.1000383107.

Rahimi-Midani, A. and Choi, T. J. (2020). Transport of phage in melon plants and inhibition of progression of bacterial fruit blotch, *Viruses* 12(4). MDPI AG. doi: 10.3390/v12040477.

Ramírez, M., Neuman, B. W. and Ramírez, C. A. (2020). Bacteriophages as promising agents for the biological control of Moko disease (Ralstonia solanacearum) of banana, *Biological Control* 149. Academic Press Inc. doi: 10.1016/j.biocontrol.2020.104238.

Rezzonico, F., Smits, T. H. M. and Duffy, B. (2011). Diversity, evolution, and functionality of clustered regularly interspaced short palindromic repeat (CRISPR) regions in the fire blight pathogen Erwinia amylovora, *Applied and Environmental Microbiology* 77(11). American Society for Microbiology, 3819–3829. doi: 10.1128/AEM.00177-11.

Rombouts, S., Volckaert, A., Venneman, S., Declercq, B., Vandenheuvel, D., Allonsius, C. N., Van Malderghem, C., Jang, H. B., Briers, Y., Noben, J. P., Klumpp, J., Van Vaerenbergh, J., Maes, M. and Lavigne, R. (2016). 'Characterization of novel bacteriophages for biocontrol of bacterial blight in leek caused by Pseudomonas syringae pv. porri', *Frontiers in Microbiology, Frontiers in Microbiology* 7(Mar), 279. doi: 10.3389/fmicb.2016.00279.

Schwartz, A. R., Potnis, N., Timilsina, S., Wilson, M., Patané, J., Martins, J., Minsavage, G. V., Dahlbeck, D., Akhunova, A., Almeida, N., Vallad, G. E., Barak, J. D., White, F. F., Miller, S. A., Ritchie, D., Goss, E., Bart, R. S., Setubal, J. C., Jones, J. B. and Staskawicz, B. J. (2015). Phylogenomics of Xanthomonas field strains infecting pepper and tomato reveals diversity in effector repertoires and identifies determinants of host specificity, *Frontiers in Microbiology* 6(Jun), 535. doi: 10.3389/fmicb.2015.00535.

Semenova, E., Nagornykh, M., Pyatnitskiy, M., Artamonova, I. I. and Severinov, K. (2009). Analysis of CRISPR system function in plant pathogen *Xanthomonas oryzae*, *FEMS Microbiology Letters* 296(1), 110–116. doi: 10.1111/j.1574-6968.2009.01626.x.

Šević, M., Gašić, K., Ignjatov, M., Mijatović, M., Prokić, A. and Obradović, A. (2019). Integration of biological and conventional treatments in control of pepper bacterial spot, *Crop Protection* 119. Elsevier Ltd, 46–51. doi: 10.1016/j.cropro.2019.01.006.

Sieiro, C., Areal-Hermida, L., Pichardo-Gallardo, Á, Almuiña-González, R., de Miguel, T., Sánchez, S., Sánchez-Pérez, Á and Villa, T. G. (2020). A hundred years of bacteriophages: can phages replace antibiotics in agriculture and aquaculture?, *Antibiotics* 9(8). MDPI AG, 493. doi: 10.3390/antibiotics9080493.

Stone, E., Campbell, K., Grant, I. and McAuliffe, O. (2019). Understanding and exploiting phage–host interactions, *Viruses* 11(6). doi: 10.3390/v11060567.

Storey, M. V. and Ashbolt, N. J. (2001). Persistence of two model enteric viruses (B40-8 and MS-2 bacteriophages) in water distribution pipe biofilms, *Water Science and Technology* 43(12). IWA Publishing, 133–138. doi: 10.2166/wst.2001.0724.

Sundin, G. W. and Wang, N. (2018). Antibiotic resistance in plant-pathogenic bacteria, *Annual Review of Phytopathology* 56. Annual Reviews Inc., 161–180. doi: 10.1146/annurev-phyto-080417-045946.

Sykes, I. K., Lanning, S. and Williams, S. T. (1981). The effect of pH on soil actinophage, *Journal of General Microbiology* 122(2). Microbiology Society, 271–280. doi: 10.1099/00221287-1224-2-271.

Tanaka, H., Negishi, H. and Maeda, H. (1990). Control of tobacco bacterial wilt by an avirulent strain of pseudomonas solanacearum M4S and its bacteriophage, *Japanese Journal of Phytopathology* 56(2). The Phytopathological Society of Japan, 243–246. doi: 10.3186/jjphytopath.56.243.

Thayer, S. (1961). A survey of Xanthomonas vesicatoria resistance to streptomycin. *Proceedings of the Florida State Horticultural Society* 75:163.

Thomas, R. C. (1935). A bacteriophage in relation to Stewart's disease of corn, *Phytopathology* 25, 371. Available at: https://scholar.google.com/scholar_lookup?title=A bacteriophage in relation to Stewart's disease of corn&author=RC. Thomas&journal=Phytopathology&volume=25&pages=371-372&publication_year=1935.

Tom, E. F., Molineux, I. J., Paff, M. L. and Bull, J. J. (2018). Experimental evolution of UV resistance in a phage *PeerJ* 6. PeerJ Inc., e5190. doi: 10.7717/peerj.5190.

Twort, F. W. (1915). An investigation on the nature of ultra-microscopic viruses, *The Lancet* 186(4814). Elsevier, 1241–1243. doi: 10.1016/S0140-6736(01)20383-3.

Vandamme, E. J. and Mortelmans, K. (2019). A century of bacteriophage research and applications: impacts on biotechnology, health, ecology and the economy!, *Journal of Chemical Technology and Biotechnology* 94(2). John Wiley and Sons Ltd, 323–342. doi: 10.1002/jctb.5810.

Vu, N. T. and Oh, C. S. (2020). Bacteriophage usage for bacterial disease management and diagnosis in plants, *Plant Pathology Journal* 36(3). Korean Society of Plant Pathology, 204–217. doi: 10.5423/PPJ.RW.04.2020.0074.

Wang, X., Wei, Z., Yang, K., Wang, J., Jousset, A., Xu, Y., Shen, Q. and Friman, V. P. (2019). Phage combination therapies for bacterial wilt disease in tomato, *Nature Biotechnology* 37(12). Nature Publishing Research, 1513–1520. doi: 10.1038/s41587-019-0328-3.

Wei, C., Liu, J., Maina, A. N., Mwaura, F. B., Yu, J., Yan, C., Zhang, R. and Wei, H. (2017). Developing a bacteriophage cocktail for biocontrol of potato bacterial wilt, *Virologica Sinica* 32(6). Science Press, 476–484. doi: 10.1007/s12250-017-3987-6.

Yamada, T. (2013). Filamentous phages of Ralstonia solanacearum: double-edged swords for pathogenic bacteria, *Frontiers in Microbiology* 4(V), 325. doi: 10.3389/fmicb.2013.00325.

Yamada, T., Kawasaki, T., Nagata, S., Fujiwara, A., Usami, S. and Fujie, M. (2007). New bacteriophages that infect the phytopathogen Ralstonia solanacearum, *Microbiology* 153(8). Microbiology Society, 2630–2639. doi: 10.1099/mic.0.2006/001453-0.

Yu, J. G., Lim, J. A., Song, Y. R., Heu, S., Kim, G. H., Koh, Y. J. and Oh, C. S. (2015). Isolation and characterization of bacteriophages against pseudomonas syringae pv. Actinidiae causing bacterial canker disease in kiwifruit, *Journal of Microbiology and Biotechnology* 26(2), 385–393. doi: 10.4014/jmb.1509.09012.

Zaccardelli, M., Saccardi, A., Gambin, E., and Mazzucchi, U. (1992). Xanthomonas campestris pv. pruni bacteriophages on peach trees and their potential use for biological control on JSTOR, *Phytopathologia Mediterranea* 31(3), 133–140.

Zaczek-Moczydłowska, M. A., Young, G. K., Trudgett, J., Fleming, C. C., Campbell, K. and O'Hanlon, R. (2020). Genomic characterization, formulation and efficacy in planta of a Siphoviridae and Podoviridae protection cocktail against the bacterial plant pathogens Pectobacterium spp., *Viruses* 12(2). MDPI AG, 150. doi: 10.3390/v12020150.

Chapter 14

The use of mild viruses for control of plant pathogenic viruses

Nelia Ortega-Parra[1], De Ceuster Meststoffen BV (DCM) and Wageningen University & Research, The Netherlands; Zafeiro Zisi[1], Scientia Terrae Research Institute VZW and Katholieke Universiteit Leuven, Belgium; and Inge M. Hanssen[2], De Ceuster Meststoffen NV (DCM), Belgium

1 Introduction

The protection of crops against pests and diseases has always been a significant aspect of agriculture and horticulture. Over the past few years, the need for efficient control of plant pests and pathogens has become more evident due to the increasing global demand for food, both in quantity and quality, and also due to the worldwide spread of pests and pathogens (Bebber et al., 2014; Myers et al., 2017; Ziebell and MacDiarmid, 2017). This increased distribution of plant pathogens is facilitated by changes in agricultural practices. The conversion of many wildlands into mono-cultivated fields has led to a reduction in biodiversity, allowing the spread of pathogens from wild ecosystems into cultivated crops (Roossinck and García-Arenal, 2015). The intensification of cultivation in modern horticulture, with overlapping cropping cycles and continuous production in greenhouses using artificial lights, has reduced the option of complete disinfection of the entire system.

1 These authors contributed equally to this work.
2 Corresponding author: Inge M. Hanssen.

http://dx.doi.org/10.19103/AS.2021.0093.19

Globalization of trade and year-round supply also contribute to the rapid spread of new diseases, through the transport of seeds, produce, and other commodities (Dombrovsky and Smith, 2017). Climate change also plays an important role in the worldwide spread of pests and pathogens (Bebber, 2015; Jones, 2016).

Each group of plant pests and pathogens represents a different challenge when it comes to the development of suitable protection strategies. The management of plant viruses focuses largely on prevention as curative treatments are yet to be developed (Ziebell and MacDiarmid, 2017). The major strategies for the management of viral diseases in plants are the implementation of hygiene measures, the control of insects and nematodes that act as viral vectors, the elimination of infected plants, the use of resistant or tolerant varieties, and in particular cases, cross-protection (Varma, 1993).

In cross-protection-based viral disease management, plants are 'vaccinated' with a mild or attenuated viral isolate in an early stage of plant development to protect them against a subsequent infection with a more aggressive isolate of the same viral species. This phenomenon was first described in 1929, and since then, it has been reported for many plant viruses as well as for some viroids (McKinney, 1929; Ziebell and MacDiarmid, 2017). An overview of protective virus isolates tested for their cross-protection potential is given in a recent review by Pechinger et al. (2019). Despite the extensive study of cross-protection in laboratory settings, only a few examples exist of cross-protection strategies being applied under field or greenhouse conditions.

This chapter discusses both the theoretical and the commercial cultivation aspects of cross-protection as a control measure for viral diseases in crops. The range of research to unravel the molecular mechanisms involved in this phenomenon is discussed. The preconditions for successful application of cross-protection in commercial plant production systems are also examined. A case study on the *Pepino mosaic virus* (PepMV) in tomato (Hanssen et al., 2009) illustrates the potential and challenges of cross-protection. Finally, the chapter examines future research related to cross-protection and the ways this technology can evolve by incorporating recent scientific advances.

2 Unraveling the molecular mechanisms of cross-protection

Cross-protection, also known as superinfection exclusion, describes the phenomenon of antagonism where a primary infection with a virus prevents a secondary infection with a challenging homologous virus (Syller and Grupa, 2016). This phenomenon can be exploited if a mild virus causes the initial infection, as it may prevent secondary infections of more aggressive viruses. Since its first description, much research has been dedicated to the characterization and unraveling of the molecular biology of this remarkable phenomenon. Despite many papers confirming the occurrence of cross-protection between viral strains and isolates (Hanssen et al., 2010; Hughes and Ollennu, 1994; Ziebell and Carr, 2010), the molecular basis of the phenomenon remains unclear. Over the years, a number of mechanisms have been described, but none of those can convincingly explain all the interactions (Ziebell and Carr, 2010; Ziebell and MacDiarmid, 2017). The following three main mechanisms have been proposed:

- coat protein-mediated protection;
- RNA interference-mediated protection;
- spatial separation-induced protection.

These will be discussed in the following subsections.

2.1 Coat protein-mediated protection

Early hypotheses for the molecular basis of cross-protection ranged from the formation of protective substances, similar to antibodies, to competition for strain-specific multiplication sites. However, these theories were not supported by experimental evidence (Chester, 1933; Kassanis et al., 1974; Price, 1940; Wallace, 1940). Zoeten and Fulton (1975) suggested a new mechanism that could explain the cross-protection between *Tobacco mosaic virus* (TMV) strains. They hypothesized that, during infection, the genomic RNA of the challenging strain was captured by the coat protein of the protecting strain that was present in the plant cells.

A later study (Sherwood and Fulton, 1982) suggested that the coat protein of the protective strain prevents the uncoating of the challenging strain, therefore providing the necessary protection. The authors used *Nicotiana sylvestris* plants infected with a mosaic-inducing strain of TMV and observed that superinfection by intact necrotizing TMV strains occurred only in dark green leaf tissue, while light green tissue was protected (Sherwood and Fulton, 1982). However, when purified RNA from the necrotizing strains was used for superinfection, necrotic symptoms were found in both tissue types. Additionally, the cross-protection effects were not observed when the purified RNAs were encapsidated in a protein from a different virus (Sherwood and Fulton, 1982). This finding led the authors to propose that the coat protein plays a crucial role in cross-protection. A study on cross-protection using *Cucumber mosaic virus* (CMV) also supported this hypothesis (Dodds et al., 1985). Horikoshi et al., (1987) observed that RNA synthesis by the replicase extracted from *Brome mosaic virus* (BMV)-infected leaves gradually decreased and was eventually completely inhibited by the addition of the BMV coat protein to the reaction mixture, in an *in vitro* study. When different proteins, such as the TMV coat protein and bovine serum albumin, were added to the reaction mixture, the RNA synthesis was not affected. Based on these findings, it was concluded that during cross-protection, the coat protein of the mild strain partly coats the RNA of the closely related aggressive strain.

The precise role of the coat protein in cross-protection remained unclear since a number of contradicting studies were published in which mutant strains that produced defective or no coat proteins induced some degree of cross-protection (Sarkar and Smitamana, 1981; Zaitlin, 1976). The hypothesis that the coat protein captures the challenging RNA before the start of the replication was further undermined by the discovery that uncoating, replication, and negative-strand synthesis occur simultaneously (Shaw et al., 1986; Wu and Shaw, 1997). In a study that used the Tobamovirus *Shunn-hemp mosaic virus* (SHMV), the wildtype isolate provided effective cross-protection against a SHMV mutant inducing necrotic lesions, as well as against purified RNA of this mutant. Additionally, wildtype SHMV conferred partial protection against the distantly related Tobamovirus (TMV). In the same study, two chimeric constructs were designed, one consisting of the RNA of TMV in the coat protein of the SHMV and one consisting of the SHMV mutant RNA encapsidated in the TMV coat protein. Wildtype SHMV infection provided protection against both constructs. Based on these results, it was concluded that although the coat protein is involved in the mechanisms, the phenomenon is affected by other factors as well (Zinnen and Fulton, 1986).

The ability to generate transgenic plants has provided a new opportunity to explore the mechanism of cross-protection. In several studies, a viral coat protein gene was introduced into the host genome, leading to a steady expression of that coat protein (Abel et al., 1986; Nelson et al., 1987; Register and Beachy, 1988; Wisniewski et al., 1990; Wu et al., 1990). These transgenic plants show variable resistance against infection with differing aggressive isolates of the respective viruses. Protection was not observed, or at least significantly lowered, when the plants were inoculated with high virus titers or with infectious RNA molecules (Abel et al., 1986; Bardonnet et al., 1994; Nelson et al., 1987; Tumer et al., 1987; Valkonen et al., 2002). These studies showed the importance of expression and aggregation of the coat protein prior to the challenge inoculation (Nelson et al., 1987; Register and Beachy, 1989, 1988; Wisniewski et al., 1990). In another study, protoplasts were isolated from transgenic *Nicotiana tabacum* cv. Xanthi plants expressing the TMV coat protein. These protoplasts, along with protoplasts isolated from non-transgenic plants, were inoculated with TMV particles with radioactive-labeled coat protein and viral RNA. The formation of 'striposomes', which are complexes of partially uncoated virions, ribosomes, and viral polypeptide molecules, was observed in the protoplast from non-transgenic plants. There was little to no formation of complexes in the plants expressing coat protein, supporting the theory that the coat protein can provide protection by inhibiting the uncoating of the challenging strain (Wu et al., 1990).

Taking this mechanistic evidence into account, a role for the coat protein of the protective strain in cross-protection is evident. The main roles that are well supported by experimental data are that the coat protein of the protective virus either prevents uncoating of the challenging virus, or that it partially coats the viral RNA, thus preventing replication. It should, however, be noted that these mechanisms cannot explain phenomena such as cross-protection induced by coat protein-deficient viral strains.

2.2 RNA interference-mediated protection

The potential role of nucleic acids in cross-protection was initially proposed in 1984, when it was suggested that replication of the protective isolate generates a large amount of positive-sense RNA (Palukaitis and Zaitlin, 1984). During replication of the challenging isolate, the positive-sense RNA of the protective isolate forms dimers with the negative-sense strands of the challenging isolate, resulting in replication and transcription inhibition. In this model, the RNA sequence homology between the two isolates is a key factor for the efficiency of protection. This theory could explain the cases where cross-protection is observed with coat protein-defective strains and with viroids which are lacking a coat protein (Palukaitis and Zaitlin, 1984). The role of RNA interference (RNAi) was further supported by later studies in which transgenic plants were used for the expression of untranslatable or truncated forms of viral coat proteins (Lindbo, 2012; Lindbo et al., 1993; Lindbo and Dougherty, 1992).

RNAi is part of the innate immune system of plants and functions as a defense against viruses. It involves RNA-dependent gene silencing and includes Post-Transcriptional Gene Silencing (PTGS) and Transcriptional Gene Silencing (TGS) (Fig. 1). RNAi is initiated by the formation of double-stranded RNA (dsRNA) molecules. In plants, the dsRNA molecules can be transported over short distances from cell to cell through plasmodesmata, and over longer distances through the vascular system to distant cells. Through transportation, dsRNA molecules can initiate RNAi in distant cells and act as gene-silencing signals, providing protection of the whole plant against secondary infection (Chitwood and Timmermans, 2010; Muthamilarasan and Prasad, 2013; Syller and Grupa, 2016).

2.2.1 PTGS

The main steps in PTGS are presented in Fig. 1 and begin with the synthesis of dsRNA molecules. In the case of RNA viruses, these molecules are replication

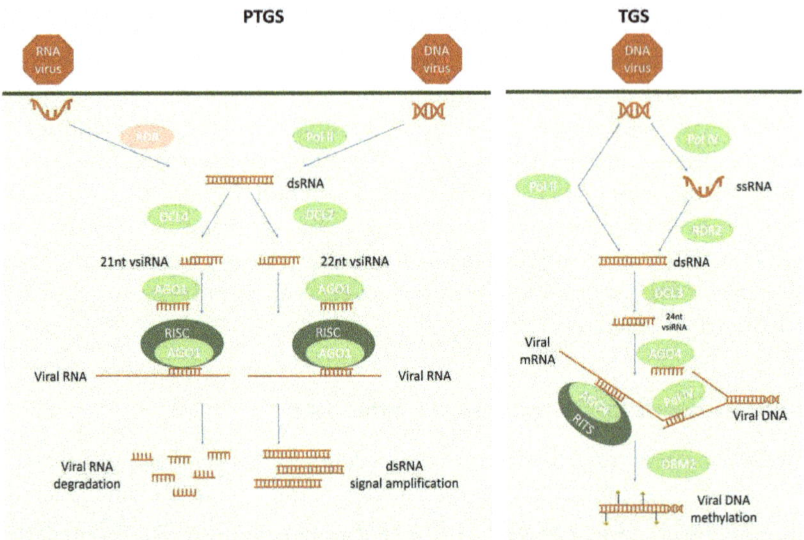

Figure 1 Mechanistic models for RNAi based on the Post-Transcriptional Gene Silencing (PTGS) and Transcriptional Gene Silencing (TGS). Adapted from: Dalakouras et al., 2020; Du et al., 2015; and Sharma et al., 2013.

intermediates formed by viral RNA-dependent RNA polymerase (RDR). In case of DNA viruses, the dsRNA molecules can result from bi-directional transcription by the host RNA polymerase II (Pol II). Molecules can also originate from self-folded hairpin precursors in viral RNA due to self-complementarity (Muthamilarasan and Prasad, 2013; Sharma et al., 2013).

As a next step, an endoribonuclease or dicer (DCL) enzyme recognizes the ends of the dsRNA molecules. The DCL enzyme consists of a PAZ (Piwi/Argonaute (AGO)/Zwille) domain on the one end and dual RNAse III domains on the other end. The ends of the dsRNA molecules are recognized by the PAZ domain and cleaved by the RNAse III domains to generate a pool of 21–24nt virus-derived small interfering RNA (vsiRNA) molecules that exhibit 3′ 2nt overhangs. The DCL enzyme has four paralogs in the model plant *Arabidopsis thaliana*. DCL1 is primarily involved in microRNA (miRNA) biogenesis and DCLs 2–4 are involved in RNAi. More specifically, DCL2 and DCL4 are involved in PTGS and DCL3 is involved in TGS (Dalakouras et al., 2020; Sharma et al., 2013).

When DCL4 cleaves dsRNA molecules, it produces 21nt vsiRNA molecules. One strand of the vsiRNA molecules (called the guide strand) is then loaded onto the endonuclease AGO. The choice of the strand depends on the 5′-terminal nucleotide; in the 21nt vsiRNAs, a 5′-U is preferred. *A. thaliana* contains ten AGO proteins, but the role of many of these remains unclear. AGO1 is primarily involved in RNAi and acts as a slicer enzyme cleaving nucleic acid, while AGO7 acts as a surrogate slicer of AGO1. Once the guide strand of the 21nt vsiRNA molecules is loaded onto AGO1, they are incorporated into the RNA-induced silencing complex (RISC), a ribonucleoprotein complex that acts as the 'effector' of the RNAi. After AGO1 loaded with the 21nt vsiRNA molecules is incorporated onto RISC, the complex uses the guide strand to scan the cytoplasm for complementary transcripts. The guide strand then forms a dimer with a matching viral mRNA molecule and AGO1 cleaves the duplex in the middle. The RNAi-mediated sliced viral mRNA is further degraded by cellular mechanisms of mRNA degradation. This process normally involves the removal of the 3′ polyA tail followed by the degradation of the remaining mRNA by the 5′-3′ exonuclease Xrnl protein (Dalakouras et al., 2020; Muthamilarasan and Prasad, 2013; Sharma et al., 2013).

When DCL2 cleaves dsRNAs, it creates 22nt vsiRNA molecules. One strand of these double-stranded vsiRNA molecules is loaded onto AGO1 and they are then incorporated onto the RISC following the same mechanism described above. The guide strand of a 22nt vsiRNA molecule changes the conformation of AGO1 so that, after the formation of the guide strand–mRNA duplex, the host RDR6 is recruited to the 3′ end of the target. This results in the transcription of the viral mRNA and the formation of new viral dsRNA molecules engaged again in PTGS, increasing the silencing effects. The viral dsRNA molecules can also be transported into distant cells and initiate PTGS systemically (Dalakouras et al., 2020; Sharma et al., 2013).

2.2.2 TGS

In case of an infection with a dsDNA virus, both PTGS and TGS can be activated (Fig. 1). First, viral dsRNA molecules are generated, either through bi-directional transcription by Pol II or by the host Pol IV, which generates single-stranded viral transcripts that are converted to dsRNA molecules by the host RDR2. DCL3 cleaves the dsRNA and produces a pool of 24nt vsiRNA molecules. In a similar way to PTGS, one strand of the 24nt vsiRNA molecules is loaded onto AGO4 and they are subsequently incorporated into the RNA-induced transcriptional silencing complex (RITS). After incorporation, the complex recognizes cognate DNA or its nascent transcript produced by the host Pol V. This recognition promotes the DNA cytosine methylation at carbon 5 of the pyrimidine ring by the Domains Rearranged Methyltransferase 2 (DRM2) in a process called RNA-directed DNA methylation (RdDM). RdDM leads to chromatin modification and translation inhibition (Dalakouras et al., 2020; Du et al., 2015; Muthamilarasan and Prasad, 2013; Sharma et al., 2013). TGS has been observed to confer resistance to viruses such as *Mungbean yellow mosaic India virus* (Yadav and Chattopadhyay, 2011), *Tomato leaf curl virus* (ToLCV) (Bian et al., 2006), and *Beet curly top virus* (Raja et al., 2008).

Many studies using transgenic plants have successfully used RNAi to trigger silencing against viral threats. These approaches can be divided into four groups:

- Sense gene-induced PTGS (S-PTGS) (Lindbo et al., 1993; Prins et al., 1997).
- Hairpin RNA-induced PTGS (hp-PTGS) (Leibman et al., 2021, in press).
- Artificial miRNA-induced PTGS (AMIR) (Vu et al., 2013).
- Trans-acting siRNA-induced PTGS (TAS-PTGS).

In S-PTGS, a gene that encodes a viral protein is introduced in the host genome. The transcription of the gene leads to the formation of dsRNA molecules by the plant RDR that can enhance the silencing signal. In hp-PTGS, a silencing vector, constructed by complementary sense and antisense viral cDNA controlled by plant promoters and terminators, is introduced into the host genome. The transcripts of this construct form hairpin dsRNA that can induce the RNAi machinery. In AMIR, mature natural miRNA sequences of the host are replaced with sequences complementary to the viral RNA in order to be favorably loaded onto RISC. Lastly, in TAS, an artificial trans-acting siRNA gene is used to express multiple siRNAs and target different viruses (Khalid et al., 2017).

The study of RNAi has revealed that plant viruses generally have defenses against host-induced antiviral silencing. Most plant viruses encode at least one viral suppressor of RNA-silencing (VSR) protein. Different VSRs can block the RNAi machinery at different stages to enhance viral replication (Sharma et al., 2013). VSRs from different viruses can have differential targets within the silencing machinery. For example, while the V2 protein of *Tomato yellow leaf curl virus* (TYLCV) binds to a cofactor of RDR6 and hampers the production of dsRNA molecules, the 2b protein of CMV binds AGO1 and inhibits cleaving activity (Muthamilarasan and Prasad, 2013). This phenomenon of differential targeting of RNAi machinery components might play a role in the synergistic interactions that are observed in mixed infections (Fukuzawa et al., 2010; Gal-On and Shiboleth, 2006; Gonzalez-Jara et al., 2004).

As with coat protein-mediated protection, RNAi cannot explain every case of cross-protection. For example, a CMV deletion mutant that does not express the 2b silencing suppressor protein is expected to be a strong cross-protection inducer, since the viral mechanism to bypass RNAi is absent, and should protect the host from subsequent challenge infection with the wildtype isolate. However, the protection induced by this mutant is limited (Ziebell et al., 2007). In addition, the cross-protection observed between genetically distant viruses cannot be explained based on RNAi (Zinnen and Fulton, 1986).

2.3 Spatial separation-induced cross-protection

Viral infection requires precise coordination of replication, translation, encapsidation, and movement mechanisms. During these processes, the viral nucleic acids compete for interaction sites with host regulatory proteins and form complexes and RNA granules which play a significant role in the establishment of infection of a plant cell (Hafrén et al., 2015; Mäkinen and Hafrén, 2014). These observations, together with results of studies conducted in plants with dual infections, led to the hypothesis that different strains of the same viruses compete for infection of the same plant cell and that the formation of nucleoprotein complexes due to prior infection can exclude a challenging virus from the same plant cell (Ziebell and MacDiarmid, 2017).

One example of a study that supports this hypothesis reports the infection of *Chenopodium quinoa* plants with *Apple latent spherical virus* (ALSV) labeled with either yellow or cyan fluorescent protein (YFP or CFP). This shows yellow and cyan fluorescence separately distributed in inoculated and systemic upper leaves. Initial inoculation with ALSV-CFP followed by the subsequent challenge with ALSV-YFP resulted in yellow fluorescence only in tissues where ALSV-CFP infection was not established. When *Nicotiana benthamiana* plants were co-inoculated with *Bean yellow mosaic virus* (BYMV)-YFP and BYMV-CFP, spatial separation was observed as well. In contrast, when *N. benthamiana* plants were infected with BYMV-CFP and ALSV-YFP, both fluorescent signals were observed in the same tissues, strengthening the spatial separation hypothesis (Takahashi et al., 2007).

In addition to these three well-studied molecular mechanisms involved in cross-protection, unknown mechanisms could also play a role. Unraveling such mechanisms could facilitate the development of additional tools for viral disease control.

3 Key issues for a safe and effective cross-protection strategy

Cross-protection has been demonstrated for many viruses and viroids in combination with various hosts (McKinney, 1929; Ziebell and Carr, 2010). However, the application of cross-protection to control viral diseases under commercial cultivation conditions has been limited to a few viruses and mostly used only in a specific area and/or for a limited period of time, such as *Tomato mosaic virus* (ToMV) in tomato (Rast, 1972); *Zucchini yellow mosaic virus* (ZYMV) in cucumber, squash, and watermelon (Lecoq et al., 1991; Wang et al., 1991); *Papaya ringspot virus* (PRSV) in papaya (Yeh, 1988); and *Citrus tristeza virus* (CTV) in citrus (Costa and Müller, 1980). In most of these cases, cross-protection was only used in a specific area and/or for a limited period of time because the results were not satisfactory (ZYMV, PRSV, CTV) or because the protection became redundant due to the availability of resistant varieties (ToMV). Cross-protection against PepMV has become a common strategy in greenhouse tomato production (Agüero et al., 2018; de Nayer et al., 2011; Hanssen et al.,

2009; Schenk et al., 2010; Vermunt and Kaarsemaker, 2017), as discussed in the case study in Section 4.

The first challenge in an efficient and successful cross-protection or vaccination suitable for cultivation practice is to identify or generate an isolate which induces no or only mild symptoms (Yeh and Gonsalves, 1994). However, having a mild isolate is only the starting point, as a cross-protection strategy for use under commercial cultivation should meet a wide range of conditions to guarantee efficient and durable protection. A number of reviews have identified the minimum conditions for long-term successful and safe cross-protection (Lecoq and Raccah, 2001; Ziebell and Carr, 2010; Ziebell and MacDiarmid, 2017). This section discusses these requirements with additional insights from more recent PepMV cross-protection studies.

First, the isolate which is intended for use in a successful cross-protection strategy should comply with the following requirements:

- **The protective isolate should induce no or only mild symptoms without having a significant impact on yield or quality of the crop or produce. These mild symptoms should be consistent with a range of commercial cultivars and in a wide range of climatic and cultivation conditions.** Previous experience, for example against ZYMV and PRSV, has shown that mild isolates can cause undesirable symptoms and/or affect yield under different circumstances. In the case of ZYMV, the mild isolate (ZYMV-WK) was originally isolated from melon in France and was used to control ZYMV damage on different cucurbit species in France (Lecoq et al., 1991), the United Kingdom (Walkey et al., 1992), Italy (Desbiez and Lecoq, 1997), Hawaii (Cho et al., 1992), Taiwan (Wang et al., 1991), the United States (Perring et al., 1995), Turkey (Yilmaz et al., 1994), and Israel (S Singer et al., 1994). This mild isolate induced mild mosaic symptoms on the leaves, but no symptoms were observed on the fruits. However, yield reductions were reported ranging from 4% to 38% in later greenhouse and field trials on zucchini and marrow, at different locations and in different seasons in the UK (Spence et al., 1996). In the case of PRSV, different mild isolates were applied in commercial papaya cultivars in different parts of the world with variable success rates (Chatchawankanphanich et al., 2000; Dias and Rezend, 2001; Gonsalves, 1998; Gottula and Fuchs, 2009; McMillan and Gonsalves, 1988; Ram et al., 2006; Rezende and Pacheco, 1998; Tripathi et al., 2008; Yeh and Gonsalves, 1994, 1984). Symptoms of the attenuated PRSV isolates varied among cultivars. In addition, mild symptoms only occurred under certain environmental conditions. For example, when the plants suffered from colder periods, lack of light and rainy conditions, severe symptoms on leaves and fruits were observed (Yeh and Gonsalves, 1994). In conclusion, experimental studies under controlled conditions provide the first step to screen and document host-virus interactions. However, trials should include a wide range of growing conditions and cultivars reflecting the conditions of the commercial crop in order to assure a successful strategy.

- **The protective isolate should be genetically stable.** Most plant viruses possess RNA genomes. RNA viruses exhibit genetic variation due to high mutation rates (Drake et al., 1998; Malpica et al., 2002). The sequence stability of the protective viral isolate which will be used as a vaccine in a cross-protection strategy should be proven. A high genetic stability of the protective viral isolate is crucial to minimize nucleotide sequence variability during the multiplication of the primary inoculum and the subsequent steps of plant vaccine production. In addition, the stability of the mild isolate should be strictly controlled throughout the vaccine production process to avoid mutations that could induce aggressive symptoms.
- **The mild isolate should infect the plants fast, systematically, and confer sufficient protection during the entire cropping season.** These factors have been identified as important criteria based on experience with, for example, ZYMV and PRSV cross-protection. For ZYMV, a period of about 14 days was needed between the initial and the challenge inoculation to obtain complete immunization (Walkey et al., 1992). However, for PRSV, a much longer interval of 26 days was required (Yeh and Gonsalves, 1994). PRSV protection was also broken down under high challenge infection pressure and after a certain period of time (Gonsalves and Garnsey, 1989; Yeh and Gonsalves, 1994; Yeh and Kung, 2007). Break-down of protection might be correlated with decreasing concentrations of the protective virus isolate in the plant and with decreasing levels of vsiRNAs in the plants, possibly influenced by temperature (Szittya, 2003). If the mild isolate does not infect all cells of a plant, the cells should at least be 'primed' by mobile vsiRNA to develop resistance against secondary infections (Gal-On and Shiboleth, 2006).
- **The vaccine should provide protection against the widest possible range of the pathogenic target virus.** The importance of a high sequence homology between the protective and challenge virus isolates has been demonstrated in studies on PRSV, ZYMV, CTV, and *Barley yellow dwarf virus*. In these cases, the efficacy provided by the mild protective isolates was highly dependent on the strain of the challenge virus and protection was only effective when challenged with related isolates (Chatchawankanphanich et al., 2000; Desbiez and Lecoq, 1997; Folimonova, 2013; Lecoq and Raccah, 2001; Tennant, 1994; Wang et al., 1991; Wen et al., 1991; Yeh, 1988). This suggests that RNA silencing is the main mechanism underlying cross-protection, as RNA silencing requires a high sequence homology between the siRNA incorporated in the RISC

complex and the corresponding sequence stretch in the challenge virus (Voinnet, 2001).

As a next step, since cross-protection implies artificial inoculation with a living virus, the following concerns need to be properly addressed to ensure safe and durable plant vaccination:

- **The risk of spreading the protective virus isolate to non-target crops should be controlled.** The application of cross-protection should be done in a controlled manner to avoid the spread of the protective virus and potential damage to non-target host plants and crops. Vaccination with a mild protective virus while implementing proper hygienic measures in confined environments such as greenhouses will help to prevent the spread of the virus to unintended host plants. Application in open fields might pose a higher risk of contamination and transmission to nearby plants, through insects, water, pollen, birds, and other animals. However, this risk is low if the mild protector virus has a narrow host range. If other crops could be hosts, symptoms induced by the mild protective isolate in these crops should be evaluated.
- **Mixed viral infections should be avoided to prevent more damage to the plants.** Co-infections with different viruses can cause synergistic effects and cause damage to the plants (Kosaka and Fukunishi, 1997; Rochow and Ross, 1955; Vance, 1991). In addition, mixed infections with different strains from the same viral species can lead to genome recombination events generating new, potentially more aggressive variants (Bergua et al., 2014; Hanssen et al., 2008; Hasiów-Jaroszewska et al., 2010a; Monci et al., 2002; Moriones et al., 2007). The risk of a synergistic interaction of the mild protective isolate with other viral strains or species potentially present in the crop should be considered if such other viruses are prevalent in the target crop. Implementation of cross-protection in a controlled environment, such as a greenhouse, is therefore preferable compared to open-field application.
- **The population dynamics of viral epidemics should be studied before and during the cross-protection application.** RNA viruses are particularly prone to mutations and, as a result, new variants can easily be generated (Drake et al., 1998; Hasiów-Jaroszewska et al., 2011; Malpica et al., 2002). Increased viral fitness results in the replacement of viral strains by biologically more fit variants. Successful cross-protection strategies require a high degree of sequence homology between the protective and the challenge virus (Yeh and Gonsalves, 1984). A thorough study of the existing viral population is required to select a suitable mild isolate belonging to the dominant viral genotype in order to obtain protection

against the virulent virus population. The evolution of the virus population needs to be investigated so that the strategy can be adapted to shifting populations if needed.

- **The vaccine production process should safeguard the mild characteristics and purity of the protective isolate.** The primary inoculum of the protective isolate should be multiplied in selected host plants under strict hygiene conditions and following a standardized protocol, to avoid variability in nucleotide sequence and viral load. The quality of the mild isolate must be controlled during the propagation of the inoculum and production phase of the vaccine to avoid decrease in infectivity, conversion to a more aggressive variant, or contamination with other pathogens. During the commercial application of the mild TMV isolate in the mid-1990s, problems in production led to compromised purity and infectivity of the mild protective isolate (Rast, 1975, 1972). The storage stability of the inoculum and the plant protection product is also important to maintain both the correct viable virus load and the mild characteristics of the isolate. In the case of PRSV, severe symptoms were obtained after long-term storage at 4°C when the mild isolate was recovered from $CaCl_{2-}$dried tissue (Yeh and Gonsalves, 1998).
- **The vaccination strategy should be optimized under commercial cultivation conditions.** Important aspects in the development of a vaccination strategy include fine-tuning the dose, the final spraying volume, the application method, and the timing of the application. A well-balanced, optimized vaccination strategy should provide rapid systemic colonization without causing undesired plant responses. Slow colonization delays protection, as systemic infection is needed to obtain good protection (see above), and increases the risk of secondary infections by more aggressive virus isolates. Such secondary infections could originate from previous crops or from neighbouring plants and crops. In addition, fast colonization eliminates the risk that mutated variants of the mild isolate, occasionally formed within a treated plant during replication, would spread to adjacent plants (cf. Section 4). For the same reason, the entire crop should be treated to obtain efficient protection of all plants.
- **The strategy should be simple to apply in cultivation practice.** Mechanically-transmitted viruses can easily be inoculated on a large scale (Yarden et al., 2000; Yeh and Gonsalves, 1994), and are thus more suitable for cross-protection strategies than vector-transmitted viruses. Vaccination based on vector-transmitted viruses is more challenging, since a fast, systemic, and homogeneous infection of the crop is more difficult to achieve. A stable and suitable formulation that can conveniently and homogeneously be applied onto the crop is crucial to achieve reliable and consistent vaccination in practical cultivation conditions.

4 Application of cross-protection under commercial cultivation conditions: overview and case study

4.1 Overview

Mild isolates used in cross-protection in the EU now need to be registered as plant protection products, under Reg. (EC) No 1107/2009. The EU plant protection database (https://ec.europa.eu/food/plant/pesticides/eu-pesticides-database, accessed on 12 October 2020) shows that only a few mild plant viruses have been registered or notified to be registered as plant protection products within the EU (Table 1).

Out of the seven mild virus isolates on the list, five belong to the species PepMV. The two non-PepMV isolates on the list are either not approved (ToMV) or not in use (ZYMV). The mild strain of ToMV was applied in tomato crops before resistant varieties became available. The mild isolate was thus used mainly in the 1980s and early 1990s, before the harmonization of the EU plant protection Directive (EC) No 91/414, in effect from 1991 and replaced by the current (EC) No 1107/2009 pesticide regulation in 2009. Since the availability of ToMV-resistant varieties in the 1990s replaced the need for a cross-protection strategy, the active substance was never approved under the harmonized EU plant protection legislation, and is thus not listed in Table 1.

Table 1 EU pesticide database - active substances under Reg. (EC) No 1107/2009

Active substance	Status	Date of approval	Expiry of the approval
Mild Pepino Mosaic Virus isolate VC 1	Approved	29/03/2017	29/03/2032
Mild Pepino Mosaic Virus isolate VX 1	Approved	29/03/2017	29/03/2032
Pepino mosaic virus (PepMV) Chilean (CH2) strain, mild isolate Abp2 (PEPMVO)	Pending	Pending	Pending
Pepino mosaic virus (PepMV) European (EU) strain, mild isolate Abp1 (PEPMVO)	Pending	Pending	Pending
Pepino mosaic virus strain CH2 isolate 1906	Approved	07/08/2015	07/08/2030
Tomato mosaic virus	Not Approved	Not Approved	Not Approved
Zucchini yellow mosaic virus (ZYMV mild strain)	Not Approved	Not Approved	Not Approved
Zucchini yellow mosaic virus - weak strain	Approved	01/06/2013	31/05/2023

In the 1990s, a mild ZYMV isolate (WK) was used in different countries to control the damage of aggressive ZYMV isolates in cucurbit plants (Lecoq et al., 1991; Perring et al., 1995; Walkey et al., 1992; Wang et al., 1991; Yilmaz et al., 1994). In Israel, the strategy was applied to control damage by ZYMV (S. Singer et al., 1994; Yarden et al., 2000) in watermelon crops until the viral population changed and begomoviruses such as *Watermelon chlorotic stunt virus* (WmCSV) became the major threat (Abudy et al., 2010) (personal communication from Dr. Amit Gal-On). To our knowledge, the approved active substance has never been used in practice, as the registration of the final product was not completed.

4.2 Case study: PepMV cross-protection

This case study focuses on PepMV-tomato interaction to understand why PepMV is the only viral plant disease for which several cross-protection-based vaccines have been developed and authorized for use in Europe. There are several reasons for the wide acceptance of cross-protection as a tool to control PepMV in tomato. The lack of resistant commercial varieties is one of the main drivers for the use of mild isolate-based plant vaccines against this damaging virus.

PepMV was first identified on pepino (*Solanum muricatum*) in 1974 in Peru (Jones et al., 1980). However, it was not until 1999 that the virus was first reported to infect commercial tomato crops (*S. lycopersicum*) (van der Vlugt et al., 2000; Wright and Mumford, 1999). PepMV has since become a major threat to greenhouse tomato production worldwide (Carmichael et al., 2011; Efthimiou et al., 2011; Hanssen and Thomma, 2010; Ling and Zhang, 2011; Werkman and Sansford, 2010).

PepMV contains a positive-sense single-stranded RNA genome and belongs to the genus *Potexvirus* of the family *Alphaflexiviridae*. Five different PepMV genotypes have currently been identified:

1) the Peruvian (LP) genotype, which was originally found in *S. muricatum* and wild *Solanum spp.*;
2) the European (EU) genotype, which was found to infect tomato crops in the Netherlands and the UK in 1999;
3) the Chilean (CH2) genotype, which has largely replaced the EU strain in Europe during the last decade (Gómez et al., 2012; Hanssen et al., 2008);
4) the American (US1) genotype, which circulates mainly in the US, Canada, and Mexico (Ling et al., 2013, 2008);
5) the PES genotype, described as infecting wild tomato species in Peru (Moreno-Perez et al., 2014).

Isolates belonging to the same genotype are very similar (99%), but the difference between genotypes is significant: the EU genotype shares about 96% sequence identity with the LP genotype (Hanssen and Thomma, 2010) and about 80% sequence identity with the US1 (Hanssen and Thomma, 2010; Maroon-Lango et al., 2005), CH2 (Hanssen and Thomma, 2010; Ling et al., 2007) and the PES genotypes (Moreno-Perez et al., 2014). In Europe, the initial infections with the EU genotype were mostly replaced by the CH2 genotype, which became dominant in most European countries (Gómez et al., 2012, 2009; Hanssen et al., 2008). Shifts in the PepMV population have been also reported in North America, where the initial US1 infections were first replaced by the EU genotype, then by the CH2 genotype, and later by the re-introduction of the US1 genotype (French et al., 2001; Ling et al., 2013, 2008). Mixed infections of the EU and CH2 genotypes within the same plant were found to result in recombinants and to lead to more severe symptoms and damage to the crop (Hanssen et al., 2008; Hasiów-Jaroszewska et al., 2010b).

PepMV is transmitted mechanically, through direct contact between adjacent plants, via handling of crops, and use of contaminated tools (Spence et al., 2006; Wright and Mumford, 1999). In addition, transmission through recirculating irrigation water (Schwarz et al., 2010), bumblebees (Plasencia et al., 2003; Shipp et al., 2008), and, to a limited extent, by contaminated seeds (Córdoba-Sellés et al., 2007; Hanssen et al., 2009; Ling et al., 2013; Salomone and Roggero, 2002) has been demonstrated.

The host range of PepMV is mainly limited to the *Solanaceae* family, although screenings of natural vegetation in the surroundings of infected tomato crops have revealed that the virus can also occur in some weed species belonging to different families (Córdoba et al., 2004; Jones et al., 1980; Jordá et al., 2001; Papayiannis et al., 2012; Salomone and Roggero, 2002). However, the epidemiological importance of these findings is unclear.

PepMV symptoms in tomato plants range from asymptomatic to severe leaf bubbling, nettle-head-like appearance in the apex, necrosis, leaf mosaics, and yellow spots. Although these symptoms can significantly reduce plant vigor and yield, the main impact of PepMV infection is discoloration, marbling, or necrosis on the tomato fruit (Hanssen et al., 2009; van der Vlugt et al., 2000). This reduces fruit quality, significantly lowering the market value (Blystad et al., 2015; de Nayer et al., 2011; Hanssen and Thomma, 2010; Spence et al., 2006).

Before PepMV vaccines became available, growers solely relied on the prevention of PepMV infection through strict hygiene measures, as resistant tomato varieties were not available. The success of these measurements was limited, especially in areas with high infection pressure due to the high-density production in tomato greenhouses.

Different research groups have evaluated the potential of cross-protection to control PepMV damage in tomato crops. In a study by Hanssen et al. (2010), a selection of potential mild PepMV isolates, belonging to different genotypes, was evaluated for crop protection against infection by aggressive isolates under greenhouse conditions. Efficient protection was obtained with the mild CH2 isolate 1906 against an aggressive CH2 infection. In contrast, pre-inoculation with mild isolates belonging to the EU and LP genotypes led to some infection, with more fruit symptoms and yields reduced by 13% compared with plants challenged with the same aggressive CH2 isolate. Molecular analyses confirmed the presence of both genotypes in these plants, suggesting that synergism or occurrence of recombinant variants could play a role in the enhanced symptom severity that was observed in these mixed infections (Hanssen et al., 2008; Hasiów-Jaroszewska et al., 2010b).

The study by Hanssen et al. (2010) underlines the importance of studying the virus population in the region in which application of cross-protection is envisaged in practice. Since PepMV infections in Europe were shown to be mainly caused by isolates of the CH2 genotype (Hanssen et al., 2008), the use of the mild CH2 isolate 1906 to protect tomato crops was further studied under greenhouse conditions (de Nayer et al., 2011). Greenhouse trials confirmed the efficacy of a formulated plant vaccine based on this isolate. A plant protection product designated as PMV®-01 (DCM N.V.), with the mild isolate 1906 as the active ingredient, is currently authorized for use in tomato greenhouses in 18 EU countries as well as in the UK, Switzerland, Georgia, Morocco, Canada, and the US. The vaccine is applied under strict hygiene measures, following a specifically designed vaccination strategy (https://pmv-01.com/en/, accessed on 14 October 2020). Growers receive detailed product guidelines. The vaccination strategy includes a sampling scheme before and after vaccination and follow-up during the growing season. The samples are analyzed with strain-specific RT-qPCR analyses to determine the absence of PepMV before application and to control the concentration of the mild isolate in the plants several weeks after vaccination.

This detailed follow-up has shown its value, for example with rapid detection and identification of other viral infections that might otherwise have been considered as PepMV, and thus not managed correctly. In one case, plants with severe discoloration on the leaves and fruits were observed in a PMV-01-vaccinated greenhouse (Fig. 2). These symptoms appeared mainly in one plant row, with a few adjacent plants showing similar symptoms. The symptoms could have been considered as PepMV symptoms. However, since the crop was vaccinated, further research was undertaken. Broad molecular analyses with genus-specific primers revealed the presence of a Potyvirus. Subsequent BLAST analyses led to the identification of the *Colombian datura virus* (CDV) (Kahn and Bartels, 1968; Zheng et al., 2010). CDV is spread by aphids and affects mainly plant species from the genus *Brugmansia* (Kahn and Bartels, 1968). The first virus identification in tomato was in the Netherlands,

Figure 2 Symptoms observed in vaccinated tomato plants infected with *Colombian datura virus* (CDV).

as an infected Brugmansia plant was kept in the same greenhouse and served as inoculum (Verhoeven et al., 1996). To our knowledge, this was the first case of an infection with CDV in a tomato greenhouse in Belgium. With the correct identification, the virus could easily be contained and eliminated through the removal of the plants and proper aphid control.

Practical experience with the PMV-01 vaccine has proven the importance of fast and systemic colonization of the mild isolate in the entire crop, in order to obtain efficient protection against aggressive isolates. Quality control during production of the vaccine guarantees the purity of the mild isolate in the final product. However, after application of the vaccine, mutations can occur during the replication of the mild isolate in plants, but mutants will in principle not be able to spread because adjacent plants are protected. Some point mutations can be detrimental, such as necrotic PepMV mutants (Hasiów-Jaroszewska et al., 2011; Hasiów-Jaroszewska and Borodynko, 2012; Sempere et al., 2016) and the yellowing PepMV mutants (Fig. 3) (Hasiów-Jaroszewska et al., 2013). If such a mutation would occur in one of the vaccinated plants, cross-protection will prevent the mutant from spreading to the surrounding plants that are also vaccinated and thus protected by the wild-type mild isolate. In practice, a necrotic or yellowing mutant remains contained in a vaccinated crop, with symptoms being limited to only one or a few plants. A correct application of the right dose, using the recommended spraying volume, onto the entire crop provides fast and efficient protection against such more aggressive mutants. A plant protection registration dossier highlights the need for precise fine-tuning of the dose to obtain optimal protection, to ensure only very mild viral symptoms for a short period of time in the first weeks after vaccination. The PMV-01 vaccine was reported not to cause any yield losses (de Nayer et al., 2011).

Recent studies have evaluated cross-protection with a mixture of genotypes. Vermunt and Kaarsemaker (2017) pre-infected tomato plants with a mixture of mild LP (VX1) and CH2 (VC1) isolates. The first mixed genotype vaccine V10, containing the mild isolates VX1 and VC1 as active substances (Table 1), has been approved in the Netherlands since 2018, and subsequently in several other EU countries. Agüero et al., (2018) used a mixture of mild EU and CH2 isolates for protection against aggressive EU and CH2 isolates. However, the mixed genotype vaccine caused a yield loss of approximately 8–12% when compared with non-treated plants, depending on the tomato cultivar. In areas where mixed infections with the EU and CH2 strains are dominant, such as Almeria and Murcia in Spain (Gómez et al., 2009), such a strategy may be appropriate despite the impact on yield, as long as the benefit to the grower and the safety of the crop and its surroundings are assured before large-scale application. In areas where the CH2 strain is dominant, using a mixed genotype vaccine poses risks for decreasing yield, spreading non-prevalent virus strains to neighboring crops, and providing opportunities for the occurrence of recombinant isolates.

An alternative to the natural mild isolates used in the studies described so far might be to generate attenuated isolates through guided mutagenesis. A mild PepMV mutant was generated using this approach by modifying non-conserved amino acid residues in the CP, between closely related potexviral CPs and subsequent inoculation of the generated mutants, to evaluate their induction of symptoms (Chewachong et al., 2015). This strategy may be appropriate in cases where natural mild isolates cannot be identified.

During the last decade, cross-protection strategies against PepMV have offered tomato growers a valuable tool to minimize the losses caused by this important pathogen. The case of PepMV vaccination shows that cross-protection can be considered as a sustainable biocontrol tool, to combat existing and emerging viral diseases.

5 Summary and future trends

Increasing population, climate change, and globalization continuously increase the pressure on agricultural production systems. Fast and easy movement of seeds and fresh products around the world hinders efficient control of plant pathogens. The spread of plant diseases has increased significantly in recent decades (Brasier, 2008; Fisher et al., 2012; McDonald and Stukenbrock, 2016), with plant viruses becoming more prevalent and challenging to control, due to their high infectivity and the lack of curative control strategies (Jones and Naidu, 2019; Rojas and Gilbertson, 2008). Viral diseases can cause important economic losses due to their impact on quality and yield. In addition, these diseases hamper the sustainability of crop production systems, due to higher

Figure 3 (a) Plant showing yellowing symptoms due to a mutation in the viral coat protein and (b) necrotic symptoms due to a mutation in the TGBp3 of PepMV.

production costs, increased use of pesticides to control mainly insects as viral vectors, and additional mitigation measures. Consequently, more and better approaches to preventively control viral pathogens are urgently required to guarantee sustainable and sufficient production of plant-based food sources.

This chapter has reviewed cross-protection, also referred to as plant vaccination, as a biocontrol tool to manage viral diseases in plant production systems. Three molecular mechanisms are potentially involved in cross-protection. Protective effects can be induced by the coat protein of the protective strain that either prevents the uncoating of the challenging strain or partially coats its RNA. The RNAi machinery and spatial separation phenomenon have also been found to be involved in cross-protection. Successful implementation of cross-protection strategies in commercial plant production systems requires the use of a genetically stable protective isolate that induces mild symptoms, that do not affect the quantity nor the quality of the produce. Infection should be fast, systematic, and lead to sufficient protection of the crop throughout the cropping season and against the widest range of viral isolates. Since cross-protection is induced by a living virus, a number of issues need to be addressed for safe application, such as the possibility of infection of non-target crops, the effects of possible mixed infections with the protective strain, the changes in viral population dynamics that could arise, the establishment of proper quality control protocols for virus inoculum multiplication and vaccine production, and straightforward protocols for large-scale application. The legal framework for plant vaccination in the EU is the plant protection regulation (EC) No 1107/2009. Amongst the currently approved active substances, there is one plant–virus combination, tomato–PepMV, that provides a good example of such a vaccination approach. This is shown by extensive studies on virus–host interaction and the current status of PepMV vaccines.

The next steps in research should focus on further deciphering the mechanisms involved in cross-protection. The use of the latest advances in molecular tools and analytical methods, such as CRISPR-CAS and Next Generation Sequencing (NGS), could be utilized to further unravel the molecular mechanisms behind cross-protection. A better understanding of the phenomenon will facilitate the improvement of selection and control of the protective isolate and minimize environmental and off-target risks following a vaccination strategy.

One example showing how the latest advances in molecular biology can provide insight into plant innate immune responses and illuminate the molecular mechanisms involved in cross-protection is a recent study of the effects of cross-protection on the host. The isobaric Tags for Relative and Absolute Quantification (iTRAQ) proteomics technique was used to identify host proteins that play an important role in cross-protection. The authors studied plant proteomics after inoculation with a mutant strain of TMV, where one of the highly conserved upstream pseudoknotted domains in the viral RNA was replaced with a poly(A) tail. The mutant TMV-43A caused mild mosaic symptoms in *N. benthamiana* plants. The authors observed that this mild strain could protect the plants against wildtype TMV. In cross-protected plants, as opposed to mock-inoculated and wildtype TMV-inoculated plants, they observed an augmentation in enzymes involved in chlorophyll biosynthesis and reactive oxygen species (ROS) scavenging, and a reduction in defense-related proteins (Das et al., 2019).

Future approaches in plant virus protection, either based on cross-protection, on resistance, or on other mechanisms, should concentrate on designing strategies that can confer stable protection, that can be developed over a short period of time to respond adequately to rapidly changing epidemiological dynamics, and that can ideally be used against multiple viral threats to compensate for the increase in emerging viruses, while not posing any threat to human health or the environment.

One approach that can facilitate the development of a cross-protection strategy is the induction of mutations in a known virus to create attenuated strains that can protect plants against the wildtype strain. This approach avoids the laborious and time-consuming process of screening viral isolates to identify a naturally occurring, stable, and mild isolate, although this approach requires careful molecular and biological characterization of the generated mutations. A recent study used site-directed mutagenesis to obtain five mutants of *Cucumber green mottle mosaic virus* (CGMMV). The viral mutants were introduced onto *N. benthamiana* plants, which were both screened for symptom development and used in a cross-protection assay to evaluate the effects of the attenuated mutants. The study resulted in the identification of two CGMMV mutants that provided protection against CGMMV with 91.7% and 100% efficiency (Liu et al., 2020).

Another approach showing promising results for efficient and flexible virus disease management is the design of strategies based on antiviral RNA interference (RNAi) and, more specifically, on the ability of dsRNA molecules to act as silencing triggers. RNAi-based strategies include genetically modified (GM) plants that express fragments of the viral sequence that could result in the generation of vsiRNAs, thus activating the RNAi machinery. Other strategies involve the design, based on the viral genome, of dsRNA molecules that can themselves be enzymatically processed into vsiRNAs. The dsRNAs are artificially constructed and externally applied onto plants, initiating the RNAi machinery once they enter the plant cell against viruses with homologous nucleotide sequences.

It is also important to mention the well-established strategy of Host-Induced Gene Silencing (HIGS). During HIGS, a DNA construct expressing antisense or hairpin RNA sequences that target the pathogen is introduced to the host genome. Transcription of the insertion results in RNA that form double-stranded molecules which trigger the RNAi pathways and finally result in the formation of a pool of siRNAs. When the insertion is designed to be complementary to a viral genome, it can activate the host RNAi pathways against infection with a virus sharing sufficient nucleotide sequence homology with the insert (Cai et al., 2018). The advantage of HIGS is that it can confer protection not only against a specific viral threat, but also against multiple viral genotypes, as recently shown for PepMV (Leibman et al., in press), against multiple viral species (Hu et al., 2011; Zhang et al., 2013) or through the insertion of multiple sequences of different pathogens in the same host, even multiple pathogen groups such as bacteria, fungi, nematodes, and viroids (Wang et al., 2016). Such a broad protection could be established due to the cross-kingdom RNA-trafficking mechanism in which small RNA molecules (sRNAs) are exchanged between interacting organisms (Cai et al., 2018).

Despite the reproducible results and the good protection levels obtained by transgenic approaches, field application of GM plants faces many obstacles due to negative public opinion and the current ban on transgenic plants in the EU. An alternative could be the use of CRISPR-CAS-mediated plant virus resistance. This technology can be used to confer either direct inhibition of plant viruses or to knock out host susceptibility factors to confer resistance. An example of direct inhibition is a study on *Bean Yellow dwarf virus* (BeYDV). The BeYDV genome was used to design single-guide RNAs (sgRNAs) to guide the Cas9 enzyme to the viral DNA. Cas9 induced double-stand breaks that were repaired by non-homologous end-joining, leading to mutations to the viral genome. When this construct was used to create transgenic plants, a reduction in viral copy numbers was observed (Baltes et al., 2015). The principle of host genome editing to knock out susceptibility factors required for viral infection was demonstrated by the knock out of the *eukaryotic translation initiation factor* 4E (*eIF*4E), which provided broad resistance to many members of the *Potyviridae* family (Chandrasekaran et al., 2016; Kalinina et al., 2020).

In addition to transgenic approaches, other studies have explored the exogenous application of RNAi-inducing signals (Taning et al., 2020). The direct application of dsRNA molecules onto hosts is termed Spray-Induced Gene Silencing (SIGS) and, much like HIGS, can target many different pathogens, depending on the sequence of the dsRNAs. This approach enables fast design, can be adapted to a multi-target strategy, and is less susceptible to target-resistance mutations (Cai et al., 2018). An example of a successful lab-scale SIGS application is that by Vadlamudi et al., (2020). Two dsRNA molecules were designed from the coat protein (CP) and the helper component-proteinase (HC-Pro) of an isolate PRSV from South India (PRSV-Tirupati). Topical application of the molecules led to 100% resistance against PRSV-Tirupati. Additionally, the molecules provided protection against an isolate from North India, PRSV-Delhi, with the CP molecules conferring 94% resistance and the HC-Pro molecules 81%.

Despite its many advantages, the uptake and systematic spread of dsRNA is challenging in commercial cultivation (Mitter et al., 2017). Establishment of systemic long-lasting protection is a major challenge in exogenous applications of naked dsRNAs for control of plant viruses (Dalakouras et al., 2020). Tenllado et al. (2003) showed that SIGS provided protection for five days post-application, but plants challenged at seven days after the application of purified dsRNAs were susceptible to infection. New methods are being developed to improve the delivery and stability of dsRNA molecules, using nanomaterials or protein binding to protect them from degradation (Taning et al., 2020). A very promising example is the use of clay nanosheets for topical application of dsRNA molecules that protect molecules from degradation and provide prolonged protection (Mitter et al., 2017). Some large agrobiotech companies have recently shown interest in R&D projects focusing on SIGS applications, mainly to control insects and nematodes. These developments could pave the way for SIGS to control plant virus diseases, reducing dsRNA production costs, and increasing knowledge of potential off-target effects.

6 Where to look for further information

6.1 Comprehensive reviews on cross-protection

- Antagonistic within-host interactions between plant viruses: molecular basis and impact on viral and host fitness by Syller and Grupa, 2016.
- Cross-protection: A century of mystery by Ziebell and Carr, 2010.
- Cross-protection by Gal-On and Shiboleth, 2006.

6.2 Relevant international projects

- COST-Action FA0806 on RNA vaccines against plant viruses (https://www.cost.eu/actions/FA0806/ accessed on 23 July 2021).

- COST-Action on NGS to study plant viruses (https://www.cost.eu/cost-action/application-of-next-generation-sequencing-for-the-study-and-diagnosis-of-plant-viral-diseases-in-agriculture/ accessed on 23 July 2021).

7 Acknowledgement

We would like to thank Prof. Dr. Bart P. H. J. Thomma for his valuable revisions of this chapter. Zafeiro Zisi is funded by a Baekeland scholarship from the Agency for Innovation and Entrepreneurship, Flanders, Belgium.

8 References

Abel, P. P., Nelson, R. S., De, B., Hoffmann, N., Rogers, S. G., Fraley, R. T. and Beachy, R. N. 1986. Delay of disease development in transgenic plants that express the tobacco mosaic virus coat protein gene. *Science* 232(4751), 738–743.

Abudy, A., Sufrin-Ringwald, T., Dayan-Glick, C., Guenoune-Gelbart, D., Livneh, O., Zaccai, M. and Lapidot, M. 2010. Watermelon chlorotic stunt and Squash leaf curl begomoviruses–new threats to cucurbit crops in the Middle East. *Israel Journal of Plant Sciences* 58(1), 33–42.

Agüero, J., Gómez-Aix, C., Sempere, R. N., García-Villalba, J., García-Núñez, J., Hernando, Y. and Aranda, M. A. 2018. Stable and broad spectrum cross-protection against pepino mosaic virus attained by mixed infection. *Frontiers in Plant Science* 9, 1810.

Baltes, N. J., Hummel, A. W., Konecna, E., Cegan, R., Bruns, A. N., Bisaro, D. M. and Voytas, D. F. 2015. Conferring resistance to geminiviruses with the CRISPR–Cas prokaryotic immune system. *Nature Plants* 1(10), 15145.

Bardonnet, N., Hans, F., Serghini, M. A. and Pinck, L. 1994. Protection against virus infection in tobacco plants expressing the coat protein of grapevine fanleaf Nepovirus. *Plant Cell Reports* 13(6), 357–360.

Bebber, D. P. 2015. Range-expanding pests and pathogens in a warming world. *Annual Review of Phytopathology* 53, 335–356.

Bebber, D. P., Holmes, T. and Gurr, S. J. 2014. The global spread of crop pests and pathogens. *Global Ecology and Biogeography* 23(12), 1398–1407.

Bergua, M., Luis-Arteaga, M. and Escriu, F. 2014. Genetic diversity, reassortment, and recombination in *alfalfa mosaic virus* population in Spain. *Phytopathology®* 104, 1241–1250.

Bian, X.-Y., Rasheed, M. S., Seemanpillai, M. J. and Rezaian, M. A. 2006. Analysis of silencing escape of *Tomato leaf curl virus*: an evaluation of the role of DNA methylation. *Molecular Plant-Microbe Interactions®* 19, 614–624.

Blystad, D.-R., van der Vlugt, R., Alfaro-Fernández, A., del Carmen Córdoba, M., Bese, G., Hristova, D., Pospieszny, H., Mehle, N., Ravnikar, M., Tomassoli, L., Varveri, C. and Nielsen, S. L. 2015. Host range and symptomatology of Pepino mosaic virus strains occurring in Europe. *European Journal of Plant Pathology* 143(1), 43–56.

Brasier, C. M. 2008. The biosecurity threat to the UK and global environment from international trade in plants. *Plant Pathology* 57(5), 792–808.

Cai, Q., He, B., Kogel, K. H. and Jin, H. 2018. Cross-kingdom RNA trafficking and environmental RNAi–nature's blueprint for modern crop protection strategies. *Current Opinion in Microbiology* 46, 58–64.

Carmichael, D. J., Rey, M. E. C., Naidoo, S., Cook, G. and van Heerden, S. W. 2011. First report of *Pepino mosaic virus* infecting tomato in South Africa. *Plant Disease* 95(6), 767–767.

Chandrasekaran, J., Brumin, M., Wolf, D., Leibman, D., Klap, C., Pearlsman, M., Sherman, A., Arazi, T. and Gal-On, A. 2016. Development of broad virus resistance in non-transgenic cucumber using CRISPR/Cas9 technology. *Molecular Plant Pathology* 17(7), 1140–1153.

Chatchawankanphanich, O., Jamboonsri, W., Kositratana, W. and Attathom, S. 2000. Screening of mild strains of papaya ringspot virus for cross protection. *Thai Journal of Agricultural Science* 33, 147–152.

Chester, K. S. 1933. The problem of acquired physiological immunity in plants. *The Quarterly Review of Biology* 8(3), 275–324.

Chewachong, G. M., Miller, S. A., Blakeslee, J. J., Francis, D. M., Morris, T. J. and Qu, F. 2015. Generation of an attenuated, cross-protective *Pepino mosaic virus* variant through alignment-guided mutagenesis of the viral capsid protein. *Phytopathology®* 105, 126–134.

Chitwood, D. H. and Timmermans, M. C. P. 2010. Small RNAs are on the move. *Nature* 467(7314), 415–419.

Cho, J. J., Ullman, D. E., Wheatley, E., Holly, J. and Gonsalves, D. 1992. Commercialization of ZYMV cross protection for zucchini production in Hawaii. *Phytopathology* 82, 1073.

Córdoba, M. C., Martínez-Priego, L. L. and Jordá, C. 2004. New natural hosts of *Pepino mosaic virus* in Spain. *Plant Disease* 88(8), 906–906.

Córdoba-Sellés, M. D. C., García-Rández, A., Alfaro-Fernández, A. and Jordá-Gutiérrez, C. 2007. Seed transmission of *Pepino mosaic virus* and efficacy of tomato seed disinfection treatments. *Plant Disease* 91(10), 1250–1254.

Costa, A. S. and Müller, G. 1980. Tristeza control by cross protection: a US-Brazil cooperative success. *Plant Disease* 64(6), 538–541.

Dalakouras, A., Wassenegger, M., Dadami, E., Ganopoulos, I., Pappas, M. L. and Papadopoulou, K. 2020. Genetically modified organism-free RNA interference: exogenous application of RNA molecules in plants. *Plant Physiology* 182(1), 38–50.

Das, P. P., Chua, G. M., Lin, Q. and Wong, S. M. 2019. iTRAQ-based analysis of leaf proteome identifies important proteins in secondary metabolite biosynthesis and defence pathways crucial to cross-protection against TMV. *Journal of Proteomics* 196, 42–56.

de Nayer, F., Goen, K., Paeleman, A., Parra, N. O., Vanachter, A. C. R. C., Hanssen, I. M., Wittemans, L. and Vandewoestijne, E. 2011. Cross-protection as a control strategy for pepino mosaic virus (PEPMV) in greenhouse tomato. *Acta Horticulturae* 914, 163–169.

Desbiez, C. and Lecoq, H. 1997. Zucchini yellow mosaic virus. *Plant Pathology* 46(6), 809–829.

Dias, P. R. P. and Rezend, J. A. M. 2001. Problemas na premunização de melancia para o controle do mosaico causado pelo papaya ringspot virus. *Fitopatologia Brasileira* 26(3), 651–654.

Dodds, J. A., Lee, S. Q. and Tiffany, M. 1985. Cross protection between strains of cucumber mosaic virus: effect of host and type of inoculum on accumulation of virions and double-stranded RNA of the challenge strain. *Virology* 144(2), 301–309.

Dombrovsky, A. and Smith, E. 2017. Seed transmission of tobamoviruses: aspects of global disease distribution. In: *Advances in Seed Biology*, Jimenez-Lopez, J. C. (Ed), IntechOpen, 234–260.

Drake, J. W., Charlesworth, B., Charlesworth, D. and Crow, J. F. 1998. Rates of spontaneous mutation. *Genetics* 148(4), 1667–1686.

Du, J., Johnson, L. M., Jacobsen, S. E. and Patel, D. J. 2015. DNA methylation pathways and their crosstalk with histone methylation. *Nature Reviews: Molecular Cell Biology* 16(9), 519–532.

Efthimiou, K. E., Gatsios, A. P., Aretakis, K. C., Papayiannis, L. C. and Katis, N. I. 2011. First report of *Pepino mosaic virus* infecting greenhouse cherry tomatoes in Greece. *Plant Disease* 95(1), 78–78.

Fisher, M. C., Henk, D. A., Briggs, C. J., Brownstein, J. S., Madoff, L. C., McCraw, S. L. and Gurr, S. J. 2012. Emerging fungal threats to animal, plant and ecosystem health. *Nature* 484, 186–194.

Folimonova, S. Y. 2013. Developing an understanding of cross-protection by citrus tristeza virus. *Frontiers in Microbiology* 4, 76.

French, C. J., Bouthillier, M., Bernardy, M., Ferguson, G., Sabourin, M., Johnson, R. C., Masters, C., Godkin, S. and Mumford, R. 2001. First report of *Pepino mosaic virus* in Canada and the United States. *Plant Disease* 85(10), 1121–1121.

Fukuzawa, N., Itchoda, N., Ishihara, T., Goto, K., Masuta, C. and Matsumura, T. 2010. HC-Pro, a Potyvirus RNA silencing suppressor, cancels cycling of Cucumber mosaic virus in Nicotiana benthamiana plants. *Virus Genes* 40(3), 440–446.

Gal-On, A. and Shiboleth, Y. M. 2006. *Natural Resistance Mechanisms of Plants to Viruses* Loebenstein and Carr (Eds). Springer, 261–288.

Gómez, P., Sempere, R. N., Aranda, M. A. and Elena, S. F. 2012. Phylodynamics of Pepino mosaic virus in Spain. *European Journal of Plant Pathology* 134(3), 445–449.

Gómez, P., Sempere, R. N., Elena, S. F. and Aranda, M. A. 2009. Mixed infections of pepino mosaic virus strains modulate the evolutionary dynamics of this emergent virus. *Journal of Virology* 83(23), 12378–12387.

Gonsalves, D. 1998. Control of papaya ringspot virus in papaya: a case study. *Annual Review of Phytopathology* 36, 415–437.

Gonsalves, D. and Garnsey, S. M. 1989. Cross-protection techniques for control of plant virus diseases in the tropics. *Plant Disease* 73(7), 592.

Gonzalez-Jara, P., Tenllado, F., Martinez-Garcia, B., Atencio, F. A., Barajas, D., Vargas, M., DiAZ-RUIZ, J. and Diaz-Ruiz, J. R. 2004. Host-dependent differences during synergistic infection by Potyviruses with potato virus X. *Molecular Plant Pathology* 5(1), 29–35.

Gottula, J. and Fuchs, M. 2009. Toward a quarter century of pathogen-derived resistance and practical approaches to plant virus disease control. *Advances in Virus Research* 75, 161–183.

Hafrén, A., Lõhmus, A. and Mäkinen, K. 2015. Formation of potato virus A-induced RNA granules and viral translation are interrelated processes required for optimal virus accumulation. *PLoS Pathogens* 11(12), e1005314.

Hanssen, I. M., Gutiérrez-Aguirre, I., Paeleman, A., Goen, K., Wittemans, L., Lievens, B., Vanachter, A. C. R. C., Ravnikar, M. and Thomma, B. P. H. J. 2010. Cross-protection or enhanced symptom display in greenhouse tomato co-infected with different *Pepino mosaic virus* isolates. *Plant Pathology* 59(1), 13–21.

Hanssen, I. M., Paeleman, A., Vandewoestijne, E., van Bergen, L., Bragard, C., Lievens, B., Vanachter, A. C. R. C. and Thomma, B. P. H. J. 2009. Pepino mosaic virus isolates and differential symptomatology in tomato. *Plant Pathology* 58(3), 450–460.

Hanssen, I. M., Paeleman, A., Wittemans, L., Goen, K., Lievens, B., Bragard, C., Vanachter, A. C. R. C. and Thomma, B. P. H. J. 2008. Genetic characterization of Pepino mosaic virus isolates from Belgian greenhouse tomatoes reveals genetic recombination. *European Journal of Plant Pathology* 121(2), 131–146.

Hanssen, I. M. and Thomma, B. P. H. J. 2010. Pepino mosaic virus: a successful pathogen that rapidly evolved from emerging to endemic in tomato crops. *Molecular Plant Pathology* 11(2), 179–189.

Hasiów-Jaroszewska, B. and Borodynko, N. 2012. Characterization of the necrosis determinant of the European genotype of pepino mosaic virus by site-specific mutagenesis of an infectious cDNA clone. *Archives of Virology* 157(2), 337–341.

Hasiów-Jaroszewska, B., Borodynko, N., Jackowiak, P., Figlerowicz, M. and Pospieszny, H. 2011. Single mutation converts mild pathotype of the Pepino mosaic virus into necrotic one. *Virus Research* 159(1), 57–61.

Hasiów-Jaroszewska, B., Jackowiak, P., Borodynko, N., Figlerowicz, M. and Pospieszny, H. 2010a. Quasispecies nature of Pepino mosaic virus and its evolutionary dynamics. *Virus Genes* 41(2), 260–267.

Hasiów-Jaroszewska, B., Kuzniar, A., Peters, S. A., Leunissen, J. A. M. and Pospieszny, H. 2010b. Evidence for RNA recombination between distinct isolates of Pepino mosaic virus. *Acta Biochimica Polonica* 57(3), 385–388.

Hasiów-Jaroszewska, B., Paeleman, A., Ortega-Parra, N., Borodynko, N., Minicka, J., Czerwoniec, A., Thomma, B. P. H. J. and Hanssen, I. M. 2013. Ratio of mutated versus wild-type coat protein sequences in *Pepino mosaic virus* determines the nature and severity of yellowing symptoms on tomato plants. *Molecular Plant Pathology* 14(9), 923–933.

Horikoshi, M., Nakayama, M., Yamaoka, N., Furusawa, I. and Shishiyama, J. 1987. Brome mosaic virus coat protein inhibits viral RNA synthesis in vitro. *Virology* 158(1), 15–19.

Hu, Q., Niu, Y., Zhang, K., Liu, Y. and Zhou, X. 2011. Virus-derived transgenes expressing hairpin RNA give immunity to tobacco mosaic virus and Cucumber mosaic virus. *Virology Journal* 8, 41.

Hughes, J. d'A. and Ollennu, L. A. A. 1994. Mild strain protection of cocoa in Ghana against cocoa swollen shoot virus–a review. *Plant Pathology* 43, 422–457.

Jones, R. A. C. 2016. Future scenarios for plant virus pathogens as climate change progresses. *Advances in Virus Research* 95, 87–147.

Jones, R. A. C., Koenig, R. and Lesemann, D. E. 1980. Pepino mosaic virus, a new Potexvirus from pepino (Solanum muricatum). *Annals of Applied Biology* 94(1), 61–68.

Jones, R. A. C. and Naidu, R. A. 2019. Global dimensions of plant virus diseases: current status and future perspectives. *Annual Review of Virology* 6(1), 387–409.

Jordá, C., Pérez, A. L., Martínez-Culebras, P., Abad, P., Lacasa, A. and Guerrero, M. M. 2001. First report of *Pepino mosaic virus* on tomato in Spain. *Plant Disease* 85(12), 1292–1292.

Kahn, R. P. and Bartels, R. 1968. The Colombian datura virus–a new virus in the potato virus Y group. *Phytopathology* 58, 587–592.

Kalinina, N. O., Khromov, A., Love, A. J. and Taliansky, M. E. 2020. CRISPR applications in plant virology: virus resistance and beyond. *Phytopathology®* 110(1), 18–28.

Kassanis, B., Gianinazzi, S. and White, R. F. 1974. A possible explanation of the resistance of virus-infected tobacco plants to second infection. *Journal of General Virology* 23(1), 11–16.

Khalid, A., Zhang, Q., Yasir, M. and Li, F. 2017. Small RNA based genetic engineering for plant viral resistance: application in crop protection. *Frontiers in Microbiology* 8, 43.

Kosaka, Y. and Fukunishi, T. 1997. Multiple inoculation with three attenuated viruses for the control of cucumber virus disease. *Plant Disease* 81(7), 733–738.

Lecoq, H., Lemaire, J. M. and Wipf-Scheibel, C. 1991. Control of zucchini yellow mosaic virus in squash by cross protection. *Plant Disease* 75(2), 208–211.

Lecoq, H. and Raccah, B. 2001. Cross-protection: interactions between strains exploited to control plant virus diseases. In: *Biotic Interactions in Plant-Pathogen Associations*. Jeger, M. J., Spence, N. J. (ED), CABI Publishing, Wallingford, Oxfordshire: England 177–192.

Leibman, D., Ortega-Parra, N., Wolf, D., Shterkman, M., Hanssen, I. and Gal-On, A. 2021. A transgenic RNAi approach for developing tomato plants immune to pepino mosaic virus. *Plant Pathology* 00, 1–10.

Lindbo, J. A. 2012. A historical overview of RNAi in plants. *Methods in Molecular Biology* 894, 1–16.

Lindbo, J. A. and Dougherty, W. G. 1992. Untranslatable transcripts of the tobacco etch virus coat protein gene sequence can interfere with tobacco etch virus replication in transgenic plants and protoplasts. *Virology* 189(2), 725–733.

Lindbo, J. A., Silva-Rosales, L., Proebsting, W. M. and Dougherty, W. G. 1993. Induction of a highly specific antiviral state in transgenic plants: implications for regulation of gene expression and virus resistance. *The Plant Cell* 5(12), 1749–1759.

Ling, K. S., Li, R. and Bledsoe, M. 2013. Pepino mosaic virus genotype shift in North America and development of a loop-mediated isothermal amplification for rapid genotype identification. *Virology Journal* 10, 117.

Ling, K. S., Wechter, W. P. and Jordan, R. 2007. Development of a one-step immunocapture real-time Taqman RT-PCR assay for the broad spectrum detection of Pepino mosaic virus. *Journal of Virological Methods* 144(1–2), 65–72.

Ling, K. S., Wintermantel, W. M. and Bledsoe, M. 2008. Genetic composition of *Pepino mosaic virus* population in north American greenhouse tomatoes. *Plant Disease* 92(12), 1683–1688.

Ling, K. S. and Zhang, W. 2011. First report of *Pepino mosaic virus* infecting tomato in Mexico. *Plant Disease* 95(8), 1035–1035.

Liu, J., Li, X. D. and Xu, S. 2020. Single amino acid substitutions in the coat protein and RNA-dependent RNA polymerase alleviated the virulence of Cucumber green mottle mosaic virus and conferred cross protection against severe infection. *Virus Genes* 56(2), 228–235.

Mäkinen, K. and Hafrén, A. 2014. Intracellular coordination of potyviral RNA functions in infection. *Frontiers in Plant Science* 5, 110.

Malpica, J. M., Fraile, A., Moreno, I., Obies, C. I., Drake, J. W. and García-Arenal, F. 2002. The rate and character of spontaneous mutation in an RNA virus. *Genetics* 162(4), 1505–1511.

Maroon-Lango, C. J., Guaragna, M. A., Jordan, R. L., Hammond, J., Bandla, M. and Marquardt, S. K. 2005. Two unique US isolates of Pepino mosaic virus from a limited source of pooled tomato tissue are distinct from a third (European-like) US isolate. *Archives of Virology* 150(6), 1187–1201.

McDonald, B. A. and Stukenbrock, E. H. 2016. Rapid emergence of pathogens in agro-ecosystems: global threats to agricultural sustainability and food security. *Philosophical Transactions of the Royal Society of London. Series B, Biological Sciences* 371(1709), 28080995.

McKinney, H. H. 1929. Mosaic diseases in the Canary Islands, West Africa and Gibraltar. *Journal of Agricultural Research* 39, 577–578.

McMillan, R. T. Jr. and Gonsalves, D. 1988. Effectiveness of cross-protection by a mild mutant of papaya ringspot virus for control of ringspot disease of papaya in Florida. Proceedings of the Annual Meeting of the Florida State Horticulture Society (USA) 100, 294–296.

Mitter, N., Worrall, E. A., Robinson, K. E., Li, P., Jain, R. G., Taochy, C., Fletcher, S. J., Carroll, B. J., Lu, G. Q. and Xu, Z. P. 2017. Clay nanosheets for topical delivery of RNAi for sustained protection against plant viruses. *Nature Plants* 3, 16207.

Monci, F., Sánchez-Campos, S., Navas-Castillo, J. and Moriones, E. 2002. A natural recombinant between the geminiviruses tomato yellow leaf curl Sardinia virus and tomato yellow leaf curl virus exhibits a novel pathogenic phenotype and is becoming prevalent in Spanish populations. *Virology* 303(2), 317–326.

Moreno-Perez, M. G., Pagan, I., Aragon-Caballero, L., Caceres, F., Fraile, A. and Garcia-Arenal, F. 2014. Ecological and genetic determinants of pepino mosaic virus emergence. *Journal of Virology* 88(6), 3359–3368.

Moriones, E., García-Andrés, S. and Navas-Castillo, J. 2007. Recombination in the TYLCV complex: a mechanism to increase genetic diversity: implications for plant resistance development. In: *Tomato Yellow Leaf Curl Virus Disease*, Czosnek, H. (ED), Springer, Dordrecht, 119–138.

Muthamilarasan, M. and Prasad, M. 2013. Plant innate immunity: an updated insight into defense mechanism. *Journal of Biosciences* 38(2), 433–449.

Myers, S. S., Smith, M. R., Guth, S., Golden, C. D., Vaitla, B., Mueller, N. D., Dangour, A. D. and Huybers, P. 2017. Climate change and global food systems: potential impacts on food security and undernutrition. *Annual Review of Public Health* 38, 259–277.

Nelson, R. S., Abel, P. P. and Beachy, R. N. 1987. Lesions and virus accumulation in inoculated transgenic tobacco plants expressing the coat protein gene of tobacco mosaic virus. *Virology* 158(1), 126–132.

Palukaitis, P. and Zaitlin, M. 1984. A model to explain the 'cross-protection' phenomenon shown by plant viruses and viroids. *Plant-Microbe Interactions* 1, 420–429.

Papayiannis, L. C., Kokkinos, C. D. and Alfaro-Fernández, A. 2012. Detection, characterization and host range studies of Pepino mosaic virus in Cyprus. *European Journal of Plant Pathology* 132(1), 1-7.

Pechinger, K., Chooi, K. M., MacDiarmid, R. M., Harper, S. J. and Ziebell, H. (2019). A New Era for Mild Strain Cross-Protection. *Viruses*. 1(7), 670. doi: 10.3390/v11070670. PMID: 31340444; PMCID: PMC6669575.

Perring, T. M., Farrar, C. A., Blua, M. J., Wang, H. L. and Gonsalves, D. 1995. Cross protection of cantaloupe with a mild strain of zucchini yellow mosaic virus: effectiveness and application. *Crop Protection* 14(7), 601-606.

Plasencia, A., Díaz, M. D. M., Hita, I., Francés, M., Gutiérrez, M., Bielza, P., Gallego, J., Alcázar, A. and Cano, A. 2003. Implicaciones de los abejorros (Bombus spp.) en la dispersión del virus del mosaico del pepino dulce (pepino mosaic virus) en cultivos de tomate. *Boletin de Sanidad Vegetal Plagas* 29(3), 393-403.

Price, W. C. 1940. Acquired immunity from plant virus diseases. *The Quarterly Review of Biology* 15(3), 338–361.

Prins, M., Kikkert, M., Ismayadi, C., Graauw, W. de, Haan, P. de and Goldbach, R. 1997. Characterization of RNA-mediated resistance to tomato spotted wilt virus in transgenic tobacco plants expressing NS(M) gene sequences. *Plant Molecular Biology* 33(2), 235–243.

Raja, P., Sanville, B. C., Buchmann, R. C. and Bisaro, D. M. 2008. Viral genome methylation as an epigenetic defense against geminiviruses. *Journal of Virology* 82(18), 8997–9007.

Ram, R. D., Verma, R., Tomer, S. P. S. and Prakash, S. 2006. Management of papaya ring spot virus through cross protection strategy. *Journal of Maharashtra Agricultural Universities* 31, 92–95.

Rast, A. Th. B. 1972. M II-16, an artificial symptomless mutant of tobacco mosaic virus for seedling inoculation of tomato crops. *Netherlands Journal of Plant Pathology* 78(3), 110–112.

Rast, A. Th. B. 1975. Variability of tobacco mosaic virus in relation to control of tomato mosaic in glasshouse tomato crops by resistance breeding and cross protection. *Agricultural Research in Reproduction*.

Register, J. C. and Beachy, R. N. 1988. Resistance to TMV in transgenic plants results from interference with an early event in infection. *Virology* 166(2), 524–532.

Register, J. C. and Beachy, R. N. 1989. Effect of protein aggregation state on coat protein-mediated protection against tobacco mosaic virus using a transient protoplast assay. *Virology* 173(2), 656–663.

Rezende, J. A. M. and Pacheco, D. A. 1998. Control of papaya ringspot virus-type W in zucchini squash by cross-protection in Brazil. *Plant Disease* 82(2), 171–175.

Rochow, W. F. and Ross, A. F. 1955. Virus multiplication in plants doubly infected by potato viruses X and Y. *Virology* 1(1), 10–27.

Rojas, M. R. and Gilbertson, R. L. 2008. *Emerging Plant Viruses: A Diversity of Mechanisms and Opportunities*. Springer, Berlin, Heidelberg.

Roossinck, M. J. and García-Arenal, F. 2015. Ecosystem simplification, biodiversity loss and plant virus emergence. *Current Opinion in Virology* 10, 56–62.

Salomone, A. and Roggero, P. 2002. Host range, seed transmission and detection by ELISA and lateral flow of an Italian isolate of Pepino mosaic virus. *Journal of Plant Pathology* 84 (1), 65–68.

Sarkar, S. and Smitamana, P. 1981. A proteinless mutant of tobacco mosaic virus: evidence against the role of a viral coat protein for interference. *Molecular and General Genetics: MGG* 184(1), 158–159.

Schenk, M. F., Hamelink, R., van der Vlugt, R. A. A., Vermunt, A. M. W., Kaarsenmaker, R. C. and Stijger, I. C. C. M. M. 2010. The use of attenuated isolates of Pepino mosaic virus for cross-protection. *European Journal of Plant Pathology* 127(2), 249–261.

Schwarz, D., Beuch, U., Bandte, M., Fakhro, A., Büttner, C. and Obermeier, C. 2010. Spread and interaction of Pepino mosaic virus (PepMV) and Pythium aphanidermatum in a closed nutrient solution recirculation system: effects on tomato growth and yield. *Plant Pathology* 59(3), 443–452.

Sempere, R. N., Gómez-Aix, C., Ruíz-Ramón, F., Gómez, P., Hasiów-Jaroszewska, B., Sánchez-Pina, M. A. and Aranda, M. A. 2016. *Pepino mosaic virus* RNA-dependent RNA polymerase POL domain is a hypersensitive response-like elicitor shared by necrotic and mild isolates. *Phytopathology®* 106(4), 395–406.

Sharma, N., Sahu, P. P., Puranik, S. and Prasad, M. 2013. Recent advances in plant-virus interaction with emphasis on small interfering RNAs (siRNAs). *Molecular Biotechnology* 55(1), 63–77.

Shaw, J. G., Plaskitt, K. A. and Wilson, T. M. A. 1986. Evidence that tobacco mosaic virus particles disassemble contranslationally in vivo. *Virology* 148(2), 326–336.

Sherwood, J. L. and Fulton, R. W. 1982. The specific involvement of coat protein in tobacco mosaic virus cross protection. *Virology* 119(1), 150–158.

Shipp, J. L., Buitenhuis, R., Stobbs, L., Wang, K., Kim, W. S. and Ferguson, G. 2008. Vectoring of Pepino mosaic virus by bumble-bees in tomato greenhouses. *Annals of Applied Biology* 153(2), 149–155.

Singer, S., Raccah, B., Lev, E. and Katz, G. 1994. Cross protection against the zucchini yellow mosaic virus using a mild strain. *Hassadeh* 74, 403–406.

Spence, N. J., Basham, J., Mumford, R. A., Hayman, G., Edmondson, R. and Jones, D. R. 2006. Effect of Pepino mosaic virus on the yield and quality of glasshouse-grown tomatoes in the UK. *Plant Pathology* 55(5), 595–606.

Spence, N. J., Mead, A., Miller, A., Shaw, E. D. and Walkey, D. G. A. 1996. The effect on yield in courgette and marrow of the mild strain of zucchini yellow mosaic virus used for cross-protection. *Annals of Applied Biology* 129(2), 247–259.

Syller, J. and Grupa, A. 2016. Antagonistic within-host interactions between plant viruses: molecular basis and impact on viral and host fitness. *Molecular Plant Pathology* 17(5), 769–782.

Szittya, G. 2003. Low temperature inhibits RNA silencing-mediated defence by the control of siRNA generation. *The EMBO Journal* 22(3), 633–640.

Takahashi, T., Sugawara, T., Yamatsuta, T., Isogai, M., Natsuaki, T. and Yoshikawa, N. 2007. Analysis of the spatial distribution of identical and two distinct virus populations differently labeled with cyan and yellow fluorescent proteins in coinfected plants. *Phytopathology®* 97(10), 1200–1206.

Taning, C. N., Arpaia, S., Christiaens, O., Dietz-Pfeilstetter, A., Jones, H., Mezzetti, B., Sabbadini, S., Sorteberg, H. G., Sweet, J., Ventura, V. and Smagghe, G. 2020. RNA-based biocontrol compounds: current status and perspectives to reach the market. *Pest Management Science* 76(3), 841–845.

Tenllado, F., Martínez-García, B., Vargas, M. and Díaz-Ruíz, J. R. 2003. Crude extracts of bacterially expressed dsRNA can be used to protect plants against virus infections. *BMC Biotechnology* 3, 3.

Tennant, P. F. 1994. Differential protection against papaya ringspot virus isolates in coat protein gene transgenic papaya and classically cross-protected papaya. *Phytopathology* 84(11), 1359.

Tripathi, S., Suzuki, J. Y., Ferreira, S. A. and Gonsalves, D. 2008. Papaya ringspot virus -P: characteristics, pathogenicity, sequence variability and control. *Molecular Plant Pathology* 9(3), 269–280.

Tumer, N. E., O'connell, K. M., Nelson, R. S., Sanders, P. R., Beachy, R. N., Fraley, R. T. and Shah, D. M. 1987. Expression of alfalfa mosaic virus coat protein gene confers cross-protection in transgenic tobacco and tomato plants. *The EMBO Journal* 6(5), 1181–1188.

Vadlamudi, T., Patil, B. L., Kaldis, A., Sai Gopal, D. V. R., Mishra, R., Berbati, M. and Voloudakis, A. 2020. DsRNA-mediated protection against two isolates of papaya ringspot virus through topical application of dsRNA in papaya. *Journal of Virological Methods* 275, 113750.

Valkonen, J. P. T., Rajamäki, M.-L. and Kekarainen, T. 2002. Mapping of viral genomic regions important in cross-protection between strains of a potyvirus. *Molecular Plant-Microbe Interactions®* 15, 683–692.

van der Vlugt, R. A. A., Stijger, C. C. M. M., Verhoeven, J. Th. J. and Lesemann, D. E. 2000. First report of *pepino mosaic virus* on tomato. *Plant Disease* 84(1), 103–103.

Vance, V. B. 1991. Replication of potato virus X RNA is altered in coinfections with potato virus Y. *Virology* 182(2), 486–494.

Varma, A. 1993. Integrated management of plant viral diseases. *Ciba Foundation Symposium* 177, 140–55; discussion 155.

Verhoeven, J. Th. J., Lesemann, D. -E. and Roenhorst, J. W. 1996. First report of Colombian datura Potyvirus in tomato. *European Journal of Plant Pathology* 102(9), 895–898.

Vermunt, A. M. W. and Kaarsemaker, R. C. 2017. Multi-genotype cross-protection against Pepino mosaic virus in tomato. *Crop Protection* 96, 116–122.

Voinnet, O. 2001. RNA silencing as a plant immune system against viruses. *Trends in Genetics : TIG* 17(8), 449–459.

Vu, T. V., Choudhury, N. R. and Mukherjee, S. K. 2013. Transgenic tomato plants expressing artificial microRNAs for silencing the pre-coat and coat proteins of a Begomovirus, Tomato leaf curl New Delhi virus, show tolerance to virus infection. *Virus Research* 172(1–2), 35–45.

Walkey, D. G. A., Lecoq, H., Collier, R. and Dobson, S. 1992. Studies on the control of zucchini yellow mosaic virus in courgettes by mild strain protection. *Plant Pathology* 41(6), 762–771.

Wallace, J. M. 1940. Evidence of passive immunization of tobacco, Nicotiana tabacum, from the virus of curly top. *Phytopathology* 30, 673–679.

Wang, H. L., Gonsalves, D., Provvidenti, R. and Lecoq, H. L. 1991. Effectiveness of cross protection by a mild strain of zucchini yellow mosaic virus in cucumber, melon, and squash. *Plant Disease* 75(2), 203.

Wang, M., Weiberg, A., Lin, F. M., Thomma, B. P. H. J., Huang, H. D. and Jin, H. 2016. Bidirectional cross-kingdom RNAi and fungal uptake of external RNAs confer plant protection. *Nature Plants* 2, 16151.

Wen, F., Lister, R. M. and Fattouh, F. A. 1991. Cross-protection among strains of barley yellow dwarf virus. *Journal of General Virology* 72(4), 791–799.

Werkman, A. W. and Sansford, C. E. 2010. Pest risk analysis for pepino mosaic virus for the EU. Deliverable Report 4.3. EU Sixth Framework project PEPEIRA. Available at: http://www.pepeira.com.

Wisniewski, L. A., Powell, P. A., Nelson, R. S. and Beachy, R. N. 1990. Local and systemic spread of tobacco mosaic virus in transgenic tobacco. *The Plant Cell* 2(6), 559–567.

Wright, D. and Mumford, R. 1999. Pepino Mosaic Potexvirus (PepMV): First records in tomato in the United Kingdom. Plant Disease Notice No. 89. York, UK: Central Science Laboratory.

Wu, X. J., Beachy, R. N., Wilson, T. M. and Shaw, J. G. 1990. Inhibition of uncoating of tobacco mosaic virus particles in protoplasts from transgenic tobacco plants that express the viral coat protein gene. *Virology* 179(2), 893–895.

Wu, X. and Shaw, J. G. 1997. Evidence that a viral replicase protein is involved in the disassembly of tobacco mosaic virus particles in vivo. *Virology* 239(2), 426–434.

Yadav, R. K. and Chattopadhyay, D. 2011. Enhanced viral intergenic region-specific short interfering RNA accumulation and DNA methylation correlates with resistance against a geminivirus. *Molecular Plant-Microbe Interactions: MPMI* 24(10), 1189–1197.

Yarden, G., Hemo, R., Livne, H., Maoz, E., Lev, E., Lecoq, H. and Raccah, B. 2000. Cross-protection of Cucurbitaceae from zucchini yellow mosaic Potyvirus. *Acta Horticulturae* (510), 349–356.

Yeh, S.-D. 1988. Control of papaya ringspot virus by cross protection. *Plant Disease* 72(5), 375.

Yeh, S.-D. and Gonsalves, D. 1984. Evaluation of induced mutants of papaya ringspot virus for control by cross protection. *Phytopathology* 74(9), 1086.

Yeh, S.-D. and Gonsalves, D. 1994. Practices and perspective of control of papaya ringspot virus by cross protection. In: *Advances in Disease Vector Research* Harris, K. F. (Ed), Vol. 10. Springer, New York, 237–257.

Yeh, S. -D. and Kung, Y. -J. 2007. The past and current approaches for control of papaya ringspot virus in Taiwan. *Acta Horticulturae* (740), 235–243.

Yilmaz, M., Abak, K., Lecoq, H., Baloglu, S., Sari, N., Kesici, S., Ozaslan, M. and Guldur, M. E. 1994. Control of zucchini yellow mosaic virus (ZYMV) in cucurbits by ZYMVWK strain. Turkish Phytopathological Society (eds). 9th Congress of the Mediterranean Phytopathological Union, 353–356.

Zaitlin, M. 1976. Viral cross protection: more understanding is needed. *Phytopathology* 66(4), 382–383.

Zhang, L., Xie, X., Song, Y., Jiang, F., Zhu, C. and Wen, F. 2013. Viral resistance mediated by shRNA depends on the sequence similarity and mismatched sites between the target sequence and siRNA. *Biologia Plantarum* 57(3), 547–554.

Zheng, L., Rodoni, B. C., Gibbs, M. J. and Gibbs, A. J. 2010. A novel pair of universal primers for the detection of potyviruses. *Plant Pathology* 59(2), 211–220.

Ziebell, H. and Carr, J. P. 2010. Cross-protection. *Advances in Virus Research*, 211–264.

Ziebell, H. and MacDiarmid, R. 2017. Prospects for engineering and improvement of cross-protective virus strains. *Current Opinion in Virology* 26, 8–14.

Ziebell, H., Payne, T., Berry, J. O., Walsh, J. A. and Carr, J. P. 2007. A cucumber mosaic virus mutant lacking the 2b counter-defence protein gene provides protection against wild-type strains. *Journal of General Virology* 88(10), 2862–2871.

Zinnen, T. M. and Fulton, R. W. 1986. Cross-protection between sunn-hemp mosaic and tobacco mosaic viruses. *Journal of General Virology* 67(8), 1679–1687.

Zoeten, G. A. and Fulton, R. W. 1975. Understanding generates possibilities. *Phytopathology* 65, 221–222.

Chapter 15

Biocontrol via mycoviruses: a neglected option for bioprotection?

Anne D. van Diepeningen, BU Biointeractions and Plant Health, Wageningen University and Research, The Netherlands

1 Introduction

'Mycoviruses' is the term used for viruses infecting fungi and fungus-like organisms. Mycoviruses have been reported in Ascomycota, Basidiomycota, Chytridiomycota, Deuteromycota and Zygomycota (Ghabrial et al., 2015). Most reported mycoviruses consist of linear double-stranded RNA (dsRNA; ~70%), while a small number are composed of negative single-stranded RNA (−ssRNA) or positive single-stranded RNA (+ssRNA; ~30%) (Ghabrial et al, 2015; King et al, 2011). More rarely, mycoviruses with circular single-stranded DNA (ssDNA) have been found (Jiang et al., 2013; Li et al., 2020; Yu et al., 2010). Like animal and plant viruses, mycoviruses require the living cells of their host to replicate. The main differences between animal and plant viruses on the one hand, and mycoviruses on the other, lie in the fact that most mycoviruses lack an extracellular route for infection and are transmitted between cells through cell division, sporulation and occasionally cell fusion, as well as lacking movement proteins (Son et al., 2015).

http://dx.doi.org/10.19103/AS.2021.0093.20

Many of these mycoviruses are considered to be cryptic, and supposedly have little to no effect on their host. However, some notable exceptions have been described, in which the viruses cause deleterious effects on their host, and where the host is a plant pathogen, may cause *hypovirulence*, reducing the infection, growth and reproduction of the fungus. Virocontrol, the biocontrol of pathogens via hypovirulence-causing mycoviruses, is a small field, but one which is now receiving more attention. This chapter discusses the characteristics of mycoviruses and the conditions that need to be met to make a successful virocontrol agent. The chapter describes one of the success stories so far, that of the virocontrol of *Cryphonectria parasitica*, the chestnut blight pathogen. The chapter ends by reviewing future trends and information on sources of more information on mycoviruses and virocontrol.

2 Mycovirus types and origins

In over 50 years of mycovirus research, an enormous diversity of mycoviruses has been found. They vary in their make-up from DNA to RNA, single-stranded or double-stranded, consisting of a single fragment or multiple fragments, linear or circular, with folds or without folds at their ends, with or without a protein coat, consisting of one or multiple protein components, multiple fragments packed singly or together, and localised in the fungal cytoplasm or the mitochondrion. Mycoviruses are classified based on their genetic organisation, their form and resemblance to other (myco)viruses. Mycovirus names are generally derived from their fungal host names, with the first letters deriving from the fungal genus and species epithet, some letters relating to virus organisation followed by a number.

The type of nucleic acids and the number of fragments forming the mycovirus genome determine the virus families they belong to. The majority of mycoviruses consist of dsRNA and are grouped in the families of the *Chryso-*, *Megabirna-*, *Quadri-*, *Partiti-*, *Reo-*, and *Totiviridae* (Ghabrial et al., 2015; King et al, 2011, 2012; Wang et al., 2016). Some dsRNA mycovirus species have also been found to belong to new, as yet unnamed virus families (*e.g.* Liu et al., 2019). Linear positive-sense +ssRNA mycoviruses can belong to the *Alphaflexi-*, *Ambigui-*, Barna-, *Beny-*, *Botourmia- Endorna-*, *Fusari-*, Gammaflexi-, *Hypo-*, Narna-, Tombus-, *Tymo-*, *Virga-* and *Yadokariviridae* families. Some of the described reverse transcribed +ssRNA mycoviruses fall in the *Meta-* and *Pseudoviridae* families. The rarer negative-sense −ssRNA viruses belong to the *Mycomononega- and Ophioviridae* (Ghabrial et al., 2015; Gilbert et al., 2019; Marzano et al., 2016) families. Finally, the ssDNA mycoviruses fall in the *Genomoviridae* or unclassified groups (Ghabrial et al., Li et al., 2020). For characteristics of the different virus families, see Table 1. Members of quite a number of diverse mycovirus families have been found in the past 50+ years, but undoubtedly many more are to be discovered.

Mycoviruses are usually located in the cytoplasm of the fungal host, but may be associated with the mitochondria (e.g. Heaton and Leslie, 2004; Hong et al., 1998, 1999; Polashock and Hillman, 1994). Many mycoviruses have a protein coat, but naked mycoviruses also occur, without a protein coat (e.g. Qiu et al., 2010). Isometric forms identified by electron microscopy are the predominant coat morphology, but rigid or flexuous rods, club-shaped particles, enveloped bacilliform particles and herpes-like particles have also been described (e.g. Ghabrial et al., 2015; Kazama and Schornstein, 1973; Varga et al., 2003; van Diepeningen et al., 2008). Wickner (1996) described a virus capsid protein that both provides protection in the form of subcellular compartmentalisation for transcription and replication, and has a catalytic function in decapping host messenger RNA (mRNA) in favour of viral mRNAs. The isometric particles are generally based on a 120-subunit T = 1 capsid with a dimer as the asymmetric unit, while insertions at the capsid's outer surface are likely linked to its enzymatic activities (Luque et al., 2018). Some ssRNA viruses code for RNA-capping methyl transferases as a mechanism to protect their genome (e.g. Howitt et al., 2001; Roossinck et al., 2011). Others use a cap-snatching mechanism and borrow from the 5′-part of mRNA of their host as protection and as a primer to synthesise viral transcripts (e.g Fujimura and Esteban, 2011). Some mycoviruses have been described where the mycovirus is associated with or enveloped in colloidal cellular components (e.g. KanhayuwaKanhayuwa et al., 2015; Kozlakidis et al., 2009). Other mycoviruses protect their RNA by folding their 5′- and 3′-UTR regions in stem-loop structures for protection (e.g. Torres-Trenas and Pérez-Artés, 2020).

Different hypotheses exist about the origin of mycoviruses. The two main hypotheses are: (1) the ancient co-evolution hypothesis and (2) the plant virus hypothesis (Ghabrial, 1998). In the case of ancient co-evolution, Koonin et al. (1991) suggested the acquisition of an ancestral ssRNA virus from a plant based on sequence similarities, and during co-evolution of fungus and virus, a subsequent loss of protein coat and a shift towards the dominant dsRNA replication type as observed in a *Cryphonectria parasitica* hypovirus. Support for the co-evolution hypothesis can also be found in the large number of chromosomal genes needed for replication of the so-called killer viruses in *Saccharomyces cerevisiae* (Wickner, 1996), the use of host-coded enzymes for capsid processing (Huang et al., 1997) and use of the mitochondrial genetic translation code in mycoviruses located in the mitochondria (Park et al., 2006). Several phylogeographic studies (e.g. Arthur et al., 2007; Voth et al., 2006) also support co-evolution.

The trade-off theory explains why a reduced virulence of a pathogen is needed when the transmission of it is limited, since it would go extinct when a host is too affected to transmit the pathogen. This theory may explain why

Table 1 Virus families of which members have been observed in fungi, and their characteristics

Genome build-up	Virus family	Number of fragments[1]	Phenotype[1]	Hosts[1]
dsRNA	*Chrysoviridae*	(3), 4, (5 or 7)	Isometric ~40 nm	Ascomycetes, Basidiomycetes, plants, (insects)
	Megabirnaviridae	2	Isometric ~52 nm	Ascomycetes Basidiomycetes
	Quadriviridae	4	Isometric ~45 nm	Filamentous fungi
	Partitiviridae	2	Isometric ~25-43 nm	Fungi, plants
	Reoviridae	9-12	Isometric ~60–80 nm, 1–3 capsid layers, with or without spikes	Fungi, animals
	Totiviridae	1	Isometric ~40 nm	Fungi, parasitic protozoa
+ssRNA	*Alphaflexiviridae*	1	Flexuous filaments,12–13 nm × 470–800 nm	Fungi, plants
	Ambiguiviridae	1	-	Fungi
	Barnaviridae	1	Bacilliform 18–20 nm × 48–53 nm	Fungi
	Benyviridae	≥ 2	Rod-shaped, 20 nm × 65–390 nm	Plants, fungi
	Botourmiaviridae	1-3	Bacilliform, 18 nm × 30–63 nm	Plants, fungi
	Endornaviridae	1	- pleomorphic membrane vesicles	Fungi, plants, oomycetes
	Fusariviridae	1	-	Fungi
	Gammaflexiviridae	1	Flexuous filaments 13 nm × 720 nm	Fungi
	Hypoviridae	1	- Pleomorphic vesicles 50–80 nm	Fungi
	Narnaviridae	1	- ribonucleoprotein complex with RNA-dependent RNA polymerase	Fungi, cytoplasm-mitochondrial
	Tombusviridae	1	Isometric ~32–35 nm	Plants, fungi

Published by Burleigh Dodds Science Publishing Limited, 2022.

	Tymoviridae	1	Variable	Plants (animal vectors)
	Virgaviridae	1-3	Rod-shaped ~20 x 300 nm	Plants, fungi
	Yadokariviridae	1	Variable	Animals, fungi
−ssRNA	*Mycomononegaviridae*	1	Filaments ~ 15–50 nm × 1000nm	Fungi, related to animal viruses
	Ophioviridae	3-4	- pleiomorphic	Plants
ssDNA	*Genomoviridae*	1, circular	Isometric ~20–22 nm	Fungi

[1] Data on the virus families are available at https://talk.ictvonline.org/ictv-reports/ictv_online_report.

so many presumably co-evolved mycoviruses are cryptic, with no apparent effects in their host and limited or no transmission. Göker and colleagues (2011) studied the codivergence of mycoviruses with their hosts. Their study was significantly influenced by the numbers of viruses sampled per family, but for well-sampled mycovirus families, gave significant support to co-evolution. However, not all mycoviruses are consistent with the first hypothesis, as they have deleterious effects on their host or do not follow the evolutionary pattern of their host (e.g. Carbone et al., 2004; Liu et al., 2003).

The plant virus hypothesis is supported by observations that some mycoviruses - both dsRNA and ssRNA mycoviruses - are more closely related to certain plant viruses than to other mycoviruses (e.g. Adams et al., 2004; Chu et al., 2002; Fauquet et al., 2005; Howitt et al., 2001, 2006; Linder-Basso et al., 2005; Martelli et al., 2007; Xie et al., 2006) or have a similar genome organisation and gene expression strategy as certain plant viruses (Kwon et al., 2007). Some fungi are known to be vectors of plant viruses, and carry those virions on the outside of the fungal structures, where a rare event may have led to internalisation (Varga et al., 2003). Besides having an evolutionary link, also the cological link between plant viruses and plant pathogenic fungi is clear (Roossinck, 2019).

Recent work has shown that transfer, replication and expression of plant viruses from plant to different (plant pathogenic) fungi and lichens is possible (Andika et al., 2017; Mascia and Gallitelli, 2016; Mascia et al., 2019; Petrzik et al., 2014). In reverse, transfer of fungus to plant has been observed where the mycoviruses of an endophytic fungus successfully replicated in plant cells (Nerva et al., 2017), suggesting that transfer of viruses between plant and fungus may occur in both directions.

A third alternative hypothesis would be the transfer of animal viruses to fungi. There is as yet less evidence for this. However, the *Reoviridae* contain mostly animal viruses and a few plant and fungal viruses, with only low to moderate levels of similarity between them (Hillman et al., 2004; Suzuki et al., 2004; Tanaka et al., 2012; Wei et al., 2003). More recently discovered fungal members of *Yadokariviridae* and the *Mycomononegaviridae* have animal pathogenic relatives. With the diversity of mycoviruses observed in the fungal kingdom, it seems likely that multiple hypotheses (ancient co-evolution, plant and animal transfer hypotheses) are needed to explain the diversity observed.

3 Mycovirus detection and incidence

The first hints of the existence of mycoviruses came from fungi showing abnormalities for which no visible pathogens could be detected. The work of Hollings (1962) can be seen as the first evidence of mycoviruses being the cause of die-back diseases such as La France disease and virus X disease in the white button mushroom *Agaricus bisporus*. The phenotypic evidence of mycoviruses can also present itself as slower growth, irregular colony rims, reduced spore production and/or different colony colour (e.g. Sivanesan and Holliday, 1981; van Diepeningen et al., 2006).

Hollings (1962) is also the first to show that infection with multiple mycoviruses in one host can occur. In Holling's electron microscopic photographs, both isometric particles and bacilliform rods were visible, while infectivity of the virus was demonstrated by injection of the tissue of the *Agaricus* sporocarp with a hypodermic needle. Electron microscopy used for the visualisation of *A. bisporus* mycoviruses is still the main technique, often with preceding (density gradient) centrifugation, to visualise mycoviruses (e.g. Sharma et al., 2011).

Many (cryptic) mycoviruses have been detected based on the presence of their (dsRNA) genomes. After viral nucleic acid isolation, determination of the (double-stranded) RNA nature of mycovirus genomes was first resolved chemically (Banks et al., 1968), and later enzymatically (e.g. Buck et al., 1973; Varga et al., 1994). The numbers and sizes of fragments were determined via agarose or acrylamide gel electrophoresis (e.g. Buck and Ratti, 1975). In many of the first (dsRNA) mycovirus detection protocols, a phenol-step to deproteinise viral fragments (Banks et al., 1968) in the nucleic acid extraction protocols proved essential for isolation. This compound was later replaced by sodium dodecyl sulfate and potassium acetate (SDS/KOAc) treatments, followed by CF-11 chromatography to separate dsRNA from other nucleic acids (e.g. DePaulo and Powell, 1995).

Direct cloning, amplification and sequencing of dsRNA genomes have advanced gradually (for an overview, see Potgieter et al., 2009), providing more insight into the diversity and functioning of mycoviruses. Many viruses are now detected *in silico* after analysis of (meta)transcriptome data (e.g. Gilbert et al., 2019; Marzano et al., 2016; Myers et al., 2020; Zoll et al., 2018).

Estimates of the numbers of mycovirus-infected strains in a population suggest that 30–80% of fungal species may be infected (Ghabrial and Suzuki, 2009), while infection rates per species vary between 0% and 100% (Table 2). Mycoviruses occur commonly both in Ascomycetes and Basidiomycetes as well as in earlier-diverging lineages (Myer et al., 2020). Multiple infections with two or more, often unrelated, mycoviruses in a host appear quite common, irrespective of the detection technique (e.g. Gilbert et al., 2019; Hao et al., 2018; Herrero and Zabalgogeazcoa, 2011; Hollings, 1962; Osaki et al., 2016; Ran et al., 2016). Defective dsRNA and/or satellite dsRNA can also be found (Ghabrial and Suzuki, 2008). Osaki and co-workers (2016) even reported a single strain of *Fusarium poae* that appeared to be infected with 16 different mycoviruses of the dsRNA and -ssRNA types.

Table 2 Examples of estimated incidences of virus infections and debilitating/hypovirulence-causing infections in different fungal and oomycete genera and species

Fungal genus/ species	% Mycovirus infected strains in population	% Debilitating/ hypovirulence phenotype[1] in population	References
Alternaria alternata (Japansese pear)	44	4	Hayashi et al., 1988; Fuke et al., 2011
Aspergillus nidulans	0	0	Coenen et al., 1997
Aspergillus section *flavus*	8-11	0	Elias and Cotty, 1996; van Diepeningen et al., 2008
Aspergillus section *nigri*	10	0.1	Ratti and Buck, 1972; van Diepeningen et al., 2006, 2008
Botrytis cinerea	72	≤1.5	Howitt et al., 1995; Hao et al., 2018; Wu et al., 2010
Chalara elegans	84	18.6	Bottacin et al., 1994
Cryphonectria parasitica	2-36	2-36	Adamčíková et al., 2012; Montenegro et al., 2008; Murolo et al., 2018; Park et al., 2008; Peever et al., 1998
Fusarium graminearum	3.7	1.5	Aminian et al, 2011; Chu et al., 2002, 2004; Darissa et al., 2011
Fusarium oxysporum	11	5	Kilic and Griffin, 1998
Fusarium oxysporum f.sp *dianthi*	≥15	≥0.4-2	Torres-Trenas et al., 2019; Torres-Trenas and Pérez-Artés, 2020
Fusarium poae	100	0	Fekete et al., 1995
Fusarium proliferatum	4	0	Heaton and Leslie, 2004
Fusarium solani	2.9	0	Nogawa et al., 1993
Monilia fructicola	63	0	Tsai et al., 2004; Tran et al., 2019
Magnaporte oryzae	19-71	≤6	Hunst et al., 1986; Moriyama et al., 2018; Okada et al., 2015
Phytophthora infestans	36	0	Tooley et al., 1989
Pythium irregulare	85	0	Gillings et al., 1993
Rosellinia necatrex	14-20		Arjona-Lopez et al., 2018; Arakawa et al., 2002
Sclerotinia sclerotiorum	20	≤20	Boland, 1992
Ustilago maydis	34-100	0	Voth et al., 2006

[1] Debilitation is defined as when the comparison of isogenic virus-containing and virus-free strains show clear phenotypic differences in their growth. Percentually small effects due to use of host resources are not taken into account. Hypovirulence - when a plant pathogenic strain shows reduced pathogenicity in bioassay.

4 Mycovirus effects

The majority of mycoviruses are cryptic or latent, with no apparent effect on their host. In theory, every mycovirus that draws on its host's resources should have at least a small negative effect on its host. The impact of this theft of resources may depend on the availability of resources from the fungal environment. In lab tests under nutrient-restricted conditions, the growth effects on a host by a mycovirus were more pronounced than under nutrient-rich conditions (van Diepeningen et al., 2006, 2008). This section summarises the different phenotypic effects caused by mycoviruses, providing an overview of what is known about the cellular mechanistic effects of the mycoviruses causing such phenotypic effects.

4.1 Debilitation and hypovirulence

A small number of mycoviruses (0–20%, depending on the host species) (Table 2) show clear debilitating effects on their host, with reduced growth rate, sporulation and/or virulence. In contrast, the reduction of pathogen virulence is called hypovirulence. Hypovirulence has been studied in some of the major plant pathogens like *Alternaria alternata, Cryphonectria parasitica* and *Magnaporthe oryzae*, and is linked to a reduced ability to infect, colonise, kill, and/or reproduce on susceptible host tissues (Table 3). In studies with green fluorescent marker strains from *Fusarium oxysporum* f.sp. *dianthi*, Torres-Trenas and co-workers (2019) showed that the initial colonisation of the carnation root was similar for virus-infected and virus-free strains, but that colonisation of internal tissues of the carnation root by mycovirus-infected strains was slower and less dense. They also showed that the infection more rarely became intracellular through the virus-infected strain, whereas virus-free strains readily infected the cells in the medulla.

Hypovirulence has been observed in all major groups of mycoviruses. This includes dsRNA mycoviruses like chrysoviruses (e.g. Moriyama et al., 2018), +ssRNA viruses like the hypoviruses (e.g. Koonin et al., 1991), a −ssRNA member of *Mycomononegaviridae* (Liu et al., 2014) and even in the ssDNA mycovirus of *Genomoviridae* (Yu et al., 2010). The strength of effects of a mycovirus on its host not only depends on the virus type but also on the virus titre in the mycelium (Aoki et al., 2009; Darissa et al., 2012; van Diepeningen et al., 2006).

Hypovirulence may be caused by the fitness effects of nutrient depletion due to mycovirus infection. Different virulence-related mechanisms of the fungus may be influenced directly or indirectly by the viral infection. Melanins, for example, are linked to microbial virulence and protection against oxidising agents. Moriyama and co-workers (2018) observed that chrysovirus infection in *M. oryzae* tends to lead to lighter-coloured colonies due to reduced melanin

Table 3 Examples of hypovirulence-causing mycoviruses in plant pathogens

Plant pathogen	Host(s)	Hypovirulence mycovirus	Virus type	Reference
Alternaria alternata f.sp. *mali*	Apple	AaHV1	*Hypoviridae*, 1 dsRNA fragment	Li et al., 2019
Alternaria alternata	Japanese pear	AaCV1	*Chrysoviridae*, 5 dsRNA fragments	Okada et al., 2018
Botryosphaeria dothidea	Fruit and wood trees	BdRV1	Unknown, 5 dsRNAs	Zhai et al., 2016
Botrytis cinerea	>200 dicotyledons and monocotyledons	Virus from *B. cinerea* CCg425	1 dsRNA	Castro et al., 2003
		BcMV1	Mitovirus 1 dsRNA	Wu et al., 2010
		BcHV1	Hypovirus 1 dsRNA	Hao et al., 2018
		BcFV1	Fusarivirus 1 dsRNA	Hao et al., 2018
		BcPV2	Partitivirus, 2 dsRNAs	Kamaruzzaman et al., 2019
Cryphonectria parasitica	Major hosts in the genus Castanea/chestnut blight	CHV1-EP713	1 dsRNA	Shapira et al., 1991
		MYRV-1	*Mycoreoviridae*, 11 dsRNAs	Enebak et al., 1994; Suzuki et al., 2004
		NB631 dsRNA	Mitochondrial dsRNA	Polashock et al., 1997
Fusarium graminearum	Cereals and other crops	FgV1-DK21	1 ssRNA	Cho et al., 2012
		FgGMTV1	*Genomoviridae*, 3 circular ssDNAs	Li et al., 2020
Fusarium oxysporum f.sp. *dianthi*	Carnation	FodV1	*Chrysoviridae*, 4 dsRNA fragments	Lemus-Minor et al., 2015; Torres-Trenas et al., 2019
Magnaporthe oryzae	Rice blast	MoCV1-A	*Chrysoviridae*, 5 dsRNAs	Moriyama et al., 2018
		MoCV1-B	*Chrysoviridae*, 4-5 dsRNAs	Moriyama et al., 2018
Ophiostoma novo-ulmi	Dutch Elm	3a	*Narnaviridae*, 1 dsRNA	Hong et al., 1998

Sclerotinia homeocarpa	Turfgrasses (dollar spot)	L-dsRNA of Sh12B	Mitochondrial dsRNA	Deng et al., 2003
Sclerotina sclerotiorum	Many herbaceous plants	SsHADV-1	Unknown, related to geminiviruses circular ssDNA	Yu et al., 2010
		SsHV1	*Hypoviridae*, 3 dsRNAs	Xie et al., 2011
		SsNSRV-1	Unknown, (−) ssRNA mycovirus, 1 fragment	Liu et al., 2014
Rhizoctonia solani	Many hosts	M2 of Rhs 1A1	dsRNA fragment	Lakshman et al., 1998
Rosellinia necatrix	Over 400 woody and herbaceous plants	RnMBV1	*Megabirnaviridae*, 2 dsRNAs	Kondo et al., 2013
		RnMPV1-W8 in combination with RnMBV2	*Partitiviridae*, 2 dsRNAs	Kondo et al., 2013
		MYRV3-W370	*Mycoreoviridae*, 12 dsRNAs	Kanematsu et al., 2004; Kondo et al., 2013

production and due to hypovirulence. Mycotoxins in many fungi are also linked to increased host pathogenicity. Virus infection decreases mycotoxin production in many fungi, for instance in different *Aspergillus* species (Elias and Cotty, 1996; Kotta-Loizou and Couuts, 2017); Nerva et al., 2019) and in *Fusarium graminearum* (Aminian et al., 2011; Chu et al., 2002; Li et al., 2016).

4.2 Hypervirulence

Hypervirulence is the opposite of hypovirulence, and has also been observed in mycovirus-infected strains, where infected strains show an increased growth rate, sporulation and/or virulence on their host. In *Chalara elegans* (a.k.a *Thielaviopsis basicola*), two different dsRNA mycoviruses have been observed, one of which caused hypovirulence while the other caused hypervirulence (Bottacin et al., 1994). In *Nectria radicicola*, a viral dsRNA upregulates host virulence (Ahn and Lee, 2001). In *A. alternata*, pathogenic to Japanese pear, both hypovirulent and hypervirulent mycoviruses have been observed. The hypervirulent virus AaCV1 caused impaired growth of its host fungus, but enhanced pathogenicity against Japanese pear plants, related to an increase in AK-toxin (Okada et al., 2018). In the US-8 lineage of *Phytophthora infestans*, most strains harbour the PiRV-2 virus that enhances its host's ecological fitness through down-regulation of ammonium and amino acid uptake, thus achieving stimulated sporulation (Cai et al., 2019). However, virus infection may also stimulate fungal growth without causing increased virulence (Tran et al., 2019).

In the human pathogen *Aspergillus fumigatus*, the AfuPmV-1 virus increases host virulence. An analysis of the nucleotide composition and codon usage showed that, in comparison to non-hypervirulent viruses, the AfuPmV-1 has a higher GC content and a preferred codon usage ending on G or C, while codons ending with A and U were not observed (Je et al., 2019).

The trade-off hypothesis is one way to study the symbiotic relationship between viruses and their hosts. In the tripartite system of mycovirus-fungal pathogen-plant host, the trade-off hypothesis can explain situations where either microbial (hypovirulence) or macrobial host (hypervirulence) has the larger cost, and also situations where everyone wins (Márquez and Roossinck, 2012).

4.3 Killer mycoviruses and other beneficial traits

In several yeasts and filamentous fungi, killer-mycoviruses have been observed that kill off any uninfected strains in the proximity of the fungus. The satellite dsRNAs of these killer-mycoviruses code both for secreted protein toxins and immunity to the toxin. Such 'killer-virus'-infected strains may have a direct selective advantage above killed or non-infected competitors, depending on environmental conditions (e.g. Czárán and Hoekstra, 2003; Wickner, 1996; Wloch-Salamon et al., 2008). This is seen as an extreme form of beneficial interaction. In *Saccharomyces cerevisiae*, several different satellite dsRNAs, called M1, M2, M28 or Mlus each encode a protein toxin, killing by a different mechanism (Schmitt and Breinig, 2006). The plant pathogenic yeast *Ustilago maydis* possesses multiple mycovirus-encoded killer systems based on different cellular mechanisms (e.g. Bruenn, 1986; Koltin, 1986; Park et al., 1996). Expression of the killer protein by transgenic maize renders it virtually immune to *U. maydis* (Allen et al., 2013). In the plant pathogenic fungus *Helminthosporium victoriae* (a.k.a *Cochliobolus victoriae*), a protein with antifungal properties has been isolated from a virus-infected strain (de Sa et al., 2010).

A different type of beneficial trait conferred by a mycovirus has been observed in Curvularia protuberata, an endophyte of panic grass, Dichanthelium lanuginosum. This fungus, its host and a mycovirus of C. protuberata have an interesting three-way symbiosis, where the dsRNA mycovirus (called Curvularia thermal tolerance virus (CThTV)) confers thermal tolerance to both the endophyte and host plant (Márquez et al., 2007).

In *Trichoderma harzianum*, the removal of Trichoderma harzianum mycovirus 1 (ThMV1) resulted in improved biocontrol of *Fusarium oxysporum* f.sp. *cucumerinum*. On the other hand, the same virus improved the growth of cucumber (Liu et al., 2019). Recent work in *Sclerotinia sclerotiorum* has identified a 2-kb mycovirus that converts this pathogen into beneficial endophytes for the *Brassica* species and increases yield (Zhang et al., 2020).

4.4 Intracellular mechanisms

Mycoviruses affect their hosts in different ways and via diverse mechanisms. With so many different mycoviruses, different cellular compartments where mycoviruses are located and different mechanisms within the fungal cell at different physiological stages, the intracellular processes affected by mycoviruses may vary over time.

RNA silencing can function as a virus defence mechanism in a diverse range of eukaryotes, and many viruses are capable of suppressing the silencing machinery targeting them. Hammond and co-workers (2008) have shown that *Aspergillus* viruses are both targets and suppressors of RNA silencing, and that *Aspergillus* virus 1816 in particular shows strong deleterious effects on host growth, with sporulation suppressing RNA silencing through a mechanism that alters the level of small interfering RNA. In *Fusarium graminearum*, viral fragments interact with different components of the RNA interference (RNAi) pathway (Yu et al., 2018). In *Rosellinia necatrix*, the mycoreovirus suppresses RNA silencing (Yaegashi et al., 2013). In Cryphonectria hypovirus1 (CHV1), p29, a papain-like protease suppresses hairpin RNA-induced, virus-induced RNA silencing, by inhibiting the expression of the RNAi genes (Choi et al., 1991; Segers et al., 2006). In addition to their roles in silencing viral replication by degrading viral RNAs, RNAi components also promote viral RNA recombination in C. parasitica (Sun et al., 2009a; Zhang et al., 2008).

C. parasitica has at least four RNA-dependent RNA polymerases (RdRPs) that may be involved in transcriptional and posttranscriptional gene silencing and antiviral defence (Zhang et al., 2014). The Aspergillus fumigatus polymycovirus-1 (AfuPmV-1) seems to have adapted its own RdRPs codon usage to that of its fungal host, and increased its virulence (Je et al., 2019).

Genome-wide studies have shown the transcriptional regulation by viruses to be very complex, for instance, in *Aspergilli* (Ejmal et al., 2018), *F. graminearum* (Cho et al., 2012), *R. necatrix* (Shimizu et al., 2018) and *S. sclerotiorum* (Li et al., 2008). The study of Shimizu et al. (2018) showed that 545 genes in the host were upregulated due to virus infection, while 615 were downregulated. With such profound effects on the expression of genes, mycoviruses also affect the secretion of specific gene products (Kazmierczak et al., 2012). Moriyama et al. (2018) suggested that mycovirus infection, with all its effects on the fungal cell and metabolism, can be a driving force behind the development of physiological diversity of strains and the evolution of pathogenic races.

Transcriptional reprogramming or usage of specific host genes has also been observed in several mycovirus–host systems. This started with studies on single genes. In *F. graminearum*, the *hex1* gene was found necessary for fungal asexual reproduction and pathogenesis as well as for efficient viral RNA accumulation (Son et al., 2013). In *C. parasitica*, the transcription factor gene *pro1* proved responsive to hypovirus, necessary for female fertility, asexual spore development and the stable maintenance of virus infection (Sun et al., 2009b). In *C. parasitica*, the *vir2* gene is also downregulated by the virus, resulting in reduced asexual sporulation and pycinidium production, as well as impaired sexual crossing ability (Zhang et al., 1993).

5 Transmission of mycoviruses

Transfer of (hypovirulence-causing) mycoviruses is important for effective biocontrol. One of the main differences between mycoviruses and viruses from plants and animals lies in the fact that mycoviruses in general have no extracellular phase. Mycoviruses depend on vertical transmission through asexual structures or sexual spores, and on horizontal transmission after hyphal contact for new infection. However, with so many different types of viruses and virus effects on their host, mycoviruses can differ in their modes and rates of transfer.

5.1 *Asexual transmission*

The cellular location of mycoviruses can vary. Some are found in the mitochondria, like the mitoviruses in *C. parasitica*, *F. proliferatum* and *Ophiostoma novo-ulmi* (e.g. Heaton and Leslie, 2004; Polashock and Hillman, 1994; Hong et al., 1998, 1999). Others are found in association with the trans-Golgi network (Jacob-Wilk et al., 2006), while for many, the exact intracellular location is not known. They may spread within the growing mycelium or between dividing yeast cells. However, mycovirus titres need not be homogenous throughout the mycelium, as variations in mycovirus-titres have been observed especially in mycovirus infections with negative effects (Aoki et al., 2009; Darissa et al., 2012; van Diepeningen et al., 2006). As a consequence of this inhomogeneous spread, (sequential) hyphal tip isolations have been a way to obtain cured or partially-cured isolates (e.g. Tran et al., 2019; van Diepeningen et al., 2006). However, the mycovirus spread throughout the growing fungal mycelium in general seems sufficient to infect the large majority of cells within the mycelium.

Whether the viral transport through the mycelium takes place by an active or a passive process is unknown. Plasma streaming in general may be a transport mechanism (Sasaki et al., 2006), but transport via the microtubules may also be possible. In the case of mitochondria-associated mycoviruses ('mitoviruses'), transport with the organelle may be likely (Wu et al., 2010).

Many fungi are capable of producing asexual reproductive structures throughout their mycelium. The asexual reproduction varies from the production of conidiospores on special conidiophores or in asexual fruiting bodies like pycnidia, to a transformation of mycelial cells to survival structures like chlamydospores and sclerotia. The transfer rates to such asexual structures seem generally high (e.g. Buck, 1998; van Diepeningen et al., 1997, 1998), but clearly can vary by mycovirus within a species and even within a multiple-infected strain (Herrero and Zabalgogeazcoa, 2011).

5.2 *Sexual transmission*

The vertical transmission of mycoviruses through sexual spores seems usually much less effective than via asexual spores in the same species. This is the case, for example, in several *Aspergillus* species, where the majority of the sexual spores prove virus-free, while asexual spores of the same species prove infected (Coenen et al., 1997; Varga et al., 2001). Exclusion of dsRNA segments from sexual spores was further observed in, for example, the ascomycetes *Gaeumannomyces graminis* strains (McFadden et al., 1983), *Ophiostoma ulmi* (Brasier, 1983), *Epichloë festucae* (Romo Vaquero et al., 2007) and *R. necatrix* and in the basidiomycete *Helicobasidium mompa* (Ikeda et al., 2004). The exclusion of mycovirus from the sexual spore is in line with the selection arena hypothesis – to exclude deleterious elements and mutations from the germ line (Bruggeman et al., 2004).

However, there are also species where the transmission to the sexual offspring is very efficient. This is the case, for example, in the basidiomycetes *Ustilago maydis* (Koltin and Day, 1976) and *Heterobasidion annosum* (Ihrmark et al., 2004) and ascomycetes like *S. cerevisiae* (Brewer and Fangman, 1980) and *F. graminearum* (Chu et al., 2004).

5.3 Heterokaryon incompatibility and horizontal transmission

Horizontal transmission is the transfer of one mycelium to another, as opposed to the vertical transmission to sexual and asexual offspring. This horizontal transmission via hyphal fusion or *anastomosis* seems under the control of the fungal self/non-self-recognition system regulated by vegetative (vic) or heterokaryon (het) incompatibility genes. This system is thought to have evolved to limit the spread of harmful organisms like mycoviruses (Caten, 1972). In many model organisms, this vegetative/heterokaryon incompatibility system is under the regulation of multiple allelic or non-allelic *vic* or *het* genes, and most species seem to involve different genes (Glass and Dementhon, 2006; Saupe, 2000).

In 1970, Lhoas described how mycoviruses could be transferred via heterokaryosis in *Aspergillus niger* (Lhoas, 1970). Since then, transmission between isolates carrying the same *vic* or *het* genes has been reported in many species, for example, in *C. parasitica* (Liu and Milgroom, 1996) and *A. nidulans* (Coenen et al., 1997). In contrast, incompatibility blocked virus transfer in *Aspergillus* and *Cryphonectria* (e.g. Coenen et al., 1997; Milgroom and Cortesi, 2004; van Diepeningen et al., 1997). However, transfer may depend on the strength of the incompatibility reaction in the fungal species as well as on the virus (Biella et al., 2002). Transmission of mycoviruses has been found to be effective between intersterility groups of *H. annosum* (Ihrmark et al., 2002), and by anastomosis between incompatible individuals of *Beauveria bassiana* (Dalzoto et al., 2006). Carbone and co-workers (2004) found that transmission may occur in one direction, while it may not be as efficient in the reciprocal pairing. The systemic disruption of multiple *vic/het* loci can be a way to engineer super mycovirus-donor strains, as has been shown in *C. parasitica* (Zhang and Nuss, 2008).

So far, one virus, the Sclerotinia sclerotiorum mycoreovirus 4 (SsMYRV4), has been found to suppress the host's self/non-self-recognition system and thus enable its own intermycelial transfer (Wu et al., 2017). Environmental conditions may also influence intermycelial transfer. Zinc compounds have, for example, been found to attenuate heterokaryon incompatibility reactions in *R. necatrix* (Ikeda et al., 2013).

5.4 Protoplast fusion and other alternative infection routes

The rigid cell wall of fungi has been thought to serve as a structural barrier against the uptake of virus particles from extracellular sources. This is because protoplasts, competent cells or cells during the sexual reproduction stage of the yeast *S. cerevisiae* could be infected by purified virus particles (el-Sherbeini and Bostian, 1987). Removal of the fungal cell wall and subsequent protoplast fusion has been a way to transfer mycoviruses within a species, though vegetative/heterokaryon-incompatible combinations of strains could still result in limited or no virus transfer when cytoplasms are mixed (e.g. van Diepeningen et al., 1998; Wu et al., 2012). The transmission of mycoviruses to protoplast particles has been a way to expand the host range of (hypovirulent) mycoviruses to other (plant pathogenic) fungi (e.g. Kanematsu et al., 2010; Lee et al., 2011; van Diepeningen et al., 1998).

An alternative way to transfer hypovirulence within a population of *C. parasitica* has been the construction of infectious cDNA clones of hypoviruses and the subsequent transgenic lines (Chen and Nuss, 1999; Choi and Nuss, 1992). Such transgenic hypovirulent *C. parasitica* strains differ from natural hypovirulent strains in having the ability to transmit hypoviruses to ascospore progeny. However, vegetative/heterokaryon incompatibility still limits the colonisation and conversion of treated cankers (Root et al., 2005).

5.5 Extracellular transmission

The majority of mycoviruses have a strictly intracellular lifestyle, but more mycoviruses are being discovered that have an extracellular stage. The first mycovirus found with an extracellular phase was a mycoreovirus (Hillman et al., 2004; Fauquet et al., 2005). A mycovirus consisting of 4 double-stranded RNA fragments from the human pathogenic fungus *A. fumigatus*, the Aspergillus fumigatus tetramycovirus-1, proved infectious both as purified particles as well as in naked dsRNA form (Kanhayuwa et al., 2015).

The first DNA mycovirus discovered, Sclerotinia sclerotiorum hypovirulence-associated DNA virus 1 (SsHADV-1), was also found to have an extracellular phase (Yu et al., 2010, 2013). Isolated virus particles of SsHADV-1 proved to be able to transfer hypovirulence under field conditions, reduce disease severity and enhance rapeseed yield. SsHADV-1 could be transmitted to sister species *Sclerotinia minor* and *S. nivalis*, but not to other species (Yu et al., 2013).

There is relatively little evidence for the use of vectors for the transmission of mycoviruses. SsHADV-1 proved able to use a mycophagous insect – *Lycoriella ingenua* – as a vector for transmission. SsHADV-1-infected fungus could suppress the production of repellent volatile substances to attract adults to lay eggs on its colony. Larvae fed on virus-infected fungus became viruliferous adults that could transmit SsHADV-1 transovarially, while stimulating the female adults to produce more eggs (Liu et al., 2016). *Cryphonectria parasitica* with mycovirus has been found in dejecta of the mite *Thyreophagus corticalis* fed with the fungus *in vitro*. The faecal pellets in turn could transmit the infection further (Bouneb et al., 2016).

In *Helicobasidium mompa*, mycophagous nematodes found in the basidiocarps of the fungus may play a role in virus–fungus interactions (Ikeda et al., 2005). In *C. parasitica*, the mycophagous nematode *Aphelenchoides hylurgi* and relatives can be vectors for hypovirulent strains (Griffin et al., 2009, 2012)

5.6 Phylogenetic studies do suggest interspecies transfer

Indirect evidence for interspecies transfer can be derived from (phylo)genetic studies. Such studies suggest that closely-related mycoviruses may occur in different host species, and that, despite interspecies barriers, interspecies transfer may occasionally be possible. Interkingdom transfer of viruses between plant and fungus and *vice versa* has been proven possible (Andika et al., 2017; Mascia and Gallitelli, 2016; Mascia et al., 2019; Nerva et al., 2017; Petrzik et al., 2014).

van Diepeningen et al. (2008) found that within species and between closely related species, mycovirus transfer via protoplast fusions seems to occur less than the transfer between more distantly related species when cell wall restrictions are removed. Mycoviruses transferred via protoplast fusion or transfection are often viable in related hosts (e.g. van Diepeningen et al., 2008; Yu et al., 2013), but transfer to other genera is also possible (e.g. van Diepeningen et al., 2000). Transferred mycoviruses even may keep their debilitating and hypovirulent effects in their new hosts (e.g. van Diepeningen et al., 1998).

In *H. mompa*, related mycovirus fragments were found in different vegetative/heterokaryon compatibility groups, suggestive of horizontal transfer (Ikeda et al., 2005). The same variants of Cryphonectria HypoVirus-1 (CHV1) have been detected in different species of the genus *Cryphonectria* (Liu et al., 2003). At least one virus, the Sclerotinia sclerotiorum mycoreovir us 4 (SsMYRV4), suppresses host non-self-recognition and facilitates horizontal transmission of itself and heterologous viruses in *S. sclerotiorum (Wu et al., 2017)*. Mycoviruses may also have as yet unknown extracellular infective stages or use an animal vector for transfer (see Section 5.5).

In their analysis of pathogenic, saprotrophic and mycorrhizal fungi inhabiting the same forest stand, Vainio et al. (2017) showed that horizontal transmission of mycoviruses between the different groups was possible. In the context of potential virocontrol applications, they did not find a major infection pressure towards the indigenous fungal community, but the ecological consequences of such putative interspecies virus transmission will require further investigation.

6 Essential traits for a virocontrol product

The principle of virocontrol is the biocontrol of fungal plant pathogens via hypovirulence-causing mycoviruses. It has mostly been studied in the Chestnut - *C. parasitica*-hypovirus model system. More host plant-fungal pathogen and mycovirus interactions are being studied such as fruit tree-white root rot fungus-mycoviruses (*e.g.* Chiba et al., 2010). From these studies of plant, pathogen and mycovirus interactions, and studies on mycoviruses in different fungal systems, the two most important essential traits needed for effective virocontrol can be identified:

- Hypovirulence; and
- Transferability.

These are discussed in the following sections.

6.1 Hypovirulence

Whereas most viruses have cryptic effects in their host, quite a number have deleterious effects that could cause hypovirulence. Occasionally, mycovirus infection may cause hypervirulence and stronger adverse effects of the pathogenic fungus on their host plant. The three-way interaction of plant, pathogen and mycovirus should be studied. Hypovirulence is not a trait limited to a particular mycovirus family, but may occur in all major types of mycoviruses. Many fungal species harbour mycoviruses themselves, but mycoviruses may be transferred to new host species either naturally or via protoplast fusions. Transfer between related species (or species with similar physiology and codon usage) may be more effective, and mycoviruses may then retain their effects on host physiology, including hypovirulence.

6.2 Transferability

The majority of mycoviruses do not have an extracellular route, and rely on intercellular transmission between the hyphae of different mycelia. The population structure of the pathogen in the field with respect to heterokaryon/vegetative incompatibility is important in the choice of a suitable and compatible vector strain or strains for introduction in the field. A pathogen with one clonal lineage will be easier to control than a pathogen with abundant incompatibility in the field that limits the effectiveness and transfer of the virocontrol agent. Selection or construction of an omni- or pluri-compatible strain could be advantageous, but may need genetic modification. For the rare mycoviruses that do not rely on intercellular transmission, but have an extracellular route, the fungal host strain is of importance for multiplication but presumably not for transmission.

6.3 Developing a commercial virocontrol product

To develop a successful commercial biological control product, Köhl et al. (2011, 2019) described and tested a useful series of steps for screening. Considering virocontrol, these steps could be adapted as shown below:

- Step 1: Assessment of targeted crops, diseases and markets;
- Step 2: Origin and isolation of candidate organisms;
- Step 3: Stepwise assessment of biomass production, safety and ecological characteristics of candidate antagonists in rapid-throughput screening systems;
- Step 4: Assessment of potential risks for use and patent positions of candidate antagonists;
- Step 5: Efficacy testing of candidate antagonistic mycovirus/fungal host combinations on infected plants or infected parts of plants;
- Step 6: Preliminary assessments of mass production;
- Step 7: Development and testing of a pilot formulation and estimation of registration costs;
- Step 8: Upscaling mass production and full-field testing; and
- Step 9: Integration into cropping systems.

These steps are discussed in more detail below.

Step 1: Assessment of targeted crops, diseases and markets for a virocontrol plant protection product. With regard to the plant, it is important to take into account crop characteristics and the affected and targeted parts of the plants. With regard to the pathogen, one must take into account the life cycle of the pathogen, heterokaryon incompatibility within the population, natural presence of a suitable hypovirulence-causing mycovirus and the potential for genetically modified organisms (e.g. when needing omni- or pluri-compatible strains). Market size, competing products and regulations should also be taken into account.

Step 2: Origin and isolation of candidate organisms. Candidate mycovirus/fungal combinations from the relevant niche and location should ideally be used. Standard growth media and growth conditions are preferable to reduce production costs. In view of the convention of biological diversity and the Nagoya Protocol, it is important to avoid restricted isolates.

Step 3: Stepwise assessment of biomass production, safety and ecological characteristics of candidate antagonists in rapid-throughput screening systems. A key criterion is purity of the colony. This is described as regular colony development and conclusive barcoding sequences. For mycovirus-containing fungal strains, the conclusive barcoding sequences would be the more important criteria, as colony morphology may be irregular due to the mycovirus infection and its effects. The sporulation criterion of 10^5 spores after 21 days on oat meal agar at 18°C may be more variable due to virus content. The criteria for safety (no germination and growth at 36°C; no mycotoxin production) and for field applicability (cold tolerance, high or low pH tolerance, draught tolerance and UV-B radiation) may be normally met. Compatibility with fungicides against the target pathogen and with pesticides against non-target pathogens as applied in the agricultural setting of the host plant is also important.

Step 4: Assessment of potential risks for use and patent positions of candidate antagonists. With regard to intellectual property rights (IPR) protection, a potential application already protected by patents or already published can affect the commercial value of the product. It is essential to carry out safety/risk assessments for human/animal pathogenicity, allergies and mycotoxins. Pathogenicity of the fungal pathogen to plants other than the target plant is an important issue. The availability of relevant registration data in dossiers or publications and toxicology profiles can make registration and marketing easier.

Step 5: Efficacy testing of candidate antagonistic mycovirus/fungal host combinations on infected plants or infected parts of plants. When assessing the efficacy of the mycovirus/host combinations, it is necessary to check for disease control *in planta* but also for possible growth effects on the host plant.

Step 6: Preliminary assessments of mass production. The mass production of the fungus and mycovirus will probably take place using either solid or liquid fermentation. Mycovirus infections may lower both mycelial growth rates as well as spore production. In general, mycovirus effects seem less severe under nutrient-rich conditions than under nutrient-restricted growth conditions (van Diepeningen et al., 2006), making it important to optimise mass production as well as minimise production costs.

Step 7: Development and testing of a pilot formulation and estimation of registration costs. Scaling up mass production, downstream processing, product formulation and shelf life are the next steps, all of which may be affected by the intracellular presence of the mycovirus. Human toxicity and environmental safety of the formulated product are important issues for the estimation of registration costs.

Step 8: Upscaling mass production and full-field testing. The best candidates for both the production process and the formulation must finally be selected. Candidates with expected low-risk profiles will probably have lower assessment costs for registration.

Step 9: Integration into cropping systems. In this last phase before registration, evaluation of the virocontrol within common plant protection schemes in target crops and other crops, and with existing or under-development non-chemical control should be evaluated. Furthermore, persistence in the environment should be evaluated.

Though quite a number of hypovirulence-inducing mycoviruses have been described, no biocontrol product based on mycovirus/fungal host combinations has been registered and/or is commercially available. Only the *Cryphonectria parasitica* virus I has been extensively used experimentally, but not commercially in the field, to reduce chestnut blight. This means that as there are no guidelines yet detailing whether a mycovirus-based product faces the same requirements as a fungal biocontrol product or whether additional requirements need to be met.

7 Success story: *Cryphonectria parasitica*

In 1904, chestnut blight caused by *C. parasitica* was first observed in American chestnut trees in New York City. Within forty years, the pathogen had killed around 3.5 billion trees in North America and destroyed much of the former oak-chestnut forest in Eastern North America, in the triangle between Ontario, Maine and Alabama, and impacted forest ecology, including insects and wildlife. In Europe, the chestnut blight was first discovered in 1938, where it was less destructive and killed fewer trees, partly because of higher blight resistance levels in the European chestnut, and partly due to the natural occurrence of hypovirulent strains (Anagnostakis, 1982). The origin of *C. parasitica* lies in Asia, and the presumably co-evolved resistance in chestnuts is observed there (Anagnostakis, 1992). The Chinese chestnut is generally considered to be highly resistant, but the pathogen can affect various other members of the genus *Castanea* in China (Qin et al., 2002). In Japan, the pathogen is sometimes destructive on the otherwise relatively blight-resistant Japanese chestnut (Uchida, 1977).

The pathogen causing chestnut blight, *Cryphonectria parasitica* (Murrill) M.E. Barr was first described as *Diaporthe parasitica* (Murrill, 1906), renamed *Endothia parasitica* by Anderson and Anderson (1912), until it was given its current name in 1978 (Barr, 1978). *C. parasitica* causes cankers on branches and stems of susceptible chestnut trees with conspicuous yellow to orange-brown stromata, containing fungal conidiomata (containing asexual spores), ascomata (containing sexual spores) as well as host cells. On resistant trees, the stromata are infrequent or inconspicuous. In natural populations, *C. parasitica* exhibits a mixed mating system and sexual offspring is formed both by outcrossing as well as selfing (Marra et al., 2004). *C. parasitica* also has a saprophytic phase on dead wood with both perithecia and pycnidia being formed (Prospero et al., 2006). After introduction of *C. parasitica* in a new area, eradication efforts by cutting and burning the infected plants/trees have mostly failed (Rigling and Prospero, 2018).

Different dsRNA mycoviruses have been found in association with *C. parasitica*, some of which cause hypovirulence of their host, which results in less virulent, non-fatal infections of the chestnut tree. The infection rate of both non-virulent and hypovirulent mycoviruses in *C. parasitica* lies in the range of between 2% and 36% of the isolates being infected (e.g. Adamčíková et al., 2012; Montenegro et al., 2008; Murolo et al., 2018; Park et al., 2008; Peever et al., 1998). The first hypovirulence virus described was CHV-1 (Anagnostakis, 1982) which is also the type virus of the *Hypoviridae*, but since then, more hypovirulence-causing viruses of *Hypoviridae* as well as other mycovirus families have been found (e.g. Hillman et al., 2004). Different subtypes of CHV-1 based on sequence variation and RFLP have been described. Phylogenetic analysis of these subtypes and their spread suggests that the subtypes diverged several hundreds of years ago, and that different subtypes were introduced in Europe and elsewhere afterwards (Gobbin et al., 2003). The mycovirus-induced hypovirulence caused by CHV-1 in *C. parasitica* has been well studied both in the field, in the lab, and back in the field again in recurrent cycles, to enhance its potential as a biocontrol method (Nuss, 2000).

The effects of CHV-1 on its host vary from transcriptional repression of specific host genes (Kazmierczak et al., 1996; Sun et al., 2009b; Zhang et al., 1993) to influencing the secretion of proteins by its host (Kazmierczak et al., 2012). The virus-induced RNA-silencing process can be suppressed by a papain-like protease of CHV-1 (Choi et al., 1991; Segers et al., 2006). In addition to their roles in silencing viral replication by degrading viral RNAs, RNAi components also promote viral RNA recombination in *C. parasitica* (Sun et al., 2009a; Zhang and Nuss, *2008). C. parasitica* has at least four RNA-dependent RNA polymerases (RdRPs) that may be involved in transcriptional and post transcriptional gene silencing and antiviral defence (Zhang et al., 2014).

Genetic variation exists in the interaction of different *C. parasitica* isolates and different CHV variants (Peever et al., 2000; Brusini et al., 2019). Some observed effects may be mitigated by environmental conditions, for instance, the typical 'white' phenotype of *C. parasitica* upon infection with a hypovirus and lower conidiation can be overcome with a high dose of light (Hillman et al., 1990).

CHV-1 does not have an extracellular life stage. Vertical transmission of mycovirus occurs to asexual offspring, where transfer rates can vary with host and mycovirus genotype (Brusini et al., 2019). The production of sexual offspring is largely inhibited by virus infection (Nuss, 2005). However, the construction of transgenic lines with cDNA clones of hypoviruses has been a way to construct hypovirulent lineages that have the ability to transmit the hypovirulence to ascospore progeny (Chen and Nuss, 1999; Choi and Nuss, 1992). In the saprophytic phase on dead wood, perithecia yield only hypovirus-free sexual spores, while pycnidia contain 5–41% hypovirulence-infected conidia (Prospero et al., 2006).

Horizontal transmission between isolates carrying the same *vic* genes occurs readily (Liu and Milgroom, 1996), but incompatibility effectively blocks virus transfer (Milgroom and Cortesi, 2004), also from virulence encoded by the transgenic lines (Root et al., 2005). The systemic disruption of multiple *vic* loci can be a way to engineer super mycovirus-donor strains (Zhang and Nuss, 2008).

In North America, the transmission of virulence-attenuating mycoviruses is severely hampered by the at least 120 vegetative incompatibility groups present in the field, and therefore the effects of virocontrol are very limited. To restore some of the former oak-chestnut forests, breeding programmes exist where some of the remaining American chestnuts, mainly existing as understory sprouts from extant root systems, are crossed with the blight-resistant Asian chestnut species and back-crossed afterwards towards producing trees with the form, phenology and growth characteristics of the American chestnut (Anagnostakis, 2012).

In Europe, less vegetative incompatibility groups occur, and hence the more ready transfer of viruses from hypovirulent strains. In Europe, however, hypovirulence was already present, caused by different viral subtypes (Anagnostakis, 1982; Bisseger et al., 1997; Gobbin et al., 2003). The European chestnut also has a higher resistance level (Anagnostakis, 1982). Bissigger and colleagues (1997) in their study in two fields in Switzerland observed 21 Vegetative Compatibility Groups (VCGs). European vegetative compatibility groups proved to have equal infection rates with the hypovirulent virus (Bissegger et al., 1997). Young chestnut sprouts were especially killed in their first year, while over the following years, the growth of cankers decreased and hypovirulent strains were observed on all surviving sprouts, suggesting natural spread of the hypovirus infection and/or of hypovirulent strains (Bissegger et al., 1997).

When looking at spreading infections throughout Europe, one can see that in general, more recently colonised areas contain lower numbers of VCGs or only one mating type may be present (e.g. Montenegro et al., 2008; Robin and Heiniger, 2001; Robin et al, 2009). The spontaneous introduction and further spread of hypovirulence in these populations may now occur to mitigate the disease (e.g. Montenegro et al., 2008; Trapiello et al., 2017). On the other hand, as the success of biocontrol with the hypovirus CHV1 is negatively correlated to the number of VCGs present, an increase in VCG diversity may jeopardise biocontrol efforts. Movement of chestnut plants and wood should be restricted and mycelial mixtures of CHV1-infested *C. parasitica* for biocontrol should only contain the local VCG and mating types (Robin and Heiniger, 2001).

In Asia, the place of origin of *C. parastica*, the numbers of VCGs appear to be highest. In Japan, at least 71 groups could be recognised in 79 isolates, while in China, two populations of 28 and 11 isolates all consisted of unique VCGs, while a third population consisted of 15 VCGs in 25 isolates (Anagnostakis et al., 1998; Liu and Milgroom, 2007). However, due to the high resistance to chestnut blight in the different chestnut species in Asia, the disease causes only minor problems.

The application of hypovirulent *C. parasitica* in the field is quite laborious. Holes are drilled around cankers on sprouts and branches, and are then filled with hypovirulent inoculum, with each canker treated separately. Heiniger and Rigling (2009) found that while treated cankers were cured, the hypovirus might slowly spread to untreated cankers. The choice of strains of the right locally present VCGs, mating type and the type of hypovirulent virus are important. Virus types that are too virulent/hypovirulent may limit host growth and spore production, thereby limiting their own spread. The spread of hypovirulent inoculum may be helped by certain arthropods, for example, mites (Bouneb et al., 2016) or mycophagous nematodes (Griffin et al., 2009, 2012)

No simple all-round product for the treatment of *C. parasitica* is available, as locally abundant VCGs have to be taken into account, as well as choice of the virus. However, *C. parasitica* is still the best-studied example of virocontrol. While *C. parasitica* eradicated many of the American chestnuts in America, the largely natural biocontrol of the pathogen in Europe by mycoviruses has proved quite successful.

8 Future trends

Mycoviruses have been reported in Ascomycetes, Basidiomycetes as well as in Deuteromycetes, and in groups formerly considered as fungi, like Chytridiomycota and Zygomycota (Ghabrial et al., 2015). As whole-genome sequencing and RNA-Seq are becoming ever more available and are integrated into routine descriptions of fungi, more viruses will be detected, even in low titres. Some of these viruses - both DNA, ss and dsRNA viruses - cause hypovirulence and debilitation in their host (Table 3). Virus and hypovirulence incidences vary per species, and sometimes per population (Table 2). For species with no known natural mycovirus infections, mycoviruses transfer via protoplast fusion has shown to be a successful way of interspecies transfer (e.g. Ejmal et al., 2018; Lee et al., 2011; van Diepeningen et al., 1998, 2000). The integration of cDNA and the construction of transgenic hypervirulent lines may also be an option to get hypovirulence out in the field (e.g. Choi and Nuss, 1992; Chen and Nuss, 1999). Hypovirulence or debilitation-causing viruses often retain this trait in a new host (e.g. Ejmal et al., 2018; Lee et al., 2011; van

Diepeningen et al., 2008). There is thus a range of potential virocontrol viruses available for the control of pathogenic fungi.

The limited vegetative transfer between different isolates of a species, due to the omnipresent heterokaryon or vegetative incompatibility genes, may be a more serious limitation to virocontrol than the availability of mycoviruses. So far, especially in natural populations with lower diversity in *het/vic* variation, virus transmission is established via donor strains compatible with the occurring pathogen(s). Engineering of super mycovirus donor strains disrupted in multiple *het/vic* loci may be an option to overcome the problem in more diverse populations (Zhang and Nuss, 2016). Alternatively, adding compounds like zinc to the virocontrol formulation to attenuate incompatibility reactions and enhance anastomosis rates between heterogenic strains, may increase virus transfer (Ikeda et al., 2013). How and where to apply virocontrol of fungi in the field is another question, and depends on the host-pathogen system and spread of hypovirulent fungal propagules via naturally present routes or via necessary additional vectors, be it human or animal.

There is also the question of whether a risk of virocontrol is that plant pathogens are used as donor strains. In principle, the plant pathogens used are hypovirulent and should not be a risk in themselves, in comparison to the virus-free pathogens. Chances of losing a mycovirus spontaneously are relatively small as shown by laboratory experiments on curing infected strains (e.g. van Diepeningen et al., 2006), while the presence of other isogenic and thus vegetative-compatible biocontrol isolates will lead to rapid re-infections and consequent hypovirulence. However, one should evaluate carefully if a hypovirulent strain is sufficiently a-virulent and that the introduction of high amounts of a mild pathogen does not cause problems, whether applied preventively or curatively.

There are currently no examples of non-pathogenic variants of pathogen species being used (only hypovirulent ones). However, with new insights into the molecular organisation of pathogenicity in many species and whole-genome sequencing techniques, one can use donor strains containing less virulence-related genes or make knock-out constructs without certain virulence or toxigenic genes. How registration regulations will deal with application of a hypovirulent pathogen or potentially of a non-pathogenic strain of this fungus as donor is not certain, as no products have been registered yet.

Present heterokaryon or vegetative incompatibility in and between species limits both needed and unwanted spontaneous transfer, and reduces the risk of accidental infection of neutral or beneficial fungal species. As most mycoviruses lack a viable extracellular phase, transfer via that route is also very limited. Some mycoviruses are related to plant viruses, most seem co-evolved with their fungal host, while a relation to animal viruses seems very limited (Göker et al., 2011). Recently, Andika et al. (2017) showed the natural occurrence of a cross-kingdom infection of pathogenic fungi with a plant virus. No research has been done to see if the reverse also may occur. That does leave the risk observed by Moriyama and co-workers (2018) in *M. oryzae* and *A. alternata*, that at least some chrysoviruses may not only confer hypovirulence but may also be the driving force behind the development of physiological diversity in the hosts, including new pathogenic races.

Virocontrol via hypovirulent fungi is not only used in plant pathology, but is also gaining interest in human and animal mycology. A lack of applicable, effective antifungals is limiting treatment, while the number of vulnerable, often immunocompromised, patients is growing. Treatment of *A. fumigatus*, the main cause of Aspergillosis, with introduced (Refos et al., 2013) or intrinsic (Kanhayuwa et al., 2015; Takahasi-Nagaguchi et al., 2020; Zoll et al., 2018) mycoviruses has been explored. In the case of the AfuTmV-1, which seems to be an intermediate between a dsRNA and positive strand ssRNA virus in build-up, the virus is infectious both as purified entity and as naked DNA (Kanhayuwa et al., 2015). One study shows such a natural mycovirus may cause hypovirulence when tested in the *Galleria mellonella* model (Özkan and Coutts, 2015). On the other hand, recent studies on mycoviruses in *Malassezia* yeasts, common inhabitants of the human skin and occasionally causing infections, show that mycovirus infection may enhance skin colonisation and enhance interferon production (Applen Clancey et al., 2020; Park et al., 2020).

9 Conclusion

Virocontrol and the use of hypovirulent mycoviruses is a small field in biocontrol research. However, advances in genomic techniques have made the screening for mycoviruses and assessing (molecular) effects on their host easier, and a plethora of mycoviruses have been found in all divisions of fungi and related phyla. Infections with mycoviruses appear quite common in most fungal species, with infection rates varying between 0% and 100%, with corresponding effects on host pathogenicity. For those pathogens that do not contain hypovirulent mycoviruses themselves in nature, alternative ways of infection are available, using viruses from other (related) species and protoplast techniques.

Fungal (population) biology has taught us a lot about the reproduction modes of fungi and vegetative (in)compatibility, and how that may affect mycovirus transmissions. Vertical transmission rates of viruses to sexual and asexual offspring can vary per species and per mycovirus, as does horizontal transfer between isolates. For successful application in the field, virus donors need to be compatible with the target field isolates and demonstrate incompatibility and species boundaries to limit further spread.

As yet, no commercial virocontrol product exists for the control of plant pathogenic fungi. However, mycovirus research so far has shown that for many of the major plant pathogens, potential candidates for virocontrol with hypovirulent mycoviruses exist.

10 Where to look for further information

10.1 Mycoviruses in general

- Ghabrial, S. A., Castón, J. R., Jiang, D., Nibert, M. L. and Suzuki, N. (2015), 50 - plus years of fungal viruses. *Virology*479, 356-368.
- Luque, D., Mata, C. P. Suzuki, N., Ghabrial, S. A. and Castón, J. R. (2018), Capsid structure of dsRNA fungal viruses. *Viruses* 10, 481.
- Son, M., Yu, J. and Kim, K. H. (2015), Five questions about mycoviruses. *Plos Pathog.* 11, e1005172.

10.2 Principles of virocontrol and biocontrol

- Chiba, S., Kondo, H., Kanematsu, S. and Suzuki, N. (2010), Mycoviruses and virocontrol. *Uirusu* 60, 163-176. [in Japanese +English abstract].
- Köhl, J., Goossen-van de Geijn, H., Groenenboom-de Haas, L., Henken, B., Hauschild, R., Hilscher, U., Lombaers-van der Plas, C., van den Bosch, T. and Wikström, M. (2019), Stepwise screening of candidate antagonists for biological control of *Blumeria graminis* f. sp. *tritici*. *Biol. Contr.* 136: 104008.
- Köhl, J., Postma, J., Nicot, P. and Ruocco, M. (2011), Stepwise screening of microorganisms for commercial use in biological control of plant pathogenic fungi and bacteria. *Biol. Contr. 57:* 1-12.
- Xie, J. and Jiang, D. (2014), New insights into mycoviruses and exploration for the biological control of crop fungal diseases. *Annu. Rev. Phytopathol.* 52: 45-68.

10.3 C. parasitica *mycoviruses and hypovirulence*

- Bisseger, M., Rigling, D. and Heiniger, U. (1997), Population structure and disease development of *Cryphonectria parasitica* in European chestnut forests in the presence of natural hypovirulence. *Phytopathology* 87, 50-59.
- Eusebio-Cope, A., Sun, L., Tanaka, T., Chiba, S., Kasahara, S. and Suzuki, N. (2015), *The chestnut blight fungus for studies on virus/host and virus/virus interactions: From a natural to a model host. Virology 477, 164-175.*
- Hillman, B. and Suzuki, N (2004), Viruses of the chestnut blight fungus, *Cryphonectria parasitica*. *Adv. Virus Res.* 63, 423-472.
- Liu, Y.-C. and Milgroom, M. G. (1996), Correlation between hypovirus transmission and the number of vegetative incompatibility (*vic*) genes

different among isolates from a natural population of *Cryphonectria parasitica*. *Phytopathology* 86, 79–86.

- Milgroom, M. G. and Cortesi, P. (2004), Biological control of chestnut blight with hypovirulence: a critical analysis. *Ann. Rev. Phytopathol.* 42, 311–338.
- Rigling, D. and Prospero, S. (2018), *Cryphonectria parasitica*, the causal agent of chestnut blight: invasion history, population biology and disease control. *Mol. Plant Pathol.* 19, 7–20.
- Zhang, D. X. and Nuss, D. L. (2016), Engineering super mycovirus donor strains of chestnut blight fungus by systematic disruption of multilocus *vic* genes. *Proc. Natl. Acad. Sci. USA*. 113, 2062–2067.

10.4 Rosellinia necatrix mycoviruses and hypovirulence

The white root rot fungus *Rosellinia necatrix* is a soil-borne pathogen that causes serious economic losses worldwide in various crops, including fruit trees. After *C. parasitica*, it probably is the best studied plant-fungal pathogen-mycovirus system studied.

- Arjona-Lopez, J. M., Telengech, P., Jamal, A., Hisano, S., Kondo, H., Yelin, M. D., Arjona-Girona, I., Kanematsu, S., Lopez-Herrera, C. J. and Suzuki, N. (2018), Novel, diverse RNA viruses from Mediterranean isolates of the phytopathogenic fungus, *Rosellinia necatrix*: insights into evolutionary biology of fungal viruses. *Environm. Microbiol.* 20, 1464–1483.
- Kondo, H., Kanematsu, S., and Suzuki, N. (2013), Chapter Seven – Viruses of the White Root Rot Fungus, *Rosellinia necatrix*. In: Ghabrial, S. A. (Ed.) *Advances in Virus Research*. Cambridge, MA: Academic Press, Volume 86, pp. 177–214.
- Shimizu, T., Kanematsu, S. and Yaegashi, H. (2018), Draft genome sequence and transcriptional analysis of *Rosellinia nacatrix* infected with a virulent mycovirus. *Phytopathol.* 108, 1206–1211.
- Yaegashi, H., Yoshikawa, N., Ito, T. and Kanematsu, S. (2013), A mycoreovirus suppresses RNA silencing in the white root rot fungus, *Rosellinia necatrix*. *Virology* 444, 409–416.

11 References

Adamčíková, K., Kobza, M., Bolvanský, M. and Ondruskova, E. (2012). Spread and population structure of *Cryphonectria parasitica* in a young chestnut orchard in Slovakia. *Eur. J. Biol.* 7(2), 267–274.

Adams, M. J., Antoniw, J. F., Bar-Joseph, M., Brunt, A. A., Candresse, T., Foster, G. D., Martelli, G. P., Milne, R. G., Zavriev, S. K. and Fauquet, C. M. (2004). The new plant virus family Flexiviridae and assessment of molecular criteria for species demarcation. *Arch. Virol.* 149(5), 1045–1060.

Ahn, I.-P. and Lee, Y.-H. (2001). A viral double-stranded RNA up regulates the fungal virulence of *Nectria radicicola*. *Mol. Plant. Micro.* 14, 496–507.

Allen, A., Islamovic, E., Kaur, J., Gold, S., Shah, D. and Smith, T. J. (2013). The virally encoded killer proteins from Ustilago maydis. *Fungal Biol. Rev.* 26(4), 166–173.

Aminian, P., Azizollah, A., Abbas, S. and Naser, S. (2011). Effect of double-stranded RNAs on virulence and deoxynivalenol production of *Fusarium graminearum* isolates. *J. Plant Prot. Res.* 51(1), 29–37.

Anagnostakis, S. L. (1982). Biological control of chestnut blight. *Science* 215(4532), 466–471.

Anagnostakis, S. L. (1992). Chestnuts and the introduction of chestnut blight. *Annu. Rep. North. Nut. Grow Assoc.* 83, 39–42.

Anagnostakis, S. L. (2012). Chestnut Breeding in the United States for disease and insect resistance. *Plant Dis.* 96(10), 1392–1403.

Anagnostakis, S. L., Chen, B., Geletka, L. M. and Nuss, D. L. (1998). *Hypovirus* transmission to ascospore progeny by field-released transgenic hypovirulent strains of *Cryphonectria parasitica*. *Phytopathology* 88(7), 598–604.

Anderson, P. J. and Anderson, H. W. (1912). *Endothia virginiana*. *Phytopathology* 2, 261–262.

Andika, I. B., Wei, S., Cao, C., Salaipeth, L., Kondo, H. and Sun, L. (2017). Phytopathogenic fungus hosts a plant virus: a naturally occurring cross-kingdom viral infection. *Proc. Natl. Acad. Sci. U. S. A.* 114(46), 12267–12272.

Aoki, N., Moriyama, H., Kodama, M., Arie, T., Teraoka, T. and Fukuhara, T. (2009). A novel mycovirus associated with four double-stranded RNAs affects host fungal growth in *Alternaria alternata*. *Virus Res.* 140(1–2), 179–187.

Applen Clancey, S., Ruchti, F., LeibundGut-Landmann, S., Heitman, J. and Ianiri, G. (2020). A novel mycovirus evokes transcriptional rewiring in the fungus *Malassezia* and stimulates beta interferon production in macrophages. *mBio* 11(5), e01534-20.

Arakawa, M., Nakamura, H., Uetake, Y. and Matsumoto, N. (2002). Presence and distribution of double-stranded RNA elements in the white root rot fungus *Rosellinia necatrex*. *Mycoscience* 43(1), 21–26.

Arjona-Lopez, J. M., Telengech, P., Jamal, A., Hisano, S., Kondo, H., Yelin, M. D., Arjona-Girona, I., Kanematsu, S., Lopez-Herrera, C. J. and Suzuki, N. (2018). Novel, diverse RNA viruses from Mediterranean isolates of the phytopathogenic fungus, *Rosellinia necatrix*: insights into evolutionary biology of fungal viruses. *Environ. Microbiol.* 20(4), 1464–1483.

Arthur, K., Pearson, M. N. and Beever, R. E. (2007). Detection and incidence of the mycoviruses BVX and BCVF. Abstract. Proceedings of the 16th Australasian Plant Pathology Conference, Adelaide, 24–27 September 2007.

Banks, G. T., Buck, K. W., Chain, E. B., Himmelweit, F., Marks, J. E., Tyler, J. M., Hollings, M., Last, F. T. and Stobe, O. M. (1968). Viruses in fungi and interferon stimulation. *Nature* 218(5141), 542–545.

Barr, M. E. (1978). The Diaporthales in North America: with emphasis on *Gnomonia* and its segregates. *Mycol. Mem.* 7, 1–232.

Biella, S., Smith, M. L., Aist, J. R., Cortesi, P. and Milgroom, M. G. (2002). Programmed cell death correlates with virus transmission in a filamentous fungus. *Proc. Biol. Sci.* 269(1506), 2269–2276.

Bisseger, M., Rigling, D. and Heiniger, U. (1997). Population structure and disease development of *Cryphonectria parasitica* in European chestnut forests in the presence of natural hypovirulence. *Phytopathology* 87(1), 50–59.

Boland, G. J. (1992). Hypovirulence and double-stranded RNA in *Sclerotinia sclerotiorum*. *Can. J. Plant Pathol.* 14(1), 10–17.

Bottacin, A. M., Levesque, C. A. and Punja, Z. K. (1994). Characterization of dsRNA in *Chalara elegans* and effects on growth and virulence. *Phytopathology* 84(3), 303–312.

Bouneb, M., Turchetti, T., Nannelli, R., Roversi, P. F., Paoli, F., Danti, R. and Simoni, S. (2016). Occurrence and transmission of mycovirus Cryphonectria Hypovirus 1 from dejecta of *Thyreophagus corticalis* (Acari, Acaridae). *Fungal Biol.* 120(3), 351–357.

Brasier, C. M. (1983). A cytoplasmically transmitted disease of *Ceratocystis ulmi*. *Nature* 305(5931), 220–223.

Brewer, B. J. and Fangman, W. L. (1980). Preferential inclusion of extrachromosomal genetic elements in yeast meiotic spores. *Proc. Natl. Acad. Sci. U. S. A.* 77(9), 5380–5384.

Bruenn, J. (1986). The killer systems of *Saccharomyces cerevisiae* and other yeasts. In: Buck, K. W. (Ed), *Fungal Virology*. Boca Raton, FL: CRC Press Press, pp. 86–108.

Bruggeman, J., Debets, A. J. and Hoekstra, R. F. (2004). Selection arena in *Aspergillus nidulans*. *Fungal Genet. Biol.* 41(2), 181–188.

Brusini, J., Wayne, M. L., Franc, A. and Robin, C. (2019). The impact of parasitism on resource allocation in a fungal host: the case of *Cryphonectria parasitica* and its mycovirus, Cryphonectria Hypovirus 1. *Ecol. Evol.* 7(15), 5967–5976.

Buck, K. (1998). Molecular variability of viruses of fungi. In: Couteaudier, P. D. Y. and Clackson, J. M. (Eds), *Molecular Variability of Fungal Pathogens Bridge*. Wallingford, UK: CAB International, pp. 53–72.

Buck, K. W., Girvan, R. F. and Ratti, G. (1973). Two serologically distinct double-stranded ribonucleic acid viruses isolated from *Aspergillus niger*. *Biochem. Soc. Trans.* 1(5), 1138–1140.

Buck, K. W. and Ratti, G. (1975). Biophysical and biochemical properties of two viruses isolated from *Aspergillus foetidus*. *J. Gen. Virol.* 27, 211–224.

Cai, G., Fry, W. E. and Hillman, B. I. (2019). PiRV-2 stimulates sporulation in *Phytophthora infestans*. *Virus Res.* 271, 197674.

Carbone, I., Liu, Y. C., Hillman, B. I. and Milgroom, M. G. (2004). Recombination and migration of Cryphonectria Hypovirus 1 as inferred from gene genealogies and the coalescent. *Genetics* 166(4), 1611–1629.

Castro, M., Kramer, K., Valdivia, L., Ortiz, S. and Castillo, A. (2003). A double-stranded RNA mycovirus confers hypovirulence-associated traits to *Botrytis cinerea*. *FEMS Microbiol. Lett.* 228(1), 87–91.

Caten, C. E. (1972). Vegetative incompatibility and cytoplasmic infection in fungi. *J. Gen. Microbiol.* 72(2), 221–229.

Chen, B. and Nuss, D. L. (1999). Infectious cDNA clone of Hypovirus CHV1-Euro7: a comparative virology approach to investigate virus-mediated hypovirulence of the chestnut blight fungus *Cryphonectria parasitica*. *J. Virol.* 73(2), 985–992.

Chiba, S., Kondo, H., Kanematsu, S. and Suzuki, N. (2010). [Mycoviruses and virocontrol]. Uirusu 60:163–176. [in Japanese + English abstract].

Cho, W. K., Yu, J., Lee, K. M., Son, M., Min, K., Lee, Y. W. and Kim, K. H. (2012). Genome-wide expression profiling shows transcriptional reprogramming in Fusarium graminearum by Fusarium graminearum virus 1-DK21 infection. *BMC Genom.* 13, 173.

Choi, G. H. and Nuss, D. L. (1992). Hypovirulence of chestnut blight fungus conferred by an infectious viral cDNA. *Science* 257(5071), 800–803.

Choi, G. H., Pawlyk, D. M. and Nuss, D. L. (1991). The autocatalytic protease p29 encoded by a hypovirulence-associated virus of the chestnut blight fungus resembles the Potyvirus-encoded protease HC-Pro. *Virology* 183(2), 747–752.

Chu, Y. M., Jeon, J. J., Yea, S. J., Kim, Y. H., Yun, S. H., Lee, Y. W. and Kim, K. H. (2002). Double-stranded RNA mycovirus from *Fusarium graminearum*. *Appl. Environ. Microbiol.* 68(5), 2529–2534.

Chu, Y. M., Lim, W. S., Yea, S. J., Cho, J. D., Lee, Y. W. and Kim, K. H. (2004). Complexity of dsRNA mycovirus isolated from *Fusarium graminearum*. *Virus Genes* 28(1), 135–143.

Coenen, A., Kevei, F. and Hoekstra, R. F. (1997). Factors affecting the spread of double-stranded RNA viruses in *Aspergillus nidulans*. *Genet. Res.* 69(1), 1–10.

Czárán, T. L. and Hoekstra, R. F. (2003). Killer-sensitive coexistence in metapopulations of micro-organisms. *Proc. Biol. Sci.* 270(1522), 1373–1378.

Dalzoto, P. R., Glienke-Blanco, C., Kava-Cordeiro, V., Ribeiro, J. Z., Kitajima, E. W. and Azevedo, J. L. (2006). Horizontal transfer and hypovirulence associated with double-stranded RNA in Beauveria bassiana. *Mycol. Res.* 110(12), 1475–1481.

Darissa, O., Adam, G. and Schäfer, W. (2012). A dsRNA mycovirus causes hypovirulence of *Fusarium graminearum* to wheat and maize. *Eur. J. Plant Pathol.* 134, 181–189.

Darissa, O., Willingmann, P., Schäfer, W. and Adam, G. (2011). A novel double-stranded RNA mycovirus from *Fusarium graminearum*: nucleic acid sequence and genomic structure. *Arch. Virol.* 156(4), 647–658.

de Sa, P. B., Havens, W. M. and Ghabrial, S. A. (2010). Characterization of a novel broad-spectrum antifungal protein from virus-infected *Hemninthosporium* (*Cochliobolus*) *victoriae*. *Phytopathology* 100(9), 880–889.

Deng, F., Xu, R. and Boland, G. J. (2003). Hypovirulence-associated double-stranded RNA from *Sclerotinia homoeocarpa* is conspecific with Ophiostoma novo-ulmi Mitovirus 3a-Ld. *Phytopathology* 93(11), 1407–1414.

DePaulo, J. J. and Powell, C. (1995). Extraction of double-stranded RNA from plant tissues without the use of organic solvents. *Plant Dis.* 79(3), 246–248.

Ejmal, M. A., Holland, D. J., MacDiarmid, R. M. and Pearson, M. N. (2018). The effect of Aspergillus Thermomutatus Chrysovirus 1 on the biology of three Aspergillus species. *Viruses* 10(10), 539.

el-Sherbeini, M. and Bostian, K. A. (1987). Viruses in fungi: infection of yeast with the K1 and K2 killer viruses. *Proc. Natl. Acad. Sci. U. S. A.* 84(12), 4293–4297.

Elias, K. S. and Cotty, P. J. (1996). Incidence and stability of infection by double-stranded RNA genetic elements in *Aspergillus* section *flavi* and effects on aflatoxigenicity. *Can. J. Bot.* 74(5), 716–725.

Enebak, S. A., Hillman, B. I. and MacDonald, W. L. (1994). A hypovirulent isolate of *Cryphonectria parasitica* with multiple, genetically unique dsRNA segments. *Mol. Plant Microbe Interact.* 7(5), 590–595 .

Fauquet, C. M., Mayo, M. A., Maniloff, J., Desselberger, U. and Ball, L. A. (Eds.) (2005). *Virus Taxonomy: Eighth Report of the International Committee on Taxonomy of Viruses*. San Diego: Elsevier Academic Press.

Fekete, C., Giczey, G., Papp, I., Szabó, L. and Hornok, L. (1995). High-frequency occurrence of virus-like particles with double-stranded RNA genome in *Fusarium poae*. *FEMS Microbiol. Lett.* 131, 295–299.

Fujimura, T. and Esteban, R. (2011). Cap-snatching mechanism in yeast L-A double-stranded RNA virus. *Proc. Natl. Acad. Sci. U. S. A.* 108(43), 17667–17671.

Fuke, K., Takeshita, K., Aoki, N., Fukuhara, T., Egusa, M., Kodama, M. and Moriyama, H. (2011). The presence of double-stranded RNAs in *Alternaria alternata* Japanese pear pathotype is associated with morphological changes. *J. Gen. Plant Pathol.* 77(4), 248–252.

Ghabrial, S. A. (1998). Origin, adaptation and evolutionary pathways of fungal viruses. *Virus Genes* 16(1), 119–131.

Ghabrial, S. A., Castón, J. R., Jiang, D., Nibert, M. L. and Suzuki, N. (2015). 50-plus years of fungal viruses. *Virology* 479, 356–368.

Ghabrial, S. A. and Suzuki, N. (2008). Fungal viruses. In: Mahy, B. W. J. and Van Regenmortel, M. H. V. (Eds), *Encyclopedia of Virology* (3rd edn.) (vol. 2). Oxford, UK: Elsevier, pp. 284–291.

Ghabrial, S. A. and Suzuki, N. (2009). Viruses of plant pathogenic fungi. *Annu. Rev. Phytopathol.* 47, 353–384.

Gilbert, K. B., Holcomb, E. E., Allscheid, R. L. and Carrington, J. C. (2019). Hiding in plain sight: new virus genomes discovered via a systematic analysis of fungal public transcriptomes. *PLoS ONE* 14(7), e0219207.

Gillings, M. R., Tesoriero, L. A. and Gunn, L. V. (1993). Detection of double-stranded RNA and virus-like particles in Australian isolates of *Pythium irregulare*. *Plant Pathol.* 42(1), 6–15.

Glass, N. L. and Dementhon, K. (2006). Non-self recognition and programmed cell death in filamentous fungi. *Curr. Opin. Microbiol.* 9(6), 553–558.

Gobbin, D., Hoegger, P. J., Heiniger, U. and Rigling, D. (2003). Sequence variation and evolution of Cryphonectria Hypovirus 1 (CHV-1) in Europe. *Virus Res.* 97(1), 39–46.

Göker, M., Scheuner, C., Klenk, H. P., Stielow, J. B. and Mezel, W. (2011). Codivergence of mycoviruses with their hosts. *PLoS ONE* 6(7), e22252.

Griffin, G. J., Eisenback, J. D., Yancey, M. M. and Templeton, J. J. (2009). *Aphelenchoides hylurgi* as a carrier of white, hypovirulent *Cryphonectria parasitica* and its possible role in hypovirulence spread on blight-controlled American chestnut trees. *J. Nematol.* 41(4), 267–273.

Griffin, G. J., Eisenback, J. D. and Oldham, K. (2012). Widespread distribution of fungivorous *Aphelenchoides* spp. in blight cankers on American Chestnut trees. *J. Nemat.* 44, 316–320.

Hammond, T. M., Andrewski, M. D., Roossinck, M. J. and Keller, N. P. (2008). Aspergillus mycoviruses are targets and suppressors of RNA silencing. *Eukaryot. Cell* 7(2), 350–357.

Hao, F., Ding, T., Wu, M., Zhang, J., Yang, L., Chen, W.. and Li, G. (2018). Two novel hypovirulence-associated mycoviruses in the phytopathogenic fungus *Botrytis cinerea*: molecular characterization and suppression of infection cushion formation. *Viruses* 10(5), 254.

Hayashi, N., Tsuge, T., Kobayashi, H. and Nishimura, S. (1988). The presence of double-stranded RNAs in *Alternaria alternata* Japanese pear pathotype and their participation in AK-toxin productivity. *Jpn. J. Phytopathol.* 54(2), 250–252.

Heaton, L. A. and Leslie, J. F. (2004). Double-stranded RNAs associated with *Fusarium proliferatum* mitochondria. *Mycol. Progr.* 3(3), 193–198.

Heiniger, U. and Rigling, D. (2009). Application of the Cryphonectria hypovirus (CHV-1) to control the chestnut blight, experience from Switzerland. *Acta Hortic.* 815, 233–246.

Herrero, N. and Zabalgogeazcoa, I. (2011). Mycoviruses infecting the endophytic and entomopathogenic fungus *Tolypocladium cylindrosporum*. *Virus Res.* 160(1-2), 409-413.

Hillman, B. I., Shapira, R. and Nuss, D. L. (1990). Hypovirulence-associated suppression of host functions can be partially relieved by high light intensity. *Phytopathology* 80, 950-956.

Hillman, B. I., Supyani, S., Kondo, H. and Suzuki, N. A. (2004). A reovirus of the fungus *Cryphonectria parasitica* that is infectious as particles and related to the *Coltivirus* genus of animal pathogens. *J. Virol.* 78(2), 892-898.

Hollings, M. (1962). Viruses associated with a die-back disease of cultivated mushroom. *Nature* 196(4858), 962-965.

Hong, Y., Cole, T. E., Brasier, C. M. and Buck, K. W. (1998). Evolutionary relationships among putative RNA-dependent RNA polymerases encoded by a mitochondrial virus-like RNA in the Dutch elm disease fungus, *Ophiostoma novo-ulmi*, by other viruses and virus-like RNAs and by the Arabidopsis mitochondrial genome. *Virology* 246(1), 158-169.

Hong, Y., Dover, S. L., Cole, T. E., Brasier, C. M. and Buck, K. W. (1999). Multiple mitochondrial viruses in an isolate of the Dutch elm disease fungus Ophiostoma novo-ulmi. *Virology* 258(1), 118-127.

Howitt, R. L. J., Beever, R. E., Pearson, M. N. and Forster, R. L. S. (1995). Presence of double-stranded RNA and virus-like particles in Botrytis cinerea. *Mycol. Res.* 99(12), 1472-1478.

Howitt, R. L. J., Beever, R. E., Pearson, M. N. and Forster, R. L. S. (2001). Genome characterization of Botrytis virus F, a flexuous rod-shaped mycovirus resembling plant "potex-like" viruses. *J. Gen. Virol.* 82(1), 67-78.

Howitt, R. L. J., Beever, R. E., Pearson, M. N. and Forster, R. L. S. (2006). Genome characterization of a flexuous rod-shaped mycovirus, Botrytis virus X, reveals high amino acid identity to genes from plant 'potex-like' viruses. *Arch. Virol.* 151(3), 563-579.

Huang, S., Soldevila, A. I., Webb, B. A. and Ghabrial, S. A. (1997). Expression, assembly and proteolytic processing of Helminthosporium victoriae 190S totivirus capsid protein in insect cells. *Virology* 234, 130-137.

Hunst, P. L., Latterell, F. M. and Rossi, A. E. (1986). Variation in double stranded RNA from isolates of Pyricularia oryzae. *Phytopathology* 76(7), 674-678.

Ihrmark, K., Johannesson, H., Stenström, E. and Stenlid, J. (2002). Transmission of double-stranded RNA in *Heterobasidion annosum*. *Fungal Genet. Biol.* 36(2), 147-154.

Ihrmark, K., Stenström, E. and Stenlid, J. (2004). Double-stranded RNA transmission through basidiospores of *Heterobasidion annosum*. *Mycol. Res.* 108(2), 149-153.

Ikeda, K., Inoue, K., Kida, C., Uwamori, T., Sasaki, A., Kanematsu, S. and Park, P. (2013). Potentiation of mycovirus transmission by zinc compounds via attenuation of heterogenic incompatibility in *Rossellinia necatrex*. *Appl. Environ. Microbiol.* 79(12), 3684-3691.

Ikeda, K., Nakamura, H., Arakawa, M. and Matsumoto, N. (2004). Diversity and vertical transmission of double-stranded RNA elements in root rot pathogens of trees, *Helicobasidium* mompa and *Rosellinia necatrix*. *Mycol. Res.* 108, 626-634.

Ikeda, K.-I., Nakamura, H., Arakawa, M. and Masumoto, N. (2005). Dynamics of double-stranded RNA segments in a *Helicobasidium mompa* clone from a tulip tree plantation. *FEMS Microbiol. Ecol.* 51(2), 293-301.

Jacob-Wilk, D., Turina, M. and Van Alfen, N. K. (2006). Mycovirus Cryphonectria Hypovirus 1 elements cofractionate with trans-Golgi network membranes of the fungal host *Cryphonectria parasitica*. *J. Virol.* 80(13), 6588–6596.

Je, M., Kim, H. and Son, H. S. (2019). Analysis of the codon usage pattern of the RdRP gene of mycovirus infecting Aspergillus spp. *Virol. J.* 16(1), 10.

Jiang, D., Fu, Y. and Ghabrial, S. A. (2013). Mycoviruses: chapter eight–viruses of the plant pathogenic fungus *Sclerotinia sclerotiorum*. *Adv. Virus Res.* 86, 215–248.

Kamaruzzaman, M., He, G., Wu, M., Zhang, J., Yang, L., Chen, W. and Li, G. (2019). A novel partitivirus in the hypovirulent isolate QT5-19 of the plant pathogenic fungus *Botrytis cinerea*. *Viruses* 11(1), 24.

Kanematsu, S., Arakawa, M., Oikawa, Y., Onoue, M., Osaki, H., Nakamura, H., Ikeda, K., Kuga-Uetaki, Y., Nitta, H., Sasaki, A., Suzaki, K., Yoshida, K. and Matsumoto, N. (2004). A reovirus causes hypovirulence of Rossellinia *necatrex*. *Phytopathology* 94(6), 561–568.

Kanematsu, S., Sasaki, A., Onoue, M., Oikawa, Y. and Ito, T. (2010). Extending the fungal host range of a partitivirus and a Mycoreovirus from Rosellinia necatrix by inoculation of protoplasts with virus particles. *Phytopathology* 100(9), 922–930.

Kanhayuwa, L., Kotta-Loizou, I., Özkan, S., Gunning, A. P. and Coutts, R. H. (2015). A novel mycovirus from *Aspergillus fumigatus* contains four unique dsRNAs as its genome and is infectious as dsRNA. *Proc. Natl. Acad. Sci. U. S. A.* 112(29), 9100 -9105.

Kazama, F. Y. and Schornstein, K. L. (1973). Ultrastructure of a fungus Herpes-type virus. *Virology* 52(2), 478–487.

Kazmierczak, P., McCabe, P., Turina, M., Jacob-Wilk, D. and Van Alfen, N. K. (2012). The mycovirus CHV1 disrupts secretion of a developmentally regulated protein in *Cryphonectria parasitica*. *J. Virol.* 86(11), 6067–6074.

Kazmierczak, P., Pfeiffer, P., Zhang, L. and Van Alfen, N. K. (1996). Transcriptional repression of specific host genes by the mycovirus Cryphonectria Hypovirus 1. *J. Virol.* 70(2), 1137–1142.

Kilic, O. and Griffin, G. J. (1998). Effect of dsRNA-containing and dsRNA-free hypovirulent isolates of *Fusarium oxysporum* on severity of *Fusarium* seedling disease of soybean in naturally infested soil. *Plant Soil* 201(1), 125–135.

King, A. M., Adams, M. J. and Lefkowitz, E. J. (2011). *Virus Taxonomy: Classification and Nomenclature of Viruses: Ninth Report of the International Committee on Taxonomy of Viruses*. Amsterdam, the Netherlands: Elsevier.

King, A. M. Q., Adams, M. J., Carstens, E. B. and Lefkowitz, E. J. (2012). *Virus Taxonomy: Classification and Nomenclature of Viruses: Ninth Report of the International Committee on Taxonomy of Viruses* (vol. 9). London, UK: Elsevier Academic Press, pp. 23–1208.

Köhl, J., de Geijn, H. G., Haas, L. G., Henken, B., Hauschild, R., Hilscher, U., der Plas, C. L., van den Bosch, T. and Wikström, M. (2019). Stepwise screening of candidate antagonists for biological control of *Blumeria graminis* f. sp. *tritici*. *Biol. Control* 136, 104008.

Köhl, J., Postma, J., Nicot, P., Ruocco, M. and Blum, B. (2011). Stepwise screening of microorganisms for commercial use in biological control of plant pathogenic fungi and bacteria. *Biol. Control* 57(1), 1–12.

Koltin, Y. (1986). Ustilago maydis virus-encoded killer system. In: Extrachromosomal elements in lower eukaryotes. *Basic Life Sci.* 40, 239–257.

Koltin, Y. and Day, P. R. (1976). Inheritance of killer phenotypes and double-stranded RNA in *Ustilago* maydis. *Proc. Natl. Acad. Sci. U. S. A.* 73(2), 594–598.

Kondo, H., Kanematsu, S. and Suzuki, N. (2013). Seven-viruses of the white root rot fungus, *Rosellinia necatrix*. In: Ghabrial, S. A. (Ed) *Advances in Virus Research* (Vol. 86). Cambridge, MA: Academic Press, 177–214.

Koonin, E. V., Choi, G. H., Nuss, D. L., Shapira, R. and Carrington, J. C. (1991). Evidence for common ancestry of a chestnut blight hypovirulence associated double-stranded RNA and a group of positive-strand RNA plant viruses. *Proc. Natl. Acad. Sci. U. S. A.* 88(23), 10647–10651.

Kotta-Loizou, I. and Coutts, R. H. A. (2017). Mycoviruses in *Aspergilli*: a comprehensive review. *Front. Microbiol.* 8, 1699. DOI: 10.3389/fmicb.2017.01699.

Kozlakidis, Z., Hacker, C. V., Bradley, D., Jamal, A., Phoon, X., Webber, J., Brasier, C. M., Buck, K. W. and Coutts, R. H. A. (2009). Molecular characterisation of two novel double-stranded RNA elements from *Phlebiopsis gigantea*. *Virus Genes* 39(1), 132–136.

Kwon, S. J., Lim, W. S., Park, S. H., Park, M. R. and Kim, K. H. (2007). Molecular characterization of a dsRNA mycovirus, Fusarium graminearum virus-DK21, which is phylogenetically related to hypoviruses but has a genome organization and gene expression strategy resembling those of plant potex-like viruses. *Mol. Cells* 23(3), 304 -315.

Lakshman, D. K., Jian, J. and Tavantzis, S. M. (1998). A double-stranded RNA element from a hypovirulent strain of Rhizoctonia solani occurs in DNA form and is genetically related to the pentafunctional AROM protein of the shikimate pathway. *Proc. Natl. Acad. Sci. U. S. A.* 95(11), 6425–6429.

Lee, K. M., Yu, J., Son, M., Lee, Y. W. and Kim, K. H. (2011). Transmission of *Fusarium boothii* mycovirus via protoplast fusion causes hypovirulence in other phytopathogenic fungi. *PLoS ONE* 6(6), e21629.

Lemus-Minor, C. G., Cañizares, M. C., García-Pedrajas, M. D. and Pérez-Artéz, E. (2015). Complete genome sequence of a novel dsRNA mycovirus isolated from the phytopathogenic fungus *Fusarium oxysporum* f. sp. *dianthi*. *Arch. Virol.* 160(9), 2375–2379.

Lhoas, P. (1970). Use of heterokaryosis to infect virus free strains of Aspergillus niger. *Aspergillus News Lett* 11, 8.

Li, H., Bian, R., Liu, Q., Yang, L., Pang, T., Salaipeth, L., Andika, I. B., Kondo, H. and Sun, L. (2019). Identification of a novel hypovirulence-inducing hypovirus from alternaria alternata. *Front. Microbiol.* 10, 1076.

Li, H., Fu, Y., Jiang, D., Li, G., Ghabrial, S. A. and Yi, X. (2008). Down-regulation of *Sclerotinia sclerotiorum* gene expression in response to infection with Sclerotinia sclerotiorum debilitation-associated RNA virus. *Virus Res.* 135(1), 95–106.

Li, P., Lin, Y., Zhang, H., Wang, S., Qiu, D.. and Guo, L. (2016). Molecular characterization of a novel mycovirus of the family Tymoviridae isolated from the plant pathogenic fungus Fusarium graminearum. *Virology* 489, 86–94.

Li, P., Wang, S., Zhang, L., Qiu, D., Zhou, X. and Guo, L. (2020). A tripartite ssDNA mycovirus from a plant pathogenic fungus is infectious as cloned DNA and purified virions. *Sci. Adv.* 6(14), eaay9634.

Linder-Basso, D., Dynek, J. N. and Hillman, B. I. (2005). Genome analysis of Chryphonectria Hypovirus 4, the most common Hypovirus species in North America. *Virology* 337(1), 192–203.

Liu, C., Li, M., Redda, E. T., Mei, J., Zhang, J., Wu, B. and Jiang, X. (2019). A novel double-stranded RNA mycovirus isolated from *Trichoderma harzianum*. *Virol. J.* 16(1), 113.

Liu, Y.-C. and Milgroom, M. G. (1996). Correlation between Hypovirus transmission and the number of vegetative incompatibility (*vic*) genes different among isolates from a natural population of *Cryphonectria parasitica*. *Phytopathol* 86, 79–86.

Liu, Y. C. and Milgroom, M. G. (2007). High diversity of vegetative compatibility types in Cryohonectria *parasitica* in Japan and China. *Mycologia* 99(2), 279–284.

Liu, L., Xie, J., Cheng, J., Fu, Y., Li, G., Yi, X. and Jiang, D. (2014). Fungal negative-stranded RNA virus that is related to *bornaviruses* and *nyaviruses*. *Proc. Natl. Acad. Sci. U. S. A.* 111(33), 12205–12210.

Liu, S., Xie, J., Cheng, J., Li, B., Chen, T., Fu, Y., Li, G., Wang, M., Jin, H., Wan, H. and Jiang, D. (2016). Fungal DNA virus infects a mycophagous insect and utilizes it as a transmission vector. *Proc. Natl. Acad. Sci. U. S. A.* 113(45), 12803–12808.

Liu, Y. C., Linder-Basso, D., Hillman, B. I., Kaneko, S. and Milgroom, M. G. (2003). Evidence for interspecies transmission of viruses in natural populations of filamentous fungi in the genus *Cryphonectria*. *Mol. Ecol.* 12(6), 1619–1628.

Luque, D., Mata, C. P., Suzuki, N., Ghabrial, S. A. and Castón, J. R. (2018). Capsid structure of dsRNA fungal viruses. *Viruses* 10(9), 481.

Márquez, L. M., Redman, R. S., Rodriguez, R. J. and Roossinck, M. J. (2007). A virus in a fungus in a plant: three-way symbiosis required for thermal tolerance. *Science* 315(5811), 513–515.

Márquez, L. M. and Roossinck, M. J. (2012). Do persistent RNA viruses fit the trade-off hypothesis of virulence evolution? *Curr. Opin. Virol.* 2, 556–560.

Marra, R. E., Cortesi, P., Bissegger, M. and Milgroom, M. G. (2004). Mixed mating in natural populations of the chestnut blight fungus, *Cryphonectria parasitica*. *Heredity (Edinb)* 93(2), 189–195.

Martelli, G. P., Adams, M. J., Kreuze, J. F. and Dolja, V. V. (2007). Family Flexiviridae: a case study in virion and genome plasticity. *Annu. Rev. Phytopathol.* 45, 73–100.

Marzano, S. L., Nelson, B. D., Ajavi-Ovetunde, O., Bradley, C. A., Hughes, T. J., Hartman, G. L., Eastburn, D. M. and Domier, L. L. (2016). Identification of diverse mycoviruses through metatranscriptomics characterization of the viromes of five major fungal plant pathogens. *J. Virol.* 90(15), 6846–6863.

Mascia, T. and Gallitelli, D. (2016). Infection, replication and expression of plant viruses in filamentous fungi. In: Gaur, R. K., Petrov, N. M., Patil, B. L. and Stovanova, M. I. (Eds) *Plant Viruses: Evolution and Management*. Singapore: Springer, pp. 31–38.

Mascia, T., Vučurović, A., Minutillo, S. A., Nigro, F., Labarile, R., Savoia, M. A., Palukaitis, P. and Gallitelli, D. (2019). Infection of *Colletotrichum acutatum* and *Phytophthora infestans* by taxonomically different plant viruses. *Eur. J. Plant Pathol.* 153(4), 1001–1017.

McFadden, J. J. P., Buck, K. W. and Rawlinson, C. J. (1983). Infrequent transmission of double-stranded RNA virus particles but absence of DNA proviruses in single ascospore cultures of *Gaeumannomyces graminis*. *J. Gen. Virol.* 64(4), 927–937.

Milgroom, M. G. and Cortesi, P. (2004). Biological control of chestnut blight with hypovirulence: a critical analysis. *Annu. Rev. Phytopathol.* 42, 311–338.

Montenegro, D., Aguín, O., Sainz, M. J., Hermida, M. and Mansilla, J. P. (2008). Diversity of vegetative compatibility types, distribution of mating types and occurrence of hypovirulence of *Cryphonectria parasitica* in chestnut stands in NW Spain. *Forest Ecol. Manag.* 256(5), 973–980.

Moriyama, H., Urayama, S. I., Higashiura, T., Le, T. M. and Komatsu, K. (2018). Chrysoviruses in *Magnaporthe oryzae*. *Viruses* 10(12), 697.

Murolo, S., De Miccolis Angelini, R. M., Faretra, F. and Romanazzi, G. (2018). Phenotypic and molecular investigation on hypovirulent *Cryohonectria parasitica* in Italy. *Plant Dis.* 102(3), 540–545.

Murrill, W. A. (1906). A new chestnut disease. *Torreya* 6, 186–189.

Myers, J. M., Bonds, A. E., Clemons, R. A., Thapa, N. A., Simmons, D. R., Carter-House, D., Ortanez, J., Liu, P., Miralles-Durán, A., Desirò, A., Longcore, J. E., Bonito, G., Stajich, J. E., Spatafora, J. W., Chang, Y., Corrochano, L. M., Gryganskyi, A., Grigoriev, I. V. and James, T. Y. (2020). Survey of early-diverging lineages of fungi reveals abundant and diverse mycoviruses. *mBio* 11(5), e02027-20.

Nerva, L., Chitarra, W., Siciliano, I., Gaiotti, F., Ciuffo, M., Forgia, M., Varese, G. C. and Turina, M. (2019). Mycoviruses mediate mycotoxin regulation in *Aspergillus ochraceus*. *Environ. Microbiol.* 21(6), 1957–1968.

Nerva, L., Varese, G. C., Falk, B. W. and Turina, M. (2017). Mycoviruses of an endophytic fungus can replicate in plant cells: evolutionary implications. *Sci. Rep.* 7(1), 1908.

Nogawa, M., Shimosaka, M., Kageyama, T. and Okazaki, M. (1993). A double-stranded RNA mycovirus from the plant pathogenic fungus *Fusarium solani* f. sp. *robiniae*. *FEMS Microbiol. Lett.* 110(2), 153–157.

Nuss, D. L. (2000). Hypovirulence and chestnut blight: from the field to the laboratory and back. In: Kronstad, J. W. (Ed) *Fungal Pathology*. Dordrecht, the Netherlands: Springer, pp. 149–170.

Nuss, D. L. (2005). Hypovirulence: mycoviruses at the fungal-plant interface. *Nat. Rev. Microbiol.* 3(8), 632–642.

Okada, R., Ichinose, S., Takeshita, K., Urayama, S. I., Fukuhara, T., Komatsu, K., Arie, T., Ishihara, A., Egusa, M., Kodama, M. and Moriyama, H. (2018). Molecular characterization of a novel mycovirus in *Alternaria alternata* manifesting two-sided effects: down-regulation of host growth and up-regulation of host plant pathogenicity. *Virology* 519, 23–32.

Okada, R., Kiyota, E., Moriyama, H., Fukuhara, T. and Natsuaki, T. (2015). A simple and rapid method to purify viral dsRNA from plant and fungal tissue. *J. Gen. Plant Pathol.* 81(2), 103–107.

Osaki, H., Sasaki, A., Nomiyama, K. and Tomioka, K. (2016). Multiple virus infection in a single strain of *Fusarium poae* shown by deep sequencing. *Virus Genes* 52(6), 835–847.

Özkan, S. and Coutts, R. H. (2015). Aspergillus fumigatus mycovirus causes mild hypervirulent effect on pathogenicity when tested on *Galleria mellonella*. *Fungal Genet. Biol.* 76, 20 –26.

Park, C. M., Banerjee, N., Koltin, Y. and Bruenn, J. A. (1996). The Ustilago maydis virally encoded KP1 killer toxin. *Mol. Microbiol.* 20(5), 957–963.

Park, M., Cho, Y. J., Kim, D., Yang, C. S., Lee, S. M., Dawson, T. L., Jr., Nakamizo, S., Kabashima, K., Lee, Y. W. and Jung, W. H. (2020). A novel virus alters gene expression and vacuolar morphology in *Malassezia* cells and induces a TLR3-mediated inflammatory immune response. *mBio* 11(5), e01521-20.

Park, S. M., Kim, J. M., Chung, H. J., Lim, J. Y., Kwon, B. R., Lim, J. G., Kim, J. A., Kim, M. J., Cha, B. J., Lee, S. H., Kim, K. H., Lee, Y. S., Yang, M. S. and Kim, D. H. (2008). Occurrence of diverse dsRNA in a Korean population of the chestnut blight fungus, *Cryphonectria parasitica*. *Mycol. Res.* 112(10), 1220–1226.

Park, Y., Chen, X. and Punja, Z. K. (2006). Diversity, complexity and transmission of double-stranded RNA elements in Chalara elegans (synanam. Thielaviopsis basicola). *Mycol. Res.* 110(6), 697–704.

Peever, T. L., Liu, Y. C., Cortesi, P. and Milgroom, M. G. (2000). Variation in tolerance and virulence in the chestnut blight fungus-Hypovirus interaction. *Appl. Environ. Microbiol.* 66(11), 4863–4869.

Peever, T. L., Liu, Y. C., Wang, K., Hillman, B. I., Foglia, R. and Milgroom, M. G. (1998). Incidence and diversity of double-stranded RNAs occurring in the chestnut blight fungus, *Cryphonectria parasitica*, in China and Japan. *Phytopathology* 88(8), 811–817.

Petrzik, K., Vondrák, J., Barták, M., Peksa, O. and Kubesová, O. (2014). Lichens-a new souces or yet unknown host of herbaceous plant viruses. *Eur. J. Plant Pathol.* 138(3), 549–559.

Polashock, J. J., Bedker, P. J. and Hillman, B. I. (1997). Movement of a small mitochondrial double-stranded RNA element of *Cryphonectria parasitica*: ascospore inheritance and implications for mitochondrial recombination. *Mol. Gen. Genet.* 256(5), 566–571.

Polashock, J. J. and Hillman, B. I. (1994). A small mitochondrial double-stranded (ds) RNA element associated with a hypovirulent strain of the chestnut blight fungus and ancestrally related to yeast cytoplasmic T-dsRNA and W-dsRNA. *Proc. Natl. Acad. Sci. U. S. A.* 91, 8680–8684.

Potgieter, A. C., Page, N. A., Liebenberg, J., Wright, I. M., Landt, O. and van Dijk, A. A. (2009). Improved strategies for sequence-independent amplification and sequencing of viral dsRNA genomes. *J. Gen. Virol.* 90(6), 1423–1432.

Prospero, S., Conedera, M., Heiniger, U. and Rigling, D. (2006). Saprophytic activity and sporulation of *Cryphonectria parasitica* on dead chestnut wood in forests with naturally established hypovirulence. *Phytopathology* 96(12), 1337–1344.

Qin, L., Gao, X., Cheng, J., Liu, S., Lhang, D. and Wang, Y. (2002). Evaluation of the resistance of Chinese chestnut cultivars to *Cryphonectria parasitica*. *J.Fruit Sci.* 19, 39–42.

Qiu, L., Li, Y., Liu, Y., Gao, Y., Qi, Y. and Shen, J. (2010). Particle and naked RNA mycoviruses in industrially cultivated mushroom *Pleurotus ostreatus* in China. *Fungal Biol.* 114(5-6), 507–513.

Ran, H., Liu, L., Li, B., Cheng, J., Fu, Y., Jiang, D. and Xie, J. (2016). Co-infection of a hypovirulent isolate of *Scherotinia sclerotiorum* with a new botyvirnavirus and a strain of a Mitovirus. *Virol. J.* 13, 92.

Ratti, G. and Buck, K. W. (1972). Virus particles in Aspergillus foetidus: a multicomponent system. *J. Gen. Virol.* 14(2), 165–175.

Refos, J. M., Vonk, A. G., Eadie, K., Lo-Ten-Foe, J. R., Verbrugh, H. A., van Diepeningen, A. D. and van de Sande, W. W. J. (2013). Double-stranded RNA mycovirus infection of *Aspergillus fumigatus* is not dependent on the genetic make-up of the host. *PLoS ONE* 8(10), e77381.

Rigling, D. and Prospero, S. (2018). *Cryphonectria parasitica*, the causal agent of chestnut blight: invasion history, population biology and disease control. *Mol. Plant Pathol.* 19(1), 7–20.

Robin, C., Capdevielle, X., Martin, M., Traverand, C. and Colinas, C. (2009). *Cryphonectria parasitica* vegetative compatibility type analysis of populations in south-western France and northern Spain. *Plant Pathol.* 58(3), 527–535.

Robin, C. and Heiniger, U. (2001). Chestnut blight in Europe: diversity of *Cryphonectria parasitica*, hyporvirulence and biocontrol. *For. Snow. Landsc. Res.* 76, 361 -367.

Romo Vaquero, M., Leuchtmann, A., García, B. and Zabalgogeazcoa, I. (2007). A Totivirus infecting the mutualistic fungal endophyte *Epichloë festucae*. *Virus Res.* 124(1-2), 38-43.

Roossinck, M. J. (2019). Evolutionary and ecological links between plant and fungal viruses. *New Phytol.* 221(1), 86-92.

Roossinck, M. J., Sabanadzovic, S., Okada, R. and Valverde, R. A. (2011). The remarkable evolutionary history of endornaviruses. *J. Gen. Virol.* 92(11), 2674-2678.

Root, C., Balbalian, C., Bierman, R., Geletka, L. M., Anagnostakis, S., Double, M., MacDonald, W. and Nuss, D. L. (2005). Multi-seasonal field release and spermatization trials of transgenic hypovirulent strains of *Cryphonectria parasitica* containing cDNA copies of Hypovirus CHV1-EP713. *Forest Pathol.* 35(4), 277-297.

Sasaki, A., Kanematsu, S., Onoue, M., Oyama, Y. and Yoshida, K. (2006). Infection of *Rossellinia necatrix* with purified viral particles of a member of the Partitiviridae (RnPV1-W8). *Arch. Virol.* 151(4), 697-707.

Saupe, S. J. (2000). Molecular genetics of heterokaryon incompatibility in filamentous ascomycetes. *Microbiol. Mol. Biol. Rev.* 64(3), 489-502.

Schmitt, M. J. and Breinig, F. (2006). Yeast viral killer toxins: lethality and self-protection. *Nat. Rev. Microbiol.* 4(3), 212-221.

Segers, G. C., van Wezel, R., Zhang, X., Hong, Y. and Nuss, D. L. (2006). Hypovirus papain-like protease p29 suppresses RNA silencing in the natural fungal host and in a heterologous plant system. *Eukaryot. Cell* 5(6), 896-904.

Shapira, R., Choi, G. H. and Nuss, D. L. (1991). Virus-like genetic organization and expression strategy for a double-stranded RNA genetic element associated with biological control of chestnut blight. *EMBO J.* 10(4), 731-739.

Sharma, S., Gupta, S. and Shrivastava, J. N. (2011). Presence of virus like particles in human pathogenic fungi: *Chrysosporium* sps and *Candida albicans*. *Indian J. Virol.* 22, 104-110.

Shimizu, T., Kanematsu, S. and Yaegashi, H. (2018). Draft genome sequence and transcriptional analysis of *Rosellinia nacatrix* infected with a virulent mycovirus. *Phytopathology* 108(10), 1206-1211.

Sivanesan, A. and Holliday, P. (1981). *Cryphonectria Parasitica. CMI Descriptions of Pathogenic Fungi and Bacteria* (Vol. 704). Wallingford, UK: CAB International.

Son, M., Lee, K. M., Yu, J., Kang, M., Park, J. M., Kwon, S. J. and Kim, K. H. (2013). The *hex1* gene of *Fusarium graminearum* is required for fungal asexual reproduction and pathogenesis and for efficient viral RNA accumulation of Fusarium graminearum virus. *J. Virol.* 87(18), 10356-10367.

Son, M., Yu, J. and Kim, K. H. (2015). Five questions about mycoviruses. *PLoS Pathog.* 11(11), e1005172.

Sun, Q., Choi, G. H. and Nuss, D. L. (2009a). A single Argonaute gene is required for induction of RNA silencing antiviral defense and promotes viral RNA recombination. *Proc. Natl. Acad. Sci. U. S. A.* 106(42), 17927-17932.

Sun, Q., Choi, G. H. and Nuss, D. L. (2009b). Hypovirus-responsive transcription factor gene *pro1* of the chestnut blight fungus *Cryphonectria parasitica* is required for female fertility, asexual spore development and stable maintenance of Hypovirus infection. *Eukaryot. Cell* 8(3), 262-270.

Suzuki, N., Supyani, S., Maruyama, K. and Hillman, B. I. (2004). Complete genome sequence of *Mycoreovirus-1*/Cp9B21, a member of a novel genus within the family *Reoviridae*, isolated from the chestnut blight fungus *Cryphonectria parasitica*. *J. Gen. Virol.* 85(11), 3437–3448.

Takahashi-Nakaguchi, A., Shishido, E., Yahara, M., Urayama, S., Sakai, K., Chibana. H., Kamei K., Moriyama, H. and Gonoi, T. (2020). Analysis of an intrinsic mycovirus associated with reduced virulence of the human pathogenic fungus *Aspergillus fumigatus*. *Front. Microbiol.* 10, 2045.

Tanaka, T., Eusebio-Cope, A., Sun, L. and Suzuki, N. (2012). Mycoreovirus genome alterations: similarities to and differences from rearrangements reported for other Reoviruses. *Front. Microbiol.* 3, 186.

Tooley, P. W., Hewings, A. D. and Falkenstein, K. F. (1989). Detection of double-stranded RNA in Phytophthora infestans. *Phytopathology* 79(4), 470–474.

Torres-Trenas, A., Ptieto, P., Cañizares M. C., García-Pedrajas M. D. and Pérez-Artés E. (2019). Mycovirus *Fusarium oxysporum* f. sp. dianthi virus 1 decreases the colonizing efficiency of its fungal host. *Front. Cell. Infect. Microbiol.* 9, 51.

Torres-Trenas, A. and Pérez-Artés, E. (2020). Characterization and incidence of the first member of the genus *Mitovirus* identified in the phytopathogenic species *Fusarium oxysporum*. *Viruses* 12(3), 279.

Tran, T. T., Li, H., Nguyen, D. Q., Jones, M. G. K. and Wylie, S. J. (2019). Co-infection with three mycovrises stimulates growth of a *Molinia fructicola* siolae on nutrient medium, but does not induce hypervirulence in a natural host. *Viruses* 11(1), 89.

Trapiello, E., Rigling, D. and González, A. J. (2017). Occurrence of Hypovirus-infected *Cryphonectria parasitica* isolates in northern Spain: an encouraging situation for biological control of chestnut blight in Asturian forests. *Eur. J. Plant Pathol.* 149(2), 503–514.

Tsai, P. F., Pearson, M. N. and Beever, R. E. (2004). Mycoviruses in *Monilinia fructicola*. *Mycol. Res.* 108(8), 907–912.

Uchida, K. (1977). Studies on Endothia canker of Japanese chestnut trees caused by *Endothia parasitica*. *Bull. Ibaraki-Ken Hortic. Exp. Stn.*, Special Issue 4, 65 pp.

Vainio, E. J., Pennanen, T., Rajala, T. and Hantula, J. (2017). Occurrence of similar mycoviruses in pathogenic, saprotrophic and mycorrhizal fungi inhabiting the same forest stand. *FEMS Microbiol. Ecol.* 93(3). DOI: 10.1093/femsec/fix003.

van Diepeningen, A. D., Debets, A. J. M. and Hoekstra, R. F. (1998). Intra-and interspecies virus transfer in Aspergilli via protoplast fusion. *Fungal Genet. Biol.* 25(3), 171 –180.

van Diepeningen, A. D., Debets, A. J. M. and Hoekstra, R. F. (1997). Heterokaryon incompatibility blocks virus transfer among natural isolates of black Aspergilli. *Curr. Genet.* 32(3), 209–217.

van Diepeningen, A. D., Debets, A. J. M., Slakhorst, S. M., Fekete, C., Hornok, L. and Hoekstra, R. F. (2000). Interspecies virus transfer via protoplast fusions between *Fusarium poae* and black *Aspergillus* strains. *Fungal Genet. Newsl.* 47(1), 99–100.

van Diepeningen, A. D., Debets, A. J. M. and Hoekstra, R. F. (2006). Dynamics of dsRNA mycoviruses in the black Aspergillus population. *Fungal Genet. Biol.* 43(6), 446–452.

van Diepeningen, A. D., Varga, J., Hoekstra, R. F. and Debets, A. J. M. (2008). Mycoviruses in aspergilli. In: Samson, R. and Varga, J. (Eds) *Aspergillus in the Genomics Era*. Wageningen, The Netherlands: Wageningen Academic Publishers, pp. 133–176.

Varga, J., Kevei, F., Vagvolgyi, C., Vriesema, A.. and Croft, J. H. (1994). Double-stranded RNA mycoviruses in section Nigri of the Aspergillus genus. *Can. J. Microbiol.* 40(4), 325–329.

Varga, J., Tóth, B., Szencz, S., Molnár, J., Fekete, C. S. and Szabó, L. (2001). Double-stranded TNA elements and virus-like particles in Aspergilli. *Acta Biol. Hung.* 52(2–3), 355–363.

Varga, J., Tóth, B. and Vágvölgyi, C. (2003). Recent advances in *mycovirus* research. *Acta Microbiol. Immunol. Hung.* 50(1), 77–94.

Voth, P. D., Mairura, L., Lockhart, B. E. and May, G. (2006). Phylogeography of Ustilago maydis virus H1 in the USA and Mexico. *J. Gen. Virol.* 87(11), 3433–3441.

Wang, L., Zhang, J., Zhang, H., Qiu, D. and Guo, L. (2016). Two novel relative double-stranded RNA mycoviruses infecting Fusarium poae Strain SX63. *Int. J. Mol. Sci.* 17(5).

Wei, C. Z., Osaki, H., Iwanami, T., Matsumoto, N. and Ohtsu, Y. (2003). Molecular characterization of dsRNA segments 2 and 5 and electron microscopy of a novel reovirus from a hypovirulent isolate, W370, of the plant pathogen *Rosellinia necatrix*. *J. Gen. Virol.* 84, 2431–2437.

Wickner, R. B. (1996). Double-stranded RNA viruses of *Saccharomyces cerevisiae* . *Microbiol. Rev.* 60(1), 250–265.

Wloch-Salamon, D. M., Gerla, D., Hoekstra, R. F. and de Visser, J. A. (2008). Effect of dispersal and nutrient availability on the competitive ability of toxin-producing yeast. *Proc. Biol. Sci.* 275(1634), 535–541.

Wu, M., Jin, F., Zhang, J., Yang, L., Jiang, D. and Li, G. (2012). Characterization of a novel bipartite double-stranded RNA mycovirus conferring hypovirulence in the phytopathogenic fungus *Botrytis porri*. *J. Virol.* 86(12), 6605–6619.

Wu, M., Zhang, L., Li, G., Jiang, D. and Ghabrial, S. A. (2010). Genome characterization of a debilitation-associated Mitovirus infecting the phytopathogenic fungus *Botrytis cinerea*. *Virology* 406(1), 117–126.

Wu, S., Cheng, J., Fu, Y., Chen, T., Jiang, D., Ghabrial, S. A., et al. (2017). Virus-mediated suppression of host non-self-recognition facilitates horizontal transmission of heterologous viruses. *PLoS Pathog.* 13(3), e1006234. DOI: 10.1371/journal.ppat.1006234.

Xie, J. and Jiang, D. (2014). New insights into mycoviruses and exploration for the biological control of crop fungal diseases. *Annu. Rev. Phytopathol.* 52, 45–68.

Xie, J., Wei, D., Jiang, D., Fu, Y., Li, G., Ghabrial, S. and Peng, Y. (2006). Characterization of debilitation-associated mycovirus infecting the plant-pathogenic fungus *Sclerotina sclerotorium*. *J. Gen. Virol.* 87(1), 241–249.

Xie, J., Xiao, X., Fu, Y., Liu, H., Cheng, J., Ghabrial, S. A., Li, G. and Jiang, D. (2011). A novel mycovirus closely related to hypoviruses that infects the plant pathogenic fungus *Sclerotinia sclerotiorum*. *Virology* 418(1), 49–56.

Yaegashi, H., Yoshikawa, N., Ito, T. and Kanematsu, S. (2013). A Mycoreovirus suppresses RNA silencing in the white root rot fungus, *Rosellinia necatrix*. *Virology* 444(1–2), 409–416.

Yu, J., Lee, K. M., Cho, W. K., Park, J. Y. and Kim, K. H. (2018). Differential contribution of RNA interference components in response to distinct Fusarium graminearum virus infections. *J. Virol.* 92(9), e01756-17.

Yu, X., Li, B., Fu, Y., Jiang, D., Ghabrial, S. A., Li, G., Peng, Y., Xie, J., Cheng, J., Huang, J. and Yi, X. (2010). A geminivirus-related DNA mycovirus that confers hypovirulence to a plant pathogenic fungus. *Proc. Natl. Acad. Sci. U. S. A.* 107(18), 8387–8392.

Yu, X., Li, B., Fu, Y., Xie, J., Cheng, J., Ghabrial, S. A., Li, G., Yi, X. and Jiang, D. (2013). Extracellular transmission of a DNA mycovirus and its use as a natural fungicide. *Proc. Natl. Acad. Sci. U. S. A.* 110(4), 1452–1457.

Zhai, L., Xiang, J., Zhang, M., Fu, M., Yang, Z., Hong, N. and Wang, G. (2016). Characterization of a novel double-stranded RNA mycovirus conferring hypovirulence from the phytopathogenic fungus *Botryosphaeria dothidea*. *Virology* 493, 75–85.

Zhang, D. X. and Nuss, D. L. (2016). Engineering super mycovirus donor strains of chestnut blight fungus by systematic disruption of multilocus *vic* genes. *Proc. Natl. Acad. Sci. U. S. A.* 113(8), 2062-2067.

Zhang, D. X., Spiering, M. J. and Nuss, D. L. (2014). Characterizing the roles of *Cryphonectria parasitica* RNA-dependent RNA polymerase-like genes in antiviral defense, viral recombination and transposon transcript accumulation. *PLoS ONE* 9(9), e108653.

Zhang, H., Xie, J., Fu, Y., Cheng, J., Qu, Z., Zhao, Z., Cheng, S., Chen, T., Li, B., Wang, Q., Liu, X., Tian, B., Collinge, D. B. and Jiang, D. (2020). A 2-kb mycovirus converts a pathogenic fungus into a beneficial endophyte for Brassica Protection and yield enhancement. *Mol. Plant* 13(10), 1420–1433.

Zhang, L., Churchill, A. C., Kazmierczak, P., Kim, D. H. and van Alfen, N. K. (1993). Hypovirulence-associated traits induced by a mycovirus of *Cryphonectria parasitica* are mimicked by targeted inactivation of a host gene. *Mol. Cell. Biol.* 13(12), 7782–7792.

Zhang, X. and Nuss, D. L. (2008). A host dicer is required for defective viral RNA production and recombinant virus vector RNA instability for a positive sense RNA virus. *Proc. Natl. Acad. Sci. U. S. A.* 105(43), 16749–16754.

Zoll, J., Verweij, P. E. and Melchers, W. J. G. (2018). Discovery and characterization of novel *Aspergillus fumigatus* mycoviruses. *PLoS ONE* 13(7), e0200511.

Part 3

Examples of use of bioprotectants

Chapter 16

Development and scale-up of bioprotectants to keep staple foods safe from aflatoxin contamination in Africa

Ranajit Bandyopadhyay, Alejandro Ortega-Beltran, Matieyedou Konlambigue, Lawrence Kaptoge and Titilayo D. O. Falade, International Institute of Tropical Agriculture, Nigeria; and Peter J. Cotty, Ocean University of China, China

1 Introduction

Recent decades have seen a transformation in agriculture in response to multiple challenges and opportunities. These include increasing demand for food/feed for an ever-growing population and raw materials for diverse industries, the dynamics of climate change and technological advances, among many other factors. Agricultural production has increased dramatically in most parts of the world. However, improvements in the safety of plant and animal products have not kept pace with production. Foodborne pathogens and contaminants cause a large burden of disease globally. It is estimated that foodborne diseases account for 600 million cases and 420 000 deaths each year. Children under 5 years of age account for 30% of all foodborne deaths.

http://dx.doi.org/10.19103/AS.2021.0093.16

Aflatoxin is an important food safety challenge across the globe but is a particular problem in sub-Saharan Africa (SSA), where its occurrence and severity are most acute. Aflatoxins affect a range of food and feed products of plant, animal and insect origins as indicated in Table 1. These food and feed products frequently contain aflatoxin concentrations above official regulatory limits. Contamination is often high in groundnuts and maize, resulting in lethal concentrations surpassing 8000 parts per billion (ppb) (CDC, 2004). Animals including fish, poultry and cattle fed with contaminated feed become sources of dietary contamination themselves.

Aflatoxin contamination is critical because of its influence on health. Aflatoxicosis may be acute or chronic. Acute aflatoxicosis is episodic and results from dietary consumption of food containing extremely high levels of aflatoxins. A notorious case is the 2004 aflatoxicosis outbreak in Kenya that led to 125 deaths out of 317 cases - a high case fatality rate of 39% (CDC, 2004). Chronic aflatoxicosis is caused by long-term sub-lethal exposure to aflatoxins and results in negative health outcomes such as hepatocellular carcinoma. About 30% of all liver cancer cases in Africa can be attributed to aflatoxin exposure (Liu and Wu, 2010). In children, aflatoxin exposure has been associated with stunting. Exposure can predate birth and continue through the first 1000 days of life through breast milk and weaning foods (Hsieh and Hsieh, 1993; Partanen et al., 2010; Anthony et al., 2016). Other health outcomes associated with aflatoxin exposure include immune system suppression, toxic hepatitis, Reye's syndrome or fatty degeneration of the viscera, bile duct degeneration, low infant birth weight and acute hepatic encephalopathy (Groopman et al., 2008). Farm animals experience similar types of health effects as humans, but some are more pronounced and

Table 1 Aflatoxin-susceptible food categories, examples and extent of contamination

Food category	Examples	Extent
Cereals	Maize, sorghum, millet, sesame, rice, fonio millet and teff	Common
Oil seeds	Groundnuts, bitter melon seeds, sunflower and palm kernel	Common
Spices, condiments	Ginger, chili, Bambara groundnuts and locust beans	Common
Tree nuts	Pistachio, figs, almond and cashew nuts	Common
Roots and tubers	Yam and cassava chips	Infrequent
Vegetables	Bitter leaf and okra	Infrequent
Aquatic animals	Dried fish and farmed fish (e.g. catfish, tilapia, trout)	Infrequent
Poultry	Poultry meat and eggs	Common
Ruminants	Dairy products, muscle and offal	Common
Insects	Caterpillars and termites	Common

Source: Adedeji et al., 2017; Kachapulula et al., 2018; Okoth, 2016.

perceptible. Aflatoxin exposure in animals leads to reduced productivity and, sometimes, death.

Aflatoxins negatively impact economic development. Indirect impacts result from chronic and acute exposure to aflatoxins on human and animal health that increases health costs. The African Union's Partnership for Aflatoxin Control in Africa (PACA) estimates that annual health costs due to aflatoxin-related cancers in Nigeria are US$1600 million (m); Tanzania US$1100 m; Uganda US$577 m; Malawi US$393 m; Senegal US$161 m; and The Gambia US$22 m (https://www.aflatoxinpartnership.org/resources-catagory/country-and-regional). In the case of aflatoxin exposure in livestock, additional feed is needed to counteract poor growth as well as the cost of veterinary treatments from reduced immunity and reduced market value from animal products due to lower weight and quality and higher mortality (Aikore et al., 2019).

Direct impacts of aflatoxins on economic development are related to costs associated with lost trade opportunities and recall of food products or their destruction by an importing country when standards are not met (Wu and Khlangwiset, 2010). Allowable levels of aflatoxins in food differ among African countries, depending on prevalence, testing regimes and the ability to manage toxin exposure (van Egmond and Jonker, 2004). In-country monitoring across Africa is minimal, leading to the risk of traders segregating higher quality produce for international trade that is more strictly monitored and retaining lower quality crops for in-country trade and consumption (Wu and Guclu, 2012). For all these reasons, aflatoxin mitigation can therefore contribute to achieving the United Nations Sustainable Development Goals (Ortega-Beltran and Bandyopadhyay, 2021).

This chapter summarizes the biology of aflatoxin-producing fungi and various factors affecting their occurrence, including climate change. Various management practices for aflatoxin mitigation are then discussed. These include biocontrol, which is increasingly being adopted by farmers in several countries. We discuss biocontrol product development and commercialization in various African countries. Subsequently, we highlight some barriers to adoption and other challenges. Much of the chapter is drawn from the experiences of the International Institute of Tropical Agriculture (IITA) and its partners that spearheaded the aflatoxin biocontrol initiative in Africa: the Aflasafe Initiative.

2 *Aspergillus* biology and aflatoxin epidemiology

2.1 Causal agents of aflatoxin contamination

Several *Aspergillus* species can produce either B or both B and G aflatoxins (Frisvad et al., 2019). The most common producer of aflatoxin is *A. flavus*,

Published by Burleigh Dodds Science Publishing Limited, 2022.

which produces B aflatoxins (Amaike and Keller, 2011; Klich, 2007). There are two major morphotypes within *A. flavus*: the L and the S morphotypes (Cotty, 1989). The former produces variable aflatoxin levels, significant numbers of spores and a few large (>400 μm) sclerotia. The latter produces consistently high aflatoxin levels, fewer spores and numerous small (<400 μm) sclerotia. Another important species is *A. parasiticus*, a B and G aflatoxin-producing species associated with groundnut in most areas where the crop is produced (Horn et al., 1995; Kachapulula et al., 2017; Schroeder, 1969). *A. parasiticus* sometimes infects maize, pistachio, almond, fig and other crops (Donner et al., 2015; Probst et al., 2014).

Many aflatoxin-producing *Aspergillus* isolates have become delineated into new species (Carvajal-Campos et al., 2017; Singh et al., 2020; Soares et al., 2012). In most cases, they are not epidemiologically significant. In some areas, species or groups of fungi occurring at relatively low frequencies are significant because of their high aflatoxin-production potential (Probst et al., 2014). For example, in Kenya, a group of fungi with S morphology are potent aflatoxin producers and have been linked to severe episodes of aflatoxicosis and death (Probst et al., 2012). Similarly, in West Africa, another group of highly toxigenic fungi with S morphology occurs at relatively low proportions but may significantly contaminate staple crops grown in that region (Agbetiameh et al., 2018; Atehnkeng et al., 2008; Cardwell and Cotty, 2002; Diedhiou et al., 2011). Examination of fungal communities in areas yet to be studied, or in those not intensively examined, will certainly result in identification of new species.

Aflatoxin-producing genotypes within a species have different toxin production potentials (Horn and Dorner, 1999; Kachapulula et al., 2017; Mehl et al., 2012; Novas and Cabral, 2002; Ortega-Beltran et al., 2015). These potentials also vary among and within diverse substrates (Mehl and Cotty, 2013; Ortega-Beltran et al., 2014; Suwarno et al., 2019). There are aflatoxin-producing species with members that have lost the ability to produce aflatoxins. Such isolates are known as atoxigenic. Natural genetic defects across the aflatoxin biosynthesis gene cluster or lack of the whole cluster itself cause atoxigenicity (Adhikari et al., 2016; Chang et al., 2005; Donner et al., 2010).

Aflatoxin-producing fungal community structures are highly dynamic. There is great variation within and among fields, in both small and large areas, within cropping systems and across years (Bayman and Cotty, 1991; Mehl et al., 2012; Ortega-Beltran and Cotty, 2018). Community structures dominated by certain genotypes or species for 1 year may change dramatically within a relatively short term (Drott et al., 2019; Ortega-Beltran et al., 2020).

There is large genetic variability within and among species. One way to determine genetic variability within a species is to assign isolates into vegetative compatibility groups (VCGs) (Leslie, 1993). Isolates belonging to a VCG are thought to descend from the same clonal lineage. There is also genetic variability within a VCG (Grubisha and Cotty, 2010; Ortega-Beltran et al., 2016). Regardless of intra-VCG variability, members of a VCG are more closely related to each other than to members of other VCGs. There may be thousands of VCGs of any single species within a given area. Members of a VCG may occur in different countries (Agbetiameh et al., 2019; Moral et al., 2020) and even in different continents (Ogunbayo et al., 2013).

The name *Aspergillus* has been used to describe the anamorph stage of several species composing this genus. The teleomorph stage for various *Aspergillus* species has been described (Horn et al., 2009a; Luis et al., 2020). Although a teleomorph stage has been described for *A. flavus* (*Petromyces flavus* B.W. Horn, I. Carbone et G.G. Moore, sp. nov.) - through laboratory manipulation - the species is considered to reproduce predominantly in an asexual manner (i.e. through spores) (Islam et al., 2018; Ortega-Beltran et al., 2020; Papa, 1986). Sexual recombination is not significant in field conditions. There have been some reports of sexual recombination in field conditions but only after subjecting isolates to specific laboratory conditions that rarely occur in nature (Horn et al., 2009a,b, 2014; Olarte et al., 2012).

2.2 Factors influencing aflatoxin contamination

Aflatoxin-producing fungi survive in the soil primarily as sclerotia and mycelium. These propagules multiply in organic matter and plant debris and produce copious amounts of spores (conidia) that are carried by wind, water, rain splashes and insects onto sites of infection. Flowers and developing grains are the primary sites of infection. Subsequently, the pathogens colonize grain tissues and produces aflatoxin in the plant substrate (Diener et al., 1987). The aflatoxin problem starts at the pre-harvest stages in the field and is accentuated when storage conditions are favorable for further fungal colonization and growth.

Multiple factors are responsible for aflatoxin contamination in crops. These include the presence of aflatoxigenic fungi, susceptibility of crops, a favorable environment for aflatoxin accumulation, agronomic practices that support contamination and insect damage to crops. Significant aflatoxin contamination results in crops when these factors occur together. It is critical to manage these risk factors for aflatoxin reduction.

The prevalence of aflatoxin-producing fungi in a fungal population is important for aflatoxin contamination. High aflatoxin accumulation occurs when aflatoxigenic fungi are present in sufficiently high quantities in the crop. The prevalence of fungal groups is influenced by founder events. Founding events may occur spontaneously or be triggered by biocontrol efforts using atoxigenic fungi (Mehl et al., 2012; Lewis et al., 2019; Ortega-Beltran et al., 2020). The latter promotes high populations of atoxigenic genotypes leading to lower aflatoxin accumulation (Bandyopadhyay et al., 2016, 2019a). Spontaneous founding populations of toxigenic strains result in high aflatoxin levels in crops (Probst et al., 2010, 2012).

Until the natural defense system of a crop is compromised to make the crop susceptible to aflatoxigenic fungi, it is difficult for the fungi to infect, colonize and synthesize aflatoxins in crops. Plant stress is an important factor that compromises the natural defense system of the crop. Stresses include heat, water and drought. These result in loss of kernel integrity such as silk cut or physiological changes such as reduction in phytoalexin levels (Odvody et al., 1997; Kebede et al., 2012). Stress to the crop can also come from pest damage (Dowd, 2003) and nutrient deficiencies (Htoon et al., 2019).

Extremes of temperatures, which may occur in either warm humid climates or hot arid and semi-arid areas, can predispose crops to aflatoxin contamination (Cotty and Jaime-Garcia, 2007; Medina et al., 2014). Dryer, hotter climates promote higher densities of *Aspergillus* spp. and can favor the prevalence of fungi with high aflatoxin production capability (Cotty and Jaime-Garcia, 2007). Dryer, hotter conditions also predispose stressed crops to both insect attack and infection by aflatoxin-producing fungi (Abbas et al., 2009; Cotty et al., 1994; Paterson and Lima, 2010). Erratic rainfall during later stages of crop maturity can provide conditions for aflatoxin formation (Cotty and Jaime-Garcia, 2007; Gilbert et al., 2016) and may make it difficult to dry crops to appropriate levels for safe storage (Bradford et al., 2018; Ndemera et al., 2020).

Improper agronomic practices such as late planting, incorrect spacing, poor irrigation and insect control practices predispose crops to aflatoxin contamination (Payne et al., 1986; Rodriguez-del-Bosque, 1996; Canavar and Kaynak, 2013; Weaver et al., 2017; Williams et al., 2017). Crop rotation practices that follow aflatoxin-susceptible crops with other susceptible crops will encourage the continued propagation of aflatoxigenic fungi (Jaime-Garcia and Cotty, 2010). Untimely harvest is another predisposing factor to aflatoxin contamination. Insect damage provides entry sites on grains while insects also carry inoculum to the entry site (Dowd, 2003; Bakoye et al., 2017). Insect control strategies are thus important in aflatoxin management.

Climate change is having and will continue to have a great negative effect on crop susceptibility to plant pathogens (Pautasso et al., 2012). In the case of aflatoxin, climate change is already causing increased contamination events in areas where the problem is perennial but also in areas that were not considered at risk (Battilani et al., 2016; Paterson and Lima, 2011). Apart from increased temperatures and disruptions of rainfall cycles, climate change is also accompanied by greater levels of CO_2 (Medina et al., 2017a). Increased CO_2 has been shown to promote greater crop stress and subsequent aflatoxin accumulation in laboratory conditions. Significant changes in CO_2 levels in real conditions may contribute to greater aflatoxin accumulation. Aflatoxin mitigation strategies must consider the ways that combinations of increased temperature, increased or decreased water availability and elevated CO_2 may reduce the effectiveness of current management strategies.

3 Aflatoxin management options

There are several aflatoxin management options for reducing the risk factors that predispose crops to aflatoxins. Aflatoxin management should begin in the field during planting and continue throughout the harvest and post-harvest stages. Several reviews have described various management tools available for mitigating aflatoxin accumulation in food and feed (Ayalew et al., 2017; Udomkun et al., 2017; Falade, 2018). These tools address risks associated with high prevalence of aflatoxigenic fungi, susceptible hosts, inadequate agronomic and post-harvest practices.

Several microbial species (e.g. fluorescent *Pseudomonas*, *Bacillus subtilis*, and *Trichoderma* spp.) have been tested as biocontrol agents for reducing aflatoxin contamination at the pre- and post-harvest stages (Desai et al., 2000; Mohammed et al., 2018; Peles et al., 2021). None of those microbes have been evaluated beyond the experimental phase. Modulating *Aspergillus* populations in favor of atoxigenic genotypes to reduce the prevalence of aflatoxigenic fungi is a way to manage populations of aflatoxigenic strains. These can be achieved via application of naturally occurring atoxigenic genotypes of *A. flavus* endemic to the region of application (Dorner and Lamb, 2006; Bandyopadhyay et al., 2016, 2019b; Ortega-Beltran et al., 2016). This biocontrol approach is discussed in detail later in this chapter.

Breeding for host plant resistance to aflatoxin is used to develop plant cultivars that are less susceptible to aflatoxin contamination (Kim et al., 2006; Brown et al., 2013). Some approaches are targeted at integrating antifungal proteins or promoting physical changes such as increased tightness of husk cover, kernel hardness and pericarp wax (Kebede et al., 2012). Biotechnological approaches using RNAi technology are highly effective in reducing aflatoxin biosynthesis (Thakare et al., 2017) and colonization by *Aspergillus* (Sharma et al., 2018). Transgenic technologies have also targeted insect pest resistance which indirectly helps aflatoxin management (Wu, 2006; ICRISAT, 2016; Weaver et al., 2017).

Multiple agronomic practices are needed to manage aflatoxin contamination. Some of these include crop rotation of susceptible plants by non-susceptible plants, use of appropriate adapted varieties, irrigation where available, proper spacing, timely harvest and insect and weed control (Rodriguez-del-Bosque, 1996; Dowd, 2003; Waliyar et al., 2003; Jaime-Garcia and Cotty, 2010).

Timely harvest to avoid rain is important to keep grain moisture under control and prevent aflatoxin accumulation. Rapid drying of grains on clean surfaces (e.g. tarp) or using dryers can reduce aflatoxin contamination (Pretari et al., 2019). Dry grains, kept under optimal storage conditions such as proper aeration, insect and pest control, are critical (Turner et al., 2005). Hermetic storage (e.g. Purdue Improved Crop Storage bags) mitigates fungal growth and insect activity, contributing to aflatoxin control (Williams et al., 2014; Maina et al., 2016).

4 Biocontrol product development and the registration process in Africa

4.1 The science of atoxigenic-based biocontrol

Atoxigenic isolates of *A. flavus* lacking the ability to produce aflatoxins were identified over 50 years ago (Joffe, 1969). The common occurrence of that trait in *A. flavus*, coupled with increased competitiveness of some genotypes, suggested that such genotypes could be used as biocontrol agents to outcompete aflatoxin producers in the field (Cotty, 1989). Studies in the United States by the U.S. Department of Agriculture-Agricultural Research Service (USDA-ARS) revealed that pre-harvest application of the atoxigenic isolate *A. flavus* AF36 reduced aflatoxin accumulation in maize when tested in field trials, both at the pre- and post-harvest stages (Brown et al., 1991). Those studies paved the way for the development of aflatoxin biocontrol using atoxigenic *A. flavus* genotypes as an active ingredient in biocontrol formulations.

The main mechanism of biocontrol is the competitive exclusion that results in a modified *A. flavus* community with a lower average aflatoxin-producing ability (Horn et al., 2011). The formulated product is composed of dead sorghum grains coated with spores of the active ingredients (0.0005% w/w of the formulated product) and other adjuvants. Farmers broadcast the product in the field at the rate of 10 kg/ha 2–3 weeks before crop flowering for maize and groundnut. The carrier sorghum grain acts as a food source on which the active ingredients, which are already present on its surface, multiply and produce a large number of spores for at least 20 days. The carrier grains thus act as '*in situ* field factories of biocontrol spores.' These spores continually disperse, colonize various niches in soil and other substrates and create a founding population of active ingredients, which further multiplies and displaces other *Aspergillus* strains. The active ingredients dominate in the treated field (and to some extent in neighboring areas) in place of other *Aspergillus* strains and ultimately become associated with the crop. Spores of the active ingredients produced on soil and other substrates are dispersed by wind, rain and insects to reach the crop's infection sites. The substrate to deliver the biocontrol is key in providing a competitive advantage to the applied fungi over aflatoxin producers present in the treated fields (Cotty and Mellon, 2006). Competitive exclusion significantly prevents aflatoxin producers from reproducing in the field and thus from becoming associated with the crop. High levels of applied atoxigenic genotypes in the treated crops are associated with low levels of aflatoxins (Agbetiameh et al., 2020; Doster et al., 2014; Mehl et al., 2012; Shenge et al., 2019). In general, treated and untreated fields and crops contain the same fungal densities (Agbetiameh et al., 2020; Atehnkeng et al., 2014; Bock et al., 2004; Dorner, 2009; Doster et al., 2014; Senghor et al., 2020).

Other indirect mechanisms that may contribute to reduced aflatoxin accumulation by atoxigenic genotypes include:

- toxin degradation by atoxigenic genotypes (Maxwell et al., 2021);
- a so-called 'touch inhibit' mechanism (Damann, 2015); and
- production of both extrolites and volatile compounds by atoxigenic genotypes which may discourage aflatoxin production (Moore et al., 2019; Sweany and Damann, 2020).

However, competitive displacement mechanisms are the predominant contributors to reduced aflatoxin content in treated crops.

Much has been achieved in the United States since the late 1980s and early 1990s. The atoxigenic biocontrol product *Aspergillus flavus* AF36 was initially registered with the United States Environmental Protection Agency (US EPA) for use in cottonseed and subsequently for use in maize, pistachio, almond and fig (Cotty et al., 2007; Doster et al., 2014; Ortega-Beltran et al., 2019). The product AF36 is produced and distributed by the Arizona Cotton Research and Protection Council (ACRPC), a farmer-run organization (Cotty et al., 2007). It is used over hundreds of thousands of hectares, and most treated crops meet stringent local and international standards. Without biocontrol, it would be difficult for farmers growing susceptible crops in contamination hot spots to produce safe crops.

AF36 belongs to VCG YV36. YV36 occurs naturally across the United States (Grubisha and Cotty, 2015), and in several states across Mexico (Ortega-Beltran et al., 2016). The natural presence of YV36 across Mexico has allowed experimental use of AF36 in maize fields in certain states (N. Palacios, CIMMYT, personal communication). Another aflatoxin biocontrol product used in the United States is Afla-guard®, which is registered with the US EPA and manufactured by Syngenta® for use in maize and groundnut (Dorner, 2009).

In the mid-1990s, USDA-ARS and IITA started a collaboration to introduce the atoxigenic biocontrol technology in SSA (Bandyopadhyay et al., 2016, 2019b). Causal agents of contamination in target African environments were characterized (Agbetiameh et al., 2018; Atehnkeng et al., 2008; Cardwell and Cotty, 2002; Cotty and Cardwell, 1999; Diedhiou et al., 2011; Donner et al., 2009). Native atoxigenic isolates belonging to widely distributed VCGs were then identified in various countries (Agbetiameh et al., 2019; Atehnkeng et al., 2016; Probst et al., 2011; Senghor et al., 2020, 2021). Native atoxigenic isolates are inherently adapted to local cropping systems and environmental conditions and have interacted with the causal agents of contamination in target areas for a long time (Mehl et al., 2012; Probst et al., 2011). Native isolates should result in better aflatoxin reduction compared to exotic genotypes. Using native isolates over introduction of exotic *Aspergillus* organisms is preferred by regulatory authorities.

4.2 Product development process

Development of efficient aflatoxin biocontrol products requires mining the fungal biodiversity across a target country. The biodiversity can be comprehensively assessed by examining large numbers (e.g. 5000 isolates) of *A. flavus* isolates associated with target crops (Agbetiameh et al., 2018, 2019; Atehnkeng et al., 2008; Probst et al., 2011). These isolates are obtained from georeferenced crop or soil samples collected from different parts of the country. Such an exercise allows detection of atoxigenic genotypes belonging to the most common atoxigenic VCGs across a country of interest with known adaptation to target agroecosystems, their cropping systems, climatic and soil conditions. Examination of small numbers of isolates (say, 100 and sometimes less) from a few areas increases the probability of selecting poor atoxigenic genotypes that may not provide adequate levels of protection or atoxigenic genotypes that may belong to VCGs containing members with aflatoxin-production abilities, since VCGs containing both atoxigenic and aflatoxin-producing members are relatively common (Ortega-Beltran and Cotty, 2018).

Applications of atoxigenic genotypes are intended to cause long-term changes to communities of *Aspergillus* section Flavi, with applied genotypes becoming prevalent thereby lowering aflatoxins in the target region. Care is therefore taken to only use VCGs that do not contain aflatoxin-producing members. As with all organisms, mutations continually occur in this predominantly haploid fungus that evolves primarily in a clonal fashion (Adhikari et al., 2016; Islam et al., 2018). Defects that result in atoxigenicity of active ingredients are sufficiently ancient that multiple mechanisms of atoxigenicity occur and atoxigenicity is fixed in the VCGs to which those active ingredients belong (Adhikari et al., 2016). However, genotypes that more recently acquired an atoxigenic trait and occur in VCGs that are predominantly high aflatoxin producers may be problematic (Ortega-Beltran and Cotty, 2018). Gene flow occurs within VCGs, and this results in the potential for aflatoxin-producing ability to be acquired by atoxigenic fungi in aflatoxin-producing VCGs through the parasexual cycle (Grubisha and Cotty, 2010, 2015). VCGs with aflatoxin-producing members are thus avoided when selecting active ingredients for biocontrol products.

Traditionally, atoxigenic VCGs were found through a laborious, time-consuming process using microbiological procedures that include vegetative compatibility assays (VCA) (Atehnkeng et al., 2016; Doster et al., 2014; Probst et al., 2011). With the advent of molecular tools, the process has been simplified. Atoxigenic isolates within a population are first identified by assessing deletions in markers spaced across the aflatoxin biosynthesis gene cluster (Callicott and Cotty, 2015). Those atoxigenic isolates can then be genotyped using simple sequence repeat (SSR) markers (Agbetiameh et al., 2019; Grubisha and Cotty, 2009; Senghor et al., 2020). Isolates showing defective aflatoxin biosynthesis genes are then examined for their lack of ability to produce aflatoxins in a crop substrate (e.g., maize kernels). Groups identified by SSR analyses are then confirmed for the absence of members with toxigenic capability using VCA. Incorporation of molecular tools allows faster detection of atoxigenic VCGs and provides genetic information to establish phylogenetic relationships within and among genotypes of *A. flavus* and other *Aspergillus* species. The number of isolates belonging to each atoxigenic VCG, the number of crop samples from which the members of each VCG are obtained and their distribution in different agroecological zones are used as traits to identify widely distributed and adapted candidate VCGs for further evaluation.

It is important to determine which atoxigenic VCGs have the best ability to prevent aflatoxin contamination when challenged with highly toxigenic fungi. Competition experiments in grains of a target crop identify VCGs most likely to outcompete toxin producers when infecting the same substrate (Ortega-Beltran et al., 2019; Probst et al., 2011). It is critical to discard poor competitors. Generally, atoxigenic fungi able to limit aflatoxin contamination by 80% or more are considered good competitors (Camiletti et al., 2018; Mauro et al., 2015; Probst et al., 2011). However, there are genotypes that can limit aflatoxin contamination by well over 95% (Agbetiameh et al., 2019; Ortega-Beltran et al., 2019), and those should receive priority if they fulfill other criteria such as adaptation in large areas, repeated detection in multiple crops and ability to persist in the environment after application.

Another important criterion for selecting aflatoxin biocontrol agents is their ability to move from the field to the treated crop (e.g. maize, groundnut, pistachio, sorghum). This is an aspect of biocontrol genotype selection that has received little attention. However, there are differences in the abilities of atoxigenic genotypes to successfully colonize substrates in the field and then move from soil to crops (Agbetiameh et al., 2019). Those differences can be revealed when candidate biocontrol genotypes are released on soil and their ability to colonize and spread is studied in multiple crops under varied agroecologies. The true test of an excellent biocontrol candidate after soil application lies in its ability to be frequently isolated from soil and grain at harvest and carry-over to the next season. The best atoxigenic genotypes can result in biocontrol products providing large aflatoxin reductions, the ultimate goal of aflatoxin biocontrol technology. For example, careful selection of the best atoxigenic genotypes native to Ghana resulted in biocontrol products with the greatest efficacy reported thus far (Agbetiameh et al., 2020). Almost all treated crops contained undetectable aflatoxin levels during two cropping seasons in multiple agroecologies, while most untreated crops contained unsafe aflatoxin levels, some of them were extremely dangerous. It is important to point out that *A. flavus*, whether toxigenic or atoxigenic, have a weak pathogenic phase, which may lead to Aspergillus ear rot (AER) in susceptible crops. However, AER is not common and often requires injuries to the ear and a high inoculum dose for symptoms to occur. No perceptible (and statistical) difference in AER has been observed between biocontrol-treated and untreated fields (Atehnkeng et al., 2014).

A major difference in aflatoxin biocontrol products developed in Africa is the use of four atoxigenic genotypes instead of one atoxigenic genotype as in the United States and Italy. Traditionally, single-genotype products have been preferred because they are easier to develop, less demanding to navigate through the registration process and less cumbersome to manufacture. In theory, the various genotypes in a multigenotype product will allow more efficient filling of microniches to which one or another genotype is best adapted. Similarly, as environmental conditions change between seasons, the active ingredient best suited to conditions will be the most successful. The design of the multigenotype products thus provides product plasticity responsive to varying conditions. However, the exact conditions favoring the various genotypes are unknown. Experiments comparing displacements achieved with multigenotype versus single genotype products are difficult to design and have not been reported. Multigenotype products are also thought to instill greater complexity to the modified *A. flavus* community resulting from treatments. Associated with this greater complexity is a greater tendency to retain the modified structure with its greatly reduced average aflatoxin-producing potential (Bandyopadhyay et al., 2016). Compositions of communities of *A. flavus* can change rapidly even in the absence of atoxigenic genotype applications and, as discussed above, the increased complexity provided by multigenotype products may support longer-term residence of modified communities (Ortega-Beltran et al., 2020; Ortega-Beltran and Cotty, 2018). However, the ability of multigenotype products to promote communities with longer-term residence than single-genotype products has not been tested empirically.

Currently, recommendations for prevention of aflatoxin contamination of several crops include one treatment of target crops in a season. However, movement of active ingredients between fields and crops and carry-over of products between seasons provide opportunities for additive benefits when fields are treated over multiple seasons. Movement and carry-over also offer the potential for cost savings through fewer applications. In some regions in the United States, movement of atoxigenic genotypes between crops and carry-over between seasons are innate components of management strategies, for example, when pistachio and almond are grown in close proximity to each other and harvest practices facilitate movement of conidia between the crops (Cotty et al., 2007). Annual treatments rely on carry-over to adequately changed communities in the canopy of these perennial tree crops and to prevent inadvertent inoculation of nearby orchards with conidia of aflatoxin producers. In some parts of Texas, most maize is treated with an atoxigenic genotype-based biocontrol product. In those areas, long-term effects may mean that when some fields are not treated in a given year, a sufficient amount of the atoxigenic *A. flavus* community remains to provide continuing control (Jaime et al., 2017). Some farmers in Texas thus reduce treatments to only a portion of the treated crop each season. However, variation in operations, rotations and climate between farms and seasons have thus far prevented development of formal recommendations to reduce application frequency to below once a season. Across Africa, density of treated fields across landscapes is not yet sufficient for carry-over to allow recommendation of reduced frequencies of application.

As mentioned earlier, there are VCGs with members known to occur in multiple countries. Many atoxigenic VCGs have been found across several SSA countries. The Aflasafe Initiative recommends the use of a product developed for one country in another country if members of the same VCG in the active ingredient coexist in the second (or third) country as well. That was the case with Aflasafe SN01, which was developed initially for Senegal (Senghor et al., 2020). Subsequently, the active ingredient genotypes of Aflasafe SN01 were found to co-occur in The Gambia paving the way for use of the product in Senegal and The Gambia (Senghor et al., 2021).

4.3 Registration with regulatory authorities

The registration of aflatoxin biocontrol products in SSA countries varies. This was difficult initially because systems to register bioprotectants (the category in which aflatoxin biocontrol falls in) were not well developed (Bandyopadhyay et al., 2016). Instead of waiting for systems to be put in place (which could have taken many years), the Aflasafe Initiative worked with regulatory authorities to develop systems, helped by the USDA Foreign Agricultural Service. The experience from registering the biocontrol product AF36 with US EPA served as a template in discussions with regulators and national policy makers.

The registration process starts before the product is developed. The Aflasafe Initiative helps to sensitize regulatory agencies and key policy makers on aflatoxin and its management using biocontrol products (Schreurs et al., 2019). The problem of aflatoxin and the use of biocontrol with atoxigenic fungi were initially poorly understood in many countries. Multiple meetings took place in each country to present evidence on aflatoxin contamination and the benefits of biocontrol and other management practices. Regulators and scientists from national institutes and IITA jointly planned research to develop and test products in target areas and crops. Regulators in many cases required particular experiments to satisfy efficacy and safety concerns (Bandyopadhyay et al., 2016).

Registration of aflatoxin biocontrol products has been granted after submitting the required evidence on the efficacy (Fig. 1), safety, quality and social benefits of the technology (Bandyopadhyay et al., 2016). In Nigeria and Kenya (where the first registrations of Aflasafe products were granted in 2014 and 2015, respectively), it was necessary to submit additional toxicological and ecotoxicological studies to demonstrate that the products and the active ingredients were safe to the environment (e.g. effects on birds, rodents, bees, earthworms and soil microbes other than aflatoxin producers). Results from those studies revealed that the use of the active ingredient fungi in biocontrol formulations do not pose risks to non-target species beyond those already taking place throughout the target environments. Only native fungi commonly interacting with target crops in target areas are used in formulations. The results from the studies conducted in Nigeria and Kenya, along with other studies conducted in the United States for registration of atoxigenic products, have been accepted by regulatory authorities in CSP-CILSS countries (Senegal, The Gambia, Burkina Faso), Ghana, Tanzania, Zambia, Malawi, and Mozambique. These countries waived repeating those toxicological and ecotoxicological studies based on the equivalence principle.

The validity period of registration of aflatoxin biocontrol product is variable. Registrations have been granted for 3 years in CILSS countries and 5 years in others. Two products developed for Ghana were initially granted an Experimental

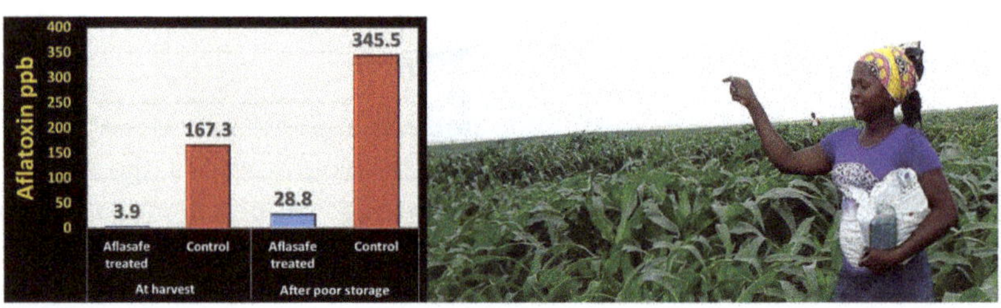

Figure 1 Aflatoxin concentration in parts per billion (ppb) in grains at harvest and after poor storage in biocontrol-treated and untreated (control) fields (left). The effectiveness data are means of over 1500 maize and groundnut fields of farmers who either treated or did not treat their crops with biocontrol products in multiple countries. A farmer broadcasting a biocontrol product in a maize field @ 10 kg/ha 2–3 weeks before flowering (right).

Use Permit for 1 year to gather additional evidence of efficacy under commercial use. In 2020, the Ghana EPA provided the regular 3-year registration usually granted to approved biopesticides. In some countries, it is necessary to register the manufacturing facilities of the biocontrol product. The factories needed to comply with environmental, structural and safety parameters.

4.4 Development of biocontrol products in different countries

There are 14 Aflasafe biocontrol products registered for use in 10 countries (product name and year of registration in parenthesis) (Moral et al., 2020):

- Nigeria (Aflasafe™; 2014);
- Kenya (Aflasafe KE01™; 2015);
- Senegal and The Gambia (Aflasafe SN01™; 2016);
- Burkina Faso (Aflasafe BF01™; 2016);
- Ghana (Aflasafe GH01™ and Aflasafe GH02™; 2018);
- Zambia (Aflasafe ZM01™ and Aflasafe ZM02™; 2018);
- Tanzania (Aflasafe TZ01™ and Aflasafe TZ02™; 2019);
- Mozambique (Aflasafe MWMZ01™ and Aflasafe MZ02™; 2019); and
- Malawi (Aflasafe MWMZ01™ and Aflasafe MW02™; 2020).

After registration, the technology has been transferred to the private sector through Technology Transfer and Licensing Agreements (TTLA) for mass manufacture and distribution. Aflasafe is now manufactured in four countries (Nigeria, Kenya, Senegal and Tanzania) and is commercially available in Nigeria, Kenya, Senegal, Tanzania, Mozambique, The Gambia, Burkina Faso, Malawi and Ghana. Product testing is currently underway in Mali, Niger, Rwanda, Burundi, Uganda and Togo. Product development is currently being done for Sudan, Benin, Cameroon, Zimbabwe and Democratic Republic of Congo. It is expected that product development will expand to Sierra Leone, Côte d'Ivoire and Chad, among other African countries.

5 Manufacturing development

Aflatoxin biocontrol products are agricultural inputs that must be produced in large quantities for use in the field. Unlike spray or seed treatment formulations generally used for biocontrol products, all formulations of commercial biocontrol products for aflatoxin control are grain-based and applied by broadcasting in the field. There are several advantages of grain-based solid formulations. These include ease of application and slow release of active ingredients in the environment for a long period, aiding competitive displacement of toxigenic strains. However, a grain-based formulation is bulky and expensive to transport over long distances.

5.1 The initial manufacturing plant

To test efficacy, products were first manufactured using a laboratory-scale adaptation of the industrial manufacturing process developed by USDA-ARS and ACRPC (Cotty et al., 2007). This method was laborious, requiring several workers to produce around 300 kg of product per week (Atehnkeng et al., 2014). The process involved sterilization of sorghum grains in an autoclave, soaking the autoclaved grains in a spore suspension, incubating the inoculated grains for 18 h at 31°C, rapid drying of the grains before sporulation begins, and bagging the product. The laboratory method allowed field efficacy trials in several hundred fields in Nigeria, Senegal, Zambia and Kenya. However, it was evident that a more efficient and cost-effective manufacturing process was needed.

In 2011, Bill Gates and senior management of the Bill & Melinda Gates Foundation (BMGF) visited IITA station in Kano, Nigeria. Aflasafe was one of the technologies showcased. The BMGF delegation recognized the potential of the technology, but the technology was out of reach for many farmers. Only industrial manufacture and effective partnerships with the private sector could realize its potential. A demonstration-scale Aflasafe manufacturing plant was thus designed and constructed in IITA-Ibadan (Nigeria) with funds granted by BMGF through the PACA (Bandyopadhyay et al., 2016). There were two main objectives for constructing the manufacturing facility. The first was to produce large quantities of biocontrol products to:

- conduct large-scale efficacy field trials in Senegal, The Gambia, Ghana and Burkina Faso; and
- allow farmers in Nigeria, Senegal, The Gambia and even Kenya to access the product to treat hundreds of thousands of hectares.

A second objective was to demonstrate that a commercially viable manufacturing facility could be set up in Africa. Without the involvement of the private sector, the technology would remain dependent on donor funding, would be unsustainable and of limited impact.

5.2 Improved manufacturing processes and facilities

The demonstration-scale Aflasafe manufacturing plant constructed at IITA-Ibadan consisted of off-the-shelf equipment: grain cleaner, feed roaster, cooling equipment, silos, seed treater and packaging machines (Bandyopadhyay et al., 2016; Schreurs et al., 2019). In contrast to the use of wheat or barley in the United States, white sorghum is used as both substrate and carrier for the atoxigenic fungi. White sorghum is cheaper, readily available and an excellent substrate for *A. flavus* reproduction. Moreover, sorghum grains are smaller than those of wheat, and a greater number of biocontrol carrier grains are dispersed across treated fields. Having more biocontrol formulation grains across a treated field can ensure a more even distribution of biocontrol fungi.

A second difference is that an improved dry process is used in African environments. In the United States, initially, the AF36 product was prepared using a wet process requiring autoclaving the wheat grain, then inoculating the grain and incubating it for 18 h. In the case of Aflasafe products, sorghum grain is cleaned and then roasted to kill both the embryo and associated microorganisms. After cooling, the grain is coated with a spore suspension containing appropriate amounts of spores of the atoxigenic genotypes, a blue food colorant to differentiate the product from regular sorghum, and a polymer to aid in the coating process (Bandyopadhyay et al., 2016). For each ton of Aflasafe product, a 10 L spore suspension (4×10^7 spores/ml) is combined with 1.5 L polymer, 2 L blue dye and 10.5 L sterile water. Roasted, sterile sorghum grain is coated with the suspension in a seed treater (Bandyopadhyay et al., 2016). After the coating, the product is automatically conveyed to a packaging system to dispense the product into bags. At full capacity, the plant can produce 5 tons of Aflasafe per hour. The Afla-guard product is also manufactured using a dry process, but the substrate type (barley) and treatment (clay coating of carrier) are different from the Aflasafe manufacturing process.

The spore suspension of the active ingredient genotypes is obtained from an inoculum production laboratory adjacent to the factory. The active ingredient genotypes are inoculated on sterile sorghum grains, incubated, diluted and mixed with the other ingredients before transfer to the factory (Agbetiameh et al., 2019; Bandyopadhyay et al., 2016). The manufacturing facility also has a quality control laboratory to check finished products. Samples are collected from each 250 kg or 500 kg batch to verify the absence of contamination, and that only atoxigenic active ingredient genotypes, at the right density, are coated on the biocontrol carrier grains. In addition, the sporulation ability of the atoxigenic genotypes on the biocontrol formulation is measured.

After the manufacturing facility in Ibadan, another three factories have been constructed. In Katumani in Kenya, IITA and partners constructed a manufacturing plant to produce Aflasafe KE01. The Katumani production process is not continuous (as in Ibadan) but modular and with significant improvements in production efficiency. It is made up of the following three modules (Fig. 2):

- Module R, where carrier grain cleaning, sterilization (using roaster) and cooling (in a cooling silo) occurs;
- Module C, for coating of the grain using a seed treater; and
- Module P, for packaging of the finished product.

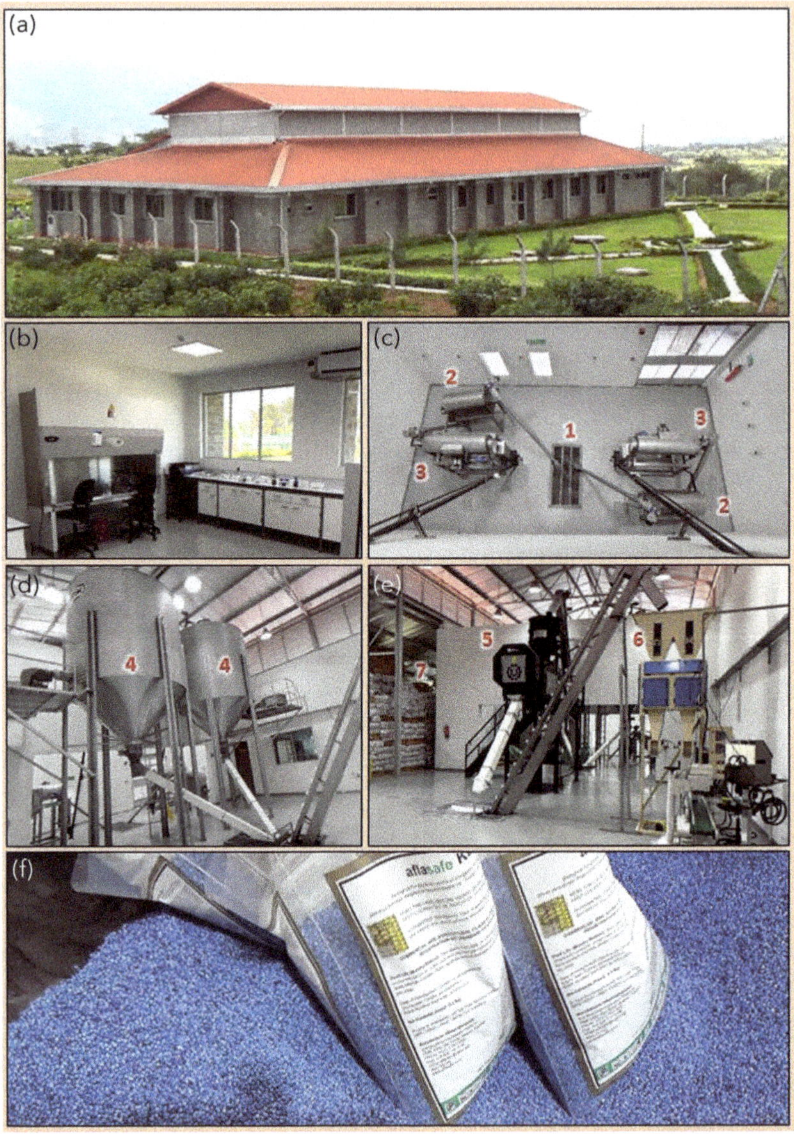

Figure 2 The KALRO/IITA Aflasafe Modular Manufacturing Plant at Kenya Agricultural and Livestock Research Organization (KALRO) Katumani Research Station in Kenya. (a) A side view of the manufacturing plant. (b) The plant consists of a laboratory where the inocula of the active ingredients are produced and combined with a blue dye and polymer sticker. (c) Carrier (sorghum) pit (1) from where sorghum is moved into a grain cleaner (2) with an auger, then transported to a roaster (3) and finally to a cooling silo for cooling and storing the roasted carrier grains. (d) From cooling silo (4), the carrier is transported via a pneumatic line into a seed treater (5) where the master solution (inoculum, dye and polymer) is coated onto the sorghum grain. (e) The formulated

The advantage of the modular facilities is that each module can be expanded when demand for the biocontrol product increases, meaning it is not necessary to make a large investment during the initial years of operation. Similar modular facilities in Kahone in Senegal and Arusha in Tanzania were constructed by the companies BAMTAARE, S.A. and A to Z Textiles Mills Ltd., respectively. In addition, a fifth modular manufacturing facility is currently being constructed by the company HarvestField Industries Ltd., the licensee of the Aflasafe product, for use in Nigeria. This new facility in Nigeria would be the largest in Africa and operational during the third quarter of 2021. New modular manufacturing facilities are currently being constructed in the Democratic Republic of Congo, Burundi, Sudan and Mozambique. Although highly efficient manufacturing facilities have been designed and constructed, investigations are continuing to develop and validate new manufacturing technologies that can simplify the process to make the products faster to manufacture and cheaper for the farmers.

6 Barriers preventing adoption and how to overcome them

For decades, many agricultural innovations have been developed for adoption by smallholder farmers to achieve higher yields. However, adoption at scale is not always successful. There are intrinsic and extrinsic factors impeding technology adoption. Some intrinsic factors are linked to limited familiarity with the innovations and lack of demonstration of effectiveness in real-life farming situations.

Other factors impeding the adoption of technologies include negative attitudes to innovation, lack of knowledge, low willingness to adopt innovations at the early stage of their development, perceived complexity and relevance to smallholder farmers as well as limited availability (Kuntosch and König, 2018; Naswem and Ejembi, 2017). Broader factors are farmers' socioeconomic characteristics, political conditions, high cost for perceived benefits, limited access to inputs, low motivation and adverse attitude toward risk and/or change (Meijer et al., 2015). In some cases, causes for low adoption are poor

Figure 2 (*Continued*)

product is then transported via an elevator to the packaging system (6) where it is bagged, sealed, baled and finally placed in a storage area (7). (f) The final formulated product (i.e. roasted sorghum coated with spores of the four active ingredient atoxigenic genotypes, a polymer and a blue food colorant) in an open bag for demonstration purpose; a bag of 5 kg of an Aflasafe product contains only 0.1 g of active ingredient atoxigenic genotypes.

communication strategies that fail to clearly convey the value and benefits of the innovation.

A particular problem has been low awareness of aflatoxin among farmers, traders, regulators and the population in general. Aflatoxins cannot be seen or smelt and do not reduce yields. Many markets do not provide incentives for production of aflatoxin-safe crops. Policies to prevent trade of contaminated crops are often poorly planned and enforced. In some cases, farmers simply do not know about and/or do not have access to products.

To deal with these challenges, scaling of aflatoxin biocontrol required developing linkages between farmers with aflatoxin-conscious buyers willing to pay a premium for high-quality, aflatoxin-safe crops. In addition, aflatoxin management requires effective practices across the value chain. This required an integrated aflatoxin management system with biocontrol as a cornerstone (Bandyopadhyay et al., 2019b).

Large-scale adoption of biocontrol was first attempted in Nigeria, through the AgResults Aflasafe Challenge (https://agresults.org/projects/nigeria), which converged several elements of aflatoxin management to address the problem. The Challenge relied on a pull mechanism in which public or private enterprises received incentives only if they delivered predetermined results. It is believed that pull mechanisms led by the private sector may be more effective than push strategies (upfront payments on the promise that results will be delivered) in overcoming constraints in scaling technologies for smallholder farmers (Kubzansky et al., 2019). The Challenge offered per-unit cash incentives to micro, small and medium enterprises handling Aflasafe-treated maize grains, referred to as implementers. The larger the quantity of Aflasafe-treated grains the implementers aggregated, the bigger were the incentives and potential income from sale of premium, aflatoxin-compliant maize. With time, the per-unit incentives were reduced.

Between them, the implementers had thousands of smallholder farmer customers. They educated farmers on the need for and benefits of aflatoxin control, sold/provided biocontrol products to them, trained them how to use both the product and other aflatoxin management practices, facilitated access to technical knowledge and inputs to increase crop yield, aggregated grains from the farmers, got the grain lots tested for aflatoxin content and marketed the biocontrol-treated maize to aflatoxin-conscious buyers who offered them a premium for the aflatoxin-reduced maize.

The Challenge began in 2013 and ended in September 2019. During this period, 41 implementers worked with nearly 76000 smallholder farmers who produced more than 315333 tons of maize earning more than US$3 m in incentives and more than US$ 5.3 m in premiums from the market. Over 95% grain lots had <20 ppb aflatoxins (US regulatory level); >90% with <4 ppb

(EU regulatory level) (https://agresults.org/projects/nigeria; Bandyopadhyay et al., 2019a). An external evaluation of the Challenge found that the annual net income from maize grown increased by US$318 (16%) per smallholder, and average consumption of Aflasafe-treated maize increased by 0.02 kg per day or 13% of daily consumption (Narayan et al., 2020). The Challenge also provided opportunities to tens of thousands of maize smallholder farmers to move from subsistence to commercial agriculture across Nigeria.

7 Scaling up aflatoxin biocontrol technology

Aflasafe products are innovations developed through collaboration between international public organizations and national agricultural systems using public funding. However, making these products available and accessible to users requires commercial investment in a manufacturing plant, distribution and marketing. Relying entirely on traditional approaches, which consist of transferring the innovation to national agricultural and extension systems to promote, may not be necessarily appropriate.

To identify the best pathway for scaling in each country, IITA and partners had to ask the right questions. Among them were the following: Is there a business case for farmers and other businesses to use the product? If so, is there an incentive for the private sector to invest in product manufacturing, distribution and market development? If not, is the public sector ready to lead the scaling of the innovation? Are there any opportunities for public-private partnerships in scaling? Responding to these questions has informed the unique three-phase technology transfer and commercialization approach employed by IITA and partners through the Aflasafe Technology Transfer and Commercialization (ATTC) project to scale up the use of Aflasafe in different African countries.

Phase 1. Developing the commercialization strategy. Countries differ widely in private sector development, cost of inputs, market structure, awareness of the risk from aflatoxin and the existence and enforcement of food-safety regulations. A one-size-fits-all approach will not work. A commercialization strategy is the first step in tracing the path to take Aflasafe to market. The strategy identifies drivers of demand, the options to meet demand and the enabling interventions required to increase product uptake in each country. It provides key market, economic and financial analyses that any potential investor would first want to know before considering an investment in manufacturing and distributing Aflasafe.

Phase 2. Selecting investors and transferring the technology. Investor selection involves an iterative solicitation and review process to identify the most qualified manufacturing and/or distribution firms in the target countries following three interrelated steps: (i) hosting the investor forum, (ii) selecting the investor and (iii) formally transferring the technology. The investor forum mobilizes stakeholders from the public and private sectors with interests in the maize, groundnut and sorghum value chains and/or agricultural industries. The forum aims to present the investment opportunity and attract investor interest. After the forum, IITA sends a call for expressions of interest to select investors. Eligible businesses must demonstrate their motivation to nurture and grow the Aflasafe business line through to delivery to farmers. Shortlisted applicants are invited to submit a full business plan with supporting documents for evaluation.

The next level of screening is the assessment of business plans and a pitch competition by investors in front of an advisory board, which selects the investor to whom the product manufacturing and distribution is licensed. The signing of a TTLA with the chosen investor is the culmination of the selection process for formally transferring the technology. This legal document sets the terms and conditions under which IITA grants limited non-transferable and non-sub-licensable rights for manufacturing, distribution and sale of the product in line with CGIAR Principles on the Management of Intellectual Assets. The TTLA provides incentives for private sector investment, balancing company profit requirements, affordability for farmers and IITA's obligation to disseminate international public goods. The TTLA also sets out manufacturing and distribution targets that the partner is expected to meet and lays out the period covered by the agreement.

Phase 3. Implementing the business plan. Once the TTLA is signed, IITA works with the selected partners to transfer the technology know-how and additionally provides technical assistance in implementing the business plan. The transfer of knowledge takes various forms depending on the partner's capacities and needs. ATTC's support package includes training technical and sales staff on the technology. To grow market demand, IITA also provides technical assistance for structured awareness-raising and demonstration of the economic and social value of the product to different market segments using business case studies. Finally, IITA supports the setting up of their factory, quality control and specialized staff training. Standard operating protocols for various manufacturing and quality control processes are also provided.

In all countries, the commercialization through a public–private partnership was the preferred pathway identified by stakeholders. Private companies lead commercial operations upstream and downstream, while the public sector is committed to investing in awareness-raising and enabling policies. IITA has licensed four private companies to manufacture and distribute Aflasafe products in Senegal and The Gambia (both countries constituting one territory), Nigeria, Tanzania and Mozambique. The manufacturers in Senegal, Tanzania and Nigeria have so far invested more than US$ 5 m in constructing factories. In Kenya, a government institution manufactures the product, while a private company has the distribution license. Four other private companies are currently distributing Aflasafe products in Ghana, Burkina Faso, Malawi, Mozambique and Mali.

Up to December 2020, manufacturers have produced over 4000 tons of Aflasafe products, enough to cover more than 400 000 hectares of maize, groundnuts and sorghum. Much of the product sale was directed at business-to-business and business-to-government clients. Usage was concentrated in the maize and groundnut value chains where awareness of the negative impacts of aflatoxin is high. Aflasafe is rarely sold at the retail level, except in Kenya. In most cases, farmers access Aflasafe as out-growers through a package of inputs provided by small and medium aggregation businesses. When they are not in such contractual relationships, farmers receive the product through Ministries of Agriculture that distribute limited quantities at a subsidized price to raise awareness about the product. For example, the Central Bank of Nigeria has included Aflasafe in the Anchor Borrowers Scheme through which farmers receive subsidized products for use in maize and groundnut fields. In addition to positive food safety, trade and income improvements, there are other positive outcomes. Two examples are shown in Boxes 1 and 2.

Box 1 Maize in Kenya

Maize is the major staple of Kenyans. The national government's food security flagship Galana-Kulalu irrigation scheme began in 2015. The objective of Galana-Kulalu was to make Kenya food secure. Unfortunately, Galana-Kulalu is in an aflatoxin-prone area. The National Irrigation Board (NIB), which was responsible for the project realized that the maize cannot be fed to people unless aflatoxin contamination is reduced below the regulatory level of 10 ppb. Therefore, the NIB treated all maize in Galana-Kulalu with the Kenya-specific biocontrol product. Nearly all (about 99%) of the scheme's 5910 tons maize was aflatoxin-compliant by Kenya's standards, while 96% met the EU standard of 4 ppb. Without treatment, much of the maize would have been lost to aflatoxin, defeating food-security goals. The scheme's aflatoxin-safe harvest was used to feed for 1 month nearly half a million (492 534) of the region's most vulnerable people (Bandyopadhyay et al. 2019b).

Box 2 Groundnut in The Gambia

For decades, people have been suffering from continuous aflatoxin exposure. In addition, the country lost the ability to produce groundnut with enough quality for premium European markets. The locally produced groundnut ends up in the local markets, creating high aflatoxin exposure. Recently, farmers working with the National Food Security, Processing and Marketing Corporation (NFSPMC) began to use biocontrol on groundnut crop and are now able to produce aflatoxin-safe crops that meet the EU standards. The re-launch of the long-lost groundnut export sector was realized due to the availability of Aflasafe coupled with other effective aflatoxin management strategies. In addition, safer groundnut is being produced for the local population leading to positive health and economic impact.

https://trade4devnews.enhancedif.org/en/impact-story/detoxifying-crops-gambia-ground.

The experience of the 18-year process of developing Aflasafe from discovery to delivery has provided a number of lessons:

- The balance between the protection of the International Public Good (IPG) and private sector investment is critical for sustainability. The TTLA has proven to be an effective tool to protect both the IPG and the private sector company's investment. Granting exclusive rights for 5 years was necessary to attract private sector investment. This has been possible by limiting sub-licensing and transfer of exclusive rights to third parties and by introducing key performance indicators, targets and pricing parameters in the TTLA.
- The right mix of policy and market incentives is required to accelerate uptake. Since food safety is not as rigorously regulated as it should be in most countries, the private sector companies interested in upgrading the agricultural value chains need to send a strong signal to farmers and intermediaries by rewarding quality. To sustain such action and thus assure safe food for all, governments should formulate and promote appropriate food-safety policies and regulations. Endorsement by Ministries of Agriculture increases the trust of users in the technology. Finally, collaborative partnerships between the public and the private sector to educate value-chain actors and food consumers might increase the demand for safe food and, thus for use of biocontrol as a mitigation tool.
- Aflasafe is not a traditional agricultural input because it does not increase yield, nor is it designed to do so. The marketing strategy for Aflasafe should

be innovative since safety improvement through aflatoxin reduction is not detectable unless verified by chemical tests. Regardless of its effectiveness, it is the businesses' ability to deliver tangible value to customers that will trigger and sustain use at scale. The use of marketing strategies that had been successful for other product lines did not work for Aflasafe. A period of experimentation with the business model is required to increase investors' confidence to finance manufacturing and distribution.
- Aflasafe scale-up should start with tailored business cases for aflatoxin-conscious market segments. The level of awareness about aflatoxin is low and highly variable within and across countries. As part of business development, companies should focus first on segments of the market that are most affected and willing to pay for a solution. By first focusing on these low-hanging fruit, companies are assured of revenue streams while investing and developing strategic partnerships to unlock the other more elusive market segments.

8 Current challenges and needs

There are several challenges in large-scale adoption of biocontrol and other technologies for aflatoxin mitigation in Africa. Multifaceted challenges are sociological, economic, regulatory, institutional, policy related and technical in nature. Some of these challenges are common to many agricultural technologies in Africa. In addition, several gaps in knowledge about biocontrol technology remain. Concerns about biocontrol have been expressed. Some of these concerns include:

- use of sorghum (a food crop) as a carrier;
- drought stress impacts on biocontrol performance;
- risk of allergies (e.g. skin, eyes);
- accumulation of other secondary metabolites; and
- influence of biocontrol on soil microbiome.

These issues have been discussed in Bandyopadhyay et al. (2016).

Substantial work is necessary to build awareness of the negative consequences of aflatoxin contamination on health and trade. Improved capacity to monitor, regulate and control aflatoxins is required in many countries in SSA. Policies to promote aflatoxin prevention and control are relatively few. There is a need to implement policies to develop standards based on dietary habits, raise awareness among key stakeholders on control strategies and regulations and reinforce food safety risk assessment, analysis and inspection. Effective legislation is needed for improving food safety in countries affected by aflatoxin contamination.

The preponderance of informal markets and lack of a price premium for aflatoxin-safe crops are major impediments for adoption of aflatoxin control technologies. Creation of demand for aflatoxin-safe products at the end market can stimulate adoption of aflatoxin management practices, including biocontrol, by farmers. Feed, poultry and food-producing companies prefer domestic aflatoxin-compliant crops and are willing to pay farmers/organizations premium prices for locally produced safe crops, which is cheaper than importing crops (Ayedun et al., 2017; Johnson et al., 2020; Migwi et al., 2020). Links between farmers/organizations producing safe crops and customers need to be facilitated.

In many countries in SSA, the human and infrastructural capacity to test crops for aflatoxins is inadequate. This has serious repercussions because most crops are commercialized and consumed without knowing if they are safe, especially those offered in local, unregulated markets. Regulators need to use appropriate sampling protocols and affordable testing systems to rapidly monitor and isolate contaminated crops to ensure that unsafe crops are not sold. Effective systems are also required to trace grain/crop lots that are contaminated to ensure appropriate disposal.

The initial emphasis of the Aflasafe Initiative was on the development of locally tailored national Aflasafe products composed of native atoxigenic active ingredients. Later, development of regional products containing active ingredients co-distributed in multiple countries in a region became a preferred approach to hasten development, testing, registration and commercialization of products. However, there are challenges for countries to mutually recognize product effectiveness, toxicological and ecotoxicological data obtained in other countries. There is a need to establish harmonized regional regulatory frameworks for biocontrol agents and to consider data already generated by neighboring countries to expedite the registration process of biocontrol agents. The current practice of naming national Aflasafe with a country suffix is also a hindrance to acceptance of regional products. Farmers, policy makers and Aflasafe marketing companies in one country (say Burundi) sometimes have reservations to use a product with a suffix of another country (e.g. Aflasafe KE01 of Kenya). Branding of products must be flexible to avoid problems associated with national pride and political considerations.

Despite the efficacy of Aflasafe products and their increased use, some basic and practical questions about the technology remain unanswered. These are related to agronomic practices, soil processes, product performance, product composition/formulation and product safety. A few of these questions are as follows:

- What is the influence of intercropping on efficacy?
- Does pesticide use influence product performance?

- Does application of Aflasafe influence soil biological processes and other toxigenic fungi?
- Is 2-3 weeks before flowering the appropriate time for application of Aflasafe?
- What is the optimum dose for field application?
- How much inoculum is carried over from one season to the next?
- Is once-a-season application necessary if inoculum is carried over from one season to the next?
- What impact will climate change have on product performance?
- What approaches should be explored to improve product performance under extended drought?
- How can loss of inoculum in the field be reduced from granivores since the carrier of Aflasafe is sorghum grain?

These and other questions demonstrate that it is necessary to critically examine various agronomic, biological and environmental factors to better understand biocontrol and improve its performance.

Several reports have suggested that the use of biocontrol products could result in superior *Aspergillus* strains with high competitiveness and with high aflatoxin production ability due to potential genetic recombination between aflatoxin producers and atoxigenic fungi in field conditions (Moore, 2015; Damann, 2015; Olarte et al., 2012; Ehrlich et al., 2015). Genetic recombination has been reported in multiple laboratory studies and in the field after exposing fungi incubated in the lab to natural populations. If genetic recombination does occur frequently in a population, we suggest that this would be beneficial in field conditions because lower aflatoxin production potentials would result in the progeny (Bandyopadhyay et al., 2016). Relatively recently, genetic recombination has been reported as beneficial, and atoxigenic genotypes with the ability to recombine are being sought in the United States (Moore, 2021; Horn et al. 2009b). Despite these challenges and the need for more information, the value of biocontrol as a component of aflatoxin mitigation strategy is largely accepted (Moore, 2021; Kagot et al., 2019; Sarrocco et al., 2019).

9 Some final thoughts

It has been a long journey toward Aflasafe development and commercialization in Africa. There are several key elements in the path to reach the current stage of aflatoxin biocontrol development in Africa. Aflatoxin is a critical food safety challenge affecting health, trade and food security. About two decades ago, aflatoxins were largely unknown to key stakeholders in agriculture and health. Meeting this challenge required education of diverse stakeholders to explain the negative consequences of aflatoxins on health, income, trade and food security

(Bandyopadhyay et al., 2016). Once the dangers of aflatoxins were known among key stakeholders, and after a solution to the contamination process was developed, champions of aflatoxin mitigation strategies (with biocontrol as a centerpiece) emerged because of their understanding of the technology and its value for public health. Such champions of aflatoxin control included leaders of farmers' organizations, program officers in the donor community, officers in the health and agricultural sectors and high-level researchers and administrators who valued aflatoxin control as a key component to achieve national public health. Aflatoxin control policies and strategies were then promoted by public sector stakeholders and organizations. This sparked the interest of the private sector for uptake and scale-up of the technology.

Reaching the current stage would not have been possible without support from the donor community and partnerships with numerous institutions for development, testing, registration and making aflatoxin biocontrol products commercially available to smallholder farmers. Institutions and governments that have funded the aflatoxin control program of IITA and partners are mentioned in the acknowledgement section. The donor community, through diverse projects, allowed the Aflasafe Initiative to create lasting partnerships with public sector institutions to develop and register products, public-private sector investors to manufacture and distribute the Aflasafe biocontrol technology and local and regional grain aggregators/exporters/industries to be able to procure protected crops from smallholder farmers with access to biocontrol.

Although the technology was initially developed by a team of plant pathologists supported by laboratory technicians, it was clear since the onset of the Aflasafe Initiative that appropriate, scalable aflatoxin management strategies needed more than the expertise of scientists. Therefore, over the years, a team was built with members from diverse disciplines, many of them unconventional for traditional agricultural systems and even less conventional for research to benefit smallholder farmers. With a team of plant pathologists, food scientists, plant breeders, industrial engineers, postharvest specialists, biocontrol experts, social scientists, commercialization specialists, administrative specialists, monitoring and evaluation personnel, communication specialists field and laboratory technicians, it has been possible to develop, test, register and commercialize aflatoxin biocontrol.

The Aflasafe multidisciplinary team had a mission-driven focus to scale up innovations beyond product development for the benefit of people with the motto: 'Safe crops, better health and higher income'. That mission is being accomplished by convergence of diverse technical, social, regulatory and structural solutions to allow moving a research technology/concept into the hands of smallholder farmers for their own benefit and the consumers, food and feed manufacturers and industries that procure aflatoxin-reduced crops.

Such a broad mission is not an easy task or one accomplished in a few years. It can take decades to develop technologies and have them available at scale and adopted by smallholder farmers.

10 Conclusion

Aflatoxin contamination of several crops continues to be a major problem across SSA. However, farmers, civil society, governments, development community and private sector, among others are now realizing that there is no food security without food safety. Failures in food safety are costly for developing countries. We have provided a general review of the aflatoxin problem in Africa and the causal agents of contamination, as well as the genesis of aflatoxin biocontrol product development in Nigeria, the beginning of the manufacturing in the laboratory and its transition to an industrial process and its subsequent expansion to >20 countries in Africa using diverse approaches. The transition required venturing into unconventional areas for a research program and for which the group had to recruit engineers, business, commercialization and communication experts. The efficacy of biocontrol and other management practices in significantly reducing aflatoxin contamination has been extensively proven. Often, technologies designed for use by smallholder farmers are developed but are not adopted or scaled up due to constraints to make them available to the end users. The IITA and partners not only improved and adapted aflatoxin biocontrol for Africa but, with the support of partners, have also taken these bioprotectants through product development, registration, manufacturing and commercialization. In the process, awareness of the aflatoxin problem was raised globally and actors across the value chain were educated on the health, economic and trade impacts of aflatoxin.

Evidence and experience are beginning to show that aflatoxin biocontrol is substantially improving the long-term health, well-being and livelihoods of millions of rural African farming families. Integrated aflatoxin management is making food supplies safer, increasing the value of harvests, improving the health and value of livestock and unlocking new markets and economic opportunities across value chains. More efforts are needed to have aflatoxin management programs used at scale in countries where these practices are currently used and to expand those practices to other susceptible crops. In addition, similar programs need to be designed, tailored and fine-tuned in countries where aflatoxin contamination is prevalent. Because of the importance of the problem, the awareness being created and the availability of efficient management tools, we are confident that the aflatoxin problem across SSA will be systematically addressed to reduce the devastating effects that the toxins cause.

11 Where to look for further information

11.1 Further reading

- For a comprehensive coverage on mycotoxins with a focus on Africa, see J. F. Leslie, R. Bandyopadhyay and A. Visconti (Ed.). (2008), *Mycotoxins: Detection Methods, Management, Public Health and Agricultural Trade*. Wallingford: CAB International, pp. 496.
- Scaling up of aflatoxin biocontrol in Africa is covered in the Strategic Brief 'Lessons Learned on Scaling Aflasafe® through Commercialization in Sub-Saharan Africa' available at https://a4nh.cgiar.org/files/2020/08/StrategicBrief_2020_A4NH_Aflasafe_web-1.pdf.
- Visit www.aflasafe.com (specially under the tabs 'Resources' and 'Multimedia') for more detailed information of aflatoxin biocontrol in Africa.
- For a good explanation of biocontrol from research to delivery in the United States, see Cotty, P. J., Antilla, L. and Wakelyn, P. J. (2007), Competitive exclusion of aflatoxin producers: Farmer driven research and development, in C. Vincent, N. Goettel and G. Lazarovits, (Eds), *Biological Control: A Global Perspective*. Wallingford: CAB International, pp. 242–253.
- World Health Organization & Joint FAO/WHO Expert Committee on Food Additives (83rd, 2017: Geneva, Switzerland). (2017), Evaluation of certain contaminants in food: eighty-third report of the Joint FAO/WHO Expert Committee on Food Additives. *WHO Technical Report Series; 1002* World Health Organization. https://apps.who.int/iris/handle/10665/254893

11.2 Key journals and conferences

- *World Mycotoxin Journal*, *Toxins*, *Food Additives and Contaminants*, and *Biological Control* are journals where aflatoxin biocontrol literature is often published.
- Gordon Conference on Mycotoxins and Phycotoxins and The World Mycotoxin Forum organize regular conferences well attended by members of the mycotoxin community.
- International Society of Microbiology and African Society of Microbiology frequently organize international conferences on various themes related to mycotoxins.
- American Phytopathological Society organizes Annual and Divisional meetings that sometimes include symposia on biocontrol.

11.3 Major international research projects

- The Africa-wide Aflasafe Initiative is composed of multiple projects on discovery to delivery of aflatoxin biocontrol products in Africa, https://aflasafe.com/.

- PACA of the African Union Commission is a collaboration that aims to protect crops, livestock and people from the effects of aflatoxins. By combating these toxins, PACA is contributing to improve food security, health and trade across the African continent. https://www.aflatoxinpartnership.org/about/.
- USDA-ARS research project on Improved Environmental and Crop Safety by Modification of the *Aspergillus flavus* Population Structure, https://www.ars.usda.gov/research/project/?accnNo=430864.
- MycoKey was a research project (under European Commission under Horizon 2020 programme) on integrated and innovative key actions for mycotoxin management in the food and feed chain, http://www.mycokey.eu/.
- MyToolBox was a research project (under European Commission under Horizon 2020 programme) on Safe Food and Feed through an Integrated ToolBox for Mycotoxin Management, https://www.mytoolbox.eu/.

12 Acknowledgement

The Aflasafe Initiative has been funded by The Bill & Melinda Gates Foundation (BMGF; OPP1007117, OPP1133356); the United States Agency for International Development (USAID); the Meridian Institute on behalf of the Partnership for Aflatoxin Control in Africa (PACA); both the Agricultural Research Service and Foreign Agricultural Service of the United States Department of Agriculture; Bundesministerium für wirtschaftliche Zusammenarbeit und Entwicklung (BMZ, German Federal Ministry for Economic Cooperation and Development); AgResults [a collaborative initiative between BMGF, Australia's Department of Foreign Affairs and Trade (DFAT), the Department for International Development of the United Kingdom (DFID), Global Affairs Canada and USAID]; MycoGlobe, MycoRed and MycoKey funded by the European Union Commission; Austrian Development Cooperation; Commercial Agriculture Development Program of the Government of Nigeria; French Development Agency; Royal Government of Norway; The World Bank, The Food and Agriculture Organization, CGIAR A4NH Research Program, CGIAR MAIZE Research Program, among others. We thank numerous officials from the public institutions and private sector for providing direct and indirect support to the Initiative. We are grateful to the thousands of farmers who graciously and willingly participated in some aspects of product development and testing.

13 References

Abbas, H. K., Wilkinson, J. R., Zablotowicz, R. M., Accinelli, C., Abel, C. A., Bruns, H. A. and Weaver, M. A. (2009). Ecology of *Aspergillus flavus*, regulation of aflatoxin production, and management strategies to reduce aflatoxin contamination of corn, *Toxin Rev.* 28(2–3), 142–53.

Adedeji, B. S., Ezeokoli, O. T., Ezekiel, C. N., Obadina, A. O., Somorin, Y. M., Sulyok, M., Adeleke, R. A., Warth, B., Nwangburuka, C. C., Omemu, A. M., Oyewole, O. B. and Krska, R. (2017). Bacterial species and mycotoxin contamination associated with locust bean, melon and their fermented products in south-western Nigeria, *Int. J. Food Microbiol.* 258, 73–80.

Adhikari, B. N., Bandyopadhyay, R. and Cotty, P. J. (2016). Degeneration of aflatoxin gene clusters in *Aspergillus flavus* from Africa and North America, *AMB Express* 6(1), 62.

Agbetiameh, D., Ortega-Beltran, A., Awuah, R. T., Atehnkeng, J., Cotty, P. J. and Bandyopadhyay, R. (2018). Prevalence of aflatoxin contamination in maize and groundnut in Ghana: population structure, distribution, and toxigenicity of the causal agents, *Plant Dis.* 102(4), 764–72.

Agbetiameh, D., Ortega-Beltran, A., Awuah, R. T., Atehnkeng, J., Islam, M. S., Callicott, K. A., Cotty, P. J. and Bandyopadhyay, R. (2019). Potential of atoxigenic *Aspergillus flavus* vegetative compatibility groups associated with maize and groundnut in Ghana as biocontrol agents for aflatoxin management, *Front. Microbiol.* 10, 2069.

Agbetiameh, D., Ortega-Beltran, A., Awuah, R. T., Atehnkeng, J., Elzein, A., Cotty, P. J. and Bandyopadhyay, R. (2020). Field efficacy of two atoxigenic biocontrol products for mitigation of aflatoxin contamination in maize and groundnut in Ghana, *Biol. Control* 150, 104351.

Aikore, M. O. S., Ortega-Beltran, A., Eruvbetine, D., Atehnkeng, J., Falade, T. D. O., Cotty, P. J. and Bandyopadhyay, R. (2019). Performance of broilers fed with maize colonized by either toxigenic or atoxigenic strains of *Aspergillus flavus* with and without an aflatoxin sequestering agent, *Toxins* 11(10), 565.

Amaike, S. and Keller, N. P. (2011). *Aspergillus flavus*, *Annu. Rev. Phytopathol.* 49, 107–33.

Anthony, M. H., Ojochenemi, A. D., Mulunda, M., Oriyomi, S. T., Jideofor, N. F., Tunde, O., Seun, E. O., Umuhani, Y. O., Robertson, O. B. and Isah, A. (2016). Aflatoxin M1 in breast milk, cow milk and milk products in Minna, Nigeria and their predisposing factors, *Biochem. Anal. Biochem.* 5, 1–6.

Atehnkeng, J., Donner, M., Ojiambo, P. S., Ikotun, B., Augusto, J., Cotty, P. J. and Bandyopadhyay, R. (2016). Environmental distribution and genetic diversity of vegetative compatibility groups determine biocontrol strategies to mitigate aflatoxin contamination of maize by *Aspergillus flavus*, *Microb. Biotechnol.* 9(1), 75–88.

Atehnkeng, J., Ojiambo, P. S., Donner, M., Ikotun, T., Sikora, R. A., Cotty, P. J. and Bandyopadhyay, R. (2008). Distribution and toxigenicity of *Aspergillus* species isolated from maize kernels from three agro-ecological zones in Nigeria, *Int. J. Food Microbiol.* 122(1–2), 74–84.

Atehnkeng, J., Ojiambo, P. S., Cotty, P. J. and Bandyopadhyay, R. (2014). Field efficacy of a mixture of atoxigenic *Aspergillus flavus* Link: FR vegetative compatibility groups in preventing aflatoxin contamination in maize (*Zea mays* L.), *Biol. Control* 72, 62–70.

Ayalew, A., Kimanya, M., Matumba, L., Bandyopadhyay, R., Menkir, A. and Cotty, P. J. (2017). Controlling aflatoxins in maize in Africa: strategies, challenges and opportunities for improvement. In: Watson, D. (Ed), *Achieving Sustainable Cultivation of Maize (Vol. 2): Cultivation Techniques, Pest and Disease Control*, Cambridge: Burleigh Dodds Science Publishing Limited, pp. 1–24.

Ayedun, B., Okpachu, G., Manyong, V., Atehnkeng, J., Akinola, A., Abu, G. A., Bandyopadhyay, R. and Abdoulaye, T. (2017). An assessment of willingness to pay by maize and groundnut farmers for aflatoxin biocontrol product in Northern Nigeria, *J. Food Prot.* 80(9), 1451–60.

Bakoye, O. N., Baoua, I. B., Seyni, H., Amadou, L., Murdock, L. L. and Baributsa, D. (2017). Quality of maize for sale in markets in Benin and Niger., *J. Stored Prod. Res.* 71, 99–105.

Bandyopadhyay, R., Atehnkeng, J., Ortega-Beltran, A., Akande, A., Falade, T. D. O. and Cotty, P. J. (2019a). "Ground-truthing" efficacy of biological control for aflatoxin mitigation in farmers' fields in Nigeria: from field trials to commercial usage, a 10-year study, *Front. Microbiol.* 10, 2528.

Bandyopadhyay, R., Cardwell, K. F., Ortega-Beltran, A., Schulthess, F., Meikle, W., Setamou, M. and Cotty, P. J. (2019b). Identifying and managing plant health risks for key African crops: maize. In: Neuenschwander, P. and Tamò, M. (Eds), *Critical Issues in Plant Health: 50 Years of Research in African Agriculture*, Cambridge: Burleigh Dodds Science Publishing Limited, pp.173–212.

Bandyopadhyay, R., Ortega-Beltran, A., Akande, A., Mutegi, C., Atehnkeng, J., Kaptoge, L., Senghor, A. L., Adhikari, B. N. and Cotty, P. J. (2016). Biological control of aflatoxins in Africa: current status and potential challenges in the face of climate change, *World Mycotoxin J.* 9(5), 771–89.

Battilani, P., Toscano, P., Van Der Fels-Klerx, H. J., Moretti, A., Camardo Leggieri, M., Brera, C., Rortais, A., Goumperis, T. and Robinson, T. (2016). Aflatoxin B_1 contamination in maize in Europe increases due to climate change, *Sci. Rep.* 6, 24328.

Bayman, P. and Cotty, P. J. (1991). Vegetative compatibility and genetic diversity in the *Aspergillus flavus* population of a single field, *Can. J. Bot.* 69(8), 1707–11.

Bock, C. H., Mackey, B. and Cotty, P. J. (2004). Population dynamics of *Aspergillus flavus* in the air of an intensively cultivated region of south-west Arizona, *Plant Pathol.* 53(4), 422–33.

Bradford, K. J., Dahal, P., Van Asbrouck, J., Kunusoth, K., Bello, P., Thompson, J. and Wu, F. (2018). The dry chain: reducing postharvest losses and improving food safety in humid climates, *Trends Food Sci. Technol.* 71, 84–93.

Brown, R. L., Cotty, P. J. and Cleveland, T. E. (1991). Reduction in aflatoxin content of maize by atoxigenic strains of *Aspergillus flavus*, *J. Food Prot.* 54(8), 623–6.

Brown, R. L., Menkir, A., Chen, Z. Y., Bhatnagar, D., Yu, J., Yao, H. and Cleveland, T. E. (2013). Breeding aflatoxin-resistant maize lines using recent advances in technologies – a review, *Food Addit. Contam. Part A Chem. Anal. Control Expo. Risk Assess.* 30(8), 1382–91.

Callicott, K. A. and Cotty, P. J. (2015). Method for monitoring deletions in the aflatoxin biosynthesis gene cluster of *Aspergillus flavus* with multiplex PCR, *Lett. Appl. Microbiol.* 60(1), 60–5.

Camiletti, B. X., Moral, J., Asensio, C. M., Torrico, A. K., Lucini, E. I., Giménez-Pecci, M. P. and Michailides, T. J. (2018) Characterization of Argentinian endemic *Aspergillus flavus* isolates and their potential use as biocontrol agents for mycotoxins in maize, *Phytopathology* 108(7), 818–28.

Canavar, Ö. and Kaynak, M. A. (2013). Prevention of pre-harvest aflatoxin production and the effect of different harvest times on peanut (*Arachis hypogaea* L.) fatty acids, *Food Addit. Contam. Part A Chem. Anal. Control Expo. Risk Assess.* 30(10), 1807–18.

Cardwell, K. F. and Cotty, P. J. (2002). Distribution of *Aspergillus* section Flavi among field soils from the four agroecological zones of the Republic of Benin, West Africa, *Plant Dis.* 79, 1039–45.

Carvajal-Campos, A., Manizan, A. L., Tadrist, S., Akaki, D. K., Koffi-Nevry, R., Moore, G. G., Fapohunda, S. O., Bailly, S., Montet, D., Oswald, I. P., Lorber, S., Brabet, C. and Puel, O. (2017). *Aspergillus korhogoensis*, a novel aflatoxin producing species from the Côte d'Ivoire, *Toxins* 9(11), 353.

CDC. (2004). Outbreak of aflatoxin poisoning – eastern and central provinces , Kenya, January-July 2004, *M.M.W.R Morb. Mortal. Wkly. Rep.* 53(34), 790–3.

Chang, P. K., Horn, B. W. and Dorner, J. W. (2005). Sequence breakpoints in the aflatoxin biosynthesis gene cluster and flanking regions in nonaflatoxigenic *Aspergillus flavus* isolates, *Fungal Genet. Biol.* 42(11), 914–23.

Cotty, P. J. (1989). Virulence and cultural characteristics of two *Aspergillus flavus* strains pathogenic on cotton, *Phytopathology* 79(7), 808–14.

Cotty, P. J., Antilla, L. and Wakelyn, P. J. (2007). Competitive exclusion of aflatoxin producers: farmer driven research and development. In: Vincent, C., Goettel, N. and Lazarovits, G. (Eds), *Biological Control: A Global Perspective*, Wallingford: CAB International, pp. 242–53.

Cotty, P. J., Bayman, P., Egel, D. S. and Elias, K. S. (1994). Agriculture, aflatoxins and *Aspergillus*. In: Powell, K. (Ed), *The Genus Aspergillus*, New York: Plenum Press, pp. 1–27.

Cotty, P. J. and Cardwell, K. F. (1999). Divergence of West African and North American communities of *Aspergillus* section Flavi, *Appl. Environ. Microbiol.* 65(5), 2264–66.

Cotty, P. J. and Jaime-Garcia, R. (2007). Influences of climate on aflatoxin producing fungi and aflatoxin contamination, *Int. J. Food Microbiol.* 119(1–2), 109–15.

Cotty, P. J. and Mellon, J. E. (2006). Ecology of aflatoxin producing fungi and biocontrol of aflatoxin contamination, *Mycotoxin Res.* 22(2), 110–17.

Damann, K. E. (2015). Atoxigenic *Aspergillus flavus* biological control of aflatoxin contamination: what is the mechanism?, *World Mycotoxin J.* 8(2), 235–44.

Desai, S., Thakur, R. P., Rao, V. P. and Anjaiah, V. (2000). Characterization of isolates of *Trichoderma* for biocontrol potential against *Aspergillus flavus* infection in groundnut, *Int. Arachis Newsl.* 20, 57–9.

Diedhiou, P. M., Bandyopadhyay, R., Atehnkeng, J. and Ojiambo, P. S. (2011). *Aspergillus* colonization and aflatoxin contamination of maize and sesame kernels in two agro-ecological zones in Senegal, *J. Phytopathol.* 159(4), 268–75.

Diener, U. L., Cole, R. J., Sanders, T. H., Payne, G. A., Lee, L. S. and Klich, M. A. (1987). Epidemiology of aflatoxin formation by *Aspergillus flavus*, *Annu. Rev. Phytopathol.* 25(1), 249–70.

Donner, M., Atehnkeng, J., Sikora, R. A., Bandyopadhyay, R. and Cotty, P. J. (2009). Distribution of *Aspergillus* section Flavi in soils of maize fields in three agroecological zones of Nigeria, *Soil Biol. Biochem.* 41(1), 37–44.

Donner, M., Atehnkeng, J., Sikora, R. A., Bandyopadhyay, R. and Cotty, P. J. (2010). Molecular characterization of atoxigenic strains for biological control of aflatoxins in Nigeria, *Food Addit. Contam. Part A Chem. Anal. Control Expo. Risk Assess.* 27(5), 576–90.

Donner, M., Lichtemberg, P. S. F., Doster, M., Picot, A., Cotty, P. J., Pucket, R. D. and Michailides, T. J. (2015). Community structure of *Aspergillus flavus* and *A. parasiticus* in major almond-producing areas of California, United States, *Plant Dis.* 99(8), 1161–9.

Dorner, J. W. and Lamb, M. C. (2006). Development and commercial use of afla-guard, an aflatoxin biocontrol agent *Aspergillus flavus*, *Mycotoxin Res.* 22(1), 33–8.

Dorner, J. W. (2009). Development of biocontrol technology to manage aflatoxin contamination in peanuts, *Pean. Sci.* 36(1), 60-7.

Doster, M. A., Cotty, P. J. and Michailides, T. J. (2014). Evaluation of the atoxigenic *Aspergillus flavus* strain AF36 in pistachio orchards, *Plant Dis.* 98(7), 948-56.

Dowd, P. F. (2003). Insect management to facilitate preharvest mycotoxin management, *J. Toxicol. Toxin Rev.* 22(2-3), 327-50.

Drott, M. T., Fessler, L. M. and Milgroom, M. G. (2019). Population subdivision and the frequency of aflatoxigenic isolates in *Aspergillus flavus* in the United States, *Phytopathology* 109(5), 878-86.

Ehrlich, K. C., Moore, G. G., Mellon, J. E. and Bhatnagar, D. (2015). Challenges facing the biological control strategy for eliminating aflatoxin contamination, *World Mycotoxin J.* 8(2), 225-33.

Falade, T. (2018). Aflatoxin management strategies in sub-Saharan Africa. In: Njobeh, P. B. (Ed), *Mycotoxins - Socio-Economic and Health Impact as Well as Pre- and Postharvest Management Strategies*, London: IntechOpen, pp. 1-14.

Frisvad, J. C., Hubka, V., Ezekiel, C. N., Hong, S. B., Nováková, A., Chen, A. J., Arzanlou, M., Larsen, T. O., Sklenář, F., Mahakarnchanakul, W., Samson, R. A. and Houbraken, J. (2019). Taxonomy of *Aspergillus* section Flavi and their production of aflatoxins, ochratoxins and other mycotoxins, *Stud. Mycol.* 93, 1-63.

Gilbert, M. K., Mack, B. M., Payne, G. A. and Bhatnagar, D. (2016). Use of functional genomics to assess the climate change impact on *Aspergillus flavus* and aflatoxin production, *World Mycotoxin J.* 9(5), 665-72.

Groopman, J. D., Kensler, T. W. and Wild, C. P. (2008). Protective interventions to prevent aflatoxin-induced carcinogenesis in developing countries, *Annu. Rev. Public Health* 29, 187-203.

Grubisha, L. C. and Cotty, P. J. (2009). Twenty-four microsatellite markers for the aflatoxin-producing fungus *Aspergillus flavus*, *Mol. Ecol. Resour.* 9(1), 264-67.

Grubisha, L. C. and Cotty, P. J. (2010). Genetic isolation among sympatric vegetative compatibility groups of the aflatoxin-producing fungus *Aspergillus flavus*, *Mol. Ecol.* 19(2), 269-80.

Grubisha, L. C. and Cotty, P. J. (2015). Genetic analysis of the *Aspergillus flavus* vegetative compatibility group to which a biological control agent that limits aflatoxin contamination in U.S. crops belongs, *Appl. Environ. Microbiol.* 81(17), 5889-99.

Horn, B. W. and Dorner, J. W. (1999). Regional differences in production of aflatoxin B1 and cyclopiazonic acid by soil isolates of *Aspergillus flavus* along a transect within the United States, *Appl. Environ. Microbiol.* 65(4), 1444-9.

Horn, B. W., Greene, R. L. and Dorner, J. W. (1995). Effect of corn and peanut cultivation on soil populations of *Aspergillus flavus* and *A. parasiticus* in southwestern Georgia, *Appl. Environ. Microbiol.* 61(7), 2472-5.

Horn, B. W., Moore, G. G. and Carbone, I. (2011). Sexual reproduction in aflatoxin-producing *Aspergillus nomius*, *Mycologia* 103, 174-83.

Horn, B. W., Ramirez-Prado, J. H. and Carbone, I. (2009a). The sexual state of *Aspergillus parasiticus*, *Mycologia* 101, 275-80.

Horn, B. W., Moore, G. G. and Carbone, I. (2009b). Sexual reproduction in *Aspergillus flavus*, *Mycologia* 101, 423-29.

Horn, B. W., Sorensen, R. B., Lamb, M. C., Sobolev, V. S., Olarte, R. A., Worthington, C. J. and Carbone, I. (2014). Sexual reproduction in *Aspergillus flavus* sclerotia naturally produced in corn. *Phytopathology*, 104: 75-85.

Hsieh, L. L. and Hsieh, T. T. (1993). Detection of aflatoxin B1-DNA adducts in human placenta and cord blood, *Cancer Res.* 53(6), 1278–80.

Htoon, W., Kaewpradit, W., Vorasoot, N., Toomsan, B., Akkasaeng, C., Puppala, N., Wongkaew, S. and Jogloy, S. (2019). Relationships between nutrient uptake and nitrogen fixation with aflatoxin contamination in peanut under terminal drought, *Agronomy* 9(8), 419.

IARC. (1972). *IARC Monographs on the Evaluation of Carcinogenic Risk of Chemicals to Man* (vol. 1), Lyon: World Health Organisation, International Agency for Research on Cancer.

IARC. (1976). *IARC Monographs on the Evaluation of Carcinogenic Risk of Chemicals to Man* (vol. 10), Lyon: World Health Organisation, International Agency for Research on Cancer.

IARC. (2002). *IARC Monographs on the Evaluation of Carcinogenic Risks to Humans* (vol. 82), Lyon, France: World Health Organisation, International Agency for Research on Cancer.

IARC. (2012). *Improving Public Health through Mycotoxin Control*, Pitt, J. I., et al. (Eds), Lyon: World Health Organisation.

ICRISAT. (2016). New study shows that groundnut immunity to aflatoxin could be within reach thanks to a double-defence approach. Available at: http://www.icrisat.org/groundnut-immunity-to-aflatoxin/ (Accessed: 5 February 2018).

Islam, M. S., Callicott, K. A., Mutegi, C., Bandyopadhyay, R. and Cotty, P. J. (2018). *Aspergillus flavus* resident in Kenya: high genetic diversity in an ancient population primarily shaped by clonal reproduction and mutation-driven evolution, *Fungal Ecol.* 35, 20–33.

Jaime-Garcia, R. and Cotty, P. J. (2010). Crop rotation and soil temperature influence the community structure of *Aspergillus flavus* in soil, *Soil Biol. Biochem.* 42(10), 1842–47.

Jaime, R., Leisner, L., Antilla, L., Mehl, H. L. and Cotty, P. J. (2017). Area-wide programs for aflatoxin mitigation: treatments to cotton can be cost effective. In: *Proceedings of the Beltwide Cotton Prod. Res. Conference*, Memphis, TN: National Cotton Council of America, pp. 243–7.

Joffe, A. Z. (1969). Aflatoxin produced by 1,626 isolates of *Aspergillus flavus* from groundnut kernels and soils in Israel, *Nature* 221(5179), 492.

Johnson, A. M., Abdoulaye, T., Ayedun, B., Fulton, J. R., Widmar, N. J. O., Adebowale, A., Bandyopadhyay, R. and Manyong, V. (2020). Willingness to pay of Nigerian poultry producers and feed millers for aflatoxin-safe maize, *Agribusiness* 36(2), 299–317.

Kachapulula, P. W., Akello, J., Bandyopadhyay, R. and Cotty, P. J. (2018). Aflatoxin contamination of dried insects and fish in Zambia, *J. Food Prot.* 81(9), 1508–18.

Kachapulula, P. W., Akello, J., Bandyopadhyay, R. and Cotty, P. J. (2017). *Aspergillus* section Flavi community structure in Zambia influences aflatoxin contamination of maize and groundnut, *Int. J. Food Microbiol.* 261, 49–56.

Kagot, V., Okoth, S., De Boevre, M. and De Saeger, S. (2019). Biocontrol of *Aspergillus* and *Fusarium* mycotoxins in Africa: benefits and limitations, *Toxins* 11(2), 109.

Kebede, H., Abbas, H. K., Fisher, D. K. and Bellaloui, N. (2012). Relationship between aflatoxin contamination and physiological responses of corn plants under drought and heat stress, *Toxins* 4(11), 1385–403.

Kim, J. H., Campbell, B. C., Molyneux, R., Mahoney, N., Chan, K. L., Yu, J., Wilkinson, J., Cary, J., Bhatnagar, D. and Cleveland, T. E. (2006). Gene targets for fungal and mycotoxin control, *Mycotoxin Res.* 22(1), 3–8.

Klich, M. A. (2007). *Aspergillus flavus*: the major producer of aflatoxin, *Mol. Plant Pathol.* 8(6), 713-22.

Kubzansky, M., Cooper, A. and Barbary, V. (2019). Promise and progress: market-based solutions to poverty in Africa, *Gates Open Res.* 3, 1-236.

Kuntosch, A. and König, B. (2018). Linking system perspectives with user perspectives to identify adoption barriers to food security innovations for smallholder farmers-evidence from rural Tanzania, *Food Secur.* 10(4), 881-96.

Leslie, J. F. (1993). Fungal vegetative compatibility, *Annu. Rev. Phytopathol.* 31, 127-50.

Lewis, M. H., Carbone, I., Luis, J. M., Payne, G. A., Bowen, K. L., Hagan, A. K., Kemerait, R., Heiniger, R. and Ojiambo, P. S. (2019). Biocontrol strains differentially shift the genetic structure of indigenous soil populations of *Aspergillus flavus*, *Front. Microbiol.* 10, 1738.

Liu, Y. and Wu, F. (2010). Global burden of aflatoxin-induced hepatocellular carcinoma: a risk assessment, *Environ. Health Perspect.* 118(6), 818-24.

Luis, J. M., Carbone, I., Payne, G. A., Bhatnagar, D., Cary, J. W., Moore, G. G., Lebar, M. D., Wei, Q., Mack, B. and Ojiambo, P. S. (2020). Characterization of morphological changes within stromata during sexual reproduction in *Aspergillus flavus*, *Mycologia* 112(5), 908-20.

Maina, A. W., Wagacha, J. M., Mwaura, F. B., Muthomi, J. W. and Woloshuk, C. P. (2016). Postharvest practices of maize farmers in Kaiti District , Kenya and the impact of hermetic storage on populations of *Aspergillus* spp. and aflatoxin contamination, *J. Food Res.* 5(6), 53-66.

Mauro, A., Battilani, P. and Cotty, P. J. (2015). Atoxigenic *Aspergillus flavus* endemic to Italy for biocontrol of aflatoxins in maize, *BioControl* 60(1), 125-34.

Maxwell, L. A., Callicott, K. A., Bandyopadhyay, R., Mehl, H. L., Orbach, M. J. and Cotty, P. J. (2021). Degradation of aflatoxin B1 by atoxigenic *Aspergillus flavus* biocontrol agents, *Plant Dis.*. https://doi.org/10.1094/PDIS-01-21-0066-RE.

Medina, A., Akbar, A., Baazeem, A., Rodriguez, A. and Magan, N. (2017a). Climate change, food security and mycotoxins: do we know enough?, *Fungal Biol. Rev.* 31(3), 143-54.

Medina, A., Gilbert, M. K., Mack, B. M., O'Brian, G. R., Rodriguez, A., Bhatnagar, D., Payne, G. and Magan, N. (2017b). Interactions between water activity and temperature on the *Aspergillus flavus* transcriptome and aflatoxin B1 production, *Int. J. Food Microbiol.* 256, 36-44.

Medina, A., Rodriguez, A. and Magan, N. (2014). Effect of climate change on *Aspergillus flavus* and aflatoxin B1 production, *Front. Microbiol.* 5, 348.

Mehl, H. L. and Cotty, P. J. (2013). Influence of plant host species on intraspecific competition during infection by *Aspergillus flavus*, *Plant Pathol.* 62(6), 1310-8.

Mehl, H. L., Jaime, R., Callicott, K. A., Probst, C., Garber, N. P., Ortega-Beltran, A., Grubisha, L. C. and Cotty, P. J. (2012). *Aspergillus flavus* diversity on crops and in the environment can be exploited to reduce aflatoxin exposure and improve health, *Ann. N. Y. Acad. Sci.* 1273, 7-17.

Meijer, S. S., Catacutan, D., Ajayi, O. C., Sileshi, G. W. and Nieuwenhuis, M. (2015). The role of knowledge, attitudes and perceptions in the uptake of agricultural and agroforestry innovations among smallholder farmers in sub-Saharan Africa, *Int. J. Agric. Sust.* 13(1), 40-54.

Migwi, B., Mutegi, C., Mburu, J., Wagacha, J. M., Cotty, P. J., Bandyopadhyay, R. and Manyong, V. M. (2020). Assessment of willingness-to-pay for Aflasafe KE01, a native biological control product for aflatoxin management in Kenya, *Food Addit. Contam. Part A Chem. Anal. Control Expo. Risk Assess.* 37(11), 1951–62.

Mohammed, A., Chala, A., Ojiewo, C. O., Dejene, M., Fininsa, C., Ayalew, A., Hoisington, D. A., Sobolev, V. S. and Arias, R. S. (2018). Integrated management of *Aspergillus* species and aflatoxin production in groundnut (*Arachis hypogaea* L.) through application of farm yard manure and seed treatments with fungicides and *Trichoderma* species, *Afr. J. Plant Sci.* 12(9), 196–207.

Moore, G. G. (2015). Heritability study of eGFP-transformed *Aspergillus flavus* strains, *World Mycotoxin J.* 8(3), 301–10.

Moore, G. G. (2021). Practical considerations will ensure the continued success of pre-harvest biocontrol using non-aflatoxigenic *Aspergillus flavus* strains, *Crit. Rev. Food Sci. Nutr.*, 1–18. https://doi.org/10.1080/10408398.2021.1873731.

Moore, G. G., Lebar, M. D. and Carter-Wientjes, C. H. (2019). The role of extrolites secreted by nonaflatoxigenic *Aspergillus flavus* in biocontrol efficacy, *J. Appl. Microbiol.* 126(4), 1257–64.

Moral, J., Garcia-Lopez, M. T., Camiletti, B. X., Jaime, R., Michailides, T. J., Bandyopadhyay, R. and Ortega-Beltran, A. (2020). Present status and perspective on the future use of aflatoxin biocontrol products, *Agronomy* 10(4), 491.

Narayan, T., Mainville, D., Geyer, J., Hausdorff, K. and Cooley, D. (2020). *AgResults Impact Evaluation Report: Nigeria Aflasafe™*, Challenge Project. Rockville, MD: Abt Associates. 83 pp.

Naswem, A. A. and Ejembi, S. A. (2017). Reviving agricultural extension for effective transition from subsistence to commercial agriculture in Nigeria, *J. Rural Soc. Sci.* 32, 2.

Ndemera, M., De Boevre, M. and De Saeger, S. (2020). Mycotoxin management in a developing country context: a critical review of strategies aimed at decreasing dietary exposure to mycotoxins in Zimbabwe, *Crit. Rev. Food Sci. Nutr.* 60(4), 529–40.

Novas, M. V. and Cabral, D. (2002). Association of mycotoxin and sclerotia production with compatibility groups in *Aspergillus flavus* from peanut in Argentina, *Plant Dis.* 86(3), 215–9.

Odvody, G. N., Spencer, N. and Remmers, J. (1997). A description of silk cut, a stress-related loss of kernel integrity in preharvest maize, *Plant Dis.* 81(5), 439–44.

Ogunbayo, T., Callicott, K. A., Probst, C. and Cotty, P. J. (2013). An atoxigenic vegetative compatibility group of *Aspergillus flavus* widely adapted to maize production in Africa and North America, *Phytopathology* 103(6), 105–6.

Okoth, S. (2016). Improving the evidence base on aflatoxin contamination and exposure in Africa, *CTA Working Paper (Agriculture and nutrition) 16/13*, pp. 1–113. Available at: https://publications.cta.int/media/publications/downloads/1975_PDF.pdf? (Accessed: 31/8/2021).

Olarte, R. A., Horn, B. W., Dorner, J. W., Monacell, J. T., Singh, R., Stone, E. A. and Carbone, I. (2012). Effect of sexual recombination on population diversity in aflatoxin production by *Aspergillus flavus* and evidence for cryptic heterokaryosis, *Mol. Ecol.* 21(6), 1453–76.

Ortega-Beltran, A. and Bandyopadhyay, R. (2021). Contributions of integrated aflatoxin management strategies to achieve the sustainable development goals in various African countries, *Glob. Food Sec.* 30, 100559.

Ortega-Beltran, A., Callicott, K. A. and Cotty, P. J. (2020). Founder events influence structures of *Aspergillus flavus* populations, *Environ. Microbiol.* 22(8), 3522–34.

Ortega-Beltran, A. and Cotty, P. J. (2018). Frequent shifts in *Aspergillus flavus* populations associated with maize production in Sonora, Mexico, *Phytopathology* 108(3), 412-20.

Ortega-Beltran, A., Grubisha, L. C., Callicott, K. A. and Cotty, P. J. (2016). The vegetative compatibility group to which the US biocontrol agent *Aspergillus flavus* AF36 belongs is also endemic to Mexico, *J. Appl. Microbiol.* 120(4), 986-98.

Ortega-Beltran, A., Guerrero-Herrera, M. D. J., Ortega-Corona, A., Vidal-Martinez, V. A. and Cotty, P. J. (2014). Susceptibility to aflatoxin contamination among maize landraces from Mexico, *J. Food Prot.* 77(9), 1554-62.

Ortega-Beltran, A., Jaime, R. and Cotty, P. J. (2015). Aflatoxin-producing fungi in maize field soils from sea level to over 2000 masl: a three year study in Sonora, Mexico, *Fungal Biol.* 119(4), 191-200.

Ortega-Beltran, A., Moral, J., Picot, A., Puckett, R. D., Cotty, P. J. and Michailides, T. J. (2019). Atoxigenic *Aspergillus flavus* isolates endemic to almond, fig, and pistachio orchards in California with potential to reduce aflatoxin contamination in these crops, *Plant Dis.* 103(5), 905-12.

Papa, K. E. (1986). Heterokaryon incompatibility in *Aspergillus flavus*, *Mycologia* 78(1), 98-101.

Partanen, H. A., El-Nezami, H. S., Leppänen, J. M., Myllynen, P. K., Woodhouse, H. J. and Vähäkangas, K. H. (2010). Aflatoxin B1 transfer and metabolism in human placenta, *Toxicol. Sci.* 113(1), 216-25.

Paterson, R. R. M. and Lima, N. (2010). How will climate change affect mycotoxins in food?, *Food Res. Int.* 43(7), 1902-14.

Paterson, R. R. M. and Lima, N. (2011). Further mycotoxin effects from climate change, *Food Res. Int.* 44(9), 2555-66.

Pautasso, M., Döring, T. F., Garbelotto, M., Pellis, L. and Jeger, M. J. (2012). Impacts of climate change on plant diseases-opinions and trends, *Eur. J. Plant Pathol.* 133(1), 295-313.

Payne, G. A., Cassel, D. K. and Adkins, C. R. (1986). Reduction of aflatoxin contamination in corn by irrigation and tillage, *Postharvest Path Mycotox* 76, 679-84.

Peles, F., Sipos, P., Kovács, S., Győri, Z., Pócsi, I. and Pusztahelyi, T. (2021). Biological control and mitigation of aflatoxin contamination in commodities, *Toxins* 13(2), 104.

Pretari, A., Hoffmann, V. and Tian, L. (2019). Post-harvest practices for aflatoxin control: evidence from Kenya, *J. Stored Prod. Res.* 82, 31-9.

Probst, C., Bandyopadhyay, R. and Cotty, P. J. (2014). Diversity of aflatoxin-producing fungi and their impact on food safety in sub-Saharan Africa, *Int. J. Food Microbiol.* 174, 113-22.

Probst, C., Bandyopadhyay, R., Price, L. E. and Cotty, P. J. (2011). Identification of atoxigenic *Aspergillus flavus* isolates to reduce aflatoxin contamination of maize in Kenya, *Plant Dis.* 95(2), 212-8.

Probst, C., Callicott, K. A. and Cotty, P. J. (2012). Deadly strains of Kenyan *Aspergillus* are distinct from other aflatoxin producers, *Eur. J. Plant Pathol.* 132(3), 419-29.

Probst, C., Schulthess, F. and Cotty, P. J. (2010). Impact of *Aspergillus* section Flavi community structure on the development of lethal levels of aflatoxins in Kenyan maize (*Zea mays*), *J. Appl. Microbiol.* 108(2), 600-10.

Rodriguez-del-Bosque, L. A. (1996). Impact of agronomic factors on aflatoxin contamination in preharvest field corn in Northeastern Mexico, *Plant Dis.* 80(9), 988-93.

Sarrocco, S., Mauro, A. and Battilani, P. (2019). Use of competitive filamentous fungi as an alternative approach for mycotoxin risk reduction in staple cereals: state of art and future perspectives, *Toxins* 11(12), 701.

Schreurs, F., Bandyopadhyay, R., Kooyman, C., Ortega-Beltran, A., Akande, A., Konlambigue, M., Kaptoge, L. and Van den Bosch, N. (2019). Commercial products promoting plant health in African agriculture. In: Neuenschwander, P. and Tamò, M. (Eds), *Critical Issues in Plant Health: 50 Years of Research in African Agriculture*, Cambridge: Burleigh Dodds Science Publishing Limited, pp. 345–363.

Schroeder, H. W. (1969). Factors influencing the development of aflatoxins in some field crops, *J. Stored Prod. Res.* 5(3), 187–92.

Senghor, A. L., Ortega-Beltran, A., Atehnkeng, J., Jarju, P., Cotty, P. J. and Bandyopadhyay, R. (2021). Aflasafe SN01 is the first biocontrol product approved for aflatoxin mitigation in two nations, Senegal and The Gambia, *Plant Dis.* 105(5), 1461–73.

Senghor, L. A., Ortega-Beltran, A., Atehnkeng, J., Callicott, K. A., Cotty, P. J. and Bandyopadhyay, R. (2020). The atoxigenic biocontrol product Aflasafe SN01 is a valuable tool to mitigate aflatoxin contamination of both maize and groundnut cultivated in Senegal, *Plant Dis.* 104(2), 510–20.

Sharma, K. K., Pothana, A., Prasad, K., Shah, D., Kaur, J., Bhatnagar, D., Chen, Z. Y., Raruang, Y., Cary, J. W., Rajasekaran, K., Sudini, H. K. and Bhatnagar-Mathur, P. (2018). Peanuts that keep aflatoxin at bay: a threshold that matters, *Plant Biotechnol. J.* 16(5), 1024–33.

Shenge, K. C., Adhikari, B. N., Akande, A., Callicott, K. A., Atehnkeng, J., Ortega-Beltran, A., Kumar, P. L., Bandyopadhyay, R. and Cotty, P. J. (2019). Monitoring *Aspergillus flavus* genotypes in a multi-genotype aflatoxin biocontrol product with quantitative pyrosequencing, *Front. Microbiol.* 10, 2529.

Singh, P., Callicott, K. A., Orbach, M. J. and Cotty, P. J. (2020). Molecular analysis of S-morphology aflatoxin producers from the United States reveals previously unknown diversity and two new taxa, *Front. Microbiol.* 11, 1236.

Soares, C., Rodrigues, P., Peterson, S. W., Lima, N. and Venâncio, A. (2012). Three new species of *Aspergillus* section Flavi isolated from almonds and maize in Portugal, *Mycologia* 104(3), 682–97.

Suwarno, W. B., Hannok, P., Palacios-Rojas, N., Windham, G., Crossa, J. and Pixley, K. V. (2019). Provitamin A carotenoids in grain reduce aflatoxin contamination of maize while combating vitamin A deficiency, *Front. Plant Sci.* 10, 30.

Sweany, R. R. and Damann Jr., K. E. (2020). Influence of neighboring clonal-colonies on aflatoxin production by *Aspergillus flavus*, *Front. Microbiol.* 10, 3038.

Thakare, D., Zhang, J., Wing, R. A., Cotty, P. J. and Schmidt, M. A. (2017). Aflatoxin-free transgenic maize using host-induced gene silencing, *Sci. Adv.* 3(3), e1602382.

Turner, P. C., Sylla, A., Gong, Y. Y., Diallo, M. S., Sutcliffe, A. E., Hall, A. J. and Wild, C. P. (2005). Reduction in exposure to carcinogenic aflatoxins by postharvest intervention measures in west Africa : a community-based intervention study, *Lancet* 365(9475), 1950–6.

Udomkun, P., Wiredu, A. N., Nagle, M., Müller, J., Vanlauwe, B. and Bandyopadhyay, R. (2017). Innovative technologies to manage aflatoxins in foods and feeds and the profitability of application–a review, *Food Control* 76, 127–38.

van Egmond, H. P. and Jonker, M. A. (2004). Current situation on regulations for mycotoxins, *JSM Mycotoxins* 2003(Suppl.3), 1–15.

Waliyar, F., Traore, A., Fatondji, D. and Ntare, B. R. (2003). Effectof irrigation interval, planting date, and cultivar on *Aspergillus flavus* and aflatoxin contamination of peanut in sandy soil of Niger, *Pean. Sci.* 30(2), 79–84.

Weaver, M. A., Abbas, H. K., Brewer, M. J., Pruter, L. S. and Little, N. S. (2017). Integration of biological control and transgenic insect protection for mitigation of mycotoxins in corn, *Crop Prot.* 98, 108–15.

Williams, S. B., Baributsa, D. and Woloshuk, C. (2014). Assessing Purdue Improved Crop Storage (PICS) bags to mitigate fungal growth and afl atoxin contamination, *J. Stored Prod. Res.* 59, 190–6.

Williams, S. B., Murdock, L. L. and Baributsa, D. (2017). Storage of maize in Purdue Improved Crop Storage (PICS) bags, *PLoS ONE* 12(1), e0168624.

Wu, F. (2006). Mycotoxin reduction in Bt corn: potential economic, health, and regulatory impacts, *Transgen. Res.* 15(3), 277–89.

Wu, F. and Guclu, H. (2012). Aflatoxin regulations in a network of global maize trade, *PLoS ONE* 7(9), e45151.

Wu, F. and Khlangwiset, P. (2010). Health economic impacts and cost-effectiveness of aflatoxin-reduction strategies in Africa: case studies in biocontrol and post-harvest interventions, *Food Addit. Contam. Part A Chem. Anal. Control Expo. Risk Assess.* 27(4), 496–509.

Chapter 17

Using *Verticillium albo-atrum* WCS850 to control Dutch elm disease

Joeke Postma, Wageningen University & Research, The Netherlands

1 Introduction

Trees are crucial for quality of life in urban areas. They limit the urban heat island effect by reducing temperature, improving air quality, contributing to water infiltration, reducing noise disturbance, supporting biodiversity and, above all, improving well-being. Given the importance of these ecosystem services in urban areas, as well as the fact that a single tree can absorb up to 150 kg of CO_2 per year, the 'Trees in Cities Challenge' was launched at the UN Climate Action Summit 2019 to increase the number of trees in cities (https://treesincities.unece.org/).

However, new pests and diseases have appeared in several tree species in recent decades. Examples of pests and diseases in trees that are commonly used in urban areas are: horse chestnut leaf miner (*Cameraria ohridella*), red palm weevil (*Rhynchophorus ferrugineus*), ash dieback (*Chalara fraxinea*) and bleeding canker in chestnut (*Pseudomonas syringae* pv. *aesculi*). Caterpillars of the oak processionary moth (*Thaumetopoea processionea*) are an increasingly serious problem since, in addition to damaging trees they also cause skin irritation and asthma in humans and animals. Given potential risks to human health, chemical pesticides to control pests and diseases in urban areas should be avoided. Biological control is an environmental-friendly alternative (Cazorla and Mercado-Blanco 2016). However, the development of effective

http://dx.doi.org/10.19103/AS.2021.0093.17

biocontrol agents for trees is more complicated due to their perennial nature than for annual crops (Cazorla and Mercado-Blanco 2016). However, 29 years of experience with biological control of Dutch elm disease using the fungus *Verticillium albo-atrum* illustrates the potential for biocontrol of tree pests and diseases. The socio-economic value of a single tree makes individual injection affordable, whereas such treatment is not realistic for individual arable crop plants. Removal of infected trees and transplanting new trees costs 1000 to 3000 Euros per tree depending on their size and the location (Veilleux et al. 2012; personal communication Ron Schraven, BTL Bomendienst). This chapter describes this successful example of biocontrol using *Verticillium albo-atrum*.

A century ago the devastating Dutch elm disease (DED) emerged in Europe (1919) and in North America (1930). Sudden wilting and dying of leaves and branches were caused by the *Ophiostoma ulmi* fungus during this first pandemic from the 1920s to 1940s. An even more aggressive strain of the *O. novo-ulmi* fungus started spreading after 1972 and is responsible for the current pandemic affecting elms. This strain has also infected large numbers of newly planted trees that were supposed to be less susceptible to DED. Since the 1920s DED has killed tens of thousands of elms annually in Europe and North America.

Transmission of the disease is due to a very effective infection cycle with bark beetles as the vector. Bark beetles of the genus *Scolytus* reproduce in weakened and dead (infected) elms. Beetles emerging from infected trees transport conidia of the fungus to healthy elms during their feeding activity on elm twigs (Scheffer et al. 2008). The fungus subsequently develops in the wood vessels, and the infected and weakened tree becomes suitable for breeding of beetles under the stem bark. Large and small elm beetles, *Scolytus scolytus* and *S. multistriatus*, respectively, are the most common and important species spreading the pathogen globally (Santini and Faccoli 2015). Other elm bark beetles are also mentioned as vectors (Anderbrant and Schlyter 1987).

Different methods to control DED have been applied with varying success since the first outbreak of the disease. Chemical control of beetles is potentially hazardous to people and harmful to the environment whilst fungicide (macro)injection in trees is laborious and injures trees (Costonis 1981). In the Netherlands, a national programme begun in the 1930s controlled the disease initially but became unaffordable after the devastating emergence of the even more aggressive strain *O. novo-ulmi*. Strict sanitation measures in some areas such as Amsterdam have reduced incidence to approximately 1% of the elms, including resistant cultivars (Hiemstra et al. 2006). The efficacy of monitoring and strict sanitation was recently evaluated in Canada, showing that rapid removal of infected trees before August reduced the infection rate to 1.5% (Veilleux et al. 2012). New elm cultivars with a high level of resistance to DED are now available on the market (Buiteveld et al. 2015; Martín et al. 2015). This is an important solution for the future but does not prevent existing trees from infection.

A biological control method for DED was developed in the Netherlands in 1980. A strain of *Verticillium albo-atrum* protecting elm trees from DED has been commercially available in several countries since 1992. The biocontrol strain, its mode of action, the product, its application, efficacy and limitations in controlling DED are described in the following sections.

2 Development of the *Verticillium albo-atrum* WCS850 strain

Research on biological control as an alternative for chemical treatments started in the 1980s. It was found that resistance towards *O. novo-ulmi* could be induced in elms that were only resistant to the less-aggressive *O. ulmi* strains by inoculating a mixture of a pathogenic and non-pathogenic strain (Scheffer et al. 1980). This promising result of reducing symptoms by inoculating a non-pathogenic strain began a process of testing antagonistic pseudomonads to control the disease (Scheffer et al. 2008; Scheffer 1983). It was found that the in vitro antagonistic activity of different bacterial isolates did not correlate with the reduction of DED symptoms after injection in trees (Scheffer 1983). Moreover, the effect of bacterial inoculation was dependent on the elm tree clone or cultivar (Scheffer 1989). Both findings supported the hypothesis that induction of resistance in the host was the underlying mechanism rather than antagonism. Unfortunately, these positive biological control effects with bacterial inoculants were only detected in the cultivar 'Commelin' and not in the other clones or cultivars tested.

Subsequent research focused on inducing resistance through different fungi. A non-aggressive *O. ulmi* strain as well as two *Verticillium* strains were tested on the basis that these vascular wilt pathogens could survive within the tree for sufficient time to elicit a resistance response (Scheffer 1990). The isolate *Verticillium albo-atrum* WCS850 (previously identified as *V. dahlia*) proved to suppress disease development effectively in 'Commelin' elms as well as in susceptible field elms (Scheffer et al. 2008). This strain was originally isolated from a potato field in the Netherlands, and is virulent on potato and tomato, but not on elm and several other tree species (Scheffer et al. 2008). It is a naturally hyaline form of *V. albo-atrum* which does not produce resting structures. Conidiospores of the fungus injected in the xylem tissue of trees induce resistance when germinating in the tree.

Efficacy of the *Verticillium albo-atrum* WCS850 strain has been tested in elm trees artificially infected with an aggressive strain of *O. ulmi*. The biocontrol strain was injected at a height of 1.2–1.4 m, and the pathogen was inoculated one or few weeks later at a height of 8 m to mimic the natural infection process. Different clones or species were tested from 1986 to 1988. Over 300 trees aged 15 years old were used in the first stage. Evaluation of disease expression was carried out after 3 months or in the following year. Spores survive during the season but remain close to the site of inoculation (Scheffer 1990). Because elms form a new layer of sapwood each year, the treatment must be repeated every spring.

3 Production, method of application, registration and use

Conidiospores of the isolate *Verticillium albo-atrum* WCS850 are produced under sterile conditions on solid media and are harvested in distilled water, obtaining a suspension with 10^7 spores per mL. Spore concentration, colony formation and absence of impurities and contaminants of each batch are checked. The spore suspensions are stored in customised plastic 40-mL vials at 4°C for a maximum period of 7 weeks.

Trees are injected (vaccinated) with the spore suspension once a year in spring, as soon as the leaves have started to sprout after flowering of the elm, and preferably before any beetle infection of DED occurs. The vaccine is injected into the tree trunk at a height of 1.3 m at every 10 cm of the tree's circumference, by pushing a chisel through the bark and releasing one drop with one pull of the trigger (Fig. 1). The closed injection system prevents spills and any impact on operators, other people, plants, trees or animals in the environment. The method is fast, taking on average 2–3 minutes per treated tree, which enables treatment of large numbers of trees in a short time frame. Only trees without symptoms are vaccinated, since the mode of action of

Figure 1 Hand-held tree injection system: the chisel is pushed through the bark; by pulling the trigger a drop of the vaccine is released which subsequently will be taken up by the elm trunk (photo by BTL Bomendienst).

the biocontrol agent is through induced resistance. A preventative treatment protects the trees against the development of the disease during one season.

All treated trees are checked for DED symptoms in summer and at the end of the year to evaluate the efficacy of the application. A distinction is made between *Ophiostoma* infection starting locally in the tree canopy and spreading downwards, which is due to beetles, and infection starting in the root system which spreads through the whole tree. A typical first sign of an infection transmitted by beetles is an upper branch of the tree with leaves starting to wither and yellow in summer, months before the normal autumnal leaf shedding. The treatment protects against infection starting at the tree canopy, but not against infection through the root system by root grafts.

The commercial name of the product is DutchTrig® with the active substance *Verticillium albo-atrum* WCS850. Although the first field experiments injecting *V. albo-atrum* WCS850 started in 1987 (Scheffer 1990), large-scale application was only initiated from 1992 when the product was registered in the Netherlands (Table 1). After extensive testing in the United States from 1995, full registration by the Environmental Protection Agency (EPA) was granted in 2005. Since 2008 the active substance has been approved for inclusion in the Annex I list of EU Directive 91/414/EEC. Germany, Sweden and Canada were the next list of countries where DutchTrig® was registered, in 2008, 2010 and 2009, respectively. In 2019, the active substance was re-evaluated by the EU under Regulation (EC) No 1107/2009 and approved as a low-risk substance (https://ec.europa.eu/food/plant/pesticides/eu-pesticides-database/public/?event=activesubstance.detail&language=EN&selectedID=2008). The development and placing on the market of low-risk substances and products is encouraged by the EU. Low-risk substances are now approved for 15 years instead of 10 years, which means that the active substance of DutchTrig® is now registered as an active substance in EU until 2034.

Table 1 DutchTrig® registration and application in different countries

	First practical application	Year of registration	Number of treated trees in 2020
The Netherlands	1992	1992	22000
The United States	1995	2005	3000
Germany	2006	2008	300
Canada	2008	2009	2500
Sweden	2008	2010	1000
UK	2016	2016	0[1]
Norway	2018	2018	700

[1] No applications due to the COVID-19 outbreak.

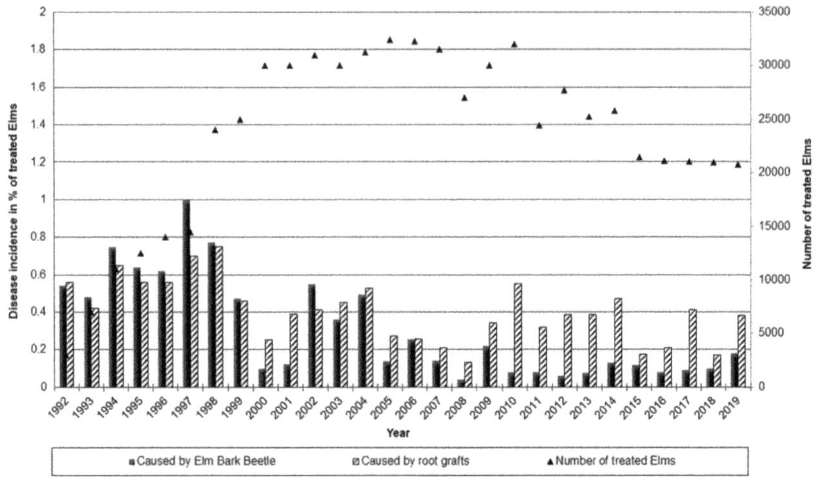

Figure 2 The number of treated elms and disease incidence among the treated trees caused by beetle infections or by root grafts in the Netherlands per year since 1992.

The Netherlands has had the largest number of elms treated with DutchTrig® where 30000 elm trees were vaccinated annually between 2000 and 2010 (Fig. 2). The number of treated trees subsequently decreased but, since 2015, 22000 elms have been vaccinated annually. The total number of treated trees in various countries was approximately 29500 in 2020 (Table 1).

The costs of treatment with DutchTrig® depend on the size and number of trees to be treated. In the Netherlands, where over 20000 trees are treated annually, the price per tree varies between 16 € and 25 €. These costs are relatively low compared to the removal of dead trees and subsequent transplantation and initial maintenance of new trees. These costs depend on the age and the size of the tree to be removed and transplanted, but are between 1000 and 3000 Euros per tree (Veilleux et al. 2012; personal communication Ron Schraven, BTL Bomendienst).

4 Testing long-term efficacy

As noted, efficacy of the biocontrol strain was first tested in elm trees artificially infected with an aggressive strain of *O. ulmi*. These experiments with 15-year-old elm trees in 1987 and 1988 demonstrated a 70–100% reduction in infected trees (Scheffer 1990). Further efficacy tests with several elm clones were performed in the United States in 2-year-old trees (Scheffer et al. 2008). The most relevant efficacy data are gained from consistently monitoring all treated trees for DED symptoms in the months after treatment (until the end of the year).

In the first years of DutchTrig® application in the Netherlands, from 1992 to 1999, approximately 0.6% of injected trees became infected with DED through beetle transmission (Fig. 2). In this period the number of treated trees increased from 3000 up to 24 000 per year. Between 2000 and 2009 efficacy ranged from 0.1% to 0.6% infection through beetle transmission. However, infection by root grafts cannot be controlled by injection with DutchTrig® and the number of trees infected via this route is, even after 2010, still 0.4% of treated trees (Fig. 2). The infection data of the last decade clearly demonstrate that the treated trees have very low infection levels by beetles, but that the infection through root grafts is often at least twice as high.

To fully assess the efficacy of the biocontrol agent, results should be compared with untreated trees. In the field experiments of 1987 and 1988 treated and untreated trees were tested in combination with artificial pathogen inoculation. In the first years of application in the Netherlands, the natural losses in untreated trees were estimated to range from 4% up to 14% per year approximately, compared to 0.6% infection through beetle transmission in the treated elm trees (Scheffer et al. 2008). However, national figures were approximate since there was no consistent monitoring of all elm trees. More precise efficacy data can be obtained from smaller areas where treated as well as untreated elms are monitored.

In The Hague (the Netherlands) where 9000–10 000 elm trees were injected annually from 1998 to 2001, the infection rate of injected trees decreased from 1% to 0.2%, whereas infection in untreated trees decreased from 5.3% to 3.5%, possibly due to the injection programme reducing disease spread (Scheffer et al. 2008). Similar results are available for Denver (USA) where between 1998 and 2003, approximately 500 out of 1300 elms were injected, resulting in a decrease in DED infection in injected trees from 7% to 0.5% (Scheffer et al. 2008). The infection of untreated trees decreased from 5% to 1%. This can be explained by a decreased number of infected trees suitable for breeding of beetles due to long-term application of the biocontrol agent. More recently, 0.86% of 3416 treated elms in Rotterdam (the Netherlands) became infected between 2007and 2013 compared to 3.47% of 647 untreated elms (Postma and Goossen-van de Geijn 2016).

5 Conclusion

Approximately 100 years after the first introduction of DED in Europe, an effective commercial biocontrol product is available as a preventive treatment to protect susceptible elm trees against DED transmission by beetles. Injection of trees with conidiospores of the fungus *V. albo-atrum* isolate WCS850, product name DutchTrig®, reduces the infection to less than 0.2% of treated trees. This product has been commercially available since 1992 and is now used in seven countries. Other biocontrol agents that have been tested for efficacy against DED are *Verticillium dahliae* isolate Vd-48 (Solla and Gil 2003) and four endophytic fungi (Díaz et al. 2013). However, none of them appeared to be effective. Selecting an effective, environmentally safe and commercially valuable biocontrol agent remains challenging (Köhl et al. 2011).

Practical limitations of the product DutchTrig® are the relatively high production costs and limited shelf-life. Living conidiospores that can germinate in the tree are needed to induce resistance and subsequently control the development of DED symptoms. The short survival period of conidiospores limits shelf life of the product and storage and transport must be at low temperature to keep the spores alive. However, this product has become a commercially viable biocontrol agent, registered and used in Europe as well as in the United States and Canada. An important advantage is that the closed injection system prevents spills, making it safe for the operatives, the public and the environment. An important limitation of DutchTrig® is that it does not protect trees already infected or trees infected via root grafts. Consequently, it is very important to avoid contact of roots between healthy and infected trees. In this respect, it is rather disadvantageous that roads are often planted with one type of tree; mixed plantings would reduce the spread of DED through root contact.

Despite its limitations, DutchTrig® is a valuable component of an integrated control strategy for DED. It can be combined with the removal of trees with DED symptoms to reduce disease pressure, together with the use of new elm cultivars with a high level of resistance to DED (Buiteveld et al. 2015; Martín et al. 2015). In the Netherlands, elm trees are now being replanted in several municipalities. This already had a positive impact on biodiversity illustrated by higher numbers of species such as the white-letter hairstreak butterfly (*Satyrium w-album*). This butterfly needs elm trees for its life cycle and has recently been observed in several Dutch cities by the Dutch butterfly foundation (https://www.vlinderstichting.nl/vlinders/overzicht-vlinders/details-vlinder/iepenpage).

Although the closed injection system is very targeted and safe with no exposure to the environment and applicator, injection of biocontrol agents is not a current practice. Most biocontrol agents are applied by spraying on plants or mixing through soil or substrate before sowing or planting the crop. Nevertheless, for perennial crops injecting biocontrol agents into the stem (or trunk) or vascular system is a viable alternative. Injection with DutchTrig® has been tested against few other pathogens in other tree species, but until now this has not resulted in new potential biocontrol applications (personal communication Ron Schraven, BTL Bomendienst). Other examples of injection are seedling vaccination by stem injecting a conidial suspension of a non-pathogenic *Fusarium oxysporum* strain to suppress *Verticillium* wilt of eggplant (Gizi et al. 2011) and injection of *Trichoderma* spp. and *Paecilomyces lilacinus* into the pseudostem of banana to control the nematode *Radopholus similis*

(Vargas et al. 2015). An example of a commercially available product used to vaccinate tomatoes is V10® containing two non-aggressive strains of PepMV (VX1 and VC1) giving cross-protection against Pepino mosaic virus (VALTO, de Lier, the Netherlands). Here the vaccination is done by rubbing the product into one leaf per plant. Although injection of plants with biocontrol agents is rarely applied, it is particularly of interest for perennial or larger sized crops (trees, grape, kiwifruit, hop, banana, tomato crop under greenhouse cultivation) when internalised pests and diseases are an issue (Rabiey et al. 2019), or when the biocontrol acts through systemic induced resistance.

6 Where to look for further information

Practical information about the product and its application, as well as details per country, can be found at: https://dutchtrig.com/.

7 Acknowledgements

We thank BTL Bomendienst for their support and for supplying data on the number of treated and DED-infected elm trees.

8 References

Anderbrant, O. and Schlyter, F. (1987). Ecology of the Dutch elm disease vectors *Scolytus laevis* and *S. Scolytus* (Coleoptera: Scolytidae) in southern Sweden. *Journal of Applied Ecology* 24(2), 539–550.

Buiteveld, J., Van Der Werf, B. and Hiemstra, J. A. (2015). Comparison of commercial elm cultivars and promising unreleased Dutch clones for resistance to *Ophiostoma novo-ulmi*. *iForest–Biogeosciences and Forestry* 8(2), 158–164.

Cazorla, F. M. and Mercado-Blanco, J. (2016). Biological control of tree and woody plant diseases: an impossible task? *BioControl* 61(3), 233–242.

Costonis, A. C. (1981). Tree infection: perspective macro-injection/micro-injection. *Journal of Arboriculture* 7(10), 275–277.

Díaz, G., Córcoles, A. I., Asencio, A. D. and Torres, M. P. (2013). In vitro antagonism of *Trichoderma* and naturally occurring fungi from elms against *Ophiostoma novo-ulmi*. *Forest Pathology* 43(1), 51–58.

Gizi, D., Stringlis, I. A., Tjamos, S. E. and Paplomatas, E. J. (2011). Seedling vaccination by stem injecting a conidial suspension of F2, a non-pathogenic Fusarium oxysporum strain, suppresses Verticillium wilt of eggplant. *Biological Control* 58(3), 387–392.

Hiemstra, J. A., Buiteveld, J., Kopinga, J., Kranenborg, K. G., Ravesloot, M. B. M., van der Sluis, B. J. and de Vries, S. M. G. (2006). Belang en toekomst van de iep in Nederland. *Praktijkonderzoek Plant & Omgeving, PPO* 421, 51 pp.

Köhl, J., Postma, J., Nicot, P., Ruocco, M. and Blum, B. (2011). Stepwise screening of microorganisms for commercial use in biological control of plant-pathogenic fungi and bacteria. *Biological Control* 57(1), 1–12.

Martín, J. A., Solla, A., Venturas, M., Collada, C., Domínguez, J., Miranda, E., Fuentes, P., Burón, M., Iglesias, S. and Gil, L. (2015). Seven *Ulmus minor* clones tolerant to *Ophiostoma novo-ulmi* registered as forest reproductive material in Spain. *iForest–Biogeosciences and Forestry* 8(2), 172–180.

Postma, J. and Goossen-van de Geijn, H. (2016). Twenty-four years of Dutch Trig® application to control Dutch elm disease. *BioControl* 61(3), 305–312.

Rabiey, M., Hailey, L. E., Roy, S. R., Grenz, K., Al-Zadjali, M. A. S., Barrett, G. A. and Jackson, R. W. (2019). Endophytes vs tree pathogens and pests: can they be used as biological control agents to improve tree health? *European Journal of Plant Pathology* 155(3), 711–729.

Santini, A. and Faccoli, M. (2015). Dutch elm disease and elm bark beetles: a century of association. *iForest–Biogeosciences and Forestry* 8(2), 126–134.

Scheffer, R. J. (1983). Biological control of Dutch elm disease by *Pseudomonas* species. *Annals of Applied Biology* 103(1), 21–30.

Scheffer, R. J. (1989). Pseudomonas for biological control of Dutch elm disease. III. Field trials at various locations in the Netherlands. *Netherlands Journal of Plant Pathology* 95(6), 305–318.

Scheffer, R. J. (1990). Mechanisms involved in biological control of Dutch elm disease. *Journal of Phytopathology* 130(4), 265–276.

Scheffer, R. J., Heybroek, H. M. and Elgersma, D. M. (1980). Symptom expression in elms after inoculation with combination of an aggressive and a non-aggressive strain of *Ophiostoma ulmi*. *Netherlands Journal of Plant Pathology* 86(6), 315–317.

Scheffer, R. J., Voeten, J. G. W. F. and Guries, R. P. (2008). Biological control of Dutch elm disease. *Plant Disease* 92(2), 192–200.

Solla, A. and Gil, L. (2003). Evaluating *Verticillium dahliae* for biological control of *Ophiostoma novo-ulmi* in *Ulmus minor*. *Plant Pathology* 52(5), 579–585.

Vargas, R., Wang, A., Obregon, M. and Araya, M. (2015). Effect of *Trichoderma* spp., *Paecilomyces* lilacinus and nematicide injection into the pseudostem, on *Radopholus similis* control and banana production. *Agronomia Costarricense* 39(2), 61–76.

Veilleux, J., Leferink, J. and Holliday, N. J. (2012). Rapid removal of symptomatic trees reduces Dutch elm disease infection rates. *Arboriculture and Urban Forestry* 38(3), 99–104.

Part 4

Future outlook on bioprotectants

Chapter 18

The role of bioprotectants for disease control in integrated crop protection approaches

Jürgen Köhl, Wageningen University & Research, The Netherlands

1 Introduction

1.1 Developments in the past

The role of bioprotectants in agriculture depends on the crop protection approaches applied in cropping systems. Approaches towards crop protection have changed during the last few decades and will continue to change. Changes in crop protection approaches allow changes in cropping systems and vice versa, and are driven by economic and societal conditions and needs as well as by the available crop protection technologies and the development of new ones. Major technological developments have determined disease control approaches during the last few decades. New breeding technologies have increased the level of resistance against pathogens in certain crops and the development and broad implementation of chemical crop protection have reduced yield losses through disease in nearly all crops (Oerke et al., 1994). Saving yields by the broad application of chemical crop protection have allowed farmers to simplify their cropping systems (Barzman et al., 2015).

http://dx.doi.org/10.19103/AS.2021.0093.09

Measures for disease prevention, such as the use of resistant varieties, crop rotation and high biodiversity at field and landscape level, have become less important. Cropping systems developed in large-scale production, simplified crop rotation or monocultures and the use of cultivars more susceptible to diseases gave higher yields in combination with chemical crop protection. Consequently, cropping systems became less resilient against losses caused by diseases and more dependent on the use of chemical crop protection.

The main drivers encouraging a change in crop protection approaches during the last few decades have been drastically increased by demand for food, feed and fibre for a growing world population, in combination with a shift in nutritional patterns of consumers towards animal protein food. The latter has resulted in a disproportional increase in feed production. An increasing awareness of the shortcomings of green revolution technologies has resulted in the adaptation of agricultural production systems, often guided by government regulations and demanded by consumer perceptions. Retailers even increased the pressure on levels of pesticide residues in food by setting more restrictive standards than legally enforced for residue levels. This has led to restricted use of chemical pesticides in advanced integrated pest management (IPM) systems (Barzman et al., 2015).

1.2 Drivers of change

Current major technological progress will further shape crop protection approaches. Better epidemiological knowledge results in reliable forecasting of certain disease epidemics, particularly concerning risk periods for infections by pathogens. This knowledge is implemented in decision support systems (DSS), guiding farmers on decisions regarding chemical crop protection, especially on the timing of applications. Developments in precision agriculture allow site-specific applications of crop protection products within fields and advanced spraying technologies allow targeting of relevant plant parts with reduced drift to the environment. This may result in reduced use of crop protection products per hectare, which results in reduced unwanted environmental side effects.

Reduced efficacies due to resistance development by pathogens against fungicides, and restrictions in the use of chemical crop protection (and subsequent withdrawal from the market) in combination with further requirements, such as those of the European Parliament (2009) to change cropping systems towards systems with IPM, is currently causing another shift towards more resilient cropping systems. In such systems, the implementation of preventative measures is preferred in combination with chemical crop protection as a last resort if other measures do not sufficiently safeguard crop yields (Barzman et al., 2015; Baker et al., 2020). This will result in the development of new complex and multi-faceted IPM systems.

Published by Burleigh Dodds Science Publishing Limited, 2022.

1.3 Future developments

Future developments in crop protection approaches are difficult to predict. Systems may slowly develop in incremental steps towards complex IPM systems with strongly guided application of crop protection products. However, the consequences of climate change and the current pressure on the environment (resulting in the degradation of soils, pollution of waterways and ground water and a decrease in biodiversity in agricultural systems and their surrounding wildlife habitats) are leading us to re-think the direction and speed of changes needed in the near future. Societal awareness of planetary boundaries and the impact of humans on ecosystems (including but not limited to climate change during the Anthropocene) may force much more rapid and drastic changes in significant transformational rather than slow incremental steps (Steffen et al., 2015). This will encourage the development of drastically changed cropping systems with closed nutrient cycles, restricted use of external resources and full protection of biodiversity (Dainese et al., 2019; van Selm et al., 2020; Willett et al., 2019; van Zanten et al., 2019). Yield may decrease per hectare compared to current high-input systems that depend on the external use of fertilizers, energy and crop protection products. Yield gaps of 8–25% have been reported by a complete switch to organic farming (Muller et al., 2017). To feed the population, the need for food production on the limited available arable land will not allow significant competing feed production for livestock industries. Such changes are only possible in combination with diet changes from animal protein to plant protein nutrition. There will be a need to produce protein food crops to fulfil future nutritional patterns instead of abundant cereal production and fodder maize for animal feed (van Selm et al., 2020; van Zanten et al., 2019). Such transformational changes in cropping systems will result in higher diversification in crops at temporal and spatial scales, a shift to other/new crops, and will have a strong emphasis on the resilience of cropping systems against the negative impacts by pests and diseases (Beillouin et al., 2019). In this context, an understanding of the use of microbiomes of soil, in and on plants and in crop residues, will be of increasing importance to maximize the impact of microbiomes on resilience and plant health (Bakker et al., 2020; Berg et al., 2020; CAST, 2020; Kerdraon et al., 2019; Syed Ab Rahman et al., 2018). Crop residue management will become an important preventative measure to reduce pathogen populations (Köhl et al., 2007). Advances in assessing microbiome components and functions are expected during the next few years so that biological information can be incorporated into the next generation of precision and digital agriculture technologies and effective microbiome transplant methods for field use will become available.

1.4 The role of bioprotectants in cropping systems

It is obvious that bioprotectants contribute in various ways towards the different crop protection approaches, ranging from the frequently used reliance on routine use of chemical crop protection products towards a possible future system relying strongly on the resilience of the cropping system with restricted use of chemical crop protection products. Currently, bioprotectants mainly complement and/or replace chemical crop protection products in spray schedules. In the envisaged future, bioprotectants will complement the functions of the microbiome and other components in the cropping system contributing to resilience against plant diseases. Multiple-strain biological control products may be designed for this purpose (Niu et al., 2020). Targeted diseases will change with changing crops and cropping systems, as well as the demands on the specificity of bioprotectants to guarantee protection of the beneficial microbiome. Consequently, different bioprotectants are needed now and in the future.

The process of development of bioprotectants - from choosing the targeted pathogens to screening candidate antagonists, product development, registration and market introduction - takes many years; often 10–15 years are needed for their broad implementation in agriculture. This chapter aims to give a brief overview of the particular roles of bioprotectants in current and potential future crop protection approaches to stimulate discussion within the biocontrol industry, amongst scientists and funding agencies on the need for new generations of bioprotectants for an agricultural industry undergoing transition.

2 The role of bioprotectants in conventional high-input cropping systems

In conventional cropping systems, high inputs of crop protection products are used to achieve maximum yields. Fungicides are intensively applied to prevent or control pathogens damaging the crop. Driving forces for crop protection measures include the avoidance of risks of losses, the expected efficacy of a measure and product and application costs. Broad-spectrum fungicides and combinations or alternations of different crop protection products are standard.

A first prerequisite for the integration of bioprotectants into such crop protection approaches is their biological compatibility. Living microorganisms as the active ingredients of bioprotectants may not survive or at least be restricted in their activity in an environment where multiple applications of different fungicides, including broad-spectrum products, are used. A possible solution is the selection of biological control strains that are resistant to one or several frequently used fungicides. However, fungicides used by growers will change over time due to changes in availability on the market and the occurrence of new compounds and mixtures of active ingredients. It will be impossible to respond to this development rapidly with the development of bioprotectants with adapted resistance to newly introduced fungicides. If bioprotectants compatible with synthetic fungicides are available and can be integrated into crop protection spray schedules, growers will compare the advantages of applying chemical fungicides versus bioprotectants. Product costs will be considered, notwithstanding the externalized costs (Pimentel and Burgess, 2014a,b; Bourguet and Guillemaud, 2016), ease of application in tank mixes including other pesticides or fertilizers, spectrum of activity from high specificity to control one disease to a broad spectrum including several diseases and direct efficacy in disease control with a rapid effect on symptom development without considering indirect, long-term effects through saving the resilience of the cropping system. Based on such considerations, the grower may decide in many cases to use synthetic fungicides instead of bioprotectants, as long as fungicides are available without any restrictions in use. Only if efficient synthetic crop protection products are not available growers will use bioprotectants as an alternative. An example is the use of bioprotectants for fire blight control in apple and pear production if the use of antibiotics such as streptomycin or copper products is restricted. The use of certain fungicides can also be restricted because of the risk of resistance development by the targeted pathogen. If no efficient fungicides are available to replace the preferred fungicides with restricted use, bioprotectants can be integrated into the spray schedule to reduce the risk of fungicide resistance development in the pathogen population. However, the most important motivation for conventional growers to integrate bioprotectants into their spray schedules may be restrictions in pesticide residue levels set by government regulations, such as the 'Maximum Residue Level' (MRL), or (often at lower levels and with further restrictions for multiple pesticide residues) by retailers (Buurma et al., 2012). Late season applications of bioprotectants that don't leave residues on the marketable product have a competitive advantage over synthetic fungicides to avoid higher residue levels of a certain fungicide or to obtain a product with a lower number of different fungicide residues.

In conclusion, there are limited niches for bioprotectants in conventional crop protection approaches where only short-term costs and benefits are valued but externalized costs and long-term benefits are not considered.

3 The role of bioprotectants in integrated pest management cropping systems

3.1 The principles of integrated pest management

Integrated pest management (IPM) is an essential component of current sustainable farm management. By avoiding routine applications of chemical pesticides, contamination of ecosystems and undesirable effects on human and animal health are reduced. In IPM, complex crop protection strategies are being developed and are applied to build on resilient crops, complex ecosystems with high buffering capacity against populations of plant pathogens and pests, and a preference for environmentally friendly crop protection products before chemical pesticides are applied as an exceptional intervention (Barzman et al., 2015).

Entomologists realised in the 1950s that broad-spectrum insecticides favoured pest outbreaks since natural enemies had been eliminated. IPM is a response to the experienced development of pesticides resistance and damage of beneficial insect populations. Important components of IPM include multiple tactics to monitor and develop and use action thresholds to keep populations under predetermined damage levels (Baker et al., 2020). Protection and enhancement of populations of natural enemies in combination with applications of selective insecticides only after pest thresholds are reached are the pillars of IPM. The original goals of IPM were to reduce dependency on pesticides and to protect the environment and health in combination with avoiding rapid resistance build-up against the pesticides by the targeted pests. Later, cultural tactics from mulch covers to tillage were added, combined with biological and chemical control (Baker et al., 2020).

Current IPM approaches cover whole crop protection, preventing damage by pests, diseases and weeds. IPM has developed a broad application in agriculture. In the European Union (EU), Directive 2009/128/EC (European Parliament, 2009) promotes the use of IPM through the development and implementation of National Action Plans in Member States since 2014, with the emphasis on growing healthy crops with the least possible disruption to agro-ecosystems and encouraging natural pest control mechanisms. The system approach is structured by eight principles (Barzman et al., 2015). Prevention and suppression of harmful organisms is always the basic first principle, for example, combining crop rotation, cultivar choice, cultivation techniques, avoiding the spread of harmful organisms and engagement of beneficial organisms, including protection of an ecological infrastructure outside production sites. Decisions regarding direct responsive crop protection measures are made only if monitoring of the crop and its pest populations and diagnostics indicate the relevant risk of crop damage by a harmful organism. Sustainable biological, physical and other non-chemical methods must be used before synthetic pesticides, if they provide satisfactory pest control (see Fig. 1; Annex III of Framework Directive 2009/128/EC; Barzman et al., 2015).

The IPM approach strengthens the position of biological control compared to conventional cropping systems where the use of bioprotectants is mainly motivated to reduce residue levels in food and feed products and to avoid the build-up of resistance to synthetic fungicides. In IPM, preferred measures

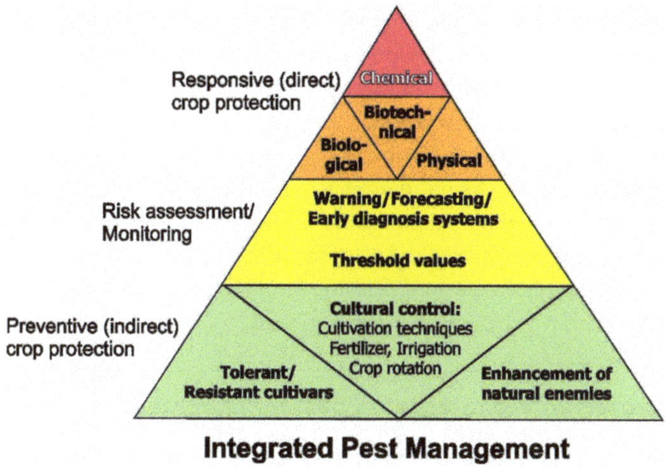

Figure 1 The principles of IPM (Meissle et al., 2011; modified by Meissle).

for preventative crop protection protect or even enhance naturally occurring biological control by natural enemies of pests and beneficial components of the microbiomes in soil, in and on plants and their residues. Resilience in the cropping system increases due to natural biological control, so that less synthetic crop protection products have to be applied and, if applied, selective products with limited negative impact on the natural biological control should be chosen in IPM. In modern IPM systems, including the framework described in the directive on sustainable use of pesticides in the EU (European Parliament, 2009), sustainable biological, physical and other non-chemical methods must be given priority over chemical methods if they provide 'satisfactory' pest control. The definition of 'biological methods' in this context is often broad and may include pheromones, plant extracts, culture extracts of microorganisms or purified metabolites of microorganisms or plants, although the scientific definition of biological control of plant diseases is the suppression of populations of plant pathogens by living organisms (Heimpel and Mills, 2017; Stenberg et al., 2021). In current IPM, the preferred selective control methods are those that are not harmful to the resilience of a cropping system before broad-spectrum methods are used that likely cause side effects on components of the natural biological control of the system. Consequently, following this way of thinking, it is important also to distinguish between selective and broad-spectrum methods in biological control. Zehnder et al. (2007) introduced such a separation for arthropod pest management. They proposed, after preventative strategies have been considered, the inundative and inoculative release of biological control agents as the first phase of a control strategy. In this phase of pest control, mass-reared live agents are released (Eilenberg et al., 2001). Only if control levels of selective biological control agents are insufficient, insecticides of biological and mineral origin and further compounds such as pheromones for mating disruption are applied in the next phase of pest management strategy (Zehnder et al., 2007). Niggli (2020) modified the pest management strategies proposed by Zehnder et al. (2007) by including disease management strategies. In the early phases of preventative measures, he included vegetation management to enhance soil microbiome inducing resistance in plants. In the next phases of disease control measures, the release of competitive antagonists against diseases is proposed, followed by the use of fungicides of biological and mineral origin and the use of chemical or synthetic fungicides (including copper) in the final phase.

3.2 Bioprotectants in integrated pest management

Compared to conventional cropping systems, the IPM approach strengthens the position of bioprotectants. The use is not complementary to the use of chemical pesticides as in conventional cropping systems, but rather biological control products with satisfactory potential in disease protection have to be used before chemicals are considered. In the EU, this principle is strongly promoted and should be implemented by the Member States (European Parliament, 2009). For the broader implementation of bioprotectants in IPM, the decision-making process for growers needs to be adapted. The use of protective measures depends on a decision-making process, often supported by sophisticated DSS developed by commercial providers. Such systems advise farmers in the timing of application of synthetic fungicides. Since the modes of action of bioprotectants are very different from those of synthetic fungicides, optimum timing of bioprotectant application may differ from the advised timing of fungicide applications. Bioprotectants may work more slowly, may need to be applied earlier during epidemics or even before a crop is established (as is the case in controlling disease outbreaks of *Sclerotinia sclerotiorum* by destroying overwintering sclerotia of the pathogen by applying hyperparasites on and in the soil (Gerlagh et al., 1999; Zeng et al., 2012)).

There is therefore the need to develop adapted DSS considering the modes of action of bioprotectants. There is also a need to evolve such DSS from the view of short-term disease to a decision-making process including systemic factors in a long-term decision strategy (Barzman et al., 2015). A change of mind is also needed regarding the understanding of the satisfactory potential in disease protection. This cannot simply consist of a comparison of the efficacy of bioprotectants and synthetic fungicides in experimental design as conventionally used for efficacy tests for crop protection products. Bioprotectants may work more slowly, and have to be applied based on different decisions as conventional fungicides because they work not only by a direct effect of the applied organism but also by the support and protection of the buffering resident microbiome, whereas fungicides may affect beneficial non-target organisms. Furthermore, due to their pathogen-specific nature, it is unlikely that the application of bioprotectants will result in new disease problems, which is often the case when broad-spectrum fungicides are used with side effects on beneficial non-target organisms. In this situation, fungicide use may encourage the additional use of other complementary fungicides to control such new disease problems (Barzman et al., 2015). As long as such possible effects (as well as other externalities causing costs for the society) are not considered in the decision-making process, fungicides may be preferred due to their faster and stronger direct control activity (Pimentel and Burgess, 2014a,b; Bourguet and Guillemaud, 2016). The strong position of bioprotectants in IPM approaches will thus become reality after a broad and long-term holistic definition of the satisfactory pest control level has been developed and applied.

4 The role of bioprotectants in organic cropping systems

4.1 Disease prevention

Organic agriculture relies on ecological processes, biodiversity and closed nutrient cycles rather than external inputs with adverse effects (http://www.ifoam.org/about_ifoam/standards/index.html). The underlying principles of organic agriculture are described by the International Federation of Organic Agricultural Movement (IFOAM) in Best Practice Guidelines and detailed organic standards and technical regulations. Governmental regulations, such as in the EU (European Commission, 2007) and the United States (United States Dept. Agric., 2011), regulate production in organic agriculture and the marketing of organic products. A worldwide certification system for organic farms and their products guarantees that organic agriculture follows the principles of the organic movement. This includes the entire chain, from organic seed production to retail.

The essential principles of organic crop production are biological cycles within the farm, closed systems regarding nutrients and organic matter, maintenance of high genetic diversity and avoidance of any form of pollution (Van Bruggen and Finckh, 2016). Farming systems should be self-regulating through interdependent natural processes in the agroecosystems with the emphasis on prevention of problems rather than solutions by use of external inputs. Typical elements in organic cropping systems are higher plant diversity in time and space due to complex rotation systems, smaller field plots, mixed crops and acceptance of certain weed levels in the crops. Soil management (including the cycling of organic matter within the farm) leads to higher organic matter in the soil and consequently higher microbial colonization levels and higher microbial diversity in the soil, as well as on above-ground parts of the crops and their residues. Organic cropping systems clearly differ from conventional cropping systems. The intrinsic properties of organic systems strongly determine the occurrence and intensity of plant diseases (Van Bruggen and Finckh, 2016), which results in disease management challenges different to those in conventional cropping systems. Consequently, the role of bioprotectants in plant disease control in organic cropping systems differs from their role in conventional cropping systems.

The importance of plant diseases in organic and in conventional cropping systems has been extensively reviewed by Van Bruggen and Finckh (2016), who analysed scientific reports on root and foot diseases in monocotyledonous and dicotyledonous arable crops, annual vegetable crops and perennial crops. Such diseases are caused by soilborne pathogens that interact with the saprophytic microorganisms active in soils. Healthy soils with sufficient organic matter and high biodiversity strongly influence populations of soilborne pathogens and thus generally lower the risk of foot and root diseases. These diseases are generally less pronounced in organic systems compared to conventional cropping systems. For foliar diseases, differences between cropping systems are harder to evaluate because airborne inocula of diseases may be transported for longer distances and thus neighbouring effects may mask system effects (Van Bruggen and Finckh, 2016). A general observation is that certain foliar diseases such as rusts and powdery mildews are favoured on plants grown at high nitrogen levels. Since nitrogen levels are generally lower in organic cropping systems, such diseases affect organic crops less severely than conventional crops. For the prevention of foliar diseases, different measures are combined in organic farming. Examples are sanitation by removal of inoculum sources, cultivar selection with an emphasis on general robustness against biotic stresses and increased habitat diversity, including use of strip crops, intercropping, mixed crops and cover crops (Ditzler et al., 2021). Van Bruggen and Finckh (2016) identified in their study the multicyclic foliar pathogens causing problems for organic growers where preventative measures are often not satisfactory and disease control measures have to be applied. Examples of such threatening diseases are late blight in potato, downy mildews in many crops and apple scab. For many other foliar diseases, differences between the cropping systems were less pronounced or diseases were less severe in organic cropping systems compared to conventional systems.

4.2 Use of crop protection products

The use of plant protection products in organic agriculture has to be authorized by general governmental regulations for plant protection products. Besides this, an additional approval for use in organic agriculture is needed (European Commission, 2008; Speiser and Tamm, 2011). Plant protection products approved for use in organic agriculture are 'mined natural products' such as copper, sulphur and silicates, salts such as bicarbonate salts, oils, extracts from plants and microorganisms, and microorganisms (European Commission, 2008; Van Bruggen et al., 2016). The use of living microorganisms (bacteria, viruses and fungi) for pest and disease control is generally allowed in organic agriculture as long as microorganisms are not genetically modified and no petroleum-based synergists or carriers are used in their formulation (Speiser and Tamm, 2021; Van Bruggen and Finckh, 2016).

Currently, copper fungicides are widely applied in organic farming to control major multicyclic diseases in organic farming such as late blight in potato, downy mildew in grapevine and apple scab (Lamichhane et al., 2018). The use of copper fungicides in the EU is restricted by regulations, allowing only gradually decreasing amounts applied per hectare due to the unwanted accumulation of copper in the soil and negative effects on soil biota. It can also be expected that broad-spectrum copper fungicides interfere with phyllosphere microbiomes so the natural microbial buffering capacity against foliar diseases might be affected. However, studies on the effect of multiple copper applications in grapevine revealed no differences in fungal microbiome composition and quantitative levels of fungal populations on grapevine leaves during a growing season (Gobbi et al., 2020). For other broad-spectrum fungicides such as lime sulphur and sulphur used in organic farming, negative effects on beneficial microorganisms can be expected but conclusive results of microbiome studies on such side effects are missing. As long as broad-spectrum fungicides are commonly used in organic farming to control the major leaf diseases, the role of bioprotectants will be limited. However, similar to the development of IPM, use of selective crop protection products should be considered before broad-spectrum products are used (Zehnder et al., 2007; Niggli, 2020), following an ecological systems approach to protect crops, relying more on larger biodiversity than in conventional cropping systems (Baker et al., 2020). It can be expected that the role of bioprotectants in organic farming will increase in the future if the use of copper fungicides is restricted and more emphasis will be given to the use of selective crop protection products. Main applications in this increasing market will be the control of problematic multicyclic foliar diseases unless fully resistant cultivars become available. A few potential candidates for the development of bioprotectants needed in future organic farming have been described, for example, *Lysobacter capsici* AZ78 for control of downy mildew in grapevine (Segarra et al., 2015) and *Cladosporium cladosporioides* H39 for control of apple scab (Köhl et al., 2015). Antagonistic effects of several bacteria and fungi on other downy mildews including *Phytophthora infestans* have been reported under controlled conditions but biological control of these pathogens under field conditions is still a challenge. Examples of successful biological control of diseases with particular importance for organic farming are limited. The specific biology of this group of pathogens and the resulting complex disease epidemiology with multiple cycles, rapid spread within the crop and high potential to cause yield losses in a short time period are all challenges facing the development of crop protection products, including bioprotectants.

5 Future integrated approaches

5.1 Food production within planetary boundaries

Future cropping systems should be built on closed nutrient cycles with limited use of external resources. Protection of biodiversity will be the other pillar to produce food within existing planetary boundaries (Steffen et al., 2015). Developing such new cropping systems will be a major challenge in the near future. A particular difficulty is the uncertainty of the impact of climate change on agricultural systems (Lamichhane et al., 2015). Crop protection approaches in future systems will rely on their intrinsic buffering capacity against pests and diseases. This robustness against damage by diseases will be based on functions of the plant microbiome (Berg et al., 2017; Berg et al., 2020; CAST, 2020) in combination with genetic diversification of crops and a variety of crop protection approaches, including the use of bioprotectants. With climate change, invasive pathogens may come, and the role of already indigenous pathogens may evolve. Knowledge gaps concerning the future importance of certain diseases hamper the choice of relevant targets for the new generation of bioprotectants needed for future cropping systems.

5.2 Fungicide effects on microbiomes

In the future, crop protection approaches relying on the buffering functions of resident microbiome and additional crop protection measures should not interfere with microbiome composition and function. The impact of insecticides on natural biological control has been studied extensively since the publication of *Silent Spring* by Rachel Carson in 1962. Similar information on potential fungicide effects on resident microbiomes is surprisingly scarce and scattered. The early work of Fokkema et al. (1975) demonstrated that the saprophytic microflora on cereal leaves has an important role in preventing leaf infection by *Cochliobolus sativus*. Fungicide application with non-target effects on the natural microflora interrupted the buffering capacity on the leaves, resulting in increased pathogen infection. It can be expected that such iatrogenic diseases, occurring after fungicide treatments, are more common in crops with high chemical crop protection input, but systematic research is missing.

Seed treatments with fungicides can change the microbiome composition of the developing crop. Such non-target effects were found by Nettles et al. (2016) in multi-year field trials with maize and soybean. Fungal communities in the rhizosphere of maize and soybean were significantly affected even one month after seeding. Seed treatments also had a significant effect on endophytic fungal populations in soybean leaves.

Recent studies of the microbiome of apple bark revealed a strong effect of orchard location, tissue age and sampling time during the growing season on the general composition of bacterial and fungal microbiomes (Arrigoni et al., 2020). Fungicide applications had no global effect on microbiome composition, but the relative abundance of specific taxa significantly differed between systems with different fungicide inputs. The relative abundance of certain potential biocontrol genera, such as *Aureobasidium*, decreased in systems with low-input disease management but other potential biocontrol genera, such as *Cryptococcus*, increased. Fungicide applications have complex effects on the microbiome on wheat leaves (Knorr et al., 2019). Certain fungal groups decreased in relative abundance, whereas other groups, especially yeasts, increased significantly. Interestingly, certain fungal groups were able to multiply after fungicide treatments and filled the available space on the leaves. Gobbi et al. (2020) compared the effect of copper fungicides and a biocontrol treatment on the grapevine phyllosphere microbiome. Bacterial populations were relatively stable whereas the fungal sample showed seasonal shifts. Significant treatment effects were limited to a few fungal taxa. In a similar study, no significant effects of a treatment with penconazole (in comparison with an untreated control) were found for indigenous bacterial and fungal populations on grapevine leaves at three experimental sites in Italy (Perazzolli et al., 2014), whereas microbiome composition significantly differed between locations. Cernava et al. (2019) reported on distinct microbiome shifts by pathogen management practices for bacterial populations on tea leaves due to strong non-target effects of the treatments. At least part of the reported differences may be due to location effects, since treated and untreated leaves were sampled at distinct locations.

5.3 Interactions between bioprotectants and microbiomes

The observed changes in microbiome composition due to fungicides used for disease management may affect the resilience of cropping systems and need further attention in future research into microbiome-pesticide interactions. For the application of biological control products based on living microorganisms, similar studies are even more essential because interactions may act in two different ways. The applied bioprotectant may affect the non-target taxa of the resident microbiome, but also the resident microbiome may affect the establishment and effectiveness of the applied bioprotectant population. Studies of potential non-target effects of biocontrol agents are limited. Brimner and Boland (2003) state that released biocontrol agents have the potential to disrupt entire ecosystems because they can interfere with the native soil communities (e.g. mycorrhizal fungi) and thus may indirectly affect plant growth. Their review on investigations of interactions between mycorrhizal fungi and biocontrol isolates of *Trichoderma* spp. revealed no consistent effects of the biocontrol strains. In different studies, mycorrhizal fungi were inhibited, unaffected or stimulated in development by *Trichoderma* spp., and even antagonism of mycorrhizal fungi against *Trichoderma* spp. has been reported. Brimner and Boland (2003) also state that the host range of several hyperparasites of fungal pathogens used in biological control is not fully known and may be broader so that potentially non-target fungi may also be affected. Deising et al. (2017) expect that applied biocontrol strains may alter resident microbial consortia and may trigger the production of microbial secondary metabolites by resident microbial consortia, causing an unpredictable risk of the use of biological control. However, wherever microbial communities interact, secondary metabolites produced in situ play a role in the communication between microbial populations, including communication with newly arriving biocontrol populations. Produced locally, at low concentrations and degraded within a short time, such secondary metabolites are present in the natural environment and no additional risk following applications of biocontrol strains is expected (Koch et al., 2018; Köhl et al., 2019). Partial effects of introduced biocontrol strains on microbiome composition have been reported, but effects were short term and often negligible, such as in applications on grapevine leaves (Perazzolli et al., 2014), strawberry leaves (Sylla et al., 2013) and lettuce rhizosphere (Scherwinski et al., 2008).

Effects of the resident microbiome on microbial biocontrol agents are more obvious since the buffering capacity of microbial communities is a general phenomenon and a rich biodiversity decreases the susceptibility of ecosystems for invasion (Köhl et al., 2019; Lugtenberg, 2018; Tilman, 1999). This general ecological principle also explains the commonly observed decrease in introduced biocontrol populations (Köhl et al., 2019). The role of the resident microbiome can be demonstrated by a 'de-coupling' experiment. Disruption of the rhizosphere microbiome by the application of a broad-spectrum fungicide led to a better establishment of an introduced antagonistic *Bacillus amyloliquefaciens* isolate and resulted in stronger suppression against *Fusarium oxysporum* f. sp. *cucumerinum* in the rhizosphere soil (Qiu et al., 2014). A similar principle is applied when antagonists are introduced following soil disinfection by steaming. Increasing the resilience of the resident microbiome in future robust cropping systems may therefore affect the establishment and effectiveness of the applied bioprotectant population. For example, the antagonist *Lysobacter capsici* AZ78 did not establish in a vineyard with rich bacterial diversity with strong antagonistic properties (Perazzolli et al., 2014). In this study, Perazzolli et al. (2014) compared the effect of the microbiome of grapevine leaves from three different vineyards on *Plasmopora viticola* by applying leaf washings on *P. viticola*-inoculated leaf discs. The foliar microbiome from one location had a strong effect on *P. viticola* in the leaf disc assay. Application of *L. capsici* AZ78 in this vineyard did not alter the microbiome structure. However, the leaf washings from another vineyard were not effective in the leaf disc assay against the pathogen. In this vineyard, application of *L. capsici* AZ78 increased the abundance of a bacterial group (Xanthomonadaceae) that includes *L. capsici*. These findings led to the hypothesis that indigenous microbiomes with high diversity and biocontrol properties as found on the first site are more resistant to exogenous microorganisms, including the pathogen *P. viticola* and the antagonist *L. capsici*. Biocontrol effects of introduced bioprotectants therefore depend on the structure and functions of the resident microbiome. The role of the bioprotectants in this regard is changing from a general crop protection product to a product complementing resident microbiomes where needed.

5.4 Shaping microbiomes

Current microbiome research is aimed at understanding the structure and function of plant microbiomes and their role in plant health and biocontrol approaches (Berg et al., 2020). Enhancing microbial diversity was identified as a new mode of action of bioprotectants (see elsewhere in this book). Insights into the functions of the microbiome in natural biological control will allow measures to enhance the resident microbiome in order to improve its biocontrol function, similar to the support of resident natural enemies of pests, such as by adapting the architecture of the landscape or strip crops. Reports on measures implemented to shape the resident microbiome towards enhanced disease suppression are still limited. A well-studied case is the increased soil suppressiveness against take-all disease in wheat monoculture (Weller et al., 2002). Recently, rare sugars have been tested to shape the microbiome of grapevine leaves (Perazzolli et al., 2020). Tagotose (TAG) application on grapevine leaves caused significant changes in microbiome composition, leading to a relative increase in potentially beneficial indigenous microorganisms. However, since amplicon sequencing revealed information on the genus level, the beneficial functions can only be hypothesized. The microbiome composition differed between the two vineyards. The hypothesized increase in beneficial microbial groups was found in one of the two vineyards, suggesting that the effect of TAG on disease development depends on the original composition of the site-specific microbiome. The disease control effect of TAG applications under field conditions of downy mildew and, more pronounced, on powdery mildew may be a combined effect of the direct control effect via the anti-nutritional effects of this rare sugar on specific taxa, including the pathogens, and the indirect effect by shaping the microbiome towards an increased relative abundance of (hypothesized) beneficial bacterial and fungal groups. Such an engineering of phyllosphere microbiomes has potential as a new biocontrol approach.

In future cropping systems with strong emphasis on the natural buffering capacity against plant diseases of resident microbiomes in the soil and on and in plants, the role of bioprotectants will change from general crop protection to specific support of and as a complement to resident microbiomes where needed. The high specificity of the bioprotectant against certain pathogens without non-target effects on the broader microbiome will be an important requirement. The interplay of the introduced bioprotectants with the resident microbiome will occur in two ways: complementing the disease suppressiveness of the system and shaping the microbiome towards increased and stable suppressiveness.

6 Case study: the role of bioprotectants in different apple scab control approaches

6.1 Apple scab epidemiology and control

Apple scab caused by *Venturia inaequalis* is the most prevalent apple disease worldwide. The pathogen affects leaves and fruits leading to losses in fruit yield and quality (MacHardy et al., 2001). The multicyclic epidemic starts in spring with the release of ascospores from overwintering apple leaves on the orchard floor. This primary inoculum is present in the orchard during just a few weeks in spring and infects young, highly susceptible developing leaves and young fruits. In a second phase of the epidemic, *V. inaequalis* produces conidia on scabbed leaf tissue during the remaining growing season, which can infect leaves and fruits in multiple infection-sporulation cycles. Young leaves formed during summer are a driving force of the summer epidemic since they are highly susceptible.

Current scab management relies on multiple applications of fungicides, often guided by sophisticated DSS, combining forecasting risks of infection periods, susceptibility of apple tissues and fungicide coverage from earlier applications. Multiple applications of copper-based fungicides or other certified crop protection products are common in organic apple production.

Two strategies for biological control of apple scab have been explored. Antagonists have been selected to interfere with the surviving mycelium and production of fruiting bodies in fallen leaves during the winter season on the orchard floor. The objective of this biocontrol strategy is to reduce the number of ascospores released from pseudothecia in spring so that the primary infections of developing leaves and fruits by ascospores are reduced. The antagonistic fungi *Athelia bombacina* (Heye and Andrews, 1983) and *Microsphaeropsis* sp. (Carisse et al., 2000) have been selected for this purpose. Carisse et al. (2000) showed that autumn applications of *Microsphaeropsis* sp. on the orchard floor reduced the number of ascospores produced by *V. inaequalis* by 70–80% so that the number of fungicide sprays needed for scab control during the growing season could be reduced. In the second biocontrol strategy, antagonists are applied during the growing season to the canopy to prevent initial infections by ascospores in spring or further spread of scab by conidia during the summer epidemic. Several screening programmes have been reported and resulted in a few candidates for biocontrol products, for example, *Chaetomium globosum* (Boudreau and Andrews, 1987), *Pseudomonas syringae* (Burr et al., 1996), epiphytic yeasts (Fiss et al., 2000) and *Cladosporium cladosporioides* (Köhl et al., 2009, 2015).

Apple fruits are produced in cropping systems with different integrated crop protection approaches (Caffi et al., 2017; Holb et al., 2017). Growers' choices depend on their economic and environmental concerns, their risk strategies concerning yield and quality losses and market prices for fruits produced under different systems (Pissonnier et al., 2016). Conventional systems mainly rely on regular applications of synthetic fungicides, whilst IPM systems use DSS before fungicides are applied and often integrate further measures to reduce apple scab risks, such as sanitation and resistant varieties – if marketing of such varieties is feasible. A limited number of fungicide applications is advised in orchards with resistant varieties to prevent the development of *V. inaequalis* populations from breaking this resistance against the pathogen. Organic cropping systems rely on applications of fungicides certified for use in organic farming, such as copper compounds and sulphur, often in combination with resistant varieties (Speiser et al., 2014). Future resilient cropping systems may be developed that fully exploit the benefits of resident microbiomes in scab-resistant varieties. Options for careful microbiome modulation may also become available to protect the crop from damage by the pathogen (Perazzolli et al., 2020). In such future apple cropping systems, interventions with crop protection products will be limited to exceptional periods of predicted high and unacceptable risks for yield and quality losses.

6.2 Bioprotectants in conventional systems

Bioprotectants with the potential to reduce ascospore production before the growing season and to control the progress of scab epidemics during the growing season could have different positioning in spray schedules in cropping systems with different crop protection approaches. In conventional approaches, regular fungicide applications are common, and often applied as calendar sprays, especially if applied by agricultural contractors (Fig. 2a).

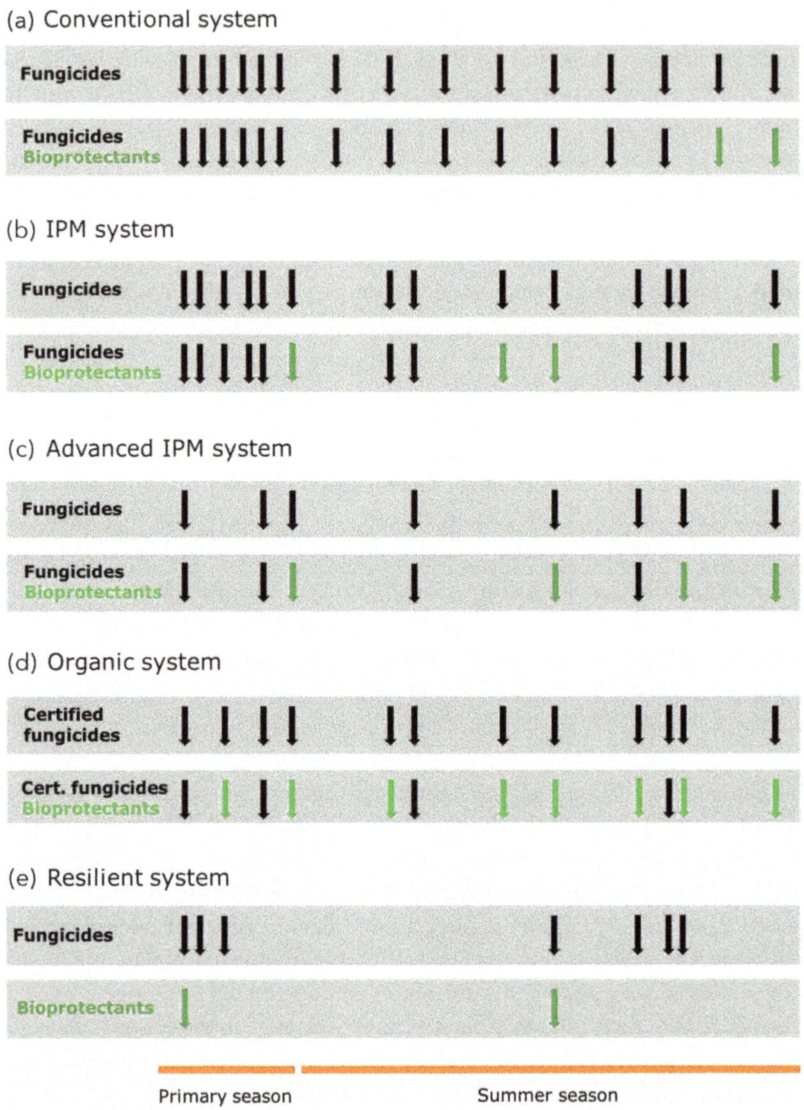

Figure 2 Integration of bioprotectants for apple scab control in different cropping systems. Application of chemical fungicides indicated by black arrows and bioprotectants by green arrows. Preventative measures and bioprotectants targeting leaf litter during the winter season in advanced IPM systems and organic systems not shown. (a) conventional cropping system; (b) IPM system; (c) advanced IPM system; (d) organic cropping system; (e) resilient cropping system.

Options for integration of biocontrol products into such crop protection approaches may be limited. Risks of selection of resistance against the key systemic fungicides within the pathogen populations may be counteracted by using different synthetic fungicides in alternation. The main reason to apply a biocontrol product is to limit the residue level on the marketed fruits below that accepted by distributors and retailers. In this situation, the biocontrol product will replace one or a few late season applications of fungicides against scab, if the biocontrol product can be integrated into spray schedules with applications of fungicides against various apple fruit rots usually applied late in the season.

6.3 Bioprotectants in integrated pest management systems

In IPM systems, fungicide applications for scab control are guided by DSS considering forecasted risks of potential infection periods and expected leaf coverage with fungicides applied during earlier treatments (Fig. 2b). Calendar applications of fungicides are avoided. The choice of a specific fungicide depends on the infection risk level and the number of applications of certain fungicides if their application frequency is limited to prevent build-up of fungicide resistance in the pathogen population. In such a management strategy, a biocontrol product - with possibly lower or more variable efficacy compared to the most effective fungicides - can be integrated during the entire growing season. The biocontrol product can be applied before forecasted moderate or low infection risk periods instead of fungicides to reduce unwanted environmental side effects of the applied spray schedule and to avoid multiple applications of fungicides at risk of resistance development. Depending on the development of the scab epidemic and the weather conditions during the season, several fungicide applications can be replaced by the biocontrol product during the primary season and the summer epidemic. Since developing young leaves are the driving force of the summer epidemic, fungicides present on older leaves from earlier applications may not interfere with the efficacy of the applied antagonist on the young leaves, allowing a flexible integration into spray schedules with alternating chemical and biological control applications. At the end of the season, applications of the biocontrol product will be preferred to achieve low residue levels on the harvested fruits.

In more advanced IPM systems, the use of DSS is complemented by various preventative measures. To reduce the risk of primary infections in the following spring, sanitation measures such as the physical removal of fallen leaves on the orchard floor or the enhancement of their decomposition by shredding or urea application (Gomez et al., 2007; Sutton et al., 2000) can be complemented or replaced by treatments with antagonists such as *Microsphaeropsis* sp. selected for their competitive ability in necrotic leaf tissues (Carisse et al., 2000). Another option is the use of scab-resistant varieties. In orchards planted with common resistant varieties, fungicide treatments applied several times during the growing season are recommended to prevent breaking of scab-resistance by local *V. inaequalis* populations (Jamar and Lateur, 2007). Such fungicide treatments can potentially be replaced by applications of bioprotectants. If scab epidemics develop in such IPM systems, their management by fungicide treatment is guided by DSS. Since the disease pressure will generally be lower, more fungicide treatments can be replaced by biocontrol products compared to less advanced IPM systems or in conventional systems (Fig. 2c).

6.4 Bioprotectants in organic systems

In organic cropping systems, the decomposition of fallen leaves is favoured by the rich biodiversity in organically managed soils. The interplay of mesofauna and microbiome reduces in this way the overwintering potential of *V. inaequalis* in leaf litter. Physical sanitation measures that enhance leaf decomposition (Gomez et al., 2007; Sutton et al., 2000) or removal of fallen leaves by specially designed leaf vacuum machinery (Benduhn et al., 2014) further lower the ascospore load at the beginning of the growing season. Examples of future biological control options in leaf litter include yeast extracts of *Saccharomyces cerivisiae* (Porsche et al., 2017) or antagonists such as *Microsphaeropsis* sp. (Carisse et al., 2000). This combination of enhanced leaf decomposition with a rich biodiversity and additional measures generally results in lower scab pressure at the beginning of the growing season compared to conventional or IPM cropping systems. During the growing season, mainly copper fungicides are applied, guided by DSS (Fig. 2d). Copper spray schedules are complemented by applications of other fungicides accepted in organic farming such as wettable sulphur, lime sulphur and potassium bicarbonates (Jamar and Lateur, 2007). Use of copper fungicides is limited due to environmental considerations, increasingly restricting the amounts applied per hectare and season due to governmental regulations, phytotoxicity and russeting caused in certain developmental stages of apple fruits. Biocontrol products as alternatives can be positioned in the primary season to complement and increasingly replace copper fungicides. Furthermore, pathogen-specific biological control products can replace a substantial number of other organic fungicides, often with broad-spectrum activities, scheduled in the summer season. In this way, the buffering capacity of an undisturbed microbiome on leaves and fruits may reduce scab severities and support the effects of specific biocontrol products.

6.5 Bioprotectants in resilient cropping systems

Future - still partly hypothetical - highly resilient cropping systems combining resistant cultivars with the full employment of microbiota in soil and in and on plants will be less susceptible to apple scab. Interference with crop protection measures will only be needed during forecasted periods of high infection risks, combining favourable weather conditions with high inoculum pressure in the orchard (Fig. 2e). In this situation, applications of broad-spectrum fungicides, as currently used in conventional and organic apple cropping systems, may have unwanted side effects on key species of the resident microbiota and may be detrimental because they may disturb the resilience of the system. If such disturbed systems do not recover before new infection periods occur, subsequent fungicide applications are needed even during periods of lower infection risks. Disturbance may continue and even progress during a longer period - sometimes even the entire growing season - due to an iatrogenic mechanism. Selective biological control products with no or limited effects on resident microbiota will have a clear advantage. Since the resilience of the systems will not be affected by the product, single sprays during periods with high infection risks will be sufficient to support the resident microbiota. The remaining resilience of the system will sufficiently suppress further scab development during low and moderate risk periods, whereas additional subsequent treatments with the biological control product are only needed again during further high-risk periods. In this way, just a few crop protection measures will be needed during a season, and scab management will mainly rely on the functions of undisturbed resident microbiota.

7 Conclusions and future trends in research

Bioprotectants have the potential to replace chemical fungicides in all agricultural cropping systems and crop protection approaches. The development of new bioprotectants in combination with more restricted use of chemical crop protection (due to environmental considerations and a lack of available new lead products) will result in bioprotectants having a stronger market position in the future.

Bioprotectants fulfil particular roles in current and future crop protection approaches, primarily reducing pesticide residues in harvested products in conventional systems, being the first and preferred control option in IPM systems and organic farming, and complementing resident microbiomes in future resilient cropping systems.

Various barriers to adoption have been identified that will slow down the broad exploitation of bioprotectants, such as a lack of biocontrol solutions for multiple diseases in crops, high registration costs, exclusion of external costs of crop protection products, a highly competitive marketplace, risk-averse customers, complex selling channels, insufficient information and training for farmers and insufficient consideration of socioeconomic factors (Marrone, 2009; Barzman et al., 2015; Lamichhane et al., 2017; Baker et al., 2020; Zaki et al., 2020). The importance of such barriers obviously also differs for the different crop protection approaches and the related roles of bioprotectants in such approaches. The main drivers encouraging the use of bioprotectants at present are the economic advantages of the individual biocontrol product over conventional products. For envisaged future crop protection approaches, relying heavily on the resilience of the cropping system against plant diseases, new drivers for using bioprotectants will evolve. A main driver will be the selectivity of bioprotectants for use against specific problematic pathogens without non-target effects and their potential to shape resident microbiomes in their functions. There will be substantial need for such bioprotectants to safeguard the transition from current cropping systems towards resilient cropping systems without unwanted interference within the core microbiome. In this situation, incentives for development and use of bioprotectants need to change from market-driven economic incentives to a more holistic view of the true costs and benefits, including currently externalized costs for the use of different external inputs in agriculture.

8 Where to look for further information

Three international journals specifically focusing on biological control (*BioControl, Biocontrol Science and Technology, Biological Control*) publish results of scientific research on biological control of pest and plant pathogens with an increasing portion of articles on plant pathogens. However, a significant number of scientific papers on biological control of plant pathogens is published in a broad range of scientific journals on plant pathology, microbial ecology and soil microbiology, and in crop-specific journals and journals on post-harvest technologies.

The International Organisation for Biological and Integrated Control (IOBC) (www.iobc-global.org) plays an important role in the communication of scientific results on biological control and their integrated use in cropping systems. The IOBC is organized globally with different regional sections. Within the West Palaearctic Regional Section of IOBC (IOBC-WPRS) (www.iobc-wprs.org) several working groups deal with the biological control of plant pathogens, such as WG 'Biological and integrated control of plant pathogens', WG 'Induced resistance in plants against insects and diseases', WG 'Integrated protection of fruit crops' with subgroups on 'Pome fruit diseases', 'Stone fruits' and 'Soft fruits', and WG 'Integrated protection in viticulture' with the subgroup 'Fungal, bacterial and physiological diseases'.

The working groups meet on a regular basis, usually bi-yearly, with international participation of non-members as well. Contributions to the workshops are published in the *IOBC-WPRS Bulletin*. These bulletins publish scientific reports on ongoing research and technical papers related to the practical use of biological control in cropping systems. IOBC-WPRS also publishes on the frameworks for integrated production and general technical guidelines and crop specific guidelines for the implementation of integrated production and IPM.

The website of the International Biocontrol Manufacturers Association (IBMA; www.ibma-global.org) informs about commercial development, registration and marketing of biocontrol products. The IBMA organizes annually a global biocontrol industry meeting to network, discover and unveil new products, market opportunities and research areas and inform policy makers and regulators of the specific needs of the biocontrol industry. The website (www.abim.ch) includes an informative archive of presentations held during the meetings since 2006.

9 References

Arrigoni, E., Albanese, D., Longa, C. M. O., Angeli, D., Donati, C., Ioriatti, C., Pertot, I. and Perazzolli, M. (2020). Tissue age, orchard location and disease management influence the composition of fungal and bacterial communities present on the bark of apple trees, *Environ. Microbiol.* 22(6), 2080–2093.

Baker, B. P., Green, T. A. and Loker, A. J. (2020). Biological control and integrated pest management in organic and conventional systems, *Biol. Control* 140, 104095.

Bakker, P. A. H. M., Berendsen, R. L., Van Pelt, J. A., Vismans, G., Yu, K., Li, E., Van Bentum, S., Poppeliers, S. W. M., Sanchez Gil, J. J., Zhang, H., Goossens, P., Stringlis, I. A., Song, Y., de Jonge, R. and Pieterse, C. M. J. (2020). The soil-borne identity and microbiome-assisted agriculture: looking back to the future, *Mol. Plant* 13(10), 1394–1401.

Barzman, M., Bàrberi, P., Birch, A. N. E., Boonekamp, P., Dachbrodt-Saaydeh, S., Graf, B., Hommel, B., Jensen, J. E., Kiss, J., Kudsk, P., Lamichhane, J. R., Messéan, A., Moonen, A.-C., Ratnadass, A., Ricci, P., Sarah, J.-L. and Sattin, M. (2015). Eight principles of integrated pest management, *Agron. Sustain. Dev.* 35(4), 1199–1215. DOI: 10.1007/s13593-015-0327-9.

Beillouin, D., Ben-Ari, T. and Makowski, D. (2019). Evidence map of crop diversification strategies at the global scale, *Environ. Res. Lett.* 14(12), 123001.

Benduhn, B., Zimmer, J. and Buchleither, S. (2014). Effect of mechanically removing of leaf litter on apple scab (Venturia inaequalis) infestation in organic apple production. In: Foerdergemeinschaft Oekologischer Obstbau e.V. (Ed), Eco-Fruit: Proc. 17th Int. Conf. Organic Fruit Growing, Hohenheim, Germany, pp. 40–44.

Berg, G., Köberl, M., Rybakova, D., Müller, H., Grosch, R. and Smalla, K. (2017). Plant microbial diversity is suggested as the key to future biocontrol and health trends, *FEMS Microbiol. Ecol.* 93(5), fix050.

Berg, G., Rybakova, D., Fischer, D., Cernava, T., Vergès, M.-C. C., Charles, T., Chen, X., Cocolin, L., Eversole, K., Corral, G. H., Kazou, M., Kinkel, L., Lange, L., Lima, N., Loy, A., Macklin, J. A., Maguin, E., Mauchline, T., McClure, R., Mitter, B., Ryan, M., Sarand, I., Smidt, H., Schelkle, B., Roume, H., Kiran, G. S., Selvin, J., Souza, R. S. C. D., Van Overbeek, L., Singh, B. K., Wagner, M., Walsh, A., Sessitsch, A. and Schloter, M. (2020). Microbiome definition re-visited: old concepts and new challenges, *Microbiome* 8(1), 103.

Boudreau, M. A. and Andrews, J. H. (1987). Factors influencing antagonism of *Chaetomium globosum* to *Venturia inaequalis*: a case study in failed biocontrol, *Phytopathology* 77(10), 1470–1475.

Bourguet, D. and Guillemaud, T. (2016). The hidden and external costs of pesticide use. In: Lichtfouse, E. (Ed), *Sustainable Agriculture Reviews* (vol. 19), Springer, pp. 35–120.

Brimner, T. A. and Boland, G. J. (2003). A review of the non-target effects of fungi used to biologically control plant diseases, *Agric. Ecosyst. Environ.* 100(1), 3–16.

Burr, T. J., Matteson, M. C., Smith, C. A., Corral-Garcia, M. R. and Huang, T.-C. (1996). Effectiveness of bacteria and yeasts from apple orchards as biological control agents of apple scab, *Biol. Control* 6(2), 151–157.

Buurma, J. S., Lamine, C. and Haynes, I. (2012). Transition to consumer driven value chains in the Netherlands, *Acta Hortic.* (930), 69–75.

Caffi, T., Helsen, H. H. M., Rossi, V., Holb, I. J., Strassemeyer, J., Buurma, J. S., Capowiez, Y., Simon, S. and Alaphilippe, A. (2017). Multicriteria evaluation of innovative IPM systems in pome fruit in Europe, *Crop Prot.* 97, 101–108.

Carisse, O., Philion, V., Rolland, D. and Bernier, J. (2000). Effect of fall application of fungal antagonist on spring ascospore production of the apple scab pathogen, *Venturia inaequalis*, *Phytopathology* 90(1), 31–37.

CAST (Council for Agricultural Science and Technology). (2020). Agriculture and the Microbiome. Issue Paper 68, CAST: Ames, IA.

Cernava, T., Chen, X., Krug, L., Li, H., Yang, M. and Berg, G. (2019). The tea leaf microbiome shows specific responses to chemical pesticides and biocontrol applications, *Sci. Total Environ.* 667, 33–40.

Dainese, M., Martin, E. A., Aizen, M. A., Albrecht, M., Bartomeus, I., Bommarco, R., Carvalheiro, L. G., Chaplin-Kramer, R., Gagic, V., Garibaldi, L. A., Ghazoul, J., Grab, H., Jonsson, M., Karp, D. S., Kennedy, C. M., Kleijn, D., Kremen, C., Landis, D. A., Letourneau, D. K., Marini, L., Poveda, K., Rader, R., Smith, H. G., Tscharntke, T., Andersson, G. K. S., Badenhausser, I., Baensch, S., Bezerra, A. D. M., Bianchi, F. J. J. A., Boreux, V., Bretagnolle, V., Caballero-Lopez, B., Cavigliasso, P., Ćetković, A., Chacoff, N. P., Classen, A., Cusser, S., da Silva E Silva, F. D., de Groot, G. A., Dudenhöffer, J. H., Ekroos, J., Fijen, T., Franck, P., Freitas, B. M., Garratt, M. P. D., Gratton, C., Hipólito, J., Holzschuh, A., Hunt, L., Iverson, A. L., Jha, S., Keasar, T., Kim, T. N., Kishinevsky, M., Klatt, B. K., Klein, A. M., Krewenka, K. M., Krishnan, S., Larsen, A. E., Lavigne, C., Liere, H., Maas, B., Mallinger, R. E., Martinez Pachon, E., Martínez-Salinas, A., Meehan, T. D., Mitchell, M. G. E., Molina, G. A. R., Nesper, M., Nilsson, L., O'Rourke, M. E., Peters, M. K., Plećaš, M., Potts, S. G., Ramos, D. L., Rosenheim, J. A., Rundlöf, M., Rusch, A., Sáez, A., Scheper, J., Schleuning, M., Schmack, J. M., Sciligo, A. R., Seymour, C., Stanley, D. A., Stewart, R., Stout, J. C., Sutter, L., Takada, M. B., Taki, H., Tamburini, G., Tschumi, M., Viana, B. F., Westphal, C., Willcox, B. K., Wratten, S. D., Yoshioka, A., Zaragoza-Trello, C., Zhang, W., Zou, Y. and Steffan-Dewenter, I. (2019). A global synthesis reveals biodiversity-mediated benefits for crop production, *Sci. Adv.* 5(10), art. no. eaax0121.

Deising, H. B., Gase, I. and Kubo, Y. (2017). The unpredictable risk imposed by microbial secondary metabolites: how safe is biological control of plant diseases?, *J. Plant Dis. Prot.* 124(5), 413–419.

Ditzler, L., van Apeldoorn, D. Fv, Schulte, R. P. O., Tittonell, P. and Rossing, W. A. H. (2021). Redefining the field to mobilize three-dimensional diversity and ecosystem services on the arable farm, *Eur. J. Agron.* 122, 126197.

Eilenberg, J., Hajek, A. and Lomer, C. (2001). Suggestions for unifying the terminology in biological control, *BioControl* 46(4), 387–400.

European Commission. (2007). Council Regulation (EC) No 834/2007 of 28 June 2007 on organic production and labelling of organic products and repealing Regulation (EEC) No 2092/91. Off. J. Eur. Union L189, 1–23.

European Commission. (2008). Commission regulation (EC) No 889/2008 of 5 September 2008, laying down detailed rules for the implementation of Council Regulation (EC) No 834/2007 on organic production and labelling of organic products with regard to organic production, labelling and control. Off. J. Eur. Commun. L 250/1, 1–84.

European Parliament. (2009). Directive 2009/128/EC of the European Parliament and of the council of 21 October 2009 establishing a framework for community action to achieve the sustainable use of pesticides, *Off. J. Eur. Union* 309, 71–86.

Fiss, M., Kucheryava, N., Schönherr, J., Kollar, A., Arnold, G. and Auling, G. (2000). Isolation and characterization of epiphytic fungi from the phyllosphere of apple as potential biocontrol agents against apple scab (*Venturia inaequalis*), *Z. Pflanzenkr. Pflanzenschutz J. Plant Dis. Prot.* 107(1), 1–11.

Fokkema, N. J., van de Laar, J. A. J., Nelis-Blomberg, A. L. and Schippers, B. (1975). The buffering capacity of the natural mycoflora of rye leaves to infection by *Cochliobolus sativus*, and its susceptibility to benomyl, *Nether. J. Plant Pathol.* 81(5), 176–186.

Gerlagh, M., Goossen-Van De Geijn, H. M., Fokkema, N. J. and Vereijken, P. F. G. (1999). Long-term biosanitation by application of *Coniothyrium minitans* on *Sclerotinia sclerotiorum*-infected crops, *Phytopathology* 89(2), 141–147.

Gobbi, A., Kyrkou, I., Filippi, E., Ellegaard-Jensen, L. and Hansen, L. H. (2020). Seasonal epiphytic microbial dynamics on grapevine leaves under biocontrol and copper fungicide treatments, *Sci. Rep.* 10(1), art. no. 681.

Gomez, C., Brun, L., Chauffour, D. and de Le Vallee, D. D. L. (2007). Effect of leaf litter management on scab development in an organic apple orchard, *Agri. Ecosyst. Environ.* 118(1-4), 249-255.

Heimpel, G. E. and Mills, N. (2017). *Biological Control–Ecology and Applications*, Cambridge University Press: Cambridge.

Heye, C. C. and Andrews, J. H. (1983). Antagonism of *Athelia bombacina* and *Chaetomium globosum* to the apple scab pathogen, *Venturia inaequalis*, *Phytopathology* 73(5), 650-654.

Holb, I. J., Abonyi, F., Buurma, J. and Heijne, B. (2017). On-farm and on-station evaluations of three orchard management approaches against apple scab and apple powdery mildew, *Crop Prot.* 97, 109–118.

Jamar, L. and Lateur, M. (2007). Strategies to reduce copper use in organic apple production, *Acta Hortic.* 737(737), 113–120.

Kerdraon, L., Laval, V. and Suffert, F. (2019). Microbiomes and pathogen survival in crop residues, an ecotone between plant and soil, *Phytobiomes J.* 3(4), 246–255.

Knorr, K., Jørgensen, L. N. and Nicolaisen, M. (2019). Fungicides have complex effects on the wheat phyllosphere mycobiome, *PLoS ONE* 14(3), art. no. e0213176.

Koch, E., Becker, J. O., Berg, G., Hauschild, R., Jehle, J., Köhl, J. and Smalla, K. (2018). Biocontrol of plant diseases is not an unsafe technology!, *J. Plant Dis. Prot.* 125(2), 121-125.

Köhl, J., Haas, B. H., Kastelein, P., Burgers, S. L. G. E. and Waalwijk, C. (2007). Population dynamics of *Fusarium* spp. and *Microdochium nivale* in crops and crop residues of winter wheat, *Phytopathology* 97(8), 971–978.

Köhl, J., Kolnaar, R. and Ravensberg, W. J. (2019). Mode of action of microbial biological control agents against plant diseases: relevance beyond efficacy, *Front. Plant Sci.* 10, 845.

Köhl, J. J., Molhoek, W. W. M. L., Groenenboom-de Haas, B. B. H. and Goossen-van de Geijn, H. H. M. (2009). Selection and orchard testing of antagonists suppressing conidia production of the apple scab pathogen *Venturia inaequalis*, *Eur. J. Plant Pathol.* 123(4), 401–414.

Köhl, J., Scheer, C., Holb, I. J., Masny, S. and Molhoek, W. M. L. (2015). Towards an integrated use of biological control by *Cladosporium cladosporioides* H39 in apple scab (*Venturia inaequalis*) management, *Plant Dis.* 99(4), 535–543.

Lamichhane, J. R., Barzman, M., Booij, K., Boonekamp, P., Desneux, N., Huber, L., Kudsk, P., Langrell, S. R. H., Ratnadass, A., Ricci, P., Sarah, J. L. and Messéan, A. (2015). Robust cropping systems to tackle pests under climate change: a review, *Agron. Sustain. Dev.* 35(2), 443–459.

Lamichhane, J. R., Bischoff-Schaefer, M., Bluemel, S., Dachbrodt-Saaydeh, S., Dreux, L., Jansen, J. P., Kiss, J., Köhl, J., Kudsk, P., Malausa, T., Messéan, A., Nicot, P. C., Ricci, P., Thibierge, J. and Villeneuve, F. (2017). Identifying obstacles and ranking common biological control research priorities for Europe to manage most economically important pests in arable, vegetable and perennial crops, *Pest Manag. Sci.* 73(1), 14–21.

Lamichhane, J. R., Osdaghi, E., Behlau, F., Köhl, J., Jeffrey, B., Jones, J. B. and Aubertot, J.-N. (2018). Thirteen decades of anti-microbial copper compounds applied in agriculture: a review, *Agron. Sustain. Dev.* 38(3), 38–28.

Lugtenberg, B. (2018). Putting concerns for caution into perspective: microbial plant protection products are safe to use in agriculture, *J. Plant Dis. Prot.* 125(2), 127–129.

MacHardy, W. E., Gadoury, D. M. and Gessler, C. (2001). Parasitic and biological fitness of *Venturia inaequalis*: relationship to disease management strategies, *Plant Dis.* 85(10), 1036–1051.

Marrone, P. G. (2009). Barriers to adoption of biological control agents and biological pesticides. In: Radcliffe, E. B., Hutchison, W. D. and Cancelado, R. E. (Eds), *Integrated Pest Management*, Cambridge University Press, pp 163–178.

Meissle, M., Romeis, J. and Bigler, F. (2011). Bt maize and integrated pest management – a European perspective, *Pest Manag. Sci.* 67(9), 1049–1058.

Muller, A., Schader, C., El-Hage Scialabba, N., Brüggemann, J., Isensee, A., Erb, K. H., Smith, P., Klocke, P., Leiber, F., Stolze, M. and Niggli, U. (2017). Strategies for feeding the world more sustainably with organic agriculture, *Nat. Commun.* 8(1), 1290. DOI: 10.1038/s41467-017-01410-w.

Nettles, R., Watkins, J., Ricks, K., Boyer, M., Licht, M., Atwood, L. W., Peoples, M., Smith, R. G., Mortensen, D. A. and Koide, R. T. (2016). Influence of pesticide seed treatments on rhizosphere fungal and bacterial communities and leaf fungal endophyte communities in maize and soybean, *Appl. Soil Ecol.* 102, 61–69.

Niggli, U. (2020). Plant protection in agroecological farming systems, *Agroscope Nachhaltigkeitstagung*, 23 January 2020. Available at: https://www.agroscope.admin.ch/agroscope/de/home/aktuell/veranstaltungen/nachhaltigkeitstagung/fruehere-praesentationen.html#-862383929 Visited 23-03-2021.

Niu, B., Wang, W., Yuan, Z., Sederoff, R. R., Sederoff, H., Chiang, V. L. and Borriss, R. (2020). Microbial interactions within multiple-strain biological control agents impact soil-borne plant disease, *Front. Microbiol.* 11, 585404.

Oerke, E.-C., Dehne, H.-W., Schönbeck, F. and Weber, A. (1994). *Crop Production and Crop Protection - Estimated Losses in Major Food and Cash Crops*, Elsevier Science: Amsterdam.

Perazzolli, M., Antonielli, L., Storari, M., Puopolo, G., Pancher, M., Giovannini, O., Pindo, M. and Pertot, I. (2014). Resilience of the natural phyllosphere microbiota of the grapevine to chemical and biological pesticides, *Appl. Environ. Microbiol.* 80(12), 3585–3596.

Perazzolli, M., Nesler, A., Giovannini, O., Antonielli, L., Puopolo, G. and Pertot, I. (2020). Ecological impact of a rare sugar on grapevine phyllosphere microbial communities, *Microbiol. Res.* 232, 126387.

Pimentel, D. and Burgess, M. (2014a). Environmental and economic costs of the application of pesticides primarily in the United States. In: Pimentel, D., and Peshin, R. (Eds), *Integrated Pest Management*, Springer Science+Business Media: Dordrecht, pp 48–71.

Pimentel, D. and Burgess, M. (2014b). Environmental and economic benefits of reducing pesticide use. In: Pimentel, D., and Peshin, R. (Eds), *Integrated Pest Management*, Springer Science+Business Media: Dordrecht, pp 128–139.

Pissonnier, S., Lavigne, C., Toubon, J.-F. and Le Gal, P.-Y. (2016). Factors driving growers' selection and implementation of an apple crop protection strategy at the farm level, *Crop Prot.* 88, 109–117.

Porsche, F. M., Pfeiffer, B. and Kollar, A. (2017). A new phytosanitary method to reduce the ascospore potential of *Venturia inaequalis*, *Plant Dis.* 101(3), 414–420.

Qiu, M., Li, S., Zhou, X., Cui, X., Vivanco, J. M., Zhang, N., Shen, Q. and Zhang, R. (2014). De-coupling of root-microbiome associations followed by antagonist inoculation improves rhizosphere soil suppressiveness, *Biol. Fertil. Soils* 50(2), 217–224.

Scherwinski, K., Grosch, R. and Berg, G. (2008). Effect of bacterial antagonists on lettuce: active biocontrol of Rhizoctonia solani and negligible, short-term effects on nontarget microorganisms, *FEMS Microbiol. Ecol.* 64(1), 106–116.

Segarra, G., Puopolo, G., Giovannini, O. and Pertot, I. (2015). Stepwise flow diagram for the development of formulations of non spore-forming bacteria against foliar pathogens: the case of Lysobacter capsici AZ78, *J. Biotechnol.* 216, 56–64.

Speiser, B. and Tamm, L. (2011). Regulation of plant protection in organic farming. In: Ehlers, R.-U. (Ed), *Regulation of Biological Control Agents*, Springer: Dordrecht Heidelberg London New York, pp 113–125.

Speiser, B., Tamm, L. and Weibel, F. P. (2014). Regulatory framework for plant protection in organic farming. In: Bellon, S. and Penvern, S. (Eds), *Organic Farming, Prototype for Sustainable Agricultures*, Springer Science+Business Media: Dordrecht, pp 65–82.

Steffen, W., Richardson, K., Rockström, J., Cornell, S. E., Fetzer, I., Bennett, E. M., Biggs, R., Carpenter, S. R., De Vries, W., De Wit, C. A., Folke, C., Gerten, D., Heinke, J., Mace, G. M., Persson, L. M., Ramanathan, V., Reyers, B. and Sörlin, S. (2015). Sustainability. Planetary boundaries: guiding human development on a changing planet, *Science* 347(6223), art. no. 1259855.

Stenberg, J. A., Sundh, I., Becher, P. G., Björkman, C., Dubey, M., Egan, P. A., Friberg, H., Gil, J. F., Jensen, D. F., Jonsson, M., Karlsson, M., Khalil, S., Ninkovic, V., Rehermann, G., Vetukuri, R. R. and Viketoft, M. (2021). When is it biological control? A framework of definitions, mechanisms, and classifications, *J. Pest Sci.* 94(3), 665–676, DOI: 10.1007/s10340-021-01354-7.

Sutton, D. K., MacHardy, W. E. and Lord, W. G. (2000). Effects of shredding or treating apple leaf litter with urea on ascospore dose of *Venturia inaequalis* and disease buildup, *Plant Dis.* 84(12), 1319–1326.

Syed Ab Rahman, S. F., Singh, E., Pieterse, C. M. J. and Schenk, P. M. (2018). Emerging microbial biocontrol strategies for plant pathogens, *Plant Sci.* 267, 102–111. DOI: 10.1016/j.plantsci.2017.11.012.

Sylla, J., Alsanius, B. W., Kruger, E., Reineke, A., Strohmeier, S. and Wohanka, W. (2013). Leaf microbiota of strawberries as affected by biological control agents, *Phytopathology* 103(10), 1001–1011.

Tilman, D. (1999). The ecological consequences of changes in biodiversity: a search for general principles, *Ecology* 80, 1455–1474.

United States Dept. Agric. (USDA). (2011). National organic program handbook, United States Department of Agriculture Agric. Mark. Serv: Washington, DC.

Van Bruggen, A. H. C. and Finckh, M. R. (2016). Plant diseases and management approaches in organic farming systems, *Annu. Rev. Phytopathol.* 54, 25–54.

Van Bruggen, A. H. C., Gamliel, A. and Finckh, M. R. (2016). Plant disease management in organic farming systems, *Pest Manag. Sci.* 72(1), 30–44.

Van Selm, B., de Boer, I. J. M., van Ittersum, M. K., Hijbeek, R. and van Zanten, H. H. E. (2020). Alternative futures of circular food systems: the environmental consequences of adopting circularity at different spatial scales. In: WIAS Annual Conference 2020: Frontiers in Animal Sciences, WIAS, pp. 57–57.

Van Zanten, H. H. E., Van Ittersum, M. K. and De Boer, I. J. M. (2019). The role of farm animals in a circular food system, *Glob. Food Sec.* 21, 18–22.

Weller, D. M., Raaijmakers, J. M., Gardener, B. B. and Thomashow, L. S. (2002). Microbial populations responsible for specific soil suppressiveness to plant pathogens, *Annu. Rev. Phytopathol.* 40, 309–348.

Willett, W., Rockström, J., Loken, B., Springmann, M., Lang, T., Vermeulen, S., Garnett, T., Tilman, D., DeClerck, F., Wood, A., Jonell, M., Clark, M., Gordon, L. J., Fanzo, J., Hawkes, C., Zurayk, R., Rivera, J. A., De Vries, W., Majele Sibanda, L., Afshin, A., Chaudhary, A., Herrero, M., Agustina, R., Branca, F., Lartey, A., Fan, S., Crona, B., Fox, E., Bignet, V., Troell, M., Lindahl, T., Singh, S., Cornell, S. E., Srinath Reddy, K., Narain, S., Nishtar, S. and Murray, C. J. L. (2019). Food in the Anthropocene: the EAT-Lancet Commission on healthy diets from sustainable food systems, *Lancet* 393(10170), 447–492.

Zaki, O., Weekers, F., Thonart, P., Tesch, E., Kuenemann, P. and Jacques, P. (2020). Limiting factors of mycopesticide development, *Biol. Control* 144, art. no. 104220.

Zehnder, G., Gurr, G. M., Kühne, S., Wade, M. R., Wratten, S. D. and Wyss, E. (2007). Arthropod pest management in organic crops, *Annu. Rev. Entomol.* 52(1), 57–80.

Zeng, W., Kirk, W. and Hao, J. (2012). Field management of Sclerotinia stem rot of soybean using biological control agents, *Biol. Control* 60(2), 141–147.

Chapter 19

Future outlook on microbial bioprotectants in agriculture

Willem J. Ravensberg, Koppert Biological Systems, The Netherlands

1 Introduction: crop protection over time

Plant diseases have had a significant influence on the history of man by their role in major famines, mortality and migration (Schumann, 1991; Zadoks, 2008; Schumann and D'Arcy, 2012). Well-known examples are ergot in grains and potato disease in Ireland. It took a long time to discover that microorganisms were the cause of crop losses and to discover methods of control, first, in the nineteenth century, with copper and arsenic, and then, after World War II, with synthetic pesticides. Although these initially worked well against pathogens, their negative effects on the environment and human health gradually became apparent. When Rachel Carlson's book *Silent Spring* (published in 1962) highlighted these issues, regulations became stricter and many toxic and persistent substances were banned. However, newer generations of chemical pesticides have continued to cause similar problems and many are being withdrawn today (Geiger et al., 2010; Zaller, 2020). Current issues include the development of pesticide resistance, environmental pollution, potentially harmful residues in food and the negative impacts on soil fertility and biodiversity, to name a few.

A recent study identified a major decline in soil invertebrates due to pesticide use, causing disruption to soil functions and soil ecosystem services (Gunstone

http://dx.doi.org/10.19103/AS.2021.0093.21

et al., 2021). Major declines in insect populations have also been attributed to the use of chemical pesticides (Hallmann et al., 2017). The negative impact of pesticides on the biodiversity of the soil microbial community has been reported by various authors (Lo, 2010; Jacobsen and Hjelmsø, 2014; Wolejko et al., 2020), although more research on non-target effects is needed. In a study comparing synthetic fungicides and a fungal biocontrol agent, it was shown that both types of control agents had significant, though contrasting, effects on the soil microbiome, with synthetic pesticides having a negative impact (Fournier et al., 2020). The compatibility of chemical pesticides needs to be tested when they are used in integrated pest management (IPM) programmes in combination with microbial bioprotectants. Synthetic fungicides can have negative effects on beneficial fungi for disease control, and their applications must be carefully timed (Ons et al., 2020).

Policies are being developed by governments to transition from the conventional means of pest control based on chemical pesticide use towards IPM programmes. The EU's Farm to Fork Strategy (https://ec.europa.eu/food/farm2fork_en; accessed 24 May 2021) aims to reduce the use of chemical pesticides by 50% by 2030 and, together with the EU Biodiversity Strategy (https://ec.europa.eu/environment/strategy/biodiversity-strategy-2030_en; accessed 25 May 2021), it seeks to promote biodiversity in agriculture. Buckwell et al. (2020) state that 'nobody is happy with the way crops are protected today, it is not sustainable and has too many negative impacts' and that a new approach that restores biodiversity and resilience in both soil and crop systems is required. Agriculture needs to transition to a new, sustainable model which respects planetary boundaries and which provides food security for a growing world population. Sustainability involves closed nutrient cycles and safe-guarding biodiversity without negative effects on human health and non-target organisms. With a changing and unpredictable climate, sustainability also means resilience to abiotic and biotic challenges, water availability, soil fertility etc. The need to change is recognised by many, and political initiatives have already started all over the world (FAO, 2014).

Alternatives to chemical protection of crops have been investigated for many years, including the use of microorganisms. The first attempts to use microorganisms in crop protection go back more than 150 years. The discovery of diseases caused by microorganisms in the silkworm industry sparked the idea of using similar microorganisms to control insect pests in crops, and can be seen as the birth of the concept of biocontrol (Lord, 2005). Biocontrol of plant diseases was first attempted in the late 1940s, using *Trichoderma* sp. to control the tobacco mosaic virus in Brazil (Bettiol and Morandi, 2009).

Microorganisms not only protect plants from pests and diseases but can also perform other valuable functions. They may also act as biostimulants in making plants more resilient to abiotic stresses (Rouphael and Colla, 2020), as soil and plant fertilisers, and in improving the shelf-life of food. The microbiomes of plants and soils are being intensively studied with the latest molecular technologies and could provide promising insights to improve agriculture and the environment (Berg et al., 2014; Trivedi et al., 2021). The enormous number of microorganisms in nature is largely uninvestigated but provides the potential for new products to replace conventional pesticides. This chapter describes the factors that promote the use of microbial bioprotectants as well as those that hamper their further adoption. A sustainable and resilient agriculture depends on the microbial interactions with plants in promoting plant growth and combatting biotic and abiotic threats. The transition to a resilient agriculture requires big changes in policy and farming practices. This chapter assesses the future outlook for the methods of controlling the plant diseases described in this book, as well as the factors determining their uptake and success.

2 Scientific developments affecting bioprotectants

2.1 Unexplored potential of microorganisms

The authors in this book have explored the huge potential of microorganisms as plant-protection tools. Many of the examples discussed are already being used by farmers all over the world. The development of a microbial bioprotectant is a complex process and takes many years of research and significant resources (Ravensberg, 2011). However, an ever increasing number of new bioprotectants are coming into the market. With new scientific advances, the number of products and applications will continue to increase. The use of microorganisms has been largely focused on a narrow range of problems, like the control of one or more diseases or a few insect pests. However, it is becoming more and more clear that microorganisms have a wide range of functions. Researchers have often categorised microorganisms as entomopathogens, or as antagonists or obligate parasites. However, new research has shown, for instance, that fungi used as entomopathogens on foliar surfaces can also act as endophytes with other modes of action that promote plant health and development (Ownley et al., 2010).

One example is the entomopathogenic fungus *Beauveria bassiana* that has been developed for the control of insects via foliar sprays. However, as an endophyte, it also contributes to the control of plant diseases as well as promoting plant growth (Barra-Bucarei et al., 2020). It also plays a role in soil remediation and may even benefit human health through some of its metabolites (Singh et al., 2015). Another example is *Trichoderma*, known for its activity against fungi, bacteria, nematodes and for plant growth promotion (Baazeem et al., 2021; Sood et al., 2020). However, it can also offer induced resistance to tomato against insect pests such as herbivorous stink bugs (Alınç et al., 2021). Plant-growth-promoting bacteria offer other functions apart from growth promotion. Some pseudomonads also produce insecticidal compounds and have potential as biocidal agents against invertebrate pests (Ruiu, 2020). Yeasts, a group of organisms with multiple functions, have broad potential as biofertilizers and biopesticides and may even have pesticide remediation properties (Hernández-Fernández et al., 2021).

These examples show that microorganisms have multiple benefits in crop protection that are still little understood, let alone utilized. The web of microorganisms and other organisms in soil can provide many benefits, through a large variety of modes of action, in areas such as crop protection, nutrient uptake, growth promotion and stress resistance. The understanding of microorganisms is still limited, and their potential, still undervalued. Ongoing scientific discoveries will surely lead to unravelling that potential for crop protection and resilient agricultural systems.

2.2 New applications of microorganisms

New applications may emerge through new discoveries such as treatment of seeds and young plant material with microorganisms, for different functions. The seed microbiome offers great potential to improve germination and growth through the whole plant life cycle (Nelson, 2018; Berg and Raaijmakers, 2018). Instead of the current practice of cleaning seeds to avoid spread of pathogens, addition of microbes to the seed could help provide protection and promote growth. As described in this book, plant microbes can also be applied in the root zone and on the canopy. Applications using bees as a vector to apply the bioprotectant in flowers have also been developed for greenhouse crops and tested for use in open fields (Mänd et al., 2020). Biocontrol products for post-harvest disease control have also been developed. The interactions between the applied microbial agents and the microbial community on the fruit surface are being investigated in order to improve their effectiveness (Sare et al., 2021). Integrated approaches such as better timing of applications and addition of nutrients to improve establishment of biocontrol strains are being investigated to address these limitations.

Microbial bioherbicides have not yet been consistently successful in the field (Triolet et al., 2020). Recent research also shows the potential for *Trichoderma koningiopsis* to cause foliar damage to the weed *Euphorbia heterophylla* (the Mexican fire plant) without damaging corn (Reichert Júnior et al., 2019), as well as the use of metabolites of *Streptomyces* strains for control of grass and broadleaf weeds (Bo et al., 2019).

During the past decade, research on endophytes for biocontrol (De Silva et al., 2019; Latz et al., 2018; Collinge et al., 2019) and as biostimulants (Omomowo and Babalola, 2019) has accelerated. A promising development is the use of the endophytic bacterium *Gluconacetobacter diazotrophicus* to fix atmospheric nitrogen in non-leguminous crops like cereals and corn, potentially reducing the need for artificial fertilisers (Dent, 2018). The use of endophytes looks like a promising area in terms of plant health benefits. However, application methods such as seed dressing, root dipping and foliar applications can be cumbersome and do not always help achieve successful long-term colonisation of the plant (Rabiey et al., 2019; Morales-Cedeño et al., 2021). There are no endophyte-based products yet registered and marketed as bioprotectants. Extremophiles, such as heat-adapted bacteria from compost heaps and arid soils, as well as cold-adapted fungi from polar habitats, are also being investigated for their potential uses and could also deliver new applications in agriculture (Mehetre et al., 2021; Torracchi et al., 2020).

2.3 Emerging technologies

Genetic and genome-assisted approaches can improve the selection of next-generation biocontrol agents (Leung et al., 2020). Gene-editing technologies have the potential to improve different properties of microorganisms (Raaijmakers et al., 2020). The target traits include virulence, speed of kill, productivity, host range and tolerance to abiotic stress (Lovett and St. Leger, 2018; Larena et al., 2020). Gene-editing technology may also improve the safety of microorganisms by knocking out antimicrobial resistance genes or genes encoding for potentially harmful metabolites. This technique can only be used in countries where a gene-edited organism is not regarded as a genetically modified organism (GMO). The European Union is currently planning to revise its GMO regulation and gene-editing may no longer be seen as a GM technique in the near future. However, it is important to consider public attitudes to gene-editing so as not to damage the positive image of biocontrol. Synthetic biology also offers the potential for designing and building synthetic microbial communities that improve plant health (Roell and Zurbriggen, 2020; Ke et al., 2021).

2.4 Amendments to improve soil resilience

Besides the use of microbial bioprotectants for control of plant diseases, soil amendments that claim to improve plant and/or soil health are available on the market. These products contain a range of ingredients, including plant extracts, organic acids, hormones, biochar and mixtures of microorganisms. They are marketed under various names such as soil improvers, bioenhancers, biofertilizers, biostimulants etc. Scientific information is often lacking to substantiate product claims. The challenge is to develop products that truly improve plant and soil health by creating an optimal microbiome for the plant.

This should then be measurable so that improvements can be monitored and followed over time. Measuring the microbiome and its effects will be difficult but necessary in order to develop the right products, dosages, frequency and interval of application. Only then will such amendments contribute to a sustainable and resilient agriculture (De Corato, 2020). The interaction between general measures to improve the resilience of the system and the targeted use of microbial bioprotectants needs much research to investigate their compatibility and whether a combined approach can strengthen the resilience of the cropping system.

2.5 Plant breeding and the microbiome

The plant breeding industry so far has not considered the relevance of the plant microbiome. Because it has focused on attributes such as yield and quality, plant breeding may even have impaired the plant's capacity to develop an optimal microbiome for its growth and health (Pérez-Jaramillo et al., 2016). As a result, new high-yielding cultivars may be more susceptible to pests and diseases, increasing the reliance on chemical crop protection. However, it is increasingly recognised that the microbiome plays a crucial role in plant health and that plants should be selected in part for their ability to co-opt microorganisms in the rhizosphere (Bakker et al., 2012; Gopal and Gupta, 2016; Compant et al., 2019). Breeding strategies are now starting to incorporate this aspect of plant behaviour. However, as breeding new cultivars is a long process, it will take many years before this will significantly contribute to more resilient crops (Nogales et al., 2016; Wei and Jousset, 2017). The use of microbial bioprotectants should be compatible with new cultivars bred with an improved microbiome, especially when bioprotectants are applied in the rhizosphere and in case of endophytic bioprotectants. This compatibility needs to be investigated by both the plant breeding and the biocontrol industry.

2.6 Microbial interactions with arthropod biocontrol agents and pests

Bacterial symbionts can interact with parasitic wasps to produce only females that parasitize pests, making mass production more efficient. Indirect effects occur through yeasts and bacteria in floral nectar and insect honeydew to produce volatiles that attract natural enemies. Bacterial symbionts can also attract pollinators (Rering et al., 2018). This area is largely unexplored (Zindel et al., 2011) and could potentially improve biocontrol using arthropods and insect-parasitic nematodes (Arora and Douglas, 2017; Machado et al., 2020). On the other hand, insect pests can benefit from defensive symbionts in strengthening their resistance against parasitoids. Selection of the most effective natural enemy population may promote pest control (Vorburger, 2018). Given the lack of understanding of their effects, chemical fungicides should be avoided, to ensure beneficial microbial interactions for a sustainable agriculture. Microbial bioprotectants should be used since these are more likely to safeguard these important microbial interactions.

2.7 Microorganisms as a cornerstone in crop protection

The potential use of microorganisms in plant health is enormous and still largely unrealised. This is true for single microorganisms, for combinations of microorganisms, and even more for consortia of microorganisms - the microbiome and the holobiont. Interventions in agriculture with single and multiple applications of microorganisms are already commonplace in certain sectors, but applying consortia of microorganisms and manipulation or engineering of the microbiome are largely untapped (Ke et al., 2021). Once we have a better understanding of all these complex interactions between plants, microorganisms, pests and diseases, it will be possible to grow crops in a much more resilient manner. Cropping systems can also be adapted to increase the positive impact of the microbiome on plant and soil health.

Instead of monocultures, crop rotation and intercropping, strip or even pixel farming (farming different crops alongside each other in small plots) may contribute to a healthier system where pests and diseases have less chance of occurring and spreading. Other measures such as no-till or reduced till, cultivation of green fertiliser crops and addition of specific organic matter and crop residues will also benefit the holobiont involving crop and soil. Improving the plant holobiont by microbiome engineering is an emerging strategy to improve pest control, crop yields and resilience (Lawson et al., 2019; Pineda et al., 2017). Indirect approaches to microbiome engineering include the use of soil amendments (Nuzzo et al., 2020) and direct approaches such as applying specific probiotic microbes and artificial microbial consortia (Arif et al., 2020; Qiu et al., 2019). Regenerative agriculture includes such measures as these and values all those ecological factors needed to maintain a resilient and healthy crop (Rhodes, 2017; Newton et al., 2020).

3 Regulatory issues

3.1 Risk assessment and registration procedures

The use of microorganisms in crop protection is heavily regulated (see elsewhere in this book), presenting a major hurdle in getting microbial bioprotectants to market. New molecular techniques are making it easier to find and identify microorganisms and research their functions and potential uses. Microbiome research opens up many new potential applications in areas such as crop protection, food, and human and animal health. The essential role of microbes in nature is being increasingly recognised, as are their potential uses. Regulations should focus on the safety of microbes using a harmonised approach. Currently, the uses of microorganisms are regulated by a variety of laws that generally focus on the separate fields of application such as feed additives, pesticides, biocides, veterinary medicines, biostimulants and fertilisers. As a new kind of product, microorganisms are often added to an existing group of substances with similar uses and regulated in the same way. A recent example is the inclusion of microbial biostimulants in the revised regulation for fertilisers in the EU (Reg.(EU) 2019/1009) (European Parliament, 2019). This usually leads to inappropriate data requirements and risk assessments. A harmonised risk assessment focusing on the safety of the microorganism rather than on their category of use should be developed, taking into account the specific characteristics of living microorganisms and their role in ecosystems. Risk assessments should also include use of consortia of microorganisms and products for manipulation of the microbiome in soils and plants. New emerging uses of microorganisms should be anticipated so that they can be included in the relevant existing legislation rather than as a new category in legislation designed to cover other uses.

3.2 Hurdles to explore microorganisms

The Nagoya Protocol on Access and Benefit Sharing of the Convention of Biological Diversity (https://www.cbd.int/abs/) hampers bioprospecting of microbes and their use, given the extensive administrative procedures required to conduct research on biocontrol agents. A key challenge is to agree on benefit-sharing between the country of origin and the commercial party beforehand. The International Organization for Biological Control (IOBC) recognises these barriers in collecting and researching beneficial organisms and is investigating a system that better facilitates the exchange of genetic resources for biocontrol purposes (Lenteren, 2021).

3.3 Human health risks associated with plants and microbes

In a number of cases, human pathogens have developed in plants and caused diseases and even deaths (Overbeek et al., 2014). Concerns over this route for human pathogens are increasing. *E. coli* and *Salmonella* strains have been reported growing in plants and causing outbreaks of foodborne disease (Deering et al., 2012). It is evident that such cases need to be studied and avoided, for example, via routine sampling for microbial contaminants. Another concern has appeared related to potential health risks from use of the biocontrol agent *Bacillus thuringiensis* for caterpillar control, despite a history of safe use of more than 50 years. *B. thuringiensis* (Bt) is closely related to *B. cereus* and is difficult to distinguish from this opportunistic human pathogen. In a few cases Bt was considered the presumed cause of a foodborne disease outbreak and the European Food Safety Authority has performed a risk assessment on Bt bacteria related to human health (EFSA, 2016). As a result, the European Commission may decide to set requirements for a Maximum Residue Limit (MRL) for Bt and a pre-harvest interval. This threatens the use of Bt in vegetables, potentially leading to more use of chemical pesticides. Issues like this do not increase confidence in the use of microbes in agriculture among the general public. There is a need for education and better communication to focus on the benefits of microbes, not just on a few cases where they can be harmful.

3.4 The real risk of microbial bioprotectants

An appropriate risk assessment of microorganisms used in agriculture is pivotal to their acceptance. The risk assessment, however, must be proportionate to the real risk that microorganisms pose to the environment and to human health. Microbiome studies show that exposure to microbes via unprocessed food is common, as shown by the plethora of microorganisms found, for example, on apples and strawberries (Abdelfattah et al., 2016a,b). There is particular concern about metabolites, though they typically occur in very low amounts, as a result of interactions with other organisms and transient events (Köhl et al., 2019a,b). Despite this low level of risk, the burden of proof to demonstrate that the risks are acceptable is unreasonable and prohibitive (Scheepmaker et al., 2019).

Microorganisms produce metabolites for their defence and survival, as demonstrated in many chapters of this book. Most metabolites are studied in a laboratory setting and the in situ analysis of their functions is extremely difficult. Do these substances pose a risk when used in agriculture? The risk assessment should consider the ecological and evolutionary characteristics of these organisms in the environment of use. Human and animal pathogens and mycotoxin-producing fungi are all well-known, and organisms with such characteristics are not selected as biocontrol agents. A new approval system should be developed, focusing on the type of microorganisms and their biology and ecology (see elsewhere in this book) to allow agriculture to benefit from these natural elements of the ecosystem.

4 Crop protection and political goals

There is a strong demand from society in general to reduce or even avoid the use of synthetic pesticides in agriculture and to produce food without residues of these chemicals. This is illustrated, for example, by a referendum in

Switzerland in 2021 where the population could vote on banning all synthetic pesticide use. Researchers in 20 EU countries in the European Research Alliance (https://www.era-pesticidefree.eu/) are anticipating such developments by creating a research agenda focusing on a chemical pesticide-free agriculture. The EU Green Deal, Farm to Fork (F2F) Strategy, the Sustainable Use Directive (SUD) (European Parliament, 2009), and many other national plans recognise public opinion and set out ambitious plans to transform agriculture. These political ambitions depend on the farmers' adoption of new crop-protection technologies.

Following implementation of the SUD, EU countries have been required to incorporate IPM into crop-protection measures since 2014. The Directive also stipulates that biological pest control measures should be given priority. However, the goals of the SUD – reducing risks and impacts on human health and environment and promoting integrated pest management – have not been achieved (ECA, 2020). The SUD is being revised to achieve a higher degree of adoption of IPM and to achieve the ambitious goal of the F2F Strategy by 2030. Farmers are willing to change as long as their income is guaranteed and appropriate tools and methods are available. The lack of information and confidence is also often mentioned as a constraint in the use of biological control measures, and training and demonstration trials are needed to increase farmers' adoption of biocontrol and IPM (Creissen et al., 2021).

5 Towards a new agriculture

5.1 System changes

Agriculture is not only required to change its crop protection methods to reduce the dependency on and the use of chemical pesticides in favour of non-chemical products and methods. It also has to change its farming systems to respond to issues such as climate change, biodiversity, carbon storage requirements, water use, nutrient cycles, soil management, changes in human diet, energy use and greenhouse gas emissions. The challenge for farmers and society is huge and extremely complex, requiring a revolutionary and holistic transition within a relatively short period, not the one that would normally need decades to achieve.

In the case of crop protection, it is not just a question of replacing a chemical tool with a biological tool. The whole system of crop protection management has to change in accordance with all eight principles of an IPM programme (Barzman et al., 2015). Standard agronomic practices, such as large fields cultivated with one crop, limited rotations, high fertiliser and pesticide use, intensive irrigation, use of susceptible cultivars bred for consumer preferences rather than pest and disease resistance, and land use unsupportive of biodiversity, need to be reconsidered. Measures to reduce pest and disease pressure include strip and even pixel farming, wildflower strips and natural hedges to increase natural enemies and pollinators, and annual crop rotation schemes. New plant varieties with an optimal microbiome-assembling capacity must be introduced. Soil management needs to build a resilient soil that is able to continue to deliver ecosystem services and has sufficient buffering capacity to keep plants healthy. Digital and precision technologies could assist the farmer in monitoring a variety of parameters so that interventions are conducted in a timely and targeted way.

The transition will take place in many aspects of farming and requires new skills from farmers and their advisors. Education and training are needed to facilitate this. A change of mindset is required for all those involved, including partners throughout the agri-food supply chain, up to the banks, investors, the consumer and, in fact, the whole of society. This kind of mindset is already present in organic agriculture, reaching from the farmer through the whole chain of organic products to the consumer. This example can be copied to create a new model, often referred to as 'regenerative agriculture'. Transition to this model is the big challenge for the coming decades, creating significant friction, since it needs major cultural and societal changes (Gosnell et al., 2019).

5.2 Microbial bioprotectants as a key factor in a new agriculture

Pest and disease management will remain necessary in any new agricultural system. Farmers need targeted and sustainable tools that protect their crop from significant losses. These should be targeted towards suppressing specific organisms without disturbing the holobiont of the crop and the resilience of the system. Microbial bioprotectants can fulfil this requirement and many already work in this way. The success of bioprotectants depends on a mix of strengths and weaknesses, opportunities and threats. A SWOT analysis is presented in Table 1.

Bioprotectants have proven to be a successful business model. Sales are growing annually by >15% (see elsewhere in this book). These figures reflect the sales of plant protection products which are still complementary to rather than an alternative for chemical pesticides. In a resilient system, more selective approaches are needed to safeguard the buffering capacity of the system. Bioprotectants are still niche products with a limited market. Lower developmental and registration costs compared to chemical pesticides have helped them enter the market. However, many potential biocontrol agents still do not make it, as the registration hurdle, in terms of costs and time, is too high, and takes too long to achieve a return on investment. To compete against the often low prices of chemicals, a radical change is needed to account for the

Table 1 Strengths, weaknesses, opportunities and threats of microbial bioprotectants

Strengths	Weaknesses
• Specific host range • Multiple modes of action • No residue (exempt of MRL) • No or short pre-harvest interval • No or short worker re-entry interval • Safe for humans and the environment • Safe for plants • Approved for organic production • Compatible with natural enemies, pollinators • Compatible with other microbial pest control agents • Probability of (cross-) resistance low • Preferred tool in IPM systems • Tool in resistance-management programmes of synthetic chemicals • No or low externalised costs	• Narrow target spectrum • Slow speed of kill • Sensitive to abiotic factors • Moderate and variable efficacy • Limited shelf life • More complicated application technology, spraying necessary, good coverage essential • Incompatibilities with chemical pesticides • Relatively high end-user price • Knowledge-intensive products that demand know-how transfer • Often only work as part of IPM programme
Opportunities	**Threats**
• Growing demand for residue-free or minimum-residue food • Withdrawal and reduction of use of chemicals • Developmental costs low to reasonable • BCAs are a natural resource and offer sustainable crop protection • BCA production using renewable resources • No or minimal impact on microbiome • Environmental benign, no environmental pollution due to production and use • Increase or maintain biodiversity • Increase of sustainable agriculture and IPM • Increase of organic production • Increasing availability of high quality bioprotectants, thereby increasing users' confidence in bioprotectants' use in general • Increasing markets where use of chemicals is forbidden, such as forestry, amenity areas, home and garden, natural environments etc.	• Novel safer chemical pesticides • Growers' scepticism based on expectation level of chemicals • Small market sizes • Transgenic crops • Increasing regulatory burden • Biodiversity and access-benefit-sharing regulations • Bioterrorism regulations • Fear of microbes among the public • Human pathogens in plants • Limited public research funding • Lack of user education and extension services • Weak economic position of farmers • Unregistered ineffective 'snake oils' • Low-quality products

Modified after Ravensberg, 2011.

externalised costs of synthetic pesticides (Bourguet and Guillemaud, 2016) in order to achieve fair competition with other means of pest control.

6 Major factors influencing the future developments of bioprotectants

To what extent, how and when microbial bioprotectants will become the most important or even the only tool for crop protection is hard to predict. Government policies can facilitate a transition since change depends on many stakeholders and market factors.

The following factors determine how crop protection will develop in the next decade:

- The demand from society for an agriculture with reduced or no synthetic pesticides - the options are a more or less conventional agriculture with slow adaptation or a more profound change ready to face the challenges of the future.
- The trend in regulating chemical use is towards greater restriction - Bioprotectants need separate regulation that takes their specific characteristics into account in a risk assessment. If the situation remains as it is today in the EU, market access for bioprotectants will increase only slowly, which will significantly delay the reduction in chemical pesticides use.
- The use of bioprotectant products can be based on a single strain or on an assembled consortium of strains. Regulations will largely determine the direction in which this will go.
- The success of biocontrol depends on breakthroughs in research in understanding microbiome functions and how they can contribute to biocontrol.
- The role of bioprotectants can either be to complement chemical plant protection products or to replace them completely and complement microbiome functions.
- Gene-editing can improve the efficacy and safety of microbial control agents if regulations are adapted and society accepts this technology.
- Shaping and complementing the microbiome will only become important tools in crop protection if they prove to be viable business models for the biocontrol industry.
- The Nagoya Protocol and ABS rules need to be facilitated for biocontrol agents to allow a rapid expansion of new products based on genetic resources from around the world.
- A rapid development of digital tools and precision agriculture will improve and accelerate the use of bioprotectants.

- The costs of agricultural production should include the currently externalised costs, for example, to account for the environmental costs of using chemical pesticides, to allow fair comparison of the costs of different crop protection approaches.

These are the pivotal factors on which the use of microbial bioprotectants in agriculture depend. They provide enormous challenges for all stakeholders. Legislation must facilitate these changes, especially in those cases where market-driven incentives are lacking. Ecological principles must be valued and ecological disturbances by external factors must be priced so that economics become a driver for change. Such a new approach not only affects farming practices but has a major impact on all stakeholders in our society. Several scenarios have to be studied to find an optimum between ecological and economic outcomes (Pretty et al., 2018).

7 Conclusion

The world population is dependent on twelve plants as main food crops, including three dominant crops: rice, wheat and corn. What will happen if an epidemic in one of these crops occurs, like the Covid-19 virus pandemic? When a devastating plant disease emerges, it may affect a crop worldwide, with dramatic consequences, as seen in the past. An example is the spread of *Fusarium* in banana which threatens this crop globally (Kema et al., 2020). A change in the way agriculture is conducted today is therefore required (OECD/FAO, 2019). The FAO estimates that up to 40 percent of global food crops are lost annually due to plant pests and diseases. In economic terms alone, plant diseases cost the global economy around $220 billion annually, and invasive insects, around $70 billion (http://www.fao.org/plant-health-2020/about/en/).

Agriculture will have to transition to a more sustainable system. Truly transformational changes are required to reach the UN Sustainable Development Goals (https://sdgs.un.org/goals). Crop protection is an essential part of growing healthy plants and maintaining resilient and fertile soils. Microbial bioprotectants can provide solutions without the negative side effects of synthetic pesticides. Their development, approval and use thus need to be prioritised and incentivised so that farmers adopt them. That this is possible is illustrated by biocontrol developments in Brazil, in the cultivation of major arable crops where microorganisms and parasitic wasps outcompete chemical use (Harris, 2021).

For the purpose of crop protection, nature potentially offers the solutions to combat pests and diseases, as shown in this book. If ecological interactions are studied and understood, they can be adjusted and engineered in new ways that will ensure a sustainable, resilient, ecologically based agriculture which requires fewer interventions. Microbial bioprotectants do not pollute and accumulate in the environment, as they have evolved in nature and are broken down as part of the natural cycle (Köhl et al., 2019b). The use of xenobiotic pesticidal substances that often upset that balance is then no longer needed. A new agricultural system will tend to go in the direction of organic agriculture. Resilient soil management will be key, and inputs will be reduced; if interventions are needed, they have to be sustainable and selective, safeguarding biodiversity and the holobiont. Microorganisms are crucial, and bioprotectants, biostimulants, biofertilizers and microbiome-enhancing amendments will be the main inputs. Besides microbes, soil invertebrates, organic matter and natural fertilisers are essential to keep the system in balance. Ecological principles must be recognised and valued. New digital technologies can assist the farmer in decision making and methods of application. With these measures, agriculture can be designed to provide food security in a sustainable agro-ecological manner within planetary boundaries.

8 Where to look for further information

The need for a transition in agriculture to a more ecological and holistic approach is described in a report of the RISE Foundation, a European think tank that provides recommendations for policy makers:

Buckwell, A., De Wachter, E., Nadeu, E., Williams, A. 2020. Crop Protection & the EU Food System. Where are they going? RISE Foundation, Brussels. https://risefoundation.eu/wp-content/uploads/2020/07/2020_RISE_CP_EU_Appendix.pdf.

The importance of biological control in sustainable agriculture is emphasised in a report of the Institute for European Environmental Policy (IEEP) which is a sustainability think tank:

Hulot, J. F. and Hiller, N. (2021) 'Exploring the benefits of biocontrol for sustainable agriculture – A literature review on biocontrol in light of the European Green Deal', Institute for European Environmental Policy. Exploring the benefits of biocontrol for sustainable agriculture (ieep.eu).

Developments in the European Union on the sustainable use of pesticides and policy actions are provided by the European Commission: Sustainable use of pesticides (europa.eu).

More research and potential applications of microorganisms can be found is recently published books:

- Yadav, A. N. (Ed). (2021). Soil Microbiomes for Sustainable Agriculture: Functional Annotation. Springer.
- Yadav, A., N, Singh, J., Rastegari, A. A. and Yadav, N. (Eds). (2020). Plant Microbiomes for Sustainable Agriculture. Springer.

A recent paper discusses the potential and the constraints of the use of microorganisms in agriculture:

- Batista, B. D. and Singh, B. K. (2021). Realities and hopes in the application of microbial tools in agriculture. *Microbial Biotechnology* (2021) 0(0), 1–11. doi:10.1111/1751-7915.13866.

Industry associations advocating the use of biocontrol, including microorganisms, for crop production and protection provide information on progress on regulations and the use of microorganism in agriculture. Examples are:

- The International Biocontrol Manufacturers Association (IBMA); Home Page IBMA-GLOBAL International Biocontrol Manufacturers.
- The Biological Products Industry Alliance (BPIA); Biological Products Industry Alliance | Advancing Knowledge About Biopesticides & Biostimulants (bpia.org).
- The European Biostimulants Industry Council (EBIC); EBIC – The European Biostimulants Industry Council.
- BioProtection Global (BPG); BioProtection Global – A Global Federation of Regional Associations Representing Biocontrols on a Global Scale.

9 References

Abdelfattah, A., Wisniewski, M., Li Destri Nicosia, M. G., Cacciola, S. O. and. Schena, L. (2016a). Metagenomic analysis of fungal diversity on strawberry plants and the effect of management practices on the fungal community structure of aerial organs, *PLoS ONE* 11(8), e0160470. https://doi.org/10.1371/journal.pone.0160470.

Abdelfattah, A., Wisniewski, M., Droby, S. and Schena, L. (2016b). Spatial and compositional variation in the fungal communities of organic and conventionally grown apple fruit at the consumer point-of-purchase, *Hortic. Res.* 3, 16047. https://doi.org/10.1038/hortres.2016.47 (accessed 5 October 2016).

Alınç, T., Cusumano, A., Peri, E., Torta, L. and Colazza, S. (2021). *Trichoderma harzianum* strain T22 modulates direct defense of tomato plants in response to *Nezara viridula* feeding activity, *J. Chem. Ecol.* 47(4–5), 455–462. https://doi.org/10.1007/s10886-021-01260-3.

Arif, I., Batool, M. and Schenk, P. M. (2020). Plant microbiome engineering: expected benefits for improved crop growth and resilience, *Trends Biotechnol.* 38(12), 1385–1396.

Arora, A. K. and Douglas, A. E. (2017). Hype or opportunity? Using microbial symbionts in novel strategies for insect pest control, *J. Insect Physiol.* 103, 10–17.

Baazeem, A., Almanea, A., Manikandan, P., Alorabi, M., Vijayaraghavan, P. and Abdel-Hadi, A. (2021). In vitro antibacterial, antifungal, nematocidal and growth-promoting activities of *trichoderma hamatum* FB10 and its secondary metabolites, *J. Fungi (Basel)* 7(5), 331. https://doi.org/10.3390/jof7050331.

Bakker, M. G., Manter, D. K., Sheflin, A. M., Weir, T. L. and Vivanco, J. M. (2012). Harnessing the rhizosphere microbiome through plant breeding and agricultural management, *Plant Soil* 360(1-2), 1-13.

Barra-Bucarei, L., González, M. G., Iglesias, A. F., Aguayo, G. S., Peñalosa, M. G. and Vera, P. V. (2020). *Beauveria bassiana* multifunction as an endophyte: growth promotion and biologic control of *Trialeurodes vaporariorum*, (Westwood) (Hemiptera: Aleyrodidae) in tomato, *Insects* 11(9), 591. https://doi.org/10.3390/insects11090591.

Barzman, M., Bàrberi, P., Birch, A. N. E., Boonekamp, P., Dachbrodt-Saaydeh, S., Graf, B., Hommel, B., Jensen, J. E., Kiss, J., Kudsk, P., Lamichhane, J. R., Messéan, A., Moonen, A., Ratnadass, A., Ricci, P., Sarah, J. and Sattin, M. (2015). Eight principles of integrated pest management, *Agron. Sustain. Dev.* 35(4), 1199-1215. https://doi.org/10.1007/s13593-015-0327-9.

Berg, G., Grube, M., Schloter, M. and Kornelia Smalla, K. (2014). The plant microbiome and its importance for plant and human health, *Front. Microbiol.* 5, 491. https://doi.org/10.3389/fmicb.2014.00491 (accessed 17 September 2014).

Berg, G. and Raaijmakers, J. M. (2018). Saving seed microbiomes, *ISME J.* 12(5), 1167-1170.

Bettiol, W. and Morandi, M. A. B. (2009). Trichoderma in Brazil: history, research, commercialization and perspectives, *IOBC WPRS Bull.* 43, 235-237.

Bo, A. B., Kim, J. D., Kim, Y. S., Sin, H. T., Kim, H. J., Khaitov, B., Ko, Y. K., Park, K. W. and Choi, J. S. (2019). Isolation, identification and characterization of *Streptomyces* metabolites as a potential bioherbicide, *PLoS ONE* 14(9), e0222933. https://doi.org/10.1371/journal.pone.0222933 (accessed 23 September 2019).

Bourguet, D. and Guillemaud, T. (2016). The hidden and external costs of pesticide use. In: Lichtfouse, E. (Ed), *Sustainable Agriculture Reviews, 19*, Cham: Springer International Publishing, pp. 35-120.

Buckwell, A., De Wachter, E., Nadeu, E. and Williams, A. (2020). *Crop Protection & the EU Food System. Where Are They Going?* Brussels: RISE Foundation.

Carlson, R. (1962). *Silent Spring*, Boston: Houghton Mifflin.

Collinge, D. B., Jørgensen, H. J. L., Latz, M. A., Manzotii, A., Ntana, F., et al. (2019). Searching for novel fungal biological control agents for plant disease control among endophytes. In: Hodkinson, T. R., Doohan, F. M., Saunders, M. J. and Murphy, B. R. (Eds), *Endophytes for a Growing World*, Cambridge: Cambridge University Press, pp. 25-51.

Compant, S., Samad, A., Faist, H. and Sessitsch, A. (2019). A review on the plant microbiome: ecology, functions, and emerging trends in microbial application, *J. Adv. Res.* 19, 29-37.

Creissen, H. E., Jones, P. J., Tranter, R. B., Girling, R. D., Jess, S., Burnett, F. J., Gaffney, M., Thorne, F. S. and Kildea, S. (2021). Identifying the drivers and constraints to adoption of IPM among arable farmers in the UK and Ireland, *Pest Manag. Sci.* 77(9), 4148-4158. https://doi.org/10.1002/ps.6452 (accessed 02 May 2021).

De Corato, U. (2020). Disease-suppressive compost enhances natural soil suppressiveness against soil-borne plant pathogens: a critical review, *Rhizosphere* 13. https://doi.org/10.1016/j.rhisph.2020.100192.

De Silva, N. I., Brooks, S., Lumyong, S. and Hyde, K. D. (2019). Use of endophytes as biocontrol agents, *Fungal Biol. Rev.* 33(2), 133-148.

Deering, A. J., Mauer, L. J. and Pruitt, R. E. (2012). Internalization of *E. coli* O157:H7 and *Salmonella* spp. in plants: a review, *Food Res. Int.* 45(2), 567-575.

Dent, D. (2018). Non-nodular endophytic bacterial symbiosis and the nitrogen fixation of *Gluconacetobacter diazotrophicus*. In: Rigobelo, E. C. (Ed), *Symbiosis*, London: Intechopen Ltd, pp. 53–82.

ECA. (2020). Sustainable use of plant protection products: limited progress in measuring and reducing risks. Special Report, European Court of Auditors. https://www.eca.europa.eu/en/Pages/DocItem.aspx?did=53001 (accessed 25 May 2021).

EFSA BIOHAZ Panel. (2016). EFSA panel on biological hazards, *EFSA J.* 14(7), 4524, 93 pp. https://doi.org/10.2903/j.efsa.2016.4524.

European Parliament. (2009). Directive 2009/128/EC of the European Parliament and of the Council of 21 October 2009 establishing a framework for Community action to achieve the sustainable use of pesticides, *OJL* 309, 71. [24.11.2009]. https://eur-lex.europa.eu/legal-content/EN/TXT/PDF/?uri=CELEX:32009L0128&from=EN, accessed 25th May 2021).

European Parliament. (2019). Regulation (EU) 2019/1009 of the European Parliament and of the Council of 5 June 2019 laying down rules on the making available on the market of EU fertilising products and amending Regulations (EC) No 1069/2009 and (EC) No 1107/2009 and repealing Regulation (EC) No 2003/2003, *OJL* 170, 1–114. [25.6.2019]. http://data.europa.eu/eli/reg/2019/1009/oj.

FAO (2014). *Building a Common Vision for Sustainable Food and Agriculture; Principles and Approaches. Rome*. ISBN 978-92-5-108471-7.

Fournier, B., Dos Santos, S. P., Gustavsen, J. A., Imfeld, G., Lamy, F., Mitchell, E. A. D., Mota, M., Noll, D., Planchamp, C. and Heger, T. J. (2020). Impact of a synthetic fungicide (fosetyl-Al and propamocarb-hydrochloride) and a biopesticide (*Clonostachys rosea*) on soil bacterial, fungal, and protist communities, *Sci. Total Environ.* 738, 139635.

Geiger, F., Bengtsson, J., Berendse, F., Weisser, W. W., Emmerson, M., Morales, M. B., Ceryngier, P., Liira, J., Tscharntke, T., Winqvist, C., Eggers, S., Bommarco, R., Pärt, T., Bretagnolle, V., Plantegenest, M., Clement, L. W., Dennis, C., Palmer, C., Oñate, J. J., Guerrero, I., Hawro, V., Aavik, T., Thies, C., Flohre, A., Hänke, S., Fischer, C., Goedhart, P. W. and Inchausti, P. (2010). Persistent negative effects of pesticides on biodiversity and biological control potential on European farmland, *Basic Appl. Ecol.* 11(2), 97–105.

Gopal, M. and Alka Gupta, A. (2016). Microbiome selection could spur next-generation plant breeding strategies, *Front. Microbiol.* 7, 1971. https://doi.org/10.3389/fmicb.2016.01971.

Gosnell, H., Gill, N. and Voyer, M. (2019). Transformational adaptation on the farm: processes of change and persistence in transitions to 'climate-smart' regenerative agriculture, *Glob. Environ. Change* 59. https://doi.org/10.1016/j.gloenvcha.2019.101965 (accessed November 2019).

Gunstone, T., Cornelisse, T., Klein, K., Dubey, A.. and Donley, N. (2021). Pesticides and soil invertebrates: a hazard assessment, *Front. Environ. Sci.* 9, Art. 643847 https://doi.org/10.3389/fenvs.2021.643847 .

Hallmann, C. A., Sorg, M., Jongejans, E., Siepel, H., Hofland, N., Schwan, H., Stenmans, W., Müller, A., Sumser, H., Hörren, T., Goulson, D. and de Kroon, H. (2017). More than 75 percent decline over 27 years in total flying insect biomass in protected areas, *PLoS ONE* 12(10), e0185809. https://doi.org/10.1371/journal.pone.0185809.

Harris, B. (2021). Biological controls in Brazil threaten crop pests–and chemical companies, *Fin. Times*. https://www.ft.com/content/d3e387a8-6350-4216-9121-98724a1059a6 (accessed 11 May 2021).

Hernández-Fernández, M., Cordero-Bueso, G., Ruiz-Muñoz, M. and Cantoral, J. M. (2021). Culturable yeasts as biofertilizers and biopesticides for a sustainable agriculture: a comprehensive review, *Plants* 10, 822. https://doi.org/10.3390.

Jacobsen, C. S. and Hjelmsø, M. H. (2014). Agricultural soils, pesticides and microbial diversity, *Curr. Opin. Biotechnol.* 27, 15–20.

Ke, J., Wang, B. and Yoshikuni, Y. (2021). Microbiome engineering: synthetic biology of plant-associated microbiomes in sustainable agriculture, *Trends Biotechnol.* 39(3), 244–261. https://doi.org/10.1016/j.tibtech.2020.07.008.

Kema, G. H. J., Drenth, A., Dita, M., Jansen, K., Vellema, S. and Stoorvogel, J. J. (2020). Editorial: Fusarium wilt of banana, a recurring threat to global banana production, *Front. Plant Sci.* 11, 628888. https://doi.org/10.3389/fpls.2020.628888.

Köhl, J., Kolnaar, R. and Ravensberg, W. J. (2019a). Mode of action of microbial biological control agents against plant diseases: relevance beyond efficacy, *Front. Plant Sci.* 10(845), 845. https://doi.org/10.3389/fpls.2019.00845.

Köhl, J., Booij, K., Kolnaar, R. and Ravensberg, W. J. (2019b). Ecological arguments to reconsider data requirements regarding the environmental fate of microbial biocontrol agents in the registration procedure in the European Union, *BioControl* 64(5), 469–487.

Larena, I., Espeso, E. A., Carreras, M., Villarino, M., De Cal, A. and Melgarejo, P. (2020). Impact of genomic resources on improving the mode of action of biocontrol agents against plant pathogens. In: De Cal, A., Melgarejo, P., Magan, N. (Eds) How Research Can Stimulate the Development of Commercial Biological Control Against Plant Diseases, Springer Nature Switzerland AG, 203–229.

Latz, M. A. C., Jensen, B., Collinge, D. B. and Jørgensen, H. J. L. (2018). Endophytic fungi as biocontrol agents: elucidating mechanisms in disease suppression, *Plant Ecol. Divers.* 11(5–6), 555–567.

Lawson, C. E., Harcombe, W. R., Hatzenpichler, R., Lindemann, S. R., Löffler, F. E., O'Malley, M. A., García Martín, H., Pfleger, B. F., Raskin, L., Venturelli, O. S., Weissbrodt, D. G., Noguera, D. R. and McMahon, K. D. (2019). Common principles and best practices for engineering microbiomes, *Nat. Rev. Microbiol.* 17(12), 725–741.

Lenteren, van, J. C. (2021). Will the "Nagoya Protocol on Access and Benefit Sharing" put an end to biological control? In: Hendrichs, J., Pereira, R. and Vreysen, M. J. B. (Eds), *Area-Wide Integrated Pest Management, Development and Field Application*, Boca Raton: CRC Press. https://doi.org/10.1201/9781003169239.

Leung, K., Ras, E., Ferguson, K. B., Ariëns, S., Babendreier, D., Bijma, P., Bourtzis, K., Brodeur, J., Bruins, M. A., Centurión, A., Chattington, S. R., Chinchilla-Ramírez, M., Dicke, M., Fatouros, N. E., González-Cabrera, J., Groot, T. V. M., Haye, T., Knapp, M., Koskinioti, P., Le Hesran, S., Lyrakis, M., Paspati, A., Pérez-Hedo, M., Plouvier, W. N., Schlötterer, C., Stahl, J. M., Thiel, A., Urbaneja, A., van de Zande, L., Verhulst, E. C., Vet, L. E. M., Visser, S., Werren, J. H., Xia, S., Zwaan, B. J., Magalhães, S., Beukeboom, L. W. and Pannebakker, B. A. (2020). Next-generation biological control: the need for integrating genetics and genomics, *Biol. Rev. Camb. Philos. Soc.* 95(6), 1838–1854. https://doi.org/10.1111/brv.12641.

Lo, C. C. (2010). Effect of pesticides on soil microbial community, *J. Environ. Sci. Health B* 45(5), 348–359.

Lord, J. C. (2005). From Metchnikoff to Monsanto and beyond: the path of microbial control, *J. Invertebr. Pathol.* 89(1), 19–29.

Lovett, B. and St. Leger, R. J. (2018). Genetically engineering better fungal biopesticides, *Pest Manag. Sci.* 74(4), 781–789. https://doi.org/10.1002/ps.4734.

Machado, R. A. R., Thönen, L., Arce, C. C. M., Theepan, V., Prada, F. F., Wüthrich, D., Robert, C. A. M., Vogiatzaki, E., Shi, Y. M., Schaeren, O. P., Notter, M., Bruggmann, R., Hapfelmeier, S., Bode, H. B. and Erb, M. (2020). Engineering bacterial symbionts of nematodes improves their biocontrol potential to counter the Western corn rootworm, *Nat. Biotechnol.* 38(5), 600–608.

Mänd, M., Karise, R. and Smagghe, G. (2020). Bumble bees and entomovectoring in open field conditions. In: Smagghe, G., Boecking, O., Maccagnani, B., Mänd, M. and Kevan, P. G. (Eds) *Entomovectoring for Precision Biocontrol and Enhanced Pollination of Crops*, Springer Nature Switzerland AG, pp. 81–93.

Mehetre, G., Leo, V. V., Singh, G., Dhawre, P., Maksimov, I., et al. (2021). Biocontrol potential and applications of extremophiles for sustainable agriculture. In: Mehetre, G., Leo, V. V., Singh, G., Dhawre, P. and Maksimov, I. (Eds). *Microbiomes of Extreme Environments*, Boca Raton: CRC Press.

Morales-Cedeño, L. R., Orozco-Mosqueda, M. D. C., Loeza-Lara, P. D., Parra-Cota, F. I., de los Santos-Villalobos, S. and Santoyo, G. (2021). Plant growth-promoting bacterial endophytes as biocontrol agents of pre- and post-harvest diseases: fundamentals, methods of application and future perspectives, *Microbiol. Res.* 242, 126612.

Nelson, E. B. (2018). The seed microbiome: origins, interactions, and impacts, *Plant Soil* 422(1–2), 7–34.

Newton, P., Civita, N., Frankel-Goldwater, L., Bartel, K. and Johns, C. (2020). What is regenerative agriculture? A review of scholar and practitioner definitions based on processes and outcomes, *Front. Sustain. Food Syst.* 4. https://doi.org/10.3389/fsufs.2020.577723 (accessed 26 October 2020).

Nogales, A., Nobre, T., Valadas, V., Ragonezi, C., Döring, M., Polidoros, A. and Arnholdt-Schmitt, B. (2016). Can functional hologenomics aid tackling current challenges in plant breeding?, *Brief. Funct. Genomics* 15(4), 288–297. https://doi.org/10.1093/bfgp/elv030.

Nuzzo, A., Satpute, A., Albrecht, U. and Strauss, S. L. (2020). Impact of soil microbial amendments on tomato rhizosphere microbiome and plant growth in field soil, *Microb. Ecol.* 80(2), 398–409.

OECD-FAO. (2019). *Agricultural Outlook 2019-2028*, Paris: OECD Publishing. https://doi.org/10.1787/agr_outlook-2019-en.

Omomowo, O. I. and Babalola, O. O. (2019) Bacterial and fungal endophytes: tiny giants with immense beneficial potential for plant growth and sustainable agricultural productivity, *Microorganisms* 7(11), 481. https://doi.org/10.3390/microorganisms7110481.

Ons, L., Bylemans, D., Thevissen, K. and Cammue, B. P. A. (2020). Combining biocontrol agents with chemical fungicides for integrated plant fungal disease control, *Microorganisms* 8(12), 1930. https://doi.org/10.3390/microorganisms8121930.

Ownley, B. H., Gwinn, K. D. and Veja, F. E. (2010). Endophytic fungal entomopathogens with activity against plant pathogens: ecology and evolution, *BioControl* 55(1), 113–128.

PérezJaramillo, J. E., Mendes, R. and Raaijmakers, J. M. (2016). Impact of plant domestication on rhizosphere microbiome assembly and functions, *Plant Mol. Biol.* 90(6), 635–644. https://doi.org/10.1007/s11103-015-0337-7.

Pineda, A., Kaplan, I. and Bezemer, T. M. (2017). Steering soil microbiomes to suppress aboveground insect pests, *Trends Plant Sci.* 22(9), 770-778.

Pretty, J., Benton, T. G., Bharucha, Z. P., Dicks, L. V., Flora, C. B., Godfray, H. C. J., Goulson, D., Hartley, S., Lampkin, N., Morris, C., Pierzynski, G., Prasad, P. V. V., Reganold, J., Rockström, J., Smith, P., Thorne, P. and Wratten, S. (2018). Global assessment of agricultural system redesign for sustainable intensification, *Nat. Sustain.* 1(8), 441-446.

Qiu, Z., Egidi, E., Liu, H., Kaur, S. and Singh, B. K. (2019). New frontiers in agriculture productivity: optimised microbial inoculants and in situ microbiome engineering, *Biotechnol. Adv.* 37(6), 107371. https://doi.org/10.1016/j.biotechadv.2019.03.010.

Rabiey, M., Hailey, L. E., Roy, S. R., Grenz, K., Al-Zadjali, M. A. S., Barrett, G. A. and Jackson, R. W. (2019). Endophytes vs tree pathogens and pests: can they be used as biological control agents to improve tree health?, *Eur. J. Plant Pathol.* 155(3), 711-729.

Raaijmakers, K., Ponet, L., Holtappels, D., Berckmans, B. and Cammue, B. P. A. (2020). Screening for novel biocontrol agents applicable in plant disease management–a review, *Biol. Control* 144. https://doi.org/10.1016/j.biocontrol.2020.104240.

Ravensberg, W. J. (2011). *A Roadmap to the Successful Development and Commercialization of Microbial Pest Control Products for Control of Arthropods.* Dordrecht: Springer.

Reichert Júnior, F. W., Scariot, M. A.,Tiago Forte, C. T., Pandolfi, L., Mara Dil, J. M., et al. (2019). New perspectives for weed control using autochthonous fungi with selective bioherbicide potential, *Heliyon* 5(5, May 2019), e01676. https://doi.org/10.1016/j.heliyon.2019.e01676.

Rering, C. C., Kessler, A., Beck, J. J., Hall, G. W., McCartney, M. M. and Vannett, R. L. (2018).Introduction to a special feature issue–new insights into plant volatiles, *New Phytol.* 220(3), 655-658.

Rhodes, C. J. (2017). The imperative for regenerative agriculture, *Sci. Prog.* 100(1), 80-129.

Roell, M. S.. and Zurbriggen, M. D. (2020). The impact of synthetic biology for future agriculture and nutrition, *Curr. Opin. Biotechnol.* 61, 102-109.

Rouphael, Y. and Colla, G. (2020). Editorial: biostimulants in agriculture, *Front. Plant Sci.* 11, 40.

Ruiu, L. (2020). Plant-growth-promoting bacteria (PGPB) against insects and other agricultural pests, *Agronomy* 10(6), 861. https://doi.org/10.3390/agronomy10060861.

Sare, A. R., Jijakli, M. H. and Massart, S. (2021). Microbial ecology to support integrative efficacy improvement of biocontrol agents for postharvest diseases management, *Postharvest Biol. Technol.* 179(September), 111572. https://doi.org/10.1016/j.postharvbio.2021.111572.

Scheepmaker, J. W. A., Busschers, M., Sundh, I., Eilenberg, J. and Butt, T. M. (2019). Sense and nonsense of the secondary metabolites data requirements in the EU for beneficial microbial control agents, *Biol. Control* 136. https://doi.org/10.1016/j.biocontrol.2019.104005.

Schumann, G. L. (1991). *Plant Diseases: Their Biology and Social Impact*, St. Paul: American Phytopathological Society.

Schumann, G. L. and D'Arcy, C. J. (2012). *Hungry Planet: Stories of Plant Diseases*, St. Paul: American Phytopathological Society.

Singh, H. B., Keswani, C., Ray, S., Yadav, S. K., Singh, S. P., et al. (2015). *Beauveria bassiana*: biocontrol beyond Lepidopteran pests'. In: Sree, K. S. and Varma, A. (Eds), *Biocontrol of Lepidopteran Pests, Soil Biology 43*, Switzerland: Springer International Publishing. https://doi.org/10.1007/978-3-319-14499-3_10219.

Sood, M., Kapoor, D., Kumar, V., Sheteiwy, M. S., Ramakrishnan, M., Landi, M., Araniti, F. and Sharma, A. (2020). *Trichoderma*: the "secrets" of a multitalented biocontrol agent, *Plants (Basel)* 9(6), 762. https://doi.org/10.3390/plants9060762.

Torracchi C, J. E., Morel, M. A., Tapia-Vázquez, I., Castro-Sowinski, S., Batista-García, R. A. and Yarzábal R, L. A. (2020). Fighting plant pathogens with cold-active microorganisms: biopesticide development and agriculture intensification in cold climates, *Appl. Microbiol. Biotechnol.* 104(19), 8243–8256.

Triolet, M., Guillemin, J. P., Andre, O. and Steinberg, C. (2020). Fungal-based bioherbicides for weed control: a myth or a reality?, *Weed Res.* 60(1), 60–77.

Trivedi, P., Mattupalli, C., Eversole, K. and Leach, J. E. (2021). Enabling sustainable agriculture through understanding and enhancement of microbiomes, *New Phytol.* 230(6), 2129–2147. https://doi.org/10.1111/nph.17319 (accessed 03 March 2021).

van Overbeek, L. S., van Doorn, J., Wichers, J. H., van Amerongen, A., van Roermund, H. J. W. and Willemsen, P. T. J. (2014). The arable ecosystem as battleground for emergence of new human pathogens, *Front. Microbiol.* 5. https://doi.org/10.3389/fmicb.2014.00104 (accessed 20 March 2014).

Vorburger, C. (2018). Symbiont-conferred resistance to parasitoids in aphids–challenges for biological control, *Biol. Control* 116, 17–26.

Wei, Z. and Jousset, A. (2017). Plant breeding goes microbial, *Trends Plant Sci.* 22(7), 555–558.

Wołejko, E., Jabłońska-Trypuća, A., Wydro, U., Butarewicz, A. and Łozowicka, B. (2020). Soil biological activity as an indicator of soil pollution with pesticides–a review, *Appl. Soil Ecol.* 147(March), 103356. https://doi.org/10.1016/j.apsoil.2019.09.006.

Zadoks, J. C. (2008). *On the Political Economy of Plant Disease Epidemics: Capita Selecta in Historical Epidemiology*, Wageningen: Wageningen Academic Publishers.

Zaller, J. G. (2020). *Daily Poison- Pesticides–An Underestimated Danger*, Cham, Switzerland: Springer Nature Switzerland AG.

Zindel, R., Gottlieb, Y. and Aebi, A. (2011). Arthropod symbioses: a neglected parameter in pest- and disease-control programmes, *J. Appl. Ecol.* 48(4), 864–872. https://doi.org/10.1111/j.1365-2664.2011.01984.x.

Index

www.ingramcontent.com/pod-product-compliance
Ingram Content Group UK Ltd.
Pitfield, Milton Keynes, MK11 3LW, UK
UKHW021823100926
8556IPUK00003B/134

9 781786 76813